Name and Dates	*Page*
26. Jacobi, Karl Gustav Jacob (1804–1851)	(528)
27. Kaprekar, D. R. (1905–1986)	(8)
28. Kronecker, Leopold (1823–1891)	(4)
29. Kummer, Ernst Eduard (1810–1893)	(594)
30. Lagrange, Joseph Louis (1736–1813)	(9)
31. Lamé, Gabriel (1795–1870)	(137)
32. Legendre, Adrien-Marie (1752–1833)	(502)
33. Lehmer, Derrick Henry (1905–1991)	(392)
34. Lucas, François-Edouard-Anatole (1842–1891)	(131)
35. Mahavira (ca. 850)	(190)
36. McCarthy, John (1827–)	(32)
37. Mersenne, Marin (1588–1648)	(382)
38. Ming, Antu (ca. 1692–ca. 1763)	(53)
39. Möbius, August Ferdinand (1790–1868)	(398)
40. Pascal, Blaise (1623–1662)	(34)
41. Pell, John (1611–1685)	(613)
42. Pythagoras (ca. 572–ca. 500)	(2)
43. Ramanujan, Srinivasa Aiyangar (1887–1920)	(567)
44. Simson, Robert (1687–1768)	(134)
45. Valleé-Poussin, Charles-Jean-Gustave-Nicholas de la (1866–1962)	(113)
46. Waring, Edward (1734–1798)	(322)
47. Wiles, Andrew J. (1953–)	(595)
48. Wilson, John (1741–1793)	(322)
49. Zeller, Christian Julius Johannes (1849–1899)	(286)

Elementary Number Theory with Applications

Second Edition

Elementary Number Theory with Applications

Second Edition

Thomas Koshy

AMSTERDAM • BOSTON • HEIDELBERG • LONDON
NEW YORK • OXFORD • PARIS • SAN DIEGO
SAN FRANCISCO • SINGAPORE • SYDNEY • TOKYO

Academic Press is an imprint of Elsevier

Academic Press is an imprint of Elsevier
30 Corporate Drive, Suite 400, Burlington, MA 01803, USA
525 B Street, Suite 1900, San Diego, California 92101-4495, USA
84 Theobald's Road, London WC1X 8RR, UK

This book is printed on acid-free paper. ♾

Library of Congress Cataloging-in-Publication Data

Koshy, Thomas.
Elementary number theory with applications / Thomas Koshy. – 2nd ed.
p. cm.
Includes bibliographical references and index.
ISBN 978-0-12-372487-8 (alk. paper)
1. Number theory. I. Title.

QA241.K67 2007
512.7–dc22

2007010165

British Library Cataloguing-in-Publication Data
A catalogue record for this book is available from the British Library.

ISBN: 978-0-12-372487-8

For information on all Academic Press publications
visit our Web site at www.books.elsevier.com

Printed in China
07 08 09 10 9 8 7 6 5 4 3 2 1

Dedicated to

*my sister, Aleyamma Zachariah, and my brother,
M. K. Tharian; and to the memory of
Professor Edwin Weiss, Professor Donald W. Blackett,
and Vice Chancellor A. V. Varughese*

Contents

Preface

Man has the faculty of becoming completely absorbed in one subject, no matter how trivial and no subject is so trivial that it will not assume infinite proportions if one's entire attention is devoted to it.

—TOLSTOY, *War and Peace*

For over two thousand years, number theory has fascinated and inspired both amateurs and mathematicians alike. A sound and fundamental body of knowledge, it has been developed by the untiring pursuits of mathematicians all over the world. Today, number theorists continue to develop some of the most sophisticated mathematical tools ever devised and advance the frontiers of knowledge.

Many number theorists, including the eminent nineteenth-century English number theorist Godfrey H. Hardy, once believed that number theory, although beautiful, had no practical relevance. However, the advent of modern technology has brought a new dimension to the power of number theory: constant practical use. Once considered the purest of pure mathematics, it is used increasingly in the rapid development of technology in a number of areas, such as art, coding theory, cryptology, and computer science. The various fascinating applications have confirmed that human ingenuity and creativity are boundless, although many years of hard work may be needed to produce more meaningful and delightful applications.

The Pursuit of a Dream

This book is the fruit of years of dreams and the author's fascination for the subject, its beauty, elegance, and historical development; the opportunities it provides for both experimentation and exploration; and, of course, its marvelous applications.

This new edition, building on the strengths of its predecessor, incorporates a number of constructive suggestions made by students, reviewers, and well-wishers. It is logically conceived, self-contained, well-organized, nonintimidating, and written with students and amateurs in mind. In clear, readable language, this book offers an overview of the historical development of the field, including major figures, as well as step-by-step development of the basic concepts and properties, leading to the more advanced exercises and discoveries.

Audience and Prerequisites

The book is designed for an undergraduate course in number theory for students majoring in mathematics and/or computer science at the sophomore/junior level and for students minoring in mathematics. No formal prerequisites are required to study the material or to enjoy its beauty except a strong background in college algebra. The main prerequisite is mathematical maturity: lots of patience, logical thinking, and the ability for symbolic manipulation. This book should enable students and number theory enthusiasts to enjoy the material with great ease.

Coverage

The text includes a detailed discussion of the traditional topics in an undergraduate number theory course, emphasizing problem-solving techniques, applications, pattern recognition, conjecturing, recursion, proof techniques, and numeric computations. It also covers figurate numbers and their geometric representations, Catalan numbers, Fibonacci and Lucas numbers, Fermat numbers, an up-to-date discussion of the various classes of prime numbers, and factoring techniques. Starred ($\star$) optional sections and optional puzzles can be omitted without losing continuity of development.

Included in this edition are new sections on Catalan numbers and the Pollard rho factoring method, a subsection on the Pollard $p-1$ factoring method, and a short chapter on continued fractions. The section on linear diophantine equations now appears in Chapter 3 to provide full prominence to congruences.

A number of well-known conjectures have been added to challenge the more ambitious students. Identified by the conjecture symbol ? in the margin, they should provide wonderful opportunities for group discussion, experimentation, and exploration.

Examples and Exercises

Each section contains a wealth of carefully prepared and well-graded examples and exercises to enhance student skills. Examples are developed in detail for easy understanding. Many exercise sets contain thought-provoking true/false problems, numeric problems to develop computational skills, and proofs to master facts and the various proof techniques. Extensive chapter-end review exercise sets provide comprehensive reviews, while chapter-end supplementary exercises provide challenging opportunities for the curious-minded to pursue.

Starred (⋆) exercises are, in general, difficult, and doubly starred (⋆⋆) ones are more difficult. Both can be omitted without losing overall understanding of the concepts under discussion. Exercises identified with a c in the margin require a knowledge of elementary calculus; they can be omitted by students with no calculus background.

Historical Comments and Biographies

Historical information, including biographical sketches of about 50 mathematicians, is woven throughout the text to enhance a historical perspective on the development of number theory. This historical dimension provides a meaningful context for prospective and in-service high school and middle school teachers in mathematics. An index of the biographies, keyed to pages in the text, can be found inside the back cover.

Applications

This book has several unique features. They include the numerous relevant and thought-provoking applications spread throughout, establishing a strong and meaningful bridge with geometry and computer science. These applications increase student interest and understanding and generate student interaction. In addition, the book shows how modular systems can be used to create beautiful designs, linking number theory with both geometry and art. The book also deals with barcodes, zip codes, International Serial Book Numbers, European Article Numbers, vehicle identification numbers, and German bank notes, emphasizing the closeness of number theory to our everyday life. Furthermore, it features Friday-the-thirteenth, the p-queens puzzle, round-robin tournaments, a perpetual calendar, the Pollard rho factoring method, and the Pollard $p-1$ factoring method.

Flexibility

The order and selection of topics offer maximum flexibility for instructors to select chapters and sections that are appropriate for student needs and course lengths. For example, Chapter 1 can be omitted or assigned as optional reading, as can the optional sections 6.2, 6.3, 7.3, 8.5, 10.4, and 11.5, without jeopardizing the core of development. Sections 2.2, 2.3, and 5.4–5.6 also can be omitted if necessary.

Foundations

All proof methods are explained and illustrated in detail in the Appendix. They provide a strong foundation in problem-solving techniques, algorithmic approach, and proof techniques.

Proofs

Most concepts, definitions, and theorems are illustrated through thoughtfully selected examples. Most of the theorems are proven, with the exception of some simple ones left as routine exercises. The proofs shed additional light on the understanding of the topic and enable students to develop their problem-solving skills. The various proof techniques are illustrated throughout the text.

Proofs Without Words

Several geometric proofs of formulas are presented without explanation. This unique feature should generate class discussion and provide opportunities for further exploration.

Pattern Recognition

An important problem-solving technique used by mathematicians is pattern recognition. Throughout the book, there are ample opportunities for experimentation and exploration: collecting data, arranging them systematically, recognizing patterns, making conjectures, and then establishing or disproving these conjectures.

Recursion

By drawing on well-selected examples, the text explains in detail this powerful strategy, which is used heavily in both mathematics and computer science. Many examples are provided to ensure that students are comfortable with this powerful problem-solving technique.

Numeric Puzzles

Several fascinating, optional number-theoretic puzzles are presented for discussion and digression. It would be a good exercise to justify each. These puzzles are useful for prospective and in-service high school and middle school teachers in mathematics.

Algorithms

A number of algorithms are given as a problem-solving technique in a straightforward fashion. They can easily be translated into computer programs in a language of your choice. These algorithms are good candidates for class discussion and are boxed in for easy identification.

Computer Assignments

Relevant and thought-provoking computer assignments are provided at the end of each chapter. They provide hands-on experience with concepts and enhance the opportunity for computational exploration and experimentation. A computer algebra system, such as *Maple* or *Mathematica*, or a language of your choice can be used.

Chapter Summary

At the end of each chapter, you will find a summary that is keyed to pages in the text. This provides a quick review and easy reference. Summaries contain the various definitions, symbols, and properties.

Enrichment Readings

Each chapter ends with a carefully prepared list of readings from various sources for further exploration of the topics and for additional enrichment.

Web Links

Relevant annotated web sites are listed in the Appendix. For instance, up-to-date information on the discovery of Mersenne primes and twin primes is available on the Internet. This enables both amateurs and professionals to access the most recent discoveries and research.

Special Symbols

The square ■ denotes the end of a proof and an example. The conjecture symbol [?] indicates an unresolved problem.

Index of Symbols

Inside the front cover, you will find, for quick reference, a list of symbols and the page numbers on which they first occur.

Odd-Numbered Solutions

The solutions to all odd-numbered exercises are given at the end of the text.

Solutions Manual for Students

The *Student's Solutions Manual* contains detailed solutions to all even-numbered exercises. It also contains valuable tips for studying mathematics, as well as for preparing and taking examinations.

Instructor's Manual

The *Instructor's Manual* contains detailed solutions to all even-numbered exercises, sample tests for each chapter, and the keys for each test. It also contains two sample final examinations and their keys.

Highlights of this Edition

They include:

- Catalan numbers (Sections 1.8, 2.5, and 8.4)
- Linear diophantine equations with Fibonacci coefficients (Section 3.5)
- Pollard rho factoring method (Section 4.3)
- Vehicle identification numbers (Section 5.3)
- German bank notes (Section 5.3)
- Factors of $2^n + 1$ (Section 7.2)
- Pollard $p - 1$ factoring method (Section 7.2)
- Pascal's binary triangle and even perfect numbers (Section 8.4)
- Continued fractions (Chapter 12)
- Well-known conjectures
- Expanded exercise sets

Acknowledgments

I am grateful to a number of people for their cooperation, support, encouragement, and thoughtful comments during the writing and revising of this book. They all have played a significant role in improving its quality.

To begin with, I am indebted to the following reviewers for their boundless enthusiasm and constructive suggestions:

Steven M. Bairos	Data Translation, Inc.
Peter Brooksbank	Bucknell University
Roger Cooke	University of Vermont
Joyce Cutler	Framingham State College
Daniel Drucker	Wayne State University
Maureen Femick	Minnesota State University at Mankato
Burton Fein	Oregon State University
Justin Wyss-Gallifent	University of Maryland
Napolean Gauthier	The Royal Military College of Canada
Richard H. Hudson	University of South Carolina
Robert Jajcay	Indiana State University
Roger W. Leezer	California State University at Sacramento

I. E. Leonard	University of Alberta
Don Redmond	Southern Illinois University
Dan Reich	Temple University
Helen Salzberg	Rhode Island College
Seung H. Son	University of Colorado at Colorado Springs
David Stone	Georgia Southern University
M. N. S. Swamy	Concordia University
Fernando Rodriguez Villegas	University of Texas at Austin
Betsey Whitman	Framingham State College
Raymond E. Whitney	Lock Haven University

Thanks also to Roger Cooke of the University of Vermont, Daniel Drucker of Wayne State University, Maureen Fenrick of Minnesota State University at Mankato, and Kevin Jackson-Mead for combing through the entire manuscript for accuracy; to Daniel Drucker of Wayne State University and Dan Reich of Temple University for class-testing the material; to the students Prasanth Kalakota of Indiana State University and Elvis Gonzalez of Temple University for their comments; to Thomas E. Moore of Bridgewater State College and Don Redmond of Southern Illinois University for preparing the solutions to all odd-numbered exercises; to Ward Heilman of Bridgewater State College and Roger Leezer of California State University at Sacramento for preparing the solutions to all even-numbered exercises; to Margarite Roumas for her superb editorial assistance; and to Madelyn Good and Ellen Keane at the Framingham State College Library, who tracked down a number of articles and books. My sincere appreciation also goes to Senior Editors Barbara Holland, who initiated the original project, Pamela Chester, and Thomas Singer; Production Editor Christie Jozwiak, Project Manager Jamey Stegmaier, Copyeditor Rachel Henriquez, and Editorial Assistant Karen Frost at Harcourt/Academic Press for their cooperation, promptness, support, encouragement, and confidence in the project.

Finally, I must confess that any errors that may yet remain are my own responsibility. However, I would appreciate hearing about any inadvertent errors, alternate solutions, or, better yet, exercises you have enjoyed.

Thomas Koshy
tkoshy@frc.mass.edu

A Word to the Student

Mathematics is music for the mind;
music is mathematics for the soul.
—ANONYMOUS

The Language of Mathematics

To learn a language, you have to know its alphabet, grammar, and syntax, and you have to develop a decent vocabulary. Likewise, mathematics is a language with its own symbols, rules, terms, definitions, and theorems. To be successful in mathematics, you must know them and be able to apply them; you must develop a working vocabulary, use it as often as you can, and speak and write in the language of math.

This book was written with you in mind, to create an introduction to number theory that is easy to understand. Each chapter is divided into short sections of approximately the same length.

Problem-Solving Techniques

Throughout, the book emphasizes problem-solving techniques such as doing experiments, collecting data, organizing them in an orderly fashion, recognizing patterns, and making conjectures. It also emphasizes recursion, an extremely powerful problem-solving strategy used heavily in both mathematics and computer science. Although you may need some practice to get used to recursion, once you know how to approach problems recursively, you will appreciate its power and beauty. So do not be turned off, even if you have to struggle a bit with it initially.

The book stresses proof techniques as well. Theorems are the bones of mathematics. So, for your convenience, the various proof methods are explained and illustrated in the Appendix. It is strongly recommended that you master them; do the worked-out examples, and then do the exercises. Keep reviewing the techniques as often as needed.

Many of the exercises use the theorems and the techniques employed in their proofs. Try to develop your own proofs. This will test your logical thinking and analytical skills. In order to fully enjoy this beautiful and elegant subject, you must feel at home with the various proof methods.

Getting Involved

Basketball players such as Michael Jordan and Larry Bird did not become superstars by reading about basketball or watching others play. Besides knowing the rules and the objects needed to play, they needed countless hours of practice, hard work, and determination to achieve their goal. Likewise, you cannot learn mathematics by simply watching your professor do it in class or by reading about it; you have to do it yourself every day, just as skill is acquired in a sport. You can learn mathematics in small, progressive steps only, building on skills you already have developed.

Suggestions for Learning

Here are a few suggestions you should find useful in your pursuit:

- Read a few sections before each class. You might not fully understand the material, but you will certainly follow it far better when your professor discusses it in class. Besides, you will be able to ask more questions in class and answer more questions.
- Always go to class well prepared. Be prepared to answer and ask questions.
- Whenever you study from the book, make sure you have a pencil and enough scrap paper next to you for writing the definitions, theorems, and proofs and for doing the exercises.
- Study the material taught in class on the same day. Do not just read it as if you were reading a novel or a newspaper. Write down the definitions, theorems, and properties in your own words without looking in your notes or the book. Rewrite the examples, proofs, and exercises done in class, all in your own words. If you cannot do them on your own, study them again and try again; continue until you succeed.
- Always study the relevant section in the text and do the examples there, then do the exercises at the end of the section. Since the exercises are graded in order of difficulty, do them in order. Do not skip steps or write over previous steps; this way you will be able to progress logically, locate your errors, and correct your mistakes. If you cannot solve a problem because it involves a new term, formula, or some property, then re-study the relevant portion of the section and try again. Do not assume that you will be able to do every problem the first time you try it. Remember, practice is the best shortcut to success.

Solutions Manual

The *Student's Solutions Manual* contains additional tips for studying mathematics, preparing for an examination in mathematics, and taking an examination in mathematics. It also contains detailed solutions to all even-numbered exercises.

A Final Word

Mathematics, especially number theory, is no more difficult than any other subject. If you have the willingness, patience, and time to sit down and do the work, then you will find number theory worth studying and this book worth studying from; you will find that number theory can be fun, and fun can be number theory. Remember that learning mathematics is a step-by-step matter. Do your work regularly and systematically; review earlier chapters every week, since things must be fresh in your mind to apply them and to build on them. In this way, you will enjoy the subject and feel confident to explore more. I look forward to hearing from you with your comments and suggestions. In the meantime, enjoy the book.

Thomas Koshy

1 Fundamentals

Tell me and I will forget.
Show me and I will remember.
Involve me and I will understand.
—CONFUCIUS

The outstanding German mathematician Karl Friedrich Gauss (1777–1855) once said, "Mathematics is the queen of the sciences and arithmetic the queen of mathematics." "Arithmetic," in the sense Gauss uses it, is number theory, which, along with geometry, is one of the two oldest branches of mathematics. Number theory, as a fundamental body of knowledge, has played a pivotal role in the development of mathematics. And as we will see in the chapters ahead, the study of number theory is elegant, beautiful, and delightful.

A remarkable feature of number theory is that many of its results are within the reach of amateurs. These results can be studied, understood, and appreciated without much mathematical sophistication. Number theory provides a fertile ground for both professionals and amateurs. We can also find throughout number theory many fascinating conjectures whose proofs have eluded some of the most brilliant mathematicians. We find a great number of unsolved problems as well as many intriguing results.

Another interesting characteristic of number theory is that although many of its results can be stated in simple and elegant terms, their proofs are sometimes long and complicated.

Generally speaking, we can define "number theory" as the study of the properties of numbers, where by "numbers" we mean integers and, more specifically, positive integers.

Studying number theory is a rewarding experience for several reasons. First, it has historic significance. Second, integers, more specifically, positive integers, are

A Greek Stamp Honoring Pythagoras

__Pythagoras__ (ca. 572–ca. 500 B.C.), a Greek philosopher and mathematician, was born on the Aegean island of Samos. After extensive travel and studies, he returned home around 529 B.C. only to find that Samos was under tyranny, so he migrated to the Greek port of Crontona, now in southern Italy. There he founded the famous Pythagorean school among the aristocrats of the city. Besides being an academy for philosophy, mathematics, and natural science, the school became the center of a closely knit brotherhood sharing arcane rites and observances. The brotherhood ascribed all its discoveries to the master.

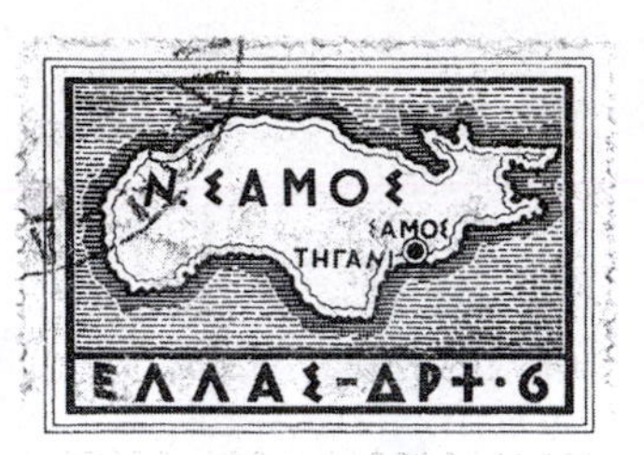

The Island of Samos

A philosopher, Pythagoras taught that number was the essence of everything, and he associated numbers with mystical powers. He also believed in the transmigration of the soul, an idea he might have borrowed from the Hindus.

Suspicions arose about the brotherhood, leading to the murder of most of its members. The school was destroyed in a political uprising. It is not known whether Pythagoras escaped death or was killed.

the building blocks of the real number system, so they merit special recognition. Third, the subject yields great beauty and offers both fun and excitement. Finally, the many unsolved problems that have been daunting mathematicians for centuries provide unlimited opportunities to expand the frontiers of mathematical knowledge. Goldbach's conjecture (Section 2.5) and the existence of odd perfect numbers (Section 8.3) are two cases in point. Modern high-speed computers have become a powerful tool in proving or disproving such conjectures.

Although number theory was originally studied for its own sake, today it has intriguing applications to such diverse fields as computer science and cryptography (the art of creating and breaking codes).

The foundations for number theory as a discipline were laid out by the Greek mathematician Pythagoras and his disciples (known as the *Pythagoreans*). The Pythagorean brotherhood believed that "everything is number" and that the central explanation of the universe lies in number. They also believed some numbers have mystical powers. The Pythagoreans have been credited with the invention of amicable numbers, perfect numbers, figurate numbers, and Pythagorean triples. They classified integers into odd and even integers, and into primes and composites.

Another Greek mathematician, Euclid (ca. 330–275 B.C.), also made significant contributions to number theory. We will find many of his results in the chapters to follow.

We begin our study of number theory with a few fundamental properties of integers.

Little is known about ***Euclid's*** *life. He was on the faculty at the University of Alexandria and founded the Alexandrian School of Mathematics. When the Egyptian ruler King Ptolemy I asked Euclid, the father of geometry, if there were an easier way to learn geometry than by studying* The Elements, *he replied, "There is no royal road to geometry."*

1.1 Fundamental Properties

The German mathematician Hermann Minkowski (1864–1909) once remarked, "Integral numbers are the fountainhead of all mathematics." We will come to appreciate how important his statement is. In fact, number theory is concerned solely with integers. The set of integers is denoted by the letter **Z**:†

$$\mathbf{Z} = \{\ldots, -3, -2, -1, 0, 1, 2, 3, \ldots\}$$

Whenever it is convenient, we write "$x \in S$" to mean "x belongs to the set S"; "$x \notin S$" means "x does not belong to S." For example, $3 \in \mathbf{Z}$, but $\sqrt{3} \notin \mathbf{Z}$.

We can represent integers geometrically on the **number line**, as in Figure 1.1.

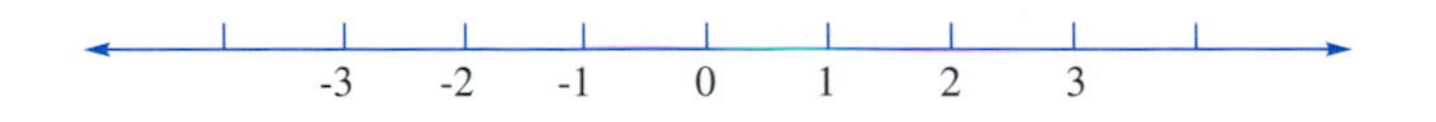

Figure 1.1

The integers $1, 2, 3, \ldots$ are **positive integers**. They are also called **natural numbers** or **counting numbers**; they lie to the right of the origin on the number line. We denote the set of positive integers by $\mathbf{Z}^+$ or **N**:

$$\mathbf{Z}^+ = \mathbf{N} = \{1, 2, 3, \ldots\}$$

† The letter **Z** comes from the German word *Zahlen* for numbers.

Leopold Kronecker *(1823–1891) was born in 1823 into a well-to-do family in Liegnitz, Prussia (now Poland). After being tutored privately at home during his early years and then attending a preparatory school, he went on to the local gymnasium, where he excelled in Greek, Latin, Hebrew, mathematics, and philosophy. There he was fortunate to have the brilliant German mathematician Ernst Eduard Kummer (1810–1893) as his teacher. Recognizing Kronecker's mathematical talents, Kummer encouraged him to pursue independent scientific work. Kummer later became his professor at the universities of Breslau and Berlin.*

In 1841, Kronecker entered the University of Berlin and also spent time at the University of Breslau. He attended lectures by Dirichlet, Jacobi, Steiner, and Kummer. Four years later he received his Ph.D. in mathematics.

Kronecker's academic life was interrupted for the next 10 years when he ran his uncle's business. Nonetheless, he managed to correspond regularly with Kummer. After becoming a member of the Berlin Academy of Sciences in 1861, Kronecker began his academic career at the University of Berlin, where he taught unpaid until 1883; he became a salaried professor when Kummer retired.

In 1891, his wife died in a fatal mountain climbing accident, and Kronecker, devastated by the loss, succumbed to bronchitis and died four months later.

Kronecker was a great lover of the arts, literature, and music, and also made profound contributions to number theory, the theory of equations, elliptic functions, algebra, and the theory of determinants. The vertical bar notation for determinants is his creation.

The German mathematician Leopold Kronecker wrote, "God created the natural numbers and all else is the work of man." The set of positive integers, together with 0, forms the set of **whole numbers W**:

$$\mathbf{W} = \{0, 1, 2, 3, \ldots\}$$

Negative integers, namely, $\ldots, -3, -2, -1$, lie to the left of the origin. Notice that 0 is neither positive nor negative.

We can employ positive integers to compare integers, as the following definition shows.

The Order Relation

Let a and b be any two integers. Then a is **less than** b, denoted by $a < b$, if there exists a positive integer x such that $a + x = b$, that is, if $b - a$ is a positive integer. When $a < b$, we also say that b is **greater than** a, and we write $b > a$.†

† The symbols $<$ and $>$ were introduced in 1631 by the English mathematician Thomas Harriet (1560–1621).

If a is not less than b, we write $a \not< b$; similarly, $a \not> b$ indicates a is not greater than b.

It follows from this definition that an integer a is positive if and only if $a > 0$.

Given any two integers a and b, there are three possibilities: either $a < b$, $a = b$, or $a > b$. This is the **law of trichotomy**. Geometrically, this means if a and b are any two points on the number line, then either point a lies to the left of point b, the two points are the same, or point a lies to the right of point b.

We can combine the less than and equality relations to define the **less than** or **equal to** relation. If $a < b$ or $a = b$, we write $a \leq b$.[†] Similarly, $a \geq b$ means either $a > b$ or $a = b$. Notice that $a \not< b$ if and only if $a \geq b$.

We will find the next result useful in Section 3.4. Its proof is fairly simple and is an application of the law of trichotomy.

THEOREM[‡] 1.1 Let $\min\{x, y\}$ denote the minimum of the integers x and y, and $\max\{x, y\}$ their maximum. Then $\min\{x, y\} + \max\{x, y\} = x + y$.[§]

PROOF **(by cases)**

case 1 Let $x \leq y$. Then $\min\{x, y\} = x$ and $\max\{x, y\} = y$, so $\min\{x, y\} + \max\{x, y\} = x + y$.

case 2 Let $x > y$. Then $\min\{x, y\} = y$ and $\max\{x, y\} = x$, so $\min\{x, y\} + \max\{x, y\} = y + x = x + y$. ■

The law of trichotomy helps us to define the absolute value of an integer.

Absolute Value

The **absolute value** of a real number x, denoted by $|x|$, is defined by

$$|x| = \begin{cases} x & \text{if } x \geq 0 \\ -x & \text{otherwise} \end{cases}$$

For example, $|5| = 5$, $|-3| = -(-3) = 3$, $|\pi| = \pi$, and $|0| = 0$.

Geometrically, the absolute value of a number indicates its distance from the origin on the number line.

Although we are interested only in properties of integers, we often need to deal with rational and real numbers also. Floor and ceiling functions are two such number-theoretic functions. They have nice applications to discrete mathematics and computer science.

[†] The symbols $\leq$ and $\geq$ were introduced in 1734 by the French mathematician P. Bouguer.

[‡] A **theorem** is a (major) result that can be proven from axioms or previously known results.

[§] Theorem 1.1 is true even if x and y are real numbers.

Floor and Ceiling Functions

The **floor** of a real number x, denoted by $\lfloor x \rfloor$, is the greatest integer $\leq x$. The **ceiling** of x, denoted by $\lceil x \rceil$, is the least integer $\geq x$.[†] The floor of x rounds down x, whereas the ceiling of x rounds up. Accordingly, if $x \notin \mathbf{Z}$, the floor of x is the nearest integer to the left of x on the number line, and the ceiling of x is the nearest integer to the right of x, as Figure 1.2 shows. The **floor function** $f(x) = \lfloor x \rfloor$ and the **ceiling function** $g(x) = \lceil x \rceil$ are also known as the **greatest integer function** and the **least integer function**, respectively.

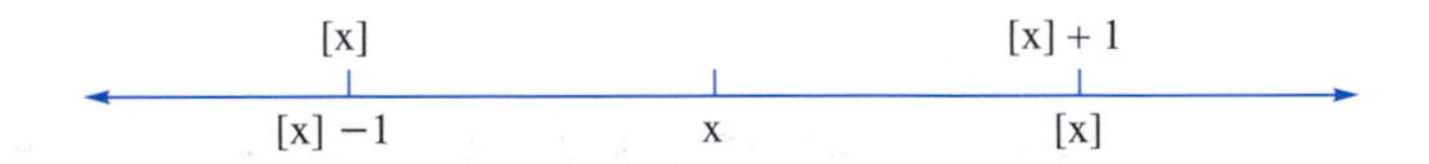

Figure 1.2

For example, $\lfloor \pi \rfloor = 3$, $\lfloor \log_{10} 3 \rfloor = 0$, $\lfloor -3.5 \rfloor = -4$, $\lfloor -2.7 \rfloor = -3$, $\lceil \pi \rceil = 4$, $\lceil \log_{10} 3 \rceil = 1$, $\lceil -3.5 \rceil = -3$, and $\lceil -2.7 \rceil = -2$.

The floor function comes in handy when real numbers are to be truncated or rounded off to a desired number of decimal places. For example, the real number $\pi = 3.1415926535\ldots$ truncated to three decimal places is given by $\lfloor 1000\pi \rfloor / 1000 = 3141/1000 = 3.141$; on the other hand, π rounded to three decimal places is $\lfloor 1000\pi + 0.5 \rfloor / 1000 = 3.142$.

There is yet another simple application of the floor function. Suppose we divide the unit interval $[0, 1)$ into 50 subintervals of equal length 0.02 and then seek to determine the subinterval that contains the number 0.4567. Since $\lfloor 0.4567/0.02 \rfloor + 1 = 23$, it lies in the 23rd subinterval. More generally, let $0 \leq x < 1$. Then x lies in the subinterval $\lfloor x/0.02 \rfloor + 1 = \lfloor 50x \rfloor + 1$.

The following example presents an application of the ceiling function to everyday life.

EXAMPLE 1.1 **(The post-office function)** In 2006, the postage rate in the United States for a first-class letter of weight x, not more than one ounce, was 39¢; the rate for each additional ounce or a fraction thereof up to 11 ounces was an additional 24¢. Thus, the postage $p(x)$ for a first-class letter can be defined as $p(x) = 0.39 + 0.24\lceil x - 1 \rceil$, $0 < x \leq 11$.

For instance, the postage for a letter weighing 7.8 ounces is $p(7.8) = 0.39 + 0.24\lceil 7.8 - 1 \rceil = \$2.07$. ■

[†] These two notations and the names, *floor* and *ceiling*, were introduced by Kenneth E. Iverson in the early 1960s. Both notations are variations of the original greatest integer notation $[x]$.

Some properties of the floor and ceiling functions are listed in the next theorem. We shall prove one of them; the others can be proved as routine exercises.

THEOREM 1.2 Let x be any real number and n any integer. Then

1. $\lfloor n \rfloor = n = \lceil n \rceil$
2. $\lceil x \rceil = \lfloor x \rfloor + 1$ $(x \notin \mathbf{Z})$
3. $\lfloor x + n \rfloor = \lfloor x \rfloor + n$
4. $\lceil x + n \rceil = \lceil x \rceil + n$
5. $\left\lfloor \frac{n}{2} \right\rfloor = \frac{n-1}{2}$ if n is odd.
6. $\left\lceil \frac{n}{2} \right\rceil = \frac{n+1}{2}$ if n is odd.

PROOF

Every real number x can be written as $x = k + x'$, where $k = \lfloor x \rfloor$ and $0 \le x' < 1$. See Figure 1.3. Then

Figure 1.3

$$x + n = k + n + x' = (k + n) + x'$$
$$\lfloor x + n \rfloor = k + n, \quad \text{since } 0 \le x' < 1$$
$$= \lfloor x \rfloor + n$$

■

EXERCISES 1.1

1. The English mathematician Augustus DeMorgan, who lived in the 19th century, once remarked that he was x years old in the year x^2. When was he born?

Evaluate each, where x is a real number.

2. $f(x) = \frac{x}{|x|}$ $(x \neq 0)$
3. $g(x) = \lfloor x \rfloor + \lfloor -x \rfloor$
4. $h(x) = \lceil x \rceil + \lceil -x \rceil$

Determine whether:

5. $-\lfloor -x \rfloor = \lfloor x \rfloor$
6. $-\lceil -x \rceil = \lceil x \rceil$
7. There are four integers between 100 and 1000 that are each equal to the sum of the cubes of its digits. Three of them are 153, 371, and 407. Find the fourth number. (Source unknown.)
8. An n-digit positive integer N is a **Kaprekar number** if the sum of the number formed by the last n digits in N^2, and the number formed by the first n (or $n-1$) digits in N^2 equals N. For example, 297 is a Kaprekar number since $297^2 = 88209$ and $88 + 209 = 297$. There are five Kaprekar numbers < 100. Find them.
9. Find the flaw in the following "proof":

   Let a and b be real numbers such that $a = b$. Then

   $$ab = b^2$$
   $$a^2 - ab = a^2 - b^2$$

   Factoring, $a(a - b) = (a + b)(a - b)$. Canceling $a - b$ from both sides, $a = a + b$. Since $a = b$, this yields $a = 2a$. Canceling a from both sides, we get $1 = 2$.

***D. R. Kaprekar** (1905–1986) was born in Dahanu, India, near Bombay. After losing his mother at the age of eight, he built a close relationship with his astrologer-father, who passed on his knowledge to his son. He attended Ferguson College in Pune, and then graduated from the University of Bombay in 1929. He was awarded the Wrangler R. P. Paranjpe prize in 1927 in recognition of his mathematical contributions. A prolific writer in recreational number theory, he worked as a schoolteacher in Devlali, India, from 1930 until his retirement in 1962.*

*Kaprekar is best known for his 1946 discovery of the **Kaprekar constant** 6174. It took him about three years to discover the number: Take a four-digit number a, not all digits being the same; let a' denote the number obtained by rearranging its digits in nondecreasing order and a'' denote the number obtained by rearranging its digits in nonincreasing order. Repeat these steps with $b = a' - a''$ and its successors. Within a maximum of eight steps, this process will terminate in 6174. It is the only integer with this property.*

10. Express 635,318,657 as the sum of two fourth powers in two different ways. (It is the smallest number with this property.)
11. The integer 1105 can be expressed as the sum of two squares in four different ways. Find them.
12. There is exactly one integer between 2 and 2×10^{14} that is a perfect square, a cube, and a fifth power. Find it. (A. J. Friedland, 1970)
13. The five-digit number $2xy89$ is the square of an integer. Find the two-digit number xy. (*Source: Mathematics Teacher*)
14. How many perfect squares can be displayed on a 15-digit calculator?
15. The number sequence 2, 3, 5, 6, 7, 10, 11, ... consists of positive integers that are neither squares nor cubes. Find the 500th term of this sequence. (*Source: Mathematics Teacher*)

Prove each, where a, b, and n are any integers, and x is a real number.

16. $|ab| = |a| \cdot |b|$
17. $|a+b| \le |a| + |b|$
18. $\left\lfloor \frac{n}{2} \right\rfloor = \frac{n-1}{2}$ if n is odd.
19. $\left\lceil \frac{n}{2} \right\rceil = \frac{n+1}{2}$ if n is odd.
20. $\left\lfloor \frac{n^2}{4} \right\rfloor = \frac{n^2-1}{4}$ if n is odd.
21. $\left\lceil \frac{n^2}{4} \right\rceil = \frac{n^2+3}{4}$ if n is odd.
22. $\left\lfloor \frac{n}{2} \right\rfloor + \left\lceil \frac{n}{2} \right\rceil = n$
23. $\lceil x \rceil = \lfloor x \rfloor + 1 \ (x \notin \mathbf{Z})$
24. $\lceil x \rceil = -\lfloor -x \rfloor$
25. $\lceil x + n \rceil = \lceil x \rceil + n$
26. $\lfloor x \rfloor + \lfloor x + 1/2 \rfloor = \lfloor 2x \rfloor$
27. $\lfloor \lfloor x \rfloor / n \rfloor = \lfloor x/n \rfloor$

The **distance** from x to y on the number line, denoted by $d(x, y)$, is defined by $d(x, y) = |y - x|$. Prove each, where x, y, and z are any integers.

28. $d(x, y) \ge 0$
29. $d(0, x) = |x|$
30. $d(x, y) = 0$ if and only if $x = y$
31. $d(x, y) = d(y, x)$
32. $d(x, y) \le d(x, z) + d(z, y)$
33. Let $\max\{x, y\}$ denote the maximum of x and y, and $\min\{x, y\}$ their minimum, where x and y are any integers. Prove that $\max\{x, y\} - \min\{x, y\} = |x - y|$.
34. A round-robin tournament has n teams, and each team plays at most once in a round. Determine the minimum number of rounds $f(n)$ needed to complete the tournament. (Romanian Olympiad, 1978)

Joseph Louis Lagrange *(1736–1813), who ranks with Leonhard Euler as one of the greatest mathematicians of the 18th century, was the eldest of eleven children in a wealthy family in Turin, Italy. His father, an influential cabinet official, became bankrupt due to unsuccessful financial speculations, which forced Lagrange to pursue a profession.*

As a young man studying the classics at the College of Turin, his interest in mathematics was kindled by an essay by astronomer Edmund Halley on the superiority of the analytical methods of calculus over geometry in the solution of optical problems. In 1754 he began corresponding with several outstanding mathematicians in Europe. The following year, Lagrange was appointed professor of mathematics at the Royal Artillery School in Turin. Three years later, he helped to found a society that later became the Turin Academy of Sciences. While at Turin, Lagrange developed revolutionary results in the calculus of variations, mechanics, sound, and probability, winning the prestigious Grand Prix of the Paris Academy of Sciences in 1764 and 1766.

In 1766, when Euler left the Berlin Academy of Sciences, Frederick the Great wrote to Lagrange that "the greatest king in Europe" would like to have "the greatest mathematician of Europe" at his court. Accepting the invitation, Lagrange moved to Berlin to head the Academy and remained there for 20 years. When Frederick died in 1786, Lagrange moved to Paris at the invitation of Louis XVI. Lagrange was appointed professor at the École Normale and then at the École Polytechnique, where he taught until 1799.

Lagrange made significant contributions to analysis, analytical mechanics, calculus, probability, and number theory, as well as helping to set up the French metric system.

1.2 The Summation and Product Notations

We will find both the summation and the product notations very useful throughout the remainder of this book. First, we turn to the summation notation.

The Summation Notation

Sums, such as $a_k + a_{k+1} + \cdots + a_m$, can be written in a compact form using the **summation symbol** $\sum$ (the Greek uppercase letter *sigma*), which denotes the word *sum*. The summation notation was introduced in 1772 by the French mathematician Joseph Louis Lagrange.

A typical term in the sum above can be denoted by a_i, so the above sum is the sum of the numbers a_i as i runs from k to m and is denoted by $\sum_{i=k}^{i=m} a_i$. Thus

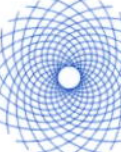

$$\sum_{i=k}^{i=m} a_i = a_k + a_{k+1} + \cdots + a_m$$

The variable i is the **summation index**. The values k and m are the **lower** and **upper limits** of the index i. The "$i =$" above the $\sum$ is usually omitted:

$$\sum_{i=k}^{i=m} a_i = \sum_{i=k}^{m} a_i$$

For example,

$$\sum_{i=-1}^{2} i(i-1) = (-1)(-1-1) + 0(0-1) + 1(1-1) + 2(2-1) = 4$$

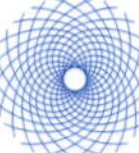

The index i is a **dummy variable**; we can use any variable as the index without affecting the value of the sum, so

$$\sum_{i=\ell}^{m} a_i = \sum_{j=\ell}^{m} a_j = \sum_{k=\ell}^{m} a_k$$

EXAMPLE 1.2 Evaluate $\sum_{j=-2}^{3} j^2$.

SOLUTION

$$\sum_{j=-2}^{3} j^2 = (-2)^2 + (-1)^2 + 0^2 + 1^2 + 2^2 + 3^2 = 19$$ ■

The following results are extremely useful in evaluating finite sums. They can be proven using mathematical induction, presented in Section 1.3.

THEOREM 1.3 Let n be any positive integer and c any real number, and $a_1, a_2, \ldots, a_n$ and $b_1, b_2, \ldots, b_n$ any two number sequences. Then

$$\sum_{i=1}^{n} c = nc \tag{1.1}$$

$$\sum_{i=1}^{n}(ca_i) = c\left(\sum_{i=1}^{n} a_i\right) \tag{1.2}$$

$$\sum_{i=1}^{n}(a_i + b_i) = \sum_{i=1}^{n} a_i + \sum_{i=1}^{n} b_i \tag{1.3}$$

(These results can be extended to any lower limit $k \in \mathbf{Z}$.) ■

The following example illustrates this theorem.

EXAMPLE 1.3 Evaluate $\sum_{j=-1}^{2} [(5j)^3 - 2j]$.

SOLUTION

$$\begin{aligned}\sum_{j=-1}^{2}[(5j)^3 - 2j] &= \sum_{j=-1}^{2}(5j)^3 - 2\left(\sum_{j=-1}^{2} j\right)\\ &= 125\left(\sum_{j=-1}^{2} j^3\right) - 2\sum_{j=-1}^{2} j\\ &= 125[(-1)^3 + 0^3 + 1^3 + 2^3] - 2(-1 + 0 + 1 + 2)\\ &= 996\end{aligned}$$

■

Indexed Summation

The summation notation can be extended to sequences with index sets I as their domains. For instance, $\sum_{i \in I} a_i$ denotes the sum of the values of a_i as i runs over the various values in I.

As an example, let $I = \{0, 1, 3, 5\}$. Then $\sum_{i \in I}(2i + 1)$ represents the sum of the values of $2i + 1$ with $i \in I$, so

$$\sum_{i \in I}(2i + 1) = (2 \cdot 0 + 1) + (2 \cdot 1 + 1) + (2 \cdot 3 + 1) + (2 \cdot 5 + 1) = 22$$

Often we need to evaluate sums of the form $\sum_{P} a_{ij}$, where the subscripts i and j satisfy certain properties P. (Such summations are used in Chapter 8.)

For example, let $I = \{1, 2, 3, 4\}$. Then $\sum_{1 \le i < j \le 4} (2i + 3j)$ denotes the sum of the values of $2i + 3j$, where $1 \le i < j \le 4$. This can be abbreviated as $\sum_{i<j}(2i + 3j)$ provided the index set is obvious from the context. To find this sum, we must consider every possible pair (i, j), where $i, j \in I$ and $i < j$. Thus,

$$\begin{aligned}\sum_{i<j}(2i+3j) &= (2 \cdot 1 + 3 \cdot 2) + (2 \cdot 1 + 3 \cdot 3) + (2 \cdot 1 + 3 \cdot 4) + (2 \cdot 2 + 3 \cdot 3) \\ &\quad + (2 \cdot 2 + 3 \cdot 4) + (2 \cdot 3 + 3 \cdot 4) \\ &= 80\end{aligned}$$

EXAMPLE 1.4 Evaluate $\sum_{\substack{d \ge 1 \\ d|6}} d$, where $d|6$ means d is a factor of 6.

SOLUTION

$$\begin{aligned}\sum_{\substack{d \ge 1 \\ d|6}} d &= \text{sum of positive integers } d, \text{ where } d \text{ is a factor of } 6 \\ &= \text{sum of positive factors of } 6 \\ &= 1 + 2 + 3 + 6 = 12\end{aligned}$$

■

Multiple summations arise often in mathematics. They are evaluated in a right-to-left fashion. For example, the double summation $\sum_i \sum_j a_{ij}$ is evaluated as $\sum_i \left(\sum_j a_{ij}\right)$, as demonstrated below.

EXAMPLE 1.5 Evaluate $\sum_{i=-1}^{1} \sum_{j=0}^{2} (2i + 3j)$.

SOLUTION

$$\begin{aligned}\sum_{i=-1}^{1} \sum_{j=0}^{2} (2i+3j) &= \sum_{i=-1}^{1} \left[\sum_{j=0}^{2} (2i+3j) \right] \\ &= \sum_{i=-1}^{1} \left[(2i + 3 \cdot 0) + (2i + 3 \cdot 1) + (2i + 3 \cdot 2) \right]\end{aligned}$$

$$= \sum_{i=-1}^{1} (6i+9)$$

$$= [6 \cdot (-1) + 9] + (6 \cdot 0 + 9) + (6 \cdot 1 + 9)$$

$$= 27$$

■

We now turn to the product notation.

The Product Notation

Just as $\sum$ is used to denote sums, the product $a_k a_{k+1} \cdots a_m$ is denoted by $\prod_{i=k}^{i=m} a_i$. The **product symbol** $\prod$ is the Greek capital letter *pi*. As in the case of the **summation notation**, the "$i =$" above the product symbol is often dropped:

$$\prod_{i=k}^{i=m} a_i = \prod_{i=k}^{m} a_i = a_k a_{k+1} \cdots a_m$$

Again, i is just a dummy variable.

The following three examples illustrate this notation.

The **factorial function**, which often arises in number theory, can be defined using the product symbol, as the following example shows.

EXAMPLE 1.6 The **factorial function** $f(n) = n!$ (read n factorial) is defined by $n! = n(n-1)\cdots 2 \cdot 1$, where $0! = 1$. Using the product notation, $f(n) = n! = \prod_{k=1}^{n} k$. ■

EXAMPLE 1.7 Evaluate $\prod_{i=2}^{5} (i^2 - 3)$.

SOLUTION

$$\prod_{i=2}^{5} (i^2 - 3) = (2^2 - 3)(3^2 - 3)(4^2 - 3)(5^2 - 3)$$

$$= 1 \cdot 6 \cdot 13 \cdot 22 = 1716$$

■

Just as we can have indexed summation, we can also have indexed multiplication, as the following example shows.

EXAMPLE 1.8 Evaluate $\prod_{\substack{i,j \in I \\ i<j}} (i+j)$, where $I = \{2, 3, 5, 7\}$.

SOLUTION

$$\begin{aligned} \text{Given product} &= \text{product of all numbers } i+j, \text{ where } i,j \in \{2,3,5,7\} \text{ and } i<j \\ &= (2+3)(2+5)(2+7)(3+5)(3+7)(5+7) \\ &= 5 \cdot 7 \cdot 9 \cdot 8 \cdot 10 \cdot 12 = 302{,}400 \end{aligned}$$ ■

The following exercises provide ample practice in both notations.

EXERCISES 1.2

Evaluate each sum.

1. $\sum_{i=1}^{6} i$
2. $\sum_{k=0}^{4} (3+k)$
3. $\sum_{j=0}^{4} (j-1)$
4. $\sum_{i=-1}^{4} 3$
5. $\sum_{n=0}^{4} (3n-2)$
6. $\sum_{j=-2}^{2} j(j-2)$
7. $\sum_{k=-2}^{4} 3k$
8. $\sum_{k=-2}^{3} 3(k^2)$
9. $\sum_{k=-1}^{3} (3k)^2$
10. $\sum_{k=1}^{5} (3-2k)k$

Rewrite each sum using the summation notation.

11. $1+3+5+\cdots+23$
12. $3^1+3^2+\cdots+3^{10}$
13. $1\cdot 2+2\cdot 3+\cdots+11\cdot 12$
14. $1(1+2)+2(2+2)+\cdots+5(5+2)$

Determine whether each is true.

15. $\sum_{i=m}^{n} i = \sum_{i=m}^{n} (n+m-i)$
16. $\sum_{i=m}^{n} x^i = \sum_{i=m}^{n} x^{n+m-i}$
17. Sums of the form $S = \sum_{i=m+1}^{n} (a_i - a_{i-1})$ are called **telescoping sums**. Show that $S = a_n - a_m$.
18. Using Exercise 17 and the identity $\frac{1}{i(i+1)} = \frac{1}{i} - \frac{1}{i+1}$, derive a formula for $\sum_{i=1}^{n} \frac{1}{i(i+1)}$.
19. Using Exercise 17 and the identity $(i+1)^2 - i^2 = 2i+1$, derive a formula for $\sum_{i=1}^{n} i$.
20. Using Exercise 17 and the identity $(i+1)^3 - i^3 = 3i^2+3i+1$, derive a formula for the sum $\sum_{i=1}^{n} i^2$.
21. Using the ideas in Exercises 19 and 20, derive a formula for $\sum_{i=1}^{n} i^3$.

Evaluate each.

22. $\sum_{i=1}^{5}\sum_{j=1}^{6} (2i+3j)$
23. $\sum_{i=1}^{3}\sum_{j=1}^{i} (i+3)$
24. $\sum_{i=1}^{5}\sum_{j=1}^{6} (i^2-j+1)$
25. $\sum_{j=1}^{6}\sum_{i=1}^{5} (i^2-j+1)$
26. $\prod_{i=0}^{3} (i+1)$
27. $\prod_{j=3}^{5} (j^2+1)$

28. $\prod_{k=0}^{50}(-1)^k$

Evaluate each, where $p \in \{2, 3, 5, 7, 11, 13\}$ and $I = \{1, 2, 3, 5\}$.

29. $\sum_{k=0}^{3} k!$

30. $\sum_{p\le 10} p$

31. $\prod_{p\le 10} p$

32. $\prod_{i\in I}(3i-1)$

33. $\sum_{\substack{d\ge 1\\ d|12}} d$

34. $\sum_{\substack{d\ge 1\\ d|12}} \binom{12}{d}$

35. $\sum_{\substack{d\ge 1\\ d|18}} 1$

36. $\sum_{p\le 10} 1$

37. $\prod_{\substack{i,j\in I\\ i<j}} (i+2j)$

38. $\prod_{\substack{i,j\in I\\ i\le j}} i^j$

39. $\sum_{\substack{i,j\in I\\ i|j}} (2^i+3^j)$

40. $\sum_{j=1}^{4}(3^j-3^{j-1})$

Expand each.

41. $\sum_{i=1}^{3}\sum_{j=1}^{2} a_{ij}$

42. $\sum_{j=1}^{2}\sum_{i=1}^{3} a_{ij}$

43. $\sum_{1\le i<j\le 3} (a_i+a_j)$

44. $\sum_{1\le i<j<k\le 3} (a_i+a_j+a_k)$

Evaluate each, where $\lg x = \log_2 x$.

45. $\sum_{n=1}^{1023} \lg(1+1/n)$

46. $\prod_{n=1}^{1023} (1+1/n)$

47. $\sum_{n=1}^{1024} \lfloor \lg(1+1/n)\rfloor$

48. $\sum_{k=1}^{n} k\cdot k!$ (*Hint*: Use Exercise 17.)

49. Find the tens digit in the sum $\sum_{k=1}^{999} k!$.

50. Find the hundreds digit in the sum $\sum_{k=1}^{999} k\cdot k!$.
(*Hint*: Use Exercise 48.)

*51. Compute $\sum_{n=0}^{\infty}\left\lfloor \dfrac{10000+2^n}{2^{n+1}}\right\rfloor$.
(*Hint*: $\lfloor x+1/2\rfloor = \lfloor 2x\rfloor - \lfloor x\rfloor$; *Source: Mathematics Teacher*, 1993.)

1.3 Mathematical Induction

The principle of mathematical induction[†] (PMI) is a powerful proof technique that we will use often in later chapters.

Many interesting results in mathematics hold true for all positive integers. For example, the following statements are true for every positive integer n and all real numbers x, y, and x_i:

- $(x\cdot y)^n = x^n \cdot y^n$
- $\log(x_1\cdots x_n) = \sum_{i=1}^{n} \log x_i$

[†] The term *mathematical induction* was coined by Augustus DeMorgan (1806–1871), although the Venetian scientist Francesco Maurocylus (1491–1575) applied it much earlier, in proofs in a book he wrote in 1575.

- $\sum_{i=1}^{n} i = \frac{n(n+1)}{2}$
- $\sum_{i=0}^{n-1} r^i = \frac{r^n - 1}{r - 1} \quad (r \neq 1)$

How do we prove that these results hold for every positive integer n? Obviously, it is impossible to substitute each positive integer for n and verify that the formula holds. The principle of induction can establish the validity of such formulas.

Before we plunge into induction, we need the well-ordering principle, which we accept as an axiom. (An **axiom** is a statement that is accepted as true; it is consistent with known facts; often it is a self-evident statement.)

The Well-Ordering Principle

Every nonempty set of positive integers has a least element.

For example, the set $\{17, 23, 5, 18, 13\}$ has a least element, namely, 5. The elements of the set can be ordered as 5, 13, 17, 18, and 23.

By virtue of the well-ordering principle, the set of positive integers is well ordered. You may notice that the set of negative integers is not well ordered.

The following example is a simple application of the well-ordering principle.

EXAMPLE 1.9 Prove that there is no positive integer between 0 and 1.

PROOF **(by contradiction)**

Suppose there is a positive integer a between 0 and 1. Let $S = \{n \in \mathbf{Z}^+ \mid 0 < n < 1\}$. Since $0 < a < 1$, $a \in S$, so S is nonempty. Therefore, by the well-ordering principle, S has a least element ℓ, where $0 < \ell < 1$. Then $0 < \ell^2 < \ell$, so $\ell^2 \in S$. But $\ell^2 < \ell$, which contradicts our assumption that ℓ is a least element of S. Thus, there are no positive integers between 0 and 1. ■

The well-ordering principle can be extended to whole numbers also, as the following example shows.

EXAMPLE 1.10 Prove that every nonempty set of nonnegative integers has a least element.

PROOF **(by cases)**

Let S be a set of nonnegative integers.

case 1 Suppose $0 \in S$. Since 0 is less than every positive integer, 0 is less than every nonzero element in S, so 0 is a least element in S.

case 2 Suppose $0 \notin S$. Then S contains only positive integers. So, by the well-ordering principle, S contains a least element.

Thus, in both cases, S contains a least element. ■

Weak Version of Induction

The following theorem is the cornerstone of the principle of induction.

THEOREM 1.4 Let S be a set of positive integers satisfying the following properties:

1. $1 \in S$.
2. If k is an arbitrary positive integer in S, then $k+1 \in S$.

Then $S = \mathbf{N}$.

PROOF **(by contradiction)**
Suppose $S \neq \mathbf{N}$. Let $S' = \{n \in \mathbf{N} \mid n \notin S\}$. Since $S' \neq \varnothing$, by the well-ordering principle, S' contains a least element ℓ'. Then $\ell' > 1$ by condition (1). Since ℓ' is the least element in S', $\ell' - 1 \notin S'$. Therefore, $\ell' - 1 \in S$. Consequently, by condition (2), $(\ell' - 1) + 1 = \ell' \in S$. This contradiction establishes the theorem. ■

This result can be generalized, as the following theorem shows. We leave its proof as an exercise.

THEOREM 1.5 Let n_0 be a fixed integer. Let S be a set of integers satisfying the following conditions:

- $n_0 \in S$.
- If k is an arbitrary integer $\geq n_0$ such that $k \in S$, then $k+1 \in S$.

Then S contains all integers $n \geq n_0$.

Before we formalize the principle of induction, let's look at a trivial example. Consider an infinite number of identical dominoes arranged in a row at varying distances from each other, as in Figure 1.4(a). Suppose we knock down the first domino. What happens to the rest of the dominoes? Do they all fall? Not necessarily. See Figures 1.4(b) and 1.4(c).

So let us assume the following: The dominoes are placed in such a way that the distance between two adjacent dominoes is less than the length of a domino; the first domino falls; and if the kth domino falls, then the $(k+1)$st domino also falls. Then they all would fall. See Figure 1.4(d).

This illustration can be expressed symbolically. Let $P(n)$ denote the statement that the nth domino falls. Assume the following statements are true:

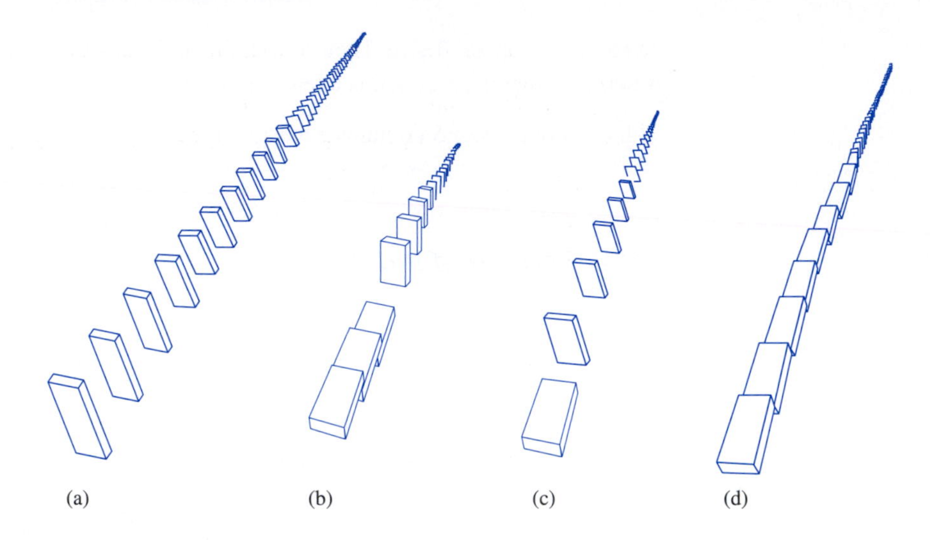

Figure 1.4

- $P(1)$.
- $P(k)$ implies $P(k+1)$ for an arbitrary positive integer k.

Then $P(n)$ is true for every positive integer n; that is, every domino would fall. This is the essence of the following weak version of the principle.

THEOREM 1.6 **(The Principle of Mathematical Induction)** Let $P(n)$ be a statement satisfying the following conditions, where $n \in \mathbf{Z}$:

1. $P(n_0)$ is true for some integer n_0.
2. If $P(k)$ is true for an arbitrary integer $k \geq n_0$, then $P(k+1)$ is also true.

Then $P(n)$ is true for every integer $n \geq n_0$.

PROOF

Let S denote the set of integers $\geq n_0$ for which $P(n)$ is true. Since $P(n_0)$ is true, $n_0 \in S$. By condition (2), whenever $k \in S$, $k+1 \in S$, so, by Theorem 1.5, S contains all integers $\geq n_0$. Consequently, $P(n)$ is true for every integer $n \geq n_0$. ■

Condition (1) in Theorem 1.6 assumes the proposition $P(n)$ is true when $n = n_0$. Look at condition (2): If $P(n)$ is true for an arbitrary integer $k \geq n_0$, it is also true for $n = k+1$. Then, by repeated application of condition (2), it follows that $P(n_0+1)$, $P(n_0+2), \ldots$ hold true. In other words, $P(n)$ holds for every $n \geq n_0$.

Theorem 1.6 can be established directly from the well-ordering principle. See Exercise 44.

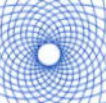

Proving a result by induction involves two key steps:

- **basis step** Verify that $P(n_0)$ is true.
- **induction step** Assume $P(k)$ is true for an arbitrary integer $k \geq n_0$ (**inductive hypothesis**).
  Then verify that $P(k+1)$ is also true.

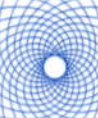

A word of caution: A question frequently asked is, "Isn't this circular reasoning? Aren't we assuming what we are asked to prove?" In fact, no. The confusion stems from misinterpreting step 2 for the conclusion. The induction step involves showing that $P(k)$ implies $P(k+1)$; that is, if $P(k)$ is true, then so is $P(k+1)$. The conclusion is "$P(n)$ is true for every $n \geq n_0$." So be careful.

Interestingly, there were television commercials for *Crest* toothpaste based on induction involving toothpastes and penguins.

Some examples will show how useful this important proof technique is.

EXAMPLE 1.11 Prove that

$$1 + 2 + 3 + \cdots + n = \frac{n(n+1)}{2} \tag{1.4}$$

for every positive integer n.

PROOF (by induction)

Let $P(n)$ be the statement that $\sum_{i=1}^{n} i = [n(n+1)]/2$.

basis step To verify that $P(1)$ is true (*note*: Here $n_0 = 1$):

When $n = 1$, RHS $= [1(1+1)]/2 = 1 = \sum_{i=1}^{1} i =$ LHS.[†] Thus, $P(1)$ is true.

[†] LHS and RHS are abbreviations of *left-hand side* and *right-hand side*, respectively.

induction step Let k be an arbitrary positive integer. We would like to show that $P(k)$ implies $P(k+1)$. Assume $P(k)$ is true; that is,

$$\sum_{i=1}^{k} i = \frac{k(k+1)}{2} \qquad \leftarrow \text{inductive hypothesis}$$

To show that $P(k)$ implies $P(k+1)$, that is, $\sum_{i=1}^{k+1} i = [(k+1)(k+2)]/2$, we start with the LHS of this equation:

$$\begin{aligned}
\text{LHS} = \sum_{i=1}^{k+1} i &= \sum_{i=1}^{k} i + (k+1) \qquad \left[\textit{Note}: \sum_{i=1}^{k+1} x_i = \left(\sum_{i=1}^{k} x_i\right) + x_{k+1}.\right] \\
&= \frac{k(k+1)}{2} + (k+1), \quad \text{by the inductive hypothesis} \\
&= \frac{(k+1)(k+2)}{2} \\
&= \text{RHS}
\end{aligned}$$

So, if $P(k)$ is true, then $P(k+1)$ is also true.

Thus, by induction, $P(n)$ is true for every integer $n \geq 1$; that is, the formula holds for every positive integer. ■

Figure 1.5 demonstrates formula (1.4) without words.

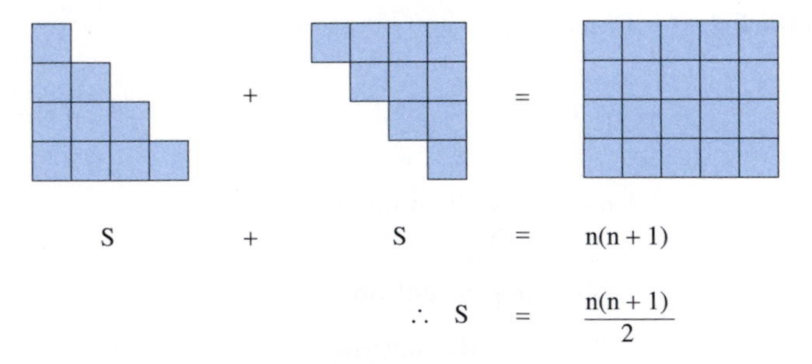

Figure 1.5

Often we arrive at a formula by studying patterns, then making a conjecture, and then establishing the formula by induction, as the following example shows.

EXAMPLE 1.12 Conjecture a formula for the sum of the first n odd positive integers and then use induction to establish the conjecture.

SOLUTION

First, we study the first five such sums, and then look for a pattern, to predict a formula for the sum of the first n odd positive integers.

The first five such sums are

$$1 = 1^2$$
$$1 + 3 = 2^2$$
$$1 + 3 + 5 = 3^2$$
$$1 + 3 + 5 + 7 = 4^2$$
$$1 + 3 + 5 + 7 + 9 = 5^2$$

There is a clear pattern here, so we conjecture that the sum of the first n odd positive integers is n^2; that is,

$$\sum_{i=1}^{n}(2i - 1) = n^2 \tag{1.5}$$

We shall now prove it by the principle of induction.

PROOF

When $n = 1$, $\sum_{i=1}^{n}(2i - 1) = \sum_{i=1}^{1}(2i - 1) = 1 = 1^2$, so the result holds when $n = 1$.

Now, assume the formula holds when $n = k$: $\sum_{i=1}^{k}(2i - 1) = k^2$. To show that it holds when $n = k + 1$, consider the sum $\sum_{i=1}^{k+1}(2i - 1)$. We have

$$\begin{aligned}\sum_{i=1}^{k+1}(2i - 1) &= \sum_{i=1}^{k}(2i - 1) + [2(k + 1) - 1]\\ &= k^2 + (2k + 1) \quad \text{by the inductive hypothesis}\\ &= (k + 1)^2\end{aligned}$$

Consequently, if the formula holds when $n = k$, it is also true when $n = k + 1$.

Thus, by induction, the formula holds for every positive integer n. ■

Figure 1.6 provides a visual illustration of formula (1.5).

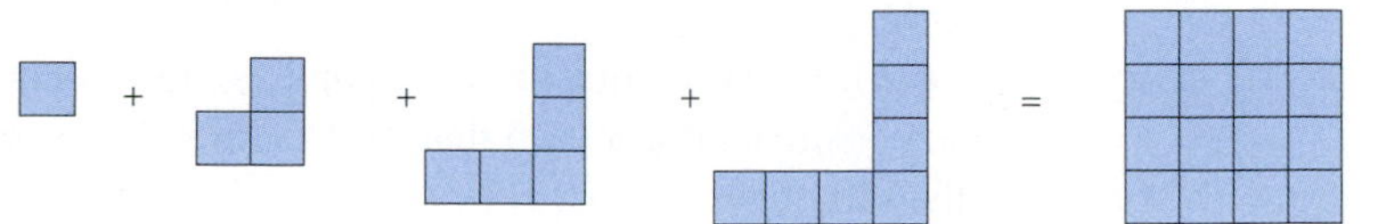

Figure 1.6

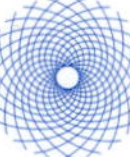

Returning to induction, we find that both the basis and the induction steps are essential in the induction proof, as the following two examples demonstrate.

EXAMPLE 1.13 Consider the "formula" $1+3+5+\cdots+(2n-1)=(n-2)^2$. Clearly it is true when $n=1$. But it is not true when $n=2$.

Conclusion? That the truth of the basis step does *not* ensure that the statement $1+3+5+\cdots+(2n-1)=(n-2)^2$ is true for every n.

The following example shows that the validity of the induction step is necessary, but not sufficient, to guarantee that $P(n)$ is true for all desired integers.

EXAMPLE 1.14 Consider the "formula" $P(n)$: $1+3+5+\cdots+(2n-1)=n^2+1$. Suppose $P(k)$ is true: $\sum_{i=1}^{k}(2i-1)=k^2+1$. Then

$$\begin{aligned}\sum_{i=1}^{k+1}(2i-1) &= \sum_{i=1}^{k}(2i-1)+[2(k+1)-1]\\ &= (k^2+1)+(2k+1)\\ &= (k+1)^2+1\end{aligned}$$

So if $P(k)$ is true, $P(k+1)$ is true. Nevertheless, the formula does not hold for any positive integer n. Try $P(1)$. ■

An interesting digression: Using induction, we "prove" in the following example that every person is of the same sex.

EXAMPLE 1.15 "Prove" that every person in a set of n people is of the same sex.

PROOF

Let $P(n)$: Everyone in a set of n people is of the same sex. Clearly, $P(1)$ is true. Let k be a positive integer such that $P(k)$ is true; that is, everyone in a set of k people is of the same sex.

To show that $P(k+1)$ is true, consider a set $A=\{a_1,a_2,\ldots,a_{k+1}\}$ of $k+1$ people. Partition A into two overlapping sets, $B=\{a_1,\ a_2,\ldots,\ a_k\}$ and $C=$

$\{a_2, \ldots, a_{k+1}\}$, as in Figure 1.7. Since B and C contain k elements, by the inductive hypothesis, everyone in B is of the same sex and everyone in C is of the same sex. Since B and C overlap, everyone in $B \cup C$† must be of the same sex; that is, everyone in A is of the same sex.

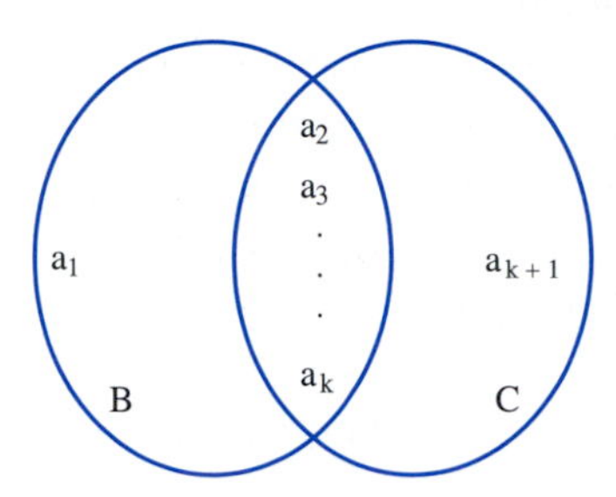

Figure 1.7

Therefore, by induction, $P(n)$ is true for every positive integer n. ■

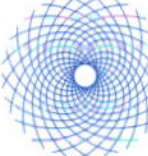

Note: Clearly the assertion that everyone is of the same sex is false. Can you find the flaw in the "proof?" See Exercise 35.

Strong Version of Induction

We now present the stronger version of induction.

Sometimes the truth of $P(k)$ might not be enough to establish that of $P(k+1)$. In other words, the truth of $P(k+1)$ may require more than that of $P(k)$. In such cases, we assume a stronger inductive hypothesis that $P(n_0), P(n_0+1), \ldots, P(k)$ are all true; then verify that $P(k+1)$ is also true. This **strong version**, which can be proven using the weak version (see Exercise 43), is stated as follows.

THEOREM 1.7 **(The Second Principle of Mathematical Induction)** Let $P(n)$ be a statement satisfying the following conditions, where $n \in \mathbf{Z}$:

1. $P(n_0)$ is true for some integer n_0.

† $B \cup C$ denotes the **union** of the sets B and C; it contains the elements in B together with those in C.

2. If k is an arbitrary integer $\geq n_0$ such that $P(n_0)$, $P(n_0+1), \ldots$, and $P(k)$ are true, then $P(k+1)$ is also true.

Then $P(n)$ is true for every integer $n \geq n_0$.

PROOF

Let $S = \{n \in \mathbf{Z} \mid P(n) \text{ is true}\}$. Since $P(n_0)$ is true by condition (1), $n_0 \in S$.

Now, assume $P(n_0), P(n_0+1), \ldots, P(k)$ are true for an arbitrary integer k. Then $n_0, n_0+1, \ldots, k$ belong to S. So, by condition (2), $k+1$ also belongs to S. Therefore, by Theorem 1.5, S contains all integers $n \geq n_0$. In other words, $P(n)$ is true for every integer $n \geq n_0$. ■

The following example illustrates this proof technique.

EXAMPLE 1.16 Prove that any postage of n (≥ 2) cents can be made with two- and three-cent stamps.

PROOF (by strong induction)

Let $P(n)$ denote the statement that any postage of n cents can be made with two- and three-cent stamps.

basis step (Notice that here $n_0 = 2$.) Since a postage of two cents can be made with one two-cent stamp, $P(2)$ is true. Likewise, $P(3)$ is also true.

induction step Assume $P(2), P(3), P(4), \ldots, P(k)$ are true; that is, any postage of two through k cents can be made with two- and three-cent stamps.

To show that $P(k+1)$ is true, consider a postage of $k+1$ cents. Since $k+1 = (k-1)+2$, a postage of $k+1$ cents can be formed with two- and three-cent stamps if a postage of $k-1$ cents can be made with two- and three-cent stamps. Since $P(k-1)$ is true by the inductive hypothesis, this implies $P(k+1)$ is also true.

Thus, by the strong version of induction, $P(n)$ is true for every $n \geq 2$; that is, any postage of n (≥ 2) cents can be made with two- and three-cent stamps. ■

The following exercises and subsequent chapters offer ample practice in both versions of induction.

EXERCISES 1.3

Determine whether each set is well ordered. If it is not, explain why.

1. Set of negative integers.
2. Set of integers.
3. $\{n \in \mathbf{N} \mid n \geq 5\}$
4. $\{n \in \mathbf{Z} \mid n \geq -3\}$

Prove each.

5. Let $a \in \mathbf{Z}$. There are no integers between a and $a+1$.

6. Let $n_0 \in \mathbf{Z}$, S a nonempty subset of the set $T = \{n \in \mathbf{Z} \mid n \geq n_0\}$, and ℓ^* be a least element of the set $T^* = \{n - n_0 + 1 \mid n \in S\}$. Then $n_0 + \ell^\star - 1$ is a least element of S.
7. **(Archimedean property)** Let a and b be any positive integers. Then there is a positive integer n such that $na \geq b$.
   (*Hint*: Use the well-ordering principle and contradiction.)
8. Every nonempty set of negative integers has a largest element.
9. Every nonempty set of integers $\leq$ a fixed integer n_0 has a largest element.

(Twelve Days of Christmas) Suppose you sent your love 1 gift on the first day of Christmas, $1 + 2$ gifts on the second day, $1 + 2 + 3$ gifts on the third day, and so on.

10. How many gifts did you send on the 12th day of Christmas?
11. How many gifts did your love receive in the 12 days of Christmas?
12. Prove that $1 + 2 + \cdots + n = [n(n + 1)]/2$ by considering the sum in the reverse order.[†] (Do not use mathematical induction.)

Using mathematical induction, prove each for every integer $n \geq 1$.

13. $\sum_{i=1}^{n} (2i - 1) = n^2$
14. $\sum_{i=1}^{n} i^2 = \dfrac{n(n+1)(2n+1)}{6}$
15. $\sum_{i=1}^{n} i^3 = \left[\dfrac{n(n+1)}{2}\right]^2$
16. $\sum_{i=1}^{n} ar^{i-1} = \dfrac{a(r^n - 1)}{r - 1}, \quad r \neq 1$

Evaluate each sum.

17. $\sum_{k=1}^{30} (3k^2 - 1)$
18. $\sum_{k=1}^{50} (k^3 + 2)$
19. $\sum_{i=1}^{n} \lfloor i/2 \rfloor$
20. $\sum_{i=1}^{n} \lceil i/2 \rceil$

Find the value of x resulting from executing each algorithm fragment, where

$$\textit{variable} \leftarrow \textit{expression}$$

means the value of *expression* is assigned to *variable*.

21. $x \leftarrow 0$
    for $i = 1$ to n do
    $x \leftarrow x + (2i - 1)$
22. $x \leftarrow 0$
    for $i = 1$ to n do
    $x \leftarrow x + i(i + 1)$
23. $x \leftarrow 0$
    for $i = 1$ to n do
    for $j = 1$ to i do
    $x \leftarrow x + 1$

Evaluate each.

24. $\sum_{i=1}^{n} \sum_{j=1}^{i} i$
25. $\sum_{i=1}^{n} \sum_{j=1}^{i} j$
26. $\sum_{i=1}^{n} \sum_{j=1}^{i} j^2$
27. $\sum_{i=1}^{n} \sum_{j=1}^{i} (2j - 1)$
28. $\prod_{i=1}^{n} 2^{2i}$
29. $\prod_{i=1}^{n} i^2$
30. $\prod_{i=1}^{n} \prod_{j=1}^{n} i^j$
31. $\prod_{i=1}^{n} \prod_{j=1}^{n} 2^{i+j}$
32. A **magic square** of order n is a square arrangement of the positive integers 1 through n^2 such that the sum of the integers along each row, column, and diagonal is a constant k, called the **magic constant**. Figure 1.8 shows two magic squares, one of order 3 and the other of order 4. Prove that the magic constant of a magic square of order n is $\dfrac{n(n^2+1)}{2}$.

[†] An interesting personal anecdote is told about Gauss. When Gauss was a fourth grader, he and his classmates were asked by his teacher to compute the sum of the first 100 positive integers. Supposedly, the teacher did so to get some time to grade papers. To the teacher's dismay, Gauss found the answer in a few moments by pairing the numbers from both ends:

$$1 + 2 + 3 + \ldots + 50 + 51 + \ldots + 98 + 99 + 100$$

The sum of each pair is 101 and there are 50 pairs. So the total sum is $50 \cdot 101 = 5050$.

8	1	6
3	5	7
4	9	2

k = 15

1	14	15	4
12	7	6	9
8	11	10	5
13	2	3	16

k = 34

Figure 1.8

According to legend, King Shirham of India was so pleased by the invention of chess that he offered to give Sissa Ben Dahir, its inventor, anything he wished. Dahir's request was a seemingly modest one: one grain of wheat on the first square of a chessboard, two on the second, four on the third, and so on. The king was delighted with this simple request but soon realized he could never fulfill it. The last square alone would take $2^{63} = 9{,}223{,}372{,}036{,}854{,}775{,}808$ grains of wheat. Find the following for an $n \times n$ chessboard.

33. The number of grains on the last square.
34. The total number of grains on the chessboard.
35. Find the flaw in the "proof" in Example 1.15.

Find the number of times the **assignment statement** $x \leftarrow x + 1$ is executed by each loop.

36. for $i = 1$ to n do
    for $j = 1$ to i do
        $x \leftarrow x + 1$
37. for $i = 1$ to n do
    for $j = 1$ to i do
        for $k = 1$ to i do
            $x \leftarrow x + 1$
38. for $i = 1$ to n do
    for $j = 1$ to i do
        for $k = 1$ to j do
            $x \leftarrow x + 1$
39. for $i = 1$ to n do
    for $j = 1$ to i do
        for $k = 1$ to i do
            for $l = 1$ to i do
                $x \leftarrow x + 1$
40. Let a_n denote the number of times the statement $x \leftarrow x + 1$ is executed in the following loop:

    for $i = 1$ to n do
        for $j = 1$ to $\lfloor i/2 \rfloor$ do
            $x \leftarrow x + 1$

    Show that $a_n = \begin{cases} \dfrac{n^2}{4} & \text{if } n \text{ is even} \\ \dfrac{n^2 - 1}{4} & \text{otherwise.} \end{cases}$

Evaluate each.

41. $\sum_{n=1}^{1024} \lfloor \lg n \rfloor$
42. $\sum_{n=1}^{1024} \lceil \lg n \rceil$

*43. Prove the strong version of induction, using the weak version.

*44. Prove the weak version of induction, using the well-ordering principle.

**45. Let S_n denote the sum of the elements in the nth set of the sequence of sets of squares $\{1\}$, $\{4, 9\}$, $\{16, 25, 36\}, \ldots$. Find a formula for S_n. (J. M. Howell, 1989)

1.4 Recursion

Recursion is one of the most elegant problem-solving techniques. It is so powerful a tool that most programming languages support it.

We begin with the well-known **handshake problem**:

There are n guests at a party. Each person shakes hands with everybody else exactly once. How many handshakes are made?

If we decide to solve a problem such as this, the solution may not be obvious. However, it is possible that the problem could be defined in terms of a simpler version of itself. Such a definition is an **inductive definition**. Consequently, the given problem can be solved provided the simpler version can be solved. This idea is pictorially represented in Figure 1.9.

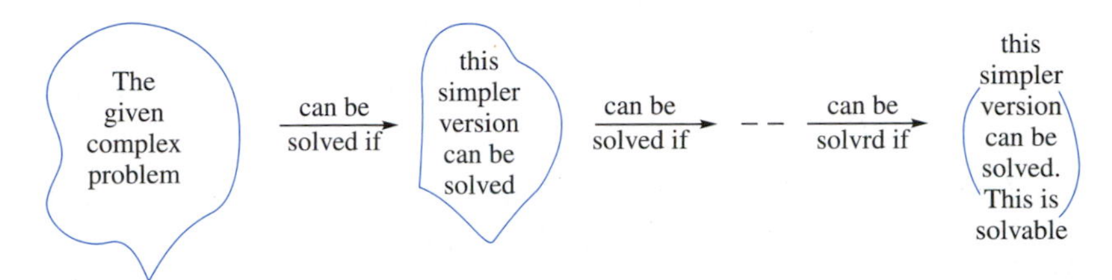

Figure 1.9

Recursive Definition of a Function

Let $a \in \mathbf{W}$ and $X = \{a, a+1, a+2, \ldots\}$. An **inductive definition** of a function f with domain X consists of three parts:

- **Basis step** A few initial values $f(a), f(a+1), \ldots, f(a+k-1)$ are specified. Equations that specify such initial values are **initial conditions**.
- **Recursive step** A formula to compute $f(n)$ from the k preceding functional values $f(n-1), f(n-2), \ldots, f(n-k)$ is made. Such a formula is a **recurrence relation** (or **recursive formula**).
- **Terminal step** Only values thus obtained are valid functional values. (For convenience, we drop this clause from the recursive definition.)

In a **recursive definition** of f, $f(n)$ may be defined using the values $f(k)$, where $k \neq n$, so *not* all recursively defined functions can be defined inductively; see Exercises 25–31.

Thus, the recursive definition of f consists of a finite number of initial conditions and a recurrence relation.

Recursion can be employed to find the minimum and maximum of three or more real numbers. For instance, $\min\{w, x, y, z\} = \min\{w, \{\min\{x, \min\{y, z\}\}\}\}$; $\max\{w, x, y, z\}$ can be evaluated similarly. For example,

$$\min\{23, 5, -6, 47, 31\} = \min\{23, \min\{5, \min\{-6, \min\{47, 31\}\}\}\} = -6$$

and

$$\max\{23, 5, -6, 47, 31\} = \max\{23, \max\{5, \max\{-6, \max\{47, 31\}\}\}\} = 47$$

The next three examples illustrate the recursive definition.

EXAMPLE 1.17 Define recursively the factorial function f.

SOLUTION

Recall that the factorial function f is defined by $f(n) = n!$, where $f(0) = 1$. Since $n! = n(n-1)!$, it can be defined recursively as follows:

$$f(0) = 1 \qquad \leftarrow \text{initial condition}$$
$$f(n) = n \cdot f(n-1), \quad n \geq 1 \qquad \leftarrow \text{recurrence relation}$$

■

Suppose we would like to compute $f(3)$ recursively. We must continue to apply the recurrence relation until the initial condition is reached, as shown below:

$$f(3) = 3 \cdot f(2) \tag{1.6}$$

return value

$$f(2) = 2 \cdot f(1) \tag{1.7}$$

return value

$$f(1) = 1 \cdot f(0) \tag{1.8}$$

return value

$$f(0) = 1 \tag{1.9}$$

Since $f(0) = 1$, 1 is substituted for $f(0)$ in equation (1.8) and $f(1)$ is computed: $f(1) = 1 \cdot f(0) = 1 \cdot 1 = 1$. This value is substituted for $f(1)$ in equation (1.7) and $f(2)$ is computed: $f(2) = 2 \cdot f(1) = 2 \cdot 1 = 2$. This value is now returned to equation (1.6) to compute $f(3)$: $f(3) = 3 \cdot f(2) = 3 \cdot 2 = 6$, as expected.

We now return to the handshake problem.

EXAMPLE 1.18 **(The handshake problem)** There are n guests at a party. Each person shakes hands with everybody else exactly once. Define recursively the number of handshakes $h(n)$ made.

SOLUTION

Clearly, $h(1) = 0$, so let $n \geq 2$. Let x be one of the guests. The number of handshakes made by the remaining $n-1$ guests among themselves, by definition, is $h(n-1)$. Now person x shakes hands with each of these $n-1$ guests, yielding $n-1$ handshakes. So the total number of handshakes made equals $h(n-1) + (n-1)$, where $n \geq 2$.

Thus, $h(n)$ can be defined recursively as follows:

$$h(1) = 0 \qquad \leftarrow \text{initial condition}$$
$$h(n) = h(n-1) + (n-1), \quad n \geq 2 \qquad \leftarrow \text{recurrence relation}$$

■

EXAMPLE 1.19 **(Tower of Brahma[†])** According to a legend, at the beginning of creation, God stacked 64 golden disks on one of three diamond pegs on a brass platform in the temple of Brahma at Benares,[‡] India (see Figure 1.10). The priests on duty were asked to move the disks from peg X to peg Z, using Y as an auxiliary peg, under the following conditions:

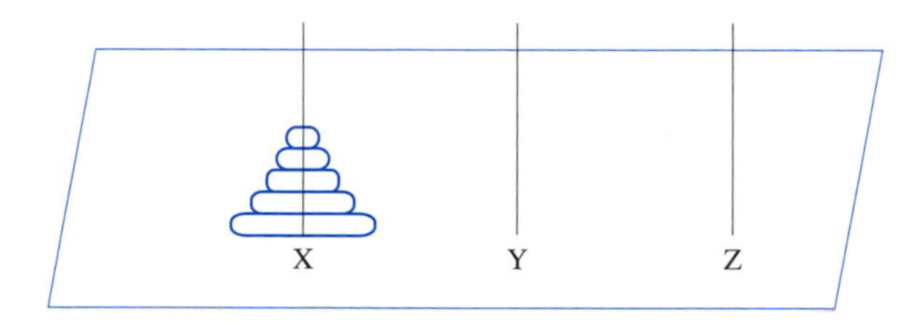

Figure 1.10

- Only one disk can be moved at a time.
- No disk can be placed on the top of a smaller disk.

The priests were told the world would end when the job was completed.

Suppose there are n disks on peg X. Let b_n denote the number of moves needed to move them from peg X to peg Z, using peg Y as an intermediary. Define b_n recursively.

SOLUTION

Clearly $b_1 = 1$. Assume $n \geq 2$. Consider the top $n-1$ disks at peg X. By definition, it takes b_{n-1} moves to transfer them from X to Y using Z as an auxiliary. That leaves the largest disk at peg X; it takes one move to transfer it from X to Z. See Figure 1.11.

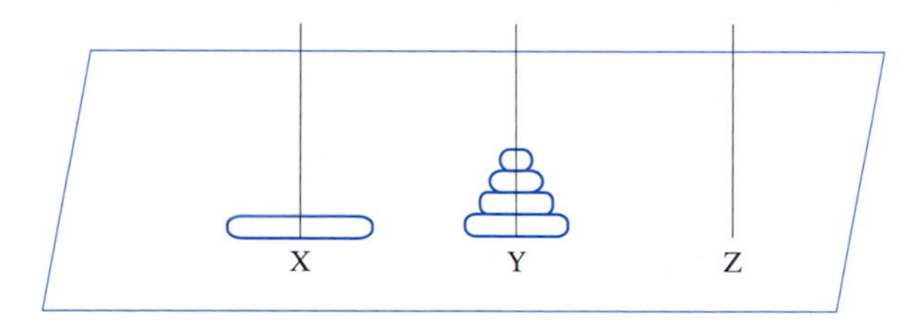

Figure 1.11

[†] A puzzle based on the Tower of Brahma was marketed by the French mathematician François-Edouard-Anatole Lucas in 1883 under the name **Tower of Hanoi**.

[‡] Benares is now known as Varanasi.

Now the $n-1$ disks at Y can be moved from Y to Z using X as an intermediary in b_{n-1} moves, so the total number of moves needed is $b_{n-1}+1+b_{n-1}=2b_{n-1}+1$. Thus b_n can be defined recursively as follows:

$$b_n = \begin{cases} 1 & \text{if } n = 1 \quad \leftarrow \text{initial condition} \\ 2b_{n-1}+1 & \text{if } n \geq 2 \quad \leftarrow \text{recurrence relation} \end{cases}$$ ■

For example,

$$\begin{aligned} b_4 &= 2b_3 + 1 &&= 2[2b_2+1]+1 \\ &= 4b_2 + 2 + 1 &&= 4[2b_1+1]+2+1 \\ &= 8b_1 + 4 + 2 + 1 &&= 8(1)+4+2+1 = 15, \end{aligned}$$

so it takes 15 moves to transfer 4 disks from X to Z.

Notice that the recursive definition of a function f does not provide us with an explicit formula for $f(n)$ but establishes a systematic procedure for finding it. The **iterative method** of finding a formula for $f(n)$ involves two steps: 1) apply the recurrence formula iteratively and look for a pattern to predict an explicit formula; 2) use induction to prove that the formula does indeed hold for every possible value of the integer n.

The following example illustrates this method.

EXAMPLE 1.20 Solve the recurrence relation in Example 1.18.

SOLUTION

Using iteration, we have:

$$\begin{aligned} h(n) &= h(n-1) + (n-1) \\ &= h(n-2) + (n-2) + (n-1) \\ &= h(n-3) + (n-3) + (n-2) + (n-1) \\ &\ \vdots \\ &= h(1) + 1 + 2 + 3 + \cdots + (n-2) + (n-1) \\ &= 0 + 1 + 2 + 3 + \cdots + (n-1) \\ &= \frac{n(n-1)}{2}, \quad \text{by Example 1.11} \end{aligned}$$

(We can verify this using induction.) ■

EXERCISES 1.4

In Exercises 1–6, compute the first four terms of the sequence defined recursively.

1. $a_1 = 1$
   $a_n = a_{n-1} + 3, n \geq 2$
2. $a_0 = 1$
   $a_n = a_{n-1} + n, n \geq 1$
3. $a_1 = 1$
   $a_n = \dfrac{n}{n-1} a_{n-1}, n \geq 2$
4. $a_1 = 1, a_2 = 2$
   $a_n = a_{n-1} + a_{n-2}, n \geq 3$
5. $a_1 = 1, a_2 = 1, a_3 = 2$
   $a_n = a_{n-1} + a_{n-2} + a_{n-3}, n \geq 4$
6. $a_1 = 1, a_2 = 2, a_3 = 3$
   $a_n = a_{n-1} + a_{n-2} + a_{n-3}, n \geq 4$

Define recursively each number sequence.
(*Hint*: Look for a pattern and define the nth term a_n recursively.)

7. $1, 4, 7, 10, 13, \ldots$
8. $3, 8, 13, 18, 23, \ldots$
9. $0, 3, 9, 21, 45, \ldots$
10. $1, 2, 5, 26, 677, \ldots$

An **arithmetic sequence** is a number sequence in which every term except the first is obtained by adding a fixed number, called the **common difference**, to the preceding term. For example, $1, 3, 5, 7, \ldots$ is an arithmetic sequence with common difference 2. Let a_n denote the nth term of the arithmetic sequence with first term a and common difference d.

11. Define a_n recursively.
12. Find an explicit formula for a_n.
13. Let S_n denote the sum of the first n terms of the sequence. Prove that

$$S_n = \frac{n}{2}\left[2a + (n-1)d\right]$$

A **geometric sequence** is a number sequence in which every term, except the first, is obtained by multiplying the previous term by a constant, called the **common ratio**. For example, $2, 6, 18, 54, \ldots$ is a geometric sequence with common ratio 3. Let a_n denote the nth term of the geometric sequence with first term a and common ratio r.

14. Define a_n recursively.
15. Find an explicit formula for a_n.
16. Let S_n denote the sum of the first n terms of the sequence. Prove that $S_n = [a(r^n - 1)]/(r - 1)$, where $r \neq 1$. Do not use induction.

Use the following triangular array of positive integers to answer Exercises 17–20.

$$\begin{array}{ccccccc} & & & 1 & & & \\ & & 2 & & 3 & & \\ & 4 & & 5 & & 6 & \\ 7 & & 8 & & 9 & & 10 \\ & & & \vdots & & & \end{array}$$

17. Let a_n denote the first term in row n, where $n \geq 1$. Define a_n recursively.
18. Find an explicit formula for a_n.
19. Find the sum of the numbers in row n.
20. Which row contains the number 2076?

Let a_n denote the number of times the assignment statement $x \leftarrow x + 1$ is executed by each nested **for** loop. Define a_n recursively.

21. for $i = 1$ to n do
    for $j = 1$ to i do
    $x \leftarrow x + 1$
22. for $i = 1$ to n do
    for $j = 1$ to i do
    for $k = 1$ to i do
    $x \leftarrow x + 1$
23. Using Example 1.19, predict an explicit formula for b_n.
24. Using induction, prove the explicit formula for b_n in Exercise 23.

The **91-function** f, invented by John McCarthy, is defined recursively on **W** as follows:

$$f(x) = \begin{cases} x - 10 & \text{if } x > 100 \\ f(f(x + 11)) & \text{if } 0 \leq x \leq 100 \end{cases}$$

Compute each:

25. $f(99)$
26. $f(98)$
27. $f(f(99))$

John McCarthy *(1927–), one of the fathers of artificial intelligence (AI), was born in Boston. He graduated in mathematics from Caltech and received his Ph.D. from Princeton in 1951. After teaching at Princeton, Stanford, Dartmouth, and MIT, he returned to Stanford as full professor. While at Princeton, he was named a Proctor Fellow and later Higgins Research Instructor in mathematics. At Stanford, he headed its Artificial Intelligence Laboratory.*

McCarthy coined the term artificial intelligence *while at Dartmouth. He developed LISP (LISt Programming), one of the most widely used programming languages in AI. In addition, he helped develop ALGOL 58 and ALGOL 60. In 1971, he received the prestigious Alan M. Turing Award for his outstanding contributions to data processing.*

28. $f(f(91))$
29. Show that $f(99) = 91$.
30. Prove that $f(x) = 91$ for $90 \le x \le 100$.
31. Prove that $f(x) = 91$ for $0 \le x < 90$.

A function of theoretical importance in the study of algorithms is **Ackermann's function**, named after the German mathematician and logician Wilhelm Ackermann (1896–1962). It is defined recursively as follows, where $m, n \in \mathbf{W}$:

$$A(m,n) = \begin{cases} n+1 & \text{if } m = 0 \\ A(m-1, 1) & \text{if } n = 0 \\ A(m-1, A(m, n-1)) & \text{otherwise} \end{cases}$$

Compute each.

32. $A(0, 7)$
33. $A(1, 1)$
*34. $A(5, 0)$
35. $A(2, 2)$

Prove each for every integer $n \ge 0$.

36. $A(1, n) = n + 2$
37. $A(2, n) = 2n + 3$
*38. Predict a formula for $A(3, n)$.
*39. Prove the formula in Exercise 38 for every integer $n \ge 0$.
*40. Let $\{u_n\}$ be a number sequence with $u_0 = 4$ and $u_n = f(u_{n-1})$, where f is a function defined by the following table and $n \ge 1$. Compute u_{9999}. (*Source: Mathematics Teacher*, 2004)

x	1	2	3	4	5
$f(x)$	4	1	3	5	2

1.5 The Binomial Theorem

Binomials are sums of two terms, and they occur often in mathematics. This section shows how to expand positive integral powers of binomials in a systematic way. The coefficients in binomial expansions have several interesting properties.

Let us begin with a discussion of binomial coefficients.

Binomial Coefficients

Let n and r be nonnegative integers. The **binomial coefficient**† $\binom{n}{r}$ is defined by $\binom{n}{r} = \dfrac{n!}{r!(n-r)!}$ if $r \le n$, and is 0 otherwise; it is also denoted by $C(n, r)$ and nCr.

For example,

$$\binom{5}{3} = \frac{5!}{3!(5-3)!} = \frac{5 \cdot 4 \cdot 3 \cdot 2 \cdot 1}{3 \cdot 2 \cdot 1 \cdot 2 \cdot 1} = 10$$

It follows from the definition that

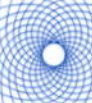

$$\binom{n}{0} = 1 = \binom{n}{n}.$$

There are many instances when we need to compute the binomial coefficients $\binom{n}{r}$ and $\binom{n}{n-r}$. Since

$$\binom{n}{n-r} = \frac{n!}{(n-r)![n-(n-r)]!} = \frac{n!}{(n-r)!r!} = \frac{n!}{r!(n-r)!} = \binom{n}{r}$$

there is no need to evaluate both; this significantly reduces our workload. For example, $\binom{25}{20} = \binom{25}{25-20} = \binom{25}{5} = 53{,}130$.

The following theorem shows an important recurrence relation satisfied by binomial coefficients. It is called **Pascal's identity**, after the outstanding French mathematician and philosopher Blaise Pascal.

---

† The term *binomial coefficient* was introduced by the German algebraist Michel Stifel (1486–1567). In his best-known work, *Arithmetica Integra* (1544), Stifel gives the binomial coefficients for $n \le 17$.

The bilevel parentheses notation for binomial coefficient was introduced by the German mathematician and physicist Baron Andreas von Ettinghausen (1796–1878). Von Ettinghausen, born in Heidelberg, attended the University of Vienna in Austria. For two years he worked as an assistant in mathematics and physics at the University. In 1821 he became professor of mathematics, and in 1835, professor of physics and director of the Physics Institute. Thirteen years later, he became the director of the Mathematical Studies and Engineering Academy in Vienna.

A pioneer in mathematical physics, von Ettinghausen worked in analysis, algebra, differential geometry, mechanics, optics, and electromagnetism.

Blaise Pascal *(1623–1662) was born in Clermont-Ferrand, France. Although he showed astounding mathematical ability at an early age, he was encouraged by his father to pursue other subjects, such as ancient languages. His father even refused to teach him any sciences and relented only when he found that Pascal by age 12 had discovered many theorems in elementary geometry. At 14, Blaise attended weekly meetings of a group of French mathematicians which later became the French Academy. At 16, he developed important results in conic sections and wrote a book on them.*

Observing that his father would spend countless hours auditing government accounts, and feeling that intelligent people should not waste their time doing mundane things, Pascal, at the age of 19, invented the first mechanical calculating machine.

THEOREM 1.8 **(Pascal's Identity)** Let n and r be positive integers, where $r \leq n$. Then $\binom{n}{r} = \binom{n-1}{r-1} + \binom{n-1}{r}$.

PROOF

We shall simplify the RHS and show that it is equal to the LHS:

$$\begin{aligned}
\binom{n-1}{r-1} + \binom{n-1}{r} &= \frac{(n-1)!}{(r-1)!(n-r)!} + \frac{(n-1)!}{r!(n-r-1)!} \\
&= \frac{r(n-1)!}{r(r-1)!(n-r)!} + \frac{(n-r)(n-1)!}{r!(n-r)(n-r-1)!} \\
&= \frac{r(n-1)!}{r!(n-r)!} + \frac{(n-r)(n-1)!}{r!(n-r)!} \\
&= \frac{(n-1)![r+(n-r)]}{r!(n-r)!} = \frac{(n-1)!n}{r!(n-r)!} = \frac{n!}{r!(n-r)!} \\
&= \binom{n}{r}
\end{aligned}$$

■

Pascal's Triangle

The various binomial coefficients $\binom{n}{r}$, where $0 \leq r \leq n$, can be arranged in the form of a triangle, called **Pascal's triangle**,† as in Figures 1.12 and 1.13.

† Although Pascal's triangle is named after Pascal, it actually appeared as early as 1303 in a work by the Chinese mathematician Chu Shi-Kie.

$$\binom{0}{0}$$

$$\binom{1}{0} \quad \binom{1}{1}$$

$$\binom{2}{0} \quad \binom{2}{1} \quad \binom{2}{2}$$

$$\binom{3}{0} \quad \binom{3}{1} \quad \binom{3}{2} \quad \binom{3}{3}$$

$$\binom{4}{0} \quad \binom{4}{1} \quad \binom{4}{2} \quad \binom{4}{3} \quad \binom{4}{4}$$

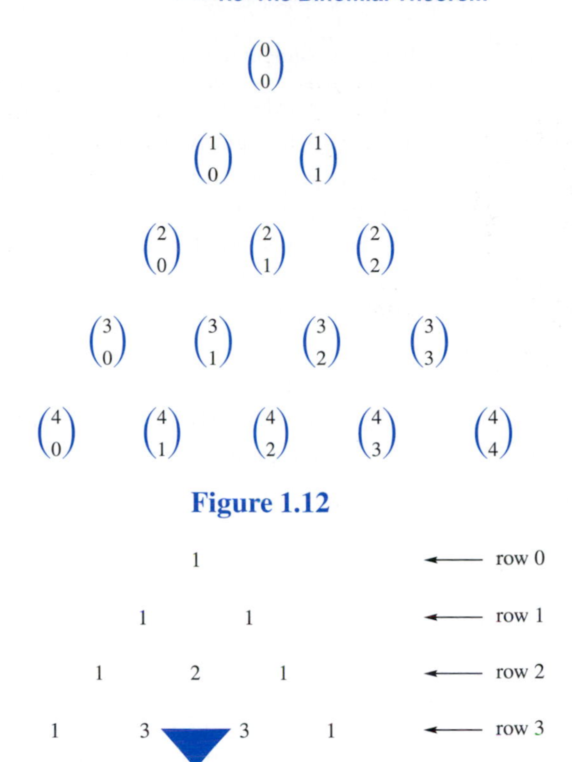

Figure 1.12

1 ⟵ row 0

1 1 ⟵ row 1

1 2 1 ⟵ row 2

1 3 ▼ 3 1 ⟵ row 3

1 4 6 4 1 ⟵ row 4

Figure 1.13

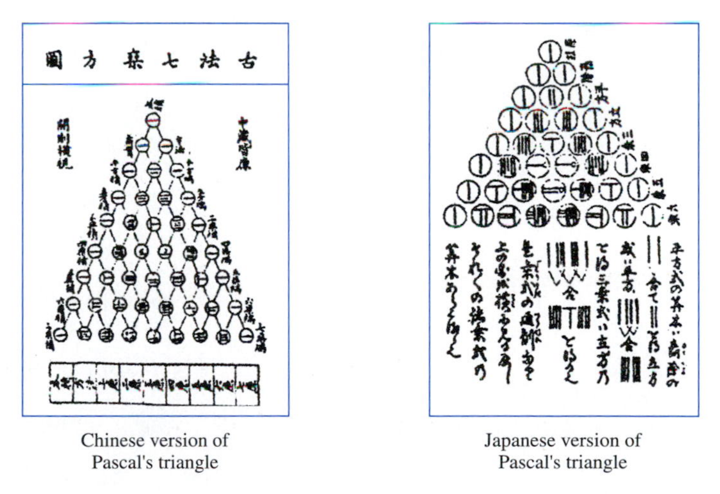

Chinese version of Pascal's triangle

Japanese version of Pascal's triangle

Figure 1.14

Figure 1.14 shows the Chinese and Japanese versions of Pascal's triangle.

Pascal's triangle has many intriguing properties:

- Every row begins with and ends in 1.
- Pascal's triangle is symmetric about a vertical line through the middle. This is so by Theorem 1.8.
- Any interior number in each row is the sum of the numbers immediately to its left and to its right in the preceding row; see Figure 1.13. This is so by virtue of Pascal's identity.
- The sum of the numbers in any row is a power of 2. Corollary 1.1 will verify this.
- The nth row can be used to determine 11^n. For example, $11^3 = 1331$ and $11^4 = 14{,}641$. To compute higher powers of 11, you should be careful since some of the numbers involve two or more digits. For instance, to compute 11^5 list row 5:

From right to left, list the single-digit numbers. When we come to a two-digit number, write the ones digit and carry the tens digit to the number on the left. Add the carry to the number to its left. Continue this process to the left. The resulting number, 161,051, is 11^5.

- Form a regular hexagon with vertices on three adjacent rows (see Figure 1.15). Find the products of numbers at alternate vertices. The two products are equal. For example, $10 \cdot 15 \cdot 4 = 6 \cdot 20 \cdot 5$. Surprised? Supplementary Exercise 10 confirms this property, known as **Hoggatt–Hansell identity**, named after V. E. Hoggatt, Jr., and W. Hansell, who discovered it in 1971; so the product of the six numbers is a square.

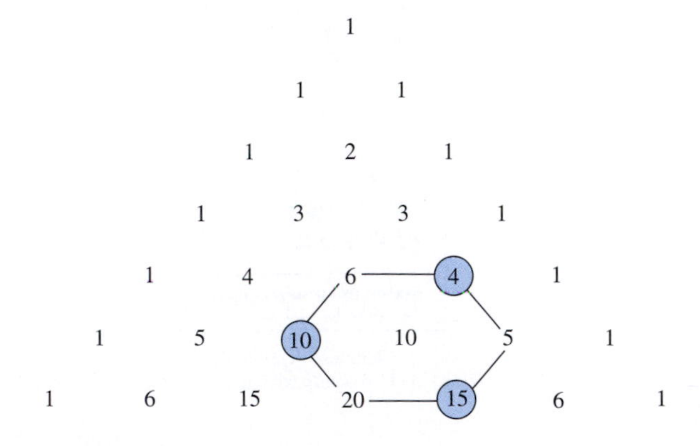

Figure 1.15

The following theorem shows how the binomial coefficients can be used to find the **binomial expansion** of $(x+y)^n$.

THEOREM 1.9 **(The Binomial Theorem)**† Let x and y be any real numbers, and n any nonnegative integer. Then $(x+y)^n = \sum_{r=0}^{n} \binom{n}{r} x^{n-r} y^r$.

PROOF (by weak induction)

When $n=0$, LHS $=(x+y)^0 = 1$ and RHS $= \sum_{r=0}^{0} \binom{r}{0} x^{0-r} y^r = x^0 y^0 = 1$, so LHS $=$ RHS.

Assume $P(k)$ is true for some $k \geq 0$:

$$(x+y)^k = \sum_{r=0}^{k} \binom{k}{r} x^{k-r} y^r \tag{1.10}$$

Then

$$\begin{aligned}
&(x+y)^{k+1} \\
&= (x+y)^k (x+y) \\
&= \left[\sum_{r=0}^{k} \binom{k}{r} x^{k-r} y^r \right] (x+y), \quad \text{by equation (1.10)} \\
&= \sum_{r=0}^{k} \binom{k}{r} x^{k+1-r} y^r + \sum_{r=0}^{k} \binom{k}{r} x^{k-r} y^{r+1} \\
&= \left[\binom{k}{0} x^{k+1} + \sum_{r=1}^{k} \binom{k}{r} x^{k+1-r} y^r \right] + \left[\sum_{r=0}^{k-1} \binom{k}{r} x^{k-r} y^{r+1} + \binom{k}{k} y^{k+1} \right] \\
&= \binom{k+1}{0} x^{k+1} + \sum_{r=1}^{k} \binom{k}{r} x^{k+1-r} y^r + \sum_{r=1}^{k} \binom{k}{r-1} x^{k+1-r} y^r + \binom{k+1}{k+1} y^{k+1} \\
&= \binom{k+1}{0} x^{k+1} + \sum_{r=1}^{k} \left[\binom{k}{r} + \binom{k}{r-1} \right] x^{k+1-r} y^r + \binom{k+1}{k+1} y^{k+1} \\
&= \binom{k+1}{0} x^{k+1} + \sum_{r=1}^{k} \binom{k+1}{r} x^{k+1-r} y^r + \binom{k+1}{k+1} x^{k+1}, \quad \text{by Theorem 1.8} \\
&= \sum_{r=0}^{k+1} \binom{k+1}{r} x^{k+1-r} y^r
\end{aligned}$$

Thus, by induction, the formula is true for every integer $n \geq 0$. ■

† The binomial theorem for $n=2$ can be found in Euclid's work (ca. 300 B.C.).

It follows from the binomial theorem that the binomial coefficients in the expansion of $(x+y)^n$ are the various numbers in row n of Pascal's triangle.

The binomial theorem can be used to establish several interesting identities involving binomial coefficients, as the following corollary shows.

COROLLARY[†] 1.1

$$\sum_{r=0}^{n} \binom{n}{r} = 2^n$$

That is, the sum of the binomial coefficients is 2^n. ■

This follows by letting $x = 1 = y$ in the binomial theorem.

The following exercises provide opportunities to explore additional relationships.

EXERCISES 1.5

(Twelve Days of Christmas) Suppose that on the first day of Christmas you sent your love 1 gift, $1+2$ gifts on the second day, $1+2+3$ gifts on the third day, and so on.

1. Show that the number of gifts sent on the nth day is $\binom{n+1}{2}$, where $1 \le n \le 12$.
2. Show that the total number of gifts sent by the nth day is $\binom{n+2}{3}$, where $1 \le n \le 12$.

Find the coefficient of each.

3. x^2y^6 in the expansion of $(2x+y)^8$.
4. x^4y^5 in the expansion of $(2x-3y)^9$.

Using the binomial theorem, expand each.

5. $(2x-1)^5$
6. $(x+2y)^6$

Find the middle term in the binomial expansion of each.

7. $\left(2x+\frac{2}{x}\right)^8$
8. $\left(x^2+\frac{1}{x^2}\right)^{10}$

Find the largest binomial coefficient in the expansion of each.

9. $(x+y)^5$
10. $(x+y)^6$
11. $(x+y)^7$
12. $(x+y)^8$
13. Using Exercises 9–12, predict the largest binomial coefficient in the binomial expansion of $(x+y)^n$.

The **Bell numbers** B_n are named after the Scottish American mathematician Eric T. Bell (1883–1960). They are used in combinatorics and are defined recursively as follows:

$$B_0 = 1$$

$$B_n = \sum_{i=0}^{n-1} \binom{n-1}{i} B_i, \quad n \ge 1$$

Compute each Bell number.

14. B_2
15. B_3
16. B_4
17. B_5
18. Verify that $\binom{n}{r} = \frac{n}{r}\binom{n-1}{r-1}$.
19. Prove that $\binom{2n}{n}$ is an even integer. (L. Moser, 1962)

Prove each.

20. $(n+1) \mid \binom{2n}{n}$, where $a|b$ means a is a factor of b and $n \ge 0$.
21. $\sum_{r=0}^{n} \binom{2n}{2r} = \sum_{r=1}^{n} \binom{2n}{2r-1}$ (*Hint*: Use Corollary 1.1.)

[†] A **corollary** is a result that follows from the previous theorem.

22. $\sum_{r=0}^{n} 2^r \binom{n}{r} = 3^n$

23. $\sum_{r=0}^{n} \binom{n}{r}\binom{n}{n-r} = \binom{2n}{n}$

(*Hint*: Consider $(1+x)^{2n} = (1+x)^n(1+x)^n$.)

24. $\sum_{i=1}^{n} \binom{n}{i-1}\binom{n}{i} = \binom{2n}{n+1}$

(*Hint*: Consider $(1+x)^{2n} = (x+1)^n(1+x)^n$.)

Evaluate each sum.

25. $1\binom{n}{1} + 2\binom{n}{2} + 3\binom{n}{3} + \cdots + n\binom{n}{n}$

(*Hint*: Let S denote the sum. Use S and the sum in the reverse order to compute $2S$.)

26. $a\binom{n}{0} + (a+d)\binom{n}{1} + (a+2d)\binom{n}{2} + \cdots + (a+nd)\binom{n}{n}$

(*Hint*: Use the same hint as in Exercise 25.)

27. Show that $C(n, r-1) < C(n, r)$ if and only if $r < \frac{n+1}{2}$, where $0 \le r < n$.

28. Using Exercise 27, prove that the largest binomial coefficient $C(n, r)$ occurs when $r = \lfloor n/2 \rfloor$.

Using induction, prove each.

29. $\binom{n}{0} + \binom{n+1}{1} + \binom{n+2}{2} + \cdots + \binom{n+r}{r} = \binom{n+r+1}{r}$

(*Hint*: Use Pascal's identity.)

30. $1\binom{n}{1} + 2\binom{n}{2} + \cdots + n\binom{n}{n} = n2^{n-1}$

31. $\binom{n}{0}^2 + \binom{n}{1}^2 + \binom{n}{2}^2 + \cdots + \binom{n}{n}^2 = \binom{2n}{n}$

(**Lagrange's identity**)

From the binomial expansion $(1+x)^n = \sum_{r=0}^{n} \binom{n}{r} x^r$, it can be shown that $n(1+x)^{n-1} = \sum_{r=1}^{n} \binom{n}{r} r x^{r-1}$.

Using this result, prove each.

32. $1\binom{n}{1} + 2\binom{n}{2} + 3\binom{n}{3} + \cdots + n\binom{n}{n} = n2^{n-1}$

33. $1\binom{n}{1} + 3\binom{n}{3} + 5\binom{n}{5} + \cdots = 2\binom{n}{2} + 4\binom{n}{4} + 6\binom{n}{6} + \cdots = n2^{n-2}$

34. Conjecture a formula for $\sum_{i=2}^{n} \binom{i}{2}$.

35. Prove the formula guessed in Exercise 34.

36. Conjecture a formula for $\sum_{i=3}^{n} \binom{i}{3}$.

37. Prove the formula guessed in Exercise 36.

38. Using Exercises 34–37, predict a formula for $\sum_{i=k}^{n} \binom{i}{k}$.

1.6 Polygonal Numbers

Figurate numbers are positive integers that can be represented by geometric patterns. They provide a fascinating link between number theory and geometry. Not surprisingly, figurate numbers are of ancient origin, and, in fact, it is believed that they were invented by the Pythagoreans. In 1665, Pascal published a book on them, *Treatise on Figurate Numbers*.

Polygonal numbers, also known as **plane figurate numbers**, are positive integers that can be represented by regular polygons in a systematic fashion. We will use four types of such numbers: triangular numbers, square numbers, pentagonal numbers, and hexagonal numbers.

If you have been to a bowling alley, you know that there are ten pins in bowling, and they are arranged initially in a triangular array. Likewise, the 15 balls in the game of pool are also initially stored in a triangular form. Both numbers, 10 and 15, are triangular numbers; likewise, the number of dots on a die is a triangular number. Accordingly, we make the following definition.

Triangular Numbers

A **triangular number** is a positive integer that can be represented in an equilateral triangular array. The *n*th triangular number is denoted by t_n, $n \geq 1$.

The first four triangular numbers are 1, 3, 6, and 10, and they are pictorially represented in Figure 1.16.

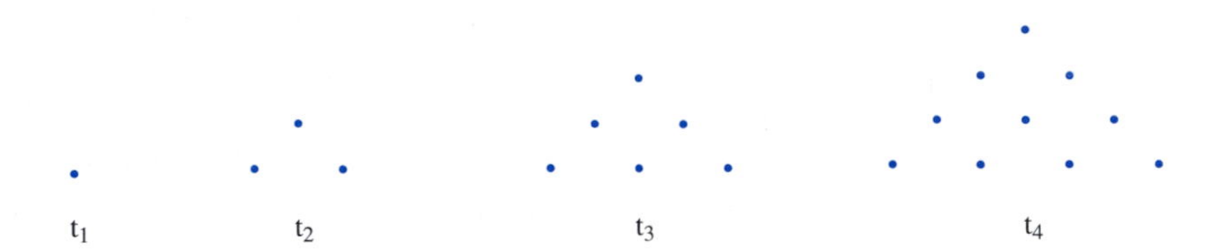

Figure 1.16

Since the *i*th row contains *i* dots, t_n equals the sum of the first *n* positive integers; that is,

$$t_n = \sum_{i=1}^{n} i = \frac{n(n+1)}{2} \quad \text{by Example 1.11}$$

For example, $t_4 = (4 \cdot 5)/2 = 10$ and $t_{36} = (36 \cdot 37)/2 = 666$.

Since $t_n = \binom{n+1}{2}$, triangular numbers can be read from Pascal's triangle.

Since each row in the triangular array contains one dot more than the previous row, t_n can be defined recursively. See Figure 1.17 and Table 1.1.

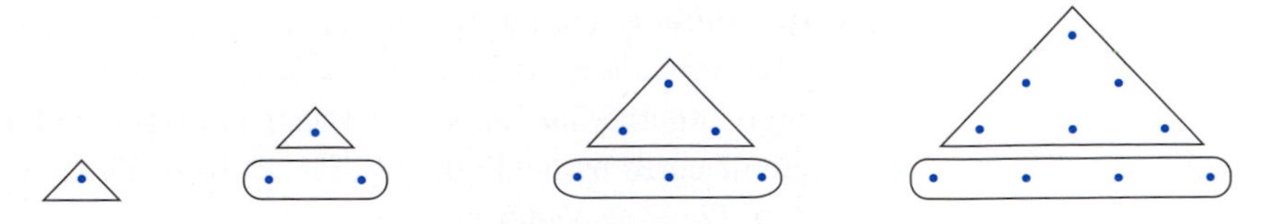

Figure 1.17

n	1	2	3	4	5	...	n
t_n	1	3	6	10	15	...	?

Table 1.1

A Recursive Definition of t_n

$$t_1 = 1$$

$$t_n = t_{n-1} + n, \quad n \geq 2$$

As an example, since $t_3 = 6$, $t_4 = t_3 + 4 = 6 + 4 = 10$ (see Figure 1.17).

We can solve the recurrence relation and obtain the explicit formula for t_n found earlier (see Exercise 1).

Now, let us take another look at *The Twelve Days of Christmas*, the traditional carol, and see how it is related to triangular numbers.

The Twelve Days of Christmas

On the first day of Christmas, my true love sent me a partridge in a pear tree. On the second day of Christmas, my true love sent me two turtle doves and a partridge in a pear tree. On the third day, my true love sent me three French hens, two turtle doves, and a partridge in a pear tree. The pattern continues until the twelfth day, on which my true love sent me twelve drummers drumming, eleven pipers piping, ten lords a-leaping, nine ladies dancing, eight maids a-milking, seven swans a-swimming, six geese a-laying, five gold rings, four calling birds, three French hens, two turtle doves, and a partridge in a pear tree.

Two interesting questions we would like to pursue:

- If the pattern in the carol continues for n days, how many gifts g_n would be sent on the nth day?
- What is the total number of gifts s_n sent in n days?

First, notice that the number of gifts sent on the nth day equals n more than the number of gifts sent on the previous day, so $g_n = g_{n-1} + n$, where $g_1 = 1$. Therefore, $g_n = t_n$, the nth triangular number. For instance, the number of gifts sent on the twelfth day is given by $t_{12} = (12 \cdot 13)/2 = 78$.

It now follows that

$$\begin{aligned} s_n &= \sum_{i=1}^{n} t_i \\ &= \sum_{i=1}^{n} \frac{i(i+1)}{2} = \frac{1}{2}\left(\sum_{i=1}^{n} i^2 + \sum_{i=1}^{n} i\right) \end{aligned}$$

$$= \frac{1}{2}\left[\frac{n(n+1)(2n+1)}{6} + \frac{n(n+1)}{2}\right]$$
$$= \frac{n(n+1)}{12}[(2n+1)+3] = \frac{n(n+1)(n+2)}{6}$$
$$= \binom{n+2}{3}$$

Figure 1.18 provides a geometric proof of this formula, developed in 1990 by M. J. Zerger of Adams State College, Alamosa, Colorado.

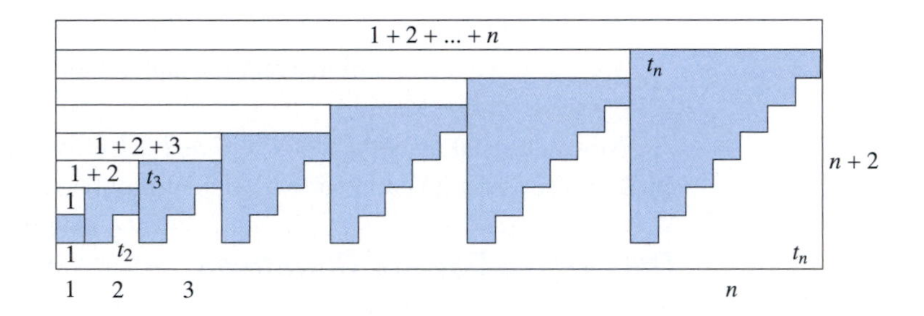

Figure 1.18

It now follows that the total number of gifts sent in 12 days is given by $s_{12} = (12 \cdot 13 \cdot 14)/6 = 364$.

The cubes $1, 8, 27, 64, 125, \ldots, n^3$ are related to triangular numbers. To see this, let c_n denote the nth cube n^3. Since $\sum_{k=1}^{n} k^3 = [n(n+1)/2]^2$, it follows by Exercise 15 in Section 1.3 that $\sum_{k=1}^{n} c_k = t_n^2$; that is, the sum of the first n cubes equals the square of the nth triangular number.

The following example shows that triangular numbers can occur in quite unexpected places. It also illustrates, step-by-step, a powerful problem-solving technique: collecting data, organizing data, conjecturing a desired formula, and then establishing the formula.

EXAMPLE 1.21 Find the number of $1 \times k$ rectangles $f(n)$ that can be formed using an array of n squares, where $1 \le k \le n$. See Figure 1.19.

Figure 1.19

SOLUTION

step 1 *Collect data by conducting a series of experiments for small values of n.*

When $n = 1$, the array looks like this: □. So only one rectangle can be formed. When $n = 2$, the array looks like this: □□. We can form two 1×1 rectangles and one 1×2 rectangle, a total of $2 + 1 = 3$ rectangles. When $n = 3$, the array consists of three squares: □□□. We then can form three 1×1 rectangles, two 1×2 rectangles, and one 1×3 rectangle, as summarized in Table 1.2.

Size of the Rectangle	*Number of Such Rectangles*
1×1	3
1×2	2
1×3	1
Total No. of Rectangles	6

Table 1.2

Continuing like this, we can find the total number of rectangles that can be formed when $n = 4$ and $n = 5$, as Tables 1.3 and 1.4 demonstrate respectively.

Size of Rectangle	*Number of Rectangles*
1×1	4
1×2	3
1×3	2
1×4	1
Total	10

Table 1.3

Size of Rectangle	*Number of Rectangles*
1×1	5
1×2	4
1×3	3
1×4	2
1×5	1
Total	15

Table 1.4

step 2 *Organize the data in a table.*

No. of squares n in the array	1	2	3	4	5	...	n
No. of rectangles $f(n)$	1	3	6	10	15	...	?

Table 1.5

step 3 *Look for a pattern and conjecture a formula for $f(n)$.*

Clearly, row 2 of Table 1.5 consists of triangular numbers. (See Table 1.1 also.) So we conjecture that $f(n) = n(n + 1)/2$.

step 4 *This formula can be established using recursion and induction.*

We now introduce the next simplest class of polygonal numbers.

Square Numbers

Positive integers that can be represented by square arrays (of dots) are **square numbers**. The nth square number is denoted by s_n. Figure 1.20 shows the first four square numbers, 1, 4, 9, and 16. In general, $s_n = n^2$, $n \geq 1$.

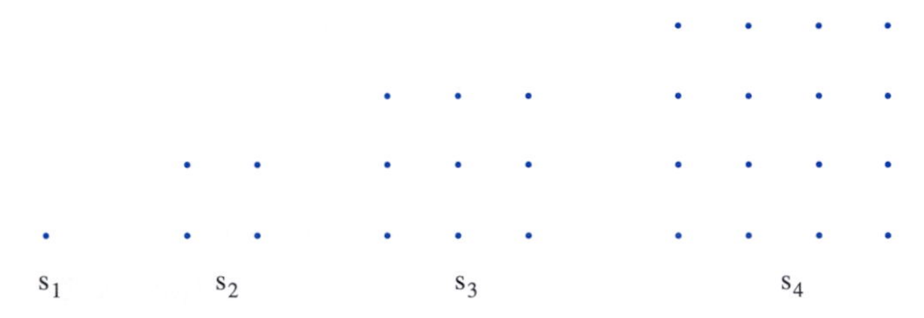

Figure 1.20

As before, s_n also can be defined recursively. To see how this can be done, consider Figure 1.21. Can we see a pattern? The number of dots in each array (except the first one) equals the number of dots in the previous array plus twice the number of dots in a row of the previous array plus one; that is,

$$\begin{aligned} s_n &= s_{n-1} + 2(n-1) + 1 \\ &= s_{n-1} + 2n - 1 \end{aligned}$$

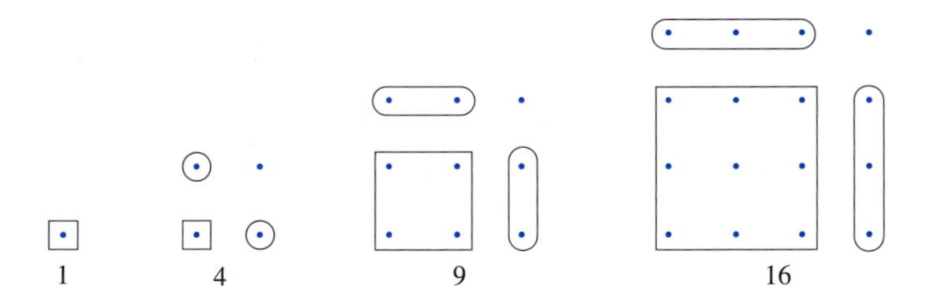

Figure 1.21

Thus, we have the following recursive definition of s_n:

A Recursive Definition of s_n

$$\begin{aligned} s_1 &= 1 \\ s_n &= s_{n-1} + 2n - 1, \quad n \geq 2 \end{aligned}$$

We now demonstrate a close relationship between t_n and s_n. To see this, it follows from Figure 1.22 that $s_5 = t_5 + t_4$. Similarly, $s_n = t_n + t_{n-1}$. The following theorem, known to the Greek mathematicians Theon of Smyrna (ca. A.D. 100) and Nicomachus, establishes this algebraically.

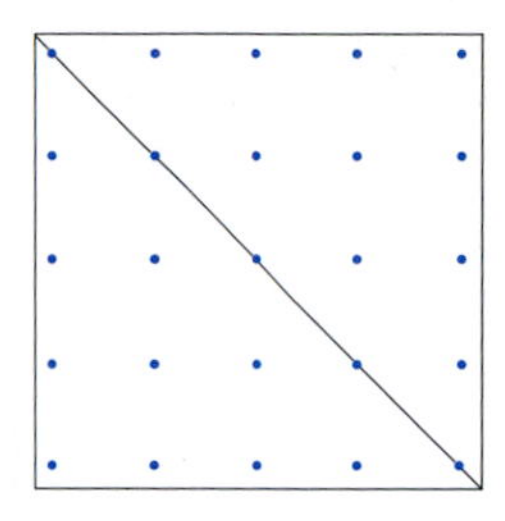

Figure 1.22

THEOREM 1.10 The sum of any two consecutive triangular numbers is a square.

PROOF

$$\begin{aligned} t_n + t_{n-1} &= \frac{n(n+1)}{2} + \frac{n(n-1)}{2} \\ &= \frac{n}{2}(n+1+n-1) = \frac{n}{2}(2n) \\ &= n^2 = s_n \end{aligned}$$ ■

Figures 1.23 and 1.24 provide a nonverbal, geometric proof of this theorem.

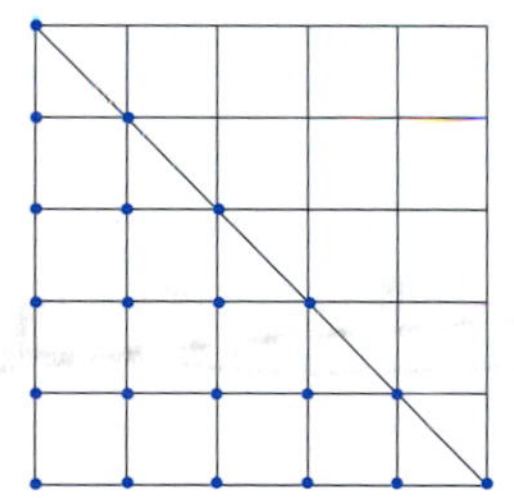

Figure 1.23

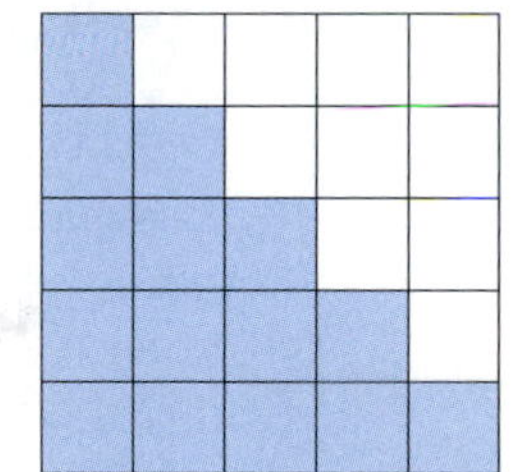

Figure 1.24

Theorem 1.10 has a companion result, which can be established algebraically. See Exercise 11.

THEOREM 1.11 $t_{n-1}^2 + t_n^2 = t_{n^2}$

Figure 1.25 provides a nonverbal, geometric proof of this result; it was developed in 1997 by R. B. Nelsen of Lewis and Clark College in Portland, Oregon.

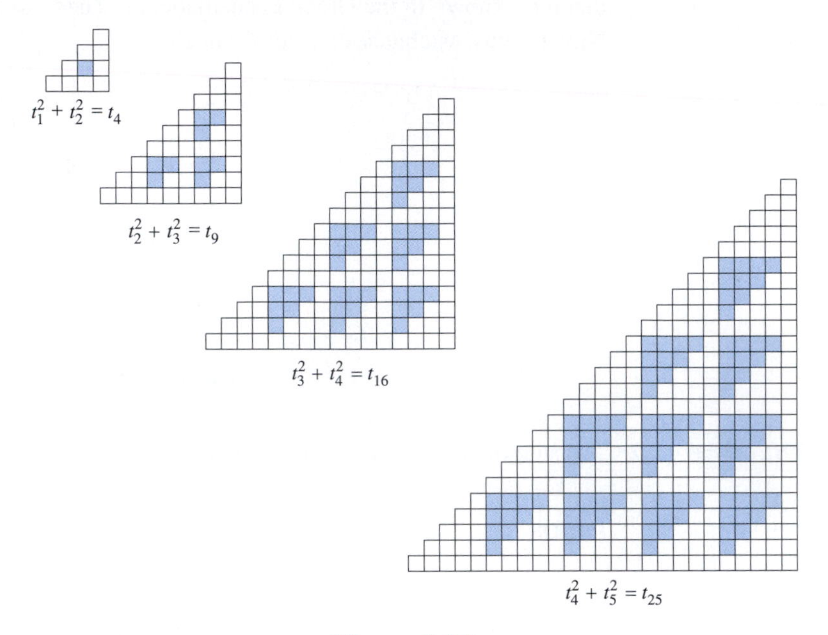

Figure 1.25

The following theorem gives two additional results. Their proofs are also simple and straightforward and can be done as routine exercises.

THEOREM 1.12

- $8t_n + 1 = (2n + 1)^2$ (Diophantus)
- $8t_{n-1} + 4n = (2n)^2$

Figure 1.26 gives a pictorial, nonverbal proof of both results. Both were developed in 1985 by E. G. Landauer of General Physics Corporation.

Next we turn to pentagonal† numbers p_n.

Pentagonal Numbers

The first four **pentagonal numbers** 1, 5, 12, and 22 are pictured in Figure 1.27. We may notice that $p_n = \frac{n(3n-1)}{2}$ (see Exercise 6).

† The Greek prefix *penta* means *five*.

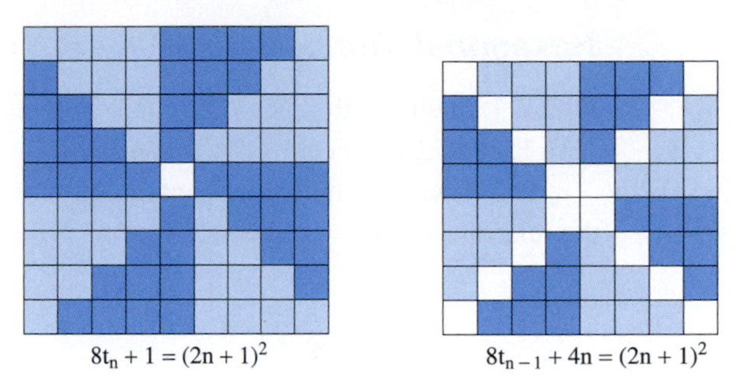

Figure 1.26

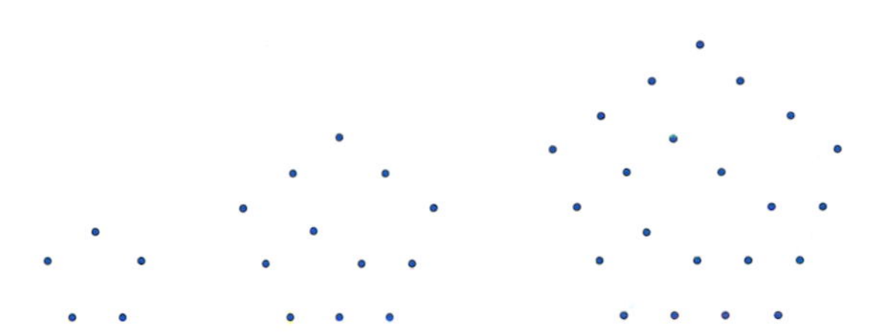

Figure 1.27

There is an interesting relationship connecting triangular numbers, square numbers, and pentagonal numbers. It follows from Figure 1.28 that $t_1 + s_2 = p_2$ and $t_2 + s_3 = p_3$. More generally, $t_{n-1} + s_n = p_n$, where $n \geq 2$. We can verify this algebraically (see Exercise 8).

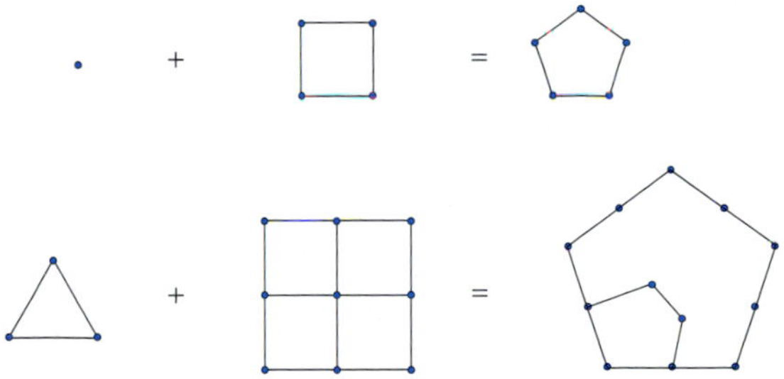

Figure 1.28

Next, we discuss hexagonal† numbers h_n.

† The Greek prefix *hexa* means *six*.

Hexagonal Numbers

Figure 1.29 shows the pictorial representations of the first four hexagonal numbers 1, 6, 15, and 28. We can verify that $h_n = n(2n-1)$, $n \geq 1$ also (see Exercise 20).

The triangular numbers, pentagonal numbers, and hexagonal numbers satisfy the relationship $p_n + t_{n-1} = h_n$. We can verify this (see Exercise 10).

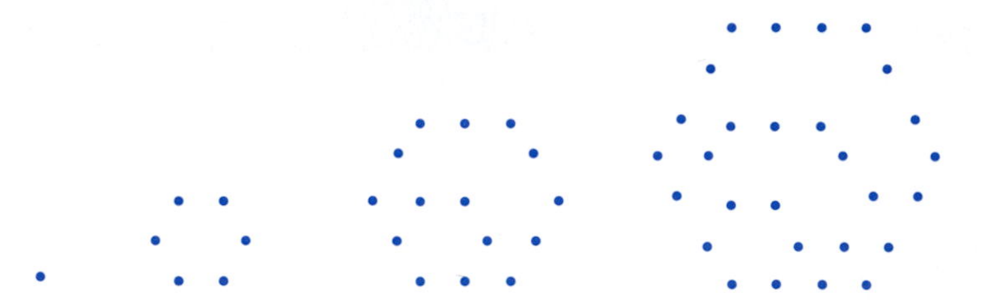

Figure 1.29

EXERCISES 1.6

1. Solve the recurrence relation satisfied by t_n.
2. Find the value of n such that $t_n = 666$. (The number 666 is called the **beastly number**.)
3. Solve the recurrence relation satisfied by s_n.
4. Show that $8t_n + 1 = s_{2n+1}$. (Diophantus)
5. Define recursively the nth pentagonal number p_n.
6. Using the recurrence relation in Exercise 5, find an explicit formula for p_n.

Prove each, where $n \geq 2$.

7. $p_n = n + 3t_{n-1}$
8. $t_{n-1} + s_n = p_n$
9. $h_n = 4t_{n-1} + n$
10. $p_n + t_{n-1} = h_n$
11. $t_{n-1}^2 + t_n^2 = t_{n^2}$
12. $8t_{n-1} + 4n = (2n)^2$
13. $t_{2n-1} - 2t_{n-1} = n^2$
14. $t_{2n} - 2t_n = n^2$
15. $t_{t_n} = t_{t_{n-1}} + t_n$
16. $t_{t_n} + t_{t_{n-1}} = t_n{}^2$
17. In 1775, Euler proved that if n is a triangular number, then so are $9n+1$, $25n+3$, and $49n+6$. Verify this.
18. Let n be a triangular number. Prove that $(2k+1)^2 n + t_k$ is also a triangular number. (Euler, 1775) (*Note*: Exercise 17 is a special case of this.)
19. Define recursively the nth hexagonal number h_n.
20. Using the recurrence relation in Exercise 19, find an explicit formula for h_n.
21. Find the first four **heptagonal**† **numbers**.
22. Define recursively the nth heptagonal number e_n.
23. Using the recurrence relation in Exercise 22, find an explicit formula for e_n.
24. Find the first four **octagonal**‡ **numbers**.
25. Define recursively the nth octagonal number o_n.
26. Using the recurrence relation in Exercise 25, find an explicit formula for o_n.
27. Find two pairs of triangular numbers whose sums and differences are also triangular.
28. Show that there are triangular numbers whose squares are also triangular.
29. There are three triangular numbers < 1000 and made up of a repeated single digit. Find them.
30. Verify that the numbers 1225, 41616, and 1413721 are both triangular and square.
31. The nth number a_n that is both triangular and square can be defined recursively as $a_n = 34a_{n-1} - a_{n-2} + 2$, where $a_1 = 1$ and $a_2 = 36$. Using this definition, compute a_4 and a_5.

† The Greek prefix *hepta* means *seven*.
‡ The Greek prefix *octa* means *eight*.

32. The nth number a_n that is both triangular and square can be computed using the formula $a_n = [(17 + 12\sqrt{2})^n + (17 - 12\sqrt{2})^n - 2]/32,\ n \geq 1$. Using this formula, compute a_2 and a_3.
33. Prove that there are infinitely many triangular numbers that are squares.

Evaluate each.

34. $\sum_{k=1}^{n} \frac{1}{t_k}$
35. (C) $\sum_{k=1}^{\infty} \frac{1}{t_k}$ (This problem, proposed by Christiaan Huygens to Baron Gottfried Wilhelm Leibniz, led to the development of the latter's harmonic triangle.)

1.7 Pyramidal Numbers

Now we pursue solid figurate numbers, which are positive integers that can be represented by pyramidal shapes. They are obtained by taking successive sums of the corresponding polygonal numbers. The number of sides in the base of a pyramid increases from three, so the various pyramidal numbers are triangular, square, pentagonal, hexagonal, and so on.

We begin with the simplest pyramidal numbers, **triangular pyramidal numbers**, also known as **tetrahedral numbers**.

Triangular Pyramidal Numbers

The nth triangular pyramidal number T_n is the sum of the first n triangular numbers t_n. The first four such numbers are: $T_1 = 1$; $T_2 = t_1 + t_2 = 1 + 3 = 4$; $T_3 = t_1 + t_2 + t_3 = 1 + 3 + 6 = 10$; and $T_4 = t_1 + t_2 + t_3 + t_4 = 1 + 3 + 6 + 10 = 20$. See Figure 1.30.

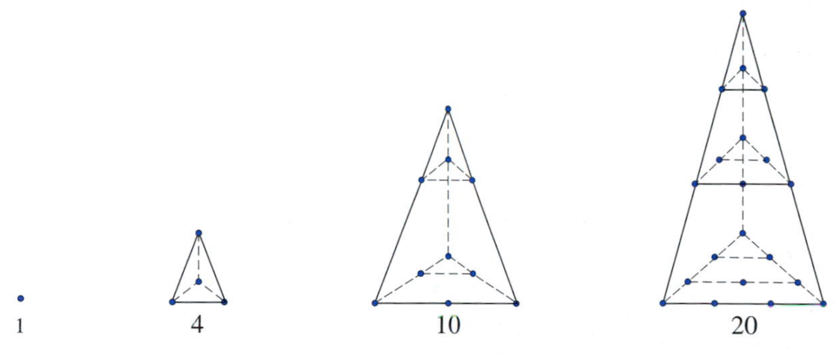

Figure 1.30

The various triangular pyramidal numbers can be constructed using Table 1.6. Just add up the numbers along the bent arrows. It follows from the table that $T_n = T_{n-1} + t_n$; that is, $T_n = T_{n-1} + [n(n+1)]/2$.

n	1	2	3	4	5	6	·	·	·	n
t_n	1	3	6	10	15	21	·	·	·	$n(n+1)/2$
T_n	1	4	10	20	35	?	·	·	·	?

Table 1.6

Since $T_n = \sum_{i=1}^{n} t_i$, it follows from the previous section that

$$T_n = \sum_{i=1}^{n} \frac{i(i+1)}{2} = \frac{n(n+1)(n+2)}{6}$$
$$= \binom{n+2}{3}$$

Consequently, T_n also can be read from Pascal's triangle.

Next, we pursue square pyramidal numbers.

Square Pyramidal Numbers

The base of the pyramid is a square, and each layer contains s_n dots. So the first four **square pyramidal numbers** are 1, 5, 14, and 30, and they are represented in Figure 1.31.

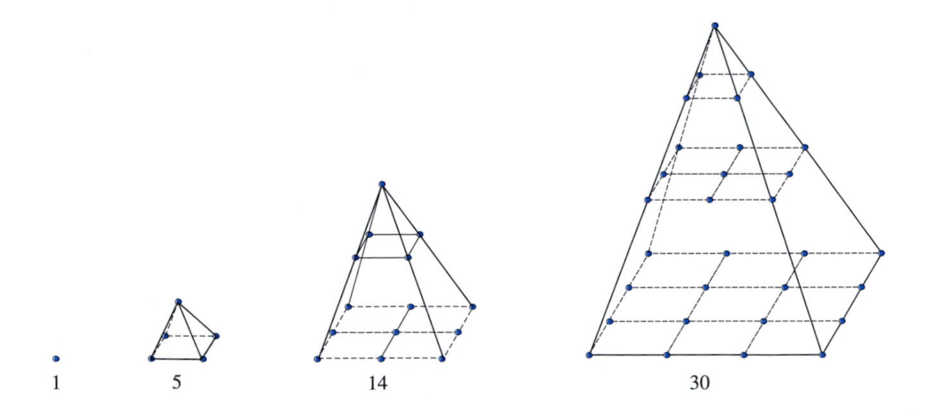

Figure 1.31

The square pyramidal numbers S_n can easily be constructed using Table 1.7, by adding the numbers along the bent arrows.

n	1	2	3	4	5	·	·	·	n
s_n	1	4	9	16	25	·	·	·	n^2
S_n	1	5	14	30	55	·	·	·	?

Table 1.7

It follows from Figure 1.31 and Table 1.7 that the nth square pyramidal number is given by

$$S_n = \sum_{k=1}^{n} s_k = \sum_{k=1}^{n} k^2$$

$$= \frac{n(n+1)(2n+1)}{6}$$

We now study pentagonal pyramidal numbers P_n.

Pentagonal Pyramidal Numbers

The nth row of a pentagonal pyramid represents the nth pentagonal number p_n, so the first five **pentagonal pyramidal numbers** are 1, 6, 18, 40, and 75. Once again, a table such as Table 1.8 comes in handy for computing them. It would be a good exercise to find an explicit formula for P_n.

n	1	2	3	4	5	·	·	·
p_n	1	5	12	22	35	·	·	·
P_n	1	6	18	40	75	·	·	·

Table 1.8

Finally, we consider the hexagonal pyramidal numbers H_n.

Hexagonal Pyramidal Numbers

The nth row of a hexagonal pyramid represents the nth hexagonal number h_n, so the first five **hexagonal pyramidal numbers** are 1, 7, 22, 50, and 95 (see Table 1.9). We can find an explicit formula for H_n as an exercise.

n	1	2	3	4	5	·	·	·	n
h_n	1	6	15	28	45	·	·	·	?
H_n	1	7	22	50	95	·	·	·	?

Table 1.9

EXERCISES 1.7

1. Find the first four triangular numbers that are squares.
2. Using the recurrence relation $T_n = T_{n-1} + \dfrac{n(n+1)}{2}$, where $T_1 = 1$, find an explicit formula for the nth triangular pyramidal number T_n.
3. Define recursively the nth square pyramidal number S_n.
4. Using Exercise 3, find an explicit formula for S_n.
5. Find a formula for the nth pentagonal pyramidal number P_n.
6. Define recursively the nth pentagonal pyramidal number P_n.
7. Using Exercise 6, find an explicit formula for P_n.
8. Find a formula for the nth hexagonal pyramidal number H_n.
9. Define recursively the nth hexagonal pyramidal number H_n.
10. Using Exercise 9, find an explicit formula for H_n.
11. Find the first five heptagonal pyramidal numbers.
12. Find a formula for the nth heptagonal pyramidal number E_n.

1.8 Catalan Numbers

Catalan numbers are both fascinating and ubiquitous. They are excellent candidates for exploration, experimentation, and conjecturing. Like Fibonacci and Lucas numbers (see Section 2.6), they have, as Martin Gardner wrote in *Scientific American*, "the same delightful propensity for popping up unexpectedly, particularly in combinatorial problems" (1976). Those unexpected places include abstract algebra, combinatorics, computer science, graph theory, and geometry.

Catalan numbers are named after the Belgian mathematician Eugene C. Catalan, who discovered them in 1838, while he was studying well-formed sequences of parentheses. Earlier, around 1751, the outstanding Swiss mathematician Leonhard Euler (see Section 7.4) found them while studying the triangulations of convex polygons. In fact, they were discussed by the Chinese mathematician Antu Ming (1692?–1763?) in 1730 through his geometric models. Since his work was available only in Chinese, his discovery was not known in the western world.

***Eugene Charles Catalan** (1814–1894) was born in Bruges, Belgium. He studied at École Polytechnique, Paris, and received his Doctor of Science in 1841. After resigning his position with the Department of Bridges and Highways, he became professor of mathematics at Collège de Chalons-sur Marne, and then at Collège Charlemagne. Catalan then taught at Lycée Saint Louis and in 1865 became professor of analysis at the University of Liège in Belgium. Besides authoring* Élements de Geometriè *(1843) and* Notions d'astronomie *(1860), he published numerous articles on multiple integrals, the theory of surfaces, mathematical analysis, calculus of probability, and geometry. He did extensive research on spherical harmonics, analysis of differential equations, transformation of variables in multiple integrals, continued fractions, series, and infinite products.*

***Antu Ming** (1692?–1763?), according to Luo, was a Zhengxianbai tribesman of Inner Mongolia and a famous scientist during the Qing Dynasty. His childhood mathematical education, specializing in astronomy and mathematics, was carefully directed by the Emperor. After mastering the scientific knowledge of the period, Ming became a mandarin, a high-ranking government official, at the national astronomical center. In 1759, he became director of the center. His work included problem solving in astronomy, meteorology, geography, surveying, and mathematics.*

Around 1730, he began to write Efficient Methods for the Precise Values of Circular Functions, *a book that clearly demonstrates his understanding of Catalan numbers. The book was completed by Ming's students before 1774, but was not published until 1839.*

Euler's Triangulation Problem

We begin our study of Catalan numbers C_n with an investigation of Euler's *triangulation problem*:

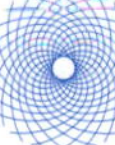

Find the number of ways A_n the interior of a convex n-gon† can be partitioned into nonoverlapping triangular areas by drawing nonintersecting diagonals, where $n \geq 3$.

There is only one way of triangulating a triangle, two different ways of triangulating a square, five different ways of triangulating a pentagon, and 14 different ways of triangulating a hexagon, as shown in Figure 1.32. Thus, we have the Catalan numbers 1, 2, 5, and 14.

† A *convex n-gon* is a polygon with n sides such that every diagonal lies entirely in the interior.

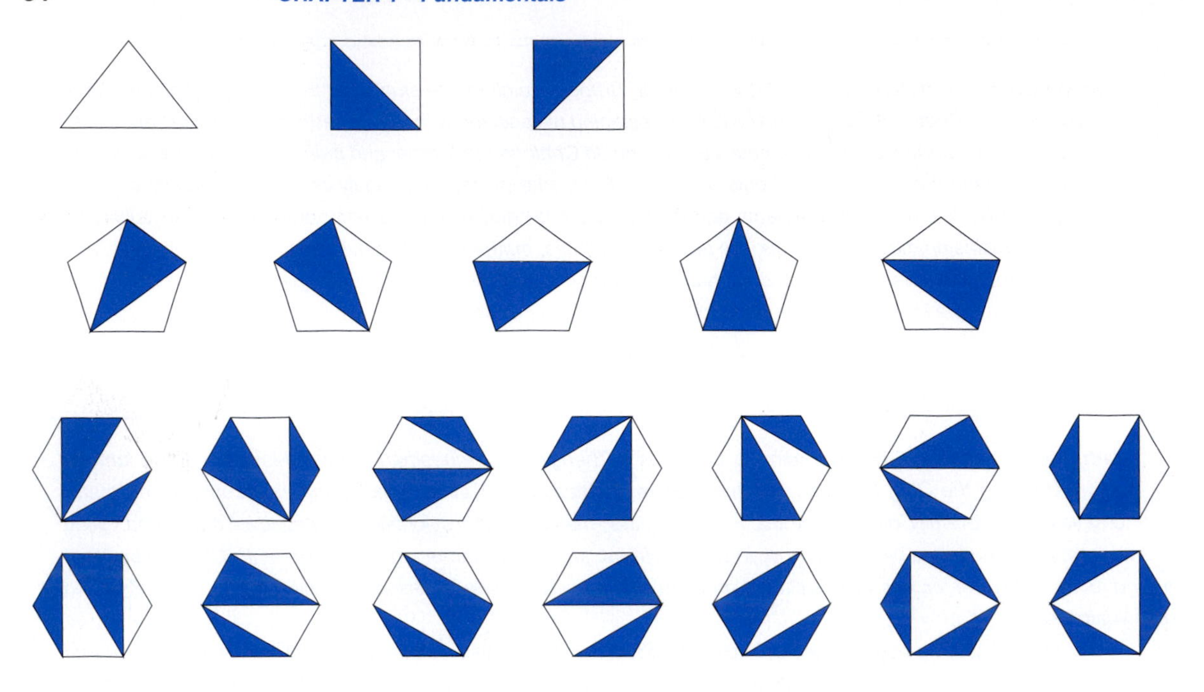

Figure 1.32 *Triangulations of an n-gon, where* $3 \leq n \leq 6$.

Euler used an inductive argument, which he called "quite laborious," to establish the formula

$$A_n = \frac{2 \cdot 6 \cdot 10 \cdots (4n-10)}{(n-1)!}, \quad n \geq 3$$

Although Euler's formula, published in 1761, makes sense only for $n \geq 3$, we can extend it to include the cases $n = 0$, 1, and 2. To this end, let $k = n - 3$. Then

$$A_{k+3} = \frac{2 \cdot 6 \cdot 10 \cdots (4k+2)}{(k+2)!}, \quad k \geq 0$$

Then $A_3 = 1$, $A_4 = 2$, and $A_5 = 5$. These are the Catalan numbers C_1, C_2, and C_3, respectively, shifted by two spaces to the right. So we define $C_n = A_{k+2}$. Thus,

$$C_n = \frac{2 \cdot 6 \cdot 10 \cdots (4n-2)}{(n+1)!}, \quad n \geq 1$$

This can be rewritten as

$$\begin{aligned} C_n &= \frac{4n-2}{n+1} \cdot \frac{2 \cdot 6 \cdot 10 \cdots (4n-6)}{n!} \\ &= \frac{4n-2}{n+1} C_{n-1} \end{aligned}$$

When $n = 1$, this yields $C_1 = C_0$. But $C_1 = 1$. So we can define $C_0 = 1$. Consequently, C_n can be defined recursively.

A Recursive Definition of C_n

$$\begin{aligned} C_0 &= 1 \\ C_n &= \frac{4n-2}{n+1} C_{n-1}, \quad n \geq 1 \end{aligned} \tag{1.11}$$

For example,

$$\begin{aligned} C_4 &= \frac{4 \cdot 4 - 2}{4+1} C_3 \\ &= \frac{14}{5} \cdot 5 = 14 \end{aligned}$$

An Explicit Formula for C_n

The recursive formula (1.11) can be employed to derive an explicit formula for C_n:

$$\begin{aligned} C_n &= \frac{4n-2}{n+1} C_{n-1} \\ &= \frac{(4n-2)(4n-6)}{(n+1)n} C_{n-2} \\ &= \frac{(4n-2)(4n-6)(4n-10)}{(n+1)n(n-1)} C_{n-3} \\ &\vdots \\ &= \frac{(4n-2)(4n-6)(4n-10)\cdots 6 \cdot 2}{(n+1)n \cdots 3 \cdot 2} C_0 \\ &= \frac{(2n-1)(2n-3)(2n-5)\cdots 3 \cdot 1}{(n+1)!} \cdot 2^n \\ &= \frac{2^n (2n)!}{2^n (n+1)! n!} = \frac{(2n)!}{(n+1)! n!} \\ &= \frac{1}{n+1} \binom{2n}{n} \end{aligned}$$

Since $(n+1)|\binom{2n}{n}$ † (see Exercise 20 in Section 1.5), it follows that every Catalan number is a positive integer. The various Catalan numbers are

$$1, 1, 2, 5, 14, 42, 132, 429, 1430, 4862, 16796, 58786, 208012, \ldots$$

It follows from the explicit formula that every Catalan number C_n can be read from Pascal's triangle: Divide each **central binomial coefficient** $\binom{2n}{n}$ by $n+1$; see Figure 1.33.

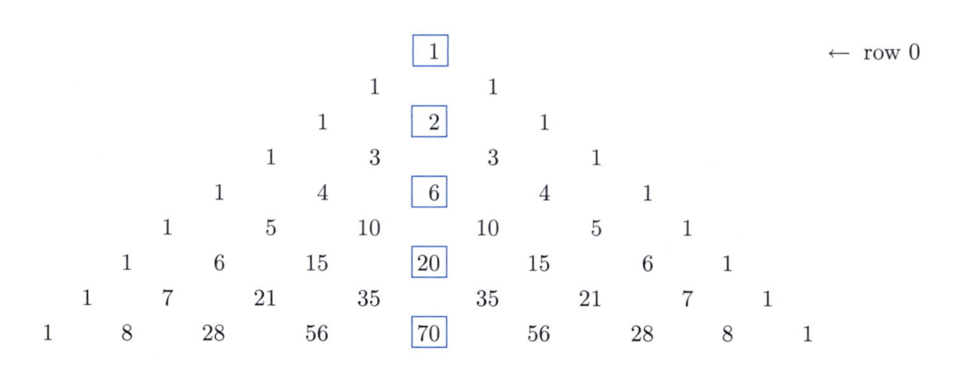

Figure 1.33 *Pascal's Triangle.*

There are several ways of reading C_n from the triangle; see Exercises 1–9.

Segner's Recursive Formula

In 1761, Johann Andreas von Segner (1704–1777), a Hungarian mathematician, physicist, and physician, developed a recursive formula for C_n using the triangulation problem:

$$C_n = C_0C_{n-1} + C_1C_{n-2} + \cdots + C_{n-2}C_1 + C_{n-1}C_0$$

where $n \geq 1$.

For example,

$$\begin{aligned} C_5 &= C_0C_4 + C_1C_3 + C_2C_2 + C_3C_1 + C_4C_0 \\ &= 1 \cdot 14 + 1 \cdot 5 + 2 \cdot 2 + 5 \cdot 1 + 14 \cdot 1 = 42 \end{aligned}$$

† $a|b$ means that a is a factor of b.

In passing, we note that by using generating functions, Segner's formula can be employed to derive the explicit formula for C_n; see Exercises 10–13.

EXERCISES 1.8

Prove each.

1. $C_n = \frac{1}{n}\binom{2n}{n-1}$
2. $C_n = \binom{2n}{n} - \binom{2n}{n-1}$
3. $C_{n+1} = \binom{2n}{n} - \binom{2n}{n-2}$
4. $C_n = \frac{1}{2n+1}\binom{2n+1}{n}$
5. $C_n = \binom{2n-1}{n-1} - \binom{2n-1}{n-2}$
6. $C_n = 2\binom{2n}{n} - \binom{2n+1}{n}$
7. $C_n = \binom{2n+1}{n+1} - 2\binom{2n}{n+1}$

Using the recursive formula

$$C_n = \sum_{r=0}^{\lfloor (n-1)/2 \rfloor} \binom{n-1}{2r} 2^{n-2r-1} C_r$$

(J. Touchard, 1928)

compute C_n for each value of n.

8. $n = 5$
9. $n = 6$

Prove each, where $C(x) = \sum_{n=0}^{\infty} C_n x^n$.

*10. $[C(x)]^2 = \frac{C(x) - C_0}{x}$

*11. $C(x) = \frac{1 - \sqrt{1-4x}}{2}$

*12. $C_n = \frac{1}{n+1}\binom{2n}{n}$

(*Hint*: $\sqrt{1-4x} = 1 - 2\sum_{n=1}^{\infty} C_{n-1}x^n$)

CHAPTER SUMMARY

This chapter presented several properties governing integers and two classes of figurate numbers—polygonal and pyramidal. The principle of induction is an extremely useful proof technique, which we will be using frequently in later chapters. Recursion is another powerful problem-solving tool.

The Order Relation

- An integer a is less than an integer b, denoted by $a < b$, if $b - a$ is a positive integer. We then also write $b > a$. If $a < b$ or $a = b$, we write $a \leq b$ or $b \geq a$. (p. 4)
- **law of trichotomy**: Given any two integers a and b, either $a < b$, $a = b$, or $a > b$. (p. 5)

Absolute Value

- The absolute value of an integer x, denoted by $|x|$, is x if $x \geq 0$ and $-x$ otherwise. (p. 5)

Floor and Ceiling Functions

- The floor of a real number x, denoted by $\lfloor x \rfloor$, is the greatest integer $\leq x$; the ceiling of x, denoted by $\lceil x \rceil$, is the least integer $\geq x$. (p. 6)

The Summation Notation

- $\sum_{i=k}^{i=m} a_i = \sum_{i=k}^{m} a_i = a_k + a_{k+1} + \cdots + a_m$ (p. 9)
- The summation notation satisfies the following properties:

$$\sum_{i=1}^{n} c = nc \quad \text{(p. 10)}$$

$$\sum_{i=1}^{n} (ca_i) = c\left(\sum_{i=1}^{n} a_i\right) \quad \text{(p. 11)}$$

$$\sum_{i=1}^{n} (a_i + b_i) = \left(\sum_{i=1}^{n} a_i\right) + \left(\sum_{i=1}^{n} b_i\right) \quad \text{(p. 11)}$$

Indexed Summation

- $\sum_{i \in I} a_i$ = sum of the values of a_i as i takes on values from the set I. (p. 11)
- $\sum_{P} a_i$ = sum of the values of a_i, where i has the properties P. (p. 11)

The Product Notation

- $\prod_{i=k}^{i=m} a_i = \prod_{i=k}^{m} a_i = a_k a_{k+1} \cdots a_m$ (p. 13)

The Factorial Function

- $n! = \begin{cases} n(n-1)\cdots 3 \cdot 2 \cdot 1 & \text{if } n \geq 1 \\ 1 & \text{if } n = 0 \end{cases}$ (p. 13)

The Well-Ordering Principle

Every nonempty set of positive integers has a least element. (p. 16)

Mathematical Induction

- *weak version* Let $P(n)$ be a statement such that
  - $P(n_0)$ is true; and
  - $P(k)$ implies $P(k+1)$ for any $k \geq n_0$.

  Then $P(n)$ is true for every $n \geq n_0$. (p. 18)

- *strong version* Let $P(n)$ be a statement such that
  - $P(n_0)$ is true; and
  - if $P(n_0), P(n_0+1), \ldots, P(k)$ are true for any $k \geq n_0$, then $P(k+1)$ is also true.

  Then $P(n)$ is true for every $n \geq n_0$. (p. 23)

Recursion

- The recursive definition of a function consists of a recurrence relation, and one or more initial conditions. (p. 27)
- A simple class of recurrence relations can be solved using iteration. (p. 30)

Binomial Coefficients

- $\binom{n}{r} = \dfrac{n!}{r!(n-r)!}$ (p. 33)
- $\binom{n}{0} = 1 = \binom{n}{n}$, $\binom{n}{r} = \binom{n}{n-r}$ (p. 33)
- $\binom{n}{r} = \binom{n-1}{r-1} + \binom{n-1}{r}$ (Pascal's identity) (p. 34)

Binomial Theorem

- $(x+y)^n = \sum_{r=0}^{n} \binom{n}{r} x^{n-r} y^r$ (p. 37)

Polygonal Numbers

- Triangular numbers

$$t_n = \frac{n(n+1)}{2} \quad \text{(p. 40)}$$

$$= t_{n-1} + n, \quad \text{where } t_1 = 1 \quad \text{(p. 41)}$$

- Square numbers

$$s_n = n^2 \quad \text{(p. 44)}$$

$$= s_{n-1} + 2n - 1, \quad \text{where } s_1 = 1 \quad \text{(p. 44)}$$

- The sum of any two consecutive triangular numbers is a square. (p. 45)
- $t_{n-1}^2 + t_n^2 = t_{n^2}$ (p. 45)
- $8t_n + 1 = (2n+1)^2$ (p. 46)
- $8t_{n-1} + 4n = (2n)^2$ (p. 46)
- Pentagonal numbers $p_n = \dfrac{n(3n-1)}{2}$ (p. 46)
- $t_{n-1} + s_n = p_n$ (p. 47)
- Hexagonal numbers $h_n = n(2n-1)$ (p. 48)

- $p_n + t_{n-1} = h_n$ (p. 48)

Pyramidal Numbers

- Triangular pyramidal numbers

$$T_n = T_{n-1} + \frac{n(n+1)}{2} \quad \text{(p. 49)}$$
$$= \frac{n(n+1)(n+2)}{6} \quad \text{(p. 50)}$$

- Square pyramidal number $S_n = [n(n+1)(2n+1)]/6$ (p. 51)
- Pentagonal pyramidal numbers P_n (p. 51)
- Hexagonal pyramidal numbers H_n (p. 51)

Catalan Numbers

$$C_n = \frac{1}{n+1}\binom{2n}{n} \quad \text{(p. 55)}$$
$$= C_0C_{n-1} + C_1C_{n-2} + \cdots + C_{n-1}C_0 \text{ (Segner's formula)} \quad \text{(p. 56)}$$

REVIEW EXERCISES

Evaluate each.

1. $\sum_{i=1}^{n} i(i+1)$
2. $\sum_{i=1}^{n}\sum_{j=1}^{n}(2i+3j)$
3. $\sum_{i=1}^{n}\sum_{j=1}^{n} 2^i 3^j$
4. $\sum_{i=1}^{n}\sum_{j=1}^{i} 2^j$
5. $\prod_{i=1}^{n}\prod_{j=1}^{n} 2^i 3^j$
6. $\prod_{i=1}^{n}\prod_{j=1}^{i} 3^{2j}$
7. $\prod_{i=1}^{n}\prod_{j=1}^{i} 2^i$

*8. $\sum_{i=1}^{n}\prod_{j=1}^{i} ij$

9. $\prod_{r=0}^{n} 2^{\binom{n}{r}}$
10. $\prod_{r=0}^{n} 2^{t_r}$

Find the value of x resulting from the execution of each algorithm fragment.

11.
```
x ← 0
for i = 1 to n do
  for j = 1 to n do
    x ← x + 1
```

12.
```
x ← 0
for i = 1 to n do
  for j = 1 to i do
    for k = 1 to j do
      x ← x + 1
```

In Exercises 13 and 14, the nth term a_n of a number sequence is defined recursively. Compute a_5.

13. $a_1 = a_2 = 1, \quad a_3 = 2$
$a_n = a_{n-1} + a_{n-2} + a_{n-3}, \quad n \geq 4$

14. $a_1 = 0, \quad a_2 = a_3 = 1$
$a_n = a_{n-1} + 2a_{n-2} + 3a_{n-3}, \quad n \geq 4$

(A modified handshake problem) Mrs. and Mr. Matrix host a party for n married couples. At the party, each person shakes hands with everyone else, except his/her spouse. Let $h(n)$ denote the total number of handshakes made.

15. Define $h(n)$ recursively.
16. Predict an explicit formula for $h(n)$.
17. Prove the formula obtained in Exercise 16 for every integer $n \geq 1$.

Using the iterative method, predict an explicit formula satisfied by each recurrence relation.

18. $a_1 = 1 \cdot 2$
$a_n = a_{n-1} + n(n+1), \quad n \geq 2$

19. $a_1 = 2 \cdot 3$
$a_n = 3a_{n-1}, \quad n \geq 2$

20. $a_1 = 1$
$a_n = a_{n-1} + 2^{n-1}, \quad n \geq 2$

21. $a_0 = 0$
$a_n = a_{n-1} + (3n-1), \quad n \geq 1$

*22. Find a formula for the number a_n of times the statement $x \leftarrow x+1$ is executed by the following loop.

for $i = 1$ to n do
  for $j = 1$ to $\lceil i/2 \rceil$ do
    $x \leftarrow x + 1$

23. Prove that one more than four times the product of any two consecutive integers is a perfect square.
24. Prove that the **arithmetic mean** $\frac{a+b}{2}$ of any two real numbers a and b is greater than or equal to their **geometric mean** $\sqrt{ab}$.
(*Hint*: Consider $(\sqrt{a} - \sqrt{b})^2$.)
25. Prove that the equation $x^2 + y^2 = z^2$ has infinitely many integral solutions.

Using induction, prove each.

26. $\sum_{i=1}^{n} (2i-1)^2 = \frac{n(4n^2-1)}{3}$

27. $\sum_{i=1}^{n} \frac{1}{(2i-1)(2i+1)} = \frac{n}{2n+1}$

28–31. Using induction, prove the formulas obtained in Exercises 18–21.

32. Prove that $\binom{2n}{n} = 2\binom{2n-1}{n}$.

33. Prove by induction that $\sum_{i=r}^{n} C(i, r) = C(n+1, r+1)$.

34. Add two lines to the following number pattern.

$$t_1 + t_2 + t_3 = t_4$$
$$t_5 + t_6 + t_7 + t_8 = t_9 + t_{10}$$
$$t_{11} + t_{12} + t_{13} + t_{14} + t_{15} = t_{16} + t_{17} + t_{18}$$ (M. N. Khatri)

35. Verify that $t_n^2 - t_{n-1}^2 = n^3$.
36. Using Exercise 35, show that $\sum_{k=1}^{n} k^3 = [n(n+1)/2]^2$.
37. A **palindrome** is a positive integer that reads the same backwards and forwards. Find the eight palindromic triangular numbers < 1000.

Prove each.

38. $\left(\sum_{k=1}^{n} k\right)^2 = \sum_{k=1}^{n} k^3$.
39. $t_n^2 = t_n + t_{n-1}t_{n+1}$
40. $2t_n t_{n-1} = t_{n^2-1}$
41. $t_{n-k} = t_n + t_k - (n+1)k$ (Casinelli, 1836)
42. $t_n t_k + t_{n-1}t_{k-1} = t_{nk}$ (R. B. Nelsen, 1997)
43. $t_{n-1}t_k + t_n t_{k-1} = t_{nk-1}$ (R. B. Nelsen, 1997)
44. $(2k+1)^2 t_n + t_k = t_{(2k+1)n+k}$ (Euler, 1775)
45. $\dfrac{(nr)!}{(r!)^n}$ is an integer. (Young, 1902)
46. $\dfrac{(nr)!}{n!(r!)^n}$ is an integer. (Feemster, 1910)

Let a_n denote the number of ways a $2 \times n$ rectangular board can be covered with 2×1 dominoes.

47. Define a_n recursively.
48. Find an explicit formula for a_n. (*Hint*: Consider $2 \times (n-1)$ and $2 \times (n-2)$ boards.)

SUPPLEMENTARY EXERCISES

1. Show that $(2mn, m^2 - n^2, m^2 + n^2)$ is a solution of the equation $x^2 + y^2 = z^2$.
2. Prove that $(a^2 + b^2)(c^2 + d^2) = (ac + bd)^2 + (ad - bc)^2$, where a, b, c, and d are any integers.

Using the number pattern in Figure 1.34, answer Exercises 3–5. (Euclides, 1949)

$$1^2 = 1$$

$$3^2 = 2 + 3 + 4$$

$$5^2 = 3 + 4 + 5 + 6 + 7$$

$$7^2 = 4 + 5 + 6 + 7 + 8 + 9 + 10$$

$$\vdots$$

Figure 1.34

3. Add the next two lines.
4. Conjecture a formula for the nth line.
5. Establish the formula in Exercise 4.
6. The array in Figure 1.35 has the property that the sum of the numbers in each band formed by two successive squares is a cube. For example, $3 + 6 + 9 + 6 + 3 = 3^3$. Using this array, establish that $\sum_{i=1}^{n} i^3 = \left(\sum_{i=1}^{n} i\right)^2$. (M. Kraitchik, 1930)

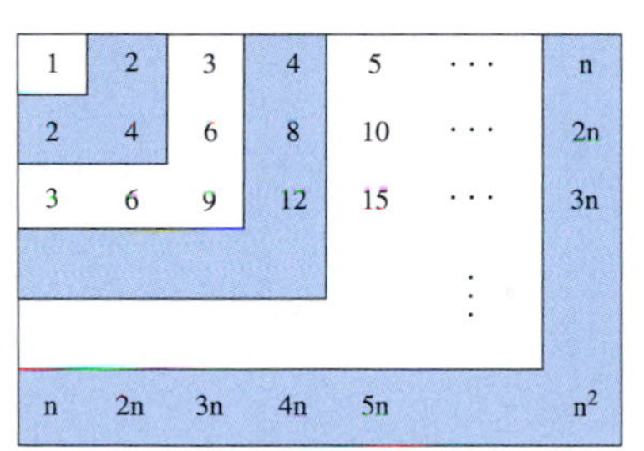

Figure 1.35

7. In 1934, the French mathematician V. Thébault studied the array in Figure 1.36. It consists of rows of arithmetic sequences and possesses several

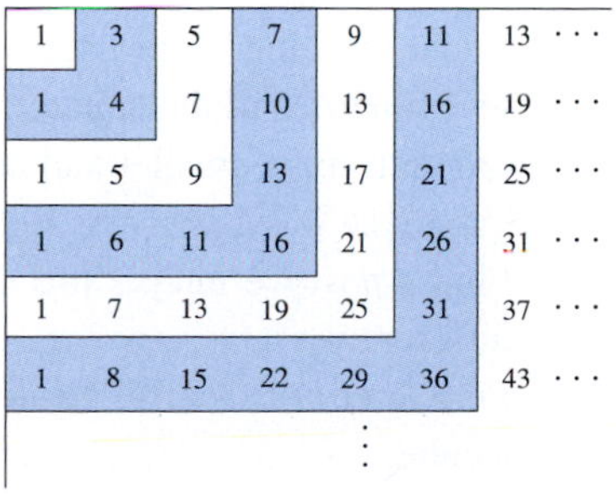

1	3	5	7	9	11	13	···
1	4	7	10	13	16	19	···
1	5	9	13	17	21	25	···
1	6	11	16	21	26	31	···
1	7	13	19	25	31	37	···
1	8	15	22	29	36	43	···
				⋮			

Figure 1.36

interesting properties. For example, the sum of the numbers in the nth band equals n^3 and the main diagonal consists of squares. Using this array, prove that $\sum_{i=1}^{n} i^3 = [n(n+1)/2]^2$.

A side of the equilateral triangle in Figure 1.37 is n units long. Let a_n denote the number of triangles pointing up.

Figure 1.37

8. Define a_n recursively.
9. Solve the recurrence relation.
10. Prove the **Hoggat–Hansell identity**

$$\binom{n-i}{r-i}\binom{n}{r+i}\binom{n+i}{r} = \binom{n-i}{r}\binom{n+i}{r+i}\binom{n}{r-i}$$

Evaluate each.

11. $\sum_{k=0}^{n} \binom{n}{k} k^2$

*12. $\sum_{k=0}^{n} \binom{n}{k} k^3$ (Kuenzi and Prielipp, 1985)

13. In 1950, P. A. Piza discovered the following formula about sums of powers of triangular numbers t_i: $3\left(\sum_{i=1}^{n} t_i\right)^3 = \sum_{i=1}^{n} t_i^3 + 2\sum_{i=1}^{n} t_i^4$. Verify it for $n = 3$ and $n = 4$.

*14. Prove that one more than the product of four consecutive integers is a perfect square, and the square root of the resulting number is the average of the product of the smaller and larger numbers, and the product of the two middle integers. (W. M. Waters, 1990)

*15. Find a positive integer that can be expressed as the sum of two cubes in two different ways.

*16. Find three consecutive positive integers such that the sum of their cubes is also a cube.

*17. Find four consecutive positive integers such that the sum of their cubes is also a cube.

**18. Let S_n denote the sum of the elements in the nth set in the sequence of sets of positive integers $\{1\}, \{3, 5\}, \{7, 9, 11\}, \{13, 15, 17, 19\}, \ldots$. Find a formula for S_n. (R. Euler, 1988)

**19. Let S denote the sum of the elements in the nth set in the sequence of positive integers $\{1\}, \{2, 3, \ldots, 8\}, \{9, 10, \ldots, 21\}, \{22, 23, \ldots, 40\}, \ldots$. Find a formula for S. (C. W. Trigg, 1980)

**20. Let S denote the sum of the numbers in the nth set of the sequence of triangular numbers $\{1\}, \{3, 6\}, \{10, 15, 21\}, \ldots$. Find a formula for S. (J. M. Howell, 1988)

**21. Redo Exercise 20 with the sets of pentagonal numbers $\{1\}, \{5, 12\}, \{22, 35, 51\}, \{70, 92, 117, 145\}, \ldots$.

**22. Three schools in each state, Alabama, Georgia, and Florida, enter one person in each of the events in a track meet. The number of events and the scoring system are unknown, but the number of points for the third place is less than that for the second place which in turn is less than the number of points for the first place. Georgia scored 22 points, and Alabama and Florida tie with 9 each. Florida wins the high jump. Who won the mile run? (M. vos Savant, 1993)

COMPUTER EXERCISES

Write a program to do each task.

1. Read in n positive integers. Find their maximum and minimum using both iteration and recursion.
2. Read in a positive integer $n \leq 20$, and compute the nth Catalan number using recursion.
3. Read in a whole number n, and print Pascal's triangle with $n + 1$ rows.
4. Print the following triangular arrays.

(a)
```
1
1 2
1 2 3
⋮
1 2 3 4 5 6 7 8 9
```

(b)
```
                1
              2 1
            3 2 1
                ⋮
9 8 7 6 5 4 3 2 1
```

5. Find the five Kaprekar numbers < 100.
6. Read in a square array of positive integers, and determine if it is a magic square. If yes, find its magic constant.
7. There are four integers between 100 and 1000, each equal to the sum of its digits. Find them.

8. The integer 1105 can be expressed as the sum of two squares in four different ways. Find them.
9. Find the smallest positive integer that can be expressed as the sum of two cubes in two different ways.
10. Find the smallest positive integer that can be expressed as the sum of two fourth powers in two different ways.
11. Read in a positive integer $n \leq 20$. Using the rules in Example 1.19, print the various moves and the number of moves needed to transfer n disks from peg X to peg Y.
12. Using Exercises 33 and 34 in Section 1.3, compute the total number of grains of wheat needed for the 8×8 chessboard.
(*Hint*: The answer is 18,446,744,073,709,551,615 grains, which may be too large for an integer variable to hold; so think of a suitable data structure.)
13. Using recursion, print the first n:

    a) Triangular numbers. b) Square numbers.
    c) Pentagonal numbers. d) Hexagonal numbers.

14. Print the triangular numbers $\leq 10^4$ that are perfect squares.
15. Print the triangular numbers $\leq 10^4$ that are prime.
16. There are 40 palindromic triangular numbers $< 10^7$. Find them.
17. Search for two triangular numbers t_n such that both t_n and n are palindromic, where $9 \leq n \leq 100$.
18. Find the first three triangular numbers consisting of the same repeated digit.
19. There are 19 palindromic pentagonal numbers $< 10^7$. Find them.
20. Find the largest three-digit integer n whose square is palindromic.
21. Find the least positive integer n such that n^3 is palindromic, but n is not.

ENRICHMENT READINGS

1. A. H. Beiler, *Recreations in the Theory of Numbers*, Dover, New York, 1966.
2. D. Birch, *The King's Chessboard*, Puffin Books, 1993.
3. P. Z. Chinn, "Inductive Patterns, Finite Differences, and a Missing Region," *Mathematics Teacher*, 81 (Sept. 1988), 446–449.
4. U. Dudley, *Mathematical Cranks*, The Math. Association of America, Washington, DC (1992), 200–204.
5. J. Dugle, "The Twelve Days of Christmas and Pascal's Triangle," *Mathematics Teacher*, 75 (Dec. 1982), 755–757.

6. M. Eng and J. Casey, "Pascal's Triangle—A Serendipitous Source for Programming Activities," *Mathematics Teacher*, 76 (Dec. 1983), 686–690.
7. M. Gardner, *Mathematics Magic and Mystery*, Dover, New York, 1956.
8. M. Gardner, "Mathematical Games," *Scientific American*, 234 (June 1976), 120–125.
9. M. Gardner, *Mathematical Puzzles and Diversions*, The University of Chicago Press, Chicago (1987), 130–140.
10. R. Honsberger, *More Mathematical Morsels*, The Math. Association of America, 1991.
11. C. Oliver, "The Twelve Days of Christmas," *Mathematics Teacher*, 70 (Dec. 1977), 752–754.
12. J. K. Smith, "The *n*th Polygonal Number," *Mathematics Teacher*, 65 (March 1972), 221–225.
13. K. B. Strangeman, "The Sum of *n* Polygonal Numbers," *Mathematics Teacher*, 67 (Nov. 1974), 655–658.
14. C. W. Trigg, "Palindromic Triangular Numbers," *J. Recreational Mathematics*, 6 (Spring 1973), 146–147.
15. T. Trotter, Jr., "Some Identities for the Triangular Numbers," *J. Recreational Mathematics*, 6 (Spring 1973), 127–135.

Divisibility

The grandest achievement of the Hindus and the one which, of all mathematical investigations, has contributed to the general progress of intelligence, is the invention of the principle of position in writing numbers.
—F. CAJORI

This chapter continues the study of properties of integers and explores five classes of positive integers: prime numbers, which are the building blocks of integers, composite numbers, Fibonacci numbers, Lucas numbers, and Fermat numbers.

2.1 The Division Algorithm

The division algorithm is a fine application of the well-ordering principle and is often employed to check the correctness of a division problem.

Suppose an integer a is divided by a positive integer b. Then we get a unique **quotient** q and a unique **remainder** r, where the remainder satisfies the condition $0 \le r < b$; a is the **dividend** and b the **divisor**. This is formally stated as follows.

THEOREM 2.1 **(The Division Algorithm)** Let a be any integer and b a positive integer. Then there exist unique integers q and r such that

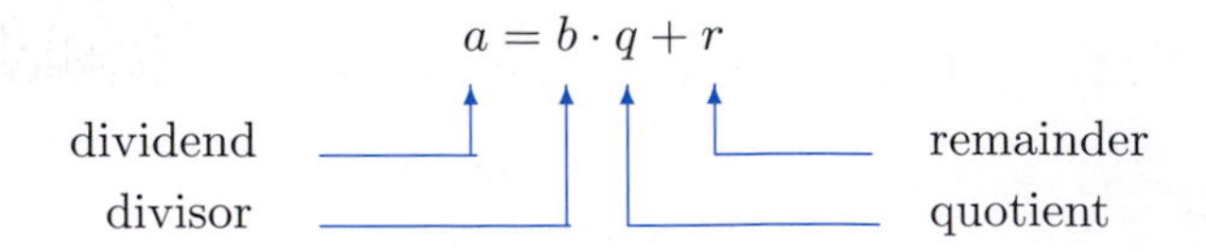

where $0 \leq r < b$.

PROOF

The proof consists of two parts. First, we must establish the existence of the integers q and r, and then we must show they are indeed unique.

1) EXISTENCE PROOF

Consider the set $S = \{a - bn \mid (n \in \mathbf{Z}) \text{ and } (a - bn \geq 0)\}$. Clearly, $S \subseteq \mathbf{W}$. We shall show that S contains a least element. To this end, first we will show that S is a nonempty subset of $\mathbf{W}$:

case 1 Suppose $a \geq 0$. Then $a = a - b \cdot 0 \in S$, so S contains an element.

case 2 Suppose $a < 0$. Since $b \in \mathbf{Z}^+$, $b \geq 1$. Then $-ba \geq -a$; that is, $a - ba \geq 0$. Consequently, $a - ba \in S$.

In both cases, S contains at least one element, so S is a nonempty subset of $\mathbf{W}$. Therefore, by the well-ordering principle, S contains a least element r.

Since $r \in S$, an integer q exists such that $r = a - bq$, where $r \geq 0$.

To show that $r < b$:

We will prove this by contradiction. Assume $r \geq b$. Then $r - b \geq 0$. But $r - b = (a - bq) - b = a - b(q + 1)$. Since $a - b(q + 1)$ is of the form $a - bn$ and is ≥ 0, $a - b(q + 1) \in S$; that is, $r - b \in S$. Since $b > 0$, $r - b < r$. Thus, $r - b$ is smaller than r and is in S. This contradicts our choice of r, so $r < b$.

Thus, there are integers q and r such that $a = bq + r$, where $0 \leq r < b$.

2) UNIQUENESS PROOF

We would like to show that the integers q and r are unique. Assume there are integers q, q', r, and r' such that $a = bq + r$ and $a = bq' + r'$, where $0 \leq r < b$ and $0 \leq r' < b$.

Assume, for convenience, that $q \geq q'$. Then $r' - r = b(q - q')$. Because $q \geq q'$, $q - q' \geq 0$ and hence $r' - r \geq 0$. But, because $r' < b$ and $r < b$, $r' - r < b$.

Suppose $q > q'$; that is, $q - q' \geq 1$. Then $b(q - q') \geq b$; that is, $r' - r \geq b$. This is a contradiction because $r' - r < b$. Therefore, $q \not> q'$; thus, $q = q'$, and hence, $r = r'$. Thus, the integers q and r are unique, completing the uniqueness proof. ■

Although this theorem has been traditionally called the division algorithm, it does not present an algorithm for finding q and r. They can be found using the familiar long division method.

EXAMPLE 2.1 Find the quotient q and the remainder r when

1. 207 is divided by 15.
2. -23 is divided by 5.

SOLUTION

1. $207 = 15 \cdot 13 + 12$; so $q = 13$ and $r = 12$.
2. Since $-23 = 5 \cdot (-4) + (-3)$, you might be tempted to say that $q = -4$ and $r = -3$. The remainder, however, can never be negative. But -23 can be written as $-23 = 5 \cdot (-5) + 2$, where $0 \leq r\ (= 2) < 5$ (see the number line in Figure 2.1). Thus, $q = -5$ and $r = 2$.

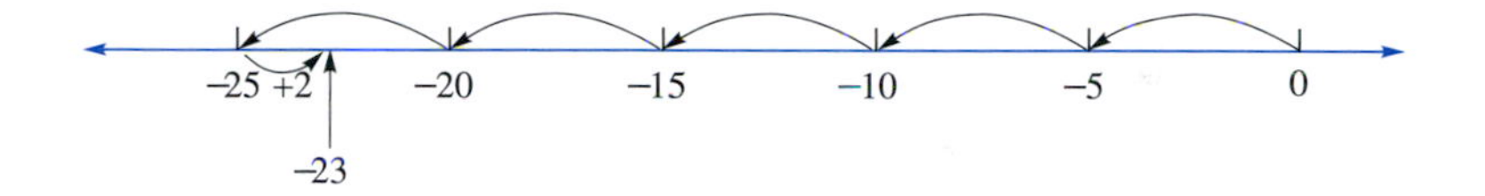

Figure 2.1

■

You may notice that the equation $a = bq + r$ can be written as

$$\frac{a}{b} = q + \frac{r}{b},$$

where $0 \leq r/b < 1$. Consequently, $q = \lfloor a/b \rfloor$ and $r = a - bq = a - b \cdot \lfloor a/b \rfloor$.

Div and Mod Operators

The binary operators, **div** and **mod**, are often used in discrete mathematics and computer science to find quotients and remainders. They are defined as follows:

$$a \textbf{ div } b = \text{quotient when } a \text{ is divided by } b$$

$$a \textbf{ mod } b = \text{remainder when } a \text{ is divided by } b$$

For example, $23 \text{ div } 5 = 4$, and $23 \text{ mod } 5 = 3$; $-23 \text{ div } 5 = -5$, and $-23 \text{ mod } 5 = 2$ (why?).

It now follows from these definitions that $q = a \text{ div } b = \lfloor a/b \rfloor$ and $r = a \text{ mod } b = a - bq = a - b \cdot \lfloor a/b \rfloor$.

The following example is a simple application of both div and mod operators.

Card Dealing (optional)

Consider a standard deck of 52 playing cards. They are originally assigned the numbers 0 through 51 in order. Use the suit labels 0 = clubs, 1 = diamonds, 2 = hearts, and 3 = spades to identify each suit, and the card labels 0 = ace, 1 = deuce, 2 = three, ..., and 12 = king to identify the cards in each suit. Suppose card x is drawn at random from a well-shuffled deck, where $0 \leq x \leq 51$. How do we identify the card?

First, we need to determine the suit to which the card belongs. It is given by x div 13. Next, we need to determine the card within the suit; this is given by x mod 13. Thus, card x is card (x mod 13) in suit (x div 13).

For example, let $x = 50$. Since 50 div 13 = 3, the card is a spade. Now 50 mod 13 = 11, so it is a queen. Thus, card 50 is the queen of spades. ■

Next, we pursue an intriguing application of the floor function and the mod operator to the game of chess.

The Two Queens Puzzle (optional)

There are two queens on an 8×8 chessboard. One can capture the other if they are on the same row, column, or diagonal. The 64 squares on the board are numbered 0 through 63. Suppose one queen is in square x and the other in square y, where $0 \leq x, y \leq 63$. Can one queen capture the other?

Because the squares are labeled 0 through 63, we can label each row with the numbers 0 through 7 and each column with the same numbers 0 through 7. In fact, each row label = $\lfloor r/8 \rfloor$ and each column label = c mod 8, where $0 \leq r, c \leq 63$ (see Figure 2.2). Thus, the queen in square x lies in row $\lfloor x/8 \rfloor$ and column x mod 8, and

	0	1	2	3	4	5	6	7
0	0	1	2	3	4	5	6	7
1	8	9	10	11	12	13	14	15
2	16	17	18	19	20	21	22	23
3	24	25	26	27	28	29	30	31
4	32	33	34	35	36	37	38	39
5	40	41	42	43	44	45	46	47
6	48	49	50	51	52	53	54	55
7	56	57	58	59	60	61	62	63

← column label

↑ row label

Figure 2.2

***Gustav Peter Lejeune Dirichlet** (1805–1859) was born in Duren, Germany. The son of a postmaster, he attended a public school and then a private school that emphasized Latin. After attending the Gymnasium in Bonn for two years, Dirichlet entered a Jesuit college in Cologne, where he received a strong background in theoretical physics under the physicist Georg Simon Ohm. In 1822, he moved to the University of Paris.*

In 1826, Dirichlet returned to Germany and taught at the University of Breslau. Three years later, he moved to the University of Berlin, where he spent the next 27 years.

Dirichlet's greatest interest in mathematics was number theory, and he was inspired by Gauss' masterpiece, Disquisitiones Arithmeticae *(1801). He established* Fermat's Last Theorem *for $n = 14$. Among the many results he discovered are the proof of a theorem presented to the Paris Academy of Sciences on algebraic number theory in 1837: The sequence $\{an + b\}$ contains infinitely many primes, where a and b are relatively prime.*

When Gauss died in 1855, Dirichlet moved to the University of Göttingen. Three years later, he went to Montreaux, Switzerland, to deliver a speech in honor of Gauss. While there, he suffered a heart attack and was barely able to return home. During his illness his wife succumbed to a stroke, and Dirichlet died soon after.

that in square y lies in row $\lfloor y/8 \rfloor$ and column y mod 8. Consequently, the two queens will be in the same row if and only if $\lfloor x/8 \rfloor = \lfloor y/8 \rfloor$, and in the same column if and only if $x \bmod 8 = y \bmod 8$. For example, if $x = 41$ and $y = 47$, the two queens lie on the same row.

How do we determine if they lie on the same diagonal? There are 15 northeast diagonals and 15 southeast diagonals. With a bit of patience, we can show that the queens lie on the same diagonal if and only if the absolute value of the difference of their row labels equals that of the difference of their column labels; that is, if and only if $|\lfloor x/8 \rfloor - \lfloor y/8 \rfloor| = |x \bmod 8 - y \bmod 8|$.

For example, let $x = 51$ and $y = 23$ (see Figure 2.2). Then $|\lfloor 51/8 \rfloor - \lfloor 23/8 \rfloor| = |6 - 2| = 4 = |3 - 7| = |51 \bmod 8 - 23 \bmod 8|$, so one queen captures the other. On the other hand, if $x = 49$ and $y = 13$, then $|\lfloor 49/8 \rfloor - \lfloor 13/8 \rfloor| \neq |49 \bmod 8 - 13 \bmod 8|$; so one queen cannot capture the other. ■

The Pigeonhole Principle and the Division Algorithm

The **pigeonhole principle** is also known as the **Dirichlet box principle** after the German mathematician Gustav Peter Lejeune Dirichlet who used it extensively in his work on number theory. It can be applied to a variety of situations.

Suppose m pigeons fly into n pigeonholes to roost, where $m > n$. What is your conclusion? Because there are more pigeons than pigeonholes, at least two pigeons must roost in the same pigeonhole; in other words, there must be a pigeonhole containing two or more pigeons (see Figure 2.3).

Figure 2.3

We now state and prove the simple version of the pigeonhole principle.

THEOREM 2.2 **(The Pigeonhole Principle)** If m pigeons are assigned to n pigeonholes, where $m > n$, then at least two pigeons must occupy the same pigeonhole.

PROOF **(by contradiction)**
Suppose the given conclusion is false; that is, no two pigeons occupy the same pigeonhole. Then every pigeon must occupy a distinct pigeonhole, so $n \geq m$, which is a contradiction. Thus, two or more pigeons must occupy some pigeonhole. ■

Next, we move on to the divisibility relation.

The Divisibility Relation

Suppose we let $r = 0$ in the division algorithm. Then $a = bq + 0 = bq$. We then say that b **divides** a, b is a **factor** of a, a is **divisible** by b, or a is a **multiple** of b, and write $b|a$. If b is *not* a factor of a, we write $b \nmid a$.

For instance, $3|12$, $5|30$, but $6 \nmid 15$.

The following example illustrates the pigeonhole principle.

EXAMPLE 2.2 Let b be an integer ≥ 2. Suppose $b + 1$ integers are randomly selected. Prove that the difference of two of them is divisible by b.

PROOF
Let q be the quotient and r the remainder when an integer a is divided by b. Then, by the division algorithm, $a = bq + r$, where $0 \leq r < b$. The $b + 1$ integers yield $b + 1$ remainders (pigeons), but there are only b possible remainders (pigeonholes). Therefore, by the pigeonhole principle, two of the remainders must be equal.

Let x and y be the corresponding integers. Then $x = bq_1 + r$ and $y = bq_2 + r$ for some quotients q_1 and q_2. Therefore,

$$\begin{aligned} x - y &= (bq_1 + r) - (bq_2 + r) \\ &= b(q_1 - q_2) \end{aligned}$$

Thus, $x - y$ is divisible by b. ■

Before we pursue divisibility properties, let us digress for a while with an interesting puzzle.

An Intriguing Puzzle (optional)

Think of a three-digit number *abc*. Multiply *abc* and the successive answers by 7, 11, and 13, respectively. Your answer is *abcabc*. Surprised? Can you explain why it works this way?

Next, we study several useful divisibility properties. We leave them as routine exercises.

THEOREM 2.3 Let a and b be positive integers such that $a|b$ and $b|a$. Then $a = b$. ■

THEOREM 2.4 Let a, b, c, α, and β be any integers.† Then

1. If $a|b$ and $b|c$, then $a|c$. (**transitive property**)
2. If $a|b$ and $a|c$, then $a|(\alpha b + \beta c)$.
3. If $a|b$, then $a|bc$. ■

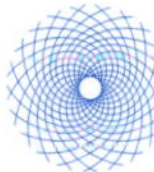

The expression $\alpha b + \beta c$ is called a **linear combination** of b and c. Thus, by part 2, if a is a factor of b and c, then a is also a factor of any linear combination of b and c. In particular, $a|(b + c)$ and $a|(b - c)$ (why?).

The floor function can be used to determine the number of positive integers less than or equal to a positive integer a and divisible by a positive integer b, as the next theorem shows.

THEOREM 2.5 Let a and b be any positive integers. Then the number of positive integers $\leq a$ and divisible by b is $\lfloor a/b \rfloor$.

† α and β are the Greek letters *alpha* and *beta*.

PROOF

Suppose there are k positive integers $\leq a$ and divisible by b. We need to show that $k = \lfloor a/b \rfloor$. The positive multiples of b less than or equal to a are $b, 2b, \ldots, kb$. Clearly, $kb \leq a$, that is, $k \leq a/b$. Further, $(k+1)b > a$. Thus, $k + 1 > a/b$ or $a/b - 1 < k$. Therefore,

$$\frac{a}{b} - 1 < k \leq \frac{a}{b}$$

Thus, k is the largest integer less than or equal to a/b, so $k = \lfloor a/b \rfloor$. ■

For example, the number of positive integers ≤ 2076 and divisible by 19 is $\lfloor 2076/19 \rfloor = \lfloor 109.26316 \rfloor = 109$.

Next, we consider some aspects of sets and the inclusion–exclusion principle.

Union, Intersection, and Complement

Let A be a finite set and $|A|$ the number of elements in A. For example, if $A = \{3, 5, 8, 17\}$, then $|A| = 4$. (In Chapter 1, we used vertical bars to denote the absolute value of a number, but here it denotes the number of elements in a set. The meaning of the notation should be clear from the context, so be a bit careful.)

Let A and B be any two sets. Their **union** $A \cup B$ is the set of elements belonging to A or B; their **intersection** $A \cap B$ consists of the common elements; A' denotes the **complement** of A, that is, the set of elements in the universal set that are *not* in A.

We now move on to the inclusion–exclusion principle. Let A and B be finite sets. Let $|A \cap B| = n$, $|A| = n + r$, and $|B| = n + s$ for some integers $n, r, s \geq 0$. Then $|A \cup B| = n + r + s = (n + r) + (n + s) - n = |A| + |B| - |A \cap B|$. Likewise, $|A \cup B \cup C| = |A| + |B| + |C| - |A \cap B| - |B \cap C| - |C \cap A| + |A \cap B \cap C|$.

More generally, we have the following result.

THEOREM 2.6 **(The Inclusion–Exclusion Principle)** Let $A_1, A_2, \ldots, A_n$ be n finite sets. Then

$$\left|\bigcup_{i=1}^{n} A_i\right| = \sum_{1 \leq i \leq n} |A_i| - \sum_{1 \leq i < j \leq n} |A_i \cap A_j| + \sum_{1 \leq i < j < k \leq n} |A_i \cap A_j \cap A_k| - \cdots + (-1)^{n+1}\left|\bigcap_{i=1}^{n} A_i\right|$$ ■

The next two examples are simple applications of this theorem.

EXAMPLE 2.3 Find the number of positive integers ≤ 2076 and divisible by neither 4 nor 5.

SOLUTION

Let $A = \{x \in \mathbf{N} \mid x \leq 2076 \text{ and divisible by } 4\}$ and $B = \{x \in \mathbf{N} \mid x \leq 2076 \text{ and divisible by } 5\}$. Then

$$\begin{aligned}|A \cup B| &= |A| + |B| - |A \cap B| \\ &= \lfloor 2076/4 \rfloor + \lfloor 2076/5 \rfloor - \lfloor 2076/20 \rfloor \\ &= 519 + 415 - 103 = 831\end{aligned}$$

Thus, among the first 2076 positive integers, there are $2076 - 831 = 1245$ integers *not* divisible by 4 or 5. ■

EXAMPLE 2.4 Find the number of positive integers ≤ 3000 and divisible by 3, 5, or 7.

SOLUTION

Let A, B, and C denote the sets of positive integers ≤ 3000 and divisible by 3, 5, or 7. By the inclusion–exclusion principle,

$$\begin{aligned}|A \cup B \cup C| &= |A| + |B| + |C| - |A \cap B| - |B \cap C| - |C \cap A| + |A \cap B \cap C| \\ &= \lfloor 3000/3 \rfloor + \lfloor 3000/5 \rfloor + \lfloor 3000/7 \rfloor - \lfloor 3000/15 \rfloor - \lfloor 3000/35 \rfloor \\ &\quad - \lfloor 3000/21 \rfloor + \lfloor 3000/105 \rfloor \\ &= 1000 + 600 + 428 - 200 - 85 - 142 + 28 = 1629\end{aligned}$$

■

In October 1582, at the request of Pope Gregory XIII, Fr. Christopher Clavius and Aloysius Giglio introduced the Gregorian calendar to rectify the errors of the Julian calendar. In the Gregorian calendar, which is now universally used, a nonleap year contains 365 days and a leap year contains 366 days. (A year is a **leap year** if it is a century divisible by 400 or if it is a noncentury and divisible by 4. For example, 1600 and 1976 were leap years, whereas 1778 and 1900 were not.) The following example shows how to derive a formula to compute the number of leap years beyond 1600 and not exceeding a given year y. (See Section 5.6 also.)

EXAMPLE 2.5 Show that the number of leap years ℓ after 1600 and not exceeding a given year y is given by $\ell = \lfloor y/4 \rfloor - \lfloor y/100 \rfloor + \lfloor y/400 \rfloor - 388$.

PROOF

Let n be a year such that $1600 < n \le y$. To derive the formula for ℓ, we proceed step by step:

step 1 *Find the number of years n in the range divisible by* 4.

Let $4n_1$ be such a year. Then $1600 < 4n_1 \le y$; that is, $400 < n_1 \le y/4$. Therefore, there are $n_1 = \lfloor y/4 \rfloor - 400$ such years.

step 2 *Find the number of centuries in the range* $1600 < n \le y$.

Let $100n_2$ be a century such that $1600 < 100n_2 \le y$. Then $16 < n_2 \le y/100$. Therefore, there are $n_2 = \lfloor y/100 \rfloor - 16$ centuries beyond 1600 and $\le y$.

step 3 *Find the number of centuries in the range divisible by* 400.

Since they are of the form $400n_3$, we have $1600 < 400n_3 \le y$. Then $4 < n_3 \le y/400$, so $n_3 = \lfloor y/400 \rfloor - 4$.

step 4 Therefore,

$$\begin{aligned}\ell &= n_1 - n_2 + n_3 \\ &= \lfloor y/4 \rfloor - 400 - \lfloor y/100 \rfloor + 16 + \lfloor y/400 \rfloor - 4 \\ &= \lfloor y/4 \rfloor - \lfloor y/100 \rfloor + \lfloor y/400 \rfloor - 388\end{aligned}$$ ■

We now return to the division algorithm and discuss some divisibility properties involving even and odd integers.

Even and Odd Integers

Suppose we let $b = 2$ in the division algorithm. Then $a = 2q + r$, where $0 \le r < 2$. So $r = 0$ or 1. When $r = 0$, $a = 2q$; such integers are **even** integers. When $r = 1$, $a = 2q + 1$; such integers are **odd** integers. It follows from this definition that every integer is either even or odd, but *not* both.

The Pythagoreans considered odd numbers male and good, and even numbers female and bad. The number 1 was considered neither male nor female. The number 5, being the sum of the first masculine and feminine numbers, was considered a symbol of marriage. Some philosophers, supported by early Christian theologians, identified the number with God.

The following properties were also known to the Pythagoreans. We shall leave them as exercises; see Exercises 40–46.

- The sum of any two even integers is even.
- The product of any two even integers is even.

- The sum of any two odd integers is even.
- The product of any two odd integers is odd.
- The sum of an even integer and an odd integer is odd.
- The product of an even integer and an odd integer is even.
- If the square of an integer is even, then the integer is even.
- If the square of an integer is odd, then the integer is odd.

EXERCISES 2.1

Find the quotient and the remainder when the first integer is divided by the second.

1. 78, 11
2. 57, 75
3. −325, 13
4. −23, 25

Let $f(n)$ denote the number of positive factors of a positive integer n. Evaluate each.

5. $f(16)$
6. $f(12)$
7. $f(15)$
8. $f(17)$

Find the number of positive integers ≤ 3076 and

9. Divisible by 19
10. Divisible by 23
11. Not divisible by 17
12. Not divisible by 24

Find the number of positive integers in the range 1976 through 3776 that are

13. Divisible by 13
14. Divisible by 15
15. Not divisible by 17
16. Not divisible by 19

Mark *true* or *false*, where a, b, and c are arbitrary integers.

17. $1|a$
18. If $a|b$, then $a|-b$.
19. $a|0$
20. If $a|b$ and $b|a$, then $a=b$.
21. If $a|b$, then $a<b$.
22. If $a<b$, then $a|b$.
23. If $a|b$ and $b|c$, then $a|c$.
24. If $a\nmid b$, then $b\nmid a$.
25. Zero is neither even nor odd.
26. There is no remainder when an even integer is divided by 2.

Prove or disprove each statement, where a, b, and c are arbitrary integers.

27. If $a^2=b^2$, then $a=b$.
28. If $a|b$ and $b|a$, then $a=b$.
29. If $a|(b+c)$, then $a|b$ and $a|c$.
30. If $a|bc$, then $a|b$ and $a|c$.

Evaluate each, where d is a positive integer.

31. $\sum_{d|12} d$
32. $\sum_{d|12} 1$
33. $\sum_{d|18} \left(\frac{1}{d}\right)$
34. $\sum_{d|18} \left(\frac{18}{d}\right)$
35. A **nude number** is a natural number n such that each of its digits is a factor of n. Find all three-digit odd nude numbers containing no repeated digits.

Let f be a function defined recursively by

$$f(n)=\begin{cases}1 & \text{if } 3|n\\ f(n+1) & \text{otherwise}\end{cases}$$

36. Find $f(16)$
37. Find an explicit formula for $f(n)$.

Prove each, where a and b are positive integers.

38. If $a|b$ and $b|a$, then $a=b$.
39. If $a|b$ and $c|d$, then $ac|bd$.
40. The sum and the product of any two even integers are even.
41. The sum of any two odd integers is even.
42. The product of any two odd integers is odd.
43. The sum of an even integer and an odd integer is odd.
44. If the square of an integer is even, then the integer must be even.
45. If the square of an integer is odd, then the integer must be odd.
46. The product of any two consecutive integers is even.

47. The sum of any two integers of the form $4k+1$ is even.
48. Every odd integer is of the form $4k+1$ or $4k+3$.
49. The product of any two integers of the form $3k+1$ is also of the same form.
50. The product of any two integers of the form $4k+1$ is also of the same form.
51. If the product of two integers is even, then at least one of them must be even.
52. If the product of two integers is odd, then both must be odd.

Prove each by cases, where n is an arbitrary integer.

53. n^2+n is an even integer.
54. $2n^3+3n^2+n$ is an even integer.
55. n^3-n is divisible by 2.
56. $30|(n^5-n)$
57. Derive the inclusion–exclusion principle for three finite sets A, B, and C.
58. Prove that the difference of the squares of two positive integers cannot be 1.
59. Prove that the product of any four consecutive positive integers cannot be a perfect square. (*Hint*: Use Exercise 58.)
60. Prove that if the sum of the cubes of three consecutive integers is a cube k^3, then $3|k$.
61. Show that the equation $n^3+(n+1)^3+(n+2)^3=(n+3)^3$ has a unique solution. (*Hint*: Use Exercise 60.)

Using induction prove each, where n is a nonnegative integer. (*Hint*: Use the binomial theorem for Exercises 64 and 65.)

62. $2n^3+3n^2+n$ is divisible by 6.
63. $n^4+2n^3+n^2$ is divisible by 4.
64. $2^{4n}+3n-1$ divisible by 9.
65. $4^{2n}+10n-1$ divisible by 25.
66. Find the largest nontrivial factor of $2^{30}-1$.

*2.2 Base-*b* Representations (optional)

The division algorithm can be used to convert a decimal integer to any other base. Furthermore, additions and multiplications can be carried out in any base, and subtraction can be accomplished using addition, as in base ten.

In everyday life, we use the decimal notation, base 10, to represent any real number. For example, $234=2(10^2)+3(10^1)+4(10^0)$, which is the **decimal expansion** of 234. Likewise, $23.45=2(10^1)+3(10^0)+4(10^{-1})+5(10^{-2})$. Computers use base two (binary); very long binary numbers are often handled by human beings using base eight (**octal**) and base sixteen (**hexadecimal**).

Actually, any positive integer $b\geq 2$ is a valid choice for a base. This is a consequence of the following fundamental result, the proof of which is a bit long but straightforward.

THEOREM 2.7 Let b be a positive integer ≥ 2. Then every positive integer N can be expressed uniquely in the form $N=a_kb^k+a_{k-1}b^{k-1}+\cdots+a_1b+a_0$, where $a_0, a_1, \ldots, a_k$ are nonnegative integers less than b, $a_k\neq 0$, and $k\geq 0$.

PROOF

The proof consists of two parts: the existence half and the uniqueness half. The existence half, applying the division algorithm, establishes the existence of such an expansion for N using powers of b; the uniqueness half shows that such an expansion is unique.

To show that N has the desired expansion:

Apply the division algorithm with N as the dividend and b as the divisor:

$$N = bq_0 + a_0, \quad 0 \le a_0 < b$$

If $q_0 \neq 0$, apply the division algorithm again with q_0 as the new dividend:

$$q_0 = bq_1 + a_1, \quad 0 \le a_1 < b$$

Continuing like this, we get a sequence of equations:

$$\begin{aligned} q_1 &= bq_2 + a_2, & 0 \le a_2 < b \\ q_2 &= bq_3 + a_3, & 0 \le a_3 < b \\ &\vdots \\ q_{k-2} &= bq_{k-1} + a_{k-1}, & 0 \le a_{k-1} < b \end{aligned}$$

where $N > q_0 > q_1 > q_2 > \cdots$. Because $q_0, q_1, q_2, \ldots$ is a decreasing sequence of nonnegative integers, this procedure must eventually terminate with the last step:

$$q_{k-1} = b \cdot 0 + a_k, \quad 0 \le a_k < b$$

To get the desired form, we begin substituting for each q_i, beginning with the first equation:

$$N = bq_0 + a_0$$

Substitute for q_0:

$$N = b(bq_1 + a_1) + a_0 = q_1 b^2 + a_1 b + a_0$$

Now substitute for q_1 and continue the procedure:

$$\begin{aligned} N &= q_2 b^3 + a_2 b^2 + a_1 b + a_0 \\ &\vdots \\ &= q_{k-1} b^k + a_{k-1} b^{k-1} + \cdots + a_2 b^2 + a_1 b + a_0 \\ &= a_k b^k + a_{k-1} b^{k-1} + \cdots + a_2 b^2 + a_1 b + a_0 \end{aligned}$$

where $0 \le a_i < b$ for every i. Also, $a_k \neq 0$, since $a_k = q_{k-1}$ is the last nonzero quotient. Thus, N has the desired expansion.

To show that the expansion of N is unique:

Suppose N has two expansions:

$$N = \sum_{i=0}^{k} a_i b^i = \sum_{i=0}^{k} c_i b^i$$

where $0 \le a_i, c_i < b$. (We can assume both expansions contain the same number of terms, since we can always add enough zero coefficients to yield the same number of terms.) Subtracting one expansion from the other yields $\sum_{i=0}^{k} (a_i - c_i) b^i = 0$. Let $d_i = a_i - c_i$. Then $\sum_{i=0}^{k} d_i b^i = 0$. If every $d_i = 0$, then $a_i = c_i$ for every i, so the two expansions are the same.

If the expansions are distinct, there must be a smallest integer j, where $0 \le j \le k$, such that $d_j \neq 0$. Then

$$\sum_{i=j}^{k} d_i b^i = 0$$

Factor out b^j:

$$b^j \left(\sum_{i=j}^{k} d_i b^{i-j} \right) = 0$$

Cancel b^j:

$$\sum_{i=j}^{k} d_i b^{i-j} = 0$$

This yields

$$d_j + b \left(\sum_{i=j+1}^{k} d_i b^{i-j-1} \right) = 0$$

$$b \left(\sum_{i=j+1}^{k} d_i b^{i-j-1} \right) = -d_j$$

Thus, $b|d_j$. But, since $0 \le a_i, c_i < b$, $-b \le a_i - c_i < b$; that is, $-b \le d_j < b$. Therefore, since $b|d_j$, $d_j = 0$, which contradicts our assumption that $d_j \neq 0$.

Thus, the two expansions are the same, establishing the uniqueness of the expansion. This concludes the proof. ■

This theorem leads us to the following definition.

Base-b Representation

The expression $a_k b^k + a_{k-1} b^{k-1} + \cdots + a_1 b + a_0$ is the **base-*b* expansion** of the integer N. Accordingly, we write $N = (a_k a_{k-1} \ldots a_1 a_0)_b$ in base b.

When the base is two, the expansion is called the **binary expansion**. When $b = 2$, each coefficient is 0 or 1; these two digits are called **bi**nary digi**ts** (or **bits**).

The number system with base ten is the **decimal system**, from the Latin word *decem*, meaning ten.[†] It was invented in India around the third century B.C., and carried to Spain in A.D. 711 by Arabs who traded with India.

The decimal system employs the ten digits 0 through 9 to represent any number. The principal reason for this choice is undoubtedly that in earlier times men and women used their fingers for counting and computing, as some still do today.

The base is omitted when it is ten. For example, $234_{\text{ten}} = 234$ and $(10110)_{\text{two}} = 22$ (see Example 2.6).

When the base is greater than ten, we use the letters $A, B, C, \ldots$ to represent the *digits* ten, eleven, twelve, ... respectively, to avoid any possible confusion. It is easy to find the decimal value of an integer from its base-b representation, as the next two examples illustrate.

EXAMPLE 2.6 Express 10110_{two} in base ten.

SOLUTION

$$\begin{aligned} 10110_{\text{two}} &= 1(2^4) + 0(2^3) + 1(2^2) + 1(2^1) + 0(2^0) \qquad \leftarrow \textit{binary expansion} \\ &= 16 + 0 + 4 + 2 + 0 = 22 \end{aligned}$$

■

EXAMPLE 2.7 Express $3ABC_{\text{sixteen}}$ in base ten.

SOLUTION

Recall that $A = 10$, $B = 11$, and $C = 12$. Therefore,

---

† December was the tenth month of the ancient Roman year. Decemvir was a member of a council of ten magistrates in ancient Rome.

$$\begin{aligned} 3ABC_{\text{sixteen}} &= 3(16^3) + 10(16^2) + 11(16^1) + 12(16^0) \\ &= 12{,}288 + 2560 + 176 + 12 = 15{,}036 \end{aligned}$$

■

Conversely, suppose we are given a decimal integer. How do we express it in another base b? By Theorem 2.9, all we have to do is express it as a sum of powers of b, then simply collect the coefficients in the correct order. Always remember to account for missing coefficients.

This method is illustrated in the following example.

EXAMPLE 2.8 Express 3014 in base eight.

SOLUTION

The largest power of 8 that is contained in 3014 is 512. Apply the division algorithm with 3014 as the dividend and 512 as the divisor:

$$3014 = 5 \cdot 512 + 454$$

Now look at 454. It lies between 64 and 512. The largest power of 8 we can now use is 64:

$$454 = 7 \cdot 64 + 6$$

Continue like this until the remainder becomes less than 8:

$$6 = 6 \cdot 1 + 0$$

Thus, we have

$$\begin{aligned} 3014 &= 5(512) + 7(64) + 6 \\ &= 5(8^3) + 7(8^2) + 0(8^1) + 6(8^0) \\ &= 5706_{\text{eight}} \end{aligned}$$

■

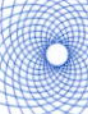

A simple algorithm expresses an integer a in any nondecimal base b: divide a and its successive quotients by b until a zero quotient is reached, then pick the remainders in the reverse order. These steps can be translated into the elegant algorithm given in Algorithm 2.1.

Algorithm nondecimal base (n, b)

(* This algorithm finds the base-b representation $(a_m a_{m-1} \ldots a_1 a_0)_b$ of a positive integer n. The variables q and r denote the quotient and the remainder of the division algorithm, and i is a subscript. *)

```
Begin (* algorithm *)
  (* initialize the variables q, r, and i *)
  q ← n
  r ← n
  i ← 0
  while q > 0 do
  begin (* while *)
    r ← q mod b
    a_i ← r
    q ← q div b
    i ← i + 1
  endwhile
End (* algorithm *)
```

Algorithm 2.1

The following example demonstrates this algorithm.

EXAMPLE 2.9 Represent 15,036 in the **hexadecimal system**, that is, in base sixteen.

SOLUTION

Applying Algorithm 2.1, we have

$$\begin{array}{rcrcrcr} 15036 & = & 939 & \cdot & 16 & + & 12 \\ 939 & = & 58 & \cdot & 16 & + & 11 \\ 58 & = & 3 & \cdot & 16 & + & 10 \\ 3 & = & 0 & \cdot & 16 & + & 3 \end{array} \quad \uparrow \text{ read up}$$

Thus, $15{,}036 = 3ABC_{\text{sixteen}}$. ■

The Egyptian Method of Multiplication

An algorithm based on Theorem 2.9 was used by the ancient Egyptians for multiplying two positive integers, say, 23 and 45. First, express one of the factors, say, 23, as a sum of powers of 2:

$$23 = 1 + 2 + 4 + 16$$

Then

$$23 \cdot 45 = 1 \cdot 45 + 2 \cdot 45 + 4 \cdot 45 + 16 \cdot 45$$

Next construct a table (Table 2.1) consisting of two rows, one headed by 1 and the other by 45; each successive column is obtained by doubling the preceding column.

1	2	4	8	16
45*	90*	180*	360	720*

Table 2.1

To find the desired result, add the starred numbers in the second row. These correspond to the terms in the binary expansion of 23:

$$\begin{aligned} 23 \cdot 45 &= 45 + 90 + 180 + 720 \\ &= 1035 \end{aligned}$$

We can use yet another algorithm for multiplication, which is a delightful application of the floor function.

The Russian Peasant Algorithm

The Russian peasant algorithm for multiplication resembles the Egyptian method. To illustrate it, suppose we want to compute $24 \cdot 43$. As before, construct a table (Table 2.2) of two rows, one headed by 24 and the other by 43. Each succeeding number in row 1 is the quotient when the number is divided by 2; continue this procedure until the quotient becomes 1. At each step, double the previous entry in row 2.

24	12	6	3	1
43	86	172	344*	688*

Table 2.2

To compute the product, just add the starred numbers that correspond to the odd numbers in row 1:

$$24 \cdot 43 = 688 + 344 = 1032$$

[Can you explain why this algorithm works? *Hint*: $ab = (a/2)(2b)$.]

The Egyptian Method of Division

The Egyptians developed a method for dividing integers that was similar to their multiplication algorithm. Suppose we would like to find the quotient and the remainder when 256 is divided by 23, by this method. Once again, build a table (Table 2.3), the first row headed by 1 and the other by the divisor 23; double each successive column until the number in the second row exceeds 256, the dividend.

1	2	4	8	16
23*	46*	92	184*	368
				↑ 256

Table 2.3

Now express 256 as a sum of the starred numbers from the second row:

$$\begin{aligned} 256 &= 184 + 72 \\ &= 184 + 46 + 26 \\ &= 184 + 46 + 23 + \boxed{3} \quad \leftarrow \text{remainder} \end{aligned}$$

Then the quotient is the sum of the numbers in row 1 that correspond to the starred numbers, namely, $1 + 2 + 8 = 11$; the remainder is the leftover, 3.

EXERCISES 2.2

Express each number in base ten.

1. 1101_{two}
2. 11011_{two}
3. 1776_{eight}
4. 1976_{sixteen}

Express each decimal number as required.

5. $1076 = (\quad)_{\text{two}}$
6. $676 = (\quad)_{\text{eight}}$
7. $1776 = (\quad)_{\text{eight}}$
8. $2076 = (\quad)_{\text{sixteen}}$

The binary representation of an integer can conveniently be used to find its octal representation. Group the bits in threes from right to left and replace each group with the corresponding octal digit. For example, $243 = 11110011_{\text{two}} = 011\ 110\ 011_{\text{two}} = 363_{\text{eight}}$. Using this shortcut, rewrite each binary number as an octal integer.

9. 1101_{two}
10. 11011_{two}
11. 111010_{two}
12. 10110101_{two}

The binary representation of an integer can also be used to find its hexadecimal representation. Group the bits in fours from right to left and then replace each group with the equivalent hexadecimal digit. For instance, $243 = 11110011_{\text{two}} = 1111\ 0011_{\text{two}} = F3_{\text{sixteen}}$. Using this method, express each binary number in base sixteen.

13. 11101_{two}
14. 110111_{two}
15. 1110101_{two}
16. 10110101_{two}

The techniques explained in Exercises 9–12 are reversible, that is, the octal and hexadecimal representations of integers can be used to find their binary representations. For example, $345_{\text{eight}} = 011\ 100\ 101_{\text{two}} = 11100101_{\text{two}}$. Using this technique, rewrite each number in base two.

17. 36_{sixteen}
18. 237_{eight}
19. 237_{sixteen}
20. $3AD_{\text{sixteen}}$

Using the Egyptian method, compute each product.

21. $19 \cdot 31$
22. $30 \cdot 43$
23. $29 \cdot 49$
24. $36 \cdot 59$

25–28. Using the Russian method, evaluate the products in Exercises 21–24.

Using the Egyptian method of division, find the quotient and the remainder when the first integer is divided by the second.

29. 243, 19
30. 1076, 31
31. 1776, 35
32. 2076, 43
33. Arrange the binary numbers 1011, 110, 11011, 10110, and 101010 in order of increasing magnitude.
34. Arrange the hexadecimal numbers 1076, 3056, 3*CAB*, 5*ABC*, and *CACB* in order of increasing magnitude.
35. What can you say about the ones bit in the binary representation of an even integer? An odd integer?

Find the value of the base b in each case.

36. $54_b = 64$
37. $1001_b = 9$
38. $1001_b = 126$
39. $144_b = 49$
40. Find the base b if $7642 = 1234_b$. (A. Dunn, 1980)
41. Find the positive integer n if the decimal values of n^3 and n^4 together contain all the digits exactly once. (A. Dunn, 1980)

Find the number of ones in the binary representations of each number.

42. $2^3 - 1$
43. $2^4 - 1$
44. $2^5 - 1$
45. $2^n - 1$
46. Suppose a space team investigating Venus sends back the picture of an addition problem scratched on a wall, as shown in Figure 2.4. The Venusian numeration system is a place value system, just like ours. The base of the system is the same as the number of fingers on a Venusian hand. Determine the base of the Venusian numeration system. (This puzzle is due to H. L. Nelson.[†])

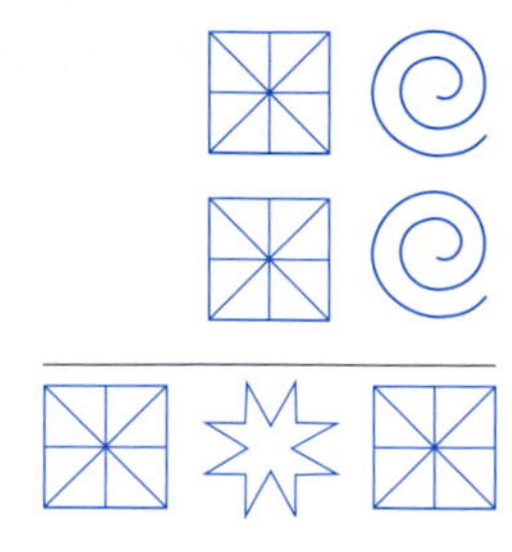

Figure 2.4

Polynomials can be evaluated efficiently using the technique of **nested multiplication**, called **Horner's method**. (This method is named after the English schoolmaster William G. Horner [1786–1837], who published it in 1819.) For instance, the polynomial $f(x) = 4x^3 + 5x^2 + 6x + 7$ can be evaluated as $f(x) = ((4x + 5)x + 6)x + 7$. Using this method, express each as a decimal integer.

47. 245_{eight}
48. 101101_{two}
49. 1100101_{two}
50. $43BC_{\text{sixteen}}$

Find the ones digit in the decimal value of each.

51. 2^{100}
52. 3^{247}

*53. Let x be a three-digit number with distinct digits in base twelve. Reverse the digits. Subtract the smaller number from the other number (save all the digits in your answer). Reverse the digits in the difference. Add this number to x. Find the sum.

*54. Redo Exercise 53 in base sixteen.

† M. Gardner, "Mathematical Games," *Scientific American*, 219 (Sept. 1968), 218–230.

*2.3 Operations in Nondecimal Bases (optional)

Before exploring how to add nondecimal numbers, let us take a close look at the familiar addition algorithm in base ten.

To find the sum of any two decimal digits a and b, first find the remainder $r = (a+b) \bmod 10$ and the quotient $q = (a+b) \text{ div } 10$. Then $a + b = (qr)_{\text{ten}}$; q is the carry resulting from the addition of a and b. Using this concept, it is possible to add any two decimal integers.

Addition in Base b

Fortunately, the addition algorithm can be extended to any nondecimal base b. For example, let $x = (x_m \dots x_0)_b$ and $y = (y_n \dots y_0)_b$, where $m \geq n$. If $m > n$, we could assume that $y_{n+1} = \cdots = y_m = 0$. We add the corresponding digits in x and y in a right-to-left fashion. Let $s_i = (x_i + y_i + c_i) \bmod b$ and $c_{i+1} = (x_i + y_i + c_i) \text{ div } b$, where $c_0 = 0$. Then $x + y = (s_{m+1}s_m \dots s_0)_b$, where s_{m+1} may be 0 or 1. (Leading zeros are deleted from the answer.)

These steps translate into a straightforward algorithm, as in Algorithm 2.2.

```
Algorithm addition (x, y, s, b)
(* This algorithm computes the sum s = (s_{m+1}s_m...s_0)_b of the integers x =
   (x_m x_{m-1}...x_0)_b and y = (y_n y_{m-1}...y_0)_b, where m ≤ n. *)
   Begin (* algorithm *)
     carry ← 0 (* initialize carry *)
     for i = 0 to n do
     begin (* for *)
       s_i ← (x_i + y_i + carry) mod b
       carry ← (x_i + y_i + carry) div b
     endfor
     for i = n + 1 to m do
     begin (* for *)
       s_i ← (x_i + carry) div b
       carry ← (x_i + carry) div b
     endfor
     if carry > 0 then
       s_{m+1} ← carry
   End (* algorithm *)
```

Algorithm 2.2

The following two examples illustrate this algorithm.

EXAMPLE 2.10 Add the binary integers 10110_{two} and 1011_{two}.

SOLUTION

First, write the integers one below the other in such a way that the corresponding bits are vertically aligned (Figure 2.5). (For convenience, the base two is not shown.)

Add the corresponding bits from right to left, beginning with the ones column: $0+1=1$. Because $1 \bmod 2 = 1$, enter 1 as the ones bit in the sum. Since $1 \text{ div } 2 = 0$, the resulting carry is 0, shown circled in Figure 2.6. (In practice when the carry is 0, it is simply ignored.) Now add the bits 0, 1, and 1 in the twos column: $0+1+1=2$. Because $2 \bmod 2 = 0$ and $2 \text{ div } 2 = 1$, enter 0 in the twos column and the new carry is 1 (Figure 2.7). Continuing like this, we get the sum 100001_{two} (Figure 2.8).

$$\begin{array}{cccccc} & 1 & 0 & 1 & 1 & 0 \\ + & & 1 & 0 & 1 & 1 \\ \hline \end{array}$$

Figure 2.5

$$\begin{array}{cccccc} & & & & (0) & \\ & 1 & 0 & 1 & 1 & 0 \\ + & & 1 & 0 & 1 & 1 \\ \hline & & & & & 1 \end{array}$$

Figure 2.6

$$\begin{array}{cccccc} & & & (1) & (0) & \\ & 1 & 0 & 1 & 1 & 0 \\ + & & 1 & 0 & 1 & 1 \\ \hline & & & & 0 & 1 \end{array}$$

Figure 2.7

$$\begin{array}{ccccccc} & (1) & (1) & (1) & (1) & (0) & \\ & & 1 & 0 & 1 & 1 & 0 \\ + & & & 1 & 0 & 1 & 1 \\ \hline & 1 & 0 & 0 & 0 & 0 & 1 \end{array}$$

Figure 2.8 ■

The addition of binary numbers can be made easy by observing that $0+0=0$, $0+1=1=1+0$, and $1+1=10$, all in base two.

The following example illustrates addition in base twelve and base sixteen.

EXAMPLE 2.11

$$\begin{array}{rcccl} & ① & ① & ① & \\ & A & 5 & 8 & B_{\text{twelve}} \\ + & & 9 & A & 3_{\text{twelve}} \\ \hline & B & 3 & 7 & 2_{\text{twelve}} \end{array} \qquad \begin{array}{rcccl} & ① & ① & ① & \\ & & A & B & C_{\text{sixteen}} \\ + & & C & B & A_{\text{sixteen}} \\ \hline & 1 & 7 & 7 & 6_{\text{sixteen}} \end{array}$$

Notice that in base twelve, $B+3=12$, $1+8+A=17$, $1+5+9=13$, and $1+A=B$; and in base sixteen, $C+A=16$, $1+B+B=17$, and $1+A+C=17$. ■

Subtraction in Base b

The following two examples illustrate nondecimal subtraction using the familiar concept of borrowing, when needed.

EXAMPLE 2.12 Evaluate $2354_{\text{seven}} - 463_{\text{seven}}$.

SOLUTION

As usual, write the numbers one below the other in such a way that the corresponding digits match vertically:

$$\begin{array}{rcccl} & 2 & 3 & 5 & 4_{\text{seven}} \\ - & & 4 & 6 & 3_{\text{seven}} \\ \hline \end{array}$$

Beginning with the ones column, $4-3=1$, so the ones digit in the answer is 1 (Figure 2.9).

$$\begin{array}{rcccl} & 2 & 3 & 5 & 4_{\text{seven}} \\ - & & 4 & 6 & 3_{\text{seven}} \\ \hline & & & & 1_{\text{seven}} \end{array}$$

Figure 2.9

$$\begin{array}{rcccl} & & ② & ⑮ & \\ & 2 & \not{3} & \not{5} & 4_{\text{seven}} \\ - & & 4 & 6 & 3_{\text{seven}} \\ \hline & & & 6 & 1_{\text{seven}} \end{array}$$

Figure 2.10

Now proceed to the sevens column. Since $5 < 6$, go to the forty-nines column and borrow a 1, leaving a 2 there. When that 1 comes to the sevens column, it becomes a 10. This yields $10+5=15$ in the sevens column. Since $15-6=6$, we get 6 as the sevens digit in the answer. (Remember, we are in base seven.) See Figure 2.10.

In the forty-nines column, $2 < 4$. So borrow a 1 from the next column, leaving a 1 there. The 1 borrowed yields $10+2 = 12$ in the forty-nines column. Since $12-4 = 5$, the forty-nines digit in the answer is 5 (Figure 2.11).

Since there are no nonzero digits left in the subtrahend, simply bring down the 1 from the minuend. This gives the final answer: 1561_{seven} (Figure 2.12).

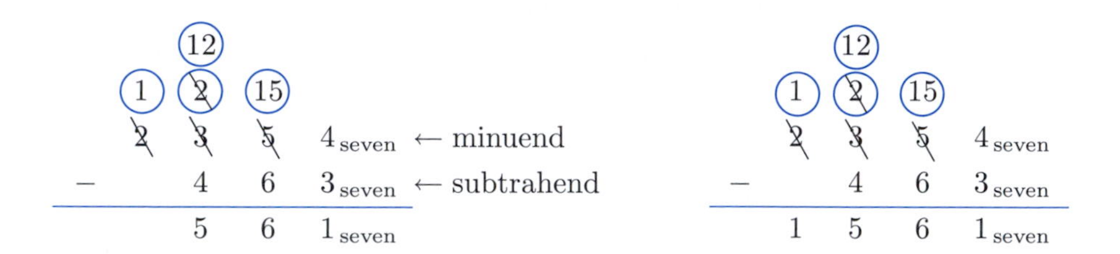

Figure 2.11 Figure 2.12

We can verify this subtraction by converting it into an addition problem: $463_{\text{seven}} + 1561_{\text{seven}} = 2354_{\text{seven}}$. ■

The following example demonstrates subtraction in bases twelve and sixteen.

EXAMPLE 2.13 Evaluate $A74_{\text{twelve}} - 39B_{\text{twelve}}$ and $2076_{\text{sixteen}} - 1777_{\text{sixteen}}$.

SOLUTION

See Figures 2.13 and 2.14. We can verify both answers as an exercise.

Figure 2.13 Figure 2.14

■

Next, we present a numeric puzzle that will test your mastery of both nondecimal addition and subtraction.

A Nondecimal Puzzle (optional)

Write down a three-digit number in base twelve, with no repetitions. Reverse its digits. Subtract the smaller number from the other (in base twelve); save all leading zeros. Reverse its digits. Add the last two numbers. Is your answer $10AB_{\text{twelve}}$? Now redo this puzzle in base sixteen. Your answer should be $10EF_{\text{sixteen}}$.

Binary Subtraction

We can subtract binary numbers without the bother of "borrows," using ones complement and addition. The **ones complement** x' of a binary number x is obtained by replacing each 0 in x with a 1 and vice versa. For example, the ones complement of 1011_{two} is 0100_{two} and that of 1001_{two} is 0110_{two}. The **twos complement** of x is $x' + 1$. For instance, the twos complement of 1011_{two} is $0100_{\text{two}} + 1 = 0101_{\text{two}}$.

The following example illustrates this new technique.

EXAMPLE 2.14 Subtract 1011_{two} from 100001_{two}.

SOLUTION

(For convenience, we shall drop the base two.)

step 1 *Find the ones complement of the subtrahend* 1011.
Since the minuend 100001 contains six bits, keep the same number of bits in the subtrahend by padding it with two 0s at the beginning. The ones complement of $1011 = 001011$ is 110100.

step 2 *Find the twos complement by adding* 1 *to the ones complement*: $110100 + 1 = 110101$.

step 3 *Add the twos complement in step* 2 *to the minuend* 100001:

$$\begin{array}{rcccccc} & & 1 & 0 & 0 & 0 & 0 & 1 \\ + & & 1 & 1 & 0 & 1 & 0 & 1 \\ \hline & \boxed{1} & 0 & 1 & 0 & 1 & 1 & 0 \end{array}$$

delete ⟶ (the boxed leading 1)

step 4 *Delete the leading carry* 1.
The resulting number $010110 = 10110$ is the desired answer.

Thus, $100001_{\text{two}} - 1011_{\text{two}} = 10110_{\text{two}}$. (To check this, you may verify that $1011_{\text{two}} + 10110_{\text{two}} = 100001_{\text{two}}$.) ■

Now we illustrate the multiplication algorithm in base b.

Multiplication in Base b

The traditional algorithm for multiplying two decimal integers x and y works for any base in an obvious way: multiply every digit in x by every digit in y as in base b and add up the partial products, as the next example shows.

EXAMPLE 2.15 Multiply 1011_{two} and 101_{two}.

SOLUTION

The various steps unfold in Figures 2.15–2.17. The product is 110111_{two}.

$$\begin{array}{rrrrl} 1 & 0 & 1 & 1 & \\ \times & 1 & 0 & 1 & \\ \hline 1 & 0 & 1 & 1 & \leftarrow \text{multiply by 1} \end{array}$$

Figure 2.15

$$\begin{array}{rrrrrl} & 1 & 0 & 1 & 1 & \\ & \times & 1 & 0 & 1 & \\ \hline & 1 & 0 & 1 & 1 & \\ 0 & 0 & 0 & 0 & & \leftarrow \text{multiply by 0} \\ 1\;0 & 1 & 1 & & & \leftarrow \text{multiply by 1} \end{array}$$

Figure 2.16

$$\begin{array}{rrrrrrl} & & 1 & 0 & 1 & 1 & \\ & \times & & 1 & 0 & 1 & \\ \hline & & 1 & 0 & 1 & 1 & \\ & 0 & 0 & 0 & 0 & & \text{add the partial products} \\ 1 & 0 & 1 & 1 & & & \\ \hline 1 & 1 & 0 & 1 & 1 & 1 & \leftarrow \text{final answer} \end{array}$$

Figure 2.17

■

Shifting and Binary Multiplication

If you were confused by this example, don't be dismayed; there is an alternative method. Most computers do binary multiplications using a technique called **shifting**, as discussed below.

Consider the binary number $x = (x_m x_{m-1} \dots x_1 x_0)_{\text{two}} = \sum_{i=0}^{m} x_i 2^i$. What is the effect of multiplying x by 2^j? Since

$$x2^j = \sum_{i=0}^{m} x_i 2^{i+j} = x_m \dots x_1 x_0 \underbrace{00\dots0}_{j \text{ zeros}}{}_{\text{two}}$$

every bit in x is shifted to the left by j columns.

More generally, let a be any bit. Then

$$x(a2^j) = \sum_{i=0}^{m}(ax_i)2^{i+j} = (ax_m)\ldots(ax_0)\underbrace{00\ldots0}_{j \text{ zeros}}{}_{\text{two}}$$

The bit ax_i equals x_i if $a = 1$, and equals 0 if $a = 0$. Thus, the effect of multiplying the number $x = (x_m \ldots x_0)_{\text{two}}$ by the bit y_j in the multiplicand $y = (y_n \ldots y_j \ldots y_0)_{\text{two}}$ is the same as multiplying each bit x_i by y_j and shifting the result to the left by j columns. Then add the partial products to get the desired product, as the following example illustrates.

EXAMPLE 2.16 Evaluate $1011_{\text{two}} \times 101_{\text{two}}$.

SOLUTION

The various steps are displayed in Figures 2.18–2.21. It follows from Figure 2.21 that the resulting product is 110111_{two}.

$$\begin{array}{rcccc}
 & 1 & 0 & 1 & 1 \\
\times & & 1 & 0 & 1 \\
\hline
 & 1 & 0 & 1 & 1
\end{array} \leftarrow \text{multiply 1011 by 1; no shifting.}$$

Figure 2.18

$$\begin{array}{ccccc}
 & 1 & 0 & 1 & 1 \\
\times & & 1 & 0 & 1 \\
\hline
 & 1 & 0 & 1 & 1 \\
0 & 0 & 0 & 0 &
\end{array} \leftarrow \text{multiply 1011 by 0; shift by one column.}$$

Figure 2.19

$$\begin{array}{cccccc}
 & & 1 & 0 & 1 & 1 \\
 & \times & & 1 & 0 & 1 \\
\hline
 & & 1 & 0 & 1 & 1 \\
 & 0 & 0 & 0 & 0 & \\
1 & 0 & 1 & 1 & &
\end{array} \leftarrow \text{multiply 1011 by 1; shift by two columns.}$$

Figure 2.20

$$\begin{array}{cccccc}
 & & 1 & 0 & 1 & 1 \\
 & \times & & 1 & 0 & 1 \\
\hline
 & & 1 & 0 & 1 & 1 \\
 & 0 & 0 & 0 & 0 & \\
1 & 0 & 1 & 1 & & \\
\hline
1 & 1 & 0 & 1 & 1 & 1
\end{array} \quad \begin{array}{l} \left.\begin{array}{c} \\ \\ \\ \end{array}\right\} \text{add the partial products} \\ \leftarrow \text{answer} \end{array}$$

Figure 2.21

■

The shifting method of multiplication leads to Algorithm 2.3 for multiplying two binary numbers.

```
Algorithm binary multiplication (x, y, p)
(* This algorithm computes the product p = (p_{m+n}p_{m+n-1}...p_0)two of the binary num-
   bers x = (x_m x_{m-1}...x_0)two and y = (y_n y_{m-1}...y_0)two, using shifting. *)
   Begin (* algorithm *)
     for j = 0 to n do
     begin (* for *)
       multiply each bit x_i by y_i
       shift the resulting binary word to the left by j columns
       w_j ← resulting binary word
     endfor
     add the partial products w_j
     p ← resulting sum
   End (* algorithm *)
```

Algorithm 2.3

Repunits

A **repunit** (**rep**eated **unit**)† is a positive integer whose decimal expansion consists of 1s. A repunit with n ones is denoted by R_n. For example, $R_2 = 11$ and $R_3 = 111$.

The following interesting problem on repunits was proposed in 1982 by L. Kuipers of Switzerland.

EXAMPLE 2.17 Show that 111 cannot be a square in any base.

PROOF **(by contradiction)**

Suppose 111 is a perfect square a^2 in some base b, so $a^2 = b^2 + b + 1 < (b+1)^2$. Then

$$(b + 1/2)^2 = b^2 + b + 1/4 < b^2 + b + 1$$

That is,

$$(b + 1/2)^2 < a^2 < (b+1)^2$$

This yields $(b + 1/2) < a < b + 1$; that is, a lies between $b + 1/2$ and $b + 1$, which is impossible. Thus, 111 cannot be a square in any base. ■

† The term **repunit** was coined by Albert H. Beiler of Brooklyn, New York.

A Brainteaser (optional)

Look at the numbers on cards A, B, C, D, and E in Figure 2.22. Assuming you are under 32 years old, if you identify the cards on which your age appears, we can easily tell your age. For example, if your age appears on cards A, B, C, and E, then you must be 23. Can you explain how this puzzle works?

A		B		C		D		E	
1	17	2	18	4	20	8	24	16	24
3	19	3	19	5	21	9	25	17	25
5	21	6	22	6	22	10	26	18	26
7	23	7	23	7	23	11	27	19	27
9	25	10	26	12	28	12	28	20	28
11	27	11	27	13	29	13	29	21	29
13	29	14	30	14	30	14	30	22	30
15	31	15	31	15	31	15	31	23	31

Figure 2.22

EXERCISES 2.3

Construct an addition table for each base.

1. Five
2. Seven

Compute $x + 1$ for each value of x.

3. 101_{two}
4. 344_{five}
5. 666_{seven}
6. $2AB_{\text{twelve}}$
7. Let b be a base such that $120_b + 211_b = 331_b$, where $b < 8$. Find the possible values(s) of b.

Perform the indicated operations.

8. $1111_{\text{two}} + 1011_{\text{two}}$
9. $1076_{\text{eight}} + 2076_{\text{eight}}$
10. $89B_{\text{twelve}} + 5A6_{\text{twelve}}$
11. $3076_{\text{sixteen}} + 5776_{\text{sixteen}}$

Compute $x - 1$ for each value of x.

12. 100_{two}
13. 210_{seven}
14. $37B_{\text{twelve}}$
15. ABC_{sixteen}

Perform the indicated operations.

16. $101101_{\text{two}} - 10011_{\text{two}}$
17. $11000_{\text{two}} - 100_{\text{two}}$
18. $2000_{\text{seven}} - 1336_{\text{seven}}$
19. $A89B_{\text{twelve}} - 65A6_{\text{twelve}}$

Construct a multiplication table for each base.

20. Five
21. Seven

Compute $x(x + 1)$ for each value of x.

22. 110_{two}
23. 243_{five}
24. 345_{seven}
25. AB_{twelve}

Compute $x(x - 1)$ for each value of x.

26. 101_{two}
27. 243_{five}
28. 343_{seven}
29. BA_{twelve}

Perform the indicated operations.

30. $10111_{\text{two}} \times 1101_{\text{two}}$
31. $1024_{\text{eight}} \times 2776_{\text{eight}}$
32. $1976_{\text{twelve}} \times 1776_{\text{twelve}}$
33. $CBA_{\text{sixteen}} \times ABC_{\text{sixteen}}$

2.4 Number Patterns

Number patterns are fun for both amateurs and professionals. Often we would like to add one or two rows to the pattern, so we must be good at pattern recognition to succeed in the art of **inductive reasoning**. It takes both skill and ingenuity. In two of the following examples, mathematical proofs establish the validity of the patterns.

The following fascinating number pattern[†] was published in 1882 by the French mathematician François-Edouard-Anatole Lucas.

EXAMPLE 2.18 Study the following number pattern and add two more lines.

$$
\begin{aligned}
1\cdot 9+2&=11\\
12\cdot 9+3&=111\\
123\cdot 9+4&=1111\\
1234\cdot 9+5&=11111\\
12345\cdot 9+6&=111111\\
123456\cdot 9+7&=1111111\\
&\vdots
\end{aligned}
$$

SOLUTION

Although the pattern here is very obvious, let us make a few observations, study them, look for a similar behavior, and apply the pattern to add two more lines:

- The LHS of each equation is a sum of two numbers. The first number is a product of the number $123\ldots n$ and 9.
- The value of n in the first equation is 1, in the second it is 2, in the third it is 3, and so on.
- Take a look at the second addends on the LHS: $2, 3, 4, 5, \ldots$. It is an increasing sequence beginning with 2, so the second addend in the nth equation is $n+1$.
- The RHS of each equation is a number made up of 1s, the nth equation containing $n+1$ ones.

Thus, a pattern emerges and we are ready to state it explicitly: The first number in the nth line is $123\ldots n$; the second number is always 9; the second addend is $n+1$; and the RHS is made up of $n+1$ ones.

[†] This curious number pattern appeared in *Mathematical Recreations* by Lucas.

So the next two lines are

$$1234567 \cdot 9 + 8 = 11111111$$

$$12345678 \cdot 9 + 9 = 111111111$$

■

The following pattern is equally charming.

EXAMPLE 2.19 Study the number pattern and add two more rows:

$$1 \cdot 8 + 1 = 9$$
$$12 \cdot 8 + 2 = 98$$
$$123 \cdot 8 + 3 = 987$$
$$1234 \cdot 8 + 4 = 9876$$
$$12345 \cdot 8 + 5 = 98765$$
$$123456 \cdot 8 + 6 = 987654$$
$$\vdots$$

SOLUTION

A close look at the various rows reveals the following pattern: The first factor of the product on the LHS of the nth equation has the form $123\ldots n$; the second factor is always 8. The second addend in the equation is n. The number on the RHS of the nth equation contains n digits, each begins with the digit 9, and the digits decrease by 1.

Thus the next two lines of the pattern are

$$1234567 \cdot 8 + 7 = 9876543$$

$$12345678 \cdot 8 + 8 = 98765432$$

■

What guarantees that these two patterns will hold? In general, conclusions reached after observing patterns do not have to be true. In other words, inductive reasoning does not necessarily lead us to true conclusions.

For instance, consider the sequence 0, 1, 2, 3, 4, 5, 6, ... Clearly, there is a pattern. So what is the next number in the sequence? Is it 7? This is certainly a possibility, but the next number could also be 0 to yield the pattern 0, 1, 2, 3, 4, 5, 6, 0, 1, 2, ...

Fortunately, it is possible to establish the validity of each pattern using mathematical proofs, as the following two examples demonstrate.

EXAMPLE 2.20 Establish the validity of the number pattern in Example 2.18.

PROOF

We would like to prove that $123\dots n \times 9 + (n+1) = \underbrace{11\dots11}_{n+1 \text{ ones}}$

$$\begin{aligned}
\text{LHS} &= 123\dots n \times 9 + (n+1) \\
&= 9(1 \cdot 10^{n-1} + 2 \cdot 10^{n-2} + 3 \cdot 10^{n-3} + \cdots + n) + (n+1) \\
&= (10-1)(1 \cdot 10^{n-1} + 2 \cdot 10^{n-2} + 3 \cdot 10^{n-3} + \cdots + n) + (n+1) \\
&= (10^n + 2 \cdot 10^{n-1} + \cdots + n \cdot 10) - (10^{n-1} + 2 \cdot 10^{n-2} + \cdots + n) + (n+1) \\
&= 10^n + 10^{n-1} + 10^{n-2} + \cdots + 10 + 1 \\
&= \underbrace{11\dots11}_{n+1 \text{ ones}} \\
&= \text{RHS}
\end{aligned}$$

(It would be interesting to see if this result holds for any positive integer n; try it.) ■

We will study one more example.

EXAMPLE 2.21 Add two more rows to the following pattern, conjecture a formula for the nth row, and prove it:

$$\begin{aligned}
9 \cdot 9 + 7 &= 88 \\
98 \cdot 9 + 6 &= 888 \\
987 \cdot 9 + 5 &= 8888 \\
9876 \cdot 9 + 4 &= 88888 \\
98765 \cdot 9 + 3 &= 888888 \\
&\vdots
\end{aligned}$$

SOLUTION

- The next two rows of the pattern are

$$\begin{aligned}
987654 \cdot 9 + 2 &= 8888888 \\
9876543 \cdot 9 + 1 &= 88888888
\end{aligned}$$

- The general pattern seems to be

$$987\ldots(10-n)\cdot 9+(8-n)=\underbrace{888\ldots888}_{n+1\text{ eighths}},\quad 1\le n\le 8$$

- To prove the conjecture:

$$\begin{aligned}
\text{LHS} &= 987\ldots(10-n)\cdot 9+(8-n)\\
&=(10-1)[9\cdot 10^{n-1}+8\cdot 10^{n-2}+7\cdot 10^{n-3}+\cdots+(11-n)10\\
&\quad +(10-n)]+(8-n)\\
&=[9\cdot 10^{n}+8\cdot 10^{n-1}+\cdots+(11-n)10^2+(10-n)10]-\\
&\quad [9\cdot 10^{n-1}+8\cdot 10^{n-2}+7\cdot 10^{n-3}+\cdots+(11-n)10+(10-n)]\\
&\quad +(8-n)\\
&=9\cdot 10^{n}-(10^{n-1}+10^{n-2}+\cdots+10)-(10-n)+(8-n)\\
&=9\cdot 10^{n}-(10^{n-1}+10^{n-2}+\cdots+10+1)-1\\
&=10\cdot 10^{n}-(10^{n}+10^{n-1}+\cdots+10+1)-1\\
&=10^{n+1}-\frac{10^{n+1}-1}{9}-1,\quad \text{since } \sum_{i=0}^{k} r^i=\frac{r^{k+1}-1}{r-1}\quad (r\ne 1)\\
&=\frac{8(10^{n+1}-1)}{9}
\end{aligned}$$

But

$$10^{n+1}-1=\underbrace{99\ldots99}_{n+1\text{ nines}},$$

so

$$\frac{10^{n+1}-1}{9}=\underbrace{11\ldots11}_{n+1\text{ ones}}$$

Therefore,

$$\text{LHS}=\frac{8(10^{n+1}-1)}{9}=\underbrace{88\ldots88}_{n+1\text{ eighths}}=\text{RHS}$$

■

EXERCISES 2.4

Find the next two elements of each sequence.

1. 1, 3, 6, 10, 15, ...
2. 1, 4, 7, 10, 13, ...
3. 1, 5, 12, 22, 35, ...
4. 1, 6, 15, 28, 45, ...
5. 1, 4, 10, 20, 35, ...
6. 1, 5, 14, 30, 55, ...
7. 1, 1, 2, 3, 5, 8, ...
8. ★ $o, t, t, f, f, s, s, \ldots$

Add two more rows to each number pattern.[†]

9.
$$\begin{aligned} 0+1 &= 1 \\ 1+3 &= 4 \\ 4+5 &= 9 \\ 9+7 &= 16 \end{aligned}$$

10.
$$\begin{aligned} 1 &= 1 \\ 1+2 &= 3 \\ 1+2+3 &= 6 \\ 1+2+3+4 &= 10 \end{aligned}$$

11.
$$\begin{aligned} 1 &= 1 \\ 1+4 &= 5 \\ 1+4+9 &= 14 \\ 1+4+9+16 &= 30 \end{aligned}$$

12.
$$\begin{aligned} 1+2 &= 3 \\ 1+2+4 &= 7 \\ 1+2+4+8 &= 15 \\ 1+2+4+8+16 &= 31 \end{aligned}$$

13.
$$\begin{aligned} 1+0\cdot 1 &= 1 \\ 1+1\cdot 3 &= 4 \\ 1+2\cdot 4 &= 9 \\ 1+3\cdot 5 &= 16 \end{aligned}$$

14.
$$\begin{aligned} 2^3-2 &= 1\cdot 2\cdot 3 \\ 3^3-3 &= 2\cdot 3\cdot 4 \\ 4^3-4 &= 3\cdot 4\cdot 5 \\ 5^3-5 &= 4\cdot 5\cdot 6 \end{aligned}$$

15.
$$\begin{aligned} 1\cdot 1 &= 1 \\ 11\cdot 11 &= 121 \\ 111\cdot 111 &= 12321 \\ 1111\cdot 1111 &= 1234321 \\ 11111\cdot 11111 &= 123454321 \end{aligned}$$

16.
$$\begin{aligned} 7\cdot 7 &= 49 \\ 67\cdot 67 &= 4489 \\ 667\cdot 667 &= 444889 \\ 6667\cdot 6667 &= 44448889 \\ 66667\cdot 66667 &= 4444488889 \end{aligned}$$

17.
$$\begin{aligned} 12345679\cdot 9 &= 111111111 \\ 12345679\cdot 18 &= 222222222 \\ 12345679\cdot 27 &= 333333333 \\ 12345679\cdot 36 &= 444444444 \\ 12345679\cdot 45 &= 555555555 \end{aligned}$$

18.
$$\begin{aligned} 4\cdot 4 &= 16 \\ 34\cdot 34 &= 1156 \\ 334\cdot 334 &= 111556 \\ 3334\cdot 3334 &= 11115556 \\ 33334\cdot 33334 &= 1111155556 \end{aligned}$$

19.
$$\begin{array}{ccccccccc} & & & & 0 & & & & \\ & & & 1 & & 1 & & & \\ & & 1 & & 0 & & 1 & & \\ & 1 & & 1 & & 1 & & 1 & \\ 1 & & 0 & & 0 & & 0 & & 1 \end{array}$$

20.
$$\begin{aligned} 10^2-10+1 &= 91 \\ 10^4-10^2+1 &= 9901 \\ 10^6-10^3+1 &= 999001 \\ 10^8-10^4+1 &= 99990001 \\ 10^{10}-10^5+1 &= 9999900001 \end{aligned}$$

21–31. Conjecture a formula for the nth row of each pattern in Exercises 10–20.

32–38. Establish the validity of your formula in Exercises 21–26 and 31.

39. Show that the formula in Example 2.18 does not hold for every integer n.

In Exercises 40–43, R_n denotes a repunit.

[†] Exercises 15–18 are based on F. B. Selkin, "Number Games Bordering on Arithmetic and Algebra," *Teachers College Record*, 13 (1912), 68. Exercise 20 is based on A. H. Beiler, *Recreations in Theory of Numbers*, Dover, New York (1966), 85.

40. Compute R_1^2, R_2^2, R_3^2, and R_4^2.
41. Using Exercise 40, predict the values of R_5^2 and R_6^2.
42. Conjecture the value of R_n^2.
43. Does the conjecture hold for R_{10}^2?

Use the following number pattern to answer Exercises 44–46.

$$1 = 1^2 - 0^2$$
$$3 = 2^2 - 1^2$$
$$5 = 3^2 - 2^2$$
$$7 = 4^2 - 3^2$$
$$9 = 5^2 - 4^2$$

44. Add two more lines.
45. Make a conjecture about row n.
46. Prove the conjecture in Exercise 45.
47. Add two more rows to the pattern in Figure 2.23.

1
2 3
4 5 6
7 8 9 10

Figure 2.23

48. Find the first and the last numbers in the nth row in Figure 2.23.
49. Conjecture a formula for the sum of the numbers in row n in Figure 2.23.

50–52. Redo Exercises 47–49 with the triangular array in Figure 2.24.

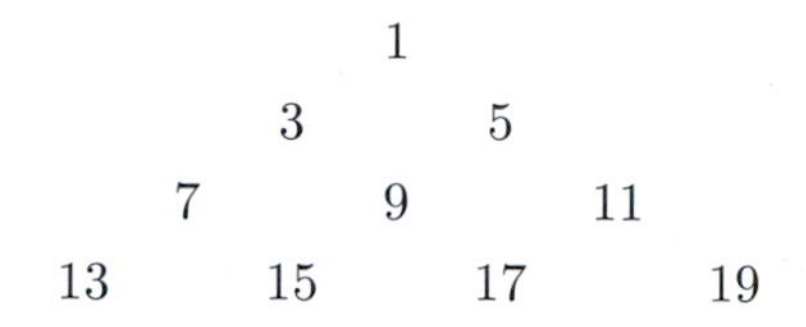

Figure 2.24

53. Show that $\dfrac{10^n - 9n - 1}{81} = \underbrace{123\ldots(n-1)}_{n-1 \text{ digits}}$, where $2 \le n \le 9$.
54. Find the value of $10^n - \dfrac{10^n - 9n - 1}{81} - 1$, where $1 \le n \le 9$.
55. Establish the validity of the pattern in Example 2.19. (*Hint*: Use Exercises 53 and 54.)
56. Prove that the numbers 49, 4489, 444889, ..., where each number, except the first, is obtained by inserting 48 in the middle, are all squares. (S. R. Conrad, 1976)

2.5 Prime and Composite Numbers

Prime numbers are the building blocks of positive integers. Two algorithms are often used to determine whether a given positive integer is a prime.

Some positive integers have exactly two positive factors and some have more than two. For example, 3 has exactly two positive factors: namely, 1 and 3; whereas 6 has four: 1, 2, 3, and 6. Accordingly, we make the following definition.

Prime and Composite Numbers

A positive integer > 1 is a **prime number** (or simply a **prime**) if its only positive factors are 1 and itself. A positive integer > 1 that is not a prime is a **composite number** (or simply a **composite**).

Notice that, by definition, 1 is neither a prime nor a composite. It is just the multiplicative identity or the **unit**.

The first ten primes are 2, 3, 5, 7, 11, 13, 17, 19, 23, and 29; the first ten composite numbers are 4, 6, 8, 9, 10, 12, 14, 15, 16, and 18.

It follows from the definition that the set of positive integers can be partitioned into three disjoint classes: the set of primes, the set of composites, and $\{1\}$.

How many primes are there? Is there a systematic way to determine whether a positive integer is a prime?

To answer the first question, we need the following lemma,[†] which we shall prove by induction. It can also be proved by contradiction (see Exercise 59).

LEMMA 2.1 Every integer $n \geq 2$ has a prime factor.

PROOF (by strong induction)

The given statement is clearly true when $n = 2$. Now assume it is true for every positive integer $n \leq k$, where $k \geq 2$. Consider the integer $k + 1$.

case 1 If $k + 1$ is a prime, then $k + 1$ is a prime factor of itself.

case 2 If $k + 1$ is not a prime, $k + 1$ must be a composite, so it must have a factor $d \leq k$. Then, by the inductive hypothesis, d has a prime factor p. So p is a factor of $k + 1$, by Theorem 2.4.

Thus, by the strong version of induction, the statement is true for every integer ≥ 2; that is, every integer ≥ 2 has a prime factor. ■

We can now prove that there is an infinite number of primes. This result, devised by Euclid, is one of the elegant results in number theory. We use essentially his technique from Book IX of *Elements* to prove it. See Theorem 3.4 and Corollary 3.8 for alternative proofs.

THEOREM 2.8 **(Euclid)** There are infinitely many primes.

PROOF (by contradiction)

Assume there is only a finite number of primes, $p_1, p_2, \ldots, p_n$. Consider the integer $N = p_1 p_2 \cdots p_n + 1$. Since $N \geq 2$, by Lemma 2.1, N is divisible by some prime p_i,

[†] A **lemma** is a minor result used to prove a theorem.

where $1 \leq i \leq n$. Since $p_i|N$ and $p_i|p_1p_2\cdots p_n$, $p_i|(N - p_1p_2\cdots p_n)$, by Theorem 2.4; that is, $p_i|1$, which is impossible.

Thus, our assumption is false, so there are infinitely many primes. ■

The proof of this theorem hinges on the choice of the number $E_n = p_1p_2\cdots p_n + 1$, where p_i denotes the ith prime and $i \geq 1$. The first five values of E_n are $E_1 = 3$, $E_2 = 7$, $E_3 = 31$, $E_4 = 211$, and $E_5 = 2311$, all primes. Unfortunately, not all values of E_n are primes; see Exercise 70.

In 1996, A. A. K. Majumdar of Jahangirnagar University, Bangladesh, established an upper bound for E_n, when $n \geq 6$: $E_n < (p_{n+1})^{n-2}$. We can establish this using induction. See Exercise 71.

*Primes and Pi (optional)

We now make an interesting digression. In 1734, the outstanding Swiss mathematician Leonhard Euler showed that the sum of the reciprocals of primes $\sum_p \frac{1}{p}$ diverges. The infinitude of primes follows from this also. However, the infinite product $\prod_p (1 - 1/p^2)$ converges to a limit ν.[†] In fact, it can be shown[‡] that $\frac{1}{\nu} = \sum_{n=1}^{\infty} \frac{1}{n^2} = \frac{1}{1^2} + \frac{1}{2^2} + \frac{1}{3^2} + \cdots$. In 1734, Euler also showed that $\sum_{n=1}^{\infty} \frac{1}{n^2} = \frac{\pi^2}{6}$, so $\nu = \frac{6}{\pi^2}$. Thus, $\prod_p \left(1 - \frac{1}{p^2}\right) = \frac{6}{\pi^2} \approx 0.6079271018$.

Now that we know there is an infinite number of primes, can we find an algorithm for determining the primality of integers ≥ 2? The great German mathematician Karl Friedrich Gauss wrote in 1801 in *Disquisitiones Arithmeticae*: "The problem of distinguishing prime numbers from composite numbers ... is known to be one of the most important and useful in arithmetic .... Further, the dignity of science itself seems to require that every possible means be explored for the solution of a problem so elegant and so celebrated." Fortunately, there is an algorithm, which is based on the following result.

THEOREM 2.9 Every composite number n has a prime factor $\leq \lfloor\sqrt{n}\rfloor$.

PROOF **(by contradiction)**

Because n is composite, there are positive integers a and b such that $n = ab$, where $1 < a < n$ and $1 < b < n$. Suppose $a > \sqrt{n}$ and $b > \sqrt{n}$. Then $n = ab > \sqrt{n} \cdot \sqrt{n} = n$,

† ν is the Greek letter *nu*.

‡ See Ogilvy and Anderson.

which is impossible. Therefore, either $a \le \sqrt{n}$ or $b \le \sqrt{n}$. Since both a and b are integers, it follows that either $a \le \lfloor\sqrt{n}\rfloor$ or $b \le \lfloor\sqrt{n}\rfloor$.

By Lemma 2.1, every positive integer ≥ 2 has a prime factor. Any such factor of a or b is also a factor of $a \cdot b = n$, so n must have a prime factor $\le \lfloor\sqrt{n}\rfloor$. ■

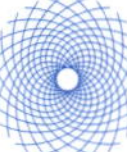

It follows from Theorem 2.11 that if n has no prime factors $\le \lfloor\sqrt{n}\rfloor$, then n is a prime; otherwise, it is a composite number.

This fact can be used to determine whether an integer $n \ge 2$ is a prime, as the next example illustrates.

EXAMPLE 2.22 Determine whether 1601 is a prime number.

SOLUTION

First list all primes $\le \lfloor\sqrt{1601}\rfloor$. They are 2, 3, 5, 7, 11, 13, 17, 19, 23, 29, 31, and 37. Since none of them is a factor of 1601 (verify), 1601 is a prime. ■

An algorithm for determining the primality of a positive integer $n \ge 2$ is given in Algorithm 2.4.

Algorithm prime number (n)
(* This algorithm using Theorem 2.9 determines whether an integer $n \le 2$ is prime or not. *)
  Begin (* algorithm *)
    list all primes $\le \lfloor\sqrt{n}\rfloor$
    if any of them is a factor of n, then n is not a prime
    else
      n is a prime
  End (* algorithm *)

Algorithm 2.4

The Sieve of Eratosthenes

Theorem 2.9 is also the basis of an ancient algorithm, the **sieve of Eratosthenes**, used for finding all primes $\le$ a positive integer n. It is an efficient algorithm for $n < 10^6$. We illustrate the sieving mechanism for $n = 100$ in Figure 2.25.

Eratosthenes *(ca. 276–ca. 194 B.C.), a Greek mathematician, was born in the ancient town of Cyrene, in present-day Libya. After spending many years at Plato's school in Athens, he went to Alexandria at the invitation of Ptolemy III to tutor his son and to serve as the chief librarian at the University. A gifted mathematician, astronomer, geographer, historian, philosopher, poet, and athlete, he was called* pentathlus *(the champion of five sports) by his students. His most important scientific achievement was the determination of the size of the earth. Around 194 B.C., he became blind and committed suicide by starvation.*

~~1~~	2	3	~~4~~	5	~~6~~	7	~~8~~	~~9~~	~~10~~
11	~~12~~	13	~~14~~	~~15~~	~~16~~	17	~~18~~	19	~~20~~
~~21~~	~~22~~	23	~~24~~	~~25~~	~~26~~	~~27~~	~~28~~	29	~~30~~
31	~~32~~	~~33~~	~~34~~	~~35~~	~~36~~	37	~~38~~	~~39~~	~~40~~
41	~~42~~	43	~~44~~	~~45~~	~~46~~	47	~~48~~	~~49~~	~~50~~
~~51~~	~~52~~	53	~~54~~	~~55~~	~~56~~	~~57~~	~~58~~	59	~~60~~
61	~~62~~	~~63~~	~~64~~	~~65~~	~~66~~	67	~~68~~	~~69~~	~~70~~
71	~~72~~	73	~~74~~	~~75~~	~~76~~	~~77~~	~~78~~	79	~~80~~
~~81~~	~~82~~	83	~~84~~	~~85~~	~~86~~	~~87~~	~~88~~	89	~~90~~
~~91~~	~~92~~	~~93~~	~~94~~	~~95~~	~~96~~	97	~~98~~	~~99~~	~~100~~

Figure 2.25

To find all primes ≤ 100, first list the positive integers 1 through 100. Then we eliminate 1 and all composite numbers ≤ 100 as follows. By Theorem 2.9, every composite number ≤ 100 must have a prime factor $\leq \lfloor\sqrt{100}\rfloor$, that is, ≤ 10. But the primes ≤ 10 are 2, 3, 5, and 7, so the composite numbers ≤ 100 are those positive integers divisible by one of them.

To eliminate the nonprimes from the list, first cross out 1 with a slash, since it is not a prime. Now cross out all multiples of 2, 3, 5, and 7, but *not* 2, 3, 5, or 7. (Why?) Numbers already eliminated need not be crossed out again. What remains are the primes ≤ 100.

There are 25 such primes: 2, 3, 5, 7, 11, 13, 17, 19, 23, 29, 31, 37, 41, 43, 47, 53, 59, 61, 67, 71, 73, 79, 83, 89, and 97.

Although the sieve looks fine, as n gets larger it becomes less efficient; the sieve is not a practical method. In fact, no simple, practical method exists for testing the primality of large numbers; see Theorem 8.14 for an efficient algorithm.

Outsider Math

Clive Thompson

Prime numbers have baffled scientists for millennia. Primes do not occur in any easily discernible order—which makes it very difficult to figure out whether a really huge number is prime or not. (And by "really huge," we're talking thousands of digits long.) Ever since the time of the ancient Greeks, finding a simple way to prove a number is prime has been the holy grail of mathematics, and the hunt has nearly driven several scientists mad. As the mathematician Karl Friedrich Gauss wrote in 1801, "The dignity of the science itself seems to require that every possible means be explored for the solution of a problem so elegant and so celebrated."

This year, it finally arrived. On Aug. 6, the Indian mathematician Manindra Agrawal distributed a nine-page paper that rocked the scientific world. He had hit upon an ingenious algorithm to prove whether a number is prime, no matter how enormous. Within weeks, stunned mathematicians had kicked the tires and pronounced it sound. Security experts were just as shocked. Encryption programs used by banks and governments rely on increasingly large primes—up to 600 digits, these days—to keep criminals and terrorists at bay. This new algorithm could guarantee primes so massive they would afford almost perfect online security.

But most astonishing of all was the simplicity of the algorithm. You can scrawl it on a single sheet of paper—double-spaced. It had been staring everyone in the face for years, like the Purloined Letter of mathematics. "When you read the paper, you slapped yourself on the forehead and asked, Why didn't *I* think of this?" says Carl Pomerance, a mathematician at Bell Labs.

Here's why. Math, like every other science, has become increasingly specialized. Prime-number theorists had been hacking away with number theory so complex and weird that barely 100 people worldwide could understand their calculations. But the new proof from India was created by a professor who isn't known as a number theorist—and his two co-authors were still undergraduates. What's more, they employed a branch of math with which any high-school student would be familiar: polynomials, like the simple expression $(A+B)^2$. "It's not really difficult at all," Agrawal says. "When you show it, it's like, Is that it?"

It is a fresh reminder of why history is riddled with innovations that came out of left field, delivered by amateurs toiling in their basements. Primes don't occur in any pattern—and sometimes, neither do discoveries.

Figure 2.26

In August 2002, M. Agrawal of the Indian Institute of Technology, Kanpur, India, and two of his undergraduate students, N. Kayal and N. Saxena, developed an efficient algorithm that is of theoretical significance. Their discovery surprised number theorists everywhere, since it runs in *polynomial time*; that is, "the number of steps (needed) is bounded by a polynomial function of the length of the input data." Two years later, H. Lenstra and C. Pomerance refined their algorithm to a theoretically more efficient one.

Number theorists often dream of finding formulas that generate primes for consecutive values of the integral variable n. Euler found one such formula in 1772: $E(n) = n^2 - n + 41$ yields a prime for every positive integer $n \leq 40$. But when $n = 41$, $E(41) = 41^2 - 41 + 41 = 41^2$ is *not* a prime.

In 1798, the eminent French mathematician Adrien-Marie Legendre (1752–1833) discovered that the formula $L(n) = n^2 + n + 41$ yields distinct primes for

$1 \le n \le 40$, but $L(41)$ is a composite. Notice that $L(n) = E(-n)$. (Several exercises based on similar formulas are included in the exercise set.)

However, no one has ever been successful in constructing a polynomial $f(n)$ that generates primes for all integers n. The reason becomes clear in the following example.

EXAMPLE 2.23 Prove that there is no polynomial $f(n)$ with integral coefficients that will produce primes for all integers n.

PROOF **(by contradiction)**

Suppose there is such a polynomial $f(n) = a_k n^k + a_{k-1}n^{k-1} + \cdots + a_1 n + a_0$, where $a_k \neq 0$. Let b be some integer. Since $f(n)$ is always a prime, $f(b)$ must be a prime p; that is,

$$f(b) = a_k b^k + a_{k-1}b^{k-1} + \cdots + a_1 b + a_0 = p \tag{2.1}$$

Let t be an arbitrary integer. Then

$$\begin{aligned} f(b+tp) &= a_k(b+tp)^k + a_{k-1}(b+tp)^{k-1} + \cdots + a_1(b+tp) + a_0 \\ &= (a_k b^k + a_{k-1}b^{k-1} + \cdots + a_1 b + a_0) + p \cdot g(t) \end{aligned}$$

where $g(t)$ is a polynomial in t. Thus,

$$\begin{aligned} f(b+tp) &= p + pg(t), \quad \text{by equation (2.1)} \\ &= p[1 + g(t)] \end{aligned}$$

So $p|f(b+tp)$. But every value of f is a prime, so $f(b+tp)$ must be a prime and hence $f(b+tp) = p$. Thus, $f(b) = p = f(b+tp)$. This implies f takes on the same value infinitely many times, since t is an arbitrary integer.

But $f(n)$ is a polynomial of degree k, so it *cannot* assume the same value more than k times, yielding a contradiction.

Thus, *no* polynomial with integral coefficients exists that will generate only primes. ■

Returning to Theorem 2.8, can we discover a way to find (or even estimate) the number of primes $\le$ a positive integer n (or a positive real number x)? This is possible, as the following theorem shows, but first we present a function.

A Number-Theoretic Function

Let x be a positive real number. Then $\pi(x)$[†] denotes the number of primes $\leq x$.

For example, $\pi(10) = 4$, $\pi(28.75) = 9$, and $\pi(100) = 25$ (see Figure 2.25). Using the summation notation, $\pi(x)$ can be defined as

$$\pi(x) = \sum_{p \leq x} 1, \quad \text{where } p \text{ denotes a prime.}$$

The following formula for $\pi(n)$, where n is a positive integer, can be established using the inclusion–exclusion principle. Its proof is a bit complicated, so we omit it.

THEOREM 2.10 Let $p_1, p_2, \ldots, p_t$ be the primes $\leq \sqrt{n}$. Then

$$\pi(n) = n - 1 + \pi(\sqrt{n}) - \sum_i \left\lfloor \frac{n}{p_i} \right\rfloor + \sum_{i<j} \left\lfloor \frac{n}{p_i p_j} \right\rfloor - \sum_{i<j<k} \left\lfloor \frac{n}{p_i p_j p_k} \right\rfloor + \cdots + (-1)^t \left\lfloor \frac{n}{p_1 p_2 \cdots p_t} \right\rfloor$$ ■

The following example illustrates this result.

EXAMPLE 2.24 Using Theorem 2.10, find the number of primes ≤ 100.

SOLUTION

Here $n = 100$. Then $\pi(\sqrt{n}) = \pi(\sqrt{100}) = \pi(10) = 4$, by Figure 2.25. The four primes ≤ 10 are 2, 3, 5, and 7; call them p_1, p_2, p_3, and p_4, respectively. Then, by Theorem 2.10,

$$\begin{aligned}\pi(100) = 100 - 1 + 4 &- \left(\left\lfloor \frac{100}{2} \right\rfloor + \left\lfloor \frac{100}{3} \right\rfloor + \left\lfloor \frac{100}{5} \right\rfloor + \left\lfloor \frac{100}{7} \right\rfloor \right) \\ &+ \left(\left\lfloor \frac{100}{2 \cdot 3} \right\rfloor + \left\lfloor \frac{100}{2 \cdot 5} \right\rfloor + \left\lfloor \frac{100}{2 \cdot 7} \right\rfloor + \left\lfloor \frac{100}{3 \cdot 5} \right\rfloor + \left\lfloor \frac{100}{3 \cdot 7} \right\rfloor + \left\lfloor \frac{100}{5 \cdot 7} \right\rfloor \right) \\ &- \left(\left\lfloor \frac{100}{2 \cdot 3 \cdot 5} \right\rfloor + \left\lfloor \frac{100}{2 \cdot 3 \cdot 7} \right\rfloor + \left\lfloor \frac{100}{2 \cdot 5 \cdot 7} \right\rfloor + \left\lfloor \frac{100}{3 \cdot 5 \cdot 7} \right\rfloor \right) \\ &+ \left\lfloor \frac{100}{2 \cdot 3 \cdot 5 \cdot 7} \right\rfloor\end{aligned}$$

[†] π is the lower case Greek letter *pi*.

$$= 103 - (50 + 33 + 20 + 14) + (16 + 10 + 7 + 6 + 4 + 2)$$
$$- (3 + 2 + 1 + 0) + 0$$
$$= 25$$

This is consistent with the sieve of Eratosthenes in Figure 2.25. ■

Although the formula for $\pi(n)$ in Theorem 2.10 is elegant in the sense that it gives the exact value of $\pi(n)$, it is not very practical when n is fairly large. This is where the prime number theorem, one of the celebrated results in number theory, becomes extremely useful. It gives an approximate value of $\pi(n)$, when n is sufficiently large.

THEOREM 2.11 **(The Prime Number Theorem)**

$$\lim_{x\to\infty} \frac{\pi(x)}{x/\ln x} = 1$$

That is, as x gets larger and larger, $\pi(x)$ approaches $x/\ln x$.[†] ■

Gauss noticed the similarity between the values of $\pi(x)$ and $x/\ln x$, as x gets larger and conjectured the theorem in 1793, but did not provide a proof. In 1850, the Russian mathematician Pafnuty Lvovich Chebychev made significant progress toward a proof; he proved that there are positive constants a and b, such that

$$a\frac{x}{\ln x} < \pi(x) < b\frac{x}{\ln x}$$

where $x \geq 2$.

In 1870, the German mathematician Ernest Meissel (1826–1895) showed that there are 5,761,455 primes less than 10^8. In 1893, one hundred years after Gauss' conjecture, the Danish mathematician N. P. Bertelsen claimed that there are 50,847,478 primes less than 10^9. In 1959, however, the American mathematician Derrick H. Lehmer (1905–1991) showed that Bertelsen's answer was incorrect and that the correct number is 50,847,534. Lehmer also showed that there are 455,052,512 primes less than 10^{10} (Table 2.4).

In 1896, the French mathematician Jacques Hadamard (1865–1963) and the Belgian mathematician Charles-Jean-Gustave-Nicholas de la Valleé-Poussin (1866–1962), working independently, proved the theorem using advanced mathematics.

[†] $\ln x$ denotes the natural logarithm of x.

***Pafnuty Lvovich Chebychev** (1821–1894), the son of an army officer, was born in Okatavo, Russia. In 1832 the family moved to Moscow, where he completed his secondary education at home. Five years later, he entered Moscow University, graduating in mathematics in 1841. As a student he published his first paper on a new method of approximating real roots of equations, for which he was awarded a silver medal, although many believed he deserved a gold one. He joined the faculty of St. Petersburg University in 1843, where he remained until 1882. His doctoral thesis,* Theory of Congruences, *submitted to Petersburg University in 1849, earned him an award from the Academy of Sciences and served as a text on number theory at Russian universities; it also dealt with the distribution of primes. With this work and a second memoir published in 1852, he became widely known in the scientific community. Besides number theory, he made significant contributions to real and numerical analysis, approximation theory, probability, and mechanics.*

Chebychev received numerous honors, and invented a calculating machine that could add and subtract and later multiply and divide; he also founded the prestigious Petersburg Mathematical School.

***Jacques Hadamard** (1865–1963) was born in Versailles, France. His father was a Latin teacher and his mother a distinguished piano teacher. After studying at the École Normale Superieure he taught at the Lycée Buffon in Paris. After receiving his doctorate in 1892 he became a lecturer at the Faculté des Sciences of Bordeaux and then at the Sorbonne. In 1909, he became professor at the Collège de France, École Polytechnique, and then at the École Centrale des Arts et Manufactures. A recipient of many honorary doctorates, in 1892 he earned the Grand Prix of the Academy of Sciences for his outstanding work in complex analysis; this in turn led to his proof of the prime number theorem in 1896.*

Nearly every branch of mathematics was influenced by the creative mind of Hadamard, especially complex analysis, functional analysis, probability, and mathematical physics.

This proof was a milestone in the development of number theory. But in 1950, the Hungarian mathematician Paul Erdös (1913–1996) and the Norwegian mathematician Alte Selberg (1917–) proved the theorem using elementary calculus.

According to the prime number theorem, when x is sufficiently large, $\pi(x)$ can be approximated by $x/\ln x$ (see columns 2 and 3 in Table 2.4). But a better approximation is the function $\mathrm{li}(x)$, defined by Gauss in 1792 at the age of 15,

Charles-Jean-Gustave-Nicholas de la Valleé-Poussin *(1866–1962), the son of a geology professor, was born in Louvain, Belgium. He attended the Jesuit College in Mons, switching his major from philosophy to engineering. After receiving his degree, however, he devoted himself to mathematics. In 1892 he joined the faculty at the University of Louvain, where he remained all his life. As the outstanding Belgian mathematician of his generation, he received many honors, including the rank of baron in 1928.*

Valleé-Poussin's most significant contribution was his proof of the prime number theorem using complex analysis. He extended his work to the distribution of primes in arithmetic progressions and primes represented by quadratic forms. He also made important contributions to approximation theory, analysis, and calculus.

Paul Erdös *(1913–1996) was born in Budapest. Both his parents were high school teachers of mathematics and physics; his father had spent six years in a Siberian prison. Young Erdös (pronounced air-dish) was home-taught, mostly by his father, except for about three years in school.*

A child prodigy, Erdös at age three discovered negative numbers for himself. In 1930, he entered Eötvös University. Three years later, he discovered a beautiful proof of Chebychev's theorem that there is a prime between a positive integer n and $2n$. In 1934, he received his Ph.D.

Erdös was one of the most prolific writers in mathematics, authoring about 1500 articles and coauthoring about 500. Ernest Straus, in a tribute in 1983, described Erdös as "the prince of problem-solvers and the absolute monarch of problem-posers." As "the Euler of our time," Erdös wrote extensively in number theory, combinatorics, function theory, complex analysis, set theory, group theory, and probability; number theory and combinatorics were his favorites.

Regarding worldly possessions "as a nuisance," he never owned a home, car, or checkbook, and never had a family or an address. "Always searching for mathematical truths," he traveled from meeting to meeting carrying a half-empty suitcase. He stayed with mathematicians wherever he went and donated the honorariums received as prizes for students.

Erdös received numerous honors. He died of a massive heart attack while attending a mathematics meeting in Warsaw, Poland.

$$\mathrm{li}(x) = \int_2^x \frac{dt}{\ln t}$$

You may notice from the table that $\dfrac{\pi(x)}{\mathrm{li}(x)}$ approaches 1 more rapidly than $\dfrac{\pi(x)}{x/\ln x}$. In fact, $\mathrm{li}(x)$ is a superior approximation for small x.

x	$\pi(x)$	$\dfrac{\pi(x)}{x/\ln x}$	$\dfrac{\pi(x)}{\mathrm{li}(x)}$
10^3	168	1.160	0.9438202
10^4	1229	1.132	0.9863563
10^5	9592	1.104	0.9960540
10^6	78498	1.085	0.9983466
10^7	664579	1.071	0.9998944
10^8	5761455	1.061	0.9998691
10^9	50847534	1.054	0.9999665
10^{10}	455052512	1.048	0.9999932

Table 2.4

In 1985, however, R. H. Hudson of the University of South Carolina showed that it is not true for arbitrary x. Four years later, C. Bays of the University of South Carolina and Hudson showed that $\pi(x) > \mathrm{li}(x)$ in the vicinity of 1.39822×10^{316}.

Although we have established the infinitude of primes, what can we say about the distribution of primes? How are they distributed among the positive integers? Are there consecutive integers that are primes? Are there consecutive odd integers that are primes?

First, there is no pattern that fits the distribution of primes. For example, 2 and 3 are the only two consecutive integers that are primes (see Exercise 45). It is also known that 3, 5, and 7 are the only three consecutive odd integers that are primes (see Exercise 46).

Although there are only two consecutive integers that are primes, we can find any number of consecutive integers that are composite numbers, as the next theorem reveals. It shows that primes occur at unpredictable intervals. Its proof is an existence proof, so we need to provide n such composite numbers.

THEOREM 2.12 For every positive integer n, there are n consecutive integers that are composite numbers.

PROOF

Consider the n consecutive integers $(n+1)!+2, (n+1)!+3, \ldots, (n+1)!+(n+1)$, where $n \geq 1$. Suppose $2 \leq k \leq n+1$, then $k|(n+1)!$, so $k|[(n+1)!+k]$, by Theorem 2.4, for every k. Therefore, each of them is a composite number.

Thus, the n consecutive integers $(n+1)!+2$, $(n+1)!+3, \ldots, (n+1)!+(n+1)$ are composites. ■

The following example illustrates the theorem.

EXAMPLE 2.25 Find six consecutive integers that are composites.

SOLUTION

By Theorem 2.12, there are six consecutive integers beginning with $(n+1)!+2 = (6+1)!+2 = 5042$, namely, 5042, 5043, 5044, 5045, 5046, and 5047. (You may notice from Figure 2.25 that the smallest consecutive chain of six composite numbers is 90, 91, 92, 93, 94, and 95.) ■

According to Theorem 2.12, we can always find arbitrarily long chains of consecutive integers that are composites. Note that the n composite numbers provided by the proof need not be the smallest consecutive composite integers that form a chain of length n. (See Supplementary Exercise 5 in Chapter 3 for constructing a considerably smaller string.)

Next we turn to some interesting classes of primes.

Cunningham Chains

A **Cunningham chain** of primes, named after the British Army officer Lt. Col. Allan J. C. Cunningham (1842–1928), is a sequence of primes $2p+1$ in which each element is one more than twice its predecessor.

The smallest five-element chain is 2–5–11–23–47 and the smallest six-element chain is 89–179–359–719–1439–2879. Lehmer discovered three chains of seven primes with the least element less than 10 million:

1122659–2245319–4490639–8981279–17962559–35925119–71850239;
2164229–4328459–8656919–17313839–34627679–69255359–138510719;
2329469–4658939–9317879–18635759–37271519–74543039–149086079

In 1965, Lehmer also found chains of length 7 of primes of the form $2p-1$. Two such chains begin with 16651 and 165901.

In 1980, Claude Lalout and Jean Meeus discovered chains of length 8 of each kind. They begin with 19099919 and 15514861, respectively.

Nine years later, Gunter Loh found many new such chains: The least elements of length 9 start with 85864769 and 857095381; those of length 10 with 26089808579 and 205528443121; those of length 11 with 665043081119 and 138912693971; and those of length 12 with 554688278429 and 216857744866621.

Until 1952, the largest known prime was the 39-digit number $2^{127}-1=$ 170,141,183,460,469,231,731,687,303,715,884,105,727, found in 1876 by Lucas. With the advent of computers, mathematicians have been able to find larger and larger primes. In 1952, mathematicians at Cambridge University, England, using EDSAC (Electronic Delay Storage Arithmetic Calculator), found a 79-digit prime given by $180(2^{127}-1)^2+1$. Since then many larger primes have been found.

In 1971, a very large prime, $2^{11213}-1$, was found at the University of Illinois, Urbana-Champaign. It contains 3376 digits. A few years later, a still larger prime, $2^{19937}-1$, was found by Bryant Tuckerman of Thomas J. Watson Research Center, International Business Machines.

The search for larger primes continues. In 1978 Noll and Nickel, two high school students from California found two still larger primes, $2^{21701}-1$ and $2^{23209}-1$. Eleven years later, an even larger prime, $2^{44497}-1$, was found by D. Slowinski of Livermore Laboratory at Livermore, California; it contains 13,395 digits.

Modern high-speed computers have certainly facilitated the pursuit of larger and larger primes. The largest known prime as of 1994 was $2^{859433}-1$, it has "only" 258,716 digits, and according to *The Boston Globe*, it "would take eight newspaper pages to print." Two years later, a still larger prime, $2^{1257787}-1$, was found by a supercomputer at Cray Research, Inc. It contains 378,632 digits.

The largest known prime in 2000 was $2^{6972593}-1$. Discovered a year earlier by the team of Nayan Hajrawala, George Woltman, and Scott Kurowski, it contains 2,098,960 digits. Hajrawala's home computer, a 350-MHz Aptiva, took 111 days of idle time to find it; the researchers estimated that it would have taken three weeks to locate it if the computer had been running full time. In 2005, two larger primes were found: $2^{25964951}-1$ with 7,816,230 digits and $2^{30402457}-1$ with 9,152,052 digits. The former was discovered on February 18 by Martin Nowak, an eye surgeon and a mathematics hobbyist in Germany, and the latter on December 15 by C. Cooper and S. R. Boone of Central Missouri State University. Table 2.5 lists the ten largest known primes. Needless to say that the hot pursuit of larger primes still continues.

For the curious minded, the largest known prime, all of whose digits are also prime, is $72323252323272325252 \times \dfrac{10^{3120}-1}{10^{20}-1}$. Discovered in 1992 by Harvey Dubner of New Jersey, it has 3120 digits.

Palindromic Primes

Interestingly, there are primes that are palindromic[†]; they are **palindromic primes**. In 1950, L. Moser of the University of North Carolina discovered 107 such primes

[†] A **palindrome** is a number that reads the same backward and forward, such as 23432.

Rank	*Prime*	*No. of Digits*	*Discoverer(s)*	*Year Discovered*
1	$2^{30402457}-1$	9,152,052	C. Cooper & S. R. Boone	2005
2	$2^{25964951}-1$	7,816,230	M. Nowak	2005
3	$2^{24036583}-1$	7,235,733	J. Findley	2004
4	$2^{20996011}-1$	6,320,430	M. Shafer	2003
5	$2^{13466917}-1$	4,053,946	M. Cameron	2001
6	$27653 \cdot 2^{9167433}+1$	2,759,677	D. Gordon	2005
7	$28433 \cdot 2^{7830457}+1$	2,357,207	S. Yates	2004
8	$2^{6972593}-1$	2,098,960	N. Hajrawala *et al.*	1999
9	$5359 \cdot 2^{5054502}+1$	1,521,561	R. Sundquist	2003
10	$4847 \cdot 2^{3321063}+1$	999,744	R. Hassler	2005

Table 2.5 *The ten largest known primes.*

$\leq 100{,}000$; 19 of them are ≤ 1000: 2, 3, 5, 7, 11, 101, 131, 151, 181, 313, 353, 373, 383, 727, 757, 787, 797, 919, and 929. The palindromic prime 16661 not only contains the embedded beast but also is the 1928th prime; it has the additional property that $1+6+6+6+1 = 1+9+2+8$, first observed by G. J. Honaker, Jr. The palindromic prime $10^{11310} + 4661644 \cdot 10^{56752} + 1$, found in 1991 by Dubner, contains 11,311 digits; it is **doubly palindromic** in the sense that the number of digits is also a palindromic prime. The largest known palindromic prime, $10^{39026} + 4538354 \cdot 10^{19510} + 1$, discovered in 2001 by Dubner, contains 39,027 digits.

Repunit Primes

The largest known **repunit prime** is R_{1031}, discovered in 1985 by Hugh C. Williams of the University of Manitoba. Repunit primes appear to be scarce, since there are only five such repunits R_n for $n < 10{,}000$. The known repunits and their discoverers are listed in Table 2.6.

n	*Discoverer*	*Year Discovered*
2	Unknown	ancient
19	O. Hoppe	1918
23	D. H. Lehmer	1929
317	H. C. Williams	1978
1031	H. C. Williams	1985

Table 2.6 *The known repunit primes.*

Twin Primes

Recall that 2 and 3 are the only two consecutive integers that are primes. Are there any primes that differ by 2? Clearly, 3 and 5, and 5 and 7 are two such pairs. Such pairs are called **twin primes**. The next two pairs are 11 and 13, and 17 and 19. (Can you find the next two pairs?)

Discovering twin primes involves essentially finding two primes; therefore, the largest known twin primes are substantially smaller than the largest known primes. Table 2.7 lists the ten largest known twin primes.

Rank	*Twin Primes*	*No. of Digits*	*Discoverer(s)*	*Year Discovered*
1, 2	$16869987339975 \cdot 2^{171960} \pm 1$	51,779	Z. Járail *et al.*	2005
3, 4	$33218925 \cdot 2^{169690} \pm 1$	51,090	D. Papp	2002
5, 6	$60194061 \cdot 2^{114689} \pm 1$	34,533	D. Underbakke	2002
7, 8	$1765199373 \cdot 2^{107520} \pm 1$	32,376	J. McElhatton	2002
9, 10	$318032361 \cdot 2^{107001} \pm 1$	32,220	D. Underbakke & P. Carmody	2001

Table 2.7 *Ten largest known twin primes.*

Although more than 100,000 twin primes are known, no one knows how many such pairs there are. This is still one of the leading mysteries in number theory.

A related conjecture is the number of pairs $z(N)$ of twin primes $n \pm 1$ that are $\leq N$:

?

$$z(N) \approx 1.3203236 \int_2^N \frac{\mathrm{d}n}{(\log n)^2}$$

where $5 \leq n+1 \leq N$.

Lehmer studied pairs of twin primes, such as 11–13–17–19 and 101–103–107–109, all lying within a decade. In his *Table of Primes*, Lehmer lists 9933611–9933613–9933617–9933619 as the largest known such quadruplet.

In 1999, B. J. Hulbert of Reading, England, investigated such quadruplets and found 1220 of them. Three of them are 22271–22273–22277–22279, 72221–72223–72227–72229, and 15222371–15222373–15222377–15222379, again all strikingly similar and lying within a decade; the latter is the largest known prime quadruplet.

Brun's Constant

In 1919, the Norwegian mathematician Viggo Brun (1885–1978) proved that the sum of the reciprocals of the twin primes $(1/3+1/5)+(1/5+1/7)+(1/11+1/13)+\cdots$ converges to a limit, called **Brun's constant**. In 1974, the American mathematicians Daniel Shanks and John Wrench, Jr., estimated Brun's constant using twin primes among the first 2 million primes. Two years later, Richard Brent of the Australian National University refined the estimate to 1.90216054 using the twin primes up to 100 billion.

Twin Primes and the Pentium Chip

When Intel Corporation, the world's largest chip manufacturer, shipped the Pentium chip to various computer manufacturers in early 1994, the chip was found to have a flaw in division involving more than five significant digits. Intel, claiming that only one in nine billion users would be affected by the error, chose not to recall the chip. Simultaneously, Thomas Nicely, a computational number theorist at Lynchburg College, Virginia, was trying to improve previous estimates of Brun's constant. Using a Pentium computer in June, he computed the constant twice, employing two different methods. One used a computer's floating point unit and the other used an extended precision arithmetic; they yielded different results. Nicely found that the Pentium was giving incorrect floating point reciprocals for the twin primes 824,633,702,441 and 824,633,702,443. After the error was made public in November, Intel offered to replace chips with the flaw (see Figure 2.27). On receiving a flurry of international

How Number Theory Got the Best of the Pentium Chip

Barry Cipra

Chalk one up for number theory. With lurid accounts of the flaw in Intel's Pentium processor making front-page and network news, users of the personal computer chip in fields ranging from science to banking are finding cases where its faulty logic sends their computations awry. But the problem might have gone undetected for much longer if the chip had not slipped up months ago during a long series of calculations in number theory, raising the suspicions of a dogged mathematics professor.

To other mathematicians, the discovery of the flaw by Thomas Nicely of Lynchburg College in Virginia emphasizes the value of number theory—the study of subtle properties of ordinary counting numbers—for providing quality control for new computer systems. By forcing a computer to perform simple operations repeatedly on many different numbers, number-theory calculations "push machines to their limits," says Peter Borwein of Simon Fraser University in Burnaby, British Columbia. Many computer makers have adopted these calculations as a shakedown test for systems intended for heavy-duty scientific computation, and although the practice has yet to spread to personal computers, Borwein and some other mathematicians think that might be a good idea.

Intel had actually found the flaw by other means after the chip had gone into production, but had decided that it was not likely to affect ordinary users. But the company hadn't counted on the use that Nicely had in mind. When he fired up a Pentium computer last March, Nicely was adding its number-crunching power to a project in computational number theory he had begun the year before. He was trying to improve on previous estimates of a number called Brun's sum, which is related to the distribution of prime numbers.

The sequence of prime numbers—2, 3, 5, 7, 11, 13, 17, 19, etc.—is a continuing source of fascination to mathematicians. Since the time of Euclid, they have known that

(continued)

Figure 2.27

there are infinitely many primes, but although primes are relatively abundant early on, they become scarce among larger numbers. For example, roughly 23% of two-digit numbers are prime (21 of 90), but the figure for ten-digit numbers is just 4%, and among hundred-digit numbers, the fraction of primes is less than half a percent. As a consequence, the gap between consecutive prime numbers tends to increase. However, every so often two odd numbers in a row turn out to be prime: 3 and 5, 41 and 43, 101 and 103, and 10,007 and 10,009, for example.

Mathematicians conjecture that such "twin primes" pop up infinitely often. But in 1919, the Norwegian mathematician Viggo Brun proved that even if there are infinitely many twin primes, the sum obtained by adding their reciprocals—the sum $(1/3 + 1/5) + (1/5 + 1/7) + (1/11 + 1/13) + \cdots$— converges to a finite value, much as the sum $1/2 + 1/4 + 1/8 + 1/16 + \cdots$ converges to 1. Brun's sum is known only to the first few digits, however—and even there, the accuracy is based on conjectures about the frequency with which twin primes occur. Number theorists think it's unlikely that clumps of twin primes are lurking among very large numbers, but they have been unable to prove it. One way to check up on this assumption is to compute better estimates for Brun's sum.

In 1974, two mathematicians working for the Navy, Daniel Shanks and John Wrench Jr., reported the first computationally intensive estimate of Brun's sum, based on the occurrence of twin primes among the first two million prime numbers.

Two years later, Richard Brent at the Australian National University calculated all twin primes up to a hundred billion (224,376,048 pairs), from which he computed an estimate of 1.90216054 for Brun's sum.

And there it sat—until Nicely entered the picture. The Lynchburg math professor decided to push Brent's work into the trillions. To be on the safe side, he computed Brun's sum twice, using two different methods: the "easy" way using a computer's built-in floating point unit, which is supposed to be accurate to 19 decimal places, and the "hard" way using an extended precision arithmetic, which he set to give 26 (and later 53) digits of accuracy. (The difference can be likened to the difference between computing $1/3 + 1/7$ as $0.33 + 0.14 = 0.47$ and computing it as $1/3 + 1/7 - 10/21 = 0.48$. The latter calculation gains accuracy by doing some exact arithmetic first.)

The comparison between the two methods is what got Intel into trouble. After Nicely added the new Pentium to his stable of computers, he found that the gap between the two results was much larger than it should have been. By trial and error and a process of elimination, he pinpointed the source of the problem: The Pentium was giving incorrect floating point reciprocals for the twin primes 824,633,702,441 and 824,633,702,443—they were wrong from the 10th digit on. Nicely still didn't know whether the error was caused by his hardware or software, in part because he'd caught an earlier error in a compiler program. "Finally, in desperation, I ran this portion of the calculation on one of the 486 [computers], rather than the Pentium," he recalls. "The error disappeared."

Even that didn't prove conclusively that it was the Pentium chip's fault; other hardware in the computer could have been responsible. But in October (4 months after he first noticed his calculations were off), Nicely nailed the culprit when he got hold of two other machines with Pentium chips and was able to reproduce the error. He notified Intel and, after getting no satisfactory answer by the end of the month, sent e-mail asking others to double-check his discovery. "I believe you are aware of events from that point on," he concludes dryly.

The Pentium's problem, as others have abundantly confirmed, lies in the way the chip does division. Although it works fine for most numbers, the chip's built-in algorithm makes mistakes in certain cases, rather like a gradeschooler who mismemorized part of a multiplication table. Nicely estimates that the chip gets roughly one in a billion reciprocals wrong. But because the work in number theory required him to compute billions of reciprocals over a wide range, he was almost bound to run into the mistake.

"We've known for a long time that number theory computations are very helpful" for turning up computer errors, notes computational number theorist Arjen Lenstra of Bellcore, in Morristown, New Jersey. "It is useful to run number theory stuff on your processor before you sell it."

Intel hasn't decided whether to make such computations a routine part of its testing procedure, says Stephen Smith, engineering manager for the Pentium processor division. But Intel was so impressed with Nicely work that it asked him to run further computations on a corrected chip. "We looked at him as the most thorough tester," says Smith.

Figure 2.27

***Sophie Germain** (1776–1831), France's great woman mathematician, was born in Paris, and educated herself at home, using her father's extensive library. At the age of thirteen, she read in J. F. Montucla's* Historie des Mathematiques *of the murder of the Greek mathematician and inventor Archimedes (ca. 287–212 B.C.) by a Roman soldier. Archimedes became her hero, and she decided to become a mathematician despite her parents' serious objections. After mastering both Latin and Greek, she studied the works of Newton and Euler.*

Because of her gender, Sophie was not allowed to attend the newly established École Central des Travaux Publics (later the École Polytechnique), but she managed to obtain the lecture notes of Lagrange and other scholars. She sent Lagrange a paper on analysis, under the pseudonym M. Leblanc. He was so impressed with the paper that he became her mathematical mentor. Germain corresponded with many mathematicians, including Legendre and Gauss.

Germain made a significant contribution toward establishing Fermat's last theorem (see Section 12.2) that the equation $x^n + y^n = z^n$ has no positive integral solutions, where $n \geq 3$. In 1825, she showed that if p is a Sophie Germain prime, then the equation $x^p + y^p = z^p$ has no positive integral solutions, where $xyz \neq 0$ and $p \nmid xyz$. She also made important contributions to the theories of acoustics and elasticity.

attention, Nicely said,[†] "Usually mathematicians have to shoot somebody to get this much publicity."

Sophie Germain Primes

Another class of primes, called **Sophie Germain primes**, played an important role in establishing Fermat's last theorem, which is discussed in Section 13.2.

Named in 1825 in honor of the French mathematician Sophie Germain, these primes have the form $2p + 1$, where p is an odd prime. The first three such primes are 7, 11, and 23. Clearly, each Sophie Germain prime belongs to a Cunningham chain. The ten largest known Sophie Germain primes are listed in Table 2.8.

It has been conjectured that there are infinitely many Sophie Germain primes. ?

Goldbach's Conjecture

?

The Prussian mathematician Christian Goldbach noticed a pattern in the following

[†] *Cincinnati Enquirer*, December 18, 1994.

Rank	Prime	No. of Digits	Discoverer(s)	Year Discovered
1	$7068555 \cdot 2^{121301} - 1$	36,523	P. Minovic	2005
2	$2540041185 \cdot 2^{114729} - 1$	34,547	D. Underbakke	2003
3	$18912879 \cdot 2^{98395} - 1$	29,628	M. Angel *et al.*	2002
4	$1213822389 \cdot 2^{81131} - 1$	24,432	M. Angel *et al.*	2002
5	$109433307 \cdot 2^{66452} - 1$	20,013	D. Underbakke	2001
6	$984798015 \cdot 2^{66444} - 1$	20,011	D. Underbakke	2001
7	$3714089895285 \cdot 2^{60000} - 1$	18,075	K. Indlekofer *et al.*	2000
8	$909004827 \cdot 2^{56789} - 1$	17,105	B. Tornberg	2005
9	$1162665081 \cdot 2^{55649} - 1$	16,762	B. Xiao	2004
10	$671383317 \cdot 2^{48345} - 1$	14,563	J. Sun	2004

Table 2.8 *The ten largest known Sophie Germain primes.*

sums:

$$\begin{array}{ccc} 4 = 2 + 2 & 6 = 3 + 3 & 8 = 3 + 5 \\ 10 = 3 + 7 & 12 = 5 + 7 & 14 = 3 + 11 \\ 16 = 5 + 11 & 18 = 7 + 11 & 20 = 7 + 13 \\ & \vdots & \end{array}$$

(Do you see a pattern here?) Based on his observations, Goldbach, in a letter to Euler in 1742, conjectured that every even integer > 2 can be expressed as the sum of two primes. Euler could not prove it, and his conjecture still remains an unsolved problem. However, Goldbach's conjecture has been shown to be true for all even integers less than 10^{10}.

The famous English mathematician Godfrey H. Hardy (1877–1947) characterized Goldbach's conjecture as one of the most difficult unsolved problems in mathematics.

In May 2000, Bloomsbury Publishing (United States) and Faber and Faber (United Kingdom) announced a million dollar prize to anyone who could provide

***Christian Goldbach** (1690–1764) was born in Königsberg, Prussia. He studied medicine and mathematics at the University of Königsberg and became professor of mathematics at the Imperial Academy of Sciences in St. Petersburg in 1725. In 1728 he moved to Moscow to tutor Tsarevich Peter II and his cousin Anna of Courland. During 1729–1963, he corresponded with Euler on number theory. He returned to the Imperial Academy in 1732 when Peter's successor Anna moved the imperial court to St. Petersburg.*

In 1742 Goldbach joined the Russian Ministry of Foreign Affairs, and later became privy councilor and established guidelines for the education of royal children.

He is also noted for his conjectures in number theory and work in analysis. Goldbach died in Moscow.

Publishers Offer Prize for Proof of Goldbach's Conjecture

Bloomsbury Publishing (USA) and Faber and Faber (UK) have announced that they are offering a one million dollar prize to any person who can prove Goldbach's Conjecture within the next two years. The prize is being offered to help promote the book *Uncle Petros and Goldbach's Conjecture*, by Apostolos Doxiadis (see the review by Keith Devlin on MAA Online's *Read This!* section, which can be found on the web at http://www.maa.org/reviews/reviews.html). To be eligible for the prize, the proof must be submitted to a journal indexed by *Mathematical Reviews* by March 15, 2002, must be published by that journal by March 15, 2004, and must be judged to be correct by a six-member judging panel whose members will be mathematicians chosen by the publisher. See Faber's web site at http://www.faber.co.uk/ (click on "Book News") for more information on the prize.

Figure 2.28

a proof of Goldbach's conjecture by March 15, 2002. See Figure 2.28. To date, it still remains a conjecture.

Bertrand's Conjecture

In 1845, Joseph Bertrand conjectured that there is a prime between n and $2n$ for every integer $n \geq 2$. For example,

3 is a prime between 2 and 4;
5 is a prime between 3 and 6;
7 is a prime between 4 and 8; and so on.

Although Bertrand could not establish the validity of his conjecture, he was able to verify it for all integers ≤ 3 million! Seven years later, Chebychev provided a successful proof. In 1944, the Indian number theorist S. S. Pillai (1901–1950) gave a simpler proof.

Using Bertrand's conjecture and induction, it can be shown that $p_{n+1} \leq 2p_n$ and hence $p_n \leq 2^n$, where p_n denotes the nth prime (see Exercise 58). For instance, $p_5 = 11$ from Figure 2.25, so clearly $p_5 \leq 2^5$.

It is worth noting that 2^n is an extremely large upper bound for p_n when n is fairly large. For instance, $p_{11} = 31$ is much smaller than $2^{11} = 2048$; nevertheless, it is true.

It is well known that $p_1 p_2 p_3 \cdots p_n > p_{n+1}$, where $n \geq 2$. In 1907, H. Bonse developed a stronger inequality, now called **Bonse's inequality**: $p_1 p_2 p_3 \cdots p_n > p_{n+1}^2$,

Joseph Louis François Bertrand *(1822–1900), the son of a writer of popular scientific articles and books, was born in Paris. At the age of 11, he unofficially began attending classes at the École Polytechnique. In 1838, at 16, he earned two degrees, one in the arts and the other in science. A year later, he received his doctorate for his work in thermomechanics and published his first paper. In 1841 he became professor at the Collège Saint-Louis. Subsequently, he taught at the Lycée Henry IV, the École Normale Supérieure, the École Polytechnique, and finally at the Collège de France until his death.*

An author of many popular textbooks, Bertrand made important contributions to applied mathematics, analysis, differential geometry, probability, and theoretical physics.

where $n \geq 4$. In 2000, M. Dalezman of Yeshiva University strengthened it even further: $p_1 p_2 p_3 \cdots p_n > p_{n+1} p_{n+2}$, where $n \geq 4$.

An interesting application of Bertrand's conjecture was proposed in 1989 by the Romanian mathematician Florentin Smarandache.† In addition to the conjecture, it uses two results: Suppose $n \geq 4$. Then $n! > 2^n$ and $1 \cdot 3 \cdot 5 \cdots (2n-1) > 2^{n+2}$. Verify both.

EXAMPLE 2.26 Prove that there are at least $3\lfloor n/2 \rfloor$ primes in the range n through $n!$, where $n \geq 4$.

PROOF

Notice that the statement is true for $4 \leq n \leq 9$. So assume $n \geq 10$.

case 1 Suppose n is even, say, $n = 2k$, where $k \geq 5$. Then

$$\begin{aligned} n! &= 1 \cdot 2 \cdot 3 \cdots (2k-2)(2k-1)n \\ &= 2^{k-1}[1 \cdot 2 \cdot 3 \cdots (k-1)][1 \cdot 3 \cdot 5 \cdots (2k-1)]n \\ &> 2^{k-1}(k-1)! 2^{k+2} n \\ &\geq 2^{k-1} \cdot 2^{k-1} \cdot 2^{k+2} n, \quad \text{since } k \geq 5 \\ &= 2^{3k} n \end{aligned}$$

A repeated application of Bertrand's conjecture shows there are at least $3k = 3(n/2) = 3\lfloor n/2 \rfloor$ primes in the range n through $2^{3k}n$, that is, between n and $n!$.

† In 1988, Smarandache escaped from the Ceausescu dictatorship, spent 2 years in a political refugee camp in Turkey, and then emigrated to the United States.

case 2 Suppose n is odd, say, $n = 2k + 1$, where $k \geq 5$. Then

$$\begin{aligned} n! &= 1 \cdot 2 \cdot 3 \cdots (2k-1)(2k)n \\ &= 2^k k![1 \cdot 3 \cdot 5 \cdots (2k-1)]n \\ &> 2^k \cdot 2^k \cdot 2^{k+2} n, \quad \text{since } k \geq 5 \\ &> 2^{3k} n \end{aligned}$$

Thus, as before, there are at least $3k = 3[(n-1)/2)] = 3\lfloor n/2 \rfloor$ primes in the range n through $2^{3k}n$, that is, between n and $n!$.

Thus, in both cases, the result is true. ■

Additional Conjectures

Are there primes of the form $n^2 + 1$? Clearly, $2 = 1^2 + 1$ and $5 = 2^2 + 1$ are two such primes. There are two more such primes ≤ 100. No one knows how many such primes exist. ?

The number of primes $p(N)$ of the form $n^2 + 1$ has been conjectured to be given by ?

$$p(N) \approx 0.6864067 \int_{n=2}^{N} \frac{\mathrm{d}n}{\log n}$$

where $2 \leq n \leq N$.

Legendre's conjecture: Is there a prime between n^2 and $(n+1)^2$? For example, ?

3 is a prime between 1 and 4; 5 is a prime between 4 and 9;
11 is a prime between 9 and 16; 19 is a prime between 16 and 25;
29 is a prime between 25 and 36.

Does this pattern hold for any positive integer n? That, too, still remains unanswered.

Bocard's conjecture: There are at least four primes between the squares of consecutive odd primes; for example, there are five primes between 3^2 and 5^2. ?

The following example singles out a unique prime.

EXAMPLE 2.27 Find the primes such that the digits in their decimal values alternate between 0s and 1s, beginning with and ending in 1.

SOLUTION

Suppose N is a prime of the desired form and it contains n ones. Then

$$\begin{aligned} N &= 10^{2n-2} + 10^{2n-4} + \cdots + 10^2 + 1 \\ &= \frac{10^{2n}-1}{10^2-1} \quad \text{since } \sum_{i=0}^{n-1} r^i = \frac{r^n-1}{r-1},\ r \neq 1 \\ &= \frac{(10^n-1)(10^n+1)}{99} \end{aligned}$$

If $n = 2$, then $N = \dfrac{(10^2-1)(10^2+1)}{99} = 101$ is a prime. If $n > 2$, $10^n - 1 > 99$ and $10^n + 1 > 99$. Then N has nontrivial factors, so N is composite. Thus, 101 is the only prime with the desired properties. ■

Primality of Catalan Numbers

Recall from Section 1.8 that the Catalan numbers C_2 and C_3 are prime. The next theorem confirms that there are no other such primes. We leave the proof as an exercise; see Exercise 75.

THEOREM 2.13 **(Koshy and Salmassi, 2004)** The only prime Catalan numbers are C_2 and C_3. ■

EXERCISES 2.5

Mark *true* or *false*, where a, b, d, and n are arbitrary positive integers.

1. A nonprime positive integer is a composite.
2. A noncomposite positive integer is a prime.
3. Every prime is odd.
4. There are no primes greater than googolplex.
5. If p is a prime, then $p+2$ is a prime.
6. If p is a prime, then p^2+1 is a prime.
7. There is an infinite number of primes.
8. There is an infinite number of composite numbers.
9. If p is a prime such that $p|ab$, then $p|a$ or $p|b$.
10. There are primes of the form $n!+1$.

Determine whether each is prime or composite.

11. 129 12. 217 13. 1001 14. 1729

Using Theorem 2.10, compute the number of primes $\leq n$ for each value of n.

15. 47 16. 61 17. 96 18. 131

19. Find five consecutive integers < 100 that are composite numbers.

Find n consecutive integers that are composites for each value of n.

20. seven 21. eight
22. nine 23. ten

24. List all twin primes ≤ 100.
25. Find all twin primes whose arithmetic mean is a triangular number.
26. List all primes of the form $n^2 + 1$ and < 100.

Find the smallest prime between n and $2n$ for each value of n.

27. 5 28. 6 29. 20 30. 47

Find the smallest prime between n^2 and $(n+1)^2$ for each value of n.

31. 6 32. 7 33. 10 34. 11

35. Prove or Disprove: $n! + 1$ is a prime for every nonnegative integer n.
36. In 1775, Lagrange conjectured that every odd integer > 5 can be written in the form $p + 2q$, where p and q are primes. Verify his conjecture for 7, 11, 15, and 23.
37. Find the flaw in the following "proof" that there are no primes greater than 101.

    Let $n > 101$. Clearly, n has to be odd. When n is odd, both $(n+1)/2$ and $(n-1)/2$ are integers. Let $x = (n+1)/2$ and $y = (n-1)/2$. Then $n = x^2 - y^2 = (x-y)(x+y)$, so n is not a prime. Thus, there are no primes > 101.

Find the positive factors of each, where p and q are distinct primes.

38. pq 39. p^2q
40. pq^2 41. p^2q^2

Let $q_1 = 2$ and $q_n = q_1q_2 \dots q_{n-1} + 1$, where $n \geq 2$.

42. Find the first four primes of the form q_n.
43. Find the smallest composite number of the form q_n.
44. Define q_n recursively.

Prove each.

45. 2 and 3 are the only two consecutive integers that are primes.
46. 3, 5, and 7 are the only three consecutive odd integers that are primes.
47. If p and $p^2 + 8$ are primes, the $p^3 + 4$ is also a prime. (D. L. Silverman, 1968)
48. If p is a prime and $1 \leq k < p$, then $p \mid \binom{p}{k}$.
49. Let p and q be successive odd primes and $p + q = 2r$. Then r is composite. (J. D. Baum, 1966)
50. The sum of two successive odd primes is the product of at least three (not necessarily distinct) prime factors. (J. D. Baum, 1967)
51. If p and $p^2 + 2$ are primes, then $p^3 + 2$ is also a prime.
52. The integral lengths of the legs of a right triangle cannot be twin primes. (J. H. Tiner, 1968)
53. If p and $p + 2$ are twin primes, then p must be odd.
54. Suppose p and q are primes such that $p - q = 3$. Then $p = 5$.
55. Every odd prime is of the form $4n + 1$ or $4n + 3$.
56. One more than the product of twin primes is a perfect square.
57. If n is composite, then $2^n - 1$ is a composite.
58. Let p_n denote the nth prime. Then $p_n < 2^n$, where $n \geq 2$.
59. Prove by contradiction that every integer ≥ 2 has a prime factor.
    (*Hint*: Use the well-ordering principle.)
60. Rewrite the proof of Euclid's theorem using p as the largest prime and $n = p! + 1$.

Let p_n denote the nth prime. Determine whether $p_{n+1} \leq p_1p_2 \cdots p_n + 1$ for each value of n.

61. 5 62. 7
63. 8 64. 10

65. Show that the repunits R_4 and R_5 are composite.
66. Find an explicit formula for R_n.
67. Prove or disprove: If n is a prime, then R_n is a prime.
68. Let $f(x) = \sum_{i=0}^{n} a_ix^i$, where a_i is an integer and $a_n \neq 0$. Suppose $f(n_0) = p$ is a prime. Prove that $f(n_0 + kp)$ is composite for any integer k.
69. The simplest consecutive prime triplet p_n–p_{n+1}–p_{n+2} such that $p_n \mid (p_{n+1}p_{n+2} + 1)$ is 2–3–5. Find two other such consecutive prime triplets. (G. L. Honaker, 1990) (*Note*: In 1991, L. Hodges of Iowa showed that there are only three such solutions below 7.263×10^{13}.)

Let $E_n = p_1p_2 \cdots p_n + 1$, where p_i denotes the ith prime and $i \geq 1$.

70. Find the least composite value of E_n.

71. Prove that $E_n < (p_{n+1})^{n-2}$, $n \geq 6$. (A. A. K. Majumdar, 1996)

*72. Establish the formula for $\pi(n)$ in Theorem 2.10.

*73. Let p_k denote the kth prime. Prove that $p_{n+1} \leq p_1 p_2 \cdots p_n + 1$, where $n \geq 1$.

*74. Let p_i denote the ith prime, where $i \geq 1$. Prove that $p_n p_{n+1} p_{n+2} > p_{n+3}^2$, where $n \geq 3$. (S. Bulman-Flemming and E. T. H. Wang, 1989)

*75. Establish Theorem 2.13.

2.6 Fibonacci and Lucas Numbers

Fibonacci numbers are one of the most intriguing number sequences, which continues to provide ample opportunities for both professional mathematicians and amateurs to make conjectures and to expand the limits of mathematical knowledge.

The Fibonacci sequence is named after Leonardo Fibonacci, the most outstanding Italian mathematician of the Middle Ages. It is so important and fascinating that there is an association of Fibonacci enthusiasts, *The Fibonacci Association*, devoted to the study of the sequence. The association, founded in 1963 by Verner E. Hoggatt, Jr. (1921–1980) of San Jose State College and Brother Alfred Brousseau (1907–1988) of St. Mary's College in California, publishes *The Fibonacci Quarterly* devoted to articles related to the Fibonacci sequence.

The following problem, proposed by Fibonacci in 1202 in his classic book, *Liber Abaci*, gave birth to the Fibonacci sequence.

The Fibonacci Problem

Suppose there are two newborn rabbits, one male and the other female. Find the number of rabbits produced in a year if

- Each pair takes one month to become mature;
- Each pair produces a mixed pair every month, from the second month; and
- All rabbits are immortal.

Suppose, for convenience, that the original pair of rabbits was born on January 1. They take a month to become mature, so there is still only one pair on February 1. On March 1, they are two months old and produce a new mixed pair, a total of two pairs. Continuing like this, there will be three pairs on April 1, five pairs on May 1, and so on. See the last row of Table 2.9.

Leonardo Fibonacci *(1170?–1250?), also known as Leonardo of Pisa, was born in the commercial city of Pisa, Italy, into the Bonacci family. His father, a customs manager, expecting Leonardo to become a merchant, took him to Bougie, Algeria, to receive advanced training in arithmetic using Indo-Arabic numerals. Leonardo's own business trips to Egypt, Syria, Greece, and Sicily gave him extensive experience with Indo-Arabic mathematics.*

In 1202, shortly after his return to Pisa, Fibonacci published his famous work, Liber Abaci, *extolling the superiority of the Indo-Arabic methods of computation. (The word* abaci *in the title does not refer to the old abacus, but to computation in general.) This book, devoted to arithmetic and elementary algebra, introduced the Hindu-Arabic notation and arithmetic algorithms to Europe.*

Fibonacci wrote three additional books: Practice Geometriae, *a collection of results in geometry and trigonometry;* Liber Quadratorum, *a major work on number theory; and* Flos, *also on number theory.*

Fibonacci's importance and usefulness to Pisa and its citizenry through his teaching and services were recognized by Emperor Frederick II.

No. of Pairs	*Jan.*	*Feb.*	*March*	*April*	*May*	*June*	*July*	*Aug.*
Adults	0	1	1	2	3	5	8	13
Babies	1	0	1	1	2	3	5	8
Total	1	1	2	3	5	8	13	21

Table 2.9

Fibonacci Numbers

The numbers 1, 1, 2, 3, 5, 8, ... in the bottom row are **Fibonacci numbers**. They have a fascinating property: Any Fibonacci number, except the first two, is the sum of the two immediately preceding Fibonacci numbers. (At the given rate, there will be 144 pairs of rabbits on December 1.)

This yields the following recursive definition of the nth Fibonacci number F_n:

$$F_1 = F_2 = 1 \qquad \leftarrow \text{initial conditions}$$
$$F_n = F_{n-1} + F_{n-2}, \quad n \geq 3 \qquad \leftarrow \text{recurrence relation}$$

Interestingly enough, Fibonacci numbers appear in quite unexpected places. They occur in nature, music, geography, and geometry. They can be found in the spiral arrangements of seeds in sunflowers, the scale patterns of pine cones, the number of petals in flowers, and the arrangement of leaves on trees. See Figure 2.29.

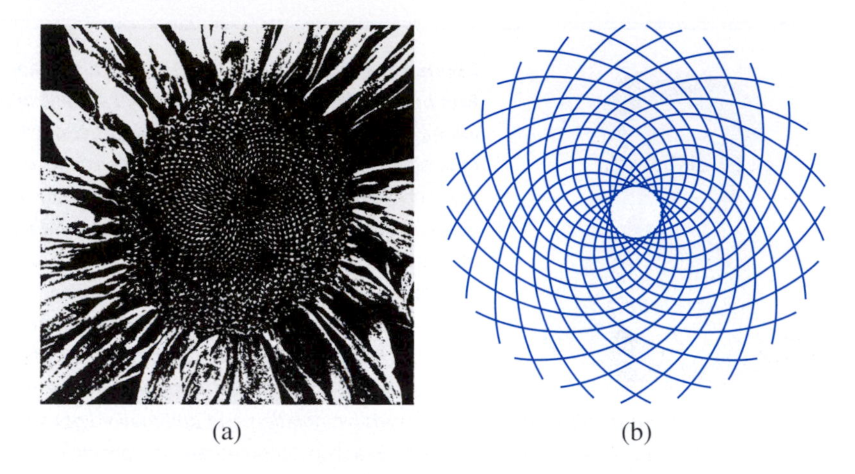

Figure 2.29

Fibonacci and Pascal's Triangle

It is surprising that Fibonacci numbers can be extracted from Pascal's triangle. Add the numbers along the northeast diagonals, as Figure 2.30 shows. Curiously enough, the sums appear to be the various Fibonacci numbers.

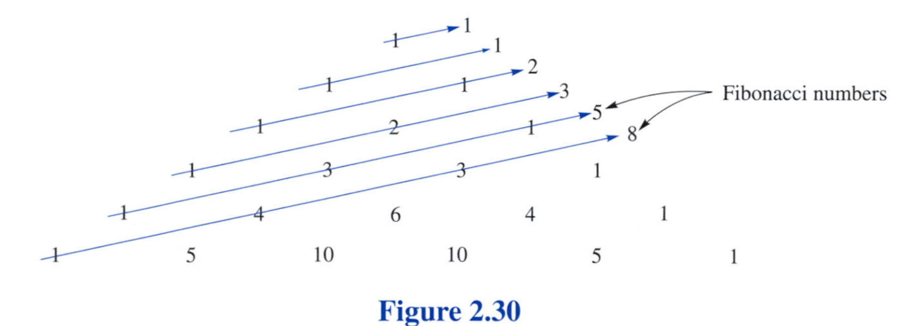

Figure 2.30

This observation is confirmed by the following theorem, discovered in 1876 by Lucas. It can be established using induction.

THEOREM 2.14 **(Lucas, 1876)**

$$F_n = \sum_{i=0}^{\lfloor (n-1)/2 \rfloor} \binom{n-i-1}{i}, \quad n \geq 1$$ ■

The recursive definition of F_n yields a straightforward method for computing it, as Algorithm 2.5 shows.

François-Edouard-Anatole Lucas *(1842–1891) was born in Amiens, France. After completing his studies at the École Normale in Amiens, he worked as an assistant at the Paris Observatory. He served as an artillery officer in the Franco-Prussian war and then became professor of mathematics at the Lycée Saint-Louis and Lycée Charlemagne, both in Paris. A gifted and entertaining teacher, Lucas died of a freak accident at a banquet; his cheek was gashed by a piece of a plate that was accidentally dropped; he died from infection within a few days.*

Lucas loved computing and developed plans for a computer that never materialized. Besides his contributions to number theory, he is known for his four-volume classic on recreational mathematics. Best known among the problems he developed is the Tower of Brahma.

```
Algorithm Fibonacci(n)
(* This algorithm computes the nth Fibonacci number using recursion. *)
  Begin (* algorithm *)
    if n = 1 or n = 2 then (* base cases *)
      Fibonacci ← 1
    else
      Fibonacci ← Fibonacci(n − 1) + Fibonacci(n − 2)
  End (* algorithm *)
```

Algorithm 2.5

The tree diagram in Figure 2.31 illustrates the recursive computing of F_5, where each dot represents an addition.

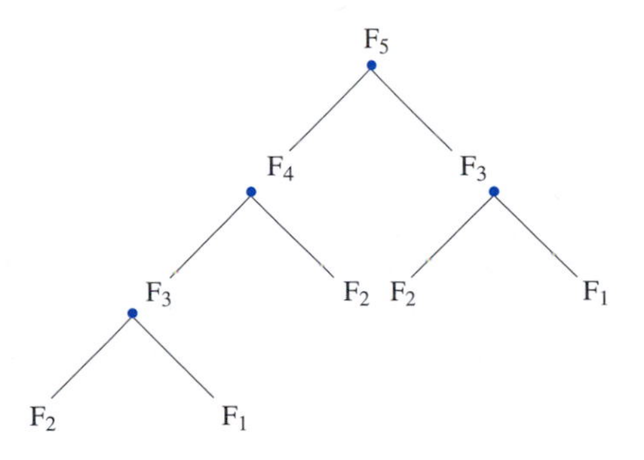

Figure 2.31

Next we pursue two interesting properties of Fibonacci numbers by way of experimentation and conjectures.

EXAMPLE 2.28 Find a formula for $\sum_{i=1}^{n} F_i$.

SOLUTION

step 1 *Collect sufficient data.*

$$\begin{aligned} F_1 &= 1 &&= 1 \\ F_1+F_2 &= 1+1 &&= 2 \\ F_1+F_2+F_3 &= 1+1+2 &&= 4 \\ F_1+F_2+F_3+F_4 &= 1+1+2+3 &&= 7 \\ F_1+F_2+F_3+F_4+F_5 &= 1+1+2+3+5 &&= 12 \end{aligned}$$

step 2 *Look for a pattern.*

These equations do not seem to manifest any pattern. So we rewrite them in such a way that a pattern emerges:

$$\begin{aligned} F_1 &= 1 = F_3 - 1 \\ F_1+F_2 &= 2 = F_4 - 1 \\ F_1+F_2+F_3 &= 4 = F_5 - 1 \\ F_1+F_2+F_3+F_4 &= 7 = F_6 - 1 \\ F_1+F_2+F_3+F_4+F_5 &= 12 = F_7 - 1 \end{aligned}$$

When we look at the subscripts on both sides, a clear pattern arises.

step 3 *Make a conjecture.*

$$\sum_{i=1}^{n} F_i = F_{n+2} - 1$$

step 4 *Establish the formula using induction.*

Since $F_1 = F_3 - 1$, the formula works for $n = 1$.

Now assume it is true for an arbitrary positive integer $k \geq 1$:

$$\sum_{i=1}^{k} F_i = F_{k+2} - 1$$

Then

$$\begin{aligned}\sum_{i=1}^{k+1} F_i &= \sum_{i=1}^{k} F_i + F_{k+1}\\ &= (F_{k+2} - 1) + F_{k+1}\\ &= (F_{k+1} + F_{k+2}) - 1\\ &= F_{k+3} - 1\end{aligned}$$

Thus, by induction, the formula is true for every positive integer n. (This formula was derived in 1876 by Lucas.) ■

For example, $\sum_{i=1}^{10} F_i = F_{12} - 1 = 144 - 1 = 143$. You may verify this by direct computation.

We now mention a Fibonacci puzzle based on this formula.

A Fibonacci Puzzle (optional)

Think of two positive integers a_1 and a_2. Add them to get a_3. Add the last two to get the next number a_4. Continue like this until you get ten numbers: $a_1, a_2, \ldots, a_{10}$. Compute their sum $s = \sum_{i=1}^{n} a_i$. Write down all ten numbers. Without adding them, we can accurately give you the sum. How does it work?

Next, we study the following Fibonacci pattern:

$$\begin{aligned}F_1F_3 - F_2^2 &= 1 \cdot 2 - 1^2 = (-1)^2\\ F_2F_4 - F_3^2 &= 1 \cdot 3 - 2^2 = (-1)^3\\ F_3F_5 - F_4^2 &= 2 \cdot 5 - 3^2 = (-1)^4\\ F_4F_6 - F_5^2 &= 3 \cdot 8 - 5^2 = (-1)^5\\ &\vdots\end{aligned}$$

Clearly, a pattern emerges. (Look at the subscripts and the power of -1 on the RHS.) Accordingly, we conjecture that $F_{n-1}F_{n+1} - F_n^2 = (-1)^n$, where $n \geq 1$. We can confirm it as an exercise.

THEOREM 2.15 **(Cassini's Formula)**

$$F_{n-1}F_{n+1} - F_n^2 = (-1)^n, \quad n \geq 1$$ ■

***Giovanni Domenico Cassini** (1625–1712) was born in a family of astronomers in Perinaldo, Imperia, Italy. He studied at Vallebone, the Jesuit College at Genoa, and then at the abbey of San Fructuoso. He manifested great enthusiasm in poetry, mathematics, and astronomy. Working at the observatory at Panzano, near Bologna, he completed his education under the tutelage of the great scientists Giovan Battista Riccioli and Francesco Maria Grimaldi, whose work influenced him a great deal. In 1650, Cassini became the principal chair of astronomy at the University of Bologna.*

Cassini left for Paris in 1669 to continue his brilliant career in planetary astronomy at the Académie Royal des Sciences. He assumed responsibility for the Academy and became a French citizen. Cassini died in Paris.

***Robert Simson** (1687–1768), son of a successful merchant, was born in West Kilbridge, Ayrshire, Scotland. After attending the University of Glasgow, he studied theology to follow the family tradition of serving in the Church of Scotland. At Glasgow he received no formal training in mathematics, but reading of George Sinclair's* Tyrocinia Mathematica in Novem Tractatus *(1661) he became interested in mathematics, and moved on to Euclid's* Elements.

During the academic year 1710–1711, he attended a mathematics school and met several prominent mathematicians, including Edmund Halley (1656–1742), the well-known astronomer and Savilian professor of geometry at Oxford. In 1711, Simson was appointed professor of mathematics at Glasgow.

He devoted most of his life to restoring the works of Greek geometers. Simson wrote on conic sections, logarithms, and the theory of limits, but by far his most influential work was the 1756 edition of Euclid's Elements, *which served as the basis of every subsequent edition of* Elements *until the beginning of the twentieth century.*

This formula was first discovered in 1680 by the Italian-born French astronomer and mathematician Giovanni Domenico Cassini, and discovered independently in 1753 by Robert Simson (1687–1768) of the University of Glasgow.

A Fibonacci Paradox (optional)

Cassini's formula is the basis of a delightful geometric paradox. This puzzle was a favorite of the famous English logician Charles Lutwidge Dodgson (1832–1898), better known as Lewis Carroll, who first published it in a mathematical periodical in Leipzig, Germany, in 1866 (666 years after Fibonacci published his rabbit problem). The brilliant American puzzlist Sam Loyd claimed that he had presented it to the American Chess Congress in 1858. Although we may never know the exact origin of the puzzle, it is nevertheless an intriguing one.

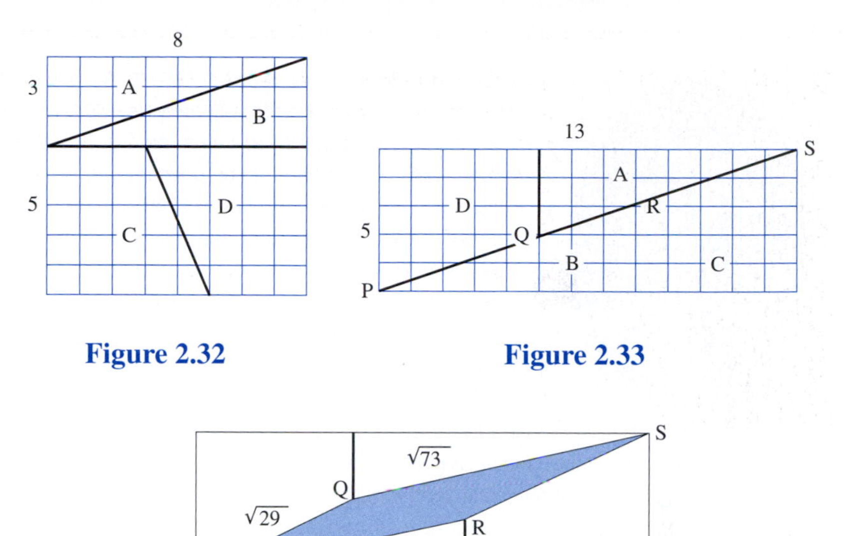

Figure 2.32

Figure 2.33

Figure 2.34

Consider an 8×8 square; cut it up into four pieces, A, B, C, and D, as in Figure 2.32. Now rearrange them to form a 5×13 rectangle, as Figure 2.33 shows. The area of the square is 64 square units, whereas that of the rectangle is 65 square units. In other words, by reassembling the four pieces of the original square, we have gained one unit. This appears to be paradoxical.

However, appearances can be deceiving. Although it appears in Figure 2.33 that the "diagonal" PQRS is a line segment, that is not in fact the case. The points P, Q, R, and S are in fact the vertices of a very narrow parallelogram, as Figure 2.34 demonstrates. The area of the parallelogram = area of the rectangle − area of the square $= 5 \cdot 13 - 8^2 = 1 = F_5F_7 - F_6^2$.

Its sides are $\sqrt{29}$ and $\sqrt{73}$ units long, and the diagonal is $\sqrt{194}$ units long. Let θ be the acute angle between the adjacent sides of the parallelogram. Then, by the law of cosines in trigonometry:

$$\cos\theta/2 = \frac{194 + 29 - 73}{2\sqrt{29 \cdot 194}}$$

$$\theta/2 \approx 0.763898460833°$$

$$\theta \approx 1°31'40''$$

This explains why it is a very narrow parallelogram.

Jacques Philippe Marie Binet *(1788–1865), a French mathematician and astronomer, was born at Rennes, Brittany. In 1804, he entered the École Polytechnique in Paris, graduated two years later, and took a job in the Department of Bridges and Roads of the French government. In 1807, Binet became a teacher at the École Polytechnique, and the following year became assistant to the professor of applied analysis and descriptive geometry. In 1814, he was appointed examiner of descriptive geometry, and professor of mechanics (1815) and then inspector general of studies (1816). In 1821, he was awarded the Chevalier de la Legion d'Honneur. Two years later, Binet was appointed chair of astronomy at the Collège de France.*

But the July 1830 revolution was not kind to him. A strong supporter of Charles X, Binet became a victim of Charles' abdication; he was dismissed from École Polytechnique by King Louis-Phillipe in November, 1830.

Binet made many contributions to mathematics, physics, and astronomy. In 1812, he discovered the rule for matrix multiplication, and in 1840, discovered the explicit formula for the nth Fibonacci number. In 1843, he was elected to the Academy of Sciences and later became its president. A devout and modest Catholic, Binet died in Paris.

In fact, there is nothing sacred about the choice of the size of the square. By virtue of Cassini's formula, the puzzle will work for any $F_{2n} \times F_{2n}$ square.

Lucas Numbers

Closely related to Fibonacci numbers are the **Lucas numbers** 1, 3, 4, 7, 11, ..., named after Lucas. Lucas numbers L_n are defined recursively as follows:

$$L_1 = 1, \qquad L_2 = 3$$
$$L_n = L_{n-1} + L_{n-2}, \quad n \geq 3$$

Binet's Formulas

Both Fibonacci numbers and Lucas numbers can be defined explicitly using **Binet's formulas**:

$$F_n = \frac{\alpha^n - \beta^n}{\alpha - \beta} \quad \text{and} \quad L_n = \alpha^n + \beta^n$$

where $\alpha = (1 + \sqrt{5})/2$ and $\beta = (1 - \sqrt{5})/2$ are the solutions of the quadratic equation $x^2 = x + 1$. See Exercises 32–37.

The explicit formula for F_n was discovered by the French mathematician Jacques-Phillipe-Marie Binet in 1843. In fact, it was first discovered in 1718 by

Gabriel Lamé *(1795–1870) was born in Tours, France. After graduating from the École Polytechnique in 1817, he continued his studies at the École des Mines from which he graduated in 1820.*

The same year, Lamé was appointed director of the School of Highways and Transportation in St. Petersburg, Russia. There he taught mathematics, physics, and chemistry, and planned roads and bridges in and around the city. In 1832 he returned to Paris to form an engineering firm. Within a few months, however, he left it to become the chair of physics at the École Polytechnique, where he remained until 1844. While teaching, he served as a consulting engineer, becoming the chief engineer of mines in 1836. He helped build the railroads from Paris to Versailles and to St. Germain.

In 1844 Lamé became graduate examiner for the University of Paris in mathematical physics and probability, and professor seven years later. In 1862 he became deaf and resigned his positions. He died in Paris in 1870.

Although Lamé made discoveries in number theory and mathematical physics, his greatest contribution was the development of the curvilinear coordinates and their applications. His work on curvilinear systems led him to number theory. In 1840 he proved Fermat's last theorem for $n = 7$.

Gauss considered Lamé the foremost French mathematician of his time. Ironically, most French mathematicians considered him too practical, and most French scientists thought him too theoretical.

the French mathematician Abraham De Moivre (1667–1754) using generating functions and arrived at independently in 1844 by the French engineer and mathematician Gabriel Lamé.

Using the recursive definitions and Binet's formulas, we can develop an array of properties of both numbers.

EXERCISES 2.6

1. Using the fact that $F_n = F_{n+1} - F_{n-1}$, derive a formula for $\sum_{i=1}^{n} F_i$.

Let a_n denote the number of additions needed to compute F_n using recursion. Compute each.

2. a_8
3. a_9
4. a_{10}
5. a_{13}
6. Using Exercises 2–5, conjecture a formula for a_n.
7. Prove the formula in Exercise 6.
8. Define a_n recursively.
9. Prove that $\sum_{i=1}^{n} a_i = a_{n+2} - n$.
10. An n-bit word containing no two consecutive ones can be constructed recursively as follows: Append a 0 to such $(n-1)$-bit words or append a 01 to such $(n-2)$-bit words. Using this procedure, construct all 5-bit words containing no two consecutive ones. There are 13 such words.
11. Compute F_{n+1}/F_n correct to eight decimal places for $1 \le n \le 10$. Compare each value to the value of $(1+\sqrt{5})/2$ places to eight decimal places.

?

12. Using Exercise 11, predict $\lim_{n\to\infty} \frac{F_{n+1}}{F_n}$.

Conjecture a formula for each.

13. $\sum_{i=1}^{n} F_{2i-1}$
14. $\sum_{i=1}^{n} F_{2i}$
15. $\sum_{i=1}^{n} L_i$
16. $\sum_{i=1}^{n} L_{2i-1}$
17. $\sum_{i=1}^{n} L_{2i}$
18. $\sum_{i=1}^{n} F_i^2$
19. $\sum_{i=1}^{n} L_i^2$

Prove each.

20. $F_n = 2F_{n-2} + F_{n-3}, \quad n \geq 4$
21. $F_{n-1}F_{n+1} - F_n^2 = (-1)^n, \quad n \geq 2$
22. F_{5n} is divisible by 5, $\quad n \geq 1$
23. $\sum_{i=1}^{n} F_{2i-1} = F_{2n}$ (E. Lucas)
24. $\sum_{i=1}^{n} F_{2i} = F_{2n+1} - 1$ (E. Lucas)
25. $\sum_{i=1}^{n} L_i = L_{n+2} - 3$
26. $\sum_{i=1}^{n} L_{2i-1} = L_{2n} - 2$
27. $\sum_{i=1}^{n} L_{2i} = L_{2n+1} - 1$
28. $\sum_{i=1}^{n} F_i^2 = F_nF_{n+1}$ (E. Lucas)
29. $\sum_{i=1}^{n} L_i^2 = L_nL_{n+1} - 2$
30. Let $A = \begin{bmatrix} 1 & 1 \\ 1 & 0 \end{bmatrix}$. Then $A^n = \begin{bmatrix} F_{n+1} & F_n \\ F_n & F_{n-1} \end{bmatrix}$, $n \geq 1$. Assume $F_0 = 0$.
31. Using Exercise 30, deduce that $F_{n-1}F_{n+1} - F_n^2 = (-1)^n$.
(*Hint*: Let A be a square matrix. Then $|A^n| = |A|^n$, where $|A|$ denotes the determinant of A.)

The nth term b_n of a number sequence is defined by $b_n = (\alpha^n - \beta^n)/(\alpha - \beta)$, where $\alpha = (1 + \sqrt{5})/2$ and $\beta = (1 - \sqrt{5})/2$ are solutions of the equation $x^2 = x + 1$. Verify each.

32. $b_1 = 1$
33. $b_2 = 1$
34. $b_n = b_{n-1} + b_{n-2}, \quad n \geq 3$

With α and β as above, let $u_n = \alpha^n + \beta^n$, $n \geq 1$. Verify each.

35. $u_1 = 1$
36. $u_2 = 3$
37. $u_n = u_{n-1} + u_{n-2}, \quad n \geq 3$

(These exercises indicate that $u_n = L_n$, the nth Lucas number.)
Using Binet's formulas, prove each.

38. $F_{2n} = F_nL_n$
39. $F_{n-1} + F_{n+1} = L_n$
40. $F_{n+2} - F_{n-2} = L_n$
41. $L_{n-1} + L_{n+1} = 5F_n$
42. $F_{n+1}^2 + F_n^2 = F_{2n+1}$ (E. Lucas)
43. $F_{n+1}^2 - F_{n-1}^2 = F_{2n}$ (E. Lucas)
44. Let a_n denote the number of rectangles that can be formed on a $1\times n$ rectangular board. Find the recurrence relation satisfied by a_n.
(*Hint*: Look for a pattern. Every square is also a rectangle.)

A subset of the set $S = \{1, 2, \ldots, n\}$ is said to be **alternating** if its elements, when arranged in increasing order, follow the pattern odd, even, odd, even, etc. For example, $\{1, 2, 5\}$ and $\{3, 4\}$ are alternating subsets of $\{1, 2, 3, 4, 5\}$, whereas $\{1, 3, 4\}$ and $\{2, 3, 4, 5\}$ are not; $\emptyset$ is considered alternating. Let a_n denote the number of alternating subsets of S. [Olry Terquem (1782–1862)]

45. Define a_n recursively.
46. Prove that $a_n = F_{n+2}$.

***Pierre de Fermat** (1601–1665), born near Toulouse, was the son of a leather merchant. A lawyer by profession, he devoted his leisure time to the pursuit of mathematics as a hobby. Although he published almost none of his brilliant discoveries, he did correspond with contemporary mathematicians.*

Fermat has contributed to several branches of mathematics, but he is best known for his work in number theory. Many of his results appear in margins of his copy of the works of the Greek mathematician Diophantus (ca. 250 A.D.). He wrote about his own famous conjecture: "I have discovered a truly wonderful proof, but the margin is too small to contain it."

2.7 Fermat Numbers

Numbers of the form $f_n = 2^{2^n} + 1$ were studied by the outstanding French mathematician Pierre de Fermat and are called **Fermat numbers**. The first five Fermat numbers are $f_0 = 3, f_1 = 5, f_2 = 17, f_3 = 257$, and $f_4 = 65537$.

The following theorem presents an interesting recurrence relation satisfied by f_n.

THEOREM 2.16 Let f_n denote the nth Fermat number. Then $f_n = f_{n-1}^2 - 2f_{n-1} + 2$, where $n \geq 1$.

PROOF

We shall substitute for f_{n-1} in the expression $f_{n-1}^2 - 2f_{n-1} + 2$, simplify it, and show that it equals f_n:

$$\begin{aligned}
f_{n-1}^2 - 2f_{n-1} + 2 &= \left(2^{2^{n-1}} + 1\right)^2 - 2\left(2^{2^{n-1}} + 1\right) + 2 \\
&= \left(2^{2^n} + 2 \cdot 2^{2^{n-1}} + 1\right) - 2 \cdot 2^{2^{n-1}} - 2 + 2 \\
&= 2^{2^n} + 1 \\
&= f_n
\end{aligned}$$

This completes the proof. ■

This theorem leads to a recursive definition of f_n.

A Recursive Definition of f_n

$$f_0 = 3$$
$$f_n = f_{n-1}^2 - 2f_{n-1} + 2, \quad n \geq 1$$

For example,

$$f_1 = f_0^2 - 2f_0 + 2 = 9 - 2 \cdot 3 + 2 = 5$$

and

$$f_2 = f_1^2 - 2f_1 + 2 = 25 - 2 \cdot 5 + 2 = 17$$

We can make an interesting observation about Fermat numbers. Notice that the numbers $f_2 = 17$, $f_3 = 257$, $f_4 = 65537$, $f_5 = 4294967297$, and $f_6 = 18446644033331951617$ all end in the same decimal digit, 7. Amazing! So what can you conjecture about Fermat numbers? Can you prove it? (See Exercises 2 and 3.)

Here is another interesting observation: The first five Fermat numbers 3, 5, 17, 257, and 65537 are primes. So Fermat conjectured that every Fermat number is a prime.

In 1732, however, Euler established the falsity of his conjecture by producing a counterexample. He showed that f_5 is divisible by 641: $f_5 = 4294967297 = 641 \cdot 6700417$. An alternate proof was given in 1926 by the Belgian mathematician M. Kraitchik (1882–1957) in his *Théorie des nombres*.

The following example furnishes a clever, elementary proof by G. T. Bennett of this result. The beauty of its proof lies in the fact that it does not involve any division.

EXAMPLE 2.29 Show that $641 | f_5$.

SOLUTION

First notice that

$$641 = 5 \cdot 2^7 + 1 \tag{2.2}$$

$$\begin{aligned}
\text{So} \quad 2^{2^5} + 1 &= 2^{32} + 1 = 2^4 \cdot 2^{28} + 1 \\
&= 16 \cdot 2^{28} + 1 = (641 - 625)2^{28} + 1 \\
&= (641 - 5^4)2^{28} + 1 = 641 \cdot 2^{28} - (5 \cdot 2^7)^4 + 1 \\
&= 641 \cdot 2^{28} - (641 - 1)^4 + 1, \quad \text{by equation (2.2)} \\
&= 641 \cdot 2^{28} - (641^4 - 4 \cdot 641^3 + 6 \cdot 641^2 - 4 \cdot 641 + 1) + 1 \\
&= 641(2^{28} - 641^3 + 4 \cdot 641^2 - 6 \cdot 641 + 4)
\end{aligned}$$

Thus, $641 | f_5$. ■

An Alternate Proof

In 1995, Stanley Peterburgsky, while studying at the New England Academy of Torah, Rhode Island, proved that f_5 is composite by showing that $\frac{f_5}{641}$ can be expressed as the sum of two squares. To see this, recall from Chapter 1 that $(a^2+b^2)(c^2+d^2) = (ac+bd)^2 + (ad-bc)^2$ for any integers a, b, c, and d. Then

$$\frac{a^2+b^2}{c^2+d^2} = \frac{(ac+bd)^2+(ad-bc)^2}{(c^2+d^2)^2}$$

Now let $a = 2^{16}$, $b = 1$, $c = 4$, and $d = 25$. Then

$$\begin{aligned}\frac{f_5}{641} &= \frac{2^{32}+1}{641}\\ &= \frac{(2^{16}\cdot 4+25)^2+(25\cdot 2^{16}-4)^2}{641^2}\\ &= 409^2+2556^2\end{aligned}$$ ■

Unfortunately, nothing is known about the infinitude of Fermat primes. It still remains an unsolved problem. In fact, no Fermat primes beyond f_4 have been found; the largest known Fermat prime continues to be f_4. The largest known Fermat composite number is $f_{2478782}$, discovered in 2003.

Is every Fermat number **square-free**, that is, free of square factors? It has been conjectured by both Lehmer and A. Schinzel that there are infinitely many square-free Fermat numbers.

The following result, derived by Lucas, is an extremely useful tool in the prime factorization of f_n. In 1747, Euler proved that every prime factor of f_n must be of the form $A \cdot 2^{n+1} + 1$. In 1879, Lucas refined Euler's work by showing that A must be an even integer $2k$. This leads us to the following theorem.

THEOREM 2.17 Every prime factor of f_n is of the form $k \cdot 2^{n+2} + 1$, where $n \geq 2$. ■

It follows by this theorem that if f_n has no prime factors of the form $k \cdot 2^{n+2} + 1$, then f_n must be a prime. The following example takes advantage of this fact.

EXAMPLE 2.30 Show that $f_4 = 65537$ is prime.

PROOF

It suffices to show that f_4 has no proper prime factors. By Theorem 2.17, every prime factor of f_4 is of the form $2^6k + 1 = 64k + 1$. By Theorem 2.9, if f_4 is composite, it must have a prime factor $\leq \lfloor\sqrt{65537}\rfloor$, that is, ≤ 256. The only prime of the form $64k+1$ and ≤ 256 is 193, but $193 \nmid 65537$; so f_4 is a prime. ■

For the curious minded, we add a bonus: In 1963, S. W. Golomb of the California Institute of Technology established that the sum of the reciprocals of Fermat numbers is an irrational number.

Finally, there is a remarkable link between Fermat primes and the ruler-and-compass construction of regular polygons, where a ruler is used as a straight edge just to draw lines, and a compass just to draw arcs. In 1796, Gauss proved the following celebrated theorem.

THEOREM 2.18 A regular polygon of n sides is constructible with a ruler and compass if and only if n is of the form $f_1 f_2 \cdots f_k$ or $2^k f_1 f_2 \cdots f_k$, where $k \geq 0$ and $f_1, f_2, \ldots, f_k$ are distinct Fermat primes. ■

The early Greeks knew the construction of regular polygons of sides 2^k, $3 \cdot 2^k$, $5 \cdot 2^k$, and $15 \cdot 2^k$. (Notice that 3 and 5 are Fermat primes.) They also knew the construction of polygons of 3, 4, 5, 6, 8, 10, 12, 15, and 16 sides, but *not* the construction of the 17-sided regular polygon. When Gauss, at the age of 19, proved that the 17-sided regular polygon is constructible, he became so elated with his discovery that he decided to devote the rest of his life to mathematics. He also requested that a 17-sided regular polygon be engraved on his tombstone. Although his wish was never fulfilled, such a polygon can be found on a monument to Gauss at his birthplace in Brunswick, Germany.

(A thorough discussion of such geometric constructions requires advanced techniques from abstract algebra, namely, Galois theory.)

EXERCISES 2.7

1. Using recursion, compute the Fermat numbers f_3 and f_4.
2. Make a conjecture about the ones digit in the decimal value of f_n.
3. Establish your conjecture in Exercise 2. (*Hint*: Use induction.)

Prove each.

4. If 2^m+1 is a prime, then m must be a power of 2.
5. If 2^m-1 is a prime, then m must be a prime.
6. Prove or disprove: If m is a prime, then 2^m-1 is a prime.
7. Prove that 3 is the only Fermat number that is also a triangular number. (S. Asadulla, 1987) (*Hint*: Use Exercises 2 and 3.)
8. Redo Exercise 7 using the fact that the product of two integers is a power of 2 if and only if both integers are powers of 2.

9. Does f_5 have a prime factor of the form $k \cdot 2^{n+2} + 1$? If yes, find such a factor.

Determine if a regular polygon of n sides is constructible with a straightedge and compass for each value of n.

10. 257 11. 36 12. 60 13. 17,476

CHAPTER SUMMARY

This chapter presented the division algorithm, one of the fundamental results in number theory. In addition, it established several divisibility properties, the pigeonhole principle, the inclusion–exclusion principle, the uniqueness of the base-b representation of a positive integer, several number patterns, and prime and composite numbers, Fibonacci and Lucas numbers, and Fermat numbers.

The Division Algorithm

- Given any integer a and any positive integer b, there exist a unique quotient q and a unique reminder r such that $a = bq + r$, where $0 \le r < b$. (p. 69)

$$q = \lfloor a/b \rfloor = a \text{ div } b \qquad \text{(p. 71)}$$

$$r = a - bq = a \bmod b \qquad \text{(p. 71)}$$

The Pigeonhole Principle

- If m pigeons are assigned to n pigeonholes, where $m > n$, then at least two pigeons must occupy the same pigeonhole. (p. 74)

Divisibility Properties

- If $a|b$ and $b|c$, then $a|c$. (p. 75)
- If $a|b$, then $a|mb$. (p. 75)
- If $a|b$ and $a|c$, then $a|(\alpha b + \beta c)$. (p. 75)
- There are $\lfloor a/b \rfloor$ positive integers $\le a$ and divisible by b. (p. 75)

The Inclusion–Exclusion Principle

- Let $A_1, A_2, \ldots, A_n$ be n finite sets. Then

$$\left|\bigcup_{i=1}^{n} A_i\right| = \sum_{1 \le i \le n} |A_i| - \sum_{1 \le i < j \le n} |A_i \cap A_j| + \sum_{1 \le i < j < k \le n} |A_i \cap A_j \cap A_k| - \cdots + (-1)^{n+1} \left|\bigcap_{i=1}^{n} A_i\right| \qquad \text{(p. 76)}$$

Odd and Even Integers

- Every even integer is of the form $2m$ and every odd integer is of the form $2n+1$. (p. 78)

Base-b Representation

- Every integer has a unique base-b representation. (p. 80)

Prime and Composite Numbers

- A prime number is a positive integer with exactly two positive factors. A positive integer ≥ 2 that is not a prime is a composite. (p. 104)
- Every positive integer ≥ 2 has a prime factor. (p. 104)
- There are infinitely many primes. (p. 104)
- Every composite number n has a prime factor $\leq \lfloor\sqrt{n}\rfloor$. (p. 105)
- $\pi(x)$ is the number of primes $\leq$ the real number x. (p. 110)
- Let $p_1, p_2, \ldots, p_t$ be the primes $\leq n$. Then

$$\pi(n) = n - 1 + \pi(\sqrt{n}) - \sum_{i} \left\lfloor \frac{n}{p_i} \right\rfloor + \sum_{i<j} \left\lfloor \frac{n}{p_i p_j} \right\rfloor - \sum_{i<j<k} \left\lfloor \frac{n}{p_i p_j p_k} \right\rfloor + \cdots + (-1)^t \left\lfloor \frac{n}{p_1 p_2 \cdots p_t} \right\rfloor \quad \text{(p. 110)}$$

Prime Number Theorem

- $\lim_{x\to\infty} \dfrac{\pi(x)}{x/\ln x} = 1$ (p. 111)
- For any positive integer n, there are n consecutive integers that are composites. (p. 114)

Cunningham Chains

- A Cunningham chain is a chain of primes $2p+1$. (p. 115)

Palindromic Primes

- A prime that is palindromic is a palindromic prime. (p. 116)

Repunit Primes

- A repunit that is prime is a repunit prime. (p. 117)

Twin Primes

- Two primes that differ by 2 are twin primes. (p. 118)

Sophie Germain Primes

- Primes of the form $2p+1$. (p. 121)
- A chain of such primes is a Cunningham chain. (p. 121)

Goldbach's Conjecture

- Every even integer > 2 can be expressed as the sum of two primes. (p. 121)

Bertrand's Conjecture

- There is a prime between n and $2n$, where $n \geq 2$. (p. 123)

Fibonacci Numbers F_n

$$F_1 = 1 = F_2$$
$$F_n = F_{n-1} + F_{n-2}, \quad n \geq 3 \quad \text{(p. 129)}$$

$$F_n = \sum_{i=0}^{\lfloor (n-1)/2 \rfloor} \binom{n-i-1}{i}, \quad n \geq 1 \quad \text{(p. 130)}$$

Lucas Numbers L_n

$$L_1 = 1, \qquad L_2 = 3$$
$$L_n = L_{n-1} + L_{n-2}, \quad n \geq 3 \quad \text{(p. 136)}$$

Binet's Formulas

$$F_n = \frac{\alpha^n - \beta^n}{\alpha - \beta} \quad \text{and} \quad L_n = \alpha^n + \beta^n$$

where $\alpha = (1 + \sqrt{5})/2$ and $\beta = (1 - \sqrt{5})/2$. (p. 136)

Fermat Numbers f_n

$$f_n = 2^{2^n} + 1 \quad \text{(p. 139)}$$
$$= f_{n-1}^2 - 2f_{n-1} + 2, \quad \text{where } f_0 = 3 \quad \text{(p. 139)}$$

- f_5 is a composite number. (p. 141)
- Every prime factor of f_n is of the form $k \cdot 2^{n+2} + 1$, where $n \geq 2$. (p. 141)
- A regular polygon of n sides is constructible with a ruler and compass if and only if n is of the form $f_1 f_2 \cdots f_k$ or $2^k f_1 f_2 \cdots f_k$, where $k \geq 0$ and $f_1, f_2, \ldots,$ and f_k are distinct Fermat primes. (p. 142)

REVIEW EXERCISES

Find the number of positive integers ≤ 2776 and

1. Divisible by 2 or 5.
2. Not divisible by 2 or 3.
3. Divisible by 2, 3, or 5.
4. Not divisible by 2, 5, or 7.

Express each number in base ten.

5. 2000_{eight}
6. 2345_{sixteen}
7. BAD_{sixteen}

*8. $BAD.CA_{\text{sixteen}}$

Rewrite each number in the indicated base b.

9. $245, b = 2$
10. $348, b = 8$
11. $1221, b = 8$
12. $1976, b = 16$

In Exercises 13–16, perform the indicated operation.

13. $\begin{array}{r} 11010_{\text{two}} \\ +\quad 111_{\text{two}} \end{array}$

14. $\begin{array}{r} 5768_{\text{sixteen}} \\ +\ 78CB_{\text{sixteen}} \end{array}$

15. $\begin{array}{r} 5AB8_{\text{sixteen}} \\ \times\ BAD_{\text{sixteen}} \end{array}$

16. $\begin{array}{r} 110110_{\text{two}} \\ -\quad 11011_{\text{two}} \end{array}$

Rewrite each binary integer in base eight.

17. 10110101
18. 1101101101
19. 100110011
20. 10011011001

21–24. Rewrite each integer in Exercises 17–20 in base sixteen.

Find the value of x resulting from the execution of each algorithm fragment.

25. $x \leftarrow 0$
    for $i = 1$ to n do
        for $j = 1$ to n do
            $x \leftarrow x + 1$

26. $x \leftarrow 0$
    for $i = 1$ to n do
        for $j = 1$ to i do
            for $k = 1$ to j do
                $x \leftarrow x + 1$

27. Find a formula for the number a_n of times the statement $x \leftarrow x + 1$ is executed by the following loops.

    for $i = 1$ to n do
        for $j = 1$ to $\lceil i/2 \rceil$ do
            $x \leftarrow x + 1$

Using induction, prove each for every positive integer n.

28. $n^2 - n$ is divisible by 2.

29. $n^3 - n$ is divisible by 3.

30. $\sum_{i=1}^{n}(2i-1)^2 = \frac{n(4n^2-1)}{3}$

31. $\sum_{i=1}^{n}\frac{1}{(2i-1)(2i+1)} = \frac{n}{2n+1}$

32. The product of any two consecutive positive integers is even.
33. Suppose you have an unlimited supply of identical black and white socks. Using induction and the pigeonhole principle, show that you must select at least $2n+1$ socks in order to ensure n matching pairs. (C. T. Long)

Add two more lines to each number pattern. (F. B. Selkin)

34.
$$\begin{aligned} 9\cdot 9 &= 81\\ 99\cdot 99 &= 9801\\ 999\cdot 999 &= 998001\\ 9999\cdot 9999 &= 99980001\\ 99999\cdot 99999 &= 9999800001 \end{aligned}$$

35.
$$\begin{aligned} 7\cdot 9 &= 63\\ 77\cdot 99 &= 7623\\ 777\cdot 999 &= 776223\\ 7777\cdot 9999 &= 77762223\\ 77777\cdot 99999 &= 7777622223 \end{aligned}$$

Determine if each is prime or composite.

36. 237
37. 327
38. 1229
39. 1997

Using Theorem 2.10, find the number of primes $\leq n$ for each value of n.

40. 129
41. 135
42. 140
43. 149

Find n consecutive integers that are composite numbers for each value of n.

44. 4
45. 6
46. 11
47. 13

48. Find all twin primes whose arithmetic mean is a square.
49. The introduction to L. Poletti's *Tavole diNumeri Primi* (Milan, 1920) contains the following statements by H. J. Scherk, where p_n denotes the nth prime, $p_0 = 1$, and $n \geq 1$:

    - p_{2n} may be expressed as the algebraic sum of all its preceding primes and p_0 each taken exactly once.
    - p_{2n-1} may be expressed in the same way, except that the last addend is to be taken twice, where $n \geq 2$.

    Verify Scherk's statement for $1 \leq n \leq 8$.

50. Euler's formula $E(n) = n^2 - n + 41$ yields a prime for $0 \leq n \leq 40$. Find 41 consecutive values of n for which $E(n)$ is composite. (S. Kravitz, 1963)

51. In 1953, J. E. Foster of Evanston, Illinois, conjectured that $2^p + 1 = 3q$, where p and q are odd primes. Show that his conjecture is false.

Prove each, where n is an arbitrary positive integer.

52. $n^3 + n$ is divisible by 2.
53. $n^4 - n^2$ is divisible by 3.
54. $H_m | R_{6m}$, where $H_m = 10^{2m} - 10^m + 1$. (Chico Problem Group, 1990)
55. The square of every odd integer is of the form $8m + 1$.
56. Thomas Greenwood claimed that if n is a prime, then one more than an even triangular number t_n or two less than an odd triangular number is a prime.
57. $b^{3n} \pm 1$ is composite, where $b \geq 2$ and $n \geq 1$.
58. $1 + 5^n + 5^{2n} + 5^{3n} + 5^{4n}$ is composite. (LSU Problem-Solving Group, 2002)
59. $346 | (365^n + 1848^n - 2021^n - 3482^n)$ (R. S. Luthar, 1970)
60. Find all triplets (a, b, c) of consecutive integers a, b, c such that $abc | (a^3 + b^3 + c^3)$. (M. J. Zerger, 2003)
61. Prove or disprove: If n is a prime, then F_n is a prime.

Determine if a regular polygon of n sides can be constructed with a straightedge and compass for each value of n.

62. 16
63. 408
64. 1275
65. 3855

Let α and β be the solutions of the equation $x^2 = x + 1$. Prove each.

66. $x^n = F_n x + F_{n-1}, \quad n \geq 2$
67. $F_n = \dfrac{\alpha^n - \beta^n}{\alpha - \beta}, \quad n \geq 1$
68. $F_1F_2 + F_2F_3 + \cdots + F_{2n-1}F_{2n} = F_{2n}^2$
69. F_n is even if and only if $3|n$.

SUPPLEMENTARY EXERCISES

1. In 1981, O. Higgins discovered that the formula $h(x) = 9x^2 - 471x + 6203$ generates a prime for 40 consecutive values of x. Give a counterexample to show that not every value of $h(x)$ is a prime.
2. The formula $g(x) = x^2 - 2999x + 2248541$ yields a prime for 80 consecutive values of x. Give a counterexample to disprove that every value of $g(x)$ is a prime.

Let n be a four-digit decimal integer, with not all digits the same. Let n' and n'' be the integers obtained by arranging the digits of n in nondecreasing and nonincreasing orders, respectively. Define $K(n) = n' - n''$. For example, $K(1995) = 9951 - 1599 = 8352$.

3. Find $K(K(1995))$.
4. Show that $K(6174) = 6174$. (The integer 6174 is the only four-digit integer that has this property. It is called **Kaprekar's constant**.)
5. Charles W. Trigg of California, who has written extensively on recreational mathematics, showed in 1968 (the year the *Journal of Recreational Mathematics* was first published) that $K^6(1968) = 6174$, where $K^n(m) = K(K^{n-1}(m))$ and $K^1 = K$. Verify this.

An **absolute prime** is a prime such that every permutation of its digits yields a prime. For example, 2, 3, and 5 are absolute primes. Every repunit prime is an absolute prime.

6. There are eight two-digit absolute primes with distinct digits. Find them.
7. There are nine three-digit absolute primes with two distinct digits. Find them.
8. Show that an absolute prime with two or more digits may contain only the digits 1, 3, 7, and 9.

A **cyclic prime** is a prime such that every cyclic permutation of its digits yields a prime. For example, 79 and 97 are cyclic primes. Every absolute prime is also a cyclic prime.

9. Find the cyclic primes that can be obtained from the cyclic prime 3779.
10. There are three three-digit cyclic primes that are not absolute primes; each consists of distinct digits. Find them.
11. Show that a cyclic prime with two or more digits may contain only the digits 1, 3, 7, and 9.
12. A **reversible prime** is a prime that yields a prime when read from right to left. For instance, 113 is a reversible prime. Determine if 199 and 733 are reversible primes.
13. Find all reversible primes < 100.

Give a reversible prime that is not:

14. An absolute prime.
15. A palindromic prime.

A sieving algorithm similar to Eratosthenes' can be employed to generate **lucky numbers**. From the list of positive integers, first strike out every other integer, leaving all odd positive integers. The smallest odd integer left after 1 is 3, so counting with 1 strike out every third integer in the new list. The next integer left is 7, so again starting at 1 cross out every seventh integer in the resulting list. Continuing like this, in step i strike out every ith integer left from step $(i-1)$, where $i > 1$. The numbers that remain are lucky numbers.

16. Find all lucky numbers < 50. (There are 13 such lucky numbers.)
17. Show that there are infinitely many lucky numbers.

Let $a, b \in \mathbf{W}$, $a = (a_n a_{n-1} \dots a_0)_{\text{two}}$, and $b = (b_n b_{n-1} \dots b_0)_{\text{two}}$. If $a_i \geq b_i$ for every i, we say a **implies** b and write $a \Rightarrow b$; otherwise $a \nRightarrow b$.

18. Determine if $43 \Rightarrow 25$ and $47 \Rightarrow 29$.
19. The binomial coefficient $C(n, r)$ is odd if and only if $n \Rightarrow r$. Using this fact, determine the **parity** (oddness or evenness) of $C(25, 18)$ and $C(29, 19)$.
20. Justify the Russian peasant algorithm.

Let b and n be integers ≥ 2. Numbers of the form $S_n = \dfrac{b^n - 1}{b - 1}$ are called **Sylvester numbers**, after the English mathematician James Joseph Sylvester (1814–1897), who investigated them in 1888.

21. Define S_n recursively.
22. If n is a composite number, prove that S_n is composite.
23. If $n > 2$ and b is a square, prove that S_n is composite.

Prove each, where n is an arbitrary positive integer.

*24. $12 | n(3n^4 + 7n^2 + 2)$
*25. $24 | n(3n^4 + 13n^2 + 8)$
*26. Guess the number of odd binomial coefficients in row n of Pascal's triangle. (*Hint*: Compare the number of odd binomial coefficients in row n and the binary expansion of n.)
*27. Find two distinct positive integers A and B such that $A + n$ is a factor of $B + n$ for every integer n, where $0 \leq n \leq 10$. (A. Friedland, 1970)
*28. Characterize all positive integers n such that $k + 1 | \binom{n}{k}$, where $0 \leq k < n$. (E. T. H. Wang, 1994)

A set of integers A is **fat** if each of its elements is $\geq$ the number of elements in A. For example, $\{5, 7, 91\}$ is a fat set, but $\{3, 7, 36, 41\}$ is not. $\varnothing$ is considered a fat set. Let g_n denote the number of fat subsets of the set $\{1, 2, \dots, n\}$. (G. F. Andrews)

*29. Define g_n recursively.
*30. Find an explicit formula for g_n.

Let $f(n, k)$ denote the number of k-element subsets of the set $S = \{1, 2, \dots, n\}$ that do not contain consecutive integers. Let g_n denote the total number of subsets of S that do not contain consecutive integers. (I. Kaplansky)

*31. Define $f(n, k)$ recursively.

*32. Find an explicit formula for g_n.

Suppose we introduce a mixed pair of 1-month-old rabbits into a large enclosure on the first day of a certain month. By the end of each month, the rabbits become mature and each pair produces $k-1$ mixed pairs of offspring at the beginning of the following month. (*Note*: $k \geq 2$.) For instance, at the beginning of the second month, there is one pair of 2-month-old rabbits and $k-1$ pairs of 0-month-olds; at the beginning of the third month, there is one pair of 3-month-olds, $k-1$ pairs of 1-month-olds, and $k(k-1)$ pairs of 0-month-olds. Assume the rabbits are immortal. Let a_n denote the average age of the rabbit-pairs at the beginning of the nth month. (P. Filipponi, 1990)

**33. Define a_n recursively.

**34. Predict an explicit formula for a_n.

**35. Prove the formula in Exercise 34.

36. Find $\lim_{n\to\infty} a_n$.

*37. Find the sum of the numbers in the nth row of the following triangular array of Fibonacci numbers.

$$
\begin{array}{ccccccccc}
 & & & & 1 & & & & \\
 & & & 1 & & 2 & & & \\
 & & 3 & & 5 & & 8 & & \\
 & 13 & & 21 & & 34 & & 55 & \\
89 & & 144 & & 233 & & 377 & & 610 \\
 & & & & \vdots & & & &
\end{array}
$$

COMPUTER EXERCISES

Write a program to do each task.

1. Read in an integer $b \geq 2$ and select $b+1$ integers at random. Find two integers in the list such that their difference is divisible by b.
2. Read in an integer $n \geq 2$ and select n positive integers at random. Find a sequence of integers from the list whose sum is divisible by n.
3. Assign the numbers 0–51 in order to the 52 playing cards in a standard deck. Read in a number x, where $0 \leq x \leq 51$. Identify the card numbered x. Use the suit labels $0 =$ clubs, $1 =$ diamonds, $2 =$ hearts, and $3 =$ spades, and the card labels $0 =$ ace, $1 =$ deuce, $2 =$ three, ..., in each suit.

4. Assign the numbers 0–63, row by row, to the various squares on an 8×8 chessboard. Read in two numbers x and y, where $0 \le x, y \le 63$. Determine if the queen at square x can capture the queen at square y.
5. Read in a sequence of pairs of integers n and b. For each integer n, determine its base-b representation and use this representation to compute the corresponding decimal value. Print each integer n, base-b, base-b representation, and its decimal value in a tabular form.
6. Print the first eight rows of the number patterns in Examples 2.18, 2.19, and 2.21.
7. Read in a positive integer n and determine if it is a prime.
8. Construct a table of values of the function $E(n) = n^2 - n + 41$, where $0 \le n \le 41$, and identify each value as prime or composite.
9. Redo program 8 with $L(n) = n^2 + n + 41$, where $0 \le n \le 41$, and identify each value as prime or composite.
10. Redo program 8 with $H(n) = 9n^2 - 471n + 6203$, where $0 \le n \le 39$, and identify each value as prime or composite.
11. Redo program 8 with $G(n) = n^2 - 2999n + 2248541$, where $1460 \le n \le 1539$, and identify each value as prime or composite.
12. Read in a positive integer n, and list all primes $\le n$ and of the form $k^2 + 1$.
13. Read in a positive integer n and find a prime between:
    a) n and $2n$
    b) n^2 and $n^2 + 1$.
14. Verify Goldbach's conjecture for all even integers ≤ 100.
15. List all twin primes ≤ 100.
16. Find all palindromic primes < 100.
17. Find all cyclic primes < 100.
18. Find all reversible primes < 100.
19. There are nine positive integers ≤ 100 for which $n! + 1$ is a prime. Find them.
20. Make a list of 12 pairs of odd primes p and q such that $2^p + 1 = 3q$.
21. Find a counterexample to show that the statement $2^p + 1 = 3q$ is false, where p and q are odd primes.
22. (**Bocard's problem**) Only three positive integers n are known for which $n! + 1$ is a square and they are < 100. Find them.
23. There are exactly two primes p for which the Fermat quotient $(2^{p-1} - 1)/p$ is a square and they are < 100. Find them.
24. Verify that R_{19} and R_{23} are primes.
25. Find all three-digit cyclic primes, each with distinct digits.
26. List all five-digit cyclic primes that can be generated from cyclic primes 11,939 and 19,937.
27. Compute the value of $\prod_p (1 - 1/p^2)$, where p is a prime < 1000.

28. Compute the sum of the reciprocals of twin primes correct to four decimal places.
29. Read in a positive integer n, and compute the first n Fibonacci numbers using recursion and iteration.
30. Verify that Fermat numbers f_0 through f_4 are primes.
31. Compute f_5 and verify that $641|f_5$.
32. Verify that both 7 and 1913 are factors of f_6.
33. Read in a positive integer n, and compute the first n Lucas numbers L_n.
34. Compute the values of F_{n+1}/F_n and L_{n+1}/L_n correct to 10 decimal places for $1 \leq n \leq 100$.

ENRICHMENT READINGS

1. P. T. Bateman *et al.*, "A Hundred Years of Prime Numbers," *The American Mathematical Monthly*, 103 (Nov. 1996), 729–741.
2. A. H. Beiler, *Recreations in the Theory of Numbers*, Dover, New York, 1966, 39–66, 83–87.
3. L. E. Card, "Patterns in Primes," *J. Recreational Mathematics*, 1 (April 1968), 93–99.
4. L. E. Card, "More Patterns in Primes," *J. Recreational Mathematics*, 2 (April 1969), 112–116.
5. D. Deutsch and B. Goldman, "Kaprekar's Constant," *Mathematics Teacher*, 98 (Nov. 2004), 234–242.
6. P. Hoffman, *The Man Who Loved Only Numbers*, Hyperion, New York, 1998.
7. T. Koshy, *Fibonacci and Lucas Numbers with Applications*, John Wiley & Sons, New York, 2001.
8. M. Křížek *et al.*, "17 Lectures on Fermat Numbers," Springer-Verlag, New York, 2001.
9. C. Oliver, "The Twelve Days of Christmas," *Mathematics Teacher*, 70 (Dec. 1977), 752–754.
10. R. Ondrejka, "Ten Extraordinary Primes," *J. Recreational Mathematics*, 18 (1985–86), 87–92.
11. C. Pomerance, "The Search for Prime Numbers," *Scientific American*, 247 (Dec. 1982), 136–147.
12. J. Varnadore, "Pascal's Triangle and Fibonacci Numbers," *Mathematics Teacher*, 84 (April 1991), 314–316, 319.

Greatest Common Divisors

What science can there be more noble, more excellent, more useful for men, more admirably high and demonstrative than this of the Mathematics.

—BENJAMIN FRANKLIN

This chapter continues to deal with the divisibility theory. We begin by exploring the common factors of two or more positive integers. We establish the *fundamental theorem of arithmetic*, the cornerstone of number theory, and then turn to the common multiples of two or more positive integers. Finally, we investigate the important class of linear diophantine equations.

3.1 Greatest Common Divisor

A positive integer can be a factor of two positive integers, a and b. Such factors are **common divisors**, or **common factors**, of a and b.

For example, 12 and 18 have four common divisors, namely, 1, 2, 3, and 6; whereas 12 and 25 have exactly one common factor, namely, 1.

Often we are not interested in all common divisors of a and b, but in the largest common divisor, so we make the following definition.

Greatest Common Divisor

The **greatest common divisor** (gcd) of two integers a and b, not both zero, is the largest positive integer that divides both a and b; it is denoted by (a, b).

For example, $(12, 18) = 6$, $(12, 25) = 1$, $(11, 19) = 1$, $(-15, 25) = 5$, and $(3, 0) = 3$.

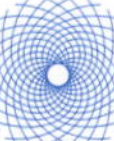

Because $(a, -b) = (-a, b) = (-a, -b) = (a, b)$, we confine our discussion of gcds to positive integers.

How do we know that the gcd of a and b always exists? Since $1|a$ and $1|b$, 1 is a common divisor of a and b, so they have a least common divisor, namely, 1. If d is a common divisor, then $d \le a$ and $d \le b$, so $d \le \min\{a, b\}$. Thus, the set of common factors is finite, so (a, b) exists.

A second important question is uniqueness: Is the gcd of a and b unique? It is, so we can talk about *the* gcd of a and b (see Exercise 46).

The preceding verbal definition of gcd, although simple and clear, is not a practical one, so we rewrite it symbolically.

A Symbolic Definition of gcd

A positive integer d is the gcd of two positive integers a and b if

- $d|a$ and $d|b$; and
- if $d'|a$ and $d'|b$, then $d' \le d$, where d' is also a positive integer.

Thus, $d = (a, b)$ if two conditions are satisfied:

- d must be a common factor of a and b.
- d must be the largest common factor of a and b; in other words, any other common factor d' must be $\le d$.

In the next section, we develop an efficient method for finding the gcd of two positive integers.

There are positive integers whose gcd is 1. For example, $(6, 35) = 1$. Accordingly, we make the following definition.

Relatively Prime Integers

Two positive integers a and b are **relatively prime** if their gcd is 1; that is, if $(a, b) = 1$.

Thus, 6 and 35 are relatively prime; so are 11 and 24.

This possible relationship between integers will be useful in our later discussions.

Cassini's formula now yields the following fascinating byproduct.

THEOREM 3.1 Any two consecutive Fibonacci numbers are relatively prime.

PROOF **(by contradiction)**

Let p be a prime factor of both F_n and F_{n+1}. Then, by Theorems 2.4 and 2.15, $p|\pm 1$, which is a contradiction. Thus, $(F_{n+1}, F_n) = 1$. ■

Interestingly, we can use Fermat numbers to reconfirm the infinitude of primes. To this end, we need the following two results.

LEMMA 3.1 Let f_i denote the ith Fermat number. Then $f_0 f_1 \cdots f_{n-1} = f_n - 2$, where $n \geq 1$.

PROOF (by weak induction)
When $n = 1$, LHS $= f_0 = 3 = 5 - 2 = f_1 - 2 =$ RHS. Thus, the result holds when $n = 1$.

Now assume the given result is true when $n = k$:

$$f_0 f_1 \cdots f_{k-1} = f_k - 2$$

Then

$$\begin{aligned} f_0 f_1 \cdots f_{k-1} f_k &= (f_0 f_1 \cdots f_{k-1}) f_k \\ &= (f_k - 2) f_k, \quad \text{by the inductive hypothesis} \\ &= \left(2^{2^k} - 1\right)\left(2^{2^k} + 1\right) \\ &= 2^{2^{k+1}} - 1 = \left(2^{2^{k+1}} + 1\right) - 2 \\ &= f_{k+1} - 2 \end{aligned}$$

So, if the result is true when $n = k$, it is also true when $n = k + 1$. Thus, by induction, the result holds for every integer $n \geq 1$. ■

The formula in this lemma, known as **Duncan's identity**, was discovered in 1964 by D. C. Duncan.

Using this result, we now show that any two distinct Fermat numbers are relatively prime; it was established in 1925 by G. Polya of Stanford University. (See Exercises 69 and 70 for an alternate proof of the lemma.)

THEOREM 3.2 **(Polya, 1925)** Let m and n be distinct nonnegative integers. Then f_m and f_n are relatively prime.

PROOF
Assume, for convenience, that $m < n$. Let $d = (f_m, f_n)$. Then $d | f_m$ and $d | f_n$. But $f_n - 2 = f_0 f_1 \cdots f_m \cdots f_{n-1}$, by Lemma 3.1. Since $d | f_m$, $d | f_0 f_1 \cdots f_m \cdots f_n$. So $d | (f_n - 2)$, but $d | f_n$; therefore, $d | 2$, by Theorem 2.4. Consequently, d must be 1 or 2. But Fermat numbers are all odd, so $d \neq 2$. Therefore, $d = 1$; that is, $(f_m, f_n) = 1$. ■

Polya's result can be generalized: Let $g_n = (2k)^{2^n} + 1$, where $k > 0$. Then $(g_m, g_n) = 1$, where $m \neq n$; see Exercise 66.

Using these two results, we can now prove again that there are infinitely many primes.

THEOREM 3.3 There is an infinitude of primes.

PROOF

By Lemma 2.1, every Fermat number has a prime factor. Therefore, by Polya's theorem, no two distinct Fermat numbers have common prime factors, meaning each has a distinct prime factor. So, since there are infinitely many Fermat numbers, there are also infinitely many primes. ■

This result can be established more formally using induction. See Exercise 71.

Next, we present an amazing confluence of number theory, probability, and analysis.

Relatively Prime Numbers and Pi (optional)

In Section 2.5, we found a close link between prime numbers and π, given by the formula $\prod_{p \in P}(1 - 1/p^2) = \pi^2/6$. Using advanced techniques, it can be shown that the infinite product represents the reciprocal of the probability that two positive integers selected at random are relatively prime.[†] Thus, the probability that two positive integers selected at random are relatively prime is given by $\prod_{p \in P} 1/(1 - 1/p^2) = 6/\pi^2$.

We now turn our attention to some interesting and useful properties of gcds.

THEOREM 3.4 Let $(a, b) = d$. Then

1. $(a/d, b/d) = 1$
2. $(a, a - b) = d$.

PROOF

1. Let $d' = (a/d, b/d)$. *To show that* $d' = 1$:
   Since d' is a common factor of a/d and b/d, $a/d = \ell d'$ and $b/d = md'$ for some integers ℓ and m. Then $a = \ell dd'$ and $b = mdd'$, so dd' is a common factor of both a and b. Then, by definition, $dd' \leq d$, so $d' \leq 1$. Thus, d' is a positive integer such that $d' \leq 1$, so $d' = 1$. Thus, if $(a, b) = d$, then a/d and b/d are relatively prime.
2. Let $d' = (a, a-b)$. To show that $d = d'$, we shall show that $d \leq d'$ and $d' \leq d$.
   To show that $d \leq d'$:
   Since d is a common divisor of a and b, $a = md$ and $b = nd$ for some integers m and n. Then $a - b = (m - n)d$. Thus $d|a$ and $d|(a - b)$; so d is a common

---

[†] See Ogilvy and Anderson.

divisor of a and $a-b$. Then, by definition, d must be less than or equal to $(a, a-b)$; that is, $d \leq d'$.

To show that $d' \leq d$:

Since d' is a common factor of a and $a-b$, $a = \alpha d'$ and $a - b = \beta d'$ for some integers α and β. Then $a-(a-b) = \alpha d' - \beta d'$; that is, $b = (\alpha - \beta)d'$. Thus, d' is a common divisor of a and b, so $d' \leq d$.

Thus, $d \leq d'$ and $d' \leq d$, so $d = d'$. ■

It follows by part (2) of this theorem that $(a, a+b) = (a, b)$. (See Exercise 50.)

Next, we prove that the gcd(a, b) can be expressed as a sum of multiples of a and b, but first we must make a definition.

Linear Combination

A **linear combination** of the integers a and b is a sum of multiples of a and b, that is, a sum of the form $\alpha a + \beta b$, where α and β are integers.

For example, $2 \cdot 3 + 5 \cdot 7$ is a linear combination of 3 and 7; so is $(-4) \cdot 3 + 0 \cdot 7$.

We now state and prove the result mentioned in the preceding paragraph. Its proof is an elegant application of the well-ordering principle.

THEOREM 3.5 **(Euler)** The gcd of the positive integers a and b is a linear combination of a and b.

PROOF

Let S be the set of positive linear combinations of a and b; that is, $S = \{ma + nb \mid ma + nb > 0, m, n \in \mathbf{Z}\}$.

To show that S has a least element:

Since $a > 0$, $a = 1 \cdot a + 0 \cdot b \in S$, so S is nonempty. So, by the well-ordering principle, S has a least positive element d.

To show that $d = (a, b)$:

Since d belongs to S, $d = \alpha a + \beta b$ for some integers α and β.

1. First we will show that $d \mid a$ and $d \mid b$:

   By the division algorithm, there exist integers q and r such that $a = dq + r$, where $0 \leq r < d$. Substituting for d,

$$\begin{aligned} r &= a - dq \\ &= a - (\alpha a + \beta b)q \\ &= (1 - \alpha q)a + (-\beta q)b \end{aligned}$$

   This shows r is a linear combination of a and b.

   If $r > 0$, then $r \in S$. Since $r < d$, r is less than the smallest element in S, which is a contradiction. So $r = 0$; thus, $a = dq$, so $d \mid a$.

   Similarly, $d \mid b$. Thus d is a common divisor of a and b.

2. *To show that any positive common divisor d' of a and b is $\leq d$:*
Since $d'|a$ and $d'|b$, $d'|(\alpha a + \beta b)$, by Theorem 2.4; that is, $d'|d$. So $d' \leq d$.
Thus, by parts (1) and (2), $d = (a, b)$. ■

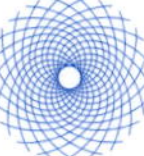

It follows by this theorem that the gcd (a, b) can always be expressed as a linear combination $\alpha a + \beta b$. In fact, it is the smallest positive such linear combination.

One way to find such a linear combination is by trial and error, especially when a and b are small, as the following example shows.

EXAMPLE 3.1 Express (28, 12) as a linear combination of 28 and 12.

SOLUTION
First, notice that $(28, 12) = 4$. Next, we need to find integers α and β such that $\alpha \cdot 28 + \beta \cdot 12 = 4$. By trial and error, $\alpha = 1$ and $\beta = -2$ works: $1 \cdot 28 + (-2) \cdot 12 = 4$. ■

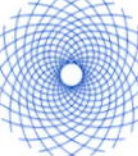

Note that the values of α and β in the linear combination need *not* be unique. For instance, in this example, you may notice that $(-5) \cdot 28 + 12 \cdot 12 = 4$.

A second way to find α and β is by using a table of multiples of a and b and then picking a right combination, as Table 3.1 shows.

28	28	56	84	112	140	...		
12	12	24	36	48	60	...	144	...

Table 3.1

The next section presents a systematic method for finding α and β.

Theorem 3.5 can be used to refine the definition of gcd and to derive several useful results about gcds.

THEOREM 3.6 If $d = (a, b)$ and d' is any common divisor of a and b, then $d'|d$.

PROOF

Since $d = (a, b)$, by Theorem 3.5, there exist α and β such that $d = \alpha a + \beta b$. Since $d'|a$ and $d'|b$, by Theorem 2.4, $d'|(\alpha a + \beta b)$; so $d'|d$. ■

Thus, every common divisor d' of a and b is a factor of their gcd d, and $d' \leq d$. Conversely, suppose that

- $d|a$ and $d|b$; and
- if $d'|a$ and $d'|b$, then $d'|d$. Then $d' \leq d$, so $d = (a, b)$.

Thus, the symbolic definition of gcd can be modified as follows.

An Alternate Definition of gcd

A positive integer d is the gcd of a and b if

- $d|a$ and $d|b$; and
- if $d'|a$ and $d'|b$, then $d'|d$, where d' is a positive integer.

THEOREM 3.7 Let a, b, and c be any positive integers. Then $(ac, bc) = c(a, b)$.

The proof of this is fairly straightforward, so we leave it as an exercise (see Exercise 51).

THEOREM 3.8 Two positive integers, a and b, are relatively prime if and only if there are integers α and β such that $\alpha a + \beta b = 1$.

PROOF

If a and b are relatively prime, then $(a, b) = 1$. Therefore, by Theorem 3.5, there are integers α and β such that $\alpha a + \beta b = 1$.

Conversely, suppose $\alpha a + \beta b = 1$. To demonstrate that $(a, b) = 1$, let $d = (a, b)$. Then, by Theorem 2.4, $d|(\alpha a + \beta b)$; that is, $d|1$, so $d = 1$. Thus, a and b are relatively prime. ■

We can deduce part (1) of Theorem 3.4 from this theorem, and it is useful to do so as an exercise (see Exercise 54).

COROLLARY 3.1 If $d = (a, b)$, then $(a/d, b/d) = 1$.

The next corollary follows by Theorem 3.5 (see Exercise 59).

COROLLARY 3.2 If $(a, b) = 1 = (a, c)$, then $(a, bc) = 1$.

Suppose $a|c$ and $b|c$. Does this mean $ab|c$? No. For example, $3|12$ and $6|12$, but $3 \cdot 6 \nmid 12$. The next corollary provides a criterion under which $ab|c$.

COROLLARY 3.3 If $a|c$ and $b|c$, and $(a, b) = 1$, then $ab|c$.

PROOF

Because $a|c$, $c = ma$ for some integer m. Similarly, $c = nb$ for some integer n. Because $(a, b) = 1$, by Theorem 3.8, $\alpha a + \beta b = 1$ for some integers α and β. Then $\alpha ac + \beta bc = c$. Now substitute nb for the first c and ma for the second:

$$\alpha a(nb) + \beta b(ma) = c$$

That is, $ab(n\alpha + m\beta) = c$, so $ab|c$. ■

Remember that $a|bc$ does not mean $a|b$ or $a|c$, although under some conditions it does. The following corollary explains when it is true.

COROLLARY 3.4 **(Euclid)** If a and b are relatively prime, and if $a|bc$, then $a|c$.

PROOF

Since a and b are relatively prime, by Theorem 3.8, there exist integers α and β such that $\alpha a + \beta b = 1$. Then $\alpha ac + \beta bc = c$. Since $a|\alpha ac$ and $a|\beta bc$, $a|\alpha ac + \beta bc$ by Theorem 2.4; that is, $a|c$. ■

The definition of gcd can be extended to three or more positive integers, as the following definition shows.

The gcd of n Positive Integers

The gcd of n (≥ 2) positive integers $a_1, a_2, \ldots, a_n$ is the largest positive integer that divides each a_i. It is denoted by $(a_1, a_2, \ldots a_n)$.

The following example illustrates this definition.

EXAMPLE 3.2 Find $(12, 18, 28)$, $(12, 36, 60, 108)$, and $(15, 28, 50)$.

SOLUTION

a) The largest positive integer that divides 12, 18, and 28 is 2, so $(12, 18, 28) = 2$.
b) 12 is the largest factor of 12, and 12 is a factor of 12, 36, 60, and 108; so $(12, 36, 60, 108) = 12$.
c) Since $(15, 28) = 1$, the largest common factor of 15, 28, and 50 is 1; that is, $(15, 28, 50) = 1$. ■

Theorem 3.5 can be extended to n integers. But first, we will extend the definition of a linear combination to n positive integers.

A Linear Combination of n Positive Integers

A **linear combination** of n positive integers $a_1, a_2, \ldots, a_n$ is a sum of the form $\alpha_1 a_1 + \alpha_2 a_2 + \cdots + \alpha_n a_n$, where $\alpha_1, \alpha_2, \ldots, \alpha_n$ are integers.

For instance, $(-1) \cdot 12 + 1 \cdot 15 + 0 \cdot 21$ is a linear combination of 12, 15, and 21; so is $3 \cdot 12 + (-2) \cdot 15 + (-5) \cdot 21$.

We now state the extension of Theorem 3.5 and leave its proof as an exercise.

THEOREM 3.9 The gcd of the positive integers $a_1, a_2, \ldots, a_n$ is the least positive integer that is a linear combination of $a_1, a_2, \ldots, a_n$. ■

The following example illustrates this theorem.

EXAMPLE 3.3 Express $(12, 15, 21)$ as a linear combination of 12, 15, and 21.

SOLUTION

First, you may notice that $(12, 15, 21) = 3$. Next, find integers α, β, and γ, by trial and error, such that $\alpha \cdot 12 + \beta \cdot 15 + \gamma \cdot 21 = 3$; $\alpha = -1$, $\beta = 1$, and $\gamma = 0$ is such a combination: $(-1) \cdot 12 + 1 \cdot 15 + 0 \cdot 21 = 3$. ■

The following theorem shows how nicely recursion can be used to find the gcd of three or more integers.

THEOREM 3.10 Let $a_1, a_2, \ldots, a_n$ be n (≥ 3) positive integers. Then $(a_1, a_2, \ldots, a_n) = ((a_1, a_2, \ldots, a_{n-1}), a_n)$.

PROOF

Let $d = (a_1, a_2, \ldots, a_n)$, $d' = (a_1, a_2, \ldots, a_{n-1})$, and $d'' = (d', a_n)$. We will show that $d = d''$:

- *To show that $d|d''$.*
  Since $d = (a_1, a_2, \ldots, a_n)$, $d|a_i$ for every i. So $d|d'$ and $d|a_n$. Then $d|(d', a_n)$; that is, $d|d''$.
- *To show that $d''|d$:*
  Since $d'' = (d', a_n)$, $d''|d'$ and $d''|a_n$. But $d''|d'$ implies $d''|a_i$ for $1 \leq i \leq n-1$. Thus, $d''|a_i$ for $1 \leq i \leq n$, so $d''|d$.

Thus, $d|d''$ and $d''|d$, so $d = d''$, by Theorem 2.3. ■

The following example illustrates this theorem.

EXAMPLE 3.4 Using recursion, evaluate $(18, 30, 60, 75, 132)$.

SOLUTION

$$\begin{aligned}(18, 30, 60, 75, 132) &= ((18, 30, 60, 75), 132) = (((18, 30, 60), 75), 132)\\ &= ((((18, 30), 60), 75), 132) = (((6, 60), 75), 132)\\ &= ((6, 75), 132) = (3, 132)\\ &= 3\end{aligned}$$

■

The following corollary follows by induction and Theorem 3.10. You can provide a proof (see Exercise 55).

COROLLARY 3.5 If $d = (a_1, a_2, \ldots, a_n)$, then $d|a_i$ for every integer i, where $1 \le i \le n$. ■

The following corollary is an extension of Corollary 3.4.

COROLLARY 3.6 If $d|a_1 a_2 \cdots a_n$ and $(d, a_i) = 1$ for $1 \le i \le n - 1$, then $d|a_n$. ■

Before we move on to another corollary, we make the following definition.

Pairwise Relatively Prime Integers

The positive integers $a_1, a_2, \ldots, a_n$ are **pairwise relatively prime** if every pair of integers is relatively prime; that is, $(a_i, a_j) = 1$, whenever $i \ne j$.

For example, the integers 8, 15, and 49 are pairwise relatively prime, whereas the integers 6, 25, 77, and 91 are *not* pairwise relatively prime.

The following result follows from Theorem 3.8.

COROLLARY 3.7 If the positive integers $a_1, a_2, \ldots, a_n$ are pairwise relatively prime, then $(a_1, a_2, \ldots, a_n) = 1$. ■

For instance, since the integers 8, 15, and 49 are pairwise relatively prime, $(8, 15, 49) = 1$.

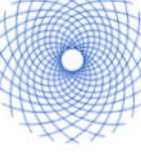

Be aware that the converse of this corollary is *not* true; that is, if $(a_1, a_2, \ldots, a_n) = 1$, then the integers $a_1, a_2, \ldots, a_n$ need not be pairwise relatively prime. For example, $(6, 15, 49) = 1$, but 6, 15, and 49 are *not* pairwise relatively prime. (Why?)

Theorem 3.1 now yields an intriguing byproduct. In 1965, M. Wunderlich of the University of Colorado employed the theorem to provide a beautiful proof that there are infinitely many primes, as the next corollary shows; it is based on the fact that $(F_m, F_n) = F_{(m,n)}$.[†]

COROLLARY 3.8 There are infinitely many primes.

PROOF

Suppose there is only a finite number of primes, $p_1, p_2, \ldots, p_k$. Consider the Fibonacci numbers $F_{p_1}, F_{p_2}, \ldots, F_{p_k}$. Clearly, they are pairwise relatively prime. Since there are only k primes, each of these Fibonacci numbers has exactly one prime factor; that is, each is a prime. This is a contradiction, since $F_{19} = 4181 = 37 \cdot 113$. Thus, our assumption that there are only finitely many primes is false. In other words, there are infinitely many prime numbers. ■

EXERCISES 3.1

Mark *true* or *false*, where a, b, and c are any positive integers, and p is an arbitrary prime.

1. $(a, b) = (b, a)$
2. $(a, b) = (a, a - b)$
3. $(a, b) = (a, a - 2b)$
4. $(a, a + 2) = 1$
5. $(p, p + 2) = 1$
6. $(ac, bc) = c(a, b)$
7. If $(a, b) = 1$, then a and b are relatively prime.
8. If a and b are relatively prime, then $(a, b) = 1$.
9. If $(a, b) = 1 = (b, c)$, then $(a, c) = 1$.
10. If $(a, b) = 2 = (b, c)$, then $(a, c) = 2$.
11. If $(a, b) = d$, then $(a + b, a - b) = d$.

Express the gcd of each pair as a linear combination of the numbers.

12. 18, 28
13. 24, 28
14. 15, 28
15. 21, 26

Let $f(n)$ denote the number of positive integers $\le n$ and relatively prime to it. For example, $f(1) = 1$, $f(2) = 1$, $f(3) = 2$, and $f(4) = 2$. Find each.

16. $f(10)$
17. $f(13)$
18. $f(18)$
19. $f(24)$
20. Evaluate $\sum_{d|n} f(d)$ for $n = 12$, 18, 19, and 25.
21. Using Exercise 20, predict a formula for $\sum_{d|n} f(d)$.
22. Find the least possible value of (a, b).

Find (a, b) if

23. $b = 1$
24. $b = a$
25. $b = a + 1$
26. $b|a$
27. $b = a^2$
28. $b = a^n$
29. $b = na$
30. $b = (b, a)$

Find the gcd of each pair, where $a > b$.

31. $a + b$, $a^2 - b^2$
32. $a^2 - b^2$, $a^3 - b^3$
33. $a^2 - b^2$, $a^4 - b^4$

Express the gcd of the given numbers as a linear combination of the numbers.

34. 12, 15, 18
35. 15, 18, 24
36. 12, 18, 20, 24
37. 15, 18, 20, 28

[†] See author's *Fibonacci and Lucas Numbers with Applications*.

Using recursion, evaluate each.

38. (12, 18, 28, 38, 44)
39. (15, 24, 28, 45)
40. (14, 18, 21, 36, 48)
41. (18, 24, 36, 63)
42. $(a^2b, ab^3, a^2b^2, a^3b^4, ab^4)$
43. $(a^2b^2, ab^3, a^2b^3, a^3b^4, a^4b^4)$

Disprove each statement.

44. If $(a, b) = 1 = (b, c)$, then $(a, c) = 1$.
45. If $(a, b) = 2 = (b, c)$, then $(a, c) = 2$.

Prove each, where a, b, c, d, k, m, and n are arbitrary positive integers, p any prime, F_n the nth Fibonacci number, f_n the nth Fermat number, and t_n the nth triangular number.

46. The gcd of any two positive integers is unique.
47. $(a, -b) = (a, b)$
48. $(-a, b) = (a, b)$
49. $(-a, -b) = (a, b)$
50. $(a, a+b) = (a, b)$
51. $(ac, bc) = c(a, b)$
52. Any two consecutive integers are relatively prime.
53. If $p \nmid a$, then p and a are relatively prime.
54. Using Theorem 3.8, prove that if $d = (a, b)$, then $(a/d, b/d) = 1$.
55. If $d = (a_1, a_2, \ldots, a_n)$, then $d|a_i$ for every integer i, where $1 \le i \le n$.
56. $(a, (a, b)) = (a, b)$
57. $(a, a-b) = 1$ if and only if $(a, b) = 1$.
58. If $(a, b) = 1$, then $(a+b, a-b) = 1$ or 2.
59. If $(a, b) = 1 = (a, c)$, then $(a, bc) = 1$.
60. Let $(a, b) = 1$. Then $(a^2 + b^2, a + 2ab) = 1$ or 5. (V. E. Hoggatt, Jr., 1972)
61. Let $(a^2 + b^2, a + 2ab) = 1$ or 5. Then $(a, b) = 1$ (V. E. Hoggatt, Jr., 1972)
62. $(n+1, n^2+1) = 1$, where n is even. (N. Schaumberger and J. Soriano, 1967)
63. $(a^n - 1, a^m + 1) = 1$ or 2. (E. Just, 1972)
64. $(t_{n-1}, t_n) \cdot (t_n, t_{n+1}) = t_n$ (T. E. Moore, 2004)
65. $a!b!|(a, b)(a+b-1)!$ (J. H. Conway, 1988)
66. Let $g_n = (2k)^{2^n} + 1$, where $n \ge 0$. Then $(g_m, g_n) = 1$, where $m \ne n$.
67. $F_m|F_n$ if and only if $m|n$. [*Hint*: $(F_m, F_n) = F_{(m,n)}$.]
68. $3|F_n$ if and only if $4|n$.
69. Let $n > m \ge 0$. Show that $f_m|(f_n - 2)$.
70. Using Exercise 69, show that $(f_m, f_n) = 1$, where $m \ne n$.
71. Using Theorem 3.2 and induction, prove that there are infinitely many primes.
*72. If $(a, b) = 1$, then $(a^2, b^2) = 1$.
*73. Let $m, n \ge 1$. Prove that $\frac{m}{(m, n)} \Big| \binom{m}{n}$ (C. Hermite)
*74. Let $m, n \ge 1$. Prove that $\frac{m-n+1}{(m, n)} \Big| \binom{m}{n}$ (C. Hermite)

3.2 The Euclidean Algorithm

Several procedures exist for finding the gcd of two positive integers. One efficient algorithm is the **euclidean algorithm**, named after Euclid, who included it in Book VII of his extraordinary work, *The Elements*. The algorithm, however, was most likely known before him. It is a fundamental tool in algorithmic number theory.

The following theorem lays the groundwork for the euclidean algorithm.

THEOREM 3.11 Let a and b be any positive integers, and r the remainder, when a is divided by b. Then $(a, b) = (b, r)$.

PROOF

Let $d = (a, b)$ and $d' = (b, r)$. To prove that $d = d'$, it suffices to show that $d|d'$ and $d'|d$. By the division algorithm, a unique quotient q exists such that

$$a = bq + r \tag{3.1}$$

To show that $d|d'$:

Since $d = (a, b)$, $d|a$ and $d|b$, so $d|bq$, by Theorem 2.4. Then $d|(a - bq)$, again by Theorem 2.4. In other words, $d|r$, by equation (3.1). Thus, $d|b$ and $d|r$, so $d|(b, r)$; that is, $d|d'$.

Similarly, it can be shown that $d'|d$ (see Exercise 17). Thus, by Theorem 2.3, $d = d'$, that is, $(a, b) = (b, r)$. ■

The following example illustrates this theorem.

EXAMPLE 3.5 Illustrate Theorem 3.11 with $a = 120$ and $b = 28$.

SOLUTION

First, you may verify that $(120, 28) = 4$.

Now, by the division algorithm, $120 = 4 \cdot 28 + 8$, so, by Theorem 3.11, $(120, 28) = (28, 8)$. But $(28, 8) = 4$. Therefore, $(120, 28) = 4$. ■

The following example illustrates how Theorem 3.11 can be used to find (a, b).

EXAMPLE 3.6 Using Theorem 3.11, evaluate (2076,1776).

SOLUTION

Apply the division algorithm with 2076 (the larger of the two numbers) as the dividend and 1776 as the divisor:

$$2076 = 1 \cdot 1776 + 300$$

Apply the division algorithm with 1776 as the dividend and 300 as the divisor:

$$1776 = 5 \cdot 300 + 276$$

Continue this procedure until a zero remainder is reached:

$$
\begin{aligned}
2076 &= 1 \cdot 1776 + 300 \\
1776 &= 5 \cdot 300 + 276 \\
300 &= 1 \cdot 276 + 24 \\
276 &= 11 \cdot 24 + \boxed{12} \quad \leftarrow \text{last nonzero remainder} \\
24 &= 2 \cdot 12 + 1
\end{aligned}
$$

By the repeated application of Theorem 3.11, we have:

$$
\begin{aligned}
(2076, 1776) &= (1776, 300) = (300, 276) \\
&= (276, 24) = (24, 12) \\
&= 12
\end{aligned}
$$

Thus, the last nonzero remainder in this procedure is the gcd. ■

We now justify this algorithm, although it is somewhat obvious.

The Euclidean Algorithm

Let a and b be any two positive integers with $a \geq b$. If $a = b$, then $(a, b) = a$, so assume $a > b$. (If this is not true, simply switch them.) Let $r_0 = b$. Then by successive application of the division algorithm, we get a sequence of equations:

$$
\begin{aligned}
a &= q_0 r_0 + r_1, \quad 0 \leq r_1 < r_0 \\
r_0 &= q_1 r_1 + r_2, \quad 0 \leq r_2 < r_1 \\
r_1 &= q_2 r_2 + r_3, \quad 0 \leq r_3 < r_2 \\
&\ \vdots
\end{aligned}
$$

Continuing like this, we get the following sequence of remainders:

$$b = r_0 > r_1 > r_2 > r_3 > \cdots \geq 0$$

Since the remainders are nonnegative and are getting smaller and smaller, this sequence should eventually terminate with remainder $r_{n+1} = 0$. Thus, the last two equations in the above procedure are

$$r_{n-2} = q_{n-1} r_{n-1} + r_n, \quad 0 \leq r_n < r_{n-1}$$

and

$$r_{n-1} = q_n r_n$$

It follows by induction that $(a, b) = (a, r_0) = (r_0, r_1) = (r_1, r_2) = \cdots = (r_{n-1}, r_n) = r_n$, the last nonzero remainder (see Exercise 18).

The following example also demonstrates the euclidean algorithm.

EXAMPLE 3.7 Apply the euclidean algorithm to find (4076, 1024).

SOLUTION

By the successive application of the division algorithm, we get:

$$\begin{aligned} 4076 &= 3 \cdot 1024 + 1004 \\ 1024 &= 1 \cdot 1004 + 20 \\ 1004 &= 50 \cdot 20 + \boxed{4} \quad \leftarrow \text{last nonzero remainder} \\ 20 &= 5 \cdot 4 + 0 \end{aligned}$$

Since the last nonzero remainder is 4, $(4076, 1024) = 4$. ■

The euclidean algorithm is purely mechanical. All we need to do is make our divisor the new dividend, and the remainder the new divisor. That is, just follow the southwest arrows in the example.

The euclidean algorithm is formally presented in Algorithm 3.1.

```
Algorithm Euclid (x, y, divisor)
(* This algorithm returns the gcd (x, y) in divisor, where x ≥ y > 0 *)
  Begin (* algorithm *)
    dividend ← x
    divisor ← y
    remainder ← dividend mod divisor
    while reminder > 0 do
      (* update dividend, divisor, and remainder *)
    begin (* while *)
      dividend ← divisor
      divisor ← dividend mod divisor
    endwhile
  End (* algorithm *)
```

Algorithm 3.1

A Jigsaw Puzzle (optional)

The euclidean algorithm has a delightful application to geometry. To this end, suppose we would like to find (23, 13). By the euclidean algorithm, we have

$$23 = 1 \cdot 13 + 10$$
$$13 = 1 \cdot 10 + 3$$
$$10 = 3 \cdot 3 + \boxed{1}$$
$$3 = 3 \cdot 1$$

So $(23, 13) = 1$.

Now consider a 23×13 rectangle; see Figure 3.1. The largest square we can place inside it is a 13×13 square, and only one such square will fit it. Now we can use one 10×10 square, three 3×3 squares, and three 1×1 squares to fit the rest of the rectangle; see Figure 3.2.

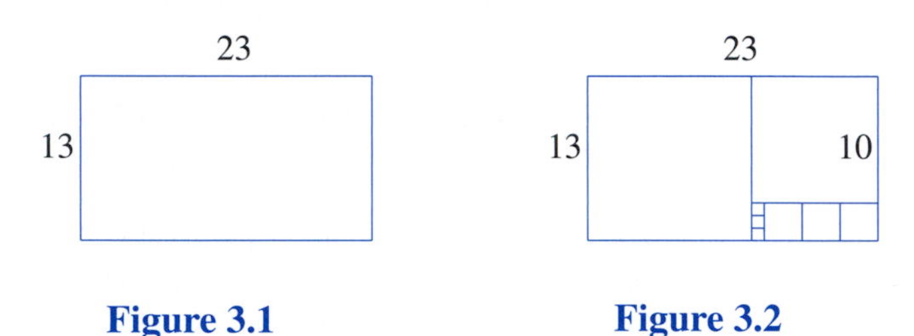

Figure 3.1 **Figure 3.2**

Each divisor d in the algorithm represents the length of the side of a $d \times d$ square, and the length of a side of the smallest square gives the gcd.

We shall revisit this jigsaw-puzzle application in Section 12.1.

As a byproduct, the euclidean algorithm provides a procedure for expressing the gcd (a, b) as a linear combination of a and b, as the following example shows.

EXAMPLE 3.8 Using the euclidean algorithm, express (4076, 1024) as a linear combination of 4076 and 1024.

SOLUTION

All we need to do is use the equations in Example 3.7 in the reverse order, each time substituting for the remainder from the previous equation:

$$
\begin{aligned}
(4076, 1024) &= 4 = \text{ last nonzero remainder} \\
&= 1004 - 50 \cdot 20 \\
&= 1004 - 50(1024 - 1 \cdot 1004) \quad \text{(substitute for 20)} \\
&= 51 \cdot 1004 - 50 \cdot 1024 \\
&= 51(4076 - 3 \cdot 1024) - 50 \cdot 1024 \quad \text{(substitute for 1004)} \\
&= 51 \cdot 4076 + (-203) \cdot 1024
\end{aligned}
$$

(We can confirm this by direct computation.) ■

Next, we shall derive an upper bound for the number of divisions needed to evaluate (a, b) by the euclidean algorithm. However, first we need to lay some groundwork in the form of a lemma that investigates yet another property of Fibonacci numbers.

LEMMA 3.2 Let $\alpha = (1 + \sqrt{5})/2$. Then $\alpha^{n-2} < F_n < \alpha^{n-1}$, where $n \geq 3$.

PROOF **(by strong induction)**
(We shall prove that $\alpha^{n-2} < F_n$ and leave the other half as an exercise.) You can verify that α is a solution of the equation $x^2 = x + 1$, so $\alpha^2 = \alpha + 1$. Let

$$P(n):\quad \alpha^{n-2} < F_n, \quad \text{where } n \geq 3$$

basis step Since the induction step below uses the recurrence relation $F_{k+1} = F_k + F_{k-1}$, the basis step involves verifying that both $P(3)$ and $P(4)$ are true.

1. *To show that $P(3)$ is true*: When $n = 3$,

$$\alpha^{n-2} = \alpha = \frac{1+\sqrt{5}}{2} < \frac{1+3}{2} = 2 = F_3,$$

   so $P(3)$ is true.

2. *To show that $P(4)$ is true*:

$$\begin{aligned}\alpha^2 &= \left(\frac{1+\sqrt{5}}{2}\right)^2 = \frac{3+\sqrt{5}}{2} \\ &< \frac{3+3}{2} = 3 = F_4\end{aligned}$$

   Therefore, $P(4)$ is also true.

induction step Assume $P(3), P(4), \ldots, P(k)$ are true; that is, assume $\alpha^{i-2} < F_i$ for $5 \leq i \leq K$. We must show that $P(k+1)$ is true; that is, $\alpha^{k-1} < F_{k+1}$. We have

$$\alpha^2 = \alpha + 1$$

Multiplying both sides by α^{k-3}, we get

$$\begin{aligned}\alpha^{k-1} &= \alpha^{k-2} + \alpha^{k-3} && (\textit{Note}: k - 3 \geq 2) \\ &< F_k + F_{k-1}, && \text{by the inductive hypothesis} \\ &= F_{k+1}, && \text{by the recurrence relation}\end{aligned}$$

So $P(k+1)$ is true.

Thus, by induction, $\alpha^{n-2} < F_n$ for every integer $n \geq 3$. ■

The irrational number α is called the **golden ratio**. It has many intriguing applications.

The following theorem, established in 1844 by Lamé, employs this result to estimate the number of divisions required by the euclidean algorithm for computing the gcd.

THEOREM 3.12 **(Lamé's Theorem)** The number of divisions needed to compute (a, b) by the euclidean algorithm is no more than five times the number of decimal digits in b, where $a \geq b \geq 2$.

PROOF

Let F_n denote the nth Fibonacci number, $a = r_0$ and $b = r_1$. By the repeated application of the division algorithm we have

$$\begin{aligned} r_0 &= r_1 q_1 + r_2, & 0 \leq r_2 < r_1 \\ r_1 &= r_2 q_2 + r_3, & 0 \leq r_3 < r_2 \\ &\vdots \\ r_{n-2} &= r_{n-1} q_{n-1} + r_n, & 0 \leq r_n < r_{n-1} \\ r_{n-1} &= r_n q_n \end{aligned}$$

Clearly, it takes n divisions to evaluate gcd $(a, b) = r_n$. Since $r_i < r_{i-1}$, $q_i \geq 1$ for $1 \leq i \leq n$. In particular, since $r_n < r_{n-1}$, $q_n \geq 2$, so $r_n \geq 1$ and $r_{n-1} \geq 2 = F_3$. Consequently, we have

$$\begin{aligned} r_{n-2} &= r_{n-1} q_{n-1} + r_n \\ &\geq r_{n-1} + r_n \\ &\geq F_3 + 1 \\ &= F_3 + F_2 = F_4 \\ r_{n-3} &= r_{n-2} q_{n-2} + r_{n-1} \\ &\geq r_{n-2} + r_{n-1} \\ &\geq F_4 + F_3 = F_5 \\ &\vdots \end{aligned}$$

Continuing like this,

$$\begin{aligned} r_1 &= r_2 q_2 + r_3 \\ &\geq r_2 + r_3 \\ &\geq F_n + F_{n-1} = F_{n+1} \end{aligned}$$

That is, $b \geq F_{n+1}$.

By Lemma 3.2, $F_{n+1} > \alpha^{n-1}$, where $\alpha = (1+\sqrt{5})/2$ and $n \geq 3$. Therefore,

$$b > \alpha^{n-1}$$

$$\log b > (n-1)\log\alpha$$

Since $\alpha = (1+\sqrt{5})/2 \approx 1.618033989$, $\log\alpha \approx 0.2089876403 > 1/5$. Therefore,

$$\log b > \frac{n-1}{5}$$

Suppose b contains k decimal digits. Then $b < 10^k$. Therefore, $\log b < k$ and hence $k > (n-1)/5$. Thus, $n < 5k+1$ or $n \leq 5k$. Thus, the number of divisions needed by the algorithm is no more than five times the number of decimal digits in n. ■

EXERCISES 3.2

Using the euclidean algorithm, find the gcd of the given integers.

1. 1024, 1000
2. 2024, 1024
3. 2076, 1076
4. 2076, 1776
5. 1976, 1776
6. 3076, 1776
7. 3076, 1976
8. 4076, 2076

9–16. Using the euclidean algorithm, express the gcd of each pair in Exercises 1–8 as a linear combination of the given numbers.

17. Let a and b be any two positive integers, and let r be the remainder when a is divided by b. Let $d = (a, b)$ and $d' = (b, r)$. Prove that $d'|d$.
18. Let a and b be any two positive integers with $a \geq b$. Using the sequence of equations in the euclidean algorithm, prove that $(a, b) = (r_{n-1}, r_n)$, where $n \geq 1$.

Prove each, where $\alpha = (1+\sqrt{5})/2$.

19. $F_n < \alpha^{n-1}$, $n \geq 2$
20. $F_n \leq 2^n$, $n \geq 1$

3.3 The Fundamental Theorem of Arithmetic

We now continue our study of primes. We can establish unequivocally the assertion that prime numbers are the building blocks of all integers. In other words, integers ≥ 2 are made up of primes; that is, every integer ≥ 2 can be decomposed into primes. This result, called the **fundamental theorem of arithmetic**, is certainly the cornerstone of number theory and one of its cardinal results. It appears in Euclid's *Elements*.

Before we state it formally and prove it, we need to lay some groundwork in the form of two lemmas. Throughout, assume all letters denote positive integers.

LEMMA 3.3 **(Euclid)** If p is a prime and $p|ab$, then $p|a$ or $p|b$.

PROOF

Suppose $p \nmid a$. Then p and a are relatively prime, so by Theorem 3.8, there are integers α and β such that $\alpha p + \beta a = 1$. Multiply both sides of this equation by b; we get $\alpha pb + \beta ab = b$. Since $p|p$ and $p|ab$, $p|(\alpha pb + \beta ab)$ by Theorem 2.4; that is, $p|b$. ∎

The following lemma extends this result to three or more factors, using induction.

LEMMA 3.4 Let p be a prime and $p|a_1 a_2 \cdots a_n$, where $a_1, a_2, \ldots, a_n$ are positive integers, then $p|a_i$ for some i, where $1 \le i \le n$.

PROOF (by weak induction)

When $n = 1$, the result follows clearly. So assume it is true for an arbitrary positive integer k: If $p|a_1 a_2 \cdots a_k$, then $p|a_i$ for some i. Suppose $p|a_1 a_2 \cdots a_{k+1}$, that is, $p|(a_1 a_2 \cdots a_k)a_{k+1}$. Then, by Lemma 3.3, $p|a_1 a_2 \cdots a_k$ or $p|a_{k+1}$. If $p|a_1 a_2 \cdots a_k$, then $p|a_i$, for some i, where $1 \le i \le k$. Thus, $p|a_i$, where $1 \le i \le k$, or $p|a_{k+1}$. In any event, $p|a_i$ for some i, where $1 \le i \le k+1$.

Thus, by induction, the result holds for every positive integer n. ∎

The following result follows nicely from this lemma.

COROLLARY 3.9 If $p, q_1, q_2, \ldots, q_n$ are primes such that $p|q_1 q_2 \cdots q_n$, then $p = q_i$ for some i, where $1 \le i \le n$.

PROOF

Since $p|q_1 q_2 \cdots q_n$, by Lemma 3.4, $p|q_i$ for some i. But p and q_i are primes, so $p = q_i$. ∎

We can now state and establish the fundamental theorem of arithmetic, the most fundamental result in number theory. The proof consists of two parts and is a bit long, so we need to follow it carefully.

THEOREM 3.13 **(The Fundamental Theorem of Arithmetic)** Every integer $n \ge 2$ either is a prime or can be expressed as a product of primes. The factorization into primes is unique except for the order of the factors.

PROOF

First, we will show by strong induction that n either is a prime or can be expressed as a product of primes. Then we will establish the uniqueness of such a factorization.

1. Let $P(n)$ denote the statement that n is a prime or can be expressed as a product of primes.

To show that $P(n)$ is true for every integer $n \geq 2$:

Since 2 is a prime, clearly $P(2)$ is true.

Now assume $P(2), P(3), \ldots, P(k)$ are true; that is, every integer 2 through k either is a prime or can be expressed as a product of primes.

If $k+1$ is a prime, then $P(k+1)$ is true. So suppose $k+1$ is composite. Then $k+1 = ab$ for some integers a and b, where $1 < a$, $b < k+1$. By the inductive hypothesis, a and b either are primes or can be expressed as products of primes; in any event, $k+1 = ab$ can be expressed as a product of primes. Thus, $P(k+1)$ is also true.

Thus, by strong induction, the result holds for every integer $n \geq 2$.

2. *To establish the uniqueness of the factorization*:

Let n be a composite number with two factorizations into primes: $n = p_1 p_2 \cdots p_r = q_1 q_2 \cdots q_S$. We will show that $r = s$ and every p_i equals some q_j, where $1 \leq i, j \leq r$; that is, the primes $q_1, q_2, \ldots, q_s$ are a permutation of the primes $p_1, p_2, \ldots, p_r$.

Assume, for convenience, that $r \leq s$. Since $p_1 p_2 \cdots p_r = q_1 q_2 \cdots q_s$, $p_1 | q_1 q_2 \cdots q_s$, by Corollary 3.9, $p_1 = q_i$ for some i. Dividing both sides by p_1, we get:

$$p_2 \cdots p_r = q_1 q_2 \cdots q_{i-1} \not{q}_i q_{i+1} \cdots q_s$$

Now p_2 divides the RHS, so again by Corollary 3.9, $p_2 = q_j$ for some j. Cancel p_2 from both sides:

$$p_3 \cdots p_r = q_1 q_2 \cdots q_{i-1} \not{q}_i q_{i+1} \cdots q_{j-1} \not{q}_j q_{j+1} q_s$$

Since $r \leq s$, continuing like this, we can cancel every p_t with some q_k. This yields a 1 on the LHS at the end. Then the RHS cannot be left with any primes, since a product of primes can never yield a 1; thus, we must have exhausted all q_ks by now. Therefore, $r = s$ and hence the primes $q_1, q_2, \ldots, q_s$ are the same as the primes $p_1, p_2, \ldots, p_r$ in some order. Thus, the factorization of n is unique, except for the order in which the primes are written. ■

It follows from this theorem that every composite number n can be factored into primes. Such a factorization is called a **prime factorization** of n.

For example, $5544 = 2 \cdot 2 \cdot 3 \cdot 7 \cdot 2 \cdot 11 \cdot 3$ is a prime factorization of 5544. Using the exponential notation, this product is often written as $5544 = 2^3 \cdot 3^2 \cdot 7 \cdot 11$. Such a product is the **prime-power decomposition** of n; if the primes occur in increasing order, then it is the **canonical decomposition**.

Canonical Decomposition

The **canonical decomposition** of a positive integer n is of the form $n = p_1^{a_1} p_2^{a_2} \cdots p_k^{a_k}$, where $p_1, p_2, \ldots, p_k$ are distinct primes with $p_1 < p_2 < \cdots < p_r$ and each exponent a_i is a positive integer.

There are two commonly used techniques for finding the canonical decomposition of a composite number. The first method involves finding all prime factors, beginning with the smallest prime, as the following example demonstrates.

EXAMPLE 3.9 Find the canonical decomposition of 2520.

SOLUTION

Beginning with the smallest prime 2, since $2|2520$, $2520 = 2 \cdot 1260$. Now 2 is a factor of 1260, so $2520 = 2 \cdot 2 \cdot 630$; $2|630$ again, so $2520 = 2 \cdot 2 \cdot 2 \cdot 315$. Now $2 \nmid 315$, but 3 does, so $2520 = 2 \cdot 2 \cdot 2 \cdot 3 \cdot 105$; 3 is a factor of 105 also, so $2520 = 2 \cdot 2 \cdot 2 \cdot 3 \cdot 3 \cdot 35$. Continuing like this we get

$$2520 = 2 \cdot 2 \cdot 2 \cdot 3 \cdot 3 \cdot 5 \cdot 7 = 2^3 \cdot 3^2 \cdot 5 \cdot 7$$

which is the desired canonical decomposition. ■

This method can be quite time consuming if the number n is fairly large. The second method, which is generally more efficient, involves splitting n as the product of two positive integers, not necessarily prime numbers, and continuing to split each factor into further factors until all factors are primes. To make this method short, look for large factors; as you will soon see, the larger the factors, the fewer the steps. The following example clarifies this fairly straightforward method.

EXAMPLE 3.10 Find the canonical decomposition of 2520 by the second method.

SOLUTION

Notice that $2520 = 40 \cdot 63$. Since none of the factors are primes, split them again: $40 = 4 \cdot 10$ and $63 = 7 \cdot 9$, so $2520 = (4 \cdot 10) \cdot (7 \cdot 9)$. Since 4, 10, and 9 are composites, split each of them: $2520 = (2 \cdot 2)(2 \cdot 5)(7)(3 \cdot 3)$. Now all the factors are primes, so the procedure stops. Rearranging them yields the canonical decomposition: $2520 = 2^3 \cdot 3^2 \cdot 5 \cdot 7$. ■

Factor Tree

This method can be illustrated in a tree diagram, called a **factor tree**. In such a diagram, if $a|b$, we connect them by a line segment. Figure 3.3 shows the factor tree

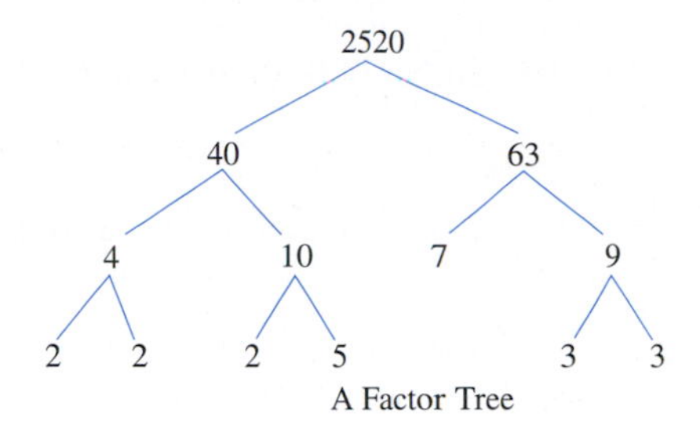

Figure 3.3

for 2520 using the above decomposition steps. To find the canonical decomposition, simply take the product of all primes at the "leaves": $2520 = 2 \cdot 2 \cdot 2 \cdot 5 \cdot 7 \cdot 3 \cdot 3 = 2^3 \cdot 3^2 \cdot 5 \cdot 7$.

The canonical decomposition of a composite number can be used to find its positive factors, as the following example shows. (It can also be used to find the number of positive factors without listing them; we will discuss this in Section 8.2.)

EXAMPLE 3.11 Find the positive factors of 60.

SOLUTION

First, notice that $60 = 2^2 \cdot 3 \cdot 5$. By the fundamental theorem of arithmetic, every factor of 60 is of the form $2^a \cdot 3^b \cdot 5^c$, where $0 \le a \le 2$, and $0 \le b, c \le 1$. Thus, the various factors are

$$\begin{array}{llll} 2^0 \cdot 3^0 \cdot 5^0 = 1 & 2^0 \cdot 3^0 \cdot 5^1 = 5 & 2^0 \cdot 3^1 \cdot 5^0 = 3 & 2^0 \cdot 3^1 \cdot 5^1 = 15 \\ 2^1 \cdot 3^0 \cdot 5^0 = 2 & 2^1 \cdot 3^0 \cdot 5^1 = 10 & 2^1 \cdot 3^1 \cdot 5^0 = 6 & 2^1 \cdot 3^1 \cdot 5^1 = 30 \\ 2^2 \cdot 3^0 \cdot 5^0 = 4 & 2^2 \cdot 3^0 \cdot 5^1 = 20 & 2^2 \cdot 3^1 \cdot 5^0 = 12 & 2^2 \cdot 3^1 \cdot 5^1 = 60 \end{array}$$

(Thus, 60 has 12 factors. Can you think of a better way to find the number of positive factors without listing them?) ■

The following example presents a beautiful application of the fundamental theorem of arithmetic and the floor function. It shows how nicely we can determine the number of trailing zeros in the decimal value of $n!$, without computing it. (For instance, $11! = 39{,}916{,}800$ has two trailing zeros.)

EXAMPLE 3.12 Find the number of trailing zeros in 234!.

SOLUTION

By the fundamental theorem of arithmetic, 234! can be factored as $2^a \cdot 5^b \cdot c$, where a and b are positive integers (why?) and c denotes the product of primes other than 2 and 5. Clearly, $a > b$ (why?). Each trailing zero in 234! corresponds to a 10 in a factorization and vice versa; each 10 is the product of a 2 and a 5.

$$\begin{pmatrix}\text{No. of trailing} \\ \text{zeros in 234!}\end{pmatrix} = \begin{pmatrix}\text{No. of products of the} \\ \text{form } 2 \cdot 5 \text{ in a prime} \\ \text{factorization of 234!}\end{pmatrix}$$

$$= \text{minimum of } a \text{ and } b \text{ (why?)}$$

$$= b$$

To find b, we proceed as follows:

No. of positive integers ≤ 234 and divisible by $5 = \lfloor 234/5 \rfloor = 46$.
Each of them contributes a 5 to the prime factorization of 234!.
No. of positive integers ≤ 234 and divisible by $25 = \lfloor 234/25 \rfloor = 9$.
Each of them contributes an additional 5 to the prime factorization of 234!.
No. of positive integers ≤ 234 and divisible by $125 = \lfloor 234/125 \rfloor = 1$.
It contributes a still additional 5 to the prime factorization.

No higher power of 5 contributes a 5 to the prime factorization of 234!, so the total number of 5s in the prime factorization equals $46 + 9 + 1 = 56$. Thus, 234! has 56 trailing zeros. (This example is pursued further in Example 4.25.) ■

It follows from this example that the highest power e of a prime p that divides $n!$ is given by

$$e = \lfloor n/p \rfloor + \lfloor n/p^2 \rfloor + \lfloor n/p^3 \rfloor + \cdots$$

Let k be the smallest integer such that $p^k > n$. Then $\lfloor n/p^k \rfloor = 0$, so the sum is a finite one.

For example, the largest power of 2 that divides 97! is

$$\begin{aligned} e &= \lfloor 97/2 \rfloor + \lfloor 97/2^2 \rfloor + \lfloor 97/2^3 \rfloor + \lfloor 97/2^4 \rfloor + \lfloor 97/2^5 \rfloor + \lfloor 97/2^6 \rfloor \\ &= 48 + 24 + 12 + 6 + 3 + 1 = 94 \end{aligned}$$

Interestingly enough, there is a close relationship between the number of ones in the binary representation of 97 and the highest power of 2 that divides 97!. To see this, notice that $97 = 1100001_{\text{two}}$, so the binary representation contains three 1s and $97 = 94 + 3$.

More generally, we have the following result due to the French mathematician Adrien-Mari Legendre.

THEOREM 3.14 Let e denote the highest power of 2 that divides $n!$ and b the number of 1s in the binary representation of n. Then $n = e + b$.

PROOF

Let $n = (a_k a_{k-1} \ldots a_1 a_0)_{\text{two}} = a_0 + a_1 \cdot 2 + \cdots + a_k \cdot 2^k$. Let $1 \le i \le k$. Then

$$\left\lfloor \frac{n}{2^i} \right\rfloor = \left\lfloor \frac{a_0 + a_1 \cdot 2 + \cdots + a_{i-1} \cdot 2^{i-1}}{2^i} \right\rfloor + a_i + a_{i+1} \cdot 2 + \cdots + a_k \cdot 2^{k-i}$$

But

$$\begin{aligned} a_0 + a_1 \cdot 2 + \cdots + a_{i-1} \cdot 2^{i-1} &\le 1 + 2 + 2^2 + \cdots + 2^{i-1} \\ &= 2^i - 1 \\ &< 2^i, \end{aligned}$$

so

$$\left\lfloor \frac{a_0 + a_1 \cdot 2 + \cdots + a_{i-1} \cdot 2^{i-1}}{2^i} \right\rfloor = 0$$

Therefore,

$$\left\lfloor \frac{n}{2^i} \right\rfloor = a_i + a_{i+1} \cdot 2 + \cdots + a_k \cdot 2^{k-i}$$

Thus,

$$\begin{aligned} \sum_{i=1}^{k} \left\lfloor \frac{n}{2^i} \right\rfloor = a_1 + a_2 \cdot 2 + a_3 \cdot 2^2 + \cdots + a_k \cdot 2^{k-1} \\ + a_2 \cdot 1 + a_3 \cdot 2 + \cdots + a_k \cdot 2^{k-2} \\ + a_3 \cdot 1 + \cdots + a_k \cdot 2^{k-3} \\ \vdots \\ + a_k \cdot 1 \end{aligned}$$

That is,

$$\begin{aligned} e &= a_1 + a_2(1+2) + a_3(1+2+2^2) + \cdots + a_k(1 + 2^2 + \cdots + 2^{k-1}) \\ &= a_1(2-1) + a_2(2^2-1) + a_3(2^3-1) + \cdots + a_k(2^k - 1) \\ &= (a_0 + a_1 \cdot 2 + a_2 \cdot 2^2 + \cdots + a_k \cdot 2^k) - (a_0 + a_1 + \cdots + a_k) \\ &= n - b \end{aligned}$$

Thus, $n = e + b$. ■

The canonical decompositions of positive integers provide a new method for finding their gcds, as the following example illustrates.

EXAMPLE 3.13 Using the canonical decompositions of 168 and 180, find their gcd.

SOLUTION

You can verify that $168 = 2^3 \cdot 3 \cdot 7$ and $180 = 2^2 \cdot 3^2 \cdot 5$. The only common prime factors are 2 and 3, so 5 or 7 cannot appear in their gcd. Since 2 appears thrice in the canonical decomposition of 168, but only twice in the canonical decomposition of 180, 2^2 is a factor in the gcd. Similarly, 3 is also a common factor, so $(168, 180) = 2^2 \cdot 3 = 12$. ■

An important observation:

$$(168, 180) = 2^2 \cdot 3 = 2^2 \cdot 3^1 \cdot 5^0 \cdot 7^0 = 2^{\min\{3,2\}} \cdot 3^{\min\{1,2\}} \cdot 5^{\min\{1,0\}} \cdot 7^{\min\{1,0\}}$$

This technique can be generalized as follows. Let a and b be positive integers with the following canonical decompositions:

$$a = p_1^{a_1} p_2^{a_2} \cdots p_n^{a_n} \quad \text{and} \quad b = p_1^{b_1} p_2^{b_2} \cdots p_n^{b_n},$$

where $a_i, b_i \geq 0$. (By letting exponents zero, we can always assume that both decompositions contain exactly the same prime bases p_i.) Then

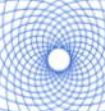

$$(a, b) = p_1^{\min\{a_1,b_1\}} p_2^{\min\{a_2,b_2\}} \cdots p_n^{\min\{a_n,b_n\}}$$

Let us look again at the distribution of primes, which we studied in the preceding chapter.

Distribution of Primes Revisited

By the division algorithm, every integer is of the form $4n + r$, where $r = 0$, 1, 2, or 3; so every odd integer is of the form $4n + 1$ or $4n + 3$. For instance, 13 and 25 are of the form $4n + 1 : 13 = 4 \cdot 3 + 1$ and $25 = 4 \cdot 6 + 1$, whereas 11 and 31 are of the form $4n + 3 : 11 = 4 \cdot 2 + 3$ and $31 = 4 \cdot 7 + 3$.

Look at positive integers of the form $4n+3$. The first eleven such numbers are 3, 7, 11, 15, 19, 23, 27, 31, 35, 39, and 43, of which seven (about 64%) are primes.

What can we reasonably conjecture from this observation? If you guessed there are infinitely many primes of the form $4n+3$, you are correct. Before we establish the validity of this educated guess, we need to lay its groundwork in the form of the following lemma.

LEMMA 3.5 The product of any two integers of the form $4n+1$ is also of the same form.

PROOF

Let a and b be any two integers of the form $4n+1$, say, $a = 4\ell + 1$ and $b = 4m+1$ for some integers ℓ and m. Then

$$\begin{aligned} ab &= (4\ell+1)(4m+1) \\ &= 16\ell m + 4\ell + 4m + 1 \\ &= 4(4\ell m + \ell + m) + 1 \\ &= 4k + 1 \quad \text{where } k = 4\ell m + \ell + m \text{ is an integer} \end{aligned}$$

Thus, ab is also of the same form. ■

This result can be extended to any finite number of such integers (see Review Exercise 60).

We are now ready to prove our conjecture. The proof looks similar to Euclid's proof, which established the infinitude of primes. See Example 11.7 also.

THEOREM 3.15 There are infinitely many primes of the form $4n+3$.

PROOF (by contradiction)

Suppose there are only finitely many primes of the form $4n+3$, say, $p_0, p_1, p_2, \dots, p_k$, where $p_0 = 3$. Consider the positive integer $N = 4p_1p_2\cdots p_k + 3$. Clearly, $N > p_k$ and is also of the same form.

case 1 If N itself is a prime, then N would be larger than the largest prime p_k of the form $4n+3$, which is a contradiction.

case 2 Suppose N is composite. Since N is odd, every factor of N is of the form $4n+1$ or $4n+3$. If every factor is of the form $4n+1$, then, by Lemma 3.5, N would be of the same form. But, since N is of the form $4n+3$, at least one of the prime factors, say, p, must be of the form $4n+3$.

subcase 1 Let $p = p_0 = 3$. Then $3|N$, so $3|(N - 3)$ by Theorem 2.4; that is, $3|4p_1p_2 \cdots p_k$. So, by Lemma 3.4, $3|2$ or $3|p_i$, where $1 \le i \le k$, but both are impossible.

subcase 2 Let $p = p_i$, where $1 \le i \le k$. Then $p|N$ and $p|4p_1p_2 \cdots p_k$, so $p|(N - 4p_1p_2 \cdots p_k)$, that is, $p|3$, again a contradiction.

Both cases lead us to a contradiction, so our assumption must be false. Thus, there is an infinite number of primes of the given form. ■

Now that we have established the infiniteness of the number of primes of the form $4n + 3$, we ask the next logical question: Are there infinitely many primes of the form $4n + 1$? Fortunately, the answer is again yes (see Example 11.7).

In fact, both results are special cases of the following remarkable result, proved by Dirichlet in 1837, but stated originally by Legendre in 1785. Its proof is extremely complicated, so we omit it.

THEOREM 3.16 **(Dirichlet's Theorem)** If a and b are relatively prime, then the arithmetic sequence $a, a + b, a + 2b, a + 3b, \ldots$ contains infinitely many primes. ■

For example, let $a = 3$ and $b = 4$; then the sequence $3, 4 \cdot 1 + 3, 4 \cdot 2 + 3, 4 \cdot 3 + 3, \ldots$ contains an infinite number of primes, namely, primes of the form $4n + 3$. Likewise, choosing $a = 1$ and $b = 4$, it follows there is an infinite number of primes of the form $4n + 1$.

For yet another example, choose $a = 7$ and $b = 100$. Then $a + nb = 100n + 7$, so the sequence $7, 107, 207, 307, \ldots$ contains an infinite number of primes, all ending in 7.

Note the crucial condition in Dirichlet's theorem that a and b be relatively prime. If they are not, then the sequence need not contain any primes at all. To verify this, choose $a = 6$ and $b = 9$; then the sequence $6, 15, 24, 33, 42, 51, \ldots$ contains only composite numbers.

EXERCISES 3.3

Find the canonical decomposition of each composite number.

1. 1947
2. 1661
3. 1863
4. 1976
5. $2^{27} + 1$
6. $2^{48} - 1$
7. 10,510,100,501
8. 1,004,006,004,001

Find the positive factors of each, where p and q are distinct primes.

9. p
10. p^2
11. pq
12. pq^2

Find the positive factors of each composite number.

13. 48 14. 90 15. 210 16. 1040

Find the number of trailing zeros in the decimal value of each.

17. 100! 18. 376! 19. 609! 20. 1010!

Find the values of n for which $n!$ contains the given number of trailing zeros.

21. 58 22. 93

Find the gcd of each pair, where p, q, and r are distinct primes.

23. $2^3 \cdot 3 \cdot 5, 2 \cdot 3^2 \cdot 5^3 \cdot 7^2$
24. $2^4 \cdot 3^2 \cdot 7^5, 3^4 \cdot 5 \cdot 11^2$
25. p^2q^3, pq^2r
26. $p^3qr^3, p^3q^4r^5$

Using canonical decompositions, find the gcd of each pair.

27. 48, 162 28. 72, 108
29. 175, 192 30. 294, 450

Find the number of trailing zeros in the binary representation of each integer.

31. 28 32. 32 33. 208 34. 235

35. Using Exercises 31–34, predict the number of trailing zeros in the binary representation of a positive integer n.

Find the highest power of each that divides 1001!

36. 2 37. 3 38. 5 39. 7

Using Theorem 3.14, find the number of ones in the binary representation of each integer.

40. 234 41. 1001 42. 1976 43. 3076

44. Using Example 3.12, conjecture the number of trailing zeros in the decimal value of $n!$

Prove each, where p is a prime, and a, b, and n are positive integers.

45. If $p|a^2$, then $p|a$.
46. If $p|a^n$, then $p|a$.
47. The product of any n integers of the form $4k+1$ is also of the same form.
48. If $(a, b) = 1$, then $(a^n, b^n) = 1$.
49. If $(a^n, b^n) = 1$, then $(a, b) = 1$.
50. There are infinitely many primes of the form $2n+3$.
51. There are infinitely many primes of the form $8n+5$.
52. Every positive integer n can be written as $n = 2^e m$, where $e \geq 0$ and m is an odd integer.
53. Every positive integer n can be written as $n = 2^a 5^b c$, where c is not divisible by 2 or 5.
54. A positive integer is a square if and only if every exponent in its canonical decomposition is an even integer.

*Find the number of positive factors of each, where p, q, and r are distinct primes.

55. pq 56. pq^2 57. p^2q^2 58. pq^2r^3

*Find the sum of the positive factors of each, where p, q, and r are distinct primes.

59. p^i 60. pq^j 61. p^iq^j 62. $p^iq^jr^k$

A positive integer is **square-free** if it is not divisible by the square of any positive integer > 1. For instance, $105 = 3 \cdot 5 \cdot 7$ is square-free.

63. An integer > 1 is square-free if and only if its prime factorization consists of distinct primes.
64. Any integer $n > 1$ can be written as the product of a square and a square-free integer.

A positive integer is said to be **powerful** if whenever a prime p is a factor of n, p^2 is also a factor. For example, 72 is a powerful number since both 3 and 3^2 are factors of 72.

65. Find the first three powerful numbers.
66. Show that every powerful number can be written in the form a^2b^3, where a and b are positive integers.

Let p be a prime, and n and a positive integers. Then p^a **exactly divides** n if $p^a|n$, but $p^{a+1} \nmid n$; we then write $p^a \| n$. Thus, $p^a \| n$ if a is the largest exponent of p such that $p^a|a$. Prove each.

67. If $p^a \| m$ and $p^b \| n$, then $p^{a+b}|mn$.
68. If $p^a \| m$, then $p^{ka} \| m^k$, where k is a positive integer.
69. If $p^a \| m$ and $p^b \| n$, then $p^{\min(a,b)} \| (m+n)$.

3.4 Least Common Multiple

The least common multiple (lcm) of two positive integers a and b is closely related to their gcd. In fact, we use the lcm every time we add and subtract fractions. Now we will explore two methods for finding the lcm of a and b. The first method employs canonical decompositions, and the second employs their gcd. We begin with a definition.

Least Common Multiple

The **least common multiple** of two positive integers a and b is the least positive integer divisible by both a and b; it is denoted by $[a, b]$.

For example, suppose we want to evaluate [18, 24]. The positive multiples of 18 are 18, 36, 54, 72, 90, ... and those of 24 are 24, 48, 72, 96, ... So their common multiples are 72, 144, 216, ... Thus, $[18, 24] =$ their lcm $= 72$.

How do we know that $[a, b]$ always exists? Since ab is a multiple of both a and b, the set of common multiples is always nonempty; so, by the well-ordering principle, the set contains a least element; thus, $[a, b]$ always exists.

Is it unique? The answer is again yes (see Exercise 32).

Next, we rewrite the previous definition of lcm symbolically.

A Symbolic Definition of lcm

The lcm of two positive integers a and b is the positive integer m such that

- $a|m$ and $b|m$; and
- if $a|m'$ and $b|m'$, then $m \le m'$, where m' is a positive integer.

Canonical decompositions of a and b can be employed to find their lcm. Suppose we want to find [90, 168]. Notice that $90 = 2 \cdot 3^2 \cdot 5$ and $168 = 2^3 \cdot 3 \cdot 7$. Looking at the prime powers, it follows that their lcm must be a multiple of 2^3, 3^2, 5, and 7; so their lcm is $2^3 \cdot 3^2 \cdot 5 \cdot 7 = 2520$.

An important observation:

$$\begin{aligned}[90, 168] &= 2^3 \cdot 3^2 \cdot 5 \cdot 7 \\ &= 2^{\max\{1,3\}} \cdot 3^{\max\{2,1\}} \cdot 5^{\max\{1,0\}} \cdot 7^{\max\{0,1\}}\end{aligned}$$

This leads us to the following generalization.

Let a and b be two positive integers with the following canonical decompositions:

$$a = p_1^{a_1} p_2^{a_2} \cdots p_n^{a_n} \quad \text{and} \quad b = p_1^{b_1} p_2^{b_2} \cdots p_n^{b_n}, \quad \text{where } a_i, b_i \geq 0.$$

(Again, we assume that both decompositions contain exactly the same prime bases p_i.) Then

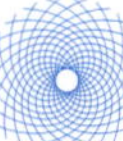

$$[a, b] = p_1^{\max\{a_1, b_1\}} p_2^{\max\{a_2, b_2\}} \cdots p_n^{\max\{a_n, b_n\}}$$

The following example illustrates this technique.

EXAMPLE 3.14 Using the canonical decompositions of 1050 and 2574, find their lcm.

SOLUTION

Notice that $1050 = 2 \cdot 3 \cdot 5^2 \cdot 7$ and $2574 = 2 \cdot 3^2 \cdot 11 \cdot 13$. Therefore,

$$\begin{aligned}[1050, 574] &= 2^{\max\{1,1\}} \cdot 3^{\max\{1,2\}} \cdot 5^{\max\{2,0\}} \cdot 7^{\max\{1,0\}} \cdot 11^{\max\{0,1\}} \cdot 13^{\max\{0,1\}} \\ &= 2^1 \cdot 3^2 \cdot 5^2 \cdot 7^1 \cdot 11^1 \cdot 13^1 = 450{,}450\end{aligned}$$ ■

Next, we derive a close relationship between the gcd and the lcm of two positive integers. But first, let us study an example and make an observation.

Notice that $(18, 24) = 6$ and $[18, 24] = 72$. Also, $6 \cdot 72 = 18 \cdot 24$; in other words, $[18, 24] = \dfrac{18 \cdot 24}{(18, 24)}$.

The following theorem shows this is not a sheer coincidence but is always the case. It is a direct application of Theorem 1.1 and canonical decompositions.

THEOREM 3.17 Let a and b be positive integers. Then $[a, b] = \dfrac{ab}{(a, b)}$.

PROOF

Let $a = p_1^{a_1} p_2^{a_2} \cdots p_n^{a_n}$ and $b = p_1^{b_1} p_2^{b_2} \cdots p_n^{b_n}$ be the canonical decompositions of a and b, respectively. Then

$$(a, b) = p_1^{\min\{a_1, b_1\}} p_2^{\min\{a_2, b_2\}} \cdots p_n^{\min\{a_n, b_n\}}$$

and

$$[a, b] = p_1^{\max\{a_1,b_1\}} p_2^{\max\{a_2,b_2\}} \cdots p_n^{\max\{a_n,b_n\}}$$

Therefore,

$$\begin{aligned}(a, b) \cdot [a, b] &= p_1^{\min\{a_1,b_1\}} \cdots p_n^{\min\{a_n,b_n\}}, p_1^{\max\{a_1,b_1\}} \cdots p_n^{\max\{a_n,b_n\}} \\ &= p_1^{\min\{a_1,b_1\}+\max\{a_1,b_1\}} \cdots p_n^{\min\{a_n,b_n\}+\max\{a_n,b_n\}} \\ &= p_1^{a_1+b_1} p_2^{a_2+b_2} \cdots p_n^{a_n+b_n} \\ &= \left(p_1^{a_1} p_2^{a_2} \cdots p_n^{a_n}\right)\left(p_1^{b_1} p_2^{b_2} \cdots p_n^{b_n}\right) \\ &= ab\end{aligned}$$

Thus,

$$[a, b] = \frac{ab}{(a, b)}$$ ■

This theorem provides a second way of computing $[a, b]$, provided (a, b) is known, as the following example illustrates.

EXAMPLE 3.15 Using (252, 360), compute [252, 360].

SOLUTION

You may notice that $252 = 2^2 \cdot 3^2 \cdot 7$ and $360 = 2^3 \cdot 3^2 \cdot 5$, so $(252, 360) = 2^2 \cdot 3^2 = 36$. Therefore, by Theorem 3.17,

$$[252, 360] = \frac{252 \cdot 360}{36} = 2520$$ ■

Returning to Theorem 3.17, suppose $(a, b) = 1$. Then $[a, b] = ab$. Accordingly, we have the following corollary.

COROLLARY 3.10 Two positive integers a and b are relatively prime if and only if $[a, b] = ab$. ■

For instance, since 15 and 28 are relatively prime, $[15, 28] = 15 \cdot 28 = 420$.

As in the case of gcd, the idea of lcm can be extended to three or more positive integers. For example, $24 = 2^3 \cdot 3$, $28 = 2^2 \cdot 7$, and $36 = 2^2 \cdot 3^2$. Therefore,

$$\begin{aligned}[24, 28, 36] &= 2^{\max\{3,2,2\}} \cdot 3^{\max\{1,0,2\}} \cdot 7^{\max\{0,1,0\}} \\ &= 2^3 \cdot 3^2 \cdot 7^1 = 504\end{aligned}$$

Again, as in the case of gcd, recursion can be applied to evaluate the lcm of three or more positive integers, as the following result shows. We leave its proof as an exercise (see Exercise 36).

THEOREM 3.18 Let $a_1, a_2, \ldots, a_n$ be n (≥ 3) positive integers. Then $[a_1, a_2, \ldots, a_n] = [[a_1, a_2, \ldots, a_{n-1}], a_n]$. ■

The following example illustrates this result.

EXAMPLE 3.16 Using recursion, evaluate [24, 28, 36, 40].

SOLUTION

$$\begin{aligned}[24, 28, 36, 40] &= [[24, 28, 36], 40] = [[[24, 28], 36], 40] \\ &= [[168, 36], 40] = [504, 40] \\ &= 2520\end{aligned}$$

(You can verify this using the canonical decompositions of 24, 28, 36, and 40.) ■

The following two results follow from Theorem 3.18.

COROLLARY 3.11 If the positive integers $a_1, a_2, \ldots, a_n$ are pairwise relatively prime, then $[a_1, a_2, \ldots, a_n] = a_1 a_2 \cdots a_{n-1} a_n$. ■

For instance, 12, 25, and 77 are pairwise relatively prime, so $[12, 25, 77] = 12 \cdot 25 \cdot 77 = 23{,}100$.

Is the converse of this corollary true? You can determine this in Exercise 38.

COROLLARY 3.12 Let $m_1, m_2, \ldots, m_k$ and a be positive integers such that $m_i | a$ for $1 \leq i \leq k$. Then $[m_1, m_2, \ldots, m_k] | a$.

***PROOF* (by strong induction on *k*)**

The statement is clearly true when $k = 1$ and $k = 2$. So assume it is true for integers 1 through t. Now let $m_i | a$ for $1 \leq i \leq t+1$. Then $[m_1, m_2, \ldots, m_t] | a$ by the inductive hypothesis and $m_{t+1} | a$; so, again by the hypothesis, $[[m_1, m_2, \ldots, m_t], m_{t+1}] | a$; that is, $[m_1, m_2, \ldots, m_{t+1}] | a$ by Theorem 3.18. Thus, by induction, the result is true for every positive integer k. ■

EXERCISES 3.4

Mark *true* or *false*, where a, b, and c are arbitrary positive integers and p is any prime.

1. The lcm of two primes is their product.
2. The lcm of two consecutive positive integers is their product.
3. The lcm of two distinct primes is their product.
4. If $(a, b) = 1$, then $[a, b] = ab$.
5. If $p \nmid a$, then $[p, a] = pa$.
6. If $[a, b] = 1$, then $a = 1 = b$.
7. If $[a, b] = b$, then $a = 1$.
8. If $[a, b] = b$, then $a|b$.
9. If $[a, b] = ab$, then $a = b$.
10. If $[a, b] = ab$ and $[b, c] = bc$, then $[a, c] = ac$.

Find the lcm of each pair of integers.

11. 110, 210
12. 65, 66

Find $[a, b]$ if

13. $a|b$
14. $b|a$
15. $a = 1$
16. $a = b$
17. a and b are distinct primes.
18. $b = a + 1$

Find $[a, b]$ if

19. $(a, b) = 3$ and $ab = 693$.
20. $ab = 156$ and a and b are relatively prime.
21. Find the positive integer a if $[a, a + 1] = 132$.
22. Find the twin primes p and q such that $[p, q] = 323$.

Find the positive integers a and b such that

23. $(a, b) = 20$ and $[a, b] = 840$
24. $(a, b) = 18$ and $[a, b] = 3780$
25. What is your conclusion if $(a, b) = [a, b]$? Why?

Using recursion, find the lcm of the given integers.

26. 12, 18, 20, 28
27. 15, 18, 24, 30
28. 10, 16, 18, 24, 28
29. 12, 15, 18, 25, 30
30. Prove or disprove: $[a, b, c] = abc/(a, b, c)$.
31. Find the smallest positive integer ≥ 2 that is a square, a cube, and a fifth power. (A. Dunn, 1983)

Prove each, where $a, b, c, k, m, a_1, a_2, \ldots, a_n, x, y$, and z are positive integers.

32. The lcm of any two integers is unique.
33. $(a, b)|[a, b]$
34. $[ka, kb] = k[a, b]$
35. Let m be any multiple of a and b. Then $[a, b]|m$.
36. Let $a_1, a_2, \ldots, a_n$ be n (≥ 3) positive integers. Then $[a_1, a_2, \ldots, a_n] = [[a_1, a_2, \ldots, a_{n-1}], a_n]$.
37. $[ka_1, ka_2, \ldots, ka_n] = k[a_1, a_2, \ldots, a_{n-1}, a_n]$.
38. If $[a_1, a_2, \ldots, a_n] = a_1 a_2 \cdots a_{n-1} a_n$, then $a_1, a_2, \ldots, a_n$ are pairwise relatively prime.
39. $\max\{x, y, z\} - \min\{x, y, z\} + \min\{x, y\} + \min\{y, z\} + \min\{z, x\} = x + y + z$
40. The sum of the twin primes p and $p + 2$ is divisible by 12, where $p > 3$. (C. Ziegenfus, 1963)
41. $(a, [b, c]) = [(a, b), (a, c)]$
42. $[a, (b, c)] = ([a, b], [a, c])$

3.5 Linear Diophantine Equations

Often we are interested in integral solutions of equations with integral coefficients. Such equations are called **diophantine equations**, after Diophantus, who wrote extensively on them. For example, when we restrict the solutions to integers, the equations $2x + 3y = 4$, $x^2 + y^2 = 1$, and $x^2 + y^2 = z^2$ are diophantine equations.

Diophantus *lived in Alexandria around* A.D. *250. Not much is known about his life or nationality, except what is found in an epigram in the* Greek Anthology: *"Diophantus passed one-sixth of his life in childhood, one-twentieth in youth, and one-seventh more as a bachelor. Five years after his marriage was born a son who died four years before his father, at half his father's age (at the time of the father's death)."*

Diophantus wrote three books, all in Greek: Arithmetica, On Polygonal Numbers, *and* Porisms. Arithmetica, *considered the earliest book on algebra, contains the first systematic use of mathematical notation for unknowns in equations. He had a symbol for subtraction and for equality.*

Six out of the thirteen copies of Arithmetica *and a portion of* On Polygonal Numbers *are still in existence;* Porisms is *lost.*

The Bishop of Laodicea, a friend of Diophantus who assumed his episcopacy around A.D. 270, *dedicated a book on Egyptian computation in his honor.*

Geometrically, such solutions of the equation $2x + 3y = 4$ are points on the line $2x + 3y = 4$ with integral coordinates. Points with integral coordinates are called **lattice points**. For example, $(-1, 2)$ is such a solution; in fact, it has infinitely many solutions $(2 + 3t, -2t)$, where t is an arbitrary integer.

The diophantine equation $x^2 + y^2 = 1$ has exactly four solutions: $(\pm 1, 0)$ and $(0, \pm 1)$, the points where the unit circle $x^2 + y^2 = 1$ intersects the axes.

The solutions of the diophantine equation $x^2 + y^2 = z^2$ represent the lengths of the sides of a right triangle; $(3, 4, 5)$ is one solution. This equation also has an infinite number of solutions, as we shall see in Section 13.1.

Linear Diophantine Equations

The simplest class of diophantine equations is the class of **linear diophantine equations** (LDEs). A **linear diophantine equation in two variables** x and y is a diophantine equation of the form $ax + by = c$. Solving such a LDE systematically involves the euclidean algorithm, as you will see shortly. First, we study LDEs in two variables.

LDEs were known in ancient China and India as applications to astronomy and riddles, so we begin our discussion with two interesting puzzles.

The first puzzle is due to the Indian mathematician Mahavira (ca. A.D. 850).

EXAMPLE 3.17 Twenty-three weary travelers entered the outskirts of a lush and beautiful forest. They found 63 equal heaps of plantains and seven single fruits, and divided them equally. Find the number of fruits in each heap.

***Mahavira**, an astronomer and mathematician at the court of King Amoghavardana Vripatunga (814–877), was born in Mysore, India. Only little is known about his life. A staunch Jain by religion, he is known for his* Ganita-Sara-Sangraha, *the most scholarly treatise of the time on Indian mathematics. Written in nine chapters, it summarizes the body of knowledge then known in India, in arithmetic, including zero, fractions, and the decimal system, and geometry. It was translated from Sanskrit into English in 1912.*

SOLUTION

Let x denote the number of plantains in a heap and y the number of plantains received by a traveler. Then we get the LDE

$$63x + 7 = 23y \tag{3.2}$$

Since both x and y must be positive, we are interested in finding only the positive integral solutions of the LDE (3.2). Solving it for y,

$$y = \frac{63x + 7}{23}$$

When $x > 0$, clearly $y > 0$. So try the values 1, 2, 3, and so on for x until the value of y becomes an integer (Table 3.2). It follows from the table that $x = 5$, $y = 14$ is a solution. We can verify that $x = 28$, $y = 77$ is yet another solution. In fact, the LDE has infinitely many solutions. See Example 3.20. ■

x	1	2	3	4	5	...	28	...
y	$\frac{70}{23}$	$\frac{133}{23}$	$\frac{196}{23}$	$\frac{252}{23}$	14	...	77	...

Table 3.2

Another ancient riddle, called the **hundred fowls puzzle**, is found in the *Mathematical Classic*, a book by the sixth-century Chinese mathematician Chang Chiu-chien.

EXAMPLE 3.18 If a cock is worth five coins, a hen three coins, and three chicks together one coin, how many cocks, hens, and chicks, totaling 100, can be bought for 100 coins?

SOLUTION

Let x, y, and z denote the number of cocks, the number of hens, and the number of chicks respectively. Clearly, $x, y, z \geq 0$. Then the given data yield two LDEs:

$$x + y + z = 100 \tag{3.3}$$

$$5x + 3y + \frac{z}{3} = 100 \tag{3.4}$$

Substituting for z [$= 100 - x - y$ from equation (3.3)] in equation (3.4) yields

$$5x + 3y + \frac{1}{3}(100 - x - y) = 100$$

That is,

$$7x + 4y = 100$$

$$y = \frac{100 - 7x}{4} = 25 - \frac{7}{4}x \tag{3.5}$$

So, for y to be an integer, $7x/4$ must be an integer; but $4 \nmid 7$, so x must be a multiple of 4: $x = 4t$, where t is an integer. Then,

$$y = 25 - \frac{7}{4}x = 25 - \frac{7(4t)}{4} = 25 - 7t$$

and

$$z = 100 - x - y = 100 - 4t - (25 - 7t) = 75 + 3t$$

Thus, every solution to the puzzle is of the form $x = 4t$, $y = 25 - 7t$, $z = 75 + 3t$, where t is an arbitrary integer.

Now, to find the possible actual solutions of the puzzle, we take the following steps: Since $x \geq 0$, $t \geq 0$. Since $y \geq 0$, $25 - 7t \geq 0$; that is, $t \leq 25/7$, so $t \leq 3$. Since $z \geq 0$, $75 + 3t \geq 0$; that is, $t \geq -25$; but this does not give us any additional information, so $0 \leq t \leq 3$.

Thus, the riddle has four possible solutions, corresponding to $t = 0$, 1, 2, and 3: $x = 0$, $y = 25$, $z = 75$; $x = 4$, $y = 18$, $z = 78$; $x = 8$, $y = 11$, $z = 81$; and $x = 12$, $y = 4$, $z = 84$. ■

Although we were able to solve successfully the LDEs in both examples, we should ask three questions:

***Aryabhata** (ca. 476–ca. 550), the first prominent Indian mathematician-astronomer, was born in Kusumapura, near Patna on the Ganges. He studied at Nalanda University, Kusumapura, and later became its head. Although he used mathematics to solve astronomical problems, he was very much interested in Diophantus' work on indeterminate equations and on the Indian astronomer Parasara's work on comets and planetary motion. Aryabhata described the earth as spherical and computed its diameter as 7980 miles. He understood the nature of eclipses and that the sun was the source of moonlight, both ideas unknown to the West until the observations of Copernicus and Galileo a thousand years later. Aryabhata's accurate astronomical calculations contributed to the development of a calendar in India. He also devised the expansions of $(x+y)^2$ and $(x+y)^3$, and formulas for extracting square roots and cube roots. Around 500, he calculated an accurate value of π as $62832/20000 = 3.1416$, more accurately than previously known.*

His masterpiece, The Aryabhatiya, *written in A.D. 499, deals with astronomy, plane and spherical trigonometry, algebra, quadratic equations, sums of powers of the first n natural numbers, and a table of sines. It was translated into Arabic around 800 and into Latin in the thirteenth century.*

In recognition of his outstanding contributions to astronomy and mathematics, India's first satellite was named Aryabhata.

- Does every LDE have a solution?
- If not, under what conditions does an LDE have a solution?
- If an LDE is solvable, what is the maximum number of solutions it can have?

The first question can be answered easily. Consider the LDE $2x + 4y = 5$. No matter what the integers x and y are, the LHS $2x + 4y$ is always even, whereas the RHS is always odd, so the LDE has no solution. Thus, *not* every LDE has a solution.

Next, we establish a necessary and sufficient condition for the LDE $ax + by = c$ to be solvable. Its proof, in fact, provides a formula for an arbitrary solution, when it is solvable.

The Indian mathematician Aryabhata provided a complete solution of the LDE in two variables. A portion of the proof of Theorem 3.19, which is long, but fairly straightforward, is a variation of his method.

THEOREM 3.19 The LDE $ax + by = c$ is solvable if and only if $d|c$, where $d = (a, b)$. If x_0, y_0 is a particular solution of the LDE, then all its solutions are given by

$$x = x_0 + \left(\frac{b}{d}\right)t \quad \text{and} \quad y = y_0 - \left(\frac{a}{d}\right)t$$

where t is an arbitrary integer.

PROOF

The proof consists of four parts:

- If the LDE is solvable, then $d|c$.

- Conversely, if $d|c$, then the LDE is solvable.
- $x = x_0 + \left(\frac{b}{d}\right)t$ and $y = y_0 - \left(\frac{a}{d}\right)t$ is a solution of the LDE.
- Every solution of the LDE is of this form.

We shall prove each part one by one in that order.

- *To prove that if the LDE is solvable, then* $d|c$:
  Suppose $x = \alpha, y = \beta$ is a solution. Then

$$a\alpha + b\beta = c \tag{3.6}$$

  Since $d = (a, b)$, $d|a$ and $d|b$, so $d|(a\alpha + b\beta)$ by Theorem 2.4; that is, $d|c$.
- *To prove that if* $d|c$, *then the LDE is solvable*:
  Suppose $d|c$. Then $c = de$ for some integer e. Since $d = (a, b)$, by Theorem 3.5, there exist integers r and s such that $ra + sb = d$. Multiplying both sides of this equation by e yields

$$rae + sbe = de$$

  That is,

$$a(re) + b(se) = c$$

  Thus, $x_0 = re$ and $y_0 = se$ is a solution of the LDE; that is, it is solvable.
- *To show that* $x = x_0 + \left(\frac{b}{d}\right)t$ and $y = y_0 - \left(\frac{a}{d}\right)t$ *is a solution*: We have

$$\begin{aligned} ax + by &= a\left[x_0 + \left(\frac{b}{d}\right)t\right] + b\left[y_0 - \left(\frac{a}{d}\right)t\right] \\ &= (ax_0 + by_0) + \frac{abt}{d} - \frac{abt}{d} \\ &= ax_0 + by_0 \\ &= c \end{aligned}$$

  Thus, $x = x_0 + \left(\frac{b}{d}\right)t$ and $y = y_0 - \left(\frac{a}{d}\right)t$ is a solution for any integer t.
- *To show that every solution* x', y' *is of the desired form*:
  Since x_0, y_0 and x', y' are solutions of the LDE, we have:

$$\begin{aligned} &ax_0 + by_0 = c \quad \text{and} \quad ax' + by' = c \\ &ax_0 + by_0 = ax' + by' \end{aligned}$$

  Therefore,

$$a(x' - x_0) = b(y_0 - y') \tag{3.7}$$

Divide both sides of this equation by d:

$$\left(\frac{a}{d}\right)(x' - x_0) = \left(\frac{b}{d}\right)(y_0 - y')$$

By Theorem 3.4, $(a/d, b/d) = 1$, so, by Corollary 3.4, $\frac{b}{d}|(x' - x_0)$ and hence $x' - x_0 = \left(\frac{b}{d}\right)t$ for some integer t.

That is,

$$x' = x_0 + \left(\frac{b}{d}\right)t$$

Now substituting for $x' - x_0$ in equation (3.7), we have

$$a\left(\frac{b}{d}\right)t = b(y_0 - y')$$

$$\left(\frac{a}{d}\right)t = y_0 - y'$$

$$y' = y_0 - \left(\frac{a}{d}\right)t$$

Thus, every solution of the LDE is of the desired form. ■

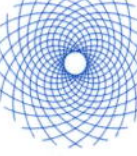

It follows by this theorem that if the LDE $ax + by = c$ is solvable, then it has infinitely many solutions. They are given by the **general solution** $x = x_0 + \left(\frac{b}{d}\right)t$ and $y = y_0 - \left(\frac{a}{d}\right)t$, t being an arbitrary integer. By giving different values to t, we can find any number of particular solutions.

This theorem has an interesting and useful corollary.

COROLLARY 3.13 If $(a, b) = 1$, then the LDE $ax + by = c$ is solvable and the general solution is given by $x = x_0 + bt$, $y = y_0 - at$, where x_0, y_0 is a particular solution. ■

The following three examples demonstrate Theorem 3.19.

EXAMPLE 3.19 Determine whether the LDEs $12x + 18y = 30$, $2x + 3y = 4$, and $6x + 8y = 25$ are solvable.

SOLUTION

- $(12, 18) = 6$ and $6|30$, so the LDE $12x + 18y = 30$ has a solution.
- $(2, 3) = 1$, so by Corollary 3.13, the LDE has a solution.
- $(6, 8) = 2$, but $2 \nmid 25$, so the LDE $6x + 8y = 25$ is not solvable. ■

The following two examples illustrate in detail how to find the general solution of an LDE in two variables.

EXAMPLE 3.20 Find the general solution to Mahavira's puzzle in Example 3.17.

SOLUTION

The LDE in Mahavira's puzzle is $63x - 23y = -7$. Since $(63, 23) = 1$, by Corollary 3.13, the LDE has a solution.

To find a particular solution x_0, y_0, first we express the gcd 1 as a linear combination of 63 and 23. To accomplish this, we apply the euclidean algorithm:

$$\begin{aligned}
63 &= 2 \cdot 23 + 17 \\
23 &= 1 \cdot 17 + 6 \\
17 &= 2 \cdot 6 + 5 \\
6 &= 1 \cdot 5 + 1 \\
5 &= 5 \cdot 1 + 0
\end{aligned}$$

Now, use the first four equations in reverse order:

$$\begin{aligned}
1 &= 6 - 1 \cdot 5 \\
&= 6 - 1(17 - 2 \cdot 6) \\
&= 3 \cdot 6 - 1 \cdot 17 \\
&= 3(23 - 1 \cdot 17) - 1 \cdot 17 \\
&= 3 \cdot 23 - 4 \cdot 17 \\
&= 3 \cdot 23 - 4(63 - 2 \cdot 23) \\
&= (-4) \cdot 63 + 11 \cdot 23
\end{aligned}$$

Multiply both sides of this equation by -7 (why?):

$$\begin{aligned}
-7 &= (-7)(-4) \cdot 63 + (-7) \cdot 11 \cdot 23 \\
&= 63 \cdot 28 - 23 \cdot 77
\end{aligned}$$

which shows $x_0 = 28$, $y_0 = 77$ is a particular solution of the LDE. [See part (2) of the proof of the theorem. Also, note that we obtained this solution in Example 3.17.]

Therefore, by Corollary 3.13, the general solution is given by $x = x_0 + bt = 28 - 23t$ and $y = y_0 - at = 77 - 63t$, where t is an arbitrary integer. ■

EXAMPLE 3.21 Using Theorem 3.19, find the general solution of the hundred fowls puzzle in Example 3.18.

SOLUTION

By Example 3.18, we have

$$x + y + z = 100 \tag{3.8}$$

$$5x + 3y + \frac{z}{3} = 100 \tag{3.9}$$

Eliminating z between these two equations, we obtain the LDE

$$7x + 4y = 100 \tag{3.10}$$

Notice that $(7, 4) = 1$ and by trial and error,

$$1 = (-1) \cdot 7 + 2 \cdot 4$$

Now multiply both sides of this equation by 100:

$$100 = (-100) \cdot 7 + 200 \cdot 4$$

Therefore, $x_0 = -100$, $y_0 = 200$ is a particular solution of the LDE (3.10). Thus, by Corollary 3.13, the general solution of the LDE (3.10) is $x = -100 + 4t'$, $y = 200 - 7t'$. Then

$$\begin{aligned} z &= 100 - x - y \\ &= 100 - (-100 + 4t') - (200 - 7t') \\ &= 3t' \end{aligned}$$

Thus, the general solution of the puzzle is $x = -100 + 4t'$, $y = 200 - 7t'$, $z = 3t'$, where t' is an arbitrary integer.

This solution can be rewritten in such a way that we can recover the general solution obtained earlier. We have

$$\begin{aligned} x &= -100 + 4t' = 4(-25 + t') \\ &= 4t, \quad \text{where } t = t' - 25 \end{aligned}$$

Then

$$y = 200 - 7(t + 25) = 25 - 7t$$

and

$$z = 3(t + 25) = 75 + 3t$$

Thus, the general solution is also given by $x = 4t$, $y = 25 - 7t$, and $z = 75 + 3t$, where t is an arbitrary integer. ■

EXAMPLE 3.22 Solve, if possible, Mahavira's puzzle if there were 24 travelers.

SOLUTION

With 24 travelers, the diophantine equation becomes $63x - 24y = -7$. Since $(63, 24) = 3$ and $3 \nmid 7$, the diophantine equation has no integral solutions, so the puzzle has no solutions. ■

We now pursue a fascinating puzzle.

The Monkey and Coconuts Puzzle

The October 9, 1926, issue of *The Saturday Evening Post* carried a fascinating puzzle by Ben Ames Williams, titled *Coconuts*. It concerned a building contractor desperate to prevent a tough competitor from getting a lucrative contract. A shrewd employee of the contractor, knowing their competitor's love for recreational mathematics, gave him a problem so intriguing that he became obsessed with solving it and forgot to enter his bid before the deadline.

Williams' problem is actually a slightly modified version of the ancient problem described in the following example. We leave his version as an exercise (see Exercise 40).

EXAMPLE 3.23 Five sailors and a monkey are marooned on a desert island. During the day they gather coconuts for food. They decide to divide them up in the morning, but first they retire for the night. While the others sleep, one sailor gets up and divides them into five equal piles, with one left over, which he throws out for the monkey. He hides his share, puts the remaining coconuts together, and goes back to sleep. Later a second sailor gets up, divides the pile into five equal shares with one coconut left over, which he discards for the monkey. One by one the remaining sailors repeat the process. In the morning, they divide the pile equally among them with one coconut left over, which they throw out for the monkey. Find the smallest possible number of coconuts in the original pile.

SOLUTION

Let n denote the number of coconuts in the original pile. Let u, v, w, x, and y denote the number of coconuts each sailor took after each division, and let z be the number of coconuts each received after the final division. Then

$$\begin{aligned} n &= 5u + 1 \\ 4u &= 5v + 1 \\ 4v &= 5w + 1 \\ 4w &= 5x + 1 \\ 4x &= 5y + 1 \\ 4y &= 5z + 1 \end{aligned}$$

These equations yield the LDE

$$15625z - 1024n = -11529 \tag{3.11}$$

Because $(15625, 1024) = 1$, the LDE has a solution. Using the euclidean algorithm, we can verify that $1 = 313 \cdot 15625 - 4776 \cdot 1024$, so

$$15625 \cdot [(-11529) \cdot 313] - 1024 \cdot [4776 \cdot (-11529)] = -11529$$

That is,

$$15625 \cdot (-3608577) - 1024 \cdot (-55062504) = -11529$$

So $z_0 = -3608577$ and $n_0 = -55062504$ is a particular solution, and the general solution is $z = -3608577 - 1024t$, $n = -55062504 - 15625t$, t being an arbitrary integer.

Because $n > 0$, $-55062504 - 15625t > 0$, so $t < -\dfrac{55062504}{15625}$; that is, $t < -3524$. Because n is a minimum when t is a maximum, $t = -3525$. Then $n = -55062504 - 15625 \cdot (-3525) = 15621$. Thus, the least number of coconuts in the original pile is 15,621. ■

We shall return to this puzzle in the next section.

Now, we make a geometric interpretation of the general solution in Theorem 3.20.

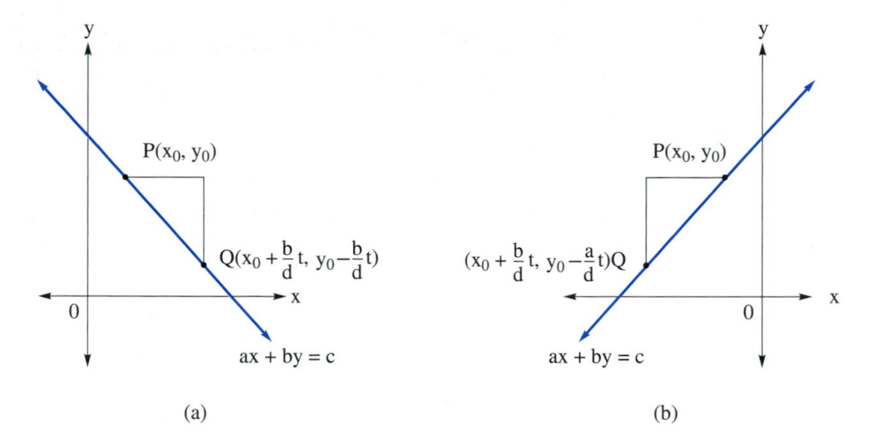

Figure 3.4

Recall that the solutions of the LDE $ax + by = c$ are the lattice points that lie on the line $ax + by = c$. (Assume $a, t > 0$ for convenience.) The slope of this line is

$$-\frac{a}{b} = -\frac{a/d}{b/d}$$

where $d = (a, b)$. Locate the point $P(x_0, y_0)$, on the line. Then move $\frac{b}{d}t$ units to the right and $\frac{a}{d}t$ units down if $b > 0$ (see Figure 3.4a), and $\frac{|b|}{d}t$ units to the left and $\frac{a}{d}t$ units down if $b < 0$ (see Figure 3.4b). In either case, the procedure determines a lattice point on the line. The cases $a < 0$ and $t < 0$ can be interpreted similarly.

Euler's Method for Solving LDEs

Euler devised a method for solving LDEs that employs the division algorithm, but not the euclidean algorithm.

EXAMPLE 3.24 Solve the LDE $1076x + 2076y = 3076$ by Euler's method.

SOLUTION

Since $(1076, 2076) = 4$ and $4|3076$, the LDE has infinitely many solutions. Euler's method involves solving the LDE for the variable with the smaller coefficient, x in this case:

$$\begin{aligned} x &= \frac{-2076y + 3076}{1076} \\ &= -y + 2 + \frac{-1000y + 924}{1076}, \quad \text{by the division algorithm} \qquad (3.12) \end{aligned}$$

Let $u = \dfrac{-1000y + 924}{1076}$. (Notice that u is an integer. Why?) This yields the LDE $1076u + 1000y = 924$. (This LDE has smaller coefficients than the original one.) Solve for y:

$$\begin{aligned} y &= \frac{-1076u + 924}{1000} \\ &= -u + \frac{-76u + 924}{1000}, \quad \text{by the division algorithm} \end{aligned} \tag{3.13}$$

Let $v = \dfrac{-76u + 924}{1000}$, so $76u + 1000v = 924$; solve for u:

$$\begin{aligned} u &= \frac{-1000v + 924}{76} \\ &= -13v + 12 + \frac{-12v + 12}{76}, \quad \text{by the division algorithm} \end{aligned} \tag{3.14}$$

Let $w = \dfrac{-12v + 12}{76}$, so $12v + 76w = 12$; solve for v:

$$v = \frac{-76w + 12}{12} = -6w + 1 - \frac{w}{3}$$

Because v is an integer, $w/3$ must be an integer, so we let $w/3 = t$.

To obtain a particular solution, we let $t = 0$; then $w = 0$ and work through the chain of equations (3.12), (3.13), and (3.14) in the reverse order:

$$\begin{aligned} v &= -6w + 1 - \frac{w}{3} &&= -6(0) + 1 - 0 &&= 1 \\ u &= \frac{-1000v + 924}{76} &&= \frac{-1000 + 924}{76} &&= -1 \\ y &= \frac{-1076u + 924}{1000} &&= \frac{1076 + 924}{1000} &&= 2 \\ x &= \frac{-2076y + 3076}{1076} &&= \frac{-4152 + 3076}{1076} &&= -1 \end{aligned}$$

You can verify that $x_0 = -1$, $y_0 = 2$ is in fact a solution of the LDE.

To find the general solution, with t as an arbitrary integer, use successive substitution, again in the reverse order:

$$\begin{aligned}
w &= 3t \\
v &= -6w + 1 - \frac{w}{3} &&= -19t + 1 \\
u &= -13v + 12 + w = 250t - 1 \\
y &= -u + v &&= -269t + 2 \\
x &= -y + 2 + u &&= 519t - 1
\end{aligned}$$

Thus, the general solution is $x = 519t - 1$, $y = -269t + 2$. (We can confirm this using Theorem 3.20.) ■

Next, we investigate LDEs with Fibonacci coefficients.

Fibonacci Numbers and LDEs

Consider the LDE $F_{n+1}x + F_n y = c$. By Theorem 3.1, $(F_{n+1}, F_n) = 1$, so the LDE is solvable.

By Cassini's formula, $F_{n+1}F_{n-1} - F_n^2 = (-1)^n$. Suppose n is even. Then $F_{n+1}F_{n-1} - F_n^2 = 1$; so $F_{n+1}(cF_{n-1}) + F_n(-cF_n) = c$. Thus, $x_0 = cF_{n-1}, y_0 = -cF_n$ is a particular solution of the LDE $F_{n+1}x + F_n y = c$.

On the other hand, let n be odd. Then $F_{n+1}(-F_{n-1}) + F_n^2 = 1$; so $F_{n+1}(-cF_{n-1}) + F_n(cF_n) = c$. Thus, $x_0 = -cF_{n-1}, y_0 = cF_n$ is a particular solution of the LDE $F_{n+1}x + F_n y = c$.

For example, consider the LDE $34x + 21y = 17$. Since $F_9F_7 - F_8^2 = 34 \cdot 13 - 21^2 = (-1)^8$ and $c = 17$, it follows that $x_0 = cF_7 = 17 \cdot 13 = 221, y_0 = -cF_8 = -17 \cdot 21 = -357$ is a particular solution. So the general solution is $x = x_0 + bt = 221 + 21t$, $y = y_0 - at = -357 - 34t$.

We now study an intriguing puzzle† whose solution involves solving LDEs.

EXAMPLE 3.25 A six-digit positive integer is cut up in the middle into two three-digit numbers. If the square of their sum yields the original number, find the number.

SOLUTION

Let N be the six-digit number, and let a and b be the two three-digit numbers. Then $N = 1000a + b$ and $N = (a + b)^2$.

† Based on A. Dunn (ed.), *Mathematical Bafflers*, Dover, New York, 1980, p. 183.

Let $a + b = c$. Then

$$(1000a + b) - (a + b) = c^2 - c$$
$$999a = c(c - 1)$$

If $c = 999$, then $999a = 999 \cdot 998$, so $a = 998$ and $b = 1$. Then $N = 998,001$ works: $998001 = (998 + 001)^2$.

If $c \neq 999$, then $999|c(c-1)$, where $999 = 27 \cdot 37$, $(27, 37) = 1 = (c, c-1)$. Therefore, $27|c$ and $37|(c-1)$, or $27|(c-1)$ and $37|c$.

case 1 Suppose $27|c$ and $37|(c-1)$. Then $c = 27x$ and $c - 1 = 37y$ for some integers x and y. These two equations yield the LDE $27x - 37y = 1$. Solving this, we get $x = 11 - 37t$ and $y = 8 - 27t$. Therefore, $c = 27(11 - 37t) = 297 - 999t$. Unfortunately, no t yields a value for c such that c^2 has the desired property.

case 2 Suppose $27|(c-1)$ and $37|c$. Then $37y - 27x = 1$. Solving this we get $y = -8 - 27t'$ and $x = -11 - 37t'$; so $c = 37(-8 - 27t') = -296 - 999t' = 703 + 999t''$, t'' being arbitrary; $t'' = 0$ yields $c = 703$. Then $999a = 703 \cdot 702$, so $a = 494$ and $b = 209$. Then $N = 494,209$ has the desired property: $494209 = (494 + 209)^2$. No other value of t'' produces such a number.

Thus, there are two six-digit positive integers satisfying the required property: 998,001 and 494,209. ■

The following theorem shows that Theorem 3.19 can be extended to LDEs containing three or more unknowns. Its proof depends on induction. See Exercises 41 and 42.

THEOREM 3.20 The LDE $a_1x_1 + a_2x_2 + \cdots + a_nx_n = c$ is solvable if and only if $(a_1, a_2, \ldots, a_n)|c$. When it is solvable, it has infinitely many solutions. ■

The following two examples illustrate this theorem.

EXAMPLE 3.26 Determine whether the LDEs $6x + 8y + 12z = 10$ and $6x + 12y + 15z = 10$ are solvable.

SOLUTION

- Since $(6, 8, 12) = 2$ and $2|10$, the LDE $6x + 8y + 12z = 10$ is solvable.
- $(6, 12, 15) = 3$, but $3 \nmid 10$, so the $6x + 12y + 15z = 10$ has no integral solutions. ■

We conclude this section with an example that demonstrates solving a LDE in three variables.

EXAMPLE 3.27 Find the general solution of the LDE $6x + 8y + 12z = 10$.

SOLUTION

By the preceding example, the LDE has infinitely many solutions. Since $8y + 12z$ is a linear combination of 8 and 12, it must be a multiple of $(8,12) = 4$; so we let

$$8y + 12z = 4u \tag{3.15}$$

This leads to a LDE in two variables: $6x + 4u = 10$. Solving this, we get $x = 5 + 2t$ and $u = -5 - 3t$, with t as an arbitrary integer (verify this).

Now substitute for u in equation (3.15):

$$8y + 12z = 4(-5 - 3t)$$

Notice that $(8, 12) = 4$ and $4 = 2 \cdot 8 + (-1) \cdot 12$. Therefore,

$$4(-5 - 3t) = (-10 - 6t) \cdot 8 + (5 - 3t) \cdot 12$$

So, by Theorem 3.19, the general solution of equation (3.15) is $y = -10 - 6t + 3t'$, $z = 5 + 3t - 2t'$. Thus, the general solution of the given linear diophantine equation is

$$x = 5 + 2t$$
$$y = -10 - 6t + 3t'$$
$$z = 5 + 3t - 2t'$$

where t and t' are arbitrary integers. ■

Obviously, this method of reducing the number of unknowns can be extended to LDEs with any finite number of unknowns. See Exercises 33–36 for additional practice.

EXERCISES 3.5

1. Using the biographical sketch of Diophantus on p. 189, determine his age at the time of his death.

Using Theorem 3.19, determine whether each LDE is solvable.

2. $12x + 16y = 18$
3. $14x + 16y = 15$
4. $12x + 13y = 14$
5. $28x + 91y = 119$
6. $1776x + 1976y = 4152$
7. $1076x + 2076y = 1155$

Find the general solution of each LDE using Theorem 3.19.

8. $2x + 3y = 4$
9. $12x + 16y = 20$
10. $12x + 13y = 14$
11. $15x + 21y = 39$
12. $28x + 91y = 119$
13. $1776x + 1976y = 4152$
14. Verify the general solution of the LDE in Example 3.24 using Theorem 3.19.

15–20. Using Euler's method, solve the LDEs in Exercises 8–13.

21. A pile of mangoes was collected. The king took one-sixth, the queen one-fifth of the remainder, the three princes one-fourth, one-third, and one-half of the successive remainders, and the youngest child took the three remaining mangoes. Find the number of mangoes in the pile. (Mahavira)
22. The total cost of nine citrons and seven fragrant wood apples is 107 coins; the cost of seven citrons and nine fragrant wood apples is 101 coins. Find the cost of a citron and a wood apple. (Mahavira)
23. A person bought some 12-cent stamps and some 15-cent stamps. The postal clerk told her the total cost was \$5.50. Is that possible?
24. A piggy bank contains nickels and dimes for a total value of \$3.15. Find the possible number of nickels and dimes.
25. A fruit basket contains apples and oranges. Each apple costs 65¢ and each orange 45¢, for a total of \$8.10. Find the minimum possible number of apples in the basket.

Solve each Fibonacci LDE.

26. $144x + 89y = 23$
27. $233x - 144y = 19$
28. Verify that $x = 5 - 4t$, $y = -10 + 12t + 3t'$, $z = 5 - 6t - 2t'$ is a solution of the LDE $6x + 8y + 12z = 10$ for any integers t and t'.

Determine whether each LDE is solvable.

29. $2x + 3y + 4z = 5$
30. $8x + 10y + 16z = 25$
31. $12x + 30y - 42z = 66$
32. $76w + 176x + 276y + 376z = 476$

Solve the following LDEs.

33. $x + 2y + 3z = 6$
34. $2x - 3y + 4z = 5$
35. $6x + 12y - 15z = 33$
36. $12x + 30y - 42z = 66$
37. A collection plate contains nickels, dimes, and quarters. The total value is \$4, and there are twice as many quarters as there are dimes. Find the possible number of combinations of each kind, if there are more quarters than nickels.

Mrs. Hall bought 10 hot dogs, 15 cheeseburgers, and 20 sandwiches for a pool party. The total bill was \$73.50.

38. Find the general solution.
39. Find the possible combinations of the numbers of hot dogs, cheeseburgers, and sandwiches she could have bought if a cheeseburger cost 65¢ more than a hot dog.
40. (Williams' version) Five sailors and a monkey are marooned on a desert island. During the day they gather coconuts for food. They decide to divide them up in the morning and retire for the night. While the others are asleep, one sailor gets up and divides them into equal piles, with one left over, which he throws out for the monkey. He hides his share, puts the remaining coconuts together, and goes back to sleep. Later a second sailor gets up, divides the pile into five equal shares with one coconut left over which he discards for the monkey. Later the remaining sailors repeat the process. Find the smallest possible number of coconuts in the original pile.

Prove each.

41. The LDE $\sum_{i=1}^{n} a_i x_i = c$ is solvable if and only if $(a_1, a_2, \ldots, a_n)|c$.
42. If the LDE $\sum_{i=1}^{n} a_i x_i = c$ is solvable, then it has infinitely many solutions.
(*Hint*: Use induction.)

CHAPTER SUMMARY

One of the most celebrated results in number theory is the fundamental theorem of arithmetic, which is indeed the cornerstone of the subject. Two concepts indispensable to its development are the gcd and the lcm. We developed a necessary and sufficient condition for an LDE to be solvable, and the general solution when it is solvable.

Greatest Common Divisor (gcd)

- The gcd (a, b) of two positive integers a and b is the largest positive integer that divides both. (p. 155)
- A positive integer d is the gcd of a and b if:
  - $d|a$ and $d|b$; and
  - if $d'|a$ and $d'|b$, then $d' \leq d$, where d' is a positive integer. (p. 156)
- Two positive integers a and b are relatively prime if $(a, b) = 1$. (p. 156)
- $f_0 f_1 \cdots f_{n-1} = f_n - 2$, where $n \geq 1$. (p. 157)
- Let m and n be distinct nonnegative integers. Then $(f_m, f_n) = 1$. (p. 157)
- If $(a, b) = d$, then $(a/d, b/d) = 1$ and $(a, a - b) = d$. (p. 158)
- $d = (a, b)$ is the least positive linear combination of a and b. (p. 159)
- If d' is a common divisor of a and b, then $d'|(a, b)$. (p. 160)
- A positive integer d is the gcd of a and b if:
  - $d|a$ and $d|b$; and
  - if $d'|a$ and $d'|b$, then $d'|d$, where d' is a positive integer. (p. 161)
- $(ac, bc) = c(a, b)$ (p. 161)
- The positive integers a and b are relatively prime if and only if $\alpha a + \beta b = 1$ for some integers α and β. (p. 161)
- If $(a, b) = 1$ and $a|bc$, then $a|c$. (p. 162)
- If $a|c$, $b|c$, and $(a, b) = 1$, then $ab|c$. (p. 162)
- $(a_1, a_2, \ldots, a_n) = ((a_1, a_2, \ldots, a_{n-1}), a_n)$ (p. 163)

Euclidean Algorithm

The euclidean algorithm for finding (a, b) is a successive application of the division algorithm and is based on the following result, where $a \geq b$:

- Let $r = a \bmod b$. Then $(a, b) = (b, r)$. (p. 166)
- The algorithm provides a systematic method for expressing (a, b) as a linear combination of a and b. (p. 170)
- The number of divisions needed to compute (a, b) by the euclidean algorithm is no more than five times the number of decimal digits in b, where $a \geq b \geq 2$. (Lamé's theorem) (p. 172)

Fundamental Theorem of Arithmetic

- If p is a prime and $p|ab$, then $p|a$ or $p|b$. (p. 174)
- If p is a prime such that $p|a_1a_2 \cdots a_n$, where $a_1, a_2, \ldots, a_n$ are positive integers, then $p|a_i$ for some i, where $1 \le i \le n$. (p. 174)
- If $p, q_1, q_2, \ldots, q_n$ are primes such that $p|q_1q_2 \cdots q_n$, then $p = q_i$ for some i, where $1 \le i \le n$. (p. 174)
- Every positive integer $n \ge 2$ either is a prime or can be expressed as a product of primes. The factorization into primes is unique except for the order of the factors. (p. 174)

Canonical Decomposition of a Positive Integer N

- $N = p_1^{a_1} p_2^{a_2} \cdots p_k^{a_k}$, where $p_1, p_2, \ldots, p_k$ are distinct primes, $p_1 < p_2 < \cdots < p_k$ and $a_i \ge 0$. (p. 176)

Dirichlet's Theorem

If a and b are relatively prime, then the arithmetic sequence $a, a+b, a+2b, a+3b, \ldots$ contains infinitely many primes. (p. 182)

Least Common Multiple (lcm)

- The lcm $[a, b]$ of two positive integers a and b is the least positive integer divisible by both. (p. 184)
- A positive integer $m = [a, b]$ if
  - $a|m$ and $b|m$; and
  - if $a|m'$ and $b|m'$, then $m \le m'$, where m' is a positive integer. (p. 184)
- $[a, b] = ab/(a, b)$ (p. 185)
- $[a_1, a_2, \ldots, a_n] = [[a_1, a_2, \ldots, a_{n-1}], a_n]$ (p. 187)
- If $a_1, a_2, \ldots, a_n$ are pairwise relatively prime, then $[a_1, a_2, \ldots, a_n] = a_1a_2 \cdots a_{n-1}a_n$. (p. 187)

Linear Diophantine Equations (LDEs)

- An LDE in two variables x and y is of the form $ax + by = c$, where a, b, and c are integers. (p. 189)
- It is solvable if and only if $d|c$, where $d = (a, b)$. (p. 192)
- If $d|c$, the general solution of the LDE is $x = x_0 + (b/d)t$, $y = y_0 - (a/d)t$, where x_0, y_0 is a particular solution. (p. 192)
- LDEs can be solved by Euler's method. (p. 199)

REVIEW EXERCISES

Find the gcd of each pair of integers.

1. $2 \cdot 3^2 \cdot 5^3, 2^5 \cdot 3^3 \cdot 5$
2. $2 \cdot 3^2 \cdot 5 \cdot 7^3, 2^3 \cdot 3 \cdot 5^2 \cdot 11^3$
3. 32, 48
4. 56, 260

Using the euclidean algorithm, find the gcd of each pair of integers.

5. 28, 12
6. 784, 48
7. 1947, 63
8. 5076, 1076

Using recursion, find the gcd of the given integers.

9. 16, 20, 36, 48
10. 20, 32, 56, 68
11. 28, 48, 68, 78
12. 24, 36, 40, 60, 88

Express the gcd of the given numbers as a linear combination of the numbers.

13. 14, 18
14. 12, 20
15. 12, 18, 20
16. 10, 12, 14, 18

Find the positive factors of each, where p and q are distinct primes.

17. 98
18. 1575
19. $p^i q$
20. $p^i q^j$

Find the canonical decomposition of each.

21. 2000
22. 3230
23. 1771
24. 4076

Find the number of trailing zeros in the decimal value of each integer.

25. 260!
26. 345!
27. 1400!
28. 1947!

Find the number of trailing zeros in the binary representation of each integer.

29. 39
30. 191
31. 243
32. 576

Find the number of trailing zeros in the **ternary** (base three) representation of each integer.

33. 45
34. 61
35. 118
36. 343

Find the lcm of each pair of integers.

37. $2 \cdot 3^2 \cdot 5, 2 \cdot 3 \cdot 7$
38. $3 \cdot 7^2 \cdot 11, 2^2 \cdot 5 \cdot 7$
39. 48, 66
40. 42, 78

41. The lcm of two consecutive positive integers is 812. Find them.
42. The lcm of twin primes is 899. Find them.

Using Theorem 3.17, find the lcm of the given integers.

43. 48, 64
44. 56, 76
45. 70, 90
46. 123, 243

47–50. Find the lcm of the integers in Exercises 9–12.

Prove each, where a, b, c, d, m, and n are positive integers.

51. Let p be a prime such that $p|a^n$. Then $p^n|a^n$.
52. If $a|m$ and $b|m$, then $[a, b]|m$.
53. The product of three consecutive integers is divisible by 6.
54. The gcd of two consecutive integers is 1.
55. The gcd of twin primes is 1.
56. If $d|ab$, $d|ac$, and $(b, c) = 1$, then $d|a$.
57. If $a|b$ and $c|d$, then $(a, c)|(b, d)$.
58. If $a|b$ and $c|d$, then $[a, c]|[b, d]$.
59. The product of two integers of the form $4n + 3$ is of the form $4n + 1$.
60. The product of n integers of the form $4k + 1$ is also of the same form.
61. A positive integer is a cube if and only if each exponent in its canonical decomposition is divisible by 3.
62. There is an infinite number of primes of the form $6n + 1$.
63. There is an infinite number of primes of the form $7n + 4$.
64. $(ca_1, ca_2, \dots, ca_n) = c(a_1, a_2, \dots, a_n)$
65. Let p be a prime such that $p|n!$. Then the exponent of p in the canonical decomposition of $n!$ is $\lfloor n/p \rfloor + \lfloor n/p^2 \rfloor + \lfloor n/p^3 \rfloor + \cdots$. (*Note*: This sum is finite since $\lfloor n/p^m \rfloor = 0$ when $p^m > n$.)
66. Let $b = a + 2$, where a is odd and $3 \nmid ab$. Then $12|(a + b)$. (M. Beiler, 1967)
67. $([a, b], [b, c], [c, a]) = [(a, b), (b, c), (c, a)]$
68. $[a, b, c] = \dfrac{abc(a, b, c)}{(a, b), (b, c), (c, a)}$

*69. $6^n n!|(3n)!$, where $n \geq 0$. (C. W. Trigg, 1968)

Using Exercise 65, find the canonical decomposition of each.

70. 12!
71. 15!
72. 18!
73. 23!

Determine whether each LDE is solvable.

74. $24x + 52y = 102$
75. $76x + 176y = 276$

Find the general solution of each LDE.

76. $12x + 20y = 28$
77. $76x + 176y = 276$

78–79. Solve the LDEs in Exercises 76 and 77 by Euler's method.
80. Solve the LDE in Example 3.23 by Euler's method.

81. Solve the LDE in Exercise 40 in Section 3.5 by Euler's method.
82. A farmer bought some calves and sheep for \$39,500, at \$475 a calf and \$275 a sheep. If she bought more calves than sheep, find the minimum number of calves she must have bought.
83. A shopper bought some apples, oranges, and pears, a total of a dozen fruits. They cost 75¢, 30¢, and 60¢ apiece respectively, for a total of \$6.30. If he bought at least one fruit of each kind, how many apples, oranges, and pears did he buy?

SUPPLEMENTARY EXERCISES

In Exercises 1–10, n is a positive integer and $n^\star = [1, 2, 3, \ldots, n]$.

1. Find $n^\star$ for $n = 5, 6, 7$, and 8.
2. Let $n = p^k$, where p is a prime and k is a positive integer. Prove that $n^\star = p(n-1)^\star$.
3. Using Exercise 2, compute $9^\star$.
4. Using the fact that $n^\star = \prod_{p \le n} p^e$, where p^e denotes the largest prime-power $\le n$, compute $9^\star$, $10^\star$, and $11^\star$.
5. Show that $(n+1)^\star + 2, (n+1)^\star + 3, \ldots, (n+1)^\star + (n+1)$ are consecutive composite numbers.
6. Using Exercise 5, find six consecutive composite numbers.
7. Compute $n^\star + 1$ for $1 \le n \le 7$ and make a conjecture.
8. Is $8^\star + 1$ a prime?
9. Compute $n^\star - 1$ for $3 \le n \le 8$ and make a conjecture.
10. Is $9^\star - 1$ a prime?
11. Let a, b, and c be positive integers such that $a + b = c$. Let $m = [a, b]$. Prove that $(c, m) = (a, b)$. (H. H. Berry, 1951)

Let a, b, m, and n be any positive integers, where $a > b$. Prove each.

*12. $(a^m - 1, a^n - 1) = a^{(m,n)} - 1$

**13. $\left(\dfrac{a^n - b^n}{a - b}, a - b\right) = (n(a,b)^{n-1}, a - b)$. (T. M. Apostol, 1980)

14. Use Exercise 13 to deduce that if p is a prime and $(a, b) = 1$, then $\left(\dfrac{a^p - b^p}{a - b}, a - b\right) = 1$ or p. (T. M. Apostol, 1980)

**15. Let m and n be any positive integers, and let a and b be relatively prime integers with $a > b$. Prove that $(a^m - b^m, a^n - b^n) = a^{(m,n)} - b^{(m,n)}$. (T. M. Apostol, 1981)

COMPUTER EXERCISES

Write a program to do each task.

1. Read in a positive integer n and determine the number of trailing zeros in each.
   (a) The decimal value of $n!$.
   (b) The binary representation of n.
   (c) The ternary expansion of n.
2. Read in two positive integers a and b, and find their gcd using the euclidean algorithm.
3. Read in an integer n (≥ 0) and a positive integer k, and find the first k primes of the form $4n+1$.
4. Read in an integer n (≥ 0) and a positive integer k, and find the first k primes of the form $4n+3$.
5. Read in a positive integer n (≥ 2) and n positive integers. Using recursion and the euclidean algorithm, find their gcd.
6. Read in a positive integer n, and print all powerful numbers $\leq n$.
7. Read in integers a, b, and c, and check if the LDE $ax+by=c$ is solvable.

ENRICHMENT READINGS

1. M. P. Cohen and W. A. Juraschek, "GCD, LCM, and Boolean Algebra," *Mathematics Teacher*, 69 (Nov. 1976), 602–605.
2. U. Dudley, *Elementary Number Theory*, W. H. Freeman, New York, 1969.
3. M. W. Kappel, "Backtracking the Euclidean Algorithm," *Mathematics Teacher*, 69 (Nov. 1976), 598–600.
4. T. Koshy, "The Euclidean Algorithm via Matrices and a Calculator," *The Mathematical Gazette*, 80 (Nov. 1996), 570–574.
5. T. Koshy, *Fibonacci and Lucas Numbers with Applications*, Wiley, New York, 2001.
6. C. S. Ogilvy and J. T. Anderson, *Excursions in Number Theory*, Dover, New York, 1988.
7. M. Polezzi, "A Geometrical Method for Finding an Explicit Formula for the Greatest Common Divisor," *The American Mathematical Monthly*, 104 (May 2000), 445–446.

Congruences

The invention of the symbol ≡ by Gauss affords a striking example of the advantage which may be derived from an appropriate notation, and marks an epoch in the development of the science of arithmetic.
—G. B. MATHEWS

This chapter investigates the congruence relation, an extremely useful and powerful number-theoretic relation used throughout number theory, and its fundamental properties. LDEs and the congruence relation are closely related, as you will see in Section 4.2. Furthermore, we discuss a practical factoring algorithm based on gcd and congruence.

4.1 Congruences

One of the most remarkable relations in number theory is the congruence relation, introduced and developed by the German mathematician Karl Friedrich Gauss, who is ranked with Archimedes (287–212 B.C.) and Isaac Newton (1642–1727) as one of the greatest mathematicians of all time. Gauss, known as the "prince of mathematics," presented the theory of congruences, a beautiful arm of divisibility theory, in his outstanding work *Disquistiones Arithmeticae*, published in 1801 when he was only 24. Gauss is believed to have submitted a major portion of the book to the French Academy for publication, but they rejected it. "It is really astonishing," writes the German mathematician Leopold Kronecker, "to think that a single man of

***Karl Friedrich Gauss** (1777–1855), the son of a laborer, was born in Brunswick, Germany. A child prodigy, he detected an error in his father's bookkeeping at the age of three. Recognizing his remarkable talents, the Duke of Brunswick sponsored his education. Gauss received his doctorate in 1799 from the University of Göttingen. In his doctoral dissertation, he gave the first rigorous proof of the fundamental theorem of algebra, which states, "Every polynomial of degree n (≥ 1) with real coefficients has at least one zero." Newton and Euler, among others, had attempted unsuccessfully to prove this.*

Gauss made significant contributions to algebra, geometry, analysis, physics, and astronomy. His Disquisitiones Arithmeticae *laid the foundation for modern number theory.*

From 1807 until his death, he was the director of the Observatory and professor of mathematics at the University of Göttingen.

such young years was able to bring to light such a wealth of results, and above all to present such a profound and well-organized treatment of an entirely new discipline."

The congruence relation, as we will see shortly, shares many interesting properties with the equality relation, so it is no accident that the congruence symbol $\equiv$, invented by Gauss around 1800, parallels the equality symbol $=$. The congruence symbol facilitates the study of divisibility theory and has many fascinating applications.

Let us begin our discussion with a definition.

Congruence Modulo m

Let m be a positive integer. Then an integer a is **congruent** to an integer b **modulo** m if $m|(a-b)$. In symbols, we then write $a \equiv b \pmod{m}$; m is the **modulus** of the **congruence relation**.

If a is not congruent to b modulo m, then a is **incongruent** to b modulo m; we then write $a \not\equiv b \pmod{m}$.

The following example illustrates these definitions.

EXAMPLE 4.1 Since $5|(23-3)$, $23 \equiv 3 \pmod{5}$; likewise, $6|(48-12)$, so $48 \equiv 12 \pmod{6}$; also $28 \equiv -4 \pmod{16}$. But $20 \not\equiv 3 \pmod{4}$, since $4 \nmid (20-3)$; likewise, $18 \not\equiv -6 \pmod{7}$. ■

Note that we use congruences in everyday life, often without realizing it. We use congruences modulo 12 to tell the time of the day and congruence modulo 7 to tell the day of the week. Odometers in automobiles use 1,000,000 as the modulus.

The following result is simple, but useful. It translates congruence into equality and enables us to characterize congruences in a different way. Throughout our discussion of congruences, assume that all letters denote integers and all **moduli** (plural of modulus) are positive integers.

THEOREM 4.1 $a \equiv b \pmod{m}$ if and only if $a = b + km$ for some integer k.

PROOF

Suppose $a \equiv b \pmod{m}$. Then $m|(a-b)$, so $a - b = km$ for some integer k; that is, $a = b + km$. Conversely, suppose $a = b + km$ for some integer k. Then $a - b = km$, so $m|(a-b)$ and consequently, $a \equiv b \pmod{m}$. ■

For example, $23 \equiv 3 \pmod{5}$ and $23 = 3 + 4 \cdot 5$; on the other hand, $49 = -5 + 9 \cdot 6$, so $49 \equiv -5 \pmod{6}$.

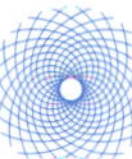

A useful observation: It follows from the definition (also from Theorem 4.1) that $a \equiv 0 \pmod{m}$ if and only if $m|a$; that is, an integer is congruent to 0 if and only if it is divisible by m. Thus, $a \equiv 0 \pmod{m}$ and $m|a$ mean exactly the same thing. For instance, $28 \equiv 0 \pmod{4}$ and $4|28$.

Using the congruence relation, Dirichlet's theorem (Theorem 3.16) can be restated as follows: There are infinitely many primes p such that $p \equiv a \pmod{b}$, where $(a, b) = 1$.

The following theorem presents three additional properties of congruence.

THEOREM 4.2

- $a \equiv a \pmod{m}$. (**Reflexive property**)
- If $a \equiv b \pmod{m}$, then $b \equiv a \pmod{m}$. (**Symmetric property**)
- If $a \equiv b \pmod{m}$ and $b \equiv c \pmod{m}$, then $a \equiv c \pmod{m}$. (**Transitive property**)

PROOF

- Since $m|(a-a)$, $a \equiv a \pmod{m}$.
- Suppose $a \equiv b \pmod{m}$. Then $m|(a-b)$; that is, $m|-(b-a)$. So $m|(b-a)$; that is, $b \equiv a \pmod{m}$.
- Suppose $a \equiv b \pmod{m}$ and $b \equiv c \pmod{m}$. Then $m|(a-b)$ and $m|(b-c)$, so, by Theorem 2.4, $m|[(a-b)+(b-c)]$; that is, $m|(a-c)$; consequently, $a \equiv c \pmod{m}$. ■

For example, $6 \equiv 6 \pmod 5$; since $3 \equiv 5 \pmod 2$, $5 \equiv 3 \pmod 2$; also, since $7 \equiv -5 \pmod 4$ and $-5 \equiv 15 \pmod 4$, $7 \equiv 15 \pmod 4$. (You may verify these congruences.)

It follows by Theorem 4.2 that the congruence relation is an equivalence relation.†

The following theorem also characterizes congruences.

THEOREM 4.3 $a \equiv b \pmod m$ if and only if a and b leave the same remainder when divided by m.

PROOF

Suppose $a \equiv b \pmod m$. Then, by Theorem 4.1, $a = b + km$ for some integer k. By the division algorithm, $b = mq + r$, where $0 \le r < m$. Then $a = b + km = (mq + r) + km = m(q + k) + r$; therefore, by the division algorithm, a leaves the same remainder r when divided by m.

Conversely, suppose both a and b leave the same remainder r when divided by m. Then, again by the division algorithm, $a = mq + r$ and $b = mq' + r$, where $0 \le r < m$. Then $a - b = (mq + r) - (mq' + r) = m(q - q')$, so $a \equiv b \pmod m$. ■

For example, $48 \equiv 28 \pmod 5$; both 48 and 28, when divided by 5, leave the same remainder 3. On the other hand, when 29 and -3 are divided by 8, the remainders are the same, 5, so $29 \equiv -3 \pmod 8$.

The next corollary follows from Theorem 4.3.

COROLLARY 4.1 The integer r is the remainder when a is divided by m if and only if $a \equiv r \pmod m$, where $0 \le r < m$. ■

By this corollary, every integer a is congruent to its remainder r modulo m; r is called the **least residue** of a modulo m. For example, the least residues of 23, 4, and -3 modulo 5 are 3, 4, and 2, respectively. Since r has exactly m choices $0, 1, 2, \dots, (m-1)$, a is congruent to exactly one of them, modulo m. Accordingly, we have the following result.

COROLLARY 4.2 Every integer is congruent to exactly one of the least residues $0, 1, 2, \dots, (m-1)$ modulo m. ■

The next example uses this result.

† An **equivalence relation** is a relation that is reflexive, symmetric, and transitive.

EXAMPLE 4.2 Prove that no prime of the form $4n + 3$ can be expressed as the sum of two squares.

PROOF (by contradiction)
Let N be a prime of the form $4n + 3$. Then $N \equiv 3 \pmod 4$.

Suppose $N = A^2 + B^2$ for some integers A and B. Since N is odd, one of the squares, say, A^2, must be odd and hence B^2 must be even. Then A must be odd and B even. Let $A = 2a + 1$ and $B = 2b$ for some integers a and b. Then

$$\begin{aligned} N &= (2a+1)^2 + (2b)^2 \\ &= 4(a^2 + b^2 + a) + 1 \\ &\equiv 1 \pmod 4 \end{aligned}$$

which is a contradiction, since $N \equiv 3 \pmod 4$. ■

Returning to Corollary 4.1, we find that it justifies the definition of the **mod** operator in Section 2.1. Thus, if $a \equiv r \pmod m$ and $0 \le r < m$, then $a \bmod m = r$; conversely, if $a \bmod m = r$, then $a \equiv r \pmod m$ and $0 \le r < m$.

We now digress briefly with an unusual application of congruence.

Friday-the-Thirteenth (optional)

Congruences can be employed to find the number of Friday-the-Thirteenths in a given year. Whether or not Friday-the-Thirteenth occurs in a given month depends on two factors: the day on which the thirteenth fell in the previous month and the number of days in the previous month.

Suppose that this is a nonleap year and that we would like to find the number of Friday-the-Thirteenths in this year. Suppose also that we know the day the thirteenth occurred in December of last year. Let M_i denote each of the months December through November in that order and D_i the number of days in month M_i. The various values of D_i are 31, 31, 28, 31, 30, 31, 30, 31, 31, 30, 31, and 30, respectively.

We label the days Sunday through Saturday by 0 through 6, respectively; so day 5 is a Friday.

Let $D_i \equiv d_i \pmod 7$, where $0 \le d_i < 7$. The corresponding values of d_i are 3, 3, 0, 3, 2, 3, 2, 3, 3, 2, 3, and 2, respectively. Each value of d_i indicates the number of days the day of the thirteenth in month M_i must be advanced to find the day the thirteenth falls in month M_{i+1}.

For example, December 13, 2000, was a Wednesday. So January 13, 2001, fell on day $(3 + 3) =$ day 6, which was a Saturday.

Let $t_i \equiv \sum_{j=1}^{i} d_j \pmod 7$, where $1 \leq i \leq 12$. Then t_i represents the total number of days the day of December 13 must be moved forward to determine the day of the thirteenth in month M_i.

For example, $t_3 \equiv d_1 + d_2 + d_3 = 3 + 3 + 0 \equiv 6 \pmod 7$. So, the day of December 13, 2000 (Wednesday), must be advanced by six days to determine the day of March 13, 2001; it is given by day $(3 + 6) =$ day $2 =$ Tuesday.

Notice that the various values of t_i modulo 7 are 3, 6, 6, 2, 4, 0, 2, 5, 1, 3, 6, and 1, respectively; they include all the least residues modulo 7. Knowing the day of December 13, we can use these least residues to determine the day of the thirteenth of each month M_i in a nonleap year.

Table 4.1 summarizes the day of the thirteenth of each month in a nonleap year, corresponding to every choice of the day of December 13 of the previous year. You can verify this. Notice from the table that there can be at most three Friday-the-Thirteenths in a nonleap year.

t_i / Dec. 13	**Jan.** **3**	**Feb.** **6**	**March** **6**	**April** **2**	**May** **4**	**June** **0**	**July** **2**	**Aug.** **5**	**Sept.** **1**	**Oct.** **3**	**Nov.** **6**	**Dec.** **1**
Sun	3	6	6	2	4	0	2	5	1	3	6	1
Mon	4	0	0	3	5	1	3	6	2	4	0	2
Tue	5	1	1	4	6	2	4	0	3	5	1	3
Wed	6	2	2	5	0	3	5	1	4	6	2	4
Thu	0	3	3	6	1	4	6	2	5	0	3	5
Fri	1	4	4	0	2	5	0	3	6	1	4	6
Sat	2	5	5	1	3	6	1	4	0	2	5	0

Table 4.1 *Day of the thirteenth in each month in a nonleap year.*

For a leap year, the various values of d_i are 3, 3, 1, 3, 2, 3, 2, 3, 3, 2, 3, and 2; and the corresponding values of t_i are 3, 6, 0, 3, 5, 1, 3, 6, 2, 4, 0, and 2. Using these, we can construct a similar table for a leap year.

We now return to additional properties of congruence.

Congruence Classes

Using least residues, the set of integers **Z** can be partitioned into m nonempty pairwise disjoint classes, called **congruence classes modulo** m. To elucidate this, let $[r]$ denote the set of integers that have r as their least residue modulo m. For example, the various congruence classes modulo 5 are

$$[0] = \{\ldots, -10, -5, 0, 5, 10, \ldots\}$$
$$[1] = \{\ldots, -9, -4, 1, 6, 11, \ldots\}$$
$$[2] = \{\ldots, -8, -3, 2, 7, 12, \ldots\}$$
$$[3] = \{\ldots, -7, -2, 3, 8, 13, \ldots\}$$
$$[4] = \{\ldots, -6, -1, 4, 9, 14, \ldots\}$$

Clearly, these classes are nonempty, pairwise disjoint, and their union is the set of integers. Accordingly, these classes form a **partitioning** of the set of integers, as Figure 4.1 shows. The least residues 0, 1, 2, 3, and 4 serve as representatives (or goodwill ambassadors) of the classes [0], [1], [2], [3], and [4], respectively.

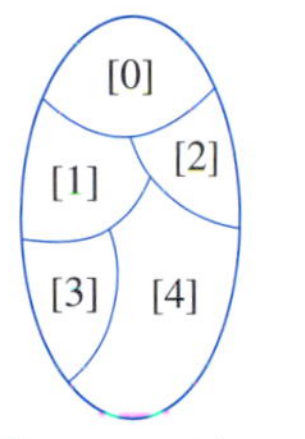

Figure 4.1

In general, we do not need to select the least residues to represent the congruence classes. By Theorem 4.3, two integers belong to the same class if and only if they leave the same remainder when divided by m; therefore, any element of the class $[r]$ can serve as a valid representative. For example, 5, 6, -3, 8, and -6 can serve as ambassadors of the classes [0], [1], [2], [3], and [4], respectively. Such a set of integers is a **complete set of residues** modulo 5.

A Complete Set of Residues Modulo m

A set of m integers is a **complete set of residues modulo** m if every integer is congruent modulo m to exactly one of them.

Thus, the set of integers $\{a_1, a_2, \ldots, a_m\}$ is a complete set of residues modulo m, if they are congruent modulo m to the least residues $0, 1, 2, \ldots, (m-1)$ in *some* order. For instance, the set $\{-12, 9, 6, 23\}$ is a complete set of residues modulo 4 since $-12 \equiv 0 \pmod 4$, $9 \equiv 1 \pmod 4$, $6 \equiv 2 \pmod 4$, and $23 \equiv 3 \pmod 4$.

The following theorem shows that two congruences with the same modulus can be added and multiplied, just as with equality.

THEOREM 4.4 Let $a \equiv b \pmod{m}$ and $c \equiv d \pmod{m}$. Then (1) $a+c \equiv b+d \pmod{m}$ and (2) $ac \equiv bd \pmod{m}$.

PROOF

Since $a \equiv b \pmod{m}$ and $c \equiv d \pmod{m}$, $a = b + \ell m$ and $c = d + km$ for some integers ℓ and m. Then

(1)
$$\begin{aligned} a+c &= (b+\ell m)+(d+km) \\ &= (b+d)+(\ell+k)m \\ &\equiv b+d \pmod{m} \end{aligned}$$

(2)
$$\begin{aligned} ac-bd &= (ac-bc)+(bc-bd) \\ &= c(a-b)+b(c-d) \\ &= c\ell m+bkm \\ &= (c\ell+bk)m \end{aligned}$$

So $ac \equiv bd \pmod{m}$. ■

EXAMPLE 4.3 We have $17 \equiv -4 \pmod{3}$ and $28 \equiv 7 \pmod{3}$. So, by Theorem 4.4, $17+28 \equiv -4+7 \pmod{3}$; that is, $45 \equiv 3 \pmod{3}$. Also, $17 \cdot 28 \equiv (-4) \cdot 7 \pmod{3}$; that is, $476 \equiv -28 \pmod{3}$. ■

The following two examples are interesting applications of Corollary 4.1 and Theorem 4.4.

EXAMPLE 4.4 Find the remainder when $1!+2!+\cdots+100!$ is divided by 15.

SOLUTION

Notice that when $k \geq 5$, $k! \equiv 0 \pmod{15}$ (why?). Therefore,

$$\begin{aligned} 1!+2!+\cdots+100! &\equiv 1!+2!+3!+4!+0+\cdots+0 \pmod{15} \\ &\equiv 1+2+6+24 \pmod{15} \\ &\equiv 1+2+0 \pmod{15} \\ &\equiv 3 \pmod{15} \end{aligned}$$

Thus, when the given sum is divided by 15, the remainder is 3. ■

EXAMPLE 4.5 Find the positive integers n for which $\sum_{k=1}^{n} k!$ is a square.

SOLUTION

Notice that when $k \geq 5$, $k! \equiv 0 \pmod{10}$ (why?), so let $n \geq 5$. Let S denote the given sum. Then

$$\begin{aligned} S &\equiv \text{ones digit in } \sum_{k=1}^{n} k! \pmod{10} \\ &\equiv (1! + 2! + 3! + 4!) \pmod{10} \\ &\equiv (1 + 2 + 6 + 24) \pmod{10} \\ &\equiv 3 \pmod{10} \end{aligned}$$

Thus, the ones digit in S is 3, if $n \geq 5$.

But $0^2 \equiv 0 \pmod{10}$, $1^2 \equiv 1 \pmod{10}$, $2^2 \equiv 4 \pmod{10}$, $3^2 \equiv 9 \pmod{10}$, $4^2 \equiv 6 \pmod{10}$, $5^2 \equiv 5 \pmod{10}$, $6^2 \equiv 6 \pmod{10}$, $7^2 \equiv 9 \pmod{10}$, $8^2 \equiv 4 \pmod{10}$, and $9^2 \equiv 1 \pmod{10}$.

Consequently, the square of every integer must end in 0, 1, 4, 5, 6, or 9. Thus, if $n \geq 5$, S cannot be a square.

When $n = 1$, $S = 1$, and when $n = 3$, $S = 9$, both squares; but S is not a square when $n = 2$ or 4.

Thus, there are exactly two positive integers n for which S is a square, namely, 1 and 3. (See Example 11.19 also.) ■

It follows from Theorem 4.4 that one congruence can be subtracted from another, provided they have the same modulus, as the following corollary states. We leave its proof as an exercise. See Exercise 70.

COROLLARY 4.3 If $a \equiv b \pmod{m}$ and $c \equiv d \pmod{m}$, then $a - c \equiv b - d \pmod{m}$. ■

For example, $23 \equiv 13 \pmod{5}$ and $30 \equiv -5 \pmod{5}$; so $23 - 30 \equiv 13 - (-5) \pmod{5}$; that is, $-7 \equiv 18 \pmod{5}$, which is true.

The following corollary also follows from Theorem 4.4. Again, we leave its proof as an exercise. See Exercises 69–72.

COROLLARY 4.4 If $a \equiv b \pmod{m}$ and c is any integer, then

- $a + c \equiv b + c \pmod{m}$
- $a - c \equiv b - c \pmod{m}$
- $ac \equiv bc \pmod{m}$
- $a^2 \equiv b^2 \pmod{m}$ ■

For example, notice that $19 \equiv 5 \pmod{7}$. So $19 + 11 \equiv 5 + 11 \pmod{7}$, $19 - 11 \equiv 5 - 11 \pmod{7}$, and $19 \cdot 11 \equiv 5 \cdot 11 \pmod{7}$.

Part (4) of Corollary 4.4 can be generalized to any positive integral exponent n, as the following theorem shows.

THEOREM 4.5 If $a \equiv b \pmod{m}$, then $a^n \equiv b^n \pmod{m}$ for any positive integer n.

PROOF (by weak induction)
The statement is clearly true when $n = 1$, so assume it is true for an arbitrary positive integer k: $a^k \equiv b^k \pmod{m}$. Then, by Theorem 4.4, $a \cdot a^k \equiv b \cdot b^k \pmod{m}$; that is, $a^{k+1} \equiv b^{k+1} \pmod{m}$. Thus, the result follows by induction. ■

The following two examples are nice applications of Corollaries 4.3 and 4.4, and Theorem 4.5. They show how congruence can be applied to a wide variety of situations.

EXAMPLE 4.6 Show that 19^{19} cannot be expressed as the sum of the cube of an integer and the fourth power of another integer.†

PROOF (by contradiction)
Notice that $19^{19} \equiv 6^{19} \pmod{13}$. But $6^2 \equiv -3 \pmod{13}$ and $6^4 \equiv -4 \pmod{13}$, so $6^6 \equiv -1 \pmod{13}$. Therefore, $19^{19} \equiv 6^{19} \equiv (6^6)^3 \cdot 6 \equiv (-1)^3 \cdot 6 \equiv -6 \equiv 7 \pmod{13}$.

Suppose 19^{19} can be expressed as $x^3 + y^4$ for some integers x and y. With a bit of patience, we can see that $x^3 \equiv 0$, 1, 5, 8, or 12 modulo 13, and $y^4 \equiv 0$, 1, 3, or 9 modulo 13. Thus, $x^3 + y^4$ can be congruent to any least residue modulo 13, *except* 7. This is a contradiction since $19^{19} \equiv 7 \pmod{13}$.

Thus, 19^{19} cannot be expressed as the sum of the cube of an integer and the fourth power of another integer. ■

EXAMPLE 4.7 Prove that no integer of the form $8n + 7$ can be expressed as a sum of three squares.

PROOF (by contradiction)
Suppose there is an integer N of the form $8n + 7$ that can be expressed as the sum $x^2 + y^2 + z^2$ of three integers x, y, and z. Then $N \equiv 7 \pmod{8}$, so $x^2 + y^2 + z^2 \equiv 7 \pmod{8}$. By Corollary 4.2, x must be congruent modulo 8 to 0, 1, 2, 3, 4, 5, 6, or 7; but $5 \equiv -3 \pmod{8}$, $6 \equiv -2 \pmod{8}$, $7 \equiv -1 \pmod{8}$; so, by Corollary 4.4,

† Based on A. Dunn (ed.), *Mathematical Bafflers*, Dover, New York, 1980, p. 187.

x^2 must be congruent modulo 8 to 0^2, 1^2, 2^2, 3^2, 4^2, $(-3)^2$, $(-2)^2$, $(-1)^2$, that is, to 0, 1, or 4. Likewise, both y^2 and z^2 must be congruent to 0, 1, or 4 modulo 8.

Therefore, by Theorem 4.4, $x^2 + y^2 + z^2$ must be congruent modulo 8 to exactly one of the sums $0 + 0 + 0$, $0 + 0 + 1$, $0 + 0 + 4$, $0 + 1 + 0$, $\ldots$, $4 + 4 + 4$, but none of them is congruent to 7 modulo 8, which is a contradiction.

Thus, no integer of the form $8n + 7$ can be expressed as the sum of three squares. ■

Theorems 4.4 and 4.5 can effectively be used to compute the remainder when an integer b^n is divided by m, as the following two examples illustrate.

EXAMPLE 4.8 Find the remainder when 16^{53} is divided by 7.

SOLUTION

First, reduce the base to its least residue: $16 \equiv 2 \pmod 7$. So, by Theorem 4.5, $16^{53} \equiv 2^{53} \pmod 7$. Now express a suitable power of 2 congruent modulo 7 to a number less than 7: $2^3 \equiv 1 \pmod 7$. Therefore,

$$\begin{aligned} 2^{53} &= 2^{3\cdot 17+2} = (2^3)^{17} \cdot 2^2 \\ &\equiv 1^{17} \cdot 4 \pmod 7 \\ &\equiv 4 \pmod 7 \end{aligned}$$

So $16^{53} \equiv 4 \pmod 7$, by the transitive property. Thus, when 16^{53} is divided by 7, the remainder is 4. ■

Notice the tremendous power of congruences in finding the remainder quickly and easily when a very large number b^n is divided by m.

EXAMPLE 4.9 Find the remainder when 3^{247} is divided by 17.

SOLUTION

Once again, we let the congruence do the job for us. We have

$$3^3 = 27 \equiv 10 \pmod{17}$$

Squaring both sides,

$$\begin{aligned}3^6 &\equiv 100 \pmod{17}\\ &\equiv -2 \pmod{17}\end{aligned}$$

Raise both sides to the fourth power:

$$\begin{aligned}3^{24} &\equiv (-2)^4 \pmod{17}\\ &\equiv -1 \pmod{17}\end{aligned}$$

Now apply the division algorithm with 24 as the divisor:

$$\begin{aligned}3^{247} = 3^{24\cdot 10+7} &= (3^{24})^{10} \cdot 3^6 \cdot 3\\ &\equiv (-1)^{10} \cdot (-2) \cdot 3 \pmod{17}\\ &\equiv -6 \pmod{17}\end{aligned}$$

Change -6 to its least residue:

$$\equiv 11 \pmod{17}$$

Thus, the remainder is 11. (Once again, appreciate the power of congruences.) ■

Modular Exponentiation

Modular exponentiation is a less efficient method for determining the remainder when b^n is divided by m. It is based on the binary representation of $n = (n_k n_{k-1} \ldots n_1 n_0)_{\text{two}}$, successive squaring, the least residue of b^{n_i}, where $0 \le i \le k$, and Theorems 4.4 and 4.5:

$$b^n = b^{n_k 2^k + n_{k-1} 2^{k-1} + \cdots + n_0} \equiv b^{n_k 2^k} \cdot b^{n_{k-1} 2^{k-1}} \cdots b^{n_0} \pmod{m}$$

The following example illustrates this method.

EXAMPLE 4.10 Compute the remainder when 3^{247} is divided by 25.

SOLUTION

First, notice that $247 = 11110111_{\text{two}}$. Now find the least residues of 3^2 and its successive squares modulo 25:

$$\begin{array}{lll} 3^2 & & \equiv 9 \pmod{25} \\ 3^8 \equiv 6^2 & & \equiv 11 \pmod{25} \\ 3^{32} \equiv 21^2 & & \equiv 16 \pmod{25} \\ 3^{128} \equiv 6^2 & & \equiv 11 \pmod{25} \end{array} \qquad \begin{array}{l} 3^4 = 9^2 \equiv 6 \pmod{25} \\ 3^{16} \equiv 11^2 \equiv 21 \pmod{25} \\ 3^{64} \equiv 16^2 \equiv 6 \pmod{25} \end{array}$$

(128 is the largest power of 2 contained in 247.)

Then

$$\begin{aligned} 3^{247} &= 3^{128+64+32+16+4+2+1} \\ &= 3^{128} \cdot 3^{64} \cdot 3^{32} \cdot 3^{16} \cdot 3^4 \cdot 3^2 \cdot 3^1 \\ &\equiv 11 \cdot 6 \cdot 16 \cdot 21 \cdot 6 \cdot 9 \cdot 3 \pmod{25} \\ &\equiv 11 \cdot (6 \cdot 16) \cdot 21 \cdot (6 \cdot 9) \cdot 3 \pmod{25} \\ &\equiv [11 \cdot (-4)] \cdot [(-4) \cdot 4] \cdot 3 \equiv 6 \cdot 9 \cdot 3 \equiv (6 \cdot 9) \cdot 3 \pmod{25} \\ &\equiv 4 \cdot 3 \equiv 12 \pmod{25} \end{aligned}$$

Thus, 12 is the desired remainder. ■

The amount of work in such a problem can be greatly reduced if we introduce negative residues, as the following example shows.

EXAMPLE 4.11 Find the remainder when 3^{181} is divided by 17.

SOLUTION

We have

$$\begin{array}{lll} 3^2 \equiv 9 \pmod{17} & 3^4 \equiv -4 \pmod{17} & 3^8 \equiv -1 \pmod{17} \\ 3^{16} \equiv 1 \pmod{17} & 3^{32} \equiv 1 \pmod{17} & 3^{64} \equiv 1 \pmod{17} \\ 3^{128} \equiv 1 \pmod{17} & & \end{array}$$

Therefore:

$$\begin{aligned} 3^{181} &= 3^{128} \cdot 3^{32} \cdot 3^{16} \cdot 3^4 \cdot 3^1 \\ &\equiv 1 \cdot 1 \cdot 1 \cdot 13 \cdot 3 \pmod{17} \\ &\equiv 5 \pmod{17} \end{aligned}$$

Thus, the desired remainder is 5. ■

Towers of Powers Modulo m

The technique of finding remainders using congruences can be extended to numbers with exponents, which are towers of powers, as the following example demonstrates.

EXAMPLE 4.12 Find the last digit in the decimal value of $1997^{1998^{1999}}$.

SOLUTION

First, notice that $a^{b^c} = a^{(b^c)}$. Let N denote the given number. The last digit in N equals the least residue of N modulo 10.

Since $1997 \equiv 7 \pmod{10}$, let us study the various powers of 7: $7^1 \equiv 7 \pmod{10}$, $7^2 \equiv 9 \pmod{10}$, $7^3 \equiv 3 \pmod{10}$, $7^4 \equiv 1 \pmod{10}$, $7^5 \equiv 7 \pmod{10}$ and clearly a pattern emerges:

$$7^a \equiv \begin{cases} 1 \pmod{10} & \text{if } a \equiv 0 \pmod{4} \\ 7 \pmod{10} & \text{if } a \equiv 1 \pmod{4} \\ 9 \pmod{10} & \text{if } a \equiv 2 \pmod{4} \\ 3 \pmod{10} & \text{if } a \equiv 3 \pmod{4} \end{cases}$$

Now let us look at 1998. Since $1998 \equiv 2 \pmod{4}$, $1998^n \equiv 2^n \pmod{4}$, so if $n \geq 2$, then $1998^n \equiv 0 \pmod{4}$. Thus, since $1999 \geq 2$, $1998^{1999} \equiv 0 \pmod{4}$, so $N \equiv 1 \pmod{10}$. In other words, the last digit in the decimal value of N is 1. ■

The following two examples also demonstrate the power of congruences.

EXAMPLE 4.13 Show that $11 \cdot 14^n + 1$ is a composite number.[†]

PROOF

Let $N = 11 \cdot 14^n + 1$. We shall show that $p|N$ for some prime p.

Suppose n is even. Since $14 \equiv -1 \pmod{3}$, $14^n \equiv 1 \pmod{3}$. Then $N \equiv 2 \cdot 1 + 1 \equiv 0 \pmod{3}$, so $3|N$.

On the other hand, let n be odd. Since $14 \equiv -1 \pmod{5}$, $14^n \equiv -1 \pmod{5}$. Then $N \equiv 1 \cdot (-1) + 1 \equiv 0 \pmod{5}$, so $5|N$.

Thus, in both cases, N is composite. ■

It is well known that $N = n^2 + n + 41$ is a prime for $0 \leq n < 41$. The following example shows how to compute the remainder when N^2 is divided by 12, for every integer n.

† Based on A. Dunn (ed.), *Mathematical Bafflers*, Dover, New York, 1980, p. 192. The elegant proof given here is due to J. N. A. Hawkins of Pacific Palisades, California.

EXAMPLE 4.14 Find the remainder when $(n^2+n+41)^2$ is divided by 12.

PROOF

First, notice that the product of four consecutive integers is divisible by 12; that is, $(n-1)n(n+1)(n+2) \equiv 0 \pmod{12}$.

We have

$$\begin{aligned}(n^2+n+41)^2 &\equiv (n^2+n+5)^2 \pmod{12}\\ &\equiv (n^4+2n^3+11n^2+10n+25) \pmod{12}\\ &\equiv (n^4+2n^3-n^2-2n)+1 \pmod{12}\\ &\equiv n(n^3+2n^2-n-2)+1 \pmod{12}\\ &\equiv n[n^2(n+2)-(n+2)]+1 \pmod{12}\\ &\equiv n(n+2)(n^2-1)+1 \pmod{12}\\ &\equiv (n-1)n(n+1)(n+2)+1 \pmod{12}\\ &\equiv 1 \pmod{12}\end{aligned}$$

Thus when $(n^2+n+41)^2$ is divided by 12, the remainder is 1. ■

In Example 2.29 we found that the Fermat number $f_5 = 2^{2^5}+1$ is divisible by 641. The next example furnishes an elegant alternate proof of this fact, using congruences.

EXAMPLE 4.15 Show that $f_5 = 2^{2^5}+1$ is divisible by 641.

PROOF

First, notice that $640 \equiv -1 \pmod{641}$; that is, $5 \cdot 2^7 \equiv -1 \pmod{641}$. Therefore,

$$5^4 \cdot 2^{28} \equiv 1 \pmod{641} \tag{4.1}$$

But $5^4 = 625 \equiv -16 \equiv -2^4 \pmod{641}$, so congruence (4.1) can be rewritten as $(-2^4)(2^{28}) \equiv 1 \pmod{641}$; that is, $2^{32} \equiv -1 \pmod{641}$. Thus, $641|f_5$. ■

We now examine some additional properties of congruences.

The cancellation property of multiplication says, if $ac = bc$ and $c \neq 0$, then $a = b$.

Does this have an analogous result for congruences? In other words, if $ac \equiv bc \pmod{m}$ and $c \not\equiv 0 \pmod{m}$, is $a \equiv b \pmod{m}$? To answer this, notice that $3 \cdot 8 \equiv 3 \cdot 4 \pmod{6}$, but $8 \not\equiv 4 \pmod{6}$, so the answer is a definite *no*.

But under some circumstances, the answer is yes, as the following theorem shows.

THEOREM 4.6 If $ac \equiv bc \pmod{m}$ and $(c, m) = 1$, then $a \equiv b \pmod{m}$.

PROOF

Suppose $ac \equiv bc \pmod{m}$, where $(c, m) = 1$. Then $m|(ac - bc)$; that is, $m|c(a - b)$. But $(m, c) = 1$, so, by Corollary 3.4, $m|(a - b)$; that is, $a \equiv b \pmod{m}$. ■

Thus, we can cancel the same number c from both sides of a congruence, provided c and m are relatively prime, as the following example demonstrates.

EXAMPLE 4.16 Notice that $78 \equiv 48 \pmod{5}$; that is, $6 \cdot 13 \equiv 6 \cdot 8 \pmod{5}$. Since $(6, 5) = 1$, we can cancel 6 from both sides:

$$\not{6} \cdot 13 \equiv \not{6} \cdot 8 \pmod{5}$$

That is,

$$13 \equiv 8 \pmod{5}$$

which is clearly true. ■

The following example, an application of Theorem 4.6, revisits the monkey and coconuts riddle we solved earlier.

The Monkey and Coconuts Puzzle Revisited

EXAMPLE 4.17 Using congruences, solve the monkey and coconuts riddle in Example 3.23.

SOLUTION

Once again, let n denote the least possible number of coconuts in the original puzzle and z each sailor's share after the final division. Then

$$\frac{1}{5}\left(\frac{4}{5}\left(\frac{4}{5}\left(\frac{4}{5}\left(\frac{4}{5}\left(\frac{4}{5}(n-1)-1\right)-1\right)-1\right)-1\right)-1\right)=z$$

With a bit of patience, we can rewrite this equation as

$$
\begin{aligned}
n\left(\frac{4}{5}\right)^5 - \left[1 + \frac{4}{5} + \left(\frac{4}{5}\right)^2 + \left(\frac{4}{5}\right)^3 + \left(\frac{4}{5}\right)^4 + \left(\frac{4}{5}\right)^5\right] &= 5z \\
n\left(\frac{4}{5}\right)^5 - \frac{1-(4/5)^6}{1-4/5} &= 5z \\
n\left(\frac{4}{5}\right)^5 - \frac{5^6-4^6}{5^5} &= 5z \\
4^5 n + 4^6 - 5^6 &= 5^6 z \\
(n+4)4^5 &= (z+1)5^6 \\
&\equiv 0 \pmod{5^6}
\end{aligned}
$$

But $(4^5, 5^6) = 1$, so $n + 4 \equiv 0 \pmod{5^6}$. Thus, for n to be a minimum, $n + 4 = 5^6 = 15{,}625$; so $n = 15{,}621$, as found earlier. ■

Returning to Theorem 4.6, we can generalize it as follows.

THEOREM 4.7 If $ac \equiv bc \pmod{m}$ and $(c, m) = d$, then $a \equiv b \pmod{m/d}$.

PROOF

Suppose $ac \equiv bc \pmod{m}$, where $(c, m) = d$. Then $m|(ac - bc)$, so $ac - bc = km$ for some integer k; that is, $c(a - b) = km$. Divide both sides by d:

$$\left(\frac{c}{d}\right)(a-b) = k\left(\frac{m}{d}\right)$$

By Theorem 3.4, $(c/d, m/d) = 1$, so $\frac{m}{d}|(a - b)$; that is, $a \equiv b \pmod{m/d}$. ■

EXAMPLE 4.18 You can verify that $8 \cdot 37 \equiv 8 \cdot 7 \pmod{12}$. Since $(8, 12) = 4$, by Theorem 4.7, we can cancel 8 from both sides:

$$\not{8} \cdot 37 \equiv \not{8} \cdot 7 \pmod{12/4}$$

That is,

$$37 \equiv 7 \pmod{3}$$

■

Now we will see how congruences of two numbers with different moduli can be combined into a single congruence.

THEOREM 4.8 If $a \equiv b \pmod{m_1}$, $a \equiv b \pmod{m_2}$, $\ldots$, $a \equiv b \pmod{m_k}$, then $a \equiv b \pmod{[m_1, m_2, \ldots, m_k]}$.

PROOF

By the given hypotheses, $m_1|(a-b)$, $m_2|(a-b), \ldots, m_k|(a-b)$, so, by Corollary 3.12, $[m_1, m_2, \ldots, m_k]|(a-b)$; that is, $a \equiv b \pmod{[m_1, m_2, \ldots, m_k]}$. ■

The following example illustrates this result.

EXAMPLE 4.19 You can verify that $197 \equiv 77 \pmod 6$, $197 \equiv 77 \pmod{10}$, and $197 \equiv 77 \pmod{15}$; so by Theorem 4.8, $197 \equiv 77 \pmod{[6, 10, 15]}$; that is, $197 \equiv 77 \pmod{30}$. ■

The following corollary follows easily from this theorem.

COROLLARY 4.5 If $a \equiv b \pmod{m_1}$, $a \equiv b \pmod{m_2}$, $\ldots$, $a \equiv b \pmod{m_k}$, where the moduli are pairwise relatively prime, then $a \equiv b \pmod{m_1 m_2 \cdots m_k}$. ■

EXERCISES 4.1

Mark *True* or *False*, where a, b, c, and d are arbitrary integers, m a positive integer, and p a prime.

1. $12 \equiv -3 \pmod 5$
2. $18 \not\equiv -2 \pmod 4$
3. $10 \equiv 1 \pmod 9$
4. $10 \equiv -1 \pmod{11}$
5. $a \equiv a \pmod m$
6. If $a \equiv b \pmod m$, then $b \equiv a \pmod m$.
7. If $a \equiv b \pmod m$ and $b \equiv c \pmod m$, then $a \equiv c \pmod m$.
8. If $a \equiv b \pmod m$, then $-a \equiv -b \pmod m$.
9. If $a \equiv b \pmod m$ and $c \equiv d \pmod m$, then $a + c \equiv b + d \pmod m$.
10. If $a + c \equiv b + c \pmod m$, then $a \equiv b \pmod m$.
11. If $a \equiv b \pmod m$ and $c \equiv d \pmod m$, then $ac \equiv bd \pmod m$.
12. If $ac \equiv bc \pmod m$, then $a \equiv b \pmod m$.
13. If $a \equiv b \pmod m$, then $a^2 \equiv b^2 \pmod m$.
14. If $a^2 \equiv b^2 \pmod m$, then $a \equiv b \pmod m$.
15. If $a \equiv b \pmod m$ and $a \equiv b \pmod n$, then $a \equiv b \pmod{m + n}$.
16. If $a \equiv b \pmod m$ and $a \equiv b \pmod n$, then $a \equiv b \pmod{mn}$.
17. If $ab \equiv 0 \pmod m$, then $a \equiv 0 \pmod m$ and $b \equiv 0 \pmod m$.
18. If $a \not\equiv b \pmod m$, then $m \nmid (a - b)$.
19. If $a \not\equiv b \pmod m$, then $b \not\equiv a \pmod m$.
20. If $a \not\equiv b \pmod m$ and $b \not\equiv c \pmod m$, then $a \not\equiv c \pmod m$.
21. If $a \not\equiv 0 \pmod m$ and $b \not\equiv 0 \pmod m$, then $ab \not\equiv 0 \pmod m$.
22. If $ac \equiv bc \pmod p$ and $p \nmid c$, then $a \equiv b \pmod p$.
23. $9^{100} - 1$ is divisible by 10.
24. $10^{2001} + 1$ is divisible by 11.

Rewrite each sentence in Exercises 25–28, using the congruence symbol.

25. n is an odd integer.
26. n is an even integer.
27. n is divisible by 5.
28. The product of any three consecutive integers is divisible by 6.
29. If today is Tuesday, what day will it be in 129 days?
30. If today is Friday, what day will it be in 1976 days?
31. If it is 9 A.M. now, what time will it be in 1900 hours?
32. If it is 3 P.M. now, what time will it be in 4334 hours?

Give a counterexample to disprove each statement.

33. If $a^2 \equiv b^2 \pmod{m}$, then $a \equiv b \pmod{m}$.
34. If $a \not\equiv 0 \pmod{m}$ and $b \not\equiv 0 \pmod{m}$, then $ab \not\equiv 0 \pmod{m}$.

Find the remainder when $1! + 2! + 3! + \cdots + 1000!$ is divided by each integer.

35. 10 36. 11 37. 12 38. 13

Find the remainder when the first integer is divided by the second.

39. $2^{35}, 7$
40. $5^{31}, 12$
41. $23^{1001}, 17$
42. $19^{1976}, 23$

Using modular exponentiation, find the remainder when the first integer is divided by the second.

43. $2^{97}, 13$
44. $4^{117}, 15$
45. $13^{218}, 17$
46. $19^{343}, 23$

Find the units digit in the decimal value of each.

47. $1776^{1777^{1778}}$
48. $1943^{1642^{1053}}$
49. $1077^{1177^{1277^{1377}}}$
50. $1089^{2089^{3089^{4089}}}$

Find the last two digits in the decimal value of each.

51. 1776^{1976}
52. 1829^{1829}
53. Let $n \equiv r \pmod{10}$, where $0 \le r < 10$. Identify the units digit in the decimal expansion of n.

Find the least residues x such that $x^2 \equiv 1 \pmod{m}$ for each value of m.

54. 5 55. 6 56. 7 57. 8

Using Exercises 54–57, conjecture the number of least residues x such that

58. $x^2 \equiv 1 \pmod{p}$, where p is a prime.
59. $x^2 \equiv 1 \pmod{m}$, where m is a positive integer.
60. Let a be a least residue modulo 5. Compute the least residue of a^5 for each a.
61. Let a be a least residue modulo 7. Compute the least residue of a^7 for each a.
62. Using Exercises 60 and 61, predict the least residue of a^p modulo p, where p is a prime.

Compute the least residue of $(p-1)!$ modulo p for each prime.

63. 3 64. 5 65. 7 66. 11

67. Using Exercises 63–65, conjecture the least residue of $(p-1)!$ modulo p.

Prove each, where a, b, c, d, and n are any integers, m is a positive integer, and p is a prime.

68. If $a \equiv b \pmod{m}$ and $c \equiv d \pmod{m}$, then $a - c \equiv b - d \pmod{m}$.

If $a \equiv b \pmod{m}$ and c is any integer, then:

69. $a + c \equiv b + c \pmod{m}$
70. $a - c \equiv b - c \pmod{m}$
71. $ac \equiv bc \pmod{m}$
72. $a^2 \equiv b^2 \pmod{m}$
73. If $ac \equiv bc \pmod{p}$ and $p \nmid c$, then $a \equiv b \pmod{p}$.
74. If $a^2 \equiv 1$, then $a \equiv \pm 1 \pmod{p}$.
75. Let $f(x)$ be a polynomial with integral coefficients and $a \equiv b \pmod{m}$. Then $f(a) \equiv f(b) \pmod{m}$.
76. The square of every even integer is congruent to 0 modulo 4.
77. Every odd integer is congruent to 1 or 3 modulo 4.
78. If $ab \equiv 0 \pmod{p}$, then $a \equiv 0 \pmod{p}$ or $b \equiv 0 \pmod{p}$.
79. The square of every odd integer is congruent to 1 modulo 4.
80. Every prime > 3 is congruent to ± 1 modulo 6.
81. If $2a \equiv 0 \pmod{p}$ and p is an odd prime, then $a \equiv 0 \pmod{p}$.
82. $n^2 + n \equiv 0 \pmod{2}$
83. $n^4 + 2n^3 + n^2 \equiv 0 \pmod{4}$
84. $2n^3 + 3n^2 + n \equiv 0 \pmod{6}$

85. Using congruences, show that the only Fermat number that is also triangular is 3. (S. Asadulla, 1987)
86. $1155 \nmid n^{7777} + 7777n + 1$, where n is a square. (A. Kumar, 2003)
87. The last $n+1$ digits of 5^{2^n} are the same as those of $5^{2^{n-1}}$, where $n \geq 3$. (P. A. Lindstrom, 2005)
88. $\binom{2p}{p} \equiv 2 \pmod{p}$, where $p > 2$. (J. M. Gandhi, 1959)
89. $p \mid \binom{2p}{r}$, where $0 < r < p$.
90. By Theorem 3.18, every prime factor of f_n is of the form $k \cdot 2^{n+2} + 1$, where $n \geq 2$. Then $k^{2^n} \equiv (-1)^n \pmod{p}$.
91. Find all primes p such that $p, p + 2d$, and $p + 4d$ are primes, where $3 \nmid d$. (M. S. Klamkin, 1967)
92. Find the number of entries in row $2p$ of Pascal's triangle that are divisible by p, where p is an odd prime.
93. Find the remainder when googolplex is divided by 7. (H. W. Kickey, 1966)

4.2 Linear Congruences

In the previous section we studied the language of congruences and some fundamental properties of congruences. Now we look at congruences containing variables, such as $3x \equiv 4 \pmod{5}$, $x^2 \equiv 1 \pmod{8}$, and $x^2 + 2 \equiv 3x \pmod{5}$. The simplest such congruence is the **linear congruence** $ax \equiv b \pmod{m}$. We will now see that linear congruences and LDEs are interlinked. We will also learn a necessary and sufficient condition for a linear congruence to be solvable.

By a **solution** of the linear congruence, we mean an integer x_0 such that $ax_0 \equiv b \pmod{m}$. For example, $3 \cdot 3 \equiv 4 \pmod{5}$, so 3 is a solution of the congruence $3x \equiv 4 \pmod{5}$. But the congruence $4x \equiv 1 \pmod{2}$ has no solutions, since $2 \nmid (4x - 1)$ for any integer x.

To see the link between linear congruences and LDEs, consider $ax \equiv b \pmod{m}$. Then, by Theorem 4.1, $ax = b + my$ for some integer y. Consequently, $ax \equiv b \pmod{m}$ is solvable if and only if the LDE $ax - my = b$ is solvable.

Suppose x_0 is a solution of the congruence $ax \equiv b \pmod{m}$; then $ax_0 \equiv b \pmod{m}$. Suppose, in addition, $x_1 \equiv x_0 \pmod{m}$. Then, by Corollary 4.4, $ax_1 \equiv ax_0 \pmod{m}$, so, by transitivity, $ax_1 \equiv b \pmod{m}$; thus, x_1 is also a solution of the congruence. But x_1 and x_0 belong to the same congruence class; so if x_0 is a solution, then every member of its class is also a solution.

For instance, since 3 is a solution of the linear congruence $3x \equiv 4 \pmod{5}$, every member of the congruence class $[3] = \{\ldots, -7, -2, 3, 8, 13, \ldots\}$ is also a solution; they are given by $x = 3 + 5t$:

$$3(3 + 5t) = 9 + 15t$$

$$\equiv 4 + 0 \pmod 5$$
$$\equiv 4 \pmod 5$$

Thus, if the congruence $ax \equiv b \pmod m$ is solvable, it has infinitely many solutions. Consequently, we are interested in its incongruent solutions only. For example, the congruence $9x \equiv 6 \pmod{12}$ has three incongruent solutions, namely, 2, 6, and 10: $9 \cdot 2 \equiv 6 \pmod{12}$, $9 \cdot 6 \equiv 6 \pmod{12}$, and $9 \cdot 10 \equiv 6 \pmod{12}$.

The following theorem provides a necessary and sufficient condition for a linear congruence to be solvable. This theorem also gives the number of incongruent solutions, and a formula for finding them when the congruence is solvable.

THEOREM 4.9 The linear congruence $ax \equiv b \pmod m$ is solvable if and only if $d|b$, where $d = (a, m)$. If $d|b$, then it has d incongruent solutions.

PROOF

The linear congruence $ax \equiv b \pmod m$ is equivalent to the LDE $ax - my = b$; so the congruence is solvable if and only if the LDE is solvable. But, by Theorem 3.19, the LDE is solvable if and only if $d|b$. Thus $ax \equiv b \pmod m$ is solvable if and only if $d|b$.

When $d|b$, the LDE has infinitely many solutions, given by

$$x = x_0 + \left(\frac{m}{d}\right)t, \qquad y = y_0 + \left(\frac{a}{d}\right)t$$

so the congruence has infinitely many solutions $x = x_0 + \left(\frac{m}{d}\right)t$, where x_0 is a particular solution.

To find the number of incongruent solutions when the congruence is solvable, suppose $x_1 = x_0 + \left(\frac{m}{d}\right)t_1$, $x_2 = x_0 + \left(\frac{m}{d}\right)t_2$, are two congruence solutions:

$$x_0 + \left(\frac{m}{d}\right)t_1 \equiv x_0 + \left(\frac{m}{d}\right)t_2 \pmod m$$

Subtracting x_0 from both sides,

$$\left(\frac{m}{d}\right)t_1 \equiv \left(\frac{m}{d}\right)t_2 \pmod m$$

Since $\left.\frac{m}{d}\right| m$, by Theorem 4.7, $t_1 \equiv t_2 \pmod d$. Thus, the solutions x_1 and x_2 are congruent if and only if $t_1 \equiv t_2 \pmod d$; that is, if and only if t_1 and t_2 belong to the same congruence class modulo d. In other words, they are incongruent solutions if and only if they belong to distinct congruence classes.

By Corollary 4.2, there are exactly d incongruent classes modulo d. Therefore, the linear congruence, when solvable, has exactly d incongruent solutions, given by $x = x_0 + \left(\frac{m}{d}\right)t$, where $0 \le t < d$. ■

Note: $x = x_0 + \left(\frac{m}{d}\right)t$, where $0 \le t < d$, is the **general solution** of the linear congruence.

This theorem has a useful corollary.

COROLLARY 4.6 The linear congruence $ax \equiv b \pmod{m}$ has a unique solution if and only if $(a, m) = 1$. ■

The following two examples illustrate these fundamental results.

EXAMPLE 4.20 Determine if the congruences $8x \equiv 10 \pmod 6$, $2x \equiv 3 \pmod 4$, and $4x \equiv 7 \pmod 5$ are solvable. Find the number of incongruent solutions when a congruence is solvable.

SOLUTION

- $(8, 6) = 2$ and $2|10$, so the congruence $8x \equiv 10 \pmod 6$ is solvable and it has two incongruent solutions modulo 6.
- $(2, 4) = 2$, but $2 \nmid 3$, so the congruence $2x \equiv 3 \pmod 4$ has no solutions.
- $(4, 7) = 1$, so by Corollary 4.6, the congruence $4x \equiv 7 \pmod 5$ has a unique solution modulo 5. ■

The following example illustrates how to find the incongruent solutions of a linear congruence.

EXAMPLE 4.21 Solve the congruence $12x \equiv 48 \pmod{18}$.

SOLUTION

Since $(12, 18) = 6$ and $6|48$, the congruence has six incongruent solutions modulo 6. They are given by $x = x_0 + \left(\frac{m}{d}\right)t = x_0 + (18/6)t = x_0 + 3t$, where x_0 is a particular solution and $0 \le t < 6$. By trial and error, $x_0 = 1$ is a solution. Thus, the six incongruent solutions modulo 18 are $1 + 3t$, where $0 \le t < 6$, that is, 1, 4, 7, 10, 13, and 16. ■

The same congruence can be solved in a slightly different way. Using Theorem 4.7, divide the congruence by 6:

$$2x \equiv 8 \pmod{3}$$

Now multiply both sides by 2 (to yield one x on the LHS):

$$2(2x) \equiv 2 \cdot 8 \pmod{3}$$
$$x \equiv 1 \pmod{3}$$

So the solutions of this congruence are of the form $x = 1 + 3t$. Now, proceeding as before, we get all the desired solutions.

The following example shows how congruences are useful in solving LDEs.

EXAMPLE 4.22 Using congruences, solve Mahavira's puzzle in Example 3.20.

SOLUTION

From Example 3.20, we have $63x - 23y = -7$. This LDE creates two linear congruences: $63x \equiv -7 \pmod{23}$ and $-23y \equiv -7 \pmod{63}$. The first one yields $-6x \equiv -7 \pmod{23}$; that is, $6x \equiv 7 \pmod{23}$, where $(6, 23) = 1$. Multiply both sides by 4,

$$4(6x) \equiv 4 \cdot 7 \pmod{23}$$
$$x \equiv 5 \pmod{23}$$

So the general solution of the congruence $63x \equiv -7 \pmod{23}$ is $x = 5 + 23t$.

Substitute for x in the LDE and solve for y:

$$63(5 + 23t) - 23y = -7$$
$$315 + 1449t - 23y = -7$$
$$y = 14 + 63t$$

Thus, the general solution of the LDE is $x = 5 + 23t$, $y = 14 + 63t$, with t an arbitrary integer. (Notice that this agrees with the solution obtained earlier.) ■

In this example, we could have solved the second congruence $-23y \equiv -7 \pmod{63}$ and obtained the same solution. Try this and convince yourself, in Exercise 20.

Modular Inverses

Consider the special case $b = 1$ in Corollary 4.6. The linear congruence $ax \equiv 1 \pmod{m}$ has a unique solution if and only if $(a, m) = 1$; in other words, when $(a, m) = 1$, there is a unique least residue x such that $ax \equiv 1 \pmod{m}$. Then a is said to be **invertible** and x is called an **inverse** of a modulo m, denoted by a^{-1}: $aa^{-1} \equiv 1 \pmod{m}$. If $a^{-1} = a$, then a is **self-invertible**.

EXAMPLE 4.23 Since $7 \cdot 8 \equiv 1 \pmod{11}$, 7 is invertible and 8 is an inverse of 7 modulo 11; that is, 7^{-1} is 8 modulo 11; 10 is its own inverse modulo 11, since $10 \cdot 10 \equiv 1 \pmod{11}$. ■

Inverses are useful in solving linear congruences. To see this, let us return to the congruence $ax \equiv b \pmod{m}$, where $(a, m) = 1$. Since $(a, m) = 1$, a has an inverse a^{-1} modulo m. Multiplying both sides of the congruence by a^{-1}, we get

$$a^{-1}(ax) \equiv a^{-1}b \pmod{m}$$
$$(a^{-1}a)x \equiv a^{-1}b \pmod{m}$$
$$1x \equiv a^{-1}b \pmod{m}$$

That is,

$$x \equiv a^{-1}b \pmod{m}$$

Accordingly, we have the following result.

THEOREM 4.10 The unique solution of the linear congruence $ax \equiv b \pmod{m}$, where $(a, m) = 1$, is the least residue of $a^{-1}b \pmod{m}$. ■

The following example employs this result.

EXAMPLE 4.24 Using Theorem 4.10, solve the hundreds fowls riddle in Example 3.18.

SOLUTION

From Example 3.21, we have

$$x + y + z = 100 \tag{4.2}$$
$$5x + 3y + \frac{z}{3} = 100 \tag{4.3}$$

Eliminating z between these equations, we get

$$7x + 4y = 100 \tag{4.4}$$

This yields

$$7x \equiv 100 \pmod 4$$
$$3x \equiv 0 \pmod 4$$

Therefore,

$$3(3x) \equiv 3 \cdot 0 \pmod 4 \quad [\textit{Note}: 3^{-1} \equiv 3 \pmod 4]$$
$$x \equiv 0 \pmod 4$$

So $x = 4t$. Substituting for x in equation (4.4), we get

$$7(4t) + 4y = 100$$
$$y = 25 - 7t$$

Now substitute for x and y in equation (4.2):

$$4t + (25 - 7t) + z = 100$$
$$z = 3t + 75$$

Thus, the general solution is $x = 4t,\ y = 25 - 7t,\ z = 75 + 3t$, exactly the same as the one obtained in Example 3.21. ■

The following example is an interesting application of Theorems 3.13 and 4.10, and is a continuation of Example 3.12.

EXAMPLE 4.25 Find the last nonzero digit (from the left) in the decimal value of 234!.

SOLUTION

First, notice that the product of the four integers between any two consecutive multiples of 5 is congruent to -1 modulo 5; that is, if $n \equiv 0 \pmod 5$, then $(n+1)(n+2)(n+3)(n+4) \equiv 1 \cdot 2 \cdot 3 \cdot 4 \equiv -1 \pmod 5$.

In Example 3.12, we found that 234! has $46 + 9 + 1 = 56$ trailing zeros. Consequently, the desired digit d is the ones digit in $234!/10^{56}$. Since the canonical decomposition of 234! contains more 2s than 5s, d must be even. Thus, $d = 2, 4, 6$, or 8. To extract the correct value of d, we compute $234!/10^{56} \pmod 5$ in seven steps:

$$231 \cdot 232 \cdot 233 \cdot 234 \equiv -1 \pmod 5$$

$$\frac{230!}{5^{46} \cdot 46!} = \frac{230!}{5 \cdot 10 \cdot 15 \cdots 230} \equiv (-1)^{46} \equiv 1 \pmod 5$$

$$\frac{46!}{5^9 \cdot 9!} = \frac{46!}{5 \cdot 10 \cdot 15 \cdots 45} \equiv (-1)^9 \equiv -1 \pmod 5$$

$$\frac{9!}{5} \equiv (-1)^2 \equiv 1 \pmod 5$$

$$\frac{230!}{5^{46} \cdot 46!} \cdot \frac{46!}{5^9 \cdot 9!} \cdot \frac{9!}{5} \equiv 1 \cdot (-1) \cdot 1 \equiv 4 \pmod 5$$

$$\frac{234!}{5^{56}} = \frac{230!(231 \cdot 232 \cdot 233 \cdot 234)}{5^{56}} \equiv (-1) \cdot 4 \equiv -4 \equiv 1 \pmod 5$$

Since $2^{56} = 4^{28} \equiv (-1)^{28} \equiv 1 \pmod 5$ and $(2, 5) = 1$, this implies

$$\frac{234!}{10^{56}} = \frac{234!}{2^{56}5^{56}} \equiv 1 \pmod 5$$

that is, $d \equiv 1 \pmod 5$, so $d = 6$. Thus, the 56 zeros in 234! follow the digit 6. (See Example 7.2 also.) ■

We now redo this example differently, using a clever notation introduced by P. M. Dunson in 1980.

Let $n^\star$ denote the product of the integers 1 through n, omitting all multiples of 5. For example, $9^\star = 1 \cdot 2 \cdot 3 \cdot 4 \cdot 6 \cdot 7 \cdot 8 \cdot 9$. Clearly, $230^\star \equiv 9^\star \equiv 6 \pmod{10}$, $45^\star \equiv 4 \pmod{10}$ and, $231 \cdot 232 \cdot 233 \cdot 234 \equiv 4 \pmod{10}$.

Notice that

$$\begin{aligned} 234! &= 234 \cdot 233 \cdot 232 \cdot 231 \cdot 230^\star \cdot 5^{46} \cdot 46! \\ &= 234 \cdot 233 \cdot 232 \cdot 231 \cdot 230^\star \cdot 5^{46} \cdot 46 \cdot 45^\star \cdot 5^9 \cdot 9! \\ &= 234 \cdot 233 \cdot 232 \cdot 231 \cdot 46 \cdot 230^\star \cdot 45^\star \cdot 5^{55} \cdot 5 \cdot 9^\star \\ &= 234 \cdot 233 \cdot 232 \cdot 231 \cdot 46 \cdot 230^\star \cdot 45^\star \cdot 9^\star \cdot 5^{56} \end{aligned}$$

Therefore,

$$\begin{aligned} \frac{234!}{5^{56}} &= (234 \cdot 233 \cdot 232 \cdot 231) \cdot 46 \cdot 230^\star \cdot 45^\star \cdot 9^\star \\ &\equiv 4 \cdot 6 \cdot 6 \cdot 4 \cdot 6 \pmod{10} \\ &\equiv 6 \pmod{10} \end{aligned}$$

Since 234! contains exactly 56 trailing zeros, $\frac{234!}{5^{56}} \pmod{10}$ yields its last nonzero digit. Thus, the last nonzero digit in 234! is 6.

EXERCISES 4.2

Using Theorem 4.9, determine whether each linear congruence is solvable.

1. $12x \equiv 18 \pmod{15}$
2. $16y \equiv 18 \pmod{12}$
3. $12x \equiv 14 \pmod{13}$
4. $28u \equiv 119 \pmod{91}$
5. $76v \equiv 50 \pmod{176}$
6. $2076y \equiv 3076 \pmod{1076}$

Determine the number of incongruent solutions of each linear congruence.

7. $12x \equiv 18 \pmod{15}$
8. $28u \equiv 119 \pmod{91}$
9. $49x \equiv 94 \pmod{36}$
10. $91y \equiv 119 \pmod{28}$
11. $48v \equiv 144 \pmod{84}$
12. $2076x \equiv 3076 \pmod{1076}$
13. Suppose x_0 is a solution of the congruence $ax \equiv b \pmod{m}$. Show that $x = x_0 + \left(\frac{m}{d}\right)t$ is also a solution of the congruence, where $d = (a, m)$.

14–19. Find the incongruent solutions of each congruence in Exercises 7–12.

20. Using the congruence $-23y \equiv -7 \pmod{63}$, solve the LDE $63x - 23y = -7$.

Using congruences, solve each LDE.

21. $3x + 4y = 5$
22. $6x + 9y = 15$
23. $15x + 21y = 39$
24. $28x + 91y = 119$
25. $48x + 84y = 144$
26. $1776x + 1976y = 4152$

Find the least residues modulo m that are invertible for each value of m.

27. five
28. six

Find the least residues modulo m that are self-invertible for each value of m.

29. seven
30. twelve

Using inverses, find the incongruent solutions of each linear congruence.

31. $5x \equiv 3 \pmod{6}$
32. $4x \equiv 11 \pmod{13}$
33. $19x \equiv 29 \pmod{16}$
34. $48x \equiv 39 \pmod{17}$
35. Suppose b is an inverse of a modulo m. Show that a is an inverse of b modulo m.
36. Let $f(n)$ denote the number of positive integers $\le n$ and relatively prime to n. Using the function f, give the number of least residues modulo m that are invertible.
37. Let p be a prime. Prove that a least residue modulo p is self-invertible if and only if $a \equiv \pm 1 \pmod{p}$.

Find the last two digits of each number.

*38. 7^{777}

*39. 19^{1991}

Find the last three digits of each number.

*40. 4^{2076}

*41. 17^{1776}

Find the last nonzero digit in the decimal value of each. (*Hint*: Use Exercises 17–20 in Section 3.3.)

*42. 100!

*43. 376!

*44. 609!

*45. 1010!

The linear congruence $ax \equiv c \pmod{b}$ is solvable if and only if $r_n | c$, and the solutions are given by $x = x_0 + \frac{(-1)^n b}{r_n} t$, where $r_n = (a, b)$. Using this fact, solve each linear congruence.

*46. $1024x \equiv 376 \pmod{1000}$

*47. $2076x \equiv 564 \pmod{1776}$

4.3 The Pollard Rho Factoring Method

Over the years, number theorists have expended considerable time and effort to develop efficient algorithms for primality and factorization. In this section, we pursue a factorization technique developed in 1974 by John M. Pollard. Although Pollard called it the *Monte Carlo method* to reflect the seemingly random nature of the numbers generated in the factorization process, it is now called the *Pollard rho method* for reasons that will become clear later. This method works remarkably well for factors with no more than 20 digits.

The composite nature of the Fermat number f_8 had been known since 1909. However, no factors were discovered until 1980, when R. P. Brent and Pollard successfully employed the rho method to find one of its two prime factors.

To describe the algorithm, consider a large odd integer n known to be composite. Choose some seed value x_0 and a diophantine polynomial $f(x)$ of degree ≥ 2, say,

$$f(x) = x^2 + a$$

where $a \neq 0, -2$. We then generate a "random" sequence $\{x_k\}$ of distinct least nonnegative residues modulo n using the recursive formula

$$x_{k+1} \equiv f(x_k) \pmod{n}$$

where $k \geq 0$.

Our goal is to find a nontrivial factor d of n. Assume it is very small compared to n. Since there are exactly d congruent classes modulo d and $d < n$, the integers x_k modulo d must become periodic; that is, there must exist residues x_i and x_j such that $x_i \equiv x_j \pmod{d}$, where $i < j$. Thus, the choice of x_0 and $f(x)$ must be such that $x_i \equiv x_j \pmod{d}$, but $x_i \not\equiv x_j \pmod{n}$. Since $d | (x_j - x_i)$ and $n \nmid (x_j - x_i)$, it follows that the gcd $(x_j - x_i, n)$ is a nontrivial factor of n, which can be found using the euclidean algorithm. Notice that the knowledge of d does not occur in the computation of $(x_j - x_i, n)$.

Thus, to find a nontrivial factor of n, we continue computing $(x_j - x_i, n)$ for every distinct pair x_j, x_i until we encounter a nontrivial gcd. Such a gcd need not be a prime or the smallest factor of n.

The following example illustrates this sophisticated algorithm.

EXAMPLE 4.26 Let $n = 7943$, $x_0 = 2$, and $f(x) = x^2 + 1$. Then

$$x_1 = 5, \quad x_2 = 26, \quad x_3 = 677, \quad x_4 = 5579, \quad x_5 = 4568, \quad x_6 = 364,$$
$$x_7 = 5409, \quad \ldots$$

We now compute the gcd $(x_j - x_i, n)$ for every distinct pair x_j, x_i until a nontrivial gcd emerges. Since $(x_6 - x_2, n) = (364 - 26, 7943) = (338, 7943) = 169$, $169|7943$. ■

The above algorithm has the disadvantage that we need to compute $(x_j - x_i, n)$ for every distinct pair x_j, x_i until a nontrivial gcd occurs. This can be time consuming. Fortunately, we can do better.

A Refined Version

Since $x_i \equiv x_j \pmod{d}$,

$$x_{i+1} \equiv f(x_i) \equiv f(x_j) \equiv x_{j+1} \pmod{d}$$

where $i < j$. Consequently, the elements of the sequence $\{x_k\}$ reduced modulo d repeat in every block of $j-i$ elements; that is, $x_r \equiv x_s \pmod{d}$, where $r \equiv s \pmod{j-i}$, and $r, s \geq i$. In fact, $\{x_k\}$ reduced modulo d is periodic with period that is a factor of $j-i$.

In particular, let t be the smallest multiple of $j-i$ that is greater than i. Then $t \equiv 0 \pmod{j-i}$; so $2t \equiv t \pmod{j-i}$. Consequently, $x_t \equiv x_{2t} \pmod{d}$. Thus, to find a nontrivial factor of n, we compute the gcd's $(x_{2k} - x_k, n)$, where $k \geq 1$, as the next example demonstrates.

EXAMPLE 4.27 Using the Pollard rho method, factor the integer 3893.

SOLUTION

We have $n = 3893$. Choosing $x_0 = 2$ and $f(x) = x^2 + 1$, we generate the sequence $\{x_k\}$:

$$5, 26, 677, 2849, 3790, 2824, 2113, 3392, 1850, 554, 3263, 3708, \ldots$$

Next, we compute $(x_{2k} - x_k, n)$ for each value of $k \geq 1$ until a nontrivial gcd appears:

$$\begin{array}{llll}
(x_2 - x_1, n) = (21, 3893) = 1 & (x_4 - x_2, n) = (2823, 3893) = 1 \\
(x_6 - x_3, n) = (2147, 3893) = 1 & (x_8 - x_4, n) = (543, 3893) = 1 \\
(x_{10} - x_5, n) = (3236, 3893) = 1 & (x_{12} - x_6, n) = (884, 3893) = 17
\end{array}$$

Thus, $17|3893$ and $3893 = 17 \cdot 229$. ■

The sequence $\{x_k\}$ in Example 4.26

$$2, 5, 26, 677, 5579, 4568, 364, 5409, \ldots$$

when reduced modulo 13, yields the periodic sequence

$$2, \underbrace{5, 0, 1, 2}, \underbrace{5, 0, 1, 2}, \underbrace{5, 0, 1, 2}, 5, 0, \ldots$$

with period 4.

This periodic behavior can be displayed pictorially, as in Figure 4.2. Since it resembles the Greek letter ρ (rho), the factoring method is now known as the **rho method**.

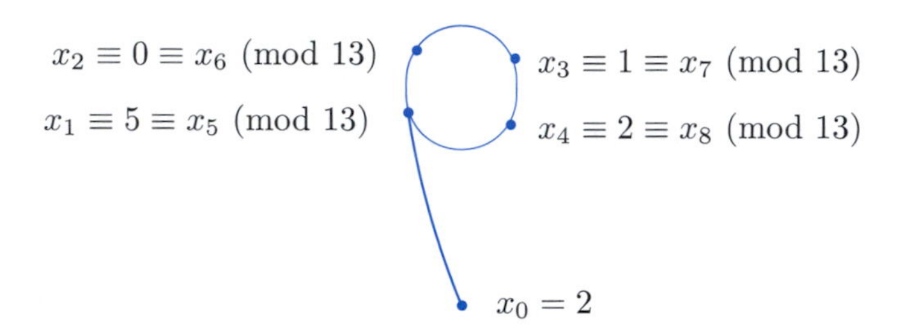

Figure 4.2

In Section 7.2, we shall describe another factoring technique developed by Pollard.

EXERCISES 4.3

Using the Pollard rho method with $x_0 = 2$ and $f(x) = x^2 + 1$, find the canonical decomposition of each integer.

1. 377
2. 3199
3. 5567
4. 9773

Find a factor of 39,997 using the Pollard rho method, the given seed x_0, and the given polynomial $f(x)$.

5. $x_0 = 1, \quad f(x) = x^2 + 1$
6. $x_0 = 2, \quad f(x) = x^2 + 1$
7. $x_0 = 2, \quad f(x) = x^2 - 1$
8. $x_0 = 3, \quad f(x) = x^2 - 1$

CHAPTER SUMMARY

We developed the language of congruences, some fundamental properties, and some simple applications, including a practical factoring technique.

Congruences

- $a \equiv b \pmod{m}$ if and only if $m|(a-b)$. (p. 212)
- $a \equiv b \pmod{m}$ if and only if $a = b + km$ for some integer k. (p. 213)
- $a \equiv a \pmod{m}$ (reflexive property) (p. 213)
- If $a \equiv b \pmod{m}$, then $b \equiv a \pmod{m}$ (symmetric property). (p. 213)
- If $a \equiv b \pmod{m}$ and $b \equiv c \pmod{m}$, then $a \equiv c \pmod{m}$ (transitive property). (p. 213)
- If $a \equiv b \pmod{m}$, then
  - $a + c \equiv b + c \pmod{m}$ (p. 219)
  - $ac \equiv bc \pmod{m}$ (p. 219)
  - $a^n \equiv b^n \pmod{m}$ (p. 220)
- If $ac \equiv bc \pmod{m}$ and $(c, m) = 1$, then $a \equiv b \pmod{m}$. (p. 226)
- If $ac \equiv bc \pmod{m}$ and $(c, m) = d$, then $a \equiv b \pmod{m/d}$. (p. 227)
- If $a \equiv b \pmod{m_i}$, where $1 \le i \le k$, then $a \equiv b \pmod{[m_1, m_2, \ldots, m_k]}$. (p. 228)

Linear Congruences

- A congruence of the form $ax \equiv b \pmod{m}$ is a linear congruence. (p. 230)
- The linear congruence $ax \equiv b \pmod{m}$ is solvable if and only if $d|b$, where $d = (a, m)$; when solvable, it has d incongruent solutions. (p. 231)
- The unique solution of $ax \equiv b \pmod{m}$, where $(a, m) = 1$, is the least residue of $a^{-1}b$ modulo m. (p. 234)

The Pollard Rho Factoring Method

- Let x_0 be a starting value and $f(x) = x^2 + a$, where $a \neq 0, -2$. Then $(x_j - x_i, n)$ is a nontrivial factor of n. (p. 238)

REVIEW EXERCISES

1. If today is Thursday, what day will it be in 1001 days?
2. If today is Wednesday, what day will it be in 4567 days?
3. If it is 11:30 A.M. now, what time will it be in 1770 hours?
4. If it is 11:30 P.M. now, what time will it be in 4455 hours?
5. Give a counterexample to show that $(a, m) = (b, m)$ does not imply that $a \equiv b \pmod{m}$.

Let p be a prime. What is your conclusion if

6. $p \equiv 2 \pmod{10}$?

7. $p \equiv 5 \pmod{10}$?

Determine whether each linear congruence is solvable.

8. $7x \equiv 10 \pmod{13}$

9. $15x \equiv 24 \pmod{20}$

Determine the number of incongruent solutions of each linear congruence.

10. $13x \equiv 14 \pmod{15}$

11. $15x \equiv 40 \pmod{25}$

Find the incongruent solutions of each linear congruence.

12. $5x \equiv 7 \pmod{8}$

13. $13x \equiv 14 \pmod{15}$

14. $15x \equiv 40 \pmod{25}$

15. $36x \equiv 96 \pmod{156}$

Using congruences, solve each LDE.

16. $15x + 25y = 40$

17. $36x + 156y = 96$

18. Find the least residues modulo 15 that are invertible.
19. Find the least residues modulo 18 that are self-invertible.

Find the remainder when

20. $1! + 2! + \cdots + 100!$ is divided by 11.
21. $1! + 2! + \cdots + 300!$ is divided by 13.
22. Find the ones digit in the sum $1! + 2! + \cdots + 100!$.
23. Find the ones digit in the ternary representation of a triangular number t_n.
24. Find the ones digit in the base-four representation of a square number s_n.

Using modular exponentiation, find the remainder when

25. 5^{103} is divided by 13.

26. 13^{1001} is divided by 17.

Find the remainder when

27. 3^{100} is divided by 91.

28. 23^{243} is divided by 17.

29. $2^{100} + 3^{123}$ is divided by 11.

30. $7^{2002} - 13^{1024}$ is divided by 19.

31. $13^{13!}$ is divided by 17.

32. $23^{18!}$ is divided by 19.

Find the last two digits in each number.

33. 3^{3434}

34. 4^{4444}

Find the last three digits in each number.

35. 3^{3003}

36. 19^{1776}

Find the units digit in the decimal value of each.

37. $1024^{1025^{1026^{1027}}}$

38. $1773^{1776^{1779^{2002}}}$

39. Find the remainder when $\sum_{k=1}^{100} k!$ is divided by 12.

40. Find the ones digit when $\sum_{k=1}^{100} k!$ is represented in base fifteen.

It is 3 P.M. now. What time will it be in

41. $\sum_{k=1}^{100} k!$ hours?
42. $\sum_{k=1}^{1000} k!$ hours?
43. Let p and q be twin primes such that $pq - 2$ is also a prime. Find the possible values of p. (J. D. Baum, 1977)

*44. Find the ones digit in $\left\lfloor \frac{10^{20000}}{10^{100}+3} \right\rfloor$. (Putnam Mathematics Competition, 1986)

Using the Pollard rho method, find the canonical decomposition of each integer.

45. 7429
46. 12121

Prove each, where p and q are distinct primes.

47. If $a^2 \equiv b^2 \pmod{p}$, $a \equiv \pm b \pmod{p}$.
48. $n^2 \equiv n \pmod{2}$
49. $n^3 \equiv n \pmod{3}$
50. $2^{4n} + 3n \equiv 1 \pmod{9}$
51. $4^{2n} + 10n \equiv 1 \pmod{25}$
52. If $a \equiv b \pmod{m}$, then $(a, m) = (b, m)$.
53. Let $a \equiv b \pmod{p}$ and $a \equiv b \pmod{q}$. Then $a \equiv b \pmod{pq}$.
54. Let p_n denote the nth prime. Then $p_1 p_2 \cdots p_n + 1$ is not a square. (L. Moser, 1951)
55. If $12 \cdot 900^n + 1$ is a prime, then it is a twin prime. (L. Marvin, 1970)
56. $99991 | \{1 + [1 + (10^{10} - 1)^{99989}](10^{999890} - 1)\}$ (F. J. Durante, 1955)
57. $2^p + 3^p$ is never a perfect power, where $p > 2$. (E. Just, 1973)
58. Let $p^2 \not\equiv p \pmod{p}$. Then $p^{2n-1} + p^{2n-3} + \cdots + p + n \equiv 0 \pmod{3}$ (R. S. Luthar and S. Wurzel, 1966)

SUPPLEMENTARY EXERCISES

1. The integer $1287xy6$ is a multiple of 72. Find the number xy. (*Mathematics Teacher*, 1986)
2. Solve: $1! + 2! + 3! + \cdots + n! = m^2$. (E. T. H. Wang, 1979)
3. Find the largest factor of $A_n = 2801^n - 2696^n - 2269^n + 169^n$, for all $n \geq 1$. (*The Mathematica Gazette*, 1995)

4. The year 1456 was the only recorded perihelion year of Halley's comet that was a multiple of 7. The most recent Halley years were 1835, 1910, and 1986, and the next one is 2061. Show that $1835^{1910} + 1986^{2061}$ is a multiple of 7.
5. Show that $1^{1999} + 2^{1999} + \cdots + 2000^{1999}$ is a multiple of 2001.

An n-digit positive integer a is an **automorphic number** if the last n digits of a^2 equals a. Clearly, 0 and 1 are automorphic.

6. Find four nontrivial automorphic numbers.
7. Prove: Every automorphic number must end in 0, 1, 5, or 6.
8. Prove: If a is automorphic, then $a^2 \equiv a \pmod{10^n}$.
9. Find all integer triplets (x, y, z) such that $xy \equiv 1 \pmod{z}$, $yz \equiv 1 \pmod{x}$, and $zx \equiv 1 \pmod{y}$, where $2 \leq x \leq y \leq z$. (G. Gilbert, 1991)
10. Let $n \geq 2$. Prove that n is a prime if and only if $\binom{n-1}{k} \equiv (-1)^k \pmod{n}$, where $0 \leq k < n$. (E. Deutsch and I. M. Gessel, 1997)

*11. Find all integer solutions (x, y, z) of the equations $xy \bmod z = yz \bmod x = zx \bmod y = 2$. (D. Knuth, 2003)

*12. Find all positive integers m and n such that $2^m + 3^n$ is a square. (E. Just, 1973)

COMPUTER EXERCISES

Write a program to perform each task.

1. Read in a positive integer n. Suppose today is day d, where $0 \leq d < 7$. Determine the day in n days.
2. Read in a certain time of the day and a positive integer n. Determine the time of the day in n hours.
3. Read in integers a, b, and m, and determine if the congruence $ax \equiv b \pmod{m}$ is solvable. Find the number of incongruent solutions when it is solvable.
4. Read in a positive integer n. Find the least residues modulo n that are
   (a) invertible.
   (b) self-invertible.
5. Verify that the sum of no combination of the integers 0, 1, and 4 is congruent to 7 modulo 8.
6. Using modular exponentiation, find the remainder when
   (a) 3^{181} is divided by 17.
   (b) 3^{247} is divided by 25.

7. Solve the original monkey and coconuts puzzle.
8. Solve Williams' version of the monkey and coconuts puzzle.
9. Construct a table of values of the function $K(n) = [(n+8d)/9]^2 - [(n+8d)/9] + 41$, where $-167 \leq n \leq 168$ and d is the least residue of n modulo 9. Identify each value as prime or composite. (T. Koshy, 1994)
10. Redo Program 9 with $K(n) = [(n+8d)/9]^2 - 79[(n+8d)/9] + 1601$, where $0 \leq n \leq 367$ and d is the least residue of n modulo 9. Identify each value as prime or composite. (T. Koshy, 1994)
11. Using the Pollard rho method, factor $2^{32} + 1$ and $2^{64} + 1$.

ENRICHMENT READINGS

1. I. G. Bashmakova, *Diophantus and Diophantine Equations*, Mathematical Association of America, Washington, DC, 1997.
2. A. H. Beiler, *Recreations in the Theory of Numbers*, Dover, New York, 1966, pp. 31–38.
3. M. Gardner, *Mathematical Puzzles and Diversions*, University of Chicago Press, Chicago, 1987.
4. T. Koshy, "Linear Diophantine Equations, Linear Congruences, and Matrices," *The Mathematics Gazette*, 82 (July 1998), 274–277.
5. C. S. Ogilvy and J. T. Anderson, *Excursions in Number Theory*, Dover, New York, 1966.
6. F. Sajdak, "The Rosberry Conjecture," *Mathematical Spectrum*, 28 (1995–1996), 33.
7. S. Singh and D. Bhattacharya, "On Dividing Coconuts," *The College Mathematics Journal*, 28 (May 1987), 203–204.

Congruence Applications

Mighty are numbers, joined with art resistless.
—EURIPIDES

Congruence applications, as we will see shortly, are part of everyday life. The applications include the standard divisibility tests, interesting puzzles, modular designs, product identification codes, German bank notes, round-robin tournaments, and a perpetual calendar.

5.1 Divisibility Tests

The theory of congruences can be used to develop simple tests for checking whether a given integer n is divisible by an integer m. This section presents a few of them.

Let $n = (n_k n_{k-1} \dots n_1 n_0)_{\text{ten}}$ be the decimal representation of n; that is, $n = n_k 10^k + n_{k-1} 10^{k-1} + \cdots + n_1 10 + n_0$. We shall use this expansion to develop divisibility tests for 10, 5, 2^i, 3, 9, and 11. We begin with the test for 10.

Divisibility Test for 10

Because $10 \equiv 0 \pmod{10}$, by Theorems 4.4 and 4.5, $n \equiv n_0 \pmod{10}$. So n is divisible by 10 if and only if n_0 is divisible by 10; that is, if and only if $n_0 = 0$. Thus, *an integer is divisible by* 10 *if and only if its units digit is zero.*

Divisibility Test for 5

Because $n \equiv n_0 \pmod{10}$, n is divisible by 5 if and only if n_0 is divisible by 5. But the only single-digit numbers divisible by 5 are 0 and 5, so *an integer is divisible by 5 if and only if it ends in a 0 or 5.*

Divisibility Test for 2^i

Because $10 \equiv 0 \pmod 2$, $10^i \equiv 0 \pmod{2^i}$ for all positive integers i. Therefore, by Theorems 4.4 and 4.5, we have

$$\begin{aligned} n &\equiv n_0 \pmod 2 \\ &\equiv n_1 n_0 \pmod{2^2} \quad (\textit{Note}: n_1 n_0 \text{ denotes a two-digit number.}) \\ &\equiv n_2 n_1 n_0 \pmod{2^3} \\ &\vdots \\ &\equiv n_{i-1} n_{i-2} \ldots n_1 n_0 \pmod{2^i} \end{aligned}$$

Thus, *an integer n is divisible by 2^i if and only if the number formed by the last i digits in n is divisible by 2^i.*

In particular, n is divisible by 2 if and only if the ones digit n_0 is divisible by 2; it is divisible by 4 if the two-digit number $n_1 n_0$ is divisible by 4; it is divisible by 8 if the three-digit number $n_2 n_1 n_0$ is divisible by 8, and so on.

For example, let $n = 343{,}506{,}076$. Since $2|6$, $2|n$; $4|76$, so $4|n$; but $8 \nmid 076$, so $8 \nmid n$.

Divisibility Tests for 3 and 9

Because $10 \equiv 1 \pmod 3$, $10^i \equiv 1 \pmod 3$, by Theorem 4.5. So by Theorem 4.4, $n \equiv n_k + n_{k-1} + \cdots + n_1 + n_0 \pmod 3$. Thus, *an integer is divisible by 3 if and only if the sum of its digits is divisible by 3.*

Likewise, since $n \equiv n_k + n_{k-1} + \cdots + n_1 + n_0 \pmod 9$, *an integer is divisible by 9 if and only if the sum of its digits is divisible by 9.*

For example, let $n = 243{,}506{,}076$. The sum of its digits is $2+4+3+5+0+6+0+7+6 = 33$. Since $3|33$, $3|n$; but $9 \nmid 33$, so $9 \nmid n$.

Next we turn to the divisibility test for 11.

Divisibility Test for 11

Notice that $10 \equiv -1 \pmod{11}$, $10^i \equiv (-1)^i \pmod{11}$, by Theorem 4.5. So again by Theorem 4.5,

$$n \equiv (-1)^k n_k + \cdots - n_3 + n_2 - n_1 + n_0 \pmod{11}$$

Thus, $11|n$ *if and only if* $(n_0 + n_2 + \cdots) - (n_1 + n_3 + \cdots)$ *is divisible by* 11; *that is, if and only if the sum of the digits in the "even" positions minus that in the "odd" positions is divisible by* 11.

For example, let $n = 243{,}506{,}076$.

$$\text{Desired difference} = (6+0+0+3+2) - (7+6+5+4)$$
$$= 11 - 22 = -11$$

Because $11|-11$, $11|n$ also.

The following theorem identifies a class of integers that are divisible by 11.

THEOREM 5.1 A palindrome with an even number of digits is divisible by 11.

PROOF

Let $n = n_{2k-1}n_{2k-2}\ldots n_1 n_0$ be a palindrome with an even number of digits. Then

$$n \equiv (n_0 + n_2 + \cdots + n_{2k-2}) - (n_1 + n_3 + \cdots + n_{2k-1}) \pmod{11}$$
$$\equiv 0 \pmod{11}$$

because n is a palindrome with an even number of digits. Thus, $11|n$. ■

For example, both palindromes 1331 and 60,588,506 contain an even number of digits, so both are divisible by 11.

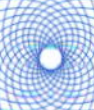

You should be aware, however, that this theorem does *not* apply to palindromes with an odd number of digits. For instance, the palindrome 131 contains an odd number of digits. However, it is not divisible by 11.

Note that these divisibility tests can be extended to nondecimal bases. See Supplementary Exercises 4 and 5.

Casting Out Nines

Next we can demonstrate a technique called **casting out nines** (in other words, canceling numbers that add up to 9). This technique can be used to detect computational errors, as the following two examples show. Casting out nines is based on the fact that *every integer is congruent to the sum of its digits modulo* 9.

EXAMPLE 5.1 Using casting out nines, check if the sum of the numbers 3569, 24,387, and 49,508 is 78,464.

SOLUTION

We have

$$\begin{aligned} 3569 &\equiv 3+5+6+9 &&\equiv 5 \pmod 9 \\ 24387 &\equiv 2+4+3+8+7 &&\equiv 6 \pmod 9 \\ 49508 &\equiv 4+9+5+0+8 &&\equiv 8 \pmod 9 \\ \hline \text{Their sum} & &&\equiv 5+6+8 \pmod 9 \\ & &&\equiv 1 \pmod 9 \end{aligned}$$

$$\begin{aligned} \text{Given answer} = 78464 &\equiv 7+8+4+6+4 \pmod 9 \\ &\equiv 2 \pmod 9 \end{aligned}$$

Thus, the given answer is not congruent to the actual sum modulo 9; consequently, the given sum is *definitely wrong*. (The correct sum is 77,464.) ■

EXAMPLE 5.2 Using casting out nines, determine whether the product of 1976 and 3458 is 6,833,080.

SOLUTION

$$\begin{aligned} 1976 &\equiv 1+9+7+6 &&\equiv 5 \pmod 9 \\ 3458 &\equiv 3+4+5+8 &&\equiv 2 \pmod 9 \\ \hline \text{Their product} & &&\equiv 1 \pmod 9 \end{aligned}$$

$$\begin{aligned} \text{Given answer} = 6{,}833{,}080 &\equiv 6+8+3+3+0+8+0 \pmod 9 \\ &\equiv 1 \pmod 9 \end{aligned}$$

Because the given answer is congruent to the actual product modulo 9, we might be tempted to say that the given answer is correct. In fact, all we can say is, it is *probably correct*. This is so because any rearrangement of the digits of an integer yields the same least residue modulo 9, an idea used by today's accountants. (The given answer is in fact wrong. The correct answer is 6,833,008.) ■

As these two examples indicate, the only answer we can provide by using casting out nines is that the given solution is either *definitely wrong* or *probably correct*.

Digital Root

Closely related to casting out nines is the concept of the **digital root** of a positive integer N. It is computed by iteration: Find the sum s of its digits; then find the sum of the digits in s; continue this procedure until a single digit d emerges; then d is the digital root of N.

For example, to find the digital root of 1976, add its digits: $1+9+7+6=23$; now add its digits: $2+3=5$; so the digital root of 1976 is 5.

Notice that $1976 \equiv 5 \pmod 9$. More generally, let $N=(a_n \ldots a_1 a_0)_{\text{ten}}$ and let d be its digital root. Then $d \equiv (a_n+\cdots+a_1+a_0) \pmod 9$. Thus, the digital root of N is the remainder when N is divided by 9, with one exception: It is 9 if the remainder is 0.

The following example identifies the possible digital roots of perfect squares.

EXAMPLE 5.3 Find the digital roots of square numbers.

SOLUTION

By the division algorithm, every integer n is of the form $9k+r$, where $0 \le r < 9$. So $n \equiv r \pmod 9$ and hence $n^2 \equiv r^2 \pmod 9$. Since $r \equiv r-9 \pmod 9$, $0^2 \equiv 0 \pmod 9$, $(\pm 1)^2 \equiv 1 \pmod 9$, $(\pm 2)^2 \equiv 4 \pmod 9$, $(\pm 3)^2 \equiv 0 \pmod 9$, and $(\pm 4)^2 \equiv 7 \pmod 9$. Thus, n^2 is congruent to 0, 1, 4, or 7, so its digital root is 1, 4, 7, or 9. ■

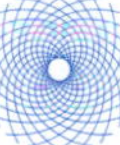

This example can serve as a test to determine whether a positive integer can be a square: *If an integer is a square, then its digital root must be* 1, 4, 7, *or* 9.

EXAMPLE 5.4 Determine whether $N = 16{,}151{,}613{,}924$ can be a square.

SOLUTION

$$\begin{aligned}\text{Digital root of } N &\equiv (1+6+1+5+1+6+1+3+9+2+4) \pmod 9\\ &\equiv 3 \pmod 9\end{aligned}$$

Because the digital root is 3, N is not a square. ■

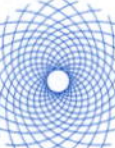

Note that the converse of the preceding statement is false; that is, *if the digital root of N is* 1, 4, 7, *or* 9, *then N need not be a square*. For instance, the digital root of 43 is 7, but 43 is not a square.

The following example identifies the digital root of the product of twin primes, except for the pair 3 and 5.

EXAMPLE 5.5 Prove that the digital root of the product of twin primes, other than 3 and 5, is 8.

PROOF

Every prime > 3 is of the form $6k-1$ or $6k+1$ (see Exercise 80 in Section 4.1), so we can take the twin primes to be $6k-1$ and $6k+1$. Their product $= (6k-1)(6k+1) = 36k^2 - 1 \equiv 0 - 1 \equiv 8 \pmod 9$. So the digital root of the product is 8. (Notice that the digital root of $3 \cdot 5$ is 6 and *not* 8.) ■

EXERCISES 5.1

Mark *True* or *False*.

1. Every integer divisible by 5 is odd.
2. Every integer divisible by 11 is odd.
3. 11 is a palindrome.
4. Every palindrome is divisible by 9.
5. $10^{1000} - 1$ is divisible by 9.
6. $10^{1000} - 1$ is divisible by 11.

Which of the following numbers are divisible by 2? By 4? By 8?

7. 427,364
8. 30,587,648
9. 800,358,816
10. 398,008,576

Which of the following numbers are divisible by 3? By 9?

11. 205,876
12. 31,876,203
13. 5,588,610,911
14. 767,767,767

Determine whether each number is divisible by 6.

15. 87,654
16. 327,723
17. 639,576
18. 2,197,584

Determine whether each number is divisible by 11.

19. 43,979
20. 548,152
21. 502,458
22. 1,928,388

Using casting out nines, identify each computation as *probably correct* or *definitely wrong*.

23. $\begin{array}{r} 35897 \\ 750971 \\ +\,908085 \\ \hline 1684953 \end{array}$

24. $\begin{array}{r} 58807 \\ 83291 \\ +\,601756 \\ \hline 748354 \end{array}$

25. $\begin{array}{r} 7958036 \\ -\,2309859 \\ \hline 5948177 \end{array}$

26. $\begin{array}{r} 8314302 \\ -\,3708594 \\ \hline 4605798 \end{array}$

27. $\begin{array}{r} 2076 \\ \times\,1076 \\ \hline 223766 \end{array}$

28. $\begin{array}{r} 4556 \\ \times\,3443 \\ \hline 15745034 \end{array}$

Using casting out nines, find the missing nonzero digit d in each computation.

29. $7961 - 1976 = 59d5$
30. $7167 - 1776 = 53d1$
31. $253 \cdot 86 = 2d758$
32. $123 \cdot 98 = 120d4$
33. Find all four-digit integers of the form $4ab8$ that are divisible by 2, 3, 4, 6, 8, and 9. (*Mathematics Teacher*, 1992)
34. The seven-digit number $21358ab$ is divisible by 99. Find a and b. (*Mathematics Teacher*, 1992)
35. Find the smallest number that leaves a remainder i when divided by $i + 1$, where $1 \leq i \leq 9$. (*Mathematics Teacher*, 1993)
36. Show that every six-digit number of the form $abcabc$ is divisible by 7, 11, and 13.
37. Develop a divisibility test for 37.
[*Hint*: $10^3 \equiv 1 \pmod{37}$.]

A procedure similar to casting out nines, called **casting out twos**, can be applied to check the accuracy of numeric computations of binary numbers. In this process, we cancel pairs of bits that add up to 0 modulo 2. Using casting out twos, determine whether each computation is *probably correct* or *definitely wrong*.

38.
$$\begin{array}{r} 10110110 \\ 1011111 \\ +\,1110011 \\ \hline 110001100 \end{array}$$

39.
$$\begin{array}{r} 110110111 \\ -\,11001101 \\ \hline 11101010 \end{array}$$

40.
$$\begin{array}{r} 1011101 \\ \times\,1011 \\ \hline 1111110011 \end{array}$$

Find the digital root of each.

41. 16,429,058
42. 17^{76}
43. 1776^{1776}
44. 2020^{9999}
45. Suppose the digital root of an integer n is 9. Show that the digital root of any multiple of n is also 9.

Determine whether each can be a square.

46. 54,893,534,046
47. 61,194,858,376
48. Find the possible values of the digital root of a cube.

Prove each.

49. The units digit of a triangular number is 0, 1, 3, 5, 6, or 8.
50. If a three-digit integer abc is divisible by 37, then its cyclic permutations are also divisible by 37.
51. Let $d|R_n$, where R_n is a repunit. If $d|a_n a_{n-1} \ldots a_0$, then d divides every cyclic permutation of $a_n a_{n-1} \ldots a_0$.
52. The digital root $\rho(f_n)$ of the nth Fermat number f_n is given by

$$\rho(f_n) = \begin{cases} 5 & \text{if } n \text{ is odd} \\ 8 & \text{otherwise} \end{cases}$$

53. $2^{2000} + 2^{2001} + 2^{2003} + 2^{2007}$ is not a square.
54. Every integer n in base b is congruent to the sum of its digits modulo $b - 1$.

5.2 Modular Designs

Modular arithmetic can be used to create beautiful designs. We will now explore three such designs: an m-pointed star, an (m, n) residue design, and quilt designs. They are really fun, so enjoy them.

m-Pointed Stars

To construct an m-pointed star, mark m equally spaced points on a large circle, and label them with the least residues 0 through $(m-1)$ modulo m. Choose a least residue

i modulo m, where $(i, m) = 1$. Join each point x with the point $x + i$ modulo m. Now color in the various regions inside the circle with some solid colors. You should get a nice m-pointed star. Figure 5.1 shows a seven-pointed star and a twelve-pointed star.

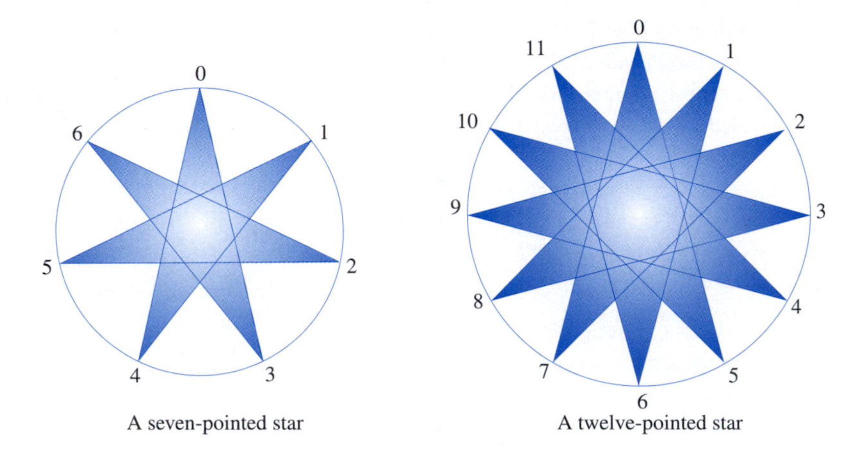

A seven-pointed star

A twelve-pointed star

Figure 5.1

(m, n) Residue Designs

To construct an (m, n) residue design, where $1 \le n < m$ and $(m, n) = 1$, select $m - 1$ equally spaced points on a large circle, label them 1 through $m - 1$, and join each point x to point nx modulo m. Then color in the various regions formed in a systematic way to create exciting designs.

For example, to construct a $(19, 9)$ residue, divide a large circle into 18 equal arcs and label the points 1 through 18. Multiply each nonzero residue modulo 19 by 9:

$9 \cdot 1 = 9$	$9 \cdot 5 = 7$	$9 \cdot 9 = 5$	$9 \cdot 13 = 3$	$9 \cdot 17 = 1$
$9 \cdot 2 = 18$	$9 \cdot 6 = 16$	$9 \cdot 10 = 14$	$9 \cdot 14 = 13$	$9 \cdot 18 = 10$
$9 \cdot 3 = 8$	$9 \cdot 7 = 6$	$9 \cdot 11 = 4$	$9 \cdot 15 = 2$	
$9 \cdot 4 = 17$	$9 \cdot 8 = 15$	$9 \cdot 12 = 13$	$9 \cdot 16 = 11$	

Then join the points 1 and 9, 2 and 18, 3 and 8, ..., and 18 and 10. Color the resulting regions systematically to obtain the beautiful design in Figure 5.2. Additional designs are shown in Figures 5.3 through 5.6.

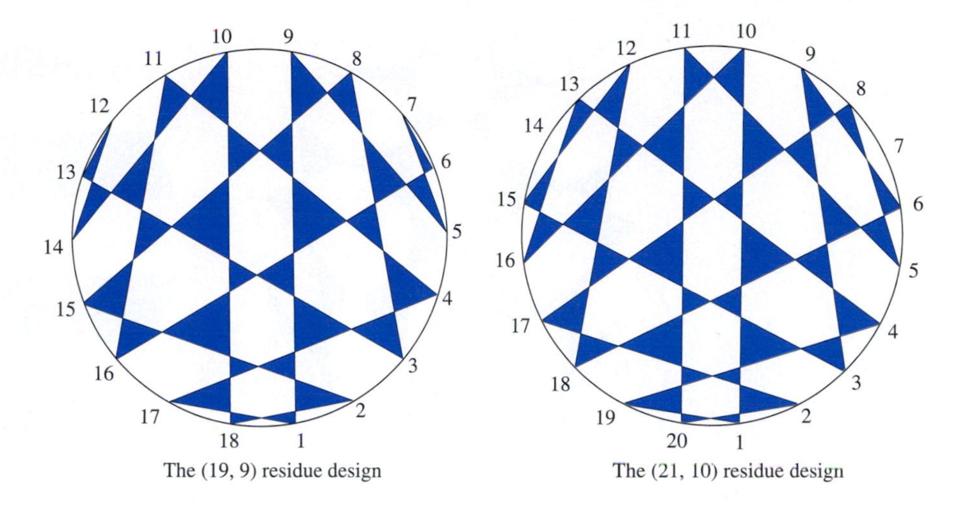

The (19, 9) residue design

The (21, 10) residue design

Figure 5.2

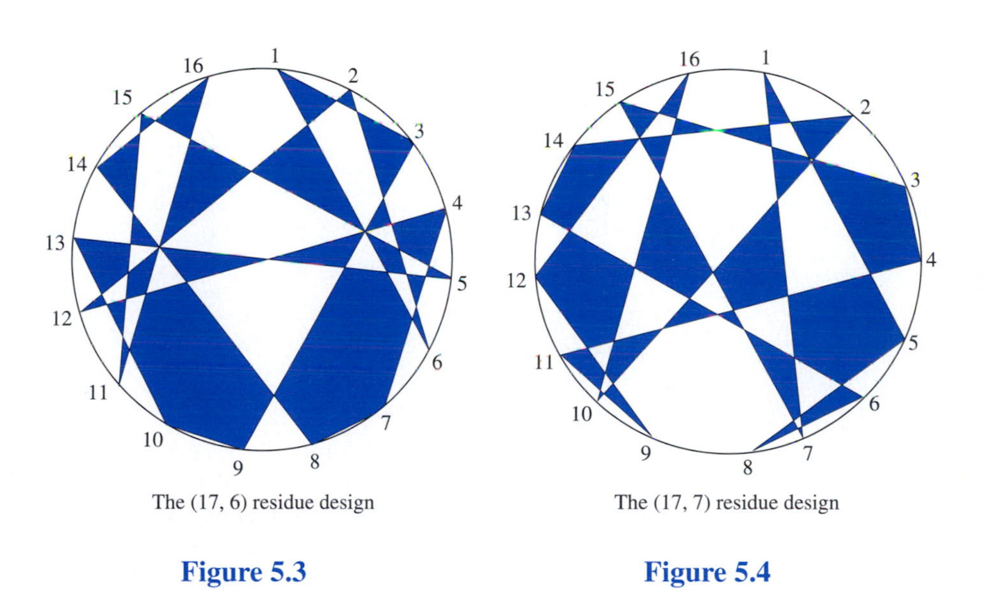

The (17, 6) residue design

The (17, 7) residue design

Figure 5.3

Figure 5.4

Quilt Designs

We can use addition and multiplication tables for least residues modulo m to generate other artistic and interesting designs. For example, choose $m = 9$. Construct the addition table for the set of least residues 0 through 8 modulo 9, as Table 5.1 shows.

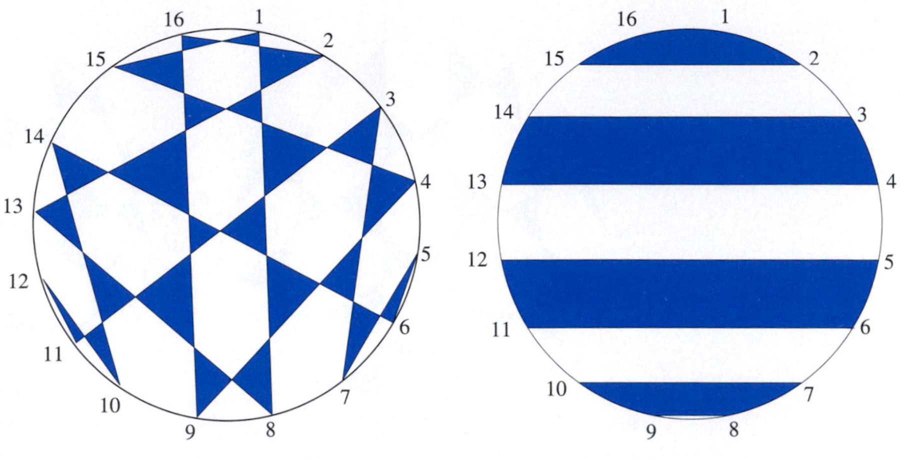

The (17, 8) residue design

Figure 5.5

The (17, 16) residue design

Figure 5.6

+	0	1	2	3	4	5	6	7	8
0	0	1	2	3	4	5	6	7	8
1	1	2	3	4	5	6	7	8	0
2	2	3	4	5	6	7	8	0	1
3	3	4	5	6	⑦	8	0	1	2
4	4	5	6	7	8	0	1	2	3
5	5	6	7	8	0	1	2	3	4
6	6	7	8	0	1	2	3	4	5
7	7	8	0	1	2	3	4	5	6
8	8	0	1	2	3	4	5	6	7

Table 5.1

In this example, the circled number 7 in row 3 and column 4 is $3 + 4$ modulo 9. Devise nine basic design elements to represent each of the numbers 0 through 8, as Figure 5.7 shows. (This design translation uses additive inverses and complements.)

Now replace each entry in the main body of Table 5.1 with the corresponding design element. Figure 5.8 shows the resulting beautiful design. This basic design can be used to generate new designs. For example, flip this design about its right-side edge and then flip the ensuing design about the bottom edge. The two flips produce the fascinating design in Figure 5.9.

Instead of a square grid, we could use a rectangular grid, like the one in Figure 5.10. Use the design elements in Figure 5.11 to develop the basic design for modulo 5. Make the two flips to produce the sensational design in Figure 5.12.

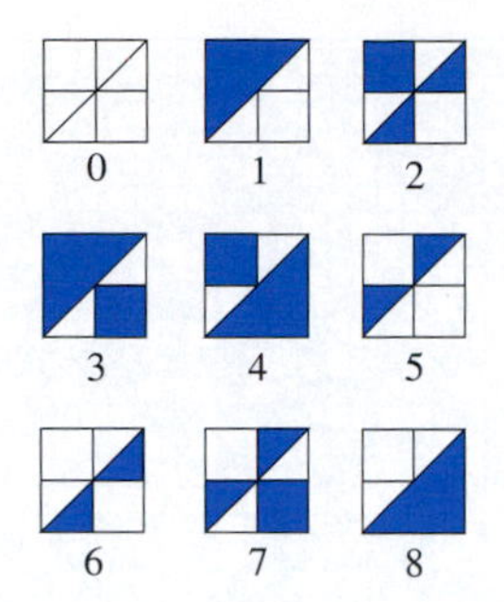

Figure 5.7 *Design elements in mod 9.*

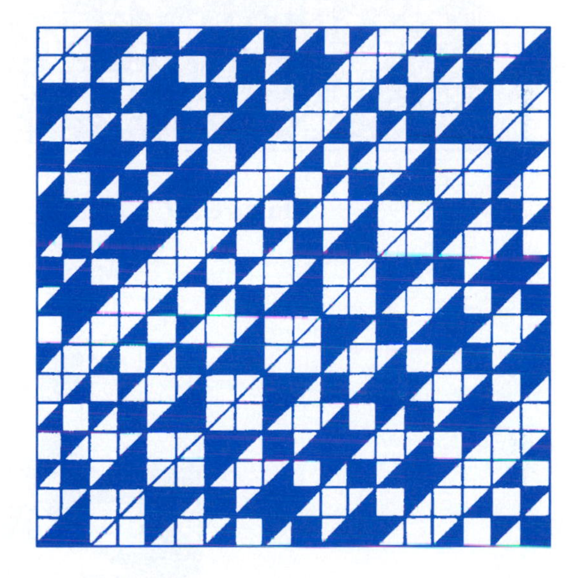

Figure 5.8 *Basic design.*

We can develop designs using different grids, design elements, and multiplication tables.

EXERCISES 5.2

Construct a seven-pointed star by joining

1. Point x to $x + 4$ modulo 7.
2. Point x to $x + 6$ modulo 7.

Construct a 12-pointed star by joining

3. Point x to $x + 7$ modulo 12.
4. Point x to $x + 11$ modulo 12.
5. Construct an 11-pointed star by joining every point x to $x + 4$ modulo 11.

Figure 5.9

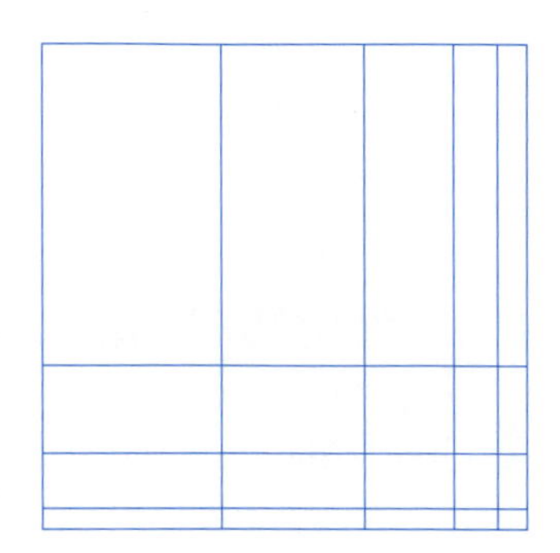

Figure 5.10

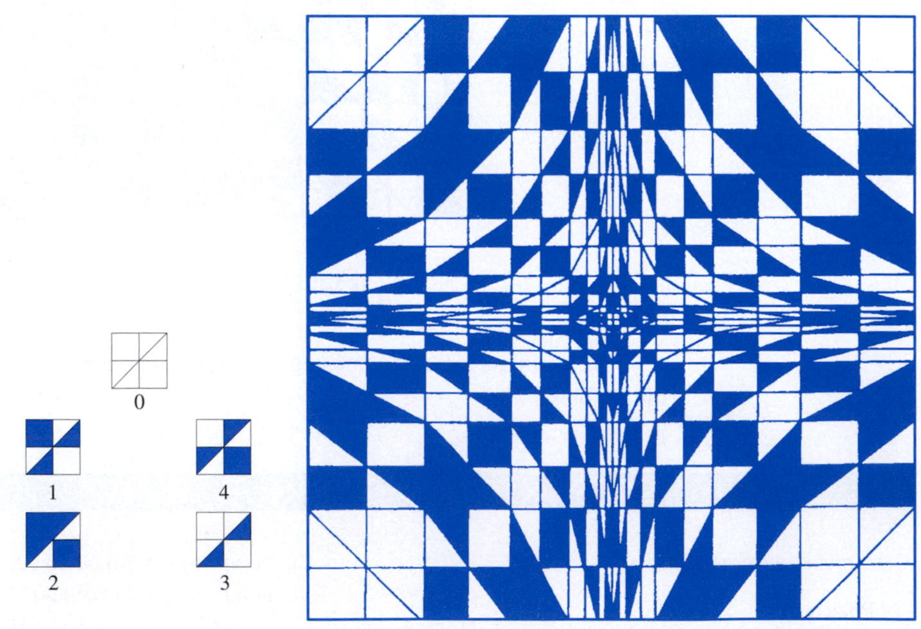

Figure 5.11 *Design elements in mod 5.*

Figure 5.12

6. Construct a 11-pointed star by joining point x to $x+7$ modulo 11.
7. Construct a 13-pointed star by joining point x to $x+4$ modulo 13.

Construct each residue design.

8. $(17, 2)$
9. $(17, 9)$
10. $(23, 11)$
11. $(23, 5)$
12. $(23, 10)$

13. Using the design elements in Figure 5.13 and the addition table in modulo 3, create the basic design in modulo 3.

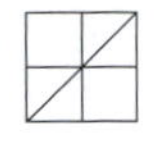

Figure 5.13

14. Flip the basic design in Exercise 13 about its right edge and the resulting design about its bottom edge.
15. Create a 12-pointed star joining point x to point $5x+3$ modulo 12.

5.3 Check Digits

Coding theory is a branch of mathematics devoted to the detection and correction of errors in codes. We will now see how congruences are used to detect and correct errors in transmitted messages.

Binary Codes

First, we turn to **binary codes**, which are messages converted (**encoded**) into bits and transmitted over a channel such as a telephone line. The receiver tries to recover the original message by **decoding** the received message. Any errors in the received message must be detected and then corrected.

The process of **casting out twos** plays a significant role in detecting and correcting errors in binary codes. Before transmission, we append a **parity check bit** x_{n+1} to each binary string $x_1x_2 \ldots x_n$, defined by $x_{n+1} \equiv x_1 + x_2 + \cdots + x_n \pmod 2$. That is, append a 1 if the number of 1s is odd and a 0 otherwise. This procedure keeps the number of 1s in the string always even.

The following example illustrates this technique.

EXAMPLE 5.6 Consider the ten-bit string 1101011101. Then $x_{11} \equiv 1+1+0+1+0+1+1+1+0+1 \equiv 1 \pmod 2$, so the check bit is 1 and the transmitted message is 11010111011. Suppose we receive the string 11010111001; since it contains an odd number of ones, an odd number of errors has occurred during transmission. If there is a single error and its location is known, by changing the bit in that location, we can recover the original message. ■

Check digits are often used to detect errors in strings of decimal digits. Banks, book publishers, libraries, and companies, such as United Parcel Service, that track large numbers of items use check digits to detect errors in their identification numbers, as the next two examples demonstrate. But first, a simple definition.

Dot Product

The dot product of the vectors $(x_1, x_2, \ldots, x_n)$ and $(y_1, y_2, \ldots, y_n)$, is defined by $(x_1, x_2, \ldots, x_n) \cdot (y_1, y_2, \ldots, y_n) = \sum_{i=1}^{n} x_i y_i$.

The following two examples employ dot products.

EXAMPLE 5.7 Every bank check has an eight-digit identification number $d_1 d_2 \ldots d_8$ followed by a check digit d, defined by $d \equiv (d_1, d_2, \ldots, d_8) \cdot (7, 3, 9, 7, 3, 9, 7, 3) \pmod{10}$. Compute the check digit for the identification number 17,761,976.

SOLUTION

$$\begin{aligned}
\text{Check digit} &\equiv (1, 7, 7, 6, 1, 9, 7, 6) \cdot (7, 3, 9, 7, 3, 9, 7, 3) \pmod{10} \\
&\equiv 1 \cdot 7 + 7 \cdot 3 + 7 \cdot 9 + 6 \cdot 7 + 1 \cdot 3 + 9 \cdot 9 + 7 \cdot 7 + 6 \cdot 3 \pmod{10} \\
&\equiv 4 \pmod{10}
\end{aligned}$$

So the nine-digit check number is 177,619,764. ■

The universal product code (UPC) found on grocery items in a supermarket contains a check digit. A UPC number consists of 12 digits $d_1, d_2, \ldots, d_{12}$, of which the first six digits identify the country and the manufacturer, the next five identify the product, and the last digit d_{12} is the check digit.

For example, the UPC number for Maxwell House Instant Coffee made by Kraft General Foods, Inc., in the United States is 043000794708. The codes for country, manufacturer, and the product are 0, 43000, and 79470, respectively:

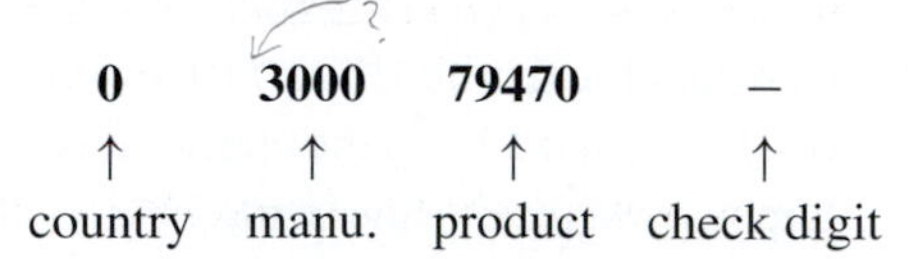

The check digit d_{12} in the UPC number must satisfy the condition $(d_1, d_2, \ldots, d_{12}) \cdot (3, 1, 3, 1, 3, 1, 3, 1, 3, 1, 3, 1) \equiv 0 \pmod{10}$; that is,

$$d_{12} \equiv -(d_1, d_2, \ldots, d_{11}) \cdot (3, 1, 3, 1, 3, 1, 3, 1, 3, 1, 3) \pmod{10}$$

The following example illustrates this method.

EXAMPLE 5.8 Compute the check digit d_{12} in the UPC number for Maxwell House Instant Coffee.

SOLUTION

$$\begin{aligned} d_{12} &\equiv -(d_1, d_2, \ldots, d_{11}) \cdot (3, 1, 3, 1, 3, 1, 3, 1, 3, 1, 3) \pmod{10} \\ &\equiv -(0, 4, 3, 0, 0, 0, 7, 9, 4, 7, 0) \cdot (3, 1, 3, 1, 3, 1, 3, 1, 3, 1, 3) \pmod{10} \\ &\equiv -(0 + 4 + 9 + 0 + 0 + 0 + 21 + 9 + 12 + 7 + 0) \pmod{10} \\ &\equiv -62 \equiv 8 \pmod{10} \end{aligned}$$

So the check digit is 8 and the UPC identification number is 0-43000-79470-8. ■

Zip Codes

The United States Postal Service† uses bar codes to encode zip code information on mail, which can be readily and rapidly read by inexpensive bar code readers. The POSTNET (**POST**al **N**umeric **E**ncoding **T**echnique) bar code may represent a five-digit zip code (32 bars), a nine-digit zip + 4 code (52 bars), or an 11-digit delivery point code (62 bars). It employs both binary numbers and check digits. Some bars are long and the others are short. See Figure 5.14.

Figure 5.14

A long bar (or full bar) represents a 1, and a short one (a half bar) a 0. The two extreme bars are always long and can be ignored. The remaining bars are grouped into blocks of five bars each, indicating that the last block represents a check digit.

† Based on **Designing Business Letter Mail**, United States Postal Service.

The scheme for converting decimal digits into binary is based on the coding scheme used by the Bell Telephone Labs (now Lucent Technologies) in the early 1940s.

There are exactly $\frac{5!}{2!3!} = 10$ arrangements of two long bars and three short bars, and they represent the 10 digits, as Table 5.2 shows.

With the exception of 0, the numeric value of each combination of five bars is found by adding the weights of the two long bars. From right to left, the bar positions are assigned the weights 0, 1, 2, 4, and 7. For example, the value of the code in Figure 5.15 is 5; and the value of the code in Figure 5.16 is 9. (A weight of 8 is not used, because then the weight of the group would be 10.)

The only exception to this rule is the combination ǀǀııı, which has a total weight of 11 but has been assigned the value 0.

Numeric Value	*Bar Position Weights* Binary 74210	Bar code 74210
1	00011	ıııǀǀ
2	00101	ııǀıǀ
3	00110	ııǀǀı
4	01001	ıǀııǀ
5	01010	ıǀıǀı
6	01100	ıǀǀıı
7	10001	ǀıııǀ
8	10010	ǀııǀı
9	10100	ǀıǀıı
0	11000	ǀǀııı

Table 5.2

Figure 5.15

Figure 5.16

Consider the five-digit zip code $z_1z_2\ldots z_5$. A check digit d is appended to it to detect errors:

$$d \equiv -\sum_{i=1}^{5} z_i \pmod{10}$$

For example, the check digit for the zip code 12345 is

$$d \equiv -(1+2+3+4+5) \pmod{10}$$
$$\equiv -5 \equiv 5 \pmod{10}$$

So $d = 5$. The bar code for the zip code is shown in Figure 5.17.

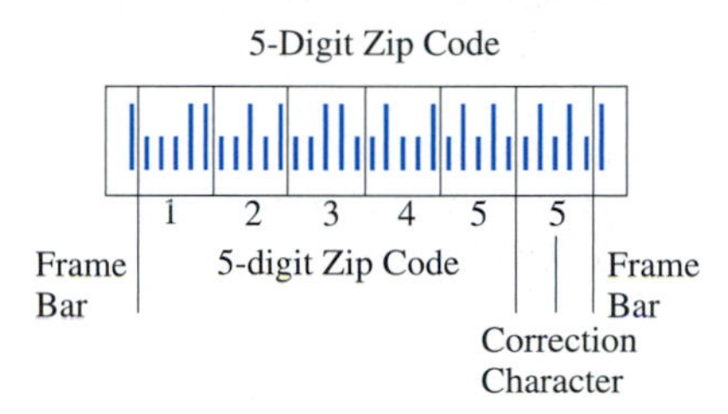

Figure 5.17

A check digit is also appended to every zip + 4 code, which was introduced by the Postal Service in 1983. For instance, consider the nine-digit zip code 12345-6789; its check digit d is given by

$$d \equiv -(1+2+3+4+5+6+7+8+9) \pmod{10}$$
$$\equiv -5 \equiv 5 \pmod{10}$$

So $d = 5$ and the corresponding bar code is shown in Figure 5.18.

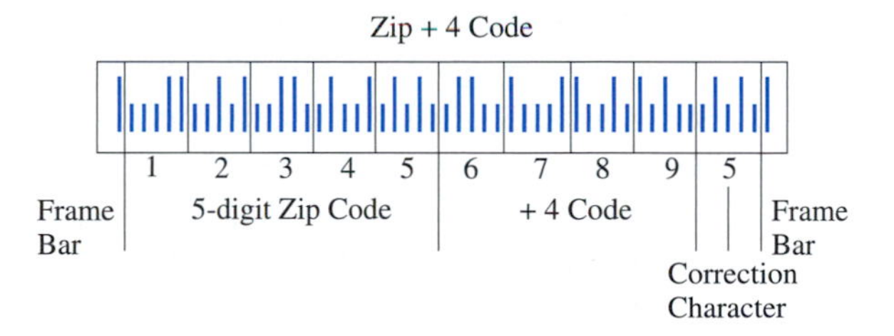

Figure 5.18

The **delivery point bar code** (**DPBC**) was introduced in 1993 by the Postal Service to uniquely identify each of the 115 million delivery points in the United States. It eliminates the need for carriers to sort mail prior to delivery. The delivery point bar code is formed by adding 10 bars to an existing zip + 4 code. The 10 bars represent two additional numbers (normally, the last two numbers of the street address, P.O. box, rural route box, or highway contract route box). See Figure 5.19.

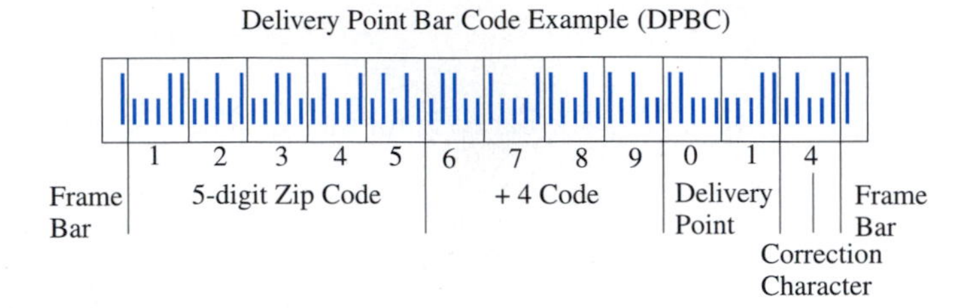

Figure 5.19

An example of the DPBC number is **12345-6789-014**, where 01 is the delivery point and 4 is the check digit. See Figure 5.19.

ISBN

Since 1972, virtually every book published anywhere in the world has an *International Standard Book Number* (**ISBN**), a 10-digit identification number. The ISBN enables computerized transmission and storage of book data. It grew out of the book numbering system introduced into the United Kingdom in 1967 by F. G. Foster of the London School of Economics. In 1968, R. R. Bowker Company introduced it into the United States.

An ISBN consists of four parts: a group code (one digit), a publisher code (two digits), a book code (six digits), and a check digit. For instance, the ISBN of a text by the author is 0-07-035471-5. The group code 0 or 1 indicates the book was published in an English-speaking country†; the publisher code 07 identifies the publisher, McGraw-Hill; and the book code 035471 is assigned by the publisher to the book. The check digit d, where $0 \le d \le 10$, and 10 is denoted by X, is defined by $d \equiv -(x_1, x_2, \ldots, x_9) \cdot (10, 9, 8, 7, 6, 5, 4, 3, 2) \pmod{11}$, where $x_1, x_2, \ldots, x_9$ denote the first nine digits in the ISBN.

The following example demonstrates this coding scheme.

EXAMPLE 5.9 Using the ISBN coding scheme, compute the check digit d if the first nine digits are 0-07-035472.

† Australia, English-speaking Canada, New Zealand, South Africa, the United Kingdom, the United States, and Zimbabwe.

SOLUTION

$$\begin{aligned} d &\equiv -(x_1, x_2, \ldots, x_9) \cdot (10, 9, 8, 7, 6, 5, 4, 3, 2) \pmod{11} \\ &\equiv -(0, 0, 7, 0, 3, 5, 4, 7, 2) \cdot (10, 9, 8, 7, 6, 5, 4, 3, 2) \pmod{11} \\ &\equiv -(0 + 0 + 56 + 0 + 18 + 25 + 16 + 21 + 4) \pmod{11} \\ &\equiv -140 \equiv 3 \pmod{11} \end{aligned}$$

Thus, the check digit is 3 and the ISBN is 0-07-035472-3. ■

EAN Bar Codes

In 1980, the International Article Numbering Association (formerly, the European Article Numbering Association, **EAN**) and the International ISBN Agency reached an agreement by which the ISBN can be translated into an EAN bar code.

All EAN bar codes begin with a national identifier (00-09 for the United States) with one exception: For books and periodicals, the national identifier is replaced with a "bookland" identifier: 978 for books and 977 for periodicals. The 978 bookland/EAN prefix is followed by the first nine ISBN digits; the ISBN check digit is dropped and replaced with a check digit computed according to the EAN rules. See Figure 5.20.

Figure 5.20

Five-Digit Add-On Code

In the United States and a few other countries, a five-digit add-on code is used to provide additional information. This code is often used for price information. The lead digit in the five-digit add-on code designates the national currency; for example, a 5 indicates the U.S. dollar and a 6 the Canadian dollar.

Figure 5.21

Publishers who do not want to indicate the price in the add-on code print the code 90090. See Figure 5.21.

Driver's License Numbers

In the United States, the method used to assign driver's license numbers varies widely from state to state. Some states use check digits when assigning driver's license numbers, in order to detect forgery or errors.

For example, Utah assigns an eight-digit number $d_1d_2\ldots d_8$ in sequential order and then appends a check digit d_9 defined by $d_9 \equiv \sum_{i=1}^{8}(10-i)d_i \pmod{10}$. The American Chemical Society uses this same system for registering chemicals, while the Canadian Province of Newfoundland uses a nearly identical scheme for driver's licenses. The following example illustrates this coding scheme.

EXAMPLE 5.10 Compute the check digit d_9 in a driver's license in Utah if the eight-digit number is 24923056.

SOLUTION

$$\begin{aligned} d_9 &\equiv (9, 8, 7, 6, 5, 4, 3, 2) \cdot (2, 4, 9, 2, 3, 0, 5, 6) \pmod{10} \\ &\equiv (18 + 32 + 63 + 12 + 15 + 0 + 15 + 12) \pmod{10} \\ &\equiv 7 \pmod{10} \end{aligned}$$

So the full license number is 249230567. ■

Some states use even more complicated coding schemes in assigning driver's license numbers. Arkansas, New Mexico, and Tennessee append a check digit d_8 to the seven-digit number $d_1d_2\ldots d_7$, determined as follows:

Let

$$x \equiv -(d_1, d_2, \ldots, d_7) \cdot (2, 7, 6, 5, 4, 3, 2) \pmod{11}$$

Then

$$d_8 = \begin{cases} 1 & \text{if } x = 0 \\ 0 & \text{if } x = 10 \\ \text{x} & \text{otherwise} \end{cases}$$

Vermont uses the same scheme, except that when $x = 0$, the letter A is used as the check symbol.

The following example illustrates this system.

EXAMPLE 5.11 Determine the check digit d_8 in a driver's license number assigned by the state of New Mexico, if the seven-digit number identification number in it is 0354729.

SOLUTION

First we compute x:

$$\begin{aligned} x &\equiv -(0, 3, 5, 4, 7, 2, 9) \cdot (2, 7, 6, 5, 4, 3, 2) \pmod{11} \\ &\equiv -(0 + 21 + 30 + 20 + 28 + 6 + 18) \pmod{11} \\ &\equiv -123 \equiv 9 \pmod{11} \end{aligned}$$

So, by definition, $d_8 = 9$ and the full license number is 03547299. ■

Exotic coding schemes are sometimes used to construct identification numbers. Norway, for instance, uses a two-check-digit scheme to assign registration numbers to its citizens. The last two digits of an eleven-digit registration number $d_1 d_2 \ldots d_{11}$ are check digits, defined as follows:

$$d_{10} \equiv -(d_1, d_2, \ldots, d_9) \cdot (3, 7, 6, 1, 8, 9, 4, 5, 2) \pmod{11}$$

$$d_{11} \equiv -(d_1, d_2, \ldots, d_{10}) \cdot (5, 4, 3, 2, 7, 6, 5, 4, 3, 2) \pmod{11}$$

Numbers for which d_{10} or d_{11} is "10" are *not* assigned.

The following example illustrates this scheme.

EXAMPLE 5.12 A registration number in Norway begins with the nine-digit number 065463334. Compute the two check digits in the identification number.

SOLUTION

We have

$$
\begin{aligned}
d_{10} &\equiv -(d_1, d_2, \ldots, d_9) \cdot (3, 7, 6, 1, 8, 9, 4, 5, 2) \pmod{11} \\
&\equiv -(0, 6, 5, 4, 6, 3, 3, 3, 4) \cdot (3, 7, 6, 1, 8, 9, 4, 5, 2) \pmod{11} \\
&\equiv -(0 + 42 + 30 + 4 + 48 + 27 + 12 + 15 + 8) \pmod{11} \\
&\equiv -186 \equiv 1 \pmod{11} \\
d_{11} &\equiv -(d_1, d_2, \ldots, d_{10}) \cdot (5, 4, 3, 2, 7, 6, 5, 4, 3, 2) \pmod{11} \\
&\equiv -(0, 6, 5, 4, 6, 3, 3, 3, 4, 1) \cdot (5, 4, 3, 2, 7, 6, 5, 4, 3, 2) \pmod{11} \\
&\equiv -(0 + 24 + 15 + 8 + 42 + 18 + 15 + 12 + 12 + 2) \pmod{11} \\
&\equiv -148 \equiv 6 \pmod{11}
\end{aligned}
$$

So the two check digits are 1 and 6, and hence the registration number is 06546333416. ■

Vehicle Identification Numbers

Automobiles and trucks built since the early 1980s have been assigned a unique vehicle identification number (VIN) by the manufacturer. A typical VIN consists of 17 alphanumeric symbols; it contains coded information for the country where the vehicle was built, manufacturer, vehicle type, body type, engine type, series, restraint system, car line, check digit, model year, plant code, and plant sequential number. See Figure 5.22.

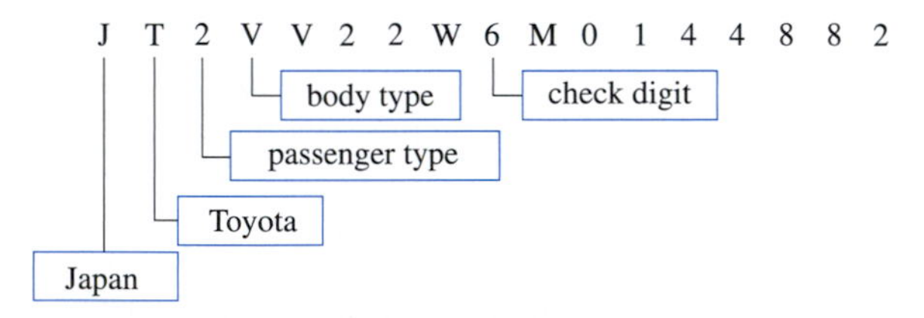

The Vehicle Identification Number for a 1991 Toyota Camry

Figure 5.22

Unlike check-digit schemes discussed earlier, the check digit in a VIN is not appended at the end, but placed in the middle.

To compute the check digit d_9, we employ the following algorithm:

- Convert the letters A through Z into the numbers 1–9, 1–9, and 2–9, respectively. This yields a 16-digit number $d_1d_2 \ldots \not{d}_9 \ldots d_{17}$.
- Assign the weights $8, 7, \ldots, 2, 10, 9, \ldots, 2$ to the positions $d_1, \ldots, \not{d}_9, \ldots, d_{17}$, respectively.
- Compute the least nonnegative residue $r \equiv (d_1, d_2, \ldots, \not{d}_9, \ldots, d_{17}) \cdot (8, 7, \ldots, 2, 10, 8, \ldots, 2) \pmod{11}$.
- Check digit $= d_9 = \begin{cases} r & \text{if } 0 \le r < 10 \\ \text{X} & \text{otherwise} \end{cases}$

The following example illustrates this algorithm.

EXAMPLE 5.13 Compute the check digit in the vehicle identification number in Figure 5.22.

SOLUTION

Replacing the letters in the VIN with their numeric codes yields the following numeric equivalents:

VIN:	J	T	2	V	V	2	2	W	–	M	0	1	4	4	8	8	2
Numeric code:	1	3	2	5	5	2	2	6	–	4	0	1	4	4	8	8	2

Vertically align each numeric code by the corresponding weight:

Numeric code:	1	3	2	5	5	2	2	6	–	4	0	1	4	4	8	8	2
Weight:	8	7	6	5	4	3	2	10	–	9	8	7	6	5	4	3	2

Now compute the weighted sum modulo 11:

$$\begin{aligned} \text{Weighted sum} &= 8 \cdot 1 + 7 \cdot 3 + 6 \cdot 2 + 5 \cdot 5 + 4 \cdot 5 + 3 \cdot 2 + 2 \cdot 2 + 10 \cdot 6 + 9 \cdot 4 \\ &\quad + 8 \cdot 0 + 7 \cdot 1 + 6 \cdot 4 + 5 \cdot 4 + 4 \cdot 8 + 3 \cdot 8 + 2 \cdot 2 \\ &\equiv 6 \pmod{11} \end{aligned}$$

Since $0 \le 6 < 10$, the check digit is 6, as desired. ■

German Bank Notes (optional)†

In 1990, the German Bundesbank adopted a mathematically sophisticated scheme based on group theory from abstract algebra to compute the check digit in the serial

† Although Germany has switched to Euro, this is still a delightful application.

number of a bank note. A typical serial number consists of ten alphanumeric symbols $s_1, s_2, \ldots, s_{10}$, and a check digit s_{11}. For example, the alphanumeric serial number of the bank note in Figure 5.23 is DD4170295U7 and the check digit is 7; the bank note features Gauss and his celebrated normal curve.

Figure 5.23 *A German bank note with serial number DD4170295U7 and check digit 7.*

To compute the check digit, we apply the following algorithm:

- Convert the letters into their numeric codes, using Table 5.3:

A	D	G	K	L	N	S	U	Y	Z
0	1	2	3	4	5	6	7	8	9

Table 5.3

- Let $f : S \to S$, defined as follows, where $S = \{0, 1, \ldots, 9\}$:

$$\begin{array}{lllll} f(0)=1 & f(1)=5 & f(2)=7 & f(3)=6 & f(4)=2 \\ f(5)=8 & f(6)=3 & f(7)=0 & f(8)=9 & f(9)=4 \end{array}$$

(f is a permutation of the elements of S.) Let $f^n = f \circ f^{n-1}$, where $\circ$ denotes the composition of functions, f^0 is the identity function, and $n \geq 1$. [For example, $f^3(5) = f(f(f(5))) = f(f(8)) = f(9) = 4$.]

- Define an operation $*$ on S using Table 5.4.

$*$	0	1	2	3	4	5	6	7	8	9
0	0	1	2	3	4	5	6	7	8	9
1	1	2	3	4	0	6	7	8	9	5
2	2	3	4	0	1	7	8	9	5	6
3	3	4	0	1	2	8	9	5	6	7
4	4	0	1	2	3	9	5	6	7	8
5	5	9	8	7	6	0	4	3	2	1
6	6	5	9	8	7	1	0	4	3	2
7	7	6	5	9	8	2	1	0	4	3
8	8	7	6	5	9	3	2	1	0	4
9	9	8	7	6	5	4	3	2	1	0

Table 5.4

(For those familiar with group theory, Table 5.4 represents the multiplication table for the *dihedral group* D_{10} of symmetries of a regular 5-gon. Notice that $*$ is a noncommutative operation; for example, $4 * 5 = 9 \neq 6 = 5 * 4$.)

- Select the check digit s_{11} such that

$$f(s_1) * f^2(s_2) * f^3(s_3) * \cdots * f^9(s_9) * f^{10}(s_{10}) * s_{11} = 0 \qquad (5.1)$$

(These steps can be stated more efficiently using group-theoretic language.)

The following example illustrates this fancy algorithm.

EXAMPLE 5.14 Compute the missing check digit in the German bank note serial number DD4170295U–.

SOLUTION

Using Table 5.3, first we convert the letters into numbers:

Serial number: DD4170295U
Numeric code: 1 1 4 1702957

Since $s_1 = 1 = s_2 = s_5$, $s_3 = 4$, $s_4 = 7 = s_{10}$, $s_6 = 0$, $s_7 = 2$, $s_8 = 9$, and $s_9 = 5$, $f(s_1) = 5$, $f^2(s_2) = 8$, $f^3(s_3) = 0$, $f^4(s_4) = 4$, $f^5(s_5) = 9$, $f^6(s_6) = 2$, $f^7(s_7) = 4$, $f^8(s_8) = 9$, $f^9(s_9) = 8$, and $f^{10}(s_{10}) = 1$ (verify these).

By equation (5.1), the check digit must satisfy the equation

$$5 * 8 * 0 * 4 * 9 * 2 * 4 * 9 * 8 * 1 * s_{11} = 0$$

Using Table 5.4, this yields:

$$\begin{aligned}(5*8)*(0*4)*(9*2)*(4*9)*(8*1)*s_{11} &= 0\\ 2*4*7*8*7*s_{11} &= 0\\ (2*4)*(7*8)*7*s_{11} &= 0\\ 1*4*7*s_{11} &= 0\\ (1*4)*7*s_{11} &= 0\\ 0*7*s_{11} &= 0\\ 7*s_{11} &= 0\\ s_{11} &= 7\end{aligned}$$

Thus, the check digit in the bank note is 7, as expected. ■

EXERCISES 5.3

Find the parity check bit that must be appended to each bit string for even parity.

1. 101101
2. 110110111
3. 10111011
4. 11011101

An n-bit string containing an even parity check bit is transmitted. What is your conclusion if the received string

5. Agrees for even parity?
6. Does not agree for even parity?

Airlines assign a check digit d to their 11-digit ticket numbers $d_1d_2\dots d_{11}$, defined by $d \equiv d_1d_2\dots d_{11} \pmod 7$. (In fact, the ticket number contains 14 digits and the check digit, but the three-digit airline code is not entered into the calculation of the check digit.) Compute the check digit if the 11-digit ticket number is

7. 20754376405
8. 17330207806

VISA traveler's checks use the negative of the least residue modulo 9 as a check digit. For instance, the check digit for the check number 1967633314327 is 8 since $1967633314327 \equiv 1 \pmod 9$ and $-1 \equiv 8 \pmod 9$. Compute the check digit for each check number.

9. 300706202013
10. 1942300317768

Using Example 5.7, compute the check digit if the eight-digit identification number is

11. 79002966
12. 88049338

Using Example 5.7, determine the missing digit d in each identification number.

13. 3313d4473
14. 78d035442

Using the ISBN coding scheme, compute the check digit if the first nine digits are

15. 0-87-620321
16. 0-201-57889

Determine whether each is a valid ISBN.

17. 0-201-57603-1
18. 0-07-095831-2

Compute the check digit in the UPC number for

19. *Cheerios* by General Mills, Inc., if the 11-digit identification number is 0-16000-66610.
20. Kellogg's *Product 19*, if the 11-digit identification number is 0-38000-01912.

Determine whether each is a valid UPC number for a grocery item.

21. 0-70734-06310-8
22. 0-16000-42080-9

Determine whether each is a valid Norway registration number.

23. 06546330708
24. 34040455642

The *International Standard Serial Number* (**ISSN**) is an internationally accepted code for identifying serial publications. It consists of two four-digit groups. The eighth digit d_8, which can be an X (for 10), is a check digit, defined by $d_8 \equiv -(d_1, d_2, \ldots, d_7) \cdot (8, 7, 6, 5, 4, 3, 2) \pmod{11}$. Compute the check digit for each seven-digit identification number.

25. 1234-567
26. 0593-303

Many European countries use check digits to detect errors in passport numbers. The check digit d_8 of the identification number $d_1 d_2 \ldots d_7$ is defined by $d_8 \equiv (d_1, d_2, \ldots, d_7) \cdot (7, 3, 1, 7, 3, 1, 7) \pmod{10}$. Determine the check digit in each case.

27. 3157406
28. 4005372

Using the Utah scheme, find the check digit in a driver's license number if the eight-digit identification number is

29. 14921994
30. 30435167

Determine the check digit d_8 in a Tennessee driver's license number if the seven-digit identification number is

31. 0243579
32. 2730373

33–34. Redo Exercises 31 and 32 using the Vermont license number scheme.

Both South Dakota and Saskatchewan employ a complex scheme developed by IBM to compute the check digit d_7 that is appended to the six-digit identification number $d_1 d_2 \ldots d_6$ in a driver's license number. It is computed as follows: Multiply d_2, d_4, and d_6 by 2; add the digits in the products; add the resulting sum to $d_1 + d_3 + d_5$ to yield s; then $d_7 \equiv -s \pmod{10}$. (This scheme is used by credit card companies, libraries, and drug stores in the United States, and by banks in Germany.)

35. Develop an algebraic formula for d_7.

Compute the check digit d_7 for each six-digit identification number.

36. 204817
37. 764076

Compute the missing check digit in each vehicle identification number.

38. 2T1BB02E–VC194572
39. 2HGES165–1H541873

Determine if each is a valid alphanumeric serial number for a German bank note.

40. GD2414993L0
41. GD3994142L0

*5.4 The p-Queens Puzzle (optional)

The n-queens puzzle, a well-known problem used in undergraduate programming courses, gives us an excellent example of **backtracking**. The goal of the puzzle is to place n queens on an $n \times n$ chessboard in such a way that no two queens can attack each other. It follows by observation that the puzzle has no solution if $n = 2$ or 3.

We can develop a formula for successfully placing p queens on a $p \times p$ chessboard, where p is a prime > 3. The following section shows how the solution yields an algorithm for constructing a schedule for a round-robin tournament with p teams.

The p-Queens Puzzle

To present a formula for solving the p-queens problem, we place the queens row by row. Let $f(i)$ denote the location (column index) of the ith queen, where $1 \le i \le p$; then $f(i)$ can be defined recursively.

A Recursive Definition of f

$$\begin{aligned} &f(0) = 0 \\ &f(i) \equiv f(i-1) + \frac{p+1}{2} \pmod{p}, \quad 1 \le i \le p-1 \\ &f(p) = p \end{aligned} \tag{5.2}$$

Using iteration, we can use this definition to find the following explicit formula for $f(i)$.

An Explicit Formula for $f(i)$

$$f(i) \equiv \left(\frac{p+1}{2}\right) i \pmod{p} \quad \text{if } 1 \le i \le p \tag{5.3}$$

Here $f(i)$ is the least residue of $(p+1)i/2$ modulo p, where the residue 0 is interpreted as p.

The following theorem singles out a property of f.

THEOREM 5.2 The function f is injective.

PROOF

Let i and j be least residues modulo p such that

$$f(i) = f(j)$$

Then

$$\left(\frac{p+1}{2}\right) i \equiv \left(\frac{p+1}{2}\right) j \pmod{p}$$

Since $((p+1)/2, p) = 1$, this implies $i \equiv j \pmod{p}$. But i and j are least residues modulo p, so $i = j$. ■

This theorem shows f assigns exactly one queen to each row and each column, as Table 5.5 shows for $p = 7$.

i \ j	1	2	3	4	5	6	7
1	.	.	.	Q	.	.	.
2	Q	.	.	.	.	.	.
3	.	.	.	.	Q	.	.
4	.	Q	.	.	.	.	.
5	.	.	.	.	.	Q	.
6	.	.	Q	.	.	.	.
7	.	.	.	.	.	.	Q

Table 5.5

Next we show that no two queens placed by the preceding assignment can attack each other.

THEOREM 5.3 No two queens placed on a $p \times p$ chessboard by the assignment f can attack each other.

PROOF

Since every row and every column contains exactly one queen, no two queens can attack each other along a row or column. So it suffices to show that they cannot attack along any southeast or northeast diagonal.

For each northeast diagonal, the sum $i + j$ of the row index i and the column index j is a constant k, where $2 \le k \le 2p$. Clearly, we need only look at the diagonals, where $3 \le k \le 2p - 1$.

Suppose there are two such queens in positions (i_1, j_1) and (i_2, j_2). Then

$$f(i_1) \equiv \left(\frac{p+1}{2}\right) i_1 \pmod{p}$$

$$f(i_2) \equiv \left(\frac{p+1}{2}\right) i_2 \pmod{p}$$

That is,

$$j_1 \equiv \left(\frac{p+1}{2}\right) i_1 \pmod{p} \quad \text{and} \quad j_2 \equiv \left(\frac{p+1}{2}\right) i_2 \pmod{p} \tag{5.4}$$

where $i_1 + j_1 = k = i_2 + j_2$. Then

$$i_1 + j_1 \equiv \left(\frac{p+3}{2}\right) i_1 \pmod{p}$$

That is,

$$k \equiv \left(\frac{p+3}{2}\right) i_1 \pmod{p}$$

Similarly,

$$k \equiv \left(\frac{p+3}{2}\right) i_2 \pmod{p}$$

These two congruences imply that $(p+3)i_1/2 \equiv (p+3)i_2/2 \pmod{p}$, so $i_1 \equiv i_2 \pmod{p}$ since $(p, (p+3)/2) = 1$. Thus, $i_1 = i_2$, since they are least residues modulo p. Then, by congruences (5.4), $j_1 = j_2$. Thus, no northeast diagonal contains two queens.

To show that no southeast diagonal contains two queens, notice that for each such diagonal, $i - j$ is a constant ℓ, where $1 - p \le \ell \le p - 1$. Clearly we can assume $\ell \ne 1 - p$ and $\ell \ne p - 1$.

Suppose a southeast diagonal contains two queens in positions (i_1, j_1) and (i_2, j_2). Then

$$f(i_1) \equiv \left(\frac{p+1}{2}\right) i_1 \pmod{p}$$

$$f(i_2) \equiv \left(\frac{p+1}{2}\right) i_2 \pmod{p}$$

That is,

$$j_1 \equiv \left(\frac{p+1}{2}\right) i_1 \pmod{p} \quad \text{and} \quad j_2 \equiv \left(\frac{p+1}{2}\right) i_2 \pmod{p} \tag{5.5}$$

where $i_1 - j_1 = \ell = i_2 - j_2$. Then

$$i_1 - j_1 \equiv i_1 - \left(\frac{p+1}{2}\right) i_1 \pmod{p}$$

$$\ell \equiv \left(\frac{1-p}{2}\right) i_1 \pmod{p}$$

$$\ell \equiv \left(\frac{p+1}{2}\right) i_1 \pmod{p}$$

Similarly,

$$\ell \equiv \left(\frac{p+1}{2}\right) i_2 \pmod{p}$$

These two congruences yield $i_1 = i_2$, since $((p+1)/2, p) = 1$, and i_1 and i_2 are least residues modulo p. Thus, by congruences (5.5), $j_1 = j_2$, so no southeast diagonal contains two queens.

Thus, no two queens on the $p \times p$ chessboard can attack each other. ■

An Algorithm for Placing p Queens on a p x p Chessboard

The recursive definition of f provides an algorithm for placing the queens row by row on the $p \times p$ chessboard:

- Place the first queen in column $(p+1)/2$. In each successive row, cyclically advance to the right by $(p+1)/2$ cells and place a queen in the resulting cell, and continue like this until a queen is placed in every row.

*5.5 Round-Robin Tournaments (optional)

In **round-robin tournaments**, every team plays every other team exactly once. Suppose there are n teams, labeled 1 through n. Then the tournament can be represented by a polygon with n vertices with every pair of vertices connected; every vertex represents a team and every line segment with endpoints i and j represents a game between teams i and j. (Such a figure is called a **complete graph** with n vertices.) For example, Figure 5.24 shows a round-robin tournament with five teams.

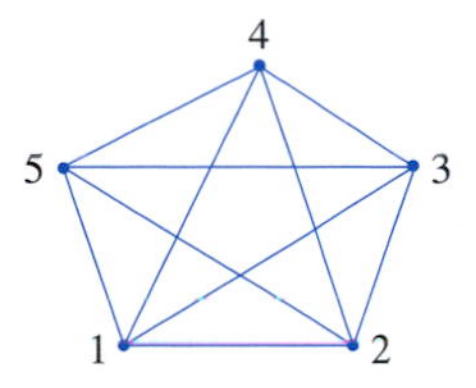

Figure 5.24 *A round-robin tournament with five teams.*

Let g_n denote the number of games by n teams in a round-robin tournament. It can be defined recursively:

$$g_1 = 0$$
$$g_n = g_{n-1} + (n-1), \quad n \geq 2$$

Solving this recurrence relation, we get

$$g_n = \frac{n(n-1)}{2} = \binom{n}{2}.$$

For example, five teams will play 10 games.

Congruences can be applied nicely to schedule round-robin tournaments. If n is even, then every team can be paired with another team; but if n is odd, not all teams can be paired, so one team gets a bye in that round. So, whenever n is odd, we add a dummy team X, so that if a team is paired with X in a certain round, it gets a bye in that round. Consequently, we assume n is even.

Solving the p-queens puzzle is closely related to constructing a schedule for a round-robin tournament with p teams.

Let $g(i,j)$ denote the team played in round i by team j. If $g(i,j) = j$, team j gets a bye in round i. We define g as

$$g(i,j) \equiv i - j \pmod{p} \tag{5.6}$$

where the least residue 0 modulo p is interpreted as p.

For example, let $p = 7$. Then $g(1,1) = 0 \pmod 7$, so $g(1,1) = 7$; similarly, $g(1,2) \equiv -1 \pmod 7$, so $g(1,2) = 6$, and so on. Table 5.6 shows a schedule for a round-robin tournament with seven teams.

Team j / *Round i*	*1*	*2*	*3*	*4*	*5*	*6*	*7*
1	7	6	5	bye	3	2	1
2	bye	7	6	5	4	3	2
3	2	1	7	6	bye	4	3
4	3	bye	1	7	6	5	4
5	4	3	2	1	7	bye	5
6	5	4	bye	2	1	7	6
7	6	5	4	3	2	1	bye

Table 5.6 *A round-robin tournament for seven teams.*

We will now show that g constructs a round-robin tournament schedule for p teams. First, we must prove the following three theorems.

THEOREM 5.4 Exactly one team draws a bye in each round.

PROOF

Suppose teams j_1 and j_2 draw byes in round i. Then

$$g(i, j_1) \equiv j_1 \pmod{p} \quad \text{and} \quad g(i, j_2) \equiv j_2 \pmod{p}$$

case 1 If $i = j_1$, then $i = j_1 = p$. Because $g(i, j_2) \equiv j_2 \pmod{p}$, $i - j_2 \equiv j_2 \pmod{p}$; that is,

$$p - j_2 \equiv j_2 \pmod{p}$$
$$2j_2 \equiv 0 \pmod{p}$$
$$j_2 \equiv 0 \pmod{p}$$

So $j_2 = p$. Thus, $j_1 = j_2$.

case 2 If $i \neq j_1$, then $g(i, j_1) \equiv i - j_1 \equiv j_1 \pmod{p}$, so $i \equiv 2j_1 \pmod{p}$.

If $i = j_2$, $g(i, j_2) \equiv i \equiv p \pmod{p}$. Then $p \equiv 2j_1 \pmod{p}$, so $2j_1 \equiv 0 \pmod{p}$; that is, $j_1 \equiv 0 \pmod{p}$ or $j_1 = p$. Then $i \equiv 2p \equiv 0 \pmod{p}$, so $i = p$. Thus, $i = j_1$, a contradiction.

So $i \neq j_2$. Therefore, $g(i, j_2) \equiv i - j_2 \equiv j_2 \pmod{p}$. This yields $i \equiv 2j_2 \pmod{p}$, so

$$2j_1 \equiv 2j_2 \pmod{p}$$
$$j_1 \equiv j_2 \pmod{p}$$

Therefore, $j_1 = j_2$, since they are least residues modulo p.

Thus, in both cases, $j_1 = j_2$, so exactly one team gets a bye in each round. ∎

The following theorem identifies the team that draws a bye in each round.

THEOREM 5.5 $g(i, j) \equiv j \pmod{p}$ if and only if $j \equiv \left(\frac{p+1}{2}\right) i \pmod{p}$.

PROOF

Assume $g(i, j) \equiv j \pmod{p}$. If $i = j$, then $g(i, j) \equiv p \pmod{p}$, so $i \equiv j \equiv p \equiv 0 \pmod{p}$. Therefore, $j \equiv (p+1)i/2 \pmod{p}$.

If $i \neq j$, then $g(i, j) \equiv i - j \pmod{p}$. Then

$$i - j \equiv j \pmod{p}$$

That is,

$$i \equiv 2j \pmod{p}$$

Therefore, $(p+1)i/2 \equiv (p+1)2j/2 \equiv pj + j \equiv j \pmod{p}$.

Thus, in both cases, team j draws a bye in round i if $j \equiv (p+1)i/2 \pmod{p}$.

Conversely, suppose $j \equiv (p+1)i/2 \pmod{p}$. Then

$$\begin{aligned} g(i,j) &\equiv i - j \pmod{p} \\ &\equiv i - (p+1)i/2 \equiv (1-p)i/2 \pmod{p} \\ &\equiv (p+1)i/2 \equiv j \pmod{p} \end{aligned}$$

Thus, team j draws a bye in round i. ■

The following theorem shows that g schedules every team exactly once in each round; that is, g outputs every value 1 through p exactly once.

THEOREM 5.6 The function g is injective for each i.

PROOF

Suppose $g(i,j_1) = g(i,j_2)$. Then $i - j_1 \equiv i - j_2 \pmod{p}$, so $j_1 \equiv j_2 \pmod{p}$; thus, $j_1 = j_2$ and g is injective. ■

It follows by Theorems 5.4 through 5.6 that the function g determines uniquely the opponent of team j in every round i, where $1 \le i, j \le p$; in round i, team j draws a bye, where $j \equiv (p+1)i/2 \pmod{p}$. Interestingly enough, this is exactly the same value (5.3) obtained earlier for placing the ith queen, where $1 \le i \le p$. Thus, a bye occurs in round i in the round-robin schedule in exactly the same cell as the one in which a Q occurs in row i of the $p \times p$ chessboard.

With this result, we can use the function g to modify the p-queens algorithm to develop an algorithm for a round-robin schedule for p teams, where $p \ge 3$.

An Algorithm for Constructing a Round-Robin Schedule for p Teams

- Place the first bye in column $(p+1)/2$; in each successive row, cyclically advance to the right by $(p+1)/2$ cells, and place a bye in the resulting cell; continue like this until a bye is placed in every row.
- Beginning with the first cell in row 1, count down the numbers p through 1 and enter them in empty cells (i.e., skip over the cell occupied by a bye), to obtain the permutation $p, p-1, \ldots$, bye, $\ldots$, 2, 1; to obtain each remaining

row, cyclically permute to the right the numbers in the preceding row. (Always skip over the byes. See Table 5.6.)

Suppose the number of teams n is not a prime. We pair the teams in round k as follows[†]: Team $i(\neq n)$ plays team $j(\neq n)$ if $i + j \equiv k \pmod{n-1}$, where $i \neq j$. This schedules all teams except teams n and i, where $2i \equiv k \pmod{n-1}$. The linear congruence $2i \equiv k \pmod{n-1}$, where $1 \leq i < n$, has a unique solution i exactly when $(2, n-1) = 1$ (by Corollary 4.6); so pair team i with team n in round k.

We can now show that this procedure pairs every team with every other team in each round. Consider team i, where $1 \leq i < n$. Since the congruence $2i \equiv k \pmod{n-1}$ has a unique solution i and team n is paired with team i, team n plays $n-1$ distinct games. Also, suppose teams i and j play in two distinct rounds k and k'; then $i+j \equiv k \pmod{n-1}$ and $i+j \equiv k' \pmod{n-1}$; this implies $k \equiv k' \pmod{n-1}$, a contradiction; in other words, teams i and j do not meet in two distinct rounds. Thus, each of the first $n-1$ teams plays $n-1$ games and no two teams play twice, so each plays exactly $n-1$ games. Team n also plays exactly $n-1$ games.

The following example illustrates this algorithm.

EXAMPLE 5.15 Develop a schedule for a round-robin tournament with seven teams.

SOLUTION

First, label the teams 1 through 7. Since the number of teams is odd, we add a dummy team X. We now prepare the schedule round by round.

To develop the schedule for round 1:

Team 1 plays team j, where $1 + j \equiv 1 \pmod 7$; then $j = 7$, so team 1 plays team 7. Team 2 plays team j, where $2 + j \equiv 1 \pmod 7$; this yields $j = 6$, so team 2 plays team 6. Similarly, team 3 plays team 5.

Because $i = 4$ is the solution of the congruence $2i \equiv 1 \pmod 7$, team 4 plays team 8; that is, team 4 gets a bye in round 1.

To develop the schedule for round 2:

Because $2i \equiv 2 \pmod 7$ implies $i = 1$, team 1 plays team 8; that is, team 1 enjoys a bye in round 2.

Team 2 plays team j, where $2 + j \equiv 2 \pmod 7$, so $j = 7$; thus, team 2 plays team 7. Similarly, team 3 plays team 6 and team 4 plays team 5.

Continuing like this, we can find the pairings in other rounds. The resulting schedule is given in Table 5.7.

[†] This method was developed in 1956 by J. E. Freund.

Round \ Team	1	2	3	4	5	6	7
1	7	6	5	bye	3	2	1
2	bye	7	6	5	4	3	2
3	2	1	7	6	bye	4	3
4	3	bye	1	7	6	5	4
5	4	3	2	1	7	bye	5
6	5	4	bye	2	1	7	6
7	6	5	4	3	2	1	bye

Table 5.7 *A schedule for a round-robin tournament for seven teams.* ■

EXERCISES 5.5

1. Solve the recurrence relation $g_n = g_{n-1} + (n-1)$, where $g_1 = 0$.

Develop a round-robin tournament schedule with

2. Five teams
3. Six teams
4. Eight teams
5. Nine teams
6. We would like to schedule a round-robin tournament with seven teams, 1 through 7. Pair team i with team j in round k, where $j \equiv i + 1 + k \pmod{7}$. Will this pairing provide a conflict-free and duplication-free schedule? If not, explain why.
7. Redo Exercise 6 if team i is paired with team j, where $j \equiv k(i+1) \pmod{7}$.
8. A company wants to schedule 1-hour meetings between every two of its six regional managers—A, B, C, D, E, and F—so each can spend an hour with each of the other five to get better acquainted. The meetings begin at 7 A.M. Find the various possible schedule-pairings. (S. W. Golomb, 1993)

*5.6 The Perpetual Calendar (optional)

In this section, we develop an interesting formula to determine the day of the week for any date in any year. Since the same day occurs every seventh day, we shall employ congruence modulo 7 to accomplish this goal, but first a few words of historical background.

Around 738 B.C., Romulus, the legendary founder of Rome, is said to have introduced a calendar consisting of 10 months, comprising a year of 304 days. His successor, Nauma, is credited with adding two months to the calendar. This new calendar was followed until Julius Caesar introduced the Julian calendar in 46 B.C., to minimize the distortions between the solar calendar and the Roman year. The Julian calendar consisted of 12 months of 30 and 31 days, except for February, which had 29 days, and every fourth year 30 days.

The first Julian year began on January 1, 45 B.C. It contained 365.25 days, was 11 minutes 14 seconds longer than the solar year, and made every fourth year a leap year of 366 days. By 1580, the Julian calendar, although the primary calendar in use, was 10 days off. It was, however, widely used until 1582.

In October 1582, astronomers Fr. Christopher Clavius and Aloysius Giglio introduced the Gregorian calendar at the request of Pope Gregory XIII, to rectify the errors of the Julian calendar. The accumulated error of 10 days was compensated by dropping 10 days in October, 1582. (October 5 became October 15.) The Gregorian calendar designates those century years divisible by 400 as **leap** years; all noncentury years divisible by 4 are also leap years. For example, 1776 and 2000 were leap years, but 1900 and 1974 were not.

The Gregorian calendar, now used throughout the world, is so accurate that it differs from the solar year only by about 24.5376 seconds. This discrepancy exists because a Gregorian year contains about 365.2425 days, whereas a solar year contains about 365.242216 days. The result is an error of 3 days every 10,000 years.

With this in mind, we can now return to our goal: *Determine the day d of the week for the rth day in a given month m of any given year y in the Gregorian calendar.* The first century leap year occurred in 1600 (18 years after the introduction of the Gregorian calendar); so we will develop the formula to hold for years beyond 1600. Also, since a leap year adds a day to February, we will count the new year beginning with March 1. For example, January 3000 is considered the eleventh month of 2999, whereas April 3000 is the second month of year 3000; also February 29 of 1976 is the last day of the 12th month of 1975.

So we assign the numbers 1 through 12 for March through February, and 0 through 6 for Sunday through Saturday, so $1 \le m \le 12$, $1 \le r \le 31$, and $0 \le d \le 6$. For example, $m = 3$ denotes May and $d = 5$ indicates Friday.

The derivation is lengthy and complicated, so we shall develop the formula in small steps.

Let d_y denote the day of the week of March 1 (the first day of the year) in year y, where $y \ge 1600$.

To Compute d from d_{1600}:

Because $365 \equiv 1 \pmod 7$, d_y is advanced from d_{y-1} by 1 if y is not a leap year and by 2 if y is a leap year:

$$d_y = \begin{cases} d_{y-1} + 1 & \text{if } y \text{ is not a leap year} \\ d_{y-1} + 2 & \text{otherwise} \end{cases}$$

To compute d_y from d_{1600}, we need to know the number of leap years ℓ since 1600. By Example 2.5,

$$\ell = \lfloor y/4 \rfloor - \lfloor y/100 \rfloor + \lfloor y/400 \rfloor - 388 \tag{5.7}$$

By the division algorithm, $y = 100C + D$, where $0 \leq D < 100$, so C denotes the number of centuries in y and D the leftover:

$$C = \lfloor y/100 \rfloor \quad \text{and} \quad D = y \pmod{100}$$

(For example, if $y = 2345$, then $C = 23$ and $D = 45$.) Then

$$\begin{aligned}
\ell &= \lfloor (100C + D)/4 \rfloor - \lfloor (100C + D)/100 \rfloor + \lfloor (100C + D)/400 \rfloor - 388 \\
&= \lfloor 25C + D/4 \rfloor - \lfloor C + D/100 \rfloor + \lfloor C/4 + D/400 \rfloor - 388 \\
&= 25C + \lfloor D/4 \rfloor - C + \lfloor C/4 \rfloor - 388, \quad \text{since } D < 100 \\
&= 24C + \lfloor D/4 \rfloor + \lfloor C/4 \rfloor - 388 \\
&\equiv 3C + \lfloor C/4 \rfloor + \lfloor D/4 \rfloor - 3 \pmod{7}
\end{aligned} \tag{5.8}$$

Therefore,

$$\begin{aligned}
d_y &\equiv d_{1600} + \begin{pmatrix} \text{one day for each} \\ \text{year since 1600} \end{pmatrix} + \begin{pmatrix} \text{one extra day for each} \\ \text{leap year since 1600} \end{pmatrix} \pmod{7} \\
&\equiv d_{1600} + (y - 1600) + \ell \pmod{7}
\end{aligned}$$

Substituting for y and ℓ,

$$\begin{aligned}
d_y &\equiv d_{1600} + (100C + D - 1600) + 3C + \lfloor C/4 \rfloor + \lfloor D/4 \rfloor - 3 \pmod{7} \\
&\equiv d_{1600} + (2C + D - 4 + 3C - 3) + \lfloor C/4 \rfloor + \lfloor D/4 \rfloor \pmod{7} \\
&\equiv d_{1600} + 5C + D + \lfloor C/4 \rfloor + \lfloor D/4 \rfloor \pmod{7} \\
&\equiv d_{1600} - 2C + D + \lfloor C/4 \rfloor + \lfloor D/4 \rfloor \pmod{7}
\end{aligned} \tag{5.9}$$

We can use this formula to identify d_y, the day of March 1 in year y, provided we know d_{1600}. In fact, we can also use it to find d_{1600} from some known value of d_y.

To Determine d_{1600}:

Because March 1, 1994, fell on a Tuesday, $d_{1994} = 2$. For $y = 1994$, $C = 19$, and $D = 94$, so, by formula (5.9),

$$\begin{aligned} d_{1600} &\equiv 2 + 2 \cdot 19 - 94 - \lfloor 19/4 \rfloor - \lfloor 94/4 \rfloor \pmod 7 \\ &\equiv 2 + 3 - 3 - 4 - 2 \pmod 7 \\ &\equiv -4 \equiv 3 \pmod 7 \end{aligned}$$

Thus, d_{1600} was a Wednesday.

Substituting for d_{1600} in formula (5.9),

$$d_y \equiv 3 - 2C + D + \lfloor C/4 \rfloor + \lfloor D/4 \rfloor \pmod 7 \tag{5.10}$$

This formula enables us to determine the day on which March 1 of any year falls. Now we extend this formula for an arbitrary day of a given month of the year.

To Extend Formula (5.10) to the rth Day of Month m in Year y:

To generalize formula (5.10), we need to know the number of days the first of the month is moved up from that of the previous month modulo 7. For this, notice that $30 \equiv 2 \pmod 7$ and $31 \equiv 3 \pmod 7$. So the day of the first of the month following a month with 30 days is advanced by 2 days, whereas that following a month with 31 days is advanced by 3 days.

For example, December 1, 1992, was a Tuesday. So January 1, 1993, fell on day $(2+3) =$ day 5, a Friday.

Thus, we have the following eleven monthly increments:

March 1 to April 1:	3 days
April 1 to May 1:	2 days
May 1 to June 1:	3 days
June 1 to July 1:	2 days
July 1 to August 1:	3 days
August 1 to September 1:	3 days
September 1 to October 1:	2 days
October 1 to November 1:	3 days
November 1 to December 1:	2 days
December 1 to January 1:	3 days
January 1 to February 1:	3 days

Next, we look for a function that yields these incremental values.

To Find a Function f That Produces These Increments:

First, notice that the sum of the increments = 29 days. So, the average number of increments $= 29/11 \approx 2.6$ days, so it was observed by Christian Zeller that the function $f(m) = \lfloor 2.6m - 0.2 \rfloor - 2$ can be employed to yield the above increments as m

> **Christian Julius Johannes Zeller** (1824–1899) was born at Mühlhausen on the Neckar. After studying theology, he became a priest at Schöckingen in 1854. From 1874 to 1898 he was the principal of Women's Elementary School Teachers' College at Markgröningen. Zeller spent his spare time studying mathematics, especially number theory. He published on remainders, Bernoulli numbers, and arithmetic progressions. His article on the calendar problem was published in a bulletin of the French Mathematical Society. Zeller died in Cannstadt.

varies from 2 to 12. For example,

$$\begin{aligned} f(3) - f(2) &= (\lfloor 7.8 - 0.2 \rfloor - 2) - (\lfloor 5.2 - 0.2 \rfloor - 2) \\ &= (7 - 2) - (5 - 2) = 2 \end{aligned}$$

so there is an increment of 2 days from month 2 (April 1) to month 3 (May 1).

Therefore, by formula (5.10), the first day d' of month m is given by $d_y + \lfloor 2.6m - 0.2 \rfloor - 2 \pmod 7$; that is,

$$\begin{aligned} d' &\equiv 3 - 2C + D + \lfloor C/4 \rfloor + \lfloor D/4 \rfloor + \lfloor 2.6m - 0.2 \rfloor - 2 \pmod 7 \\ &\equiv 1 + \lfloor 2.6m - 0.2 \rfloor - 2C + D + \lfloor C/4 \rfloor + \lfloor D/4 \rfloor \pmod 7 \end{aligned}$$

To Find the Formula for the rth Day of Month m:
The day d of the week for the rth day of month m is given by $d' + (r - 1) \pmod 7$; that is,

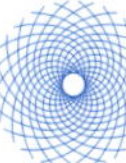

$$d \equiv r + \lfloor 2.6m - 0.2 \rfloor - 2C + D + \lfloor C/4 \rfloor + \lfloor D/4 \rfloor \pmod 7 \qquad (5.11)$$

This formula enables us to determine the day of the week of any given day in the Gregorian calendar, as the next example demonstrates.

EXAMPLE 5.16 Determine the day of the week on which January 13, 2020, falls.

SOLUTION

Notice that January 2020 is the eleventh month of year 2019, so here $y = 2019$, $C = 20$, $D = 19$, $m = 11$, and $r = 13$. Therefore, by formula (5.11),

$$\begin{aligned} d &\equiv 13 + \lfloor 2.6 \times 11 - 0.2 \rfloor - 2 \times 20 + 19 + \lfloor 20/4 \rfloor + \lfloor 19/4 \rfloor \pmod 7 \\ &\equiv 13 + 28 - 40 + 19 + 5 + 4 \pmod 7 \\ &\equiv 1 \pmod 7 \end{aligned}$$

Thus, January 13, 2020, falls on a Monday. ■

EXERCISES 5.6

Find the day of the week in each case.

1. 234 days from Monday.
2. 365 days from Friday.
3. 1776 days from Wednesday.
4. 2076 days from Saturday.

Let $S = \{\text{true, false}\}$. Define a **boolean function** $f: \mathbf{N} \to S$ by $f(n) =$ true if year n is a leap year and false otherwise. Find $f(n)$ for each year n.

5. 1996
6. 2020
7. 2076
8. 3000
9. January 1, 2000, falls on a Saturday. What day of the week will January 1, 2020, be?
   (*Hint*: Look for leap years.)
10. January 1, 1990, was a Monday. What day of the week was January 1, 1976?
    (*Hint*: Again, look for leap years.)

Determine the day of the week of each historical date.

11. January 17, 1706 (Benjamin Franklin's date of birth)
12. February 22, 1732 (George Washington's date of birth)
13. July 4, 1776 (U.S. Declaration of Independence)
14. November 19, 1863 (Gettysburg Address)
15. March 7, 1876 (first telephone patent issued to Alexander Graham Bell)
16. December 17, 1903 (world's first flight)
17. July 16, 1945 (first atomic bomb detonation)
18. October 24, 1945 (United Nations established)
19. April 12, 1961 (first human travel in space)
20. July 20, 1969 (first lunar landing)

The day of January 1 of any year y can be determined using the formula

$$\begin{aligned} x = y + \lfloor (y-1)/4 \rfloor - \lfloor (y-1)/100 \rfloor \\ + \lfloor (y-1)/400 \rfloor \pmod 7 \end{aligned} \tag{5.12}$$

where $0 \le x \le 6$. (G. L. Ritter, 1977) Using this formula, determine the first day in each year.

21. 2000
22. 2020
23. 2076
24. 3000

The number of Friday-the-thirteenths in a given year y can be computed using formula (5.12) and Table 5.8. For example, suppose that January 1 of a year falls on a Sunday (0). If it is not a leap year, there will be two Friday-the-thirteenths: January 13 and October 13; if it is a leap year, there will be three: January 13, April 13, and July 13. Compute the number of Friday-the-thirteenths in each year.

Code, x	*January 1*	*Nonleap Year, y*	*Leap Year, y*
0	Sunday	January, October	January, April, July
1	Monday	April, July	September, December
2	Tuesday	September, December	June
3	Wednesday	June	March, November
4	Thursday	February, March, November	February, August
5	Friday	August	May
6	Saturday	May	October

Table 5.8

25. 2000
26. 2020
27. 2076
28. 3076

(**Easter Sunday**) The date for Easter Sunday in any year y can be computed as follows. Let $a = y \bmod 19$, $b = y \bmod 4$, $c = y \bmod 7$, $d = (19a + 24) \bmod 30$, $e = (2b + 4c + 6d + 5) \bmod 7$, and $r = 22 + d + e$. If $r \leq 31$, then Easter Sunday is March r; otherwise, it is April $[r \pmod{31}]$. Compute the date for Easter Sunday in each year.

29. 1996
30. 2000
31. 2076
32. 3000

We can calculate the day of the week for the rth day of an arbitrary month m in year y in the Gregorian calendar by a different formula

$$d \equiv \lfloor 23m/9 \rfloor + r + 4 + y + \lfloor z/4 \rfloor - \lfloor z/100 \rfloor + \lfloor z/400 \rfloor - k \pmod{7}$$

where $z = y - 1$ and $k = 0$ if $m < 3$; and $z = y$ and $k = 2$ if $m \geq 3$. (M. Keith, 1990)

33–42. Using this formula, compute the days of each date in Exercises 11–20.

43. Show that $\lfloor C/4 + D/400 \rfloor = \lfloor C/4 \rfloor$, where $0 \leq D < 100$.

CHAPTER SUMMARY

This chapter explored the power of congruences in several applications.

Divisibility Tests

- An integer is divisible by 10 if and only if it ends in 0. (p. 247)
- An integer is divisible by 5 if and only if it ends in 0 or 5. (p. 248)
- An integer is divisible by 2^i if and only if the number formed by the last i digits is divisible by 2^i. (p. 248)
- An integer is divisible by 3 if and only if the sum of its digits is divisible by 3. (p. 248)
- An integer is divisible by 9 if and only if the sum of its digits is divisible by 9. (p. 248)
- An integer is divisible by 11 if and only if the sum of its digits in the even positions minus that of its digits in the odd positions is divisible by 11. (p. 249)
- Every palindrome with an even number of digits is divisible by 11. (p. 249)

Casting Out Nines

- This technique can detect computational errors. (p. 249)
- The digital root of a positive integer is its least residue modulo 9; if it is zero, the digital root is 9. (p. 251)

Dot Product

- The dot product of two vectors $(x_1, x_2, \ldots, x_n)$ and $(y_1, y_2, \ldots, y_n)$ is $\sum_{i=1}^{n} x_i y_i$. (p. 260)

REVIEW EXERCISES

Find the largest exponent i such that 2^i divides each integer.

1. 24,050,856
2. 300,472,336

Determine whether each number is divisible by 6.

3. 140,736
4. 3,041,079

Determine whether each number is divisible by 11.

5. 9,240,622
6. 85,140,643

Find the missing nonzero digit d in each case.

7. $645d56 \equiv 4 \pmod 8$
8. $29d224 \equiv 3 \pmod 3$
9. $889d849 \equiv 2 \pmod{11}$
10. $29992d5 \equiv 4 \pmod{11}$

Using casting out nines, identify each computation as *probably correct* or *definitely wrong*.

11. $$\begin{array}{r} 53467 \\ 498773 \\ +\,877008 \\ \hline 1439248 \end{array}$$
12. $$\begin{array}{r} 8700579 \\ -\,4099247 \\ \hline 4610332 \end{array}$$
13. $$\begin{array}{r} 780086 \\ \times\,27753 \\ \hline 21549726758 \end{array}$$

Find the parity check bit that must be appended to each bit string for even parity.

14. 110110110
15. 11101101101
16. 10101011011
17. 101011011111

Find the digital root of each.

18. 999,000,555
19. 888,777,666
20. 2323^{3232}
21. 5555^{1122}

Determine whether each can be a square.

22. 1,161,310,084
23. 3,656,973,729

Determine whether each can be a cube.

24. 15,064,223
25. 95,506,664,111

Determine whether each is a valid UPC number.

26. 0-49200-05100-9
27. 0-13130-03622-4
28. *Prego* spaghetti sauce, made by Campbell Soup Company, has a country code, manufacturer code, and product code of 0, 51000, and 02549, respectively. Compute its UPC number.
29. Compute the UPC number for *Classic Ovaltine* by Sandoz Nutrition Corporation if the country code, manufacturer code, and product code are 7, 51746, and 03361, respectively.
30. The United Parcel Service assigns to each parcel an identification number of nine digits and a check digit. The check digit is the least residue modulo 9 of the nine-digit number. Find the check digit for 038575447.
31. Libraries use a sophisticated **code-a-bar system** to assign each book a 13-digit identification number $d_1 d_2 \dots d_{13}$ and a check digit d_{14}. The check digit is computed as $d_{14} \equiv [-(d_1, d_2, \dots, d_{13}) \cdot (2,1,2,1,2,1,2,1,2,1,2,1,2) - k]$ (mod 10), where k denotes the number of digits among d_1, d_3, d_5, d_7, d_9, d_{11}, and d_{13} greater than or equal to 5. Compute the check digit for 3-3014-00099-073.

A MasterCard number contains 16 digits d_1 through d_{16}, with d_{16} being the check digit. It is computed as

$$d_{16} \equiv -\left[\sum_{i=1}^{8} \rho(2d_{2i-1}) + \sum_{i=1}^{7} d_{2i}\right] \pmod{10}$$

where $\rho(m)$ denotes the digital root of m. (ρ is the Greek letter *rho*.) Compute the check digit if the 15-digit identification number is

32. 5300-7402-4001-638
33. 5329-0419-4253-736

Determine the day of the week of each historical date.

34. December 21, 1620 (the Pilgrims landed at Plymouth, Massachusetts)
35. February 1, 1790 (the first meeting of the U.S. Supreme Court)

Compute the missing check digit in each VIN.

36. JT2DB02T–T0049506
37. 1B4GP44R–XB582510

Compute the missing check digit in each German bank note serial number.

38. YA8174491A–
39. DG6244129Y–

SUPPLEMENTARY EXERCISES

Let $\rho(n)$ denote the digital root of a positive integer n. Prove each, where m and n are positive integers.

1. $\rho(\rho(n)) = \rho(n)$
2. $\rho(m+n) = \rho(\rho(m) + \rho(n))$
3. $\rho(mn) = \rho(\rho(m)\rho(n))$

Prove each.

4. An integer $n = (n_k n_{k-1} \ldots n_1 n_0)_b$ is divisible by d^i if and only if the integer $(n_{i-1} \ldots n_1 n_0)_b$ is divisible by d^i, where $d|b$ and $i > 0$.
5. An integer $n = (n_k n_{k-1} \ldots n_1 n_0)_b$ is divisible by d if and only if the integer $n_k + \cdots + n_1 + n_0$ is divisible by d, where $d|(b-1)$.
6. Let a and d be positive integers such that $d|(10a-1)$. Then $d|10q + r$ if and only if $d|q + ar$. (C. F. Liljevalch, 1838)

Using **Liljevalch's theorem** in Exercise 6, deduce a divisibility test for each number.

7. 3
8. 7
9. 9
10. 11
11. 13
12. 17
13. 19

Using the divisibility tests in Exercises 7–13, determine whether each integer is divisible by the corresponding second integer.

14. 1953, 7
15. 28303, 11
16. 51814, 17
17. 61731, 19
18. Suppose a two-digit number N is divisible by 7. Reverse its digits; add the tens digit of N to it. Prove that the resulting number N' is also divisible by 7. (A. P. Stevens, 1951)
19. Suppose a three-digit number N is divisible by 7. Reverse its digits; subtract the difference of its end digits from the reverse. Prove that the resulting number N' is also divisible by 7. (A. P. Stevens, 1951)

*20. Find the least positive integer that equals eleven times the sum of its digits.

*21. Find the digital root of the integer $n = 2^{p-1}(2^p - 1)$, where p and $2^p - 1$ are primes.
[*Hint*: $2 \equiv -1 \pmod 3$.]

COMPUTER EXERCISES

Write a program to perform each task.

1. Read in a positive integer n and find the largest exponent i such that $2^i | n$.

2. Read in a positive integer n. Using divisibility tests, determine if it is divisible by 3, 5, 6, 9, or 11.
3. Read in a bit string and append a bit for even parity.
4. Read in the first nine digits of an ISBN and print the corresponding ISBN.
5. Read in the first eight digits in a Utah driver's license number and print the nine-digit license number.
6. Read in the first seven digits in a New Mexico driver's license number and print the eight-digit license number.
7. Redo program 6 if the driver lives in Vermont.
8. Read in a 12-digit number and determine whether is a valid UPC number.
9. Read in the country code, manufacturer code, and the product code of a grocery item. Compute its UPC number.
10. Read in an 11-digit number and determine if it is a valid Norwegian registration number.
11. Read in the nine-digit identification number in a Norwegian registration number and determine the 11-digit registration number.
12. Read in a prime $p > 3$ and place p queens on a $p \times p$ chessboard.
13. Read in a positive integer $n > 2$ and print a schedule for a round-robin tournament with n teams.
14. Read in a date in the form month/date/year in the Gregorian calendar. Print the day of the week corresponding to the date using

    (a) Formula (5.11) in Section 5.6.
    (b) Formula in problems 33–42 of Exercises 5.6.

ENRICHMENT READINGS

1. J. A. Gallian, "The Zip Code Bar Code," *UMAP J.*, 7 (1986), 191–194.
2. J. A. Gallian, "Assigning Driver's License Numbers," *Mathematics Magazine*, 64 (Feb. 1991), 13–22.
3. J. A. Gallian, "The Mathematics of Identification Numbers," *College Mathematics J.*, 22 (May 1991), 194–202.
4. M. Keith, "A Mental Perpetual Calendar," *J. Recreational Mathematics*, 8 (1975–1976), 242–245.
5. M. Keith and T. Craver, "The Ultimate Perpetual Calendar?" *J. Recreational Mathematics*, 22 (1990), 280–282.
6. J. Kirkland, "Identification Numbers and Check Digit Schemes," Math. Association of America, Washington, DC, 2001.
7. P. M. Tuchinsky, "International Standard Book Numbers," *UMAP J.*, 5 (1985), 41–54.

8. S. J. Winters, "Error Detecting Schemes Using Dihedral Groups," *UMAP J.*, 11 (1990), 299–308.
9. E. F. Wood, "Self-Checking Codes—An Application of Modular Arithmetic," *Mathematics Teacher*, 80 (1987), 312–316.

Systems of Linear Congruences

Time as he grows old teaches many lessons.
—AESCHYLUS

We have studied linear congruences and seen how to solve such congruences when they are solvable. We now turn to solving a set of two or more linear congruences in the same number of variables. Such a set is called a **system of linear congruences**.

To begin with, we shall study systems of linear congruences in a single variable x with pairwise relatively prime moduli. We then turn to systems in x with moduli that are not necessarily relatively prime, and finally to systems in two variables x and y with the same modulus.

6.1 The Chinese Remainder Theorem

Systems of linear congruences in a single variable were known in ancient China, India, and Greece; originally, they were used by astronomers for calendar making. The following puzzle, for example, is due to the Chinese mathematician Sun-Tsu, and appears in *Master Sun's Mathematical Manual*, written between 287 A.D. and 473 A.D.:

> *Find a number that leaves a remainder of 1 when divided by 3, a remainder of 2 when divided by 5, and a remainder of 3 when divided by 7.*

Using the congruence language, the riddle is to find an integer x such that $x \equiv 1 \pmod 3$, $x \equiv 2 \pmod 5$, and $x \equiv 3 \pmod 7$.

A **solution** of a linear system is a number that satisfies every member of the system. For example, since $52 \equiv 1 \pmod{3}$, $52 \equiv 2 \pmod{5}$, and $52 \equiv 3 \pmod{7}$, 52 is a solution of the above Chinese riddle; so are -53 and 157; in fact, the system has infinitely many solutions (see Example 6.1). You may verify that 22 is *not* a solution of the system.

A straightforward method for solving such a linear system is **iteration**: successive substitution for x until the last congruence is used, as the following example demonstrates.

EXAMPLE 6.1 Solve Sun-Tsu's puzzle by iteration.

SOLUTION

We have $x \equiv 1 \pmod{3}$, $x \equiv 2 \pmod{5}$, and $x \equiv 3 \pmod{7}$. Since $x \equiv 1 \pmod{3}$, by Theorem 4.3, $x = 1 + 3t_1$, where t_1 is an arbitrary integer.

Substitute for x in the second congruence $x \equiv 2 \pmod{5}$:

$$\begin{aligned} 1 + 3t_1 &\equiv 2 \pmod{5} \\ 3t_1 &\equiv 1 \pmod{5} \\ t_1 &\equiv 2 \pmod{5} \end{aligned}$$

That is, $t_1 = 2 + 5t_2$, with t_2 being an arbitrary integer. Therefore,

$$\begin{aligned} x &= 1 + 3t_1 = 1 + 3(2 + 5t_2) \\ &= 7 + 15t_2 \end{aligned}$$

Now substitute this value of x in the third congruence $x \equiv 3 \pmod{7}$:

$$\begin{aligned} 7 + 15t_2 &\equiv 3 \pmod{7} \\ 15t_2 &\equiv 3 \pmod{7} \\ t_2 &\equiv 3 \pmod{7} \end{aligned}$$

So $t_2 = 3 + 7t$, with t being arbitrary. Therefore,

$$\begin{aligned} x &= 7 + 15t_2 = 7 + 15(3 + 7t) \\ &= 52 + 105t \end{aligned}$$

Thus, any integer of the form $x = 52 + 105t$ is a solution of the linear system; it is the **general solution** of the system. (*Note*: $105 = 3 \cdot 5 \cdot 7$.) ■

In this example, 52 is the unique solution to the puzzle modulo 105, where the moduli are pairwise relatively prime. The following theorem is called the **Chinese Remainder Theorem** (CRT) in honor of early contributions by Chinese mathematicians to the theory of congruences. It shows that every linear system in the same single variable with pairwise relatively prime moduli has a unique solution. Sun-Tsu's puzzle is the earliest known instance of the CRT. A general method for solving such linear systems can be found in *Mathematical Treatise in Nine Sections*, written by Ch'in Chiu-Shao (1202–1261) in 1247.

THEOREM 6.1 **(The Chinese Remainder Theorem)** The linear system of congruences $x \equiv a_i \pmod{m_i}$, where the moduli are pairwise relatively prime and $1 \le i \le k$, has a unique solution modulo $m_1 m_2 \cdots m_k$.

PROOF

The proof consists of two parts. First, we will construct a solution and then show that it is unique modulo $m_1 m_2 \cdots m_k$.

Let $M = m_1 m_2 \cdots m_k$ and $M_i = M/m_i$, $1 \le i \le k$. Since the moduli are pairwise relatively prime, $(M_i, m_i) = 1$ for every i. Also, $M_i \equiv 0 \pmod{m_j}$ whenever $i \ne j$.

To construct a solution to the linear system:
Since $(M_i, m_i) = 1$, by Corollary 4.6, the congruence $M_i y_i \equiv 1 \pmod{m_i}$ has a unique solution y_i. (y_i is in fact the inverse of M_i modulo m_i.) Let $x = a_1 M_1 y_1 + a_2 M_2 y_2 + \cdots + a_k M_k y_k$.

To show that x is a solution of the linear system, we have

$$
\begin{aligned}
x &= \sum_{\substack{i=1 \\ i \ne j}}^{k} a_i M_i y_i + a_j M_j y_j \\
&\equiv \sum_{i \ne j} a_i \cdot 0 \cdot y_i + a_j \cdot 1 \pmod{m_j} \\
&\equiv 0 + a_j \pmod{m_j} \\
&\equiv a_j \pmod{m_j}, \qquad 1 \le j \le k
\end{aligned}
$$

Thus, x satisfies every congruence in the system, so x is a solution of the linear system.

To show that the solution is unique modulo M:
Let x_0 and x_1 be two solutions of the system. We shall show that $x_0 \equiv x_1 \pmod{M}$.

Since $x_0 \equiv a_j \pmod{m_j}$ and $x_1 \equiv a_j \pmod{m_j}$ for $1 \le j \le k$, $x_1 - x_0 \equiv 0 \pmod{m_j}$; that is, $m_j | (x_1 - x_0)$ for every j. By Corollary 3.12, $[m_1, m_2, \ldots, m_k] | (x_1 - x_0)$. But, by Corollary 3.11, $[m_1, m_2, \ldots, m_k] = M$. Therefore, $M | (x_1 - x_0)$, so $x_1 - x_0 \equiv$

0 (mod M); that is, $x_1 \equiv x_0 \pmod{M}$. Thus, any two solutions of the linear system are congruent modulo M, so the solution is unique modulo M. ■

The following examples illustrate this theorem.

EXAMPLE 6.2 Using the CRT, solve Sun-Tsu's puzzle:

$$x \equiv 1 \pmod{3}, \quad x \equiv 2 \pmod{5}, \quad \text{and} \quad x \equiv 3 \pmod{7}$$

SOLUTION

Since the moduli $m_1 = 3$, $m_2 = 5$, and $m_3 = 7$ are pairwise relatively prime, by the CRT, the linear system has a unique solution. To find it, first we find the M_1, M_2, M_3, y_1, y_2, and y_3 in the proof of the theorem.

To find M_1, M_2, and M_3:

$$M_1 = \frac{M}{m_1} = \frac{3 \cdot 5 \cdot 7}{3} = 35$$

$$M_2 = \frac{M}{m_2} = \frac{3 \cdot 5 \cdot 7}{5} = 21$$

$$M_3 = \frac{M}{m_3} = \frac{3 \cdot 5 \cdot 7}{7} = 15$$

To find y_1, y_2, and y_3:
y_1 is the solution of the congruence $M_1 y_1 \equiv 1 \pmod{m_1}$; that is,

$$35y_1 \equiv 1 \pmod{3}$$

$$(-1)y_1 \equiv 1 \pmod{3}$$

$$y_1 \equiv 2 \pmod{3}$$

Similarly, $M_2 y_2 \equiv 1 \pmod{m_2}$ implies

$$21y_2 \equiv 1 \pmod{5}$$

$$y_2 \equiv 1 \pmod{5}$$

Finally, $M_3 y_3 \equiv 1 \pmod{m_3}$ yields

$$15y_3 \equiv 1 \pmod{7}$$

$$y_3 \equiv 1 \pmod{7}$$

Thus, by the CRT,

$$\begin{aligned} x &\equiv \sum_{i=1}^{3} a_i M_i y_i \pmod{M} \\ &\equiv 1 \cdot 35 \cdot 2 + 2 \cdot 21 \cdot 1 + 3 \cdot 15 \cdot 1 \pmod{105} \\ &\equiv 52 \pmod{105} \end{aligned}$$

Therefore, 52 is the unique solution of the linear system modulo 105. Thus, the general solution is $x = 52 + 105t$. ■

The following example also demonstrates the CRT, but we leave the details for you to fill in, so follow the steps carefully.

EXAMPLE 6.3 Solve the linear system

$$x \equiv 1 \pmod{3}, \quad x \equiv 2 \pmod{4}, \quad \text{and} \quad x \equiv 3 \pmod{5}$$

SOLUTION

Here $M = 3 \cdot 4 \cdot 5 = 60$, $M_1 = M/3 = 20$, $M_2 = M/4 = 15$, and $M_3 = M/5 = 12$. The unique solutions of the congruences $M_1 y_1 \equiv 1 \pmod{m_1}$, $M_2 y_2 \equiv 1 \pmod{m_2}$, and $M_3 y_3 \equiv 1 \pmod{m_3}$, that is, $20y_1 \equiv 1 \pmod{3}$, $15y_2 \equiv 1 \pmod{4}$, and $12y_3 \equiv 1 \pmod{5}$ are 2, 3, and 3, respectively.

Thus, by the CRT,

$$\begin{aligned} x &\equiv \sum_{i=1}^{3} a_i M_i y_i \pmod{M} \\ &\equiv 1 \cdot 20 \cdot 2 + 2 \cdot 15 \cdot 3 + 3 \cdot 12 \cdot 3 \pmod{60} \\ &\equiv 58 \pmod{60} \end{aligned}$$

■

We close this section with the following example, which presents an interesting application of the CRT.

EXAMPLE 6.4 The largest integer the scientific calculator Casio fx 330A can handle is the eight-digit number 99,999,999. Compute the exact value of 2^{31} using this calculator and the CRT.

SOLUTION

To compute the value of $x = 2^{31}$, we select k pairwise relatively prime numbers $m_1, m_2, \ldots, m_k$, where $M = m_1 m_2 \cdots m_k > x$, and then compute the least residue r

of x modulo M. Since $x \equiv r \pmod{M}$ and $0 < r < M$, r would be the exact value of x.

The largest power of 2 the eight-digit calculator can handle is $2^{26} = 67{,}108{,}864$, whereas $2^{31} \approx 2.1474836 \times 10^9$. So we select four pairwise relatively prime numbers $m_1 = 300$, $m_2 = 301$, $m_3 = 307$, and $m_4 = 311$ such that $M = m_1 m_2 m_3 m_4 = 300 \cdot 301 \cdot 307 \cdot 311 > x$; to check this, we have

$$\begin{aligned} M &> 3^4 \times 10^8 \\ &= 81 \times 10^8 \\ &> 8 \times 10^9 \end{aligned}$$

so $M > x$. (We do not need to know the exact value of M.)

Notice that

$$\begin{aligned} 2^{10} &= 1024 \\ &\equiv 124 \pmod{300} \\ &\equiv 121 \pmod{301} \\ &\equiv 103 \pmod{307} \\ &\equiv 91 \pmod{311} \end{aligned}$$

Since $2^{31} = 2 \cdot 2^{10} \cdot 2^{10} \cdot 2^{10}$,

$$\begin{aligned} 2^{31} &\equiv 2 \cdot 124 \cdot 124 \cdot 124 \pmod{300} \\ &\equiv -52 \pmod{300} \end{aligned}$$

Similarly, $2^{31} \equiv 51 \pmod{301}$, $2^{31} \equiv 228 \pmod{307}$, and $2^{31} \equiv 36 \pmod{311}$. (Verify these.)

Thus, $x = 2^{31}$ satisfies the linear system:

$$\begin{aligned} x &\equiv -52 \pmod{300} \\ x &\equiv 51 \pmod{301} \\ x &\equiv 228 \pmod{307} \\ x &\equiv 36 \pmod{311} \end{aligned}$$

To apply the CRT, we have

$$M_1 = M/m_1 = 301 \cdot 307 \cdot 311, \qquad M_2 = M/m_2 = 300 \cdot 307 \cdot 311$$
$$M_3 = M/m_3 = 300 \cdot 301 \cdot 311, \quad \text{and} \quad M_4 = M/m_4 = 300 \cdot 301 \cdot 307$$

To find the values of y_1, y_2, y_3, *and* y_4 *in the CRT*:

The congruence $M_1 y_1 \equiv 1 \pmod{m_1}$ yields

$$\begin{aligned} 301 \cdot 307 \cdot 311 y_1 &\equiv 1 \pmod{300} \\ 1 \cdot 7 \cdot 11 y_1 &\equiv 1 \pmod{300} \\ 77 y_1 &\equiv 1 \pmod{300} \\ y_1 &\equiv 113 \pmod{300} \end{aligned}$$

The congruence $M_2 y_2 \equiv 1 \pmod{m_2}$ yields

$$\begin{aligned} 300 \cdot 307 \cdot 311 y_2 &\equiv 1 \pmod{301} \\ (-1) \cdot 6 \cdot 10 y_2 &\equiv 1 \pmod{301} \\ 60 y_2 &\equiv -1 \pmod{301} \\ y_2 &\equiv 5 \pmod{301} \end{aligned}$$

Similarly, $y_3 \equiv 53 \pmod{307}$ and $y_4 \equiv 135 \pmod{311}$ (verify these).

Therefore, by the CRT,

$$\begin{aligned} x &\equiv a_1 M_1 y_1 + a_2 M_2 y_2 + a_3 M_3 y_3 + a_4 M_4 y_4 \pmod{M} \\ &\equiv (-52) \cdot 301 \cdot 307 \cdot 311 \cdot 113 + 51 \cdot 300 \cdot 307 \cdot 311 \cdot 5 \\ &\quad + 228 \cdot 300 \cdot 301 \cdot 311 \cdot 53 + 36 \cdot 300 \cdot 301 \cdot 307 \cdot 135 \pmod{M} \end{aligned}$$

Each individual term in this sum contains 12 digits, so we reduce each modulo M as follows.

$$\begin{aligned} (-52) \cdot 113 &\equiv 124 \pmod{300} \\ (-52) \cdot 301 \cdot 307 \cdot 311 \cdot 113 &\equiv 124 \cdot 301 \cdot 307 \cdot 311 \pmod{M} \\ 51 \cdot 5 &\equiv -46 \pmod{301} \\ 51 \cdot 300 \cdot 307 \cdot 311 \cdot 5 &\equiv -46 \cdot 300 \cdot 307 \cdot 311 \pmod{M} \end{aligned}$$

Similarly,

$$228 \cdot 300 \cdot 301 \cdot 311 \cdot 53 \equiv 111 \cdot 300 \cdot 301 \cdot 311 \pmod{M}$$

and

$$36 \cdot 300 \cdot 301 \cdot 307 \cdot 135 \equiv -116 \cdot 300 \cdot 301 \cdot 307 \pmod{M}$$

Therefore,

$$\begin{aligned}x &\equiv 124 \cdot 301 \cdot 307 \cdot 311 - 46 \cdot 300 \cdot 307 \cdot 311 + 111 \cdot 300 \cdot 301 \cdot 311 \\ &\quad - 116 \cdot 300 \cdot 301 \cdot 307 \pmod{M} \\ &\equiv (124 \cdot 301 - 46 \cdot 300) \cdot 307 \cdot 311 + (111 \cdot 311 - 116 \cdot 307) \cdot 300 \cdot 301 \pmod{M} \\ &\equiv 23524 \cdot 307 \cdot 311 - 1091 \cdot 300 \cdot 301 \pmod{M} \\ &\equiv 23000 \cdot 307 \cdot 311 + 524 \cdot 307 \cdot 311 - 1091 \cdot 300 \cdot 301 \pmod{M} \\ &\equiv 2{,}195{,}971{,}000 + 50{,}029{,}948 - 98{,}517{,}300 \pmod{M} \\ &\equiv 2{,}147{,}483{,}648 \pmod{M}\end{aligned}$$

Thus, $2^{31} = 2{,}147{,}483{,}648$.

(Fortunately, a calculator such as the TI-86 will find this answer in seconds. Nevertheless, the solution exemplifies the power of the CRT.) ■

EXERCISES 6.1

Determine whether the given number is a solution of the corresponding system of linear congruences.

1. 52; $x \equiv 2 \pmod 5$, $x \equiv 3 \pmod 7$
2. 43; $x \equiv 1 \pmod 2$, $x \equiv 2 \pmod 3$, $x \equiv 3 \pmod 5$
3. 14; $x \equiv 2 \pmod 3$, $x \equiv 4 \pmod 5$, $x \equiv 5 \pmod 7$
4. 67; $x \equiv 1 \pmod 3$, $x \equiv 5 \pmod 4$, $x \equiv 4 \pmod 7$, $x \equiv 7 \pmod{11}$

Solve the following linear systems using iteration.

5. $x \equiv 2 \pmod 5$, $x \equiv 3 \pmod 7$
6. $x \equiv 3 \pmod 4$, $x \equiv 5 \pmod 9$
7. $x \equiv 1 \pmod 2$, $x \equiv 2 \pmod 3$, $x \equiv 3 \pmod 5$
8. $x \equiv 2 \pmod 3$, $x \equiv 4 \pmod 5$, $x \equiv 5 \pmod 7$
9. $x \equiv 1 \pmod 3$, $x \equiv 3 \pmod 4$, $x \equiv 4 \pmod 7$, $x \equiv 7 \pmod{11}$
10. $x \equiv 2 \pmod 4$, $x \equiv 3 \pmod 5$, $x \equiv 4 \pmod 9$, $x \equiv 5 \pmod{13}$

11–16. Using the CRT, solve the linear systems in Exercises 5–10.

Show that the following linear systems are *not* solvable.

17. $x \equiv 2 \pmod 4$, $x \equiv 3 \pmod 6$
18. $x \equiv 3 \pmod 4$, $x \equiv 4 \pmod 5$, $x \equiv 4 \pmod 6$

Find the least positive integer that leaves

19. The remainder 3 when divided by 7, 4 when divided by 9, and 8 when divided by 11.
20. The remainder 2 when divided by 5, 4 when divided by 6, and 5 when divided by 11, and 6 when divided by 13.
21. Find the least positive multiple of 7 that leaves the remainder 2 when divided by 5, 3 when divided by 6, and 5 when divided by 11.
22. Find the least positive multiple of 12 that leaves the remainder 4 when divided by 8, 6 when divided by 9, and 8 when divided by 14.
23. Find the smallest positive integer n such that $2|n$, $3|n+1$, $5|n+2$, $7|n+3$, and $11|n+4$.
24. Find the smallest integer > 10000 such that $3|n$, $4|n+3$, $5|n+4$, $7|n+5$, and $11|n+7$.
25. Find the smallest positive integer n such that $3^2|n$, $4^2|n+1$, and $5^2|n+2$.

26. A child has some marbles in a box. If the marbles are grouped in sevens, there will be five left over; if they are grouped in elevens, there will be six left over; if they are grouped in thirteens, eight will be left over. Determine the least number of marbles in the box.
27. Find the largest integer < 6000 that leaves the remainders 0, 2, 3, and 5 when divided by 3, 5, 7, and 13, respectively.
28. Find the largest integer < 4000 that leaves the remainders 1, 2, 3, and 4 when divided by 5, 6, 7, and 11, respectively.
29. Find the smallest positive integer n, if it exists, such that $2^3|n$, $3^3|n+1$, and $4^3|n+2$.
30. Find the smallest positive integer n such that $3^3|n$ and $5^3|n+1$.

*6.2 General Linear Systems (optional)

The proof of the CRT establishes a solution to a linear system with pairwise relatively prime moduli and shows the solution is unique. It does not, however, indicate anything about a system where the moduli are not necessarily pairwise relatively prime. We will establish a necessary and sufficient condition for such a system to be solvable.

We begin with the following theorem for a system consisting of two linear congruences.

THEOREM 6.2 The linear system

$$x \equiv a \pmod{m}$$

$$x \equiv b \pmod{n}$$

is solvable if and only if $(m, n)|(a-b)$. When it is solvable, the solution is unique modulo $[m, n]$.

PROOF

The proof consists of two parts. We will show that the linear system is solvable if and only if $(m, n)|(a-b)$; and when the system is solvable, the solution is unique modulo $[m, n]$.

- *To show that the linear system is solvable if and only if $(m, n)|(a-b)$:*

Suppose x_0 is a solution of the linear system. Then $x_0 \equiv a \pmod{m}$ and $x_0 \equiv b \pmod{n}$. The first congruence implies $x_0 = a + km$ for some integer k. So, $a + km \equiv b \pmod{n}$; that is, $mk \equiv b - a \pmod{n}$. This congruence, by Theorem 4.9, has a solution (for k) if and only if $(m, n)|(b-a)$; that is, if and only if $(m, n)|(a-b)$.

- *To show that the linear system has a unique solution when the system is solvable*:

Suppose $(m, n)|(a-b)$ and x_0 is a solution of the linear system. Let x_1 be an arbitrary solution of the system. We will show that $x_1 \equiv x_0 \pmod{[m, n]}$.

Because x_1 and x_0 are solutions of the linear system, $x_1 \equiv a \pmod m$, $x_1 \equiv b \pmod n$, $x_0 \equiv a \pmod m$, and $x_0 \equiv b \pmod n$.

Therefore, by symmetry and transitivity, $x_1 \equiv x_0 \pmod m$ and $x_1 \equiv x_0 \pmod n$. Then $m|(x_1 - x_0)$ and $n|(x_1 - x_0)$, so by Corollary 3.11, $[m, n]|(x_1 - x_0)$; that is, $x_1 \equiv x_0 \pmod{[m, n]}$.

Thus, every solution is congruent to x_0 mod$[m, n]$; that is, the solution is unique modulo $[m, n]$. ■

Unlike the CRT, this theorem does not supply a formula for the solution to the system. However, if we can find one solution x_0, then the **general solution** is $x = x_0 + [m, n]t$, with t being an arbitrary integer.

The following example illustrates the theorem.

EXAMPLE 6.5 Determine whether the following linear systems are solvable.

1. $x \equiv 3 \pmod 6$
   $x \equiv 5 \pmod 8$
2. $x \equiv 7 \pmod 9$
   $x \equiv 11 \pmod{12}$

SOLUTION

1. Since $(6, 8) = 2$ and $2|(3-5)$, the first linear system has a solution.
2. We have $(9, 12) = 3$, but $3 \nmid (7-11)$, so the second system is not solvable. ■

The following example illustrates once again how we can employ iteration to solve linear systems of congruences.

EXAMPLE 6.6 Solve the linear system

$$x \equiv 3 \pmod 6$$

$$x \equiv 5 \pmod 8$$

SOLUTION

By the preceding example, the system has a unique solution. Because $x \equiv 3 \pmod 6$, $x = 3 + 6t'$, with t' being an arbitrary integer. Now substitute for x in the second congruence:

$$3 + 6t' \equiv 5 \pmod 8$$
$$6t' \equiv 2 \pmod 8$$

Dividing both sides by 2, using Theorem 4.9,

$$3t' \equiv 1 \pmod 4$$
$$t' \equiv 3 \pmod 4$$

so $t' = 3 + 4t$, with t being an arbitrary integer. Then $x = 3 + 6(3 + 4t) = 21 + 24t$. Thus, $x = 21$ is the unique solution modulo $[6, 8] = 24$. ■

Theorem 6.2 can be generalized to any system of linear congruences, as the following theorem shows. We leave its proof as an exercise.

THEOREM 6.3 The linear system $x \equiv a_i \pmod{m_i}$ is solvable if and only if $(m_i, m_j)|(a_i - a_j)$ for every i and j, where $1 \leq i < j \leq k$. When it is solvable, the solution is unique modulo $[m_1, m_2, \ldots, m_k]$. ■

The following two examples demonstrate this theorem.

EXAMPLE 6.7 Determine whether the following linear systems are solvable:

1. $x \equiv 4 \pmod 6$
   $x \equiv 2 \pmod 8$
   $x \equiv 1 \pmod 9$
2. $x \equiv 3 \pmod 4$
   $x \equiv 5 \pmod 9$
   $x \equiv 7 \pmod{12}$

SOLUTION

1. Since $(6, 8)|(4 - 2)$, $(8, 9)|(2 - 1)$, and $(6, 9)|(4 - 1)$, the first linear system has a solution.
2. For the second congruence, $(4, 9)|(3 - 5)$, and $(9, 12)|(5 - 8)$, but $(4, 12) = 4$ and $4 \nmid (3 - 8)$; so the second system is *not* solvable. ■

The following example shows how to solve a linear system using iteration.

EXAMPLE 6.8 Solve the linear system (1) in Example 6.7.

SOLUTION

By the preceding example, we know the system has a unique solution. To find it, the first congruence implies $x = 4 + 6w$, with w being arbitrary.

Now substitute for x in the second congruence:

$$4+6w \equiv 2 \pmod 8$$
$$6w \equiv -2 \pmod 8$$
$$3w \equiv -1 \pmod 4, \quad \text{by Theorem 4.7}$$
$$w \equiv 1 \pmod 4$$

that is, $w = 1+4v$, so $x = 4+6w = 10+24v$.

Substitute for x in the third congruence:

$$10+24v \equiv 1 \pmod 9$$
$$6v \equiv 0 \pmod 9$$
$$2v \equiv 0 \pmod 3$$
$$v \equiv 0 \pmod 3$$

that is, $v = 3t$. Therefore, $x = 10+24v = 10+72t$, where $72 = [6,8,9]$. So the unique solution is 10 modulo 72. ■

Suppose the moduli of the linear system in Theorem 6.3 are pairwise relatively prime. Then $(m_i, m_j) = 1$ for every pair of i and j, so $(m_i, m_j) | (a_i - a_j)$ for $i \neq j$. Thus, the system is solvable; further, since $[m_1, m_2, \ldots, m_k] = m_1 m_2 \cdots m_k$, the solution is unique modulo $m_1 m_2 \cdots m_k$. Thus, the CRT follows from Theorem 6.3 in the form of the following corollary.

COROLLARY 6.1 **(The Chinese Remainder Theorem)** The linear system $x \equiv a_i \pmod{m_i}$, where $1 \le i \le k$ and the moduli are pairwise relatively prime, is solvable and has a unique solution modulo $m_1 m_2 \cdots m_k$. ■

EXERCISES 6.2

Determine whether each linear system is solvable.

1. $x \equiv 2 \pmod{10}$
   $x \equiv 7 \pmod{15}$

2. $x \equiv 5 \pmod 9$
   $x \equiv 8 \pmod{12}$

3. $x \equiv 4 \pmod 9$
   $x \equiv 10 \pmod{12}$
   $x \equiv -2 \pmod{18}$

4. $x \equiv 7 \pmod 8$
   $x \equiv 3 \pmod{10}$
   $x \equiv 2 \pmod{15}$

Check whether the given value is a solution of the corresponding linear system.

5. $22+30t$; $x \equiv 2 \pmod{10}$
   $x \equiv 7 \pmod{15}$

6. $7+24t$; $x \equiv -1 \pmod 8$
   $x \equiv 7 \pmod{12}$

7. 426; $x \equiv 2 \pmod{8}$
   $x \equiv 3 \pmod{9}$
   $x \equiv 6 \pmod{10}$

8. 170; $x \equiv 3 \pmod{12}$
   $x \equiv 6 \pmod{15}$
   $x \equiv 11 \pmod{20}$

Using the given solution, find the general solution of each linear system.

9. 66; $x \equiv 2 \pmod{8}$
   $x \equiv 3 \pmod{9}$
   $x \equiv 6 \pmod{10}$

10. 51; $x \equiv 3 \pmod{12}$
    $x \equiv 6 \pmod{15}$
    $x \equiv 11 \pmod{20}$

Solve each linear system using iteration.

11. $x \equiv 10 \pmod{12}$
    $x \equiv 4 \pmod{15}$

12. $x \equiv 17 \pmod{20}$
    $x \equiv 5 \pmod{28}$

13. $x \equiv 1 \pmod{10}$
    $x \equiv 5 \pmod{12}$
    $x \equiv -4 \pmod{15}$

14. $x \equiv 7 \pmod{12}$
    $x \equiv 7 \pmod{15}$
    $x \equiv 7 \pmod{18}$

15. $x \equiv 2 \pmod{6}$
    $x \equiv 5 \pmod{9}$
    $x \equiv 8 \pmod{11}$
    $x \equiv 11 \pmod{15}$

16. $x \equiv 2 \pmod{6}$
    $x \equiv 5 \pmod{7}$
    $x \equiv 6 \pmod{8}$
    $x \equiv 8 \pmod{9}$

17. Assuming the linear system $x \equiv a \pmod{m_i}$, where $1 \le i \le k$, is solvable, find a formula for its general solution.

18. A piggy bank contains no more than 300 coins. When the coins are grouped in stacks of sixes, three coins are left; when they are grouped in eights, five are left; when they are grouped in twelves, nine are left. Find the maximum number of coins possible in the piggy bank.

19. A person has more than 500 fruits. If they are arranged in piles of 12, 16, and 18, then 5, 9, and 11 fruits are left over. Find the least number of fruits he has.

20. Find the smallest integer $n \ge 3$ such that $2|n$, $3|n+1$, $4|n+2$, $5|n+3$, and $6|n+4$.

21. Find the smallest integer $n \ge 4$ such that $3|n$, $4|n+1$, $5|n+2$, $6|n+3$, and $7|n+4$.

22. Find an integer n, if it exists, such that $2^2|n$, $3^2|n+1$, and $4^2|n+2$.

23. When eggs in a basket are removed 2, 3, 4, 5, or 6 at a time, there remain 1, 2, 3, 4, and 5 eggs, respectively. However, when 7 eggs are removed at a time, no eggs are left. Find the least number of eggs that could have been in the basket.†

24. Determine if there is an integer n such that $2^2|n$, $3^2|n+1$, $4^2|n+2$, and $5^2|n+3$.

25. Find the smallest integer n such that $3^2|n$, $4^2|n+1$, $5^2|n+2$, and $7^2|n+3$.

† This problem was proposed by Brahmagupta (ca. 628), the most prominent Indian mathematician of the seventh century.

*6.3 2 x 2 Linear Systems (optional)

In the two preceding sections, we demonstrated in detail how to solve systems of linear congruences involving a single variable. Now we turn to systems of two linear congruences in two variables with the same modulus m. Such a linear system is a 2×2 linear system.

2 x 2 Linear Systems

A **2 x 2 linear system** is a system of linear congruences of the form

$$ax + by \equiv e \pmod{m}$$
$$cx + dy \equiv f \pmod{m}$$

A **solution** of the linear system is a pair $x \equiv x_0 \pmod{m}$, $y \equiv y_0 \pmod{m}$ that satisfies both congruences.

EXAMPLE 6.9 Show that $x \equiv 12 \pmod{13}$ and $y \equiv 2 \pmod{13}$ is a solution of the 2×2 linear system

$$2x + 3y \equiv 4 \pmod{13}$$
$$3x + 4y \equiv 5 \pmod{13}$$

SOLUTION

When $x \equiv 12 \pmod{13}$ and $y \equiv 2 \pmod{13}$,

$$2x + 3y \equiv 2(12) + 3(2) \equiv 4 \pmod{13}$$
$$3x + 4y \equiv 3(12) + 4(2) \equiv 5 \pmod{13}$$

Therefore, every pair $x \equiv 12 \pmod{13}$, $y \equiv 2 \pmod{13}$ is a solution of the system. (The **general solution** of the system is $x = 12 + 13t$, $y = 2 + 13t$, with t being an arbitrary integer.) ■

We will now study two methods for solving 2×2 linear systems. One is the **method of elimination** and the other is a rule that resembles the well-known **Cramer's rule** for solving linear systems of equations.

The following example illustrates the method of elimination, which involves eliminating one of the variables and solving the resulting linear congruences.

EXAMPLE 6.10 Using the method of elimination, solve the linear system

$$2x + 3y \equiv 4 \pmod{13} \quad (6.1)$$
$$3x + 4y \equiv 5 \pmod{13} \quad (6.2)$$

SOLUTION

To eliminate y, multiply congruence (6.1) by 4 and congruence (6.2) by 3:

$$8x + 12y \equiv 3 \pmod{13}$$
$$9x + 12y \equiv 2 \pmod{13}$$

Subtracting,

$$-x \equiv 1 \pmod{13}$$
$$x \equiv 12 \pmod{13}$$

To find y, substitute for x in congruence (6.1):

$$2 \cdot 12 + 3y \equiv 4 \pmod{13}$$
$$3y \equiv -7 \pmod{13}$$
$$y \equiv 2 \pmod{13}$$

Thus, the solution is given by $x \equiv 12 \pmod{13}$, $y \equiv 2 \pmod{13}$. (Notice that this agrees with Example 6.9.) ■

The following theorem provides a necessary and sufficient condition for a 2×2 linear system to have a unique solution.

THEOREM 6.4 The linear system

$$ax + by \equiv e \pmod{m}$$
$$cx + dy \equiv f \pmod{m}$$

has a unique solution if and only if $(\Delta, m) = 1$, where $\Delta \equiv ad - bc \pmod{m}$.

PROOF

Suppose the system has a solution $x \equiv x_0 \pmod{m}$ and $y \equiv y_0 \pmod{m}$:

$$ax_0 + by_0 \equiv e \pmod{m} \tag{6.3}$$
$$cx_0 + dy_0 \equiv f \pmod{m} \tag{6.4}$$

Multiply congruence (6.3) by d and congruence (6.4) by b:

$$adx_0 + bdy_0 \equiv ed \pmod{m}$$
$$bcx_0 + bdy_0 \equiv bf \pmod{m}$$

Subtracting,

$$(ad - bc)x_0 \equiv (ed - bf) \pmod{m}$$

By Corollary 4.6, x_0 has a unique value modulo m if and only if $(\Delta, m) = 1$. Similarly, y_0 has a unique value modulo m if and only if $(\Delta, m) = 1$.

Thus, the system has a unique solution modulo m if and only if $(\Delta, m) = 1$. ■

The following example demonstrates this theorem.

EXAMPLE 6.11 Verify that the linear system

$$2x + 3y \equiv 4 \pmod{13}$$

$$3x + 4y \equiv 5 \pmod{13}$$

has a unique solution modulo 13.

SOLUTION

By Theorem 6.4, all we need to check is whether $(\Delta, 13) = 1$ for the linear system: $\Delta \equiv ad - bc \equiv 2 \cdot 4 - 3 \cdot 3 \equiv -1 \equiv 12 \pmod{13}$. Since $(12, 13) = 1$, by Theorem 6.4, the system has a unique solution modulo 13. ■

Although Theorem 6.4 can be used to determine whether a system has a unique solution, it does not furnish us with the solution when it is solvable. However, the following theorem does.

THEOREM 6.5 When the linear system

$$ax + by \equiv e \pmod{m}$$

$$cx + dy \equiv f \pmod{m}$$

has a unique solution modulo m, it is given by $x_0 \equiv \Delta^{-1}(ed - bf) \pmod{m}$ and $y_0 \equiv \Delta^{-1}(af - ce) \pmod{m}$, where $\Delta \equiv ad - bc \pmod{m}$ and Δ^{-1} is an inverse of Δ modulo m.

PROOF

By Theorem 6.4, since the system has a unique solution modulo m, $(\Delta, m) = 1$; so, by Corollary 4.6, Δ is invertible.

Because the linear system has a unique solution, it suffices to show that x_0, y_0 satisfies the system:

$$\begin{aligned}
ax_0 + by_0 &\equiv a\Delta^{-1}(de - bf) + b\Delta^{-1}(af - ce) \pmod{m} \\
&\equiv (ad - bc)\Delta^{-1}e + \Delta^{-1}(abf - abf) \pmod{m} \\
&\equiv \Delta\Delta^{-1}e + 0 \pmod{m} \\
&\equiv e \pmod{m}, \quad \text{since } \Delta\,\Delta^{-1} \equiv 1 \pmod{m}
\end{aligned}$$

Also,

$$\begin{aligned}
cx_0 + dy_0 &\equiv c\Delta^{-1}(de - bf) + d\Delta^{-1}(af - ce) \pmod{m} \\
&\equiv (ad - bc)\Delta^{-1}f + \Delta^{-1}(cde - cde) \pmod{m} \\
&\equiv \Delta\Delta^{-1}f + 0 \pmod{m} \\
&\equiv f \pmod{m}, \quad \text{because } \Delta\,\Delta^{-1} \equiv 1 \pmod{m}
\end{aligned}$$

Thus, $x \equiv x_0 \pmod{m}$, $y \equiv y_0 \pmod{m}$ is the unique solution of the linear system. ■

The formulas for $x_0 \pmod{m}$ and $y_0 \pmod{m}$ closely resemble those for x and y in Cramer's rule for a linear system of equations. To see this, we can rewrite the values of Δ, x_0, and y_0 in terms of determinants:

$$\Delta \equiv ad - bc \equiv \begin{vmatrix} a & b \\ c & d \end{vmatrix} \pmod{m}$$

$$x_0 \equiv \Delta^{-1}(ed - bf) \equiv \Delta^{-1} \begin{vmatrix} e & b \\ f & d \end{vmatrix} \pmod{m}$$

$$y_0 \equiv \Delta^{-1}(af - ce) \equiv \Delta^{-1} \begin{vmatrix} a & e \\ c & f \end{vmatrix} \pmod{m}$$

The following example illustrates Theorem 6.5.

EXAMPLE 6.12 Solve the linear system

$$\begin{aligned}
3x + 13y &\equiv 8 \pmod{55} \\
5x + 21y &\equiv 34 \pmod{55}
\end{aligned}$$

SOLUTION

First notice that $\Delta \equiv 3 \cdot 21 - 13 \cdot 5 \equiv 53 \pmod{55}$ and $(53, 55) = 1$, so the system has a unique solution modulo 55. Also, $\Delta^{-1} \equiv 27 \pmod{55}$. Therefore,

$$x_0 \equiv \Delta^{-1}(de - bf) \equiv 27(21 \cdot 8 - 13 \cdot 34) \equiv 27 \pmod{55}$$
$$y_0 \equiv \Delta^{-1}(af - ce) \equiv 27(3 \cdot 34 - 5 \cdot 8) \equiv 24 \pmod{55}$$

Thus, $x \equiv 27 \pmod{55}$ and $y \equiv 24 \pmod{55}$ is the unique solution to the given system. ■

Note that the techniques employed to solve linear systems of equations using matrices and determinants can be adapted nicely to solve higher order linear systems of congruences. Some are explored in the following exercises.

EXERCISES 6.3

Determine whether each linear system is solvable.

1. $3x + 4y \equiv 5 \pmod 7$
   $4x + 5y \equiv 6 \pmod 7$

2. $4x + 5y \equiv 5 \pmod 8$
   $3x - 6y \equiv 3 \pmod 8$

3. $5x + 6y \equiv 10 \pmod{13}$
   $6x - 7y \equiv 2 \pmod{13}$

4. $7x + 8y \equiv 10 \pmod{15}$
   $5x - 9y \equiv 10 \pmod{15}$

5. $x + 3y \equiv 3 \pmod{11}$
   $5x + y \equiv 5 \pmod{11}$

6. $6x - 7y \equiv 15 \pmod{17}$
   $11x - 9y \equiv 13 \pmod{17}$

Solve the following linear systems using elimination.

7. $3x + 4y \equiv 5 \pmod 7$
   $4x + 5y \equiv 6 \pmod 7$

8. $4x + 5y \equiv 5 \pmod 8$
   $3x - 6y \equiv 3 \pmod 8$

9. $5x + 6y \equiv 10 \pmod{13}$
   $6x - 7y \equiv 2 \pmod{13}$

10. $7x + 8y \equiv 11 \pmod{15}$
    $5x - 9y \equiv 10 \pmod{15}$

11. $x + 3y \equiv 3 \pmod{11}$
    $5x + y \equiv 5 \pmod{11}$

12. $6x + 11y \equiv 9 \pmod{16}$
    $7x + 8y \equiv 9 \pmod{16}$

13–18. Solve the linear systems in Exercises 7–12 using Theorem 6.5.

Solve each linear system using Theorem 6.5.

19. $5x + 11y \equiv 8 \pmod{13}$
    $11x + 5y \equiv 9 \pmod{13}$

20. $4x - 6y \equiv 2 \pmod{14}$
    $7x + 11y \equiv 11 \pmod{14}$

21. $7x - 11y \equiv 12 \pmod{18}$
    $11x - 12y \equiv 2 \pmod{18}$

Solve each **3 × 3 linear system** using elimination.

22. $x + y + z \equiv 6 \pmod 7$
    $x + 2y + 3z \equiv 6 \pmod 7$
    $2x + 3y + 4z \equiv 5 \pmod 7$

23. $x - 2y - z \equiv 6 \pmod{11}$
    $2x + 3y + z \equiv 5 \pmod{11}$
    $3x + y + 2z \equiv 2 \pmod{11}$

The 3 × 3 linear system

$$\begin{aligned} a_1x + b_1y + c_1z &\equiv d_1 \pmod m \\ a_2x + b_2y + c_2z &\equiv d_2 \pmod m \\ a_3x + b_3y + c_3z &\equiv d_3 \pmod m \end{aligned} \tag{6.5}$$

has a unique solution modulo m if and only if $(\Delta, m) = 1$, where

$$\Delta \equiv \begin{vmatrix} a_1 & b_1 & c_1 \\ a_2 & b_2 & c_2 \\ a_3 & b_3 & c_3 \end{vmatrix} \pmod{m}$$

Using this fact, determine whether each system has a unique solution.

24. $x + y + z \equiv 6 \pmod{7}$
$x + 2y + 3z \equiv 6 \pmod{7}$
$2x + 3y + 4z \equiv 5 \pmod{7}$

25. $x - 2y - z \equiv 6 \pmod{11}$
$2x + 3y + z \equiv 5 \pmod{11}$
$3x + y + 2z \equiv 2 \pmod{11}$

26. $x - y + 2z \equiv 7 \pmod{8}$
$2x + y - z \equiv 7 \pmod{8}$
$3x + 2y + z \equiv 2 \pmod{8}$

27. $2x - 3y + z \equiv 12 \pmod{13}$
$x + 2y - z \equiv 6 \pmod{13}$
$3x - y + 2z \equiv 2 \pmod{13}$

28–31. When the linear system (6.5) has a unique solution, it is given by

$$x \equiv \Delta^{-1} \begin{vmatrix} d_1 & b_1 & c_1 \\ d_2 & b_2 & c_2 \\ d_3 & b_3 & c_3 \end{vmatrix} \pmod{m}$$

$$y \equiv \Delta^{-1} \begin{vmatrix} a_1 & d_1 & c_1 \\ a_2 & d_2 & c_2 \\ a_3 & d_3 & c_3 \end{vmatrix} \pmod{m}$$

$$z \equiv \Delta^{-1} \begin{vmatrix} a_1 & b_1 & d_1 \\ a_2 & b_2 & d_2 \\ a_3 & b_3 & d_3 \end{vmatrix} \pmod{m}$$

Using this result, solve linear systems in Exercises 24–27.

CHAPTER SUMMARY

We have established the solvability of systems of linear congruences $x \equiv a_i \pmod{m_i}$, where $1 \le i \le k$ and the moduli m_i are pairwise relatively prime. Also, we have shown that the solution is unique modulo $m_1 m_2 \cdots m_k$. We have obtained necessary and sufficient conditions for the solvability of such systems when the moduli are not necessarily pairwise relatively prime and have proved that the solution is unique modulo $[m_1, m_2, \ldots, m_k]$. We have solved linear systems in a single variable using iteration and the CRT, and 2×2 linear systems using elimination and determinants.

The Chinese Remainder Theorem

- The linear system $x \equiv a_i \pmod{m_i}$, where $1 \le i \le k$ and the moduli are pairwise relatively prime, has a unique solution modulo $m_1 m_2 \cdots m_k$. (p. 297)
- The solution of the linear system is given by $x \equiv a_1 M_1 y_1 + a_2 M_2 y_2 + \cdots + a_k M_k y_k \pmod{M}$, where $M = m_1 m_2 \cdots m_k$, $M_i = M/m_i$, and y_i is an inverse of M_i modulo m_i. (p. 297)

- The linear system $x \equiv a_i \pmod{m_i}$ is solvable if and only if $(m_i, m_j)|(a_i - a_j)$ for every i and j, where $1 \le i < j \le k$; when it is solvable, the system has a unique solution modulo $[m_1, m_2, \ldots, m_k]$. (p. 305)

2×2 Linear Systems

- The 2×2 linear system

$$ax + by \equiv e \pmod{m}$$

$$cx + dy \equiv f \pmod{m}$$

has a unique solution modulo m if and only if $(\Delta, m) = 1$, where $\Delta \equiv ad - bc \pmod{m}$. (p. 309)

- The solution is given by $x_0 \equiv \Delta^{-1}(de - bf) \pmod{m}$, $y_0 \equiv \Delta^{-1}(af - ce) \pmod{m}$, where Δ^{-1} denotes an inverse of Δ modulo m. (p. 310)

REVIEW EXERCISES

Solve each linear system using iteration.

1. $x \equiv 3 \pmod{7}$
   $x \equiv 5 \pmod{10}$
2. $x \equiv 6 \pmod{8}$
   $x \equiv -2 \pmod{12}$
3. $x \equiv 2 \pmod{5}$
   $x \equiv 3 \pmod{7}$
   $x \equiv 5 \pmod{8}$
4. $x \equiv 4 \pmod{6}$
   $x \equiv -2 \pmod{10}$
   $x \equiv -2 \pmod{15}$

Using the CRT, solve each linear system.

5. $x \equiv 3 \pmod{7}$
   $x \equiv 5 \pmod{10}$
6. $x \equiv 5 \pmod{9}$
   $x \equiv 8 \pmod{16}$
7. $x \equiv 2 \pmod{5}$
   $x \equiv 3 \pmod{7}$
   $x \equiv 5 \pmod{8}$
8. $x \equiv 3 \pmod{5}$
   $x \equiv 5 \pmod{7}$
   $x \equiv 8 \pmod{12}$

Find the smallest positive integer that leaves

9. The remainders 8, 7, and 11 when divided by 7, 11, and 15, respectively.
10. The remainders 8, 5, and 14 when divided by 11, 12, and 15, respectively.

Find the largest integer < 15,000 that leaves:

11. The remainders 2, 5, 0, and 1 when divided by 5, 11, 12, and 13, respectively.
12. The remainders 3, 7, 9, and 11 when divided by 8, 12, 14, and 17, respectively.

13. Find the smallest positive integer n such that $3|n$, $4|n+1$, $5|n+2$, $7|n+3$, and $11|n+4$.
14. Find the smallest positive integer n such that $2^2|n$, $3^2|n+2$, $5^2|n+3$, and $11^2|n+5$.
15. A jar contains at least 300 pennies. If they are grouped in fives, sixes, sevens, and elevens, there will be three, five, four, and eight pennies left over. Find the least possible number of coins in the jar.
16. A fruit basket contains not more than 3000 plums. When they are grouped in piles of sixes, nines, elevens, and fifteens, there will be two, eight, seven, and fourteen plums left over, respectively. Find the maximum possible number of plums in the basket.

Determine whether each linear system is solvable.

17. $5x + 7y \equiv 3 \pmod 9$
$6x + 5y \equiv 4 \pmod 9$

18. $6x + 5y \equiv 7 \pmod{12}$
$3x + 11y \equiv 8 \pmod{12}$

19. $8x + 11y \equiv 5 \pmod{13}$
$7x + 9y \equiv 10 \pmod{13}$

20. $3x + 8y \equiv 11 \pmod{15}$
$7x + 12y \equiv 13 \pmod{15}$

Solve each linear system using elimination.

21. $5x + 7y \equiv 3 \pmod 9$
$6x + 5y \equiv 4 \pmod 9$

22. $8x + 11y \equiv 5 \pmod{13}$
$7x + 9y \equiv 10 \pmod{13}$

23. $8x + 5y \equiv 4 \pmod{15}$
$3x + 11y \equiv 7 \pmod{15}$

24. $x + y - z \equiv 8 \pmod{11}$
$x - y + z \equiv 5 \pmod{11}$
$x - y - z \equiv 10 \pmod{11}$

25. $x - y - z \equiv 5 \pmod{17}$
$x + 2y + z \equiv 2 \pmod{17}$
$2x - 3y - z \equiv 0 \pmod{17}$

26. $x + 2y - 3z \equiv 3 \pmod{19}$
$2x - y + 4z \equiv 10 \pmod{19}$
$3x + 4y + 5z \equiv 9 \pmod{19}$

27–32. Using determinants, solve the linear systems in Exercises 21–26.

33. An apartment complex contains one- and two-bedroom apartments. They are rented for \$675 and \$975 a month, respectively. If all apartments are rented, the total monthly revenue would leave a remainder of \$54 when it is divided by 101. But if the rents are lowered by \$100 apiece, and if all apartments are rented, then the total monthly revenue would leave a remainder of \$53 when it is divided by 101. Find the number of one- and two-bedroom apartments if the total income does not exceed \$100,000.
34. Judy bought some 29-cent and 35-cent stamps. The total value of the stamps (in cents) leaves a remainder 1 when it is divided by 23. But when the stamps' costs are increased by 10¢ each, the total value leaves a remainder of 6 when divided

by 23. Find the number of 29-cent and 35-cent stamps she bought if the total cost is no more than $18.

SUPPLEMENTARY EXERCISES

To do these exercises, you will need knowledge of matrices and the following definitions:

Let $A = (a_{ij})_{k\times l}$ and $B = (b_{ij})_{k\times l}$ be two matrices with integral entries. Then A is **congruent to** B modulo m if $a_{ij} \equiv b_{ij} \pmod{m}$ for every i and j, and we then write $A \equiv B \pmod{m}$.

For example,

$$\begin{bmatrix} 8 & -5 \\ 17 & 6 \end{bmatrix} \equiv \begin{bmatrix} 2 & 1 \\ -1 & 0 \end{bmatrix} \pmod{6}$$

The matrix $A = (a_{ij})_{n\times n}$ is the **identity matrix of order** n if $a_{ij} = 1$ when $i = j$ and 0 otherwise; it is denoted by I_n or simply I. For instance, the identity matrix of order 2 is

$$\begin{bmatrix} 1 & 0 \\ 0 & 1 \end{bmatrix}$$

A matrix A^{-1} is an **inverse** of matrix A modulo m if $AA^{-1} \equiv I \equiv A^{-1}A \pmod{m}$. For example, let

$$A = \begin{bmatrix} 2 & 3 \\ 4 & 5 \end{bmatrix}$$

Then

$$A^{-1} \equiv \begin{bmatrix} 1 & 5 \\ 2 & 6 \end{bmatrix} \pmod{7},$$

because

$$AA^{-1} \equiv \begin{bmatrix} 2 & 3 \\ 4 & 5 \end{bmatrix}\begin{bmatrix} 1 & 5 \\ 2 & 6 \end{bmatrix} \equiv \begin{bmatrix} 1 & 0 \\ 0 & 1 \end{bmatrix} \equiv I \pmod{7};$$

similarly, $A^{-1}A \equiv I \pmod{7}$.

Verify that the given matrices are inverses of each other for the indicated modulus.

1. $\begin{bmatrix} 3 & 5 \\ 4 & 7 \end{bmatrix}$, $\begin{bmatrix} 7 & 6 \\ 7 & 3 \end{bmatrix}$; $m = 11$

2. $\begin{bmatrix} 6 & 8 \\ -7 & 4 \end{bmatrix}$, $\begin{bmatrix} 2 & 9 \\ 10 & 3 \end{bmatrix}$; $m = 13$

The 2×2 linear system

$$ax + by \equiv e \pmod{m}$$

$$cx + dy \equiv f \pmod{m}$$

can be written as the matrix congruence $AX \equiv B \pmod{m}$, where

$$A = \begin{bmatrix} a & b \\ c & d \end{bmatrix}, \quad X = \begin{bmatrix} x \\ y \end{bmatrix}, \quad \text{and} \quad B = \begin{bmatrix} e \\ f \end{bmatrix}$$

Rewrite the following linear systems as matrix congruences.

3. $3x + 5y \equiv 7 \pmod{11}$
   $4x + 7y \equiv 9 \pmod{11}$

4. $6x + 8y \equiv 5 \pmod{13}$
   $-7x + 4y \equiv 8 \pmod{13}$

5–6. The coefficient matrix A in the congruence $AX \equiv B \pmod{m}$ has an inverse A^{-1} modulo m if and only if $(\Delta, m) = 1$, where $\Delta \equiv ad - bc \pmod{m}$. Then $X \equiv A^{-1}B \pmod{m}$. Using these facts, and Exercises 1 and 2, solve the linear systems in Exercises 3 and 4.

7. Let $A = \begin{bmatrix} a & b \\ c & d \end{bmatrix}$, where a, b, c, and d are integers. Let $(\Delta, m) = 1$, where $\Delta \equiv ad - bc \pmod{m}$. Prove that

$$A^{-1} \equiv \Delta^{-1} \begin{bmatrix} d & -b \\ -c & a \end{bmatrix} \pmod{m}.$$

Using Exercise 7, find an inverse of each matrix for the indicated modulus m.

8. $\begin{bmatrix} 3 & 4 \\ 5 & 6 \end{bmatrix}$, $m = 7$

9. $\begin{bmatrix} 5 & 8 \\ 3 & 7 \end{bmatrix}$, $m = 13$

10. $\begin{bmatrix} 8 & 13 \\ 10 & 11 \end{bmatrix}$, $m = 17$

Solve each linear system using matrices.

11. $3x + 4y \equiv 2 \pmod{7}$
    $5x + 6y \equiv 3 \pmod{7}$

12. $5x + 8y \equiv 3 \pmod{13}$
    $3x + 7y \equiv 5 \pmod{13}$

13. $8x + 13y \equiv 9 \pmod{17}$
    $10x + 11y \equiv 8 \pmod{17}$

14. $4x + 7y \equiv 3 \pmod{16}$
    $11x + 8y \equiv 7 \pmod{16}$

COMPUTER EXERCISES

Write a program to perform each task.

1. Solve Sun-Tsu's puzzle using the CRT.
2. Find the smallest integer $n \geq 3$ such that $2|n$, $3|n+1, 4|n+2, 5|n+3$, and $6|n+4$.
3. Find the smallest integer $n \geq 4$ such that $3|n$, $4|n+1, 5|n+2, 6|n+3$, and $7|n+4$.
4. Find the smallest positive integer n such that $3^2|n$, $4^2|n+1$, and $5^2|n+2$.
5. Find the smallest positive integer n such that $3^2|n$, $4^2|n+1$, $5^2|n+2$, and $7^2|n+3$.

Determine whether each linear system is solvable.

6. $x \equiv 4 \pmod{6}$
   $x \equiv 2 \pmod{8}$
   $x \equiv 1 \pmod{9}$

7. $x \equiv 3 \pmod{4}$
   $x \equiv 5 \pmod{9}$
   $x \equiv 8 \pmod{12}$

Solve each linear system.

8. $x \equiv 7 \pmod{12}$
   $x \equiv 7 \pmod{15}$
   $x \equiv 7 \pmod{18}$

9. $x \equiv 2 \pmod{6}$
   $x \equiv 5 \pmod{9}$
   $x \equiv 8 \pmod{11}$
   $x \equiv 11 \pmod{15}$

10. $x \equiv 2 \pmod{6}$
    $x \equiv 5 \pmod{7}$
    $x \equiv 6 \pmod{8}$
    $x \equiv 8 \pmod{9}$

11. Let $m_1 = 400$, $m_2 = 401$, $m_3 = 403$, $M = m_1m_2m_3$, and $M_i = M/m_i$, where $1 \leq i \leq 3$. Solve each congruence $M_iy_i \equiv 1 \pmod{m_i}$; give the least residue in each case.
12. Let $m_1 = 300$, $m_2 = 301$, $m_3 = 307$, $m_4 = 311$, $M = m_1m_2m_3m_4$, and $M_i = M/m_i$, where $1 \leq i \leq 4$. Solve each congruence $M_iy_i \equiv 1 \pmod{m_i}$; give the least residue in each case.
13. Read in a 2×2 matrix with integral elements and a modulus m. Determine if it is invertible; if it is, find an inverse modulo m.
14. Read in a 2×2 linear system of congruences. Solve it using matrix congruences.
15. Read in a 3×3 matrix with integral elements and a modulus m. Determine if it is invertible.

ENRICHMENT READINGS

1. H. Eves, *An Introduction to the History of Mathematics*, 3rd ed., Holt, Rinehart and Winston, New York, 1969, 197–202.

2. F. T. Howard, "A Generalized Chinese Remainder Theorem," *College Math. J.*, 33 (Sept. 2002), 279–282.
3. S. Kangsheng, "Historical Development of the Chinese Remainder Theorem," *Archive for History of Exact Sciences*, 38 (1988), 285–305.
4. C. S. Ogilvy and J. T. Anderson, *Excursions in Number Theory*, Dover, New York, 1966.
5. O. Ore, *Invitation to Number Theory*, Math. Association of America, Washington, DC, 1967.
6. A. Rothbart, *The Theory of Remainders*, Janson Publications, Dedham, MA, 1995.

Three Classical Milestones

Euler calculated without effort, as men breathe, or as eagles sustain themselves in the wind.
—FRANÇOIS ARAGO

Three classical results—Wilson's theorem, Fermat's little theorem, and Euler's theorem—have played a significant role in the development of the theory of congruences. All three theorems illustrate the power of congruences and the congruence notation.

We begin our discussion with Wilson's theorem, which involves the factorial function.

7.1 Wilson's Theorem

In 1770, the English mathematician Edward Waring described in his *Meditationes Algebraicae* the following conjecture by John Wilson, one of his former students: "If p is a prime, then $p|[(p-1)!+1]$." Wilson is likely to have guessed this by using some pattern recognition. In any case, neither he nor Waring could furnish a proof of the result.

Three years after the conjecture was announced, Lagrange provided the first proof. He observed that its converse is also true.

Wilson, in fact, was not the first mathematician to discover the theorem, although it bears his name. There is evidence that the outstanding German mathematician

Edward Waring *(1734–1798) was born in Shrewsbury, England. Little is known about his early life. In 1753 he entered Magdalene College, Cambridge, where his mathematical talent blossomed. He graduated four years later, received his masters in 1760, and then became the sixth Lucasian professor of mathematics at Cambridge University, although some opposed his appointment because of his young age. His* Miscellanea Analytics, *published in 1762, silenced his critics and proved him a first-rate mathematician. He wrote five more treatises, the most important of them being* Meditationes Algebraicae. *A fellow of the Royal Society, he received the Copley Medal in 1784.*

Not all his activities, however, were mathematical. Concurrent with the writing of books, he pursued medicine and received his M.D. from Cambridge in 1767. He gave up his practice in 1770, but continued to serve as the Lucasian professor until his death.

John Wilson *(1741–1793) was born in Applethwaite, Westmoreland, England. After completing undergraduate work at Cambridge University and being a private tutor there for a brief period, he was called to the bar in 1766 and acquired considerable practice on the northern circuit. In 1786 he was elevated to the bench of the Court of Common Pleas. A fellow of the Royal Society, Wilson died in Kendal.*

Baron Gottfried Wilhelm Leibniz (1646–1716) knew it as early as 1682, although he did not publish it.

Recall from Corollary 4.6 that the congruence $ax \equiv 1 \pmod{m}$ has a unique solution if and only if $(a, m) = 1$. Further, the solution is an inverse a^{-1} of a modulo m. In particular, suppose the modulus is a prime p. Then positive least residues modulo p, that is, integers 1 through $p-1$, are invertible. For example, let $p = 7$. Then the positive least residues 1 through 6 are invertible: $1 \cdot 1 \equiv 2 \cdot 4 \equiv 3 \cdot 5 \equiv 6 \equiv 6 \equiv 1 \pmod{7}$.

The following lemma shows that exactly two of them are self-invertible.

LEMMA 7.1 A positive integer a is self-invertible modulo p if and only if $a \equiv \pm 1 \pmod{p}$.

PROOF

Suppose a is self-invertible. Then $a^2 \equiv 1 \pmod{p}$; that is, $p|(a^2-1)$; so $p|(a-1)(a+1)$. Then, by Lemma 3.3, $p|a-1$ or $p|a+1$; thus, either $a \equiv 1 \pmod{p}$ or $a \equiv -1 \pmod{p}$.

Conversely, suppose $a \equiv 1 \pmod{p}$ or $a \equiv -1 \pmod{p}$. In either case, $a^2 \equiv 1 \pmod{p}$, so a is self-invertible modulo p. ■

It follows by this lemma that exactly two least residues modulo p are self-invertible; they are 1 and $p-1$. Thus, the congruence $x^2 \equiv 1 \pmod{p}$ has exactly two solutions, 1 and $p-1$ modulo p.

For example, the self-invertible least residues modulo 13 are 1 and 12: $1^2 \equiv 1 \pmod{13}$ and $12^2 \equiv 1 \pmod{13}$. In other words, the solutions of the congruence $x^2 \equiv 1 \pmod{13}$ are 1 and 12 modulo 13.

Before formally stating Wilson's theorem, we study an example that will facilitate its proof.

EXAMPLE 7.1 Let $p = 11$. Then $(p-1)! = 10! = 1 \cdot 2 \cdot 3 \cdot 4 \cdot 5 \cdot 6 \cdot 7 \cdot 8 \cdot 9 \cdot 10$. The least residues modulo 11 that are self-invertible are 1 and 10; rearrange the remaining factors into pairs in such a way that the residues in each pair are inverses of each other modulo 11:

$$\begin{aligned} 10! &= 1 \cdot (2 \cdot 6) \cdot (3 \cdot 4) \cdot (5 \cdot 9) \cdot (7 \cdot 8) \cdot 10 \\ &\equiv 1 \cdot 1 \cdot 1 \cdot 1 \cdot 1 \cdot 10 \pmod{11} \\ &\equiv 10 \pmod{11} \\ &\equiv -1 \pmod{11} \end{aligned}$$

Thus, $(p-1)! \equiv -1 \pmod{11}$, illustrating Wilson's theorem. ■

The technique used in this example is essentially the same one employed in the proof of the theorem: Arrange into $(p-3)/2 = (11-3)/2 = 4$ pairs the positive least residues modulo $p\ (= 11)$ that are not self-invertible. We can now state and prove the first feature theorem.

THEOREM 7.1 **(Wilson's Theorem)** If p is a prime, then $(p-1)! \equiv -1 \pmod{p}$.

PROOF

When $p = 2$, $(p-1)! = 1 \equiv -1 \pmod{2}$; thus, the theorem is true when $p = 2$.

So, let $p > 2$. By Corollary 4.6, the least positive residues 1 through $p-1$ are invertible modulo p. But, by Lemma 7.1, two of them, 1 and $p-1$, are their own inverses. So we can group the remaining $p-3$ residues, 2 through $p-2$, into $(p-3)/2$ pairs of inverses a and $b = a^{-1}$ such that $ab \equiv 1 \pmod{p}$ for every pair a and b. Thus,

$$2 \cdot 3 \cdots (p-2) \equiv 1 \pmod{p}$$

$$\begin{aligned} (p-1)! &= 1 \cdot [2 \cdot 3 \cdots (p-2)] \cdot (p-1) \\ &\equiv 1 \cdot 1 \cdot (p-1) \pmod{p} \\ &\equiv -1 \pmod{p} \end{aligned}$$

■

The following example shows an interesting application of Wilson's theorem.

EXAMPLE 7.2 Let p be a prime and n any positive integer. Prove that

$$\frac{(np)!}{n!p^n} \equiv (-1)^n \pmod{p}$$

SOLUTION

First, we can make an observation. Let a be any positive integer congruent to 1 modulo p. Then, by Wilson's theorem,

$$a(a+1)\cdots[a+(p-2)] \equiv (p-1)! \equiv -1 \pmod{p}.$$

In other words, the product of the $p-1$ integers between any two consecutive multiples of p is congruent to -1 modulo p. Then

$$\begin{aligned}
\frac{(np)!}{n!p^n} &= \frac{(np)!}{p \cdot 2p \cdot 3p \cdots (np)} \\
&= \prod_{r=1}^{n} [(r-1)p+1]\cdots[(r-1)p+(p-1)] \\
&\equiv \prod_{r=1}^{n} (p-1)! \pmod{p} \\
&\equiv \prod_{r=1}^{n} (-1) \pmod{p} \\
&\equiv (-1)^n \pmod{p}
\end{aligned}$$

■

In particular, let $p = 5$ and $n = 46$. Then

$$\frac{(np)!}{n!p^n} = \frac{230!}{46!5^{46}} \equiv (-1)^{46} \equiv 1 \pmod{5}.$$

In 1957, F. G. Elston of New York generalized Wilson's theorem: Let p be a prime and $0 \le r \le p-1$. Then $r!(p-1-r)! + (-1)^r \equiv 0 \pmod{p}$. See Exercise 17.

We now turn to the converse of Wilson's theorem.

THEOREM 7.2 If n is a positive integer such that $(n-1)! \equiv -1 \pmod{n}$, then n is a prime.

PROOF (by contradiction)

Suppose n is composite, say, $n = ab$, where $1 < a, b < n$. Since $a|n$ and $n|[(n-1)! + 1]$, $a|[(n-1)! + 1]$. Since $1 < a < n$, a is one of the integers 2 through $n-1$, so

$a|(n-1)!$. Therefore, by Theorem 2.4, $a|[(n-1)!+1-(n-1)!]$; that is, $a|1$. So $a=1$, a contradiction. Thus, n must be a prime. ■

Theorems 7.1 and 7.2 together furnish a necessary and sufficient condition for a positive integer to be a prime: A positive integer $n \geq 2$ is a prime if and only if $(n-1)! \equiv -1 \pmod{n}$. This condition provides a seemingly simple test for primality. To check if n is a prime, all we need is to determine whether $(n-1)! \equiv -1 \pmod{n}$.

For example, $(7-1)! = 720 \equiv -1 \pmod{7}$, so 7 is a prime. On the other hand, $(12-1)! = 39{,}916{,}800 \equiv 0 \pmod{12}$, so $(12-1)! \not\equiv -1 \pmod{12}$, showing that 12 is *not* a prime.

Unfortunately, this test has *no* practical significance, because $(n-1)!$ becomes extremely large as n gets large.

Factorial, Multifactorial, and Primorial Primes

Theorem 7.2 naturally prompts several questions in the minds of the curious: Are there primes of the form $m!+1$? If yes, how many such primes are there? ?

Since $1!+1=2, 2!+1=3, 3!+1=7$ are primes, there do exist primes of the form $m!+1$. There are in fact nine such primes for $m \leq 100$. The largest known such prime, as of 2005, was $32659!+1$, discovered by Steven L. Harvey; it contains 44,416 digits.

On the other hand, $n!-1$ is a prime for $n=3$, 4, and 6. The largest known such prime is $974!-1$, discovered by Harvey Dubner of New Jersey; $n!-1$ is composite for $975 \leq n \leq 1155$. It remains unresolved as to whether there is an infinitude of primes of the form $n! \pm 1$; such primes are **factorial primes**. ?

In 1930, S. S. Pillai asked if every prime factor of $n!+1$ is congruent to 1 modulo n. For example, $11|(5!+1)$ and $11 \equiv 1 \pmod{5}$. In the same year, the Indian number theorist S. Chowla discovered two exceptions: $14!+1 \equiv 0 \pmod{23}$ and $18!+1 \equiv 0 \pmod{23}$, where $23 \not\equiv 1 \pmod{14}$ and $23 \not\equiv 1 \pmod{18}$. In fact, the smallest such counterexample is $8!+1 \equiv 0 \pmod{61}$, where $61 \not\equiv 1 \pmod{8}$.

In 1993, Erdös and M. V. Subbarao of the University of Alberta independently proved that there are infinitely many primes p for which there is an integer n such that $n!+1 \equiv 0 \pmod{p}$, where $p \not\equiv 1 \pmod{n}$.

Primes of the form $n!_k \pm 1$ are **multifactorial primes**, where $n!_k = n(n-k)(n-2k)(n-3k)\cdots(n-rk)$, k is a positive integer, and r is the largest positive integer such that $n-rk \geq 1$. (Notice that $n!_1 = n!$.) For example, $7!_3+1 = 7(7-3)(7-6)+1 = 29$ and $7!_5-1 = 7(7-5)-1 = 13$ are multifactorial primes.

Primes of the form $n\# \pm 1$ are **primorial primes**, where $n\#$ denotes the product of all primes $\leq n$; for example, $10\#+1 = 2\cdot 3\cdot 5\cdot 7+1 = 211$ is a primorial prime. Establishing the infinitude of primorial primes remains unresolved. ?

EXERCISES 7.1

Find the self-invertible least residues modulo each prime p.

1. 7
2. 13
3. 19
4. 23

Solve the congruence $x^2 \equiv 1 \pmod{m}$ for each modulus m.

5. 6
6. 8
7. 12
8. 15
9. Prove or disprove: If the congruence $x^2 \equiv 1 \pmod{m}$ has exactly two solutions, then m is a prime.
10. If $x^2 \equiv 1 \pmod{p}$ and $x^2 \equiv 1 \pmod{q}$, does it follow that $x^2 \equiv 1 \pmod{pq}$, where p and q are distinct primes?
11. Let a be a solution of the congruence $x^2 \equiv 1 \pmod{m}$. Show that $m - a$ is also a solution.

Without using Wilson's theorem, verify that $(p-1)! \equiv -1 \pmod{p}$ for each p.

12. 3
13. 5
14. 7
15. 13

Prove each, where p is a prime.

16. Let p be odd. Then $2(p-3)! \equiv -1 \pmod{p}$.
17. $(p-1)(p-2)\cdots(p-k) \equiv (-1)^k k! \pmod{p}$, where $1 \le k < p$.
18. Let p be odd. Then $1^2 \cdot 3^2 \cdots (p-2)^2 \equiv (-1)^{(p+1)/2} \pmod{p}$.
19. Let p be odd. Then $2^2 \cdot 4^2 \cdots (p-1)^2 \equiv (-1)^{(p+1)/2} \pmod{p}$.
20. A positive integer $n \ge 2$ is a prime if and only if $(n-2)! \equiv 1 \pmod{n}$.
21. Let r be a positive integer $< p$ such that $r! \equiv (-1)^r \pmod{p}$. Then $(p-r-1)! \equiv -1 \pmod{p}$.
22. $\dfrac{1 \cdot 3 \cdot 5 \cdots (p-2)}{2 \cdot 4 \cdot 6 \cdots (p-1)} \equiv (-1)^{(p-1)/2} \pmod{p}$, where $p > 2$. (P. S. Bruckman, 1975)
23. Let $0 \le r \le p-1$. Then $r!(p-1-r)! + (-1)^r \equiv 0 \pmod{p}$ (F. G. Elston, 1957)
24. $\dbinom{np}{p} \equiv n \pmod{p}$ (J. H. Hodges, 1959)
25. $\dbinom{np-1}{p-1} \equiv -1 \pmod{p}$, where $p \nmid n$. (This result, true even if $p|n$, was established in 1874 by Catalan.)
26. $\dbinom{p-1}{r} \equiv (-1)^r \pmod{p}$, where $0 \le r < p$.
27. *Let $p = m + n + 3$, where $m, n \ge 0$. Then $[m! + (m+1)!][n! + (n+1)!] \equiv (-1)^m \pmod{p}$ (A. Cusumano, 2005)
28. Using Exercise 21, show that $63! \equiv -1 \pmod{71}$.

Verify that $[((p-1)/2)!]^2 \equiv -1 \pmod{p}$ for each prime p.

29. 5
30. 13
31. 17
32. 29
33. Make a conjecture using Exercises 29–32.
34. Does your conjecture hold for $p = 2$, 3, or 7?
35. *Establish the conjecture in Exercise 33.
36. *Prove that $\prod_{n=1}^{p-1} (1 + p/n) \equiv 1 \pmod{p}$. (L. Talbot, 1995)

7.2 Fermat's Little Theorem

On October 18, 1640, Fermat wrote a letter to Bernhard Frenicle de Bessy (1605–1675), an official at the French mint who was a gifted student of number theory. In his letter, Fermat communicated the following result: *If p is a prime and $p \nmid a$, then $p | a^{p-1} - 1$.* Fermat did not provide a proof of this result but enclosed a note promising that he would send along a proof, provided it was not too long. This result is known as **Fermat's little theorem** or simply **Fermat's theorem**, to distinguish it from **Fermat's last theorem**, which is presented in Chapter 13. Incidentally, the special case of Fermat's little theorem for $a = 2$ was known to the Chinese as early as 500 B.C.

The first proof of Fermat's little theorem was given by Euler in 1736, almost a century after Fermat's announcement. Leibniz had given an identical proof in an unpublished work about 50 years prior to Euler's, but once again Leibniz did not receive his share of credit.

We need the following lemma for the proof of Fermat's little theorem, but before we turn to the lemma, let us study a special case.

EXAMPLE 7.3 Let $p = 7$ and $a = 12$. Clearly, $p \nmid a$. Then,

$$1 \cdot 12 \equiv 5 \pmod{7} \qquad 2 \cdot 12 \equiv 3 \pmod{7} \qquad 3 \cdot 12 \equiv 1 \pmod{7}$$
$$4 \cdot 12 \equiv 6 \pmod{7} \qquad 5 \cdot 12 \equiv 4 \pmod{7} \qquad 6 \cdot 12 \equiv 2 \pmod{7}$$

Thus, the least residues of $1 \cdot 12$, $2 \cdot 12$, $3 \cdot 12$, $4 \cdot 12$, $5 \cdot 12$, and $6 \cdot 12$ are the same as the integers 1, 2, 3, 4, 5, and 6 in *some* order. ■

More generally, we have the following result.

LEMMA 7.2 Let p be a prime and a any integer such that $p \nmid a$. Then the least residues of the integers $a, 2a, 3a, \ldots, (p-1)a$ modulo p are a permutation of the integers $1, 2, 3, \ldots, (p-1)$.

PROOF

The proof consists of two parts. [First, we will show that $ia \not\equiv 0 \pmod{p}$, where $1 \le i \le p-1$. Then we will show that the least residues of ia and ja modulo p are distinct if $i \ne j$, where $1 \le j \le p-1$.]

To show that $ia \not\equiv 0 \pmod{p}$, *where* $1 \le i \le p-1$:

Suppose $ia \equiv 0 \pmod{p}$. Then $p|ia$. But $(p, a) = 1$, so $p|i$, which is impossible since $i < p$. Therefore, $ia \not\equiv 0 \pmod{p}$.

To show that if $ia \equiv ja \pmod{p}$, *where* $1 \le i, j \le p-1$ *then* $i = j$:

Suppose $ia \equiv ja \pmod{p}$, where $1 \le i, j \le p-1$. Since $(p, a) = 1$, by Theorem 4.6, $i \equiv j \pmod{p}$. But both i and j are least residues modulo p, so $i = j$. Thus, if $ia \equiv ja \pmod{p}$, where $1 \le i, j \le p-1$, then $i = j$. In other words, no two least residues of $a, 2a, 3a, \ldots, (p-1)a$ are congruent modulo p. ■

We are now ready to present Fermat's little theorem formally and to prove it. Using the preceding lemma, we will find that the proof is short and clear.

THEOREM 7.3 **(Fermat's Little Theorem)** Let p be a prime and a any integer such that $p \nmid a$. Then $a^{p-1} \equiv 1 \pmod{p}$.

PROOF

By Lemma 7.2, the least residues of the integers $a, 2a, 3a, \ldots, (p-1)a$ modulo p are the same as the integers $1, 2, 3, \ldots, (p-1)$ in some order, so their products are congruent modulo p; that is, $a \cdot 2a \cdot 3a \cdots (p-1)a \equiv 1 \cdot 2 \cdot 3 \cdots (p-1) \pmod{p}$. In other words, $(p-1)!a^{p-1} \equiv (p-1)! \pmod{p}$. But $((p-1)!, p) = 1$, so by Theorem 4.6, $a^{p-1} \equiv 1 \pmod{p}$, as desired. ■

The following example illustrates this proof.

EXAMPLE 7.4 Let $p = 7$ and $a = 12$. By Lemma 7.2, the least residues of $1 \cdot 12$, $2 \cdot 12$, $3 \cdot 12$, $4 \cdot 12$, $5 \cdot 12$, $6 \cdot 12$ modulo 7 are a permutation of the integers 1 through 6, so $(1 \cdot 12)(2 \cdot 12)(3 \cdot 12)(4 \cdot 12)(5 \cdot 12)(6 \cdot 12) \equiv 1 \cdot 2 \cdot 3 \cdot 4 \cdot 5 \cdot 6 \pmod{7}$. That is, $6!12^6 \equiv 6! \pmod{7}$. Since $(6!, 7) = 1$, this yields $12^6 \equiv 1 \pmod{7}$. ■

Fermat's little theorem, coupled with the congruence properties we studied in Chapter 4, provides an efficient recipe to evaluate the remainder when a^n is divided by p, where $p \nmid a$ and $n \geq p - 1$, as the following example demonstrates.

EXAMPLE 7.5 Find the remainder when 24^{1947} is divided by 17.

SOLUTION

$$24 \equiv 7 \pmod{17}$$

Therefore,

$$24^{1947} \equiv 7^{1947} \pmod{17}$$

But, by Fermat's little theorem, $7^{16} \equiv 1 \pmod{17}$. So

$$\begin{aligned} 7^{1947} &= 7^{16 \cdot 121 + 11} = (7^{16})^{121} \cdot 7^{11} \\ &\equiv 1^{121} \cdot 7^{11} \equiv 7^{11} \pmod{17} \end{aligned}$$

But $7^2 \equiv -2 \pmod{17}$, so $7^{11} \equiv (7^2)^5 \cdot 7 \equiv (-2)^5 \cdot 7 \equiv -32 \cdot 7 \equiv 2 \cdot 7 \equiv 14 \pmod{17}$.

Thus, when 24^{1947} is divided by 17, the remainder is 14. ■

By Fermat's little theorem, $p|a^{p-1} - 1$, so $(a^{p-1} - 1)/p$ is an integer. The following example, a delightful application of the theorem, identifies the primes p for which $(2^{p-1} - 1)/p$ is a square.

EXAMPLE 7.6 Find the primes p for which $\dfrac{2^{p-1} - 1}{p}$ is a square.

SOLUTION

Suppose $\dfrac{2^{p-1} - 1}{p} = n^2$ for some positive integer n. Then $2^{p-1} - 1 = pn^2$. Clearly, both p and n must be odd. Let $p = 2k + 1$ for some positive integer k. Then $2^{2k} - 1 = pn^2$; that is, $(2^k - 1)(2^k + 1) = pn^2$. Since $2^k - 1$ and $2^k + 1$ are consecutive odd integers, they are relatively prime. Consequently, either $2^k - 1$ or $2^k + 1$ must be a perfect square.

Suppose $2^k - 1$ is a perfect square r^2:

$$2^k - 1 = r^2$$

$$2^k = r^2 + 1$$

That is,

$$2^{p-1} = \left(r^2 + 1\right)^2$$

Since $r \geq 1$ and is odd, $r = 2i + 1$ for some integer ≥ 0. Then $2^k = (2i + 1)^2 = 2(2i^2 + 2i + 1)$; this is possible if and only if $i = 0$. Then $r = 1$, so $2^{p-1} = (1^2 + 1)^2 = 4$, and hence $p = 3$.

Suppose $2^k + 1$ is a perfect square s^2:

$$2^k + 1 = s^2$$

$$2^k = s^2 - 1$$

That is,

$$2^{p-1} = (s - 1)^2(s + 1)^2$$

Since $s \geq 3$ and is odd, $s = 2i + 1$ for some $i \geq 1$. Then $2^k = (2i + 1)^2 - 1 = 4i(i + 1)$; that is, $2^{k-2} = i(i + 1)$. This is possible if and only if $i = 1$. Then $s = 3$ and hence $2^{p-1} = 2^2 \cdot 4^2 = 2^6$; so $p = 7$.

Thus, p must be 3 or 7. ■

An Alternate Proof of Wilson's Theorem (optional)

Lagrange developed a delightful proof of Wilson's theorem as an application of Fermat's little theorem and Euler's formula in Theorem 7.4. In the interest of brevity, we omit the proof of Theorem 7.4, which can be proved by induction. In 1996, S. M. Ruiz of Spain rediscovered the same proof.

THEOREM 7.4 **(Euler's formula)** Let $n \geq 0$ and x any real number. Then

$$\sum_{i=0}^{n}(-1)^i\binom{n}{i}(x-i)^n = n! \quad \blacksquare$$

COROLLARY 7.1 **(Wilson's Theorem)** Let p be a prime. Then $(p-1)! \equiv -1 \pmod{p}$.

PROOF

Since the result is true when $p = 2$, assume that $p > 2$. Letting $n = p - 1$ and $x = 0$, Euler's formula yields

$$\sum_{i=0}^{p-1}(-1)^i\binom{p-1}{i}(-i)^{p-1} = (p-1)!$$

Since p is odd, by Fermat's little theorem, this implies

$$\begin{aligned}
(p-1)! &\equiv \sum_{i=1}^{p-1}(-1)^i\binom{p-1}{i} \pmod{p}\\
&\equiv \sum_{i=0}^{p-1}(-1)^i\binom{p-1}{i} - 1 \pmod{p}\\
&\equiv (1-1)^{p-1} - 1 \pmod{p}\\
&\equiv 0 - 1 \pmod{p}\\
&\equiv -1 \pmod{p}
\end{aligned}$$

as desired. $\blacksquare$

The following theorem, another useful application of Fermat's little theorem, identifies an inverse of a modulo p, when $p \nmid a$.

THEOREM 7.5 Let p be a prime and a any integer such that $p \nmid a$. Then a^{p-2} is an inverse of a modulo p.

PROOF

By Fermat's little theorem, $a^{p-1} \equiv 1 \pmod{p}$. That is, $a \cdot a^{p-2} \equiv 1 \pmod{p}$, so a^{p-2} is an inverse of a modulo p. ■

The following example illustrates this theorem.

EXAMPLE 7.7 Let $p = 7$ and $a = 12$. Then, by Theorem 7.5 (see Example 7.4 also), 12^5 is an inverse of 12 modulo 7. Since $12 \equiv -2 \pmod{7}$, $12^5 \equiv (-2)^5 \equiv -2^2 \cdot 2^3 \equiv -4 \cdot 1 \equiv 3 \pmod{7}$. Thus, 3 is an inverse of 12 modulo 7: $12 \cdot 3 \equiv 1 \pmod{7}$, as expected! ■

Theorem 7.5 can be used to derive a formula for solving linear congruences with prime moduli.

THEOREM 7.6 Let p be a prime and a any integer such that $p \nmid a$. Then the solution of the linear congruence $ax \equiv b \pmod{p}$ is given by $x \equiv a^{p-2}b \pmod{p}$.

PROOF

Since $p \nmid a$, by Corollary 4.6, the congruence $ax \equiv b \pmod{p}$ has a unique solution. Since, by Theorem 7.5, a^{p-2} is an inverse of a modulo p, multiplying both sides of the congruence by a^{p-2}, we have

$$\begin{aligned} a^{p-2}(ax) &\equiv a^{p-2}b \pmod{p} \\ a^{p-1}x &\equiv a^{p-2}b \pmod{p} \\ x &\equiv a^{p-2}b \pmod{p}, \quad \text{by Fermat's little theorem} \end{aligned}$$

■

The following two examples employ this theorem.

EXAMPLE 7.8 Solve the linear congruence $12x \equiv 6 \pmod{7}$.

SOLUTION

By Example 7.7, $12^5 \equiv 3 \pmod{7}$ is an inverse of 12 modulo 7. Multiply both sides of the congruence by 3:

$$\begin{aligned} 3(12x) &\equiv 3 \cdot 6 \pmod{7} \\ x &\equiv 4 \pmod{7} \end{aligned}$$

■

EXAMPLE 7.9 Solve the congruence $24x \equiv 11 \pmod{17}$.

SOLUTION

$$24x \equiv 11 \pmod{17}$$
$$7x \equiv 11 \pmod{17}$$

So, by Theorem 7.6, $x \equiv 7^{15} \cdot 11 \pmod{17}$.

Now, we need to find the least residue of $7^{15} \cdot 11 \pmod{17}$. To this end, notice that $7^2 \equiv -2 \pmod{17}$, $7^4 \equiv 4 \pmod{17}$, and $7^8 \equiv -1 \pmod{17}$. Therefore, $7^{15} \equiv 7^8 \cdot 7^4 \cdot 7^2 \cdot 7 \equiv (-1) \cdot 4 \cdot (-2) \cdot 7 \equiv 5 \pmod{17}$. Thus, $x \equiv 5 \cdot 11 \equiv 4 \pmod{17}$. ■

We now discuss an interesting application of Fermat's little theorem, the pigeonhole principle, and the well-ordering principle.

Factors of $2^n + 1$ (optional)

Consider the prime factorization of $N = 2^n + 1$ for various values of n. It appears from Table 7.1 that when a prime p makes its debut in the prime factorization of N, $p \equiv 1 \pmod{n}$. This phenomenon was first observed in 2001 by J. E. Parkes of Staffordshire, England, and then pursued by K. R. McLean of the University of Liverpool, where $n > 1$. For example, $p = 43$ occurs first when $n = 7$ and $43 \equiv 1 \pmod{7}$.

n	$2^n + 1$	n	$2^n + 1$	n	$2^n + 1$	n	$2^n + 1$
1	$\mathbf{3}$	5	$3 \cdot \mathbf{11}$	9	$3^3 \cdot \mathbf{19}$	13	$3 \cdot \mathbf{2731}$
2	$\mathbf{5}$	6	$5 \cdot \mathbf{13}$	10	$5^2 \cdot \mathbf{41}$	14	$5 \cdot 29 \cdot \mathbf{113}$
3	3^2	7	$3 \cdot \mathbf{43}$	11	$3 \cdot \mathbf{683}$	15	$3^2 \cdot 11 \cdot \mathbf{331}$
4	$\mathbf{17}$	8	$\mathbf{257}$	12	$17 \cdot \mathbf{241}$	16	$\mathbf{65537}$

Table 7.1

To confirm this observation, notice that p first appears in the prime factorization of N corresponding to the least exponent n, $2^n \equiv -1 \pmod{p}$. By the pigeonhole principle, the least residues of $2^1, 2^2, 2^3, \ldots, 2^{p+1}$ modulo p cannot all be distinct. Therefore, there exist positive integers a and b such that $2^a \equiv 2^b \pmod{p}$, where $a < b$. Then $2^{b-a} \equiv 1 \pmod{p}$. Consequently, by the well-ordering principle, there is a least positive integer d such that $2^d \equiv 1 \pmod{p}$.

Let t be any positive integer such that $2^t \equiv 1 \pmod{p}$. By the division algorithm, $t = qd + r$, where $0 \le r < d$. Then

$$1 \equiv 2^t \equiv 2^{qd+r} \equiv \left(2^d\right)^q \cdot 2^r \equiv 1^q \cdot 2^r \equiv 2^r \pmod{p}$$

Since $r < d$, this implies that $r = 0$. Then $t = qd$ and $d|t$.

Since $2^n \equiv -1 \pmod{p}$, $2^{2n} \equiv 1 \pmod{p}$. Consequently, $d|2n$. Since $2^n \equiv -1 \pmod{p}$ and $2^d \equiv 1 \pmod{p}$, it follows that $d \ne n$. Suppose $d < n$. Then $2^{n-d} \equiv -1 \pmod{p}$, which contradicts the choice of n. Thus, $d > n$ and $d|2n$; so $d = 2n$.

By Fermat's little theorem, $2^{p-1} \equiv 1 \pmod{p}$. So $d|p-1$; thus, $p \equiv 1 \pmod{d}$ and hence $p \equiv 1 \pmod{n}$, as desired.

The following theorem shows Fermat's little theorem can be extended to all positive integers a.

THEOREM 7.7 Let p be a prime and a any positive integer. Then $a^p \equiv a \pmod{p}$.

PROOF (by cases)

case 1 Suppose $p \nmid a$. Then, by Fermat's little theorem, $a^{p-1} \equiv 1 \pmod{p}$, so $a^p \equiv a \pmod{p}$.

case 2 Suppose $p|a$. Then $p \equiv a \equiv 0 \pmod{p}$, so $a^p \equiv 0 \pmod{p}$, by Theorem 4.5. Therefore, by Theorem 4.2, $a^p \equiv a \pmod{p}$.

Thus, in both cases, $a^p \equiv a \pmod{p}$. ■

The following example illustrates this theorem.

EXAMPLE 7.10 Let $p = 7$. If $a = 12$, then by Example 7.4, $12^6 \equiv 1 \pmod{7}$, so $12^7 \equiv 12 \pmod{7}$. On the other hand, if $a = 28$, then $28 \equiv 0 \pmod{7}$, so $28^7 \equiv 0 \pmod{7}$. But $0 \equiv 8 \pmod{7}$, so $28^7 \equiv 28 \pmod{7}$. ■

Next, we present an alternate and elegant proof of Theorem 7.7, based on the binomial theorem and induction, developed in 1989 by R. J. Hendel of Dowling College, New York.

An Alternate Proof of Theorem 7.7 (optional)

Let x be any integer. Because $p|\binom{p}{k}$, where $0 < k < p$, by the binomial theorem, it can be shown that $(x+1)^p \equiv x^p + 1 \pmod{p}$. (See Exercise 33.)

Let $g(t) = t^p - t$. Then

$$\begin{aligned} g(x+1) - g(x) &= (x+1)^p - (x+1) - x^p + x \\ &= (x+1)^p - x^p - 1 \\ &\equiv 0 \pmod{p} \end{aligned}$$

Therefore, $g(x+1) \equiv g(x) \pmod{p}$. Replacing x with $x+1$, this yields $g(x+2) \equiv g(x+1) \equiv g(x) \pmod{p}$.

More generally, it can be shown that $g(x+a) \equiv g(x)$ for every positive integer a. (See Exercise 36.)

Letting $x = 0$ in this result yields $g(a) \equiv g(0) \pmod{p}$; that is, $a^p - a \equiv 0 \pmod{p}$. In other words, $a^p \equiv a \pmod{p}$. ■

In 1970, J. E. Phythian of Tanzania extended Theorem 7.7 to a finite number of distinct primes, as the following theorem shows.

THEOREM 7.8 Let $p_1, p_2, \ldots, p_k$ be any distinct primes, a any positive integer, and $\ell = [p_1 - 1, p_2 - 1, \ldots, p_k - 1]$. Then $a^{\ell+1} \equiv a \pmod{p_1 p_2 \cdots p_k}$.

PROOF

By Fermat's little theorem, $a^{p_i - 1} \equiv 1 \pmod{p_i}$, where $1 \le i \le k$. Since $p_i - 1 | \ell$, this implies $(a^{p_i} - 1)^{\ell/(p_i - 1)} \equiv 1 \pmod{p_i}$; that is, $a^\ell \equiv 1 \pmod{p_i}$. Thus, $a^{\ell+1} \equiv a \pmod{p_i}$. Consequently, $a^{\ell+1} \equiv a \pmod{[p_1, p_2, \ldots, p_k]}$; that is, $a^{\ell+1} \equiv a \pmod{p_1 p_2 \cdots p_k}$. ■

For example, let $p_1 = 3$, $p_2 = 7$, and $p_3 = 11$. Then $\ell = [2, 6, 10] = 30$ and $p_1 p_2 p_3 = 3 \cdot 7 \cdot 11 = 231$. So, by Theorem 7.8, $a^{31} \equiv a \pmod{231}$ for any positive integer a. In particular, $43^{31} \equiv 43 \pmod{231}$.

Theorem 7.8 yields an interesting byproduct. This observation was made independently in 1985 by G. Duckworth of England and J. Suck of Germany.

COROLLARY 7.2 Let a be any integer and p any prime > 3. Then $a^p \equiv a \pmod{6p}$.

PROOF

Let $p_1 = 2$, $p_2 = 3$, and $p_3 = p$ in Theorem 7.8. Since $2 \cdot 3 \cdot p = 6p$ and $[p_1 - 1, p_2 - 1, p_3 - 1] = [1, 2, p - 1] = p - 1$, the result follows by the theorem. ■

For example, let $a = 20$ and $p = 13$. Then $20^{13} = (20^2)^6 \cdot 20 \equiv 10^6 \cdot 20 = (10^3)^2 \cdot 20 \equiv (-14)^2 \cdot 20 \equiv 20 \pmod{78}$, as expected by this corollary.

This corollary can be proved without using Theorem 7.8. See Exercise 31.

We now present an application of Fermat's little theorem to the factoring of large composite numbers.

The Pollard p − 1 Factoring Method

In 1974, Pollard developed a factoring method based on Fermat's little theorem, called the *Pollard* $p-1$ *method*. Suppose n is an odd integer known to be composite. Let p be a prime factor of n such that the prime factors of $p-1$ are relatively small. Let k be a large enough positive integer such that $(p-1)|k!$. For example, let $p = 2393$. Then $p-1 = 2392 = 2^3 \cdot 13 \cdot 23$ and $(p-1)|23!$.

Let $k! = m(p-1)$ for some integer m. Since $2 \nmid p$, by Fermat's little theorem, $2^{p-1} \equiv 1 \pmod{p}$. Then

$$2^{k!} = 2^{m(p-1)} = \left(2^{p-1}\right)^m \equiv 1^m \equiv 1 \pmod{p}$$

So $p|(2^{k!}-1)$.

Let r be the least positive residue of $2^{k!}$ modulo n; so $r-1 \equiv 2^{k!}-1 \pmod{n}$. Since $p|n$ and $p|(2^{k!}-1)$, $p|(r-1)$. So the gcd $(r-1, n)$ is a nontrivial factor of n. Notice that, as in the case of the rho method, the choice of p does not occur in the computation of $(r-1, n)$.

To implement this technique, first notice that $2^{k!} = ((((2^1)^2)^3)\cdots)^k$; then compute the least residue r of $2^{k!}$ modulo n; and at each step, compute $(r-1, n)$ until a nontrivial factor emerges, as the following example illustrates.

EXAMPLE 7.11 Using the Pollard $p-1$ method, find a nontrivial factor of $n = 2813$.

SOLUTION

Using the fact that $2^{k!} = (2^{(k-1)!})^k$, we continue computing the least positive residue $r \equiv 2^{k!} \pmod{2813}$ and the gcd $(r-1, n)$ until a nontrivial factor of n appears, where $k \geq 1$:

$$\begin{array}{llll}
2^{1!} = & 2 \equiv & 2 \pmod{2813} & (1, 2813) = 1 \\
2^{2!} = & 2^2 \equiv & 4 \pmod{2813} & (3, 2813) = 1 \\
2^{3!} = & 4^3 \equiv & 64 \pmod{2813} & (63, 2813) = 1 \\
2^{4!} \equiv & 64^4 \equiv & 484 \pmod{2813} & (483, 2813) = 1 \\
2^{5!} \equiv & 484^5 \equiv & 1648 \pmod{2813} & (1647, 2813) = 1 \\
2^{6!} \equiv & 1648^6 \equiv & 777 \pmod{2813} & (776, 2813) = 97
\end{array}$$

Thus, $97|2813$. ■

The $p-1$ method fails if $2^{k!} \equiv 1 \pmod{n}$. For example, when $n = 3277$, $2^{7!} \equiv 1 \pmod{3277}$ and $(0, 3277) = 3277$. Interestingly, we can use any base b in lieu of 2 in the algorithm, where $1 < b < p$.

Returning to Fermat's little theorem, we would like to examine its converse: *If* $a^{n-1} \equiv$ (mod n) *and* $n \nmid a$, *then* n *is a prime.* Is this true or false? If it is true, can we prove it? If it is not, can we produce a counterexample? We will continue this discussion in the next section.

EXERCISES 7.2

Compute the remainder when the first integer is divided by the second.

1. $7^{1001}, 17$
2. $30^{2020}, 19$
3. $15^{1976}, 23$
4. $43^{5555}, 31$

Find the ones digit in the base-seven expansion of each decimal number.

5. 5^{101}
6. 12^{1111}
7. 29^{2076}
8. 37^{3434}

Solve each linear congruence.

9. $8x \equiv 3 \pmod{11}$
10. $15x \equiv 7 \pmod{13}$
11. $26x \equiv 12 \pmod{17}$
12. $43x \equiv 17 \pmod{23}$

Compute the least residue of each.

13. $2^{340} \pmod{341}$
14. $11^{16} + 17^{10} \pmod{187}$
15. $13^{18} + 19^{12} \pmod{247}$

Verify each.

16. $(12+15)^{17} \equiv 12^{17} + 15^{17} \pmod{17}$
17. $(16+21)^{23} \equiv 16^{23} + 21^{23} \pmod{23}$
18. Find the primes p such that $(2^{p-1} - 1)/p$ is a perfect cube.

Let p and q be distinct primes, and a, b, and n arbitrary positive integers. Prove each.

19. Let $a^p \equiv a \pmod q$ and $a^q \equiv a \pmod p$. Then $a^{pq} \equiv a \pmod{pq}$.
20. $a^{pq} - a^p - a^q + a \equiv 0 \pmod{pq}$
21. If $a^p \equiv b^p \pmod p$, then $a \equiv b \pmod p$.
22. If $a^p \equiv b^p \pmod p$, then $a^p \equiv b^p \pmod{p^2}$.
23. $p^{q-1} + q^{p-1} \equiv 1 \pmod{pq}$
24. $p^q + q^p \equiv p + q \pmod{pq}$
25. $30|(n^5 - n)$ (R. S. Hatcher, 1970)
26. There are infinitely many values of n such that $p|2^{n+1} + 3^n - 17$. (E. Just, 1976)

Let p be any odd prime and a any nonnegative integer. Prove the following.

27. $1^{p-1} + 2^{p-1} + \cdots + (p-1)^{p-1} \equiv -1 \pmod p$
28. $1^p + 2^p + \cdots + (p-1)^p \equiv 0 \pmod p$
29. $(a+1)^p + (a+2)^p + \cdots + (a+p-1)^p \equiv -a \pmod p$

Let p be any prime, a any positive integer, and x an arbitrary integer ≥ 0. Prove each.

30. $\binom{p}{k} \equiv 0 \pmod p$, where $0 < k < p$.
31. Let $p > 3$. Then $a^p \equiv a \pmod{6p}$. Do not use Theorem 7.8. (G. Duckworth, 1985)
32. Using induction, prove that $a^p \equiv a \pmod p$.
33. Using the binomial theorem, prove that $(a+b)^p \equiv a^p + b^p \pmod p$.
34. Using Fermat's little theorem, prove that $(a+b)^p \equiv a^p + b^p \pmod p$.
35. Using induction, prove that $(x+1)^p \equiv x^p + 1 \pmod p$.
36. Let $g(t) = t^p - t$. Then $g(x+a) \equiv g(x) \pmod p$ for every positive integer a.
37. Let p be a prime, and a and b be any integers such that $a \geq b \geq 0$. Prove that

$$\binom{pa}{pb} \equiv \binom{a}{b} \pmod p.$$

(Putnam Mathematics Competition, 1977)

Using the Pollard $p-1$ method, find a nontrivial factor of each.

38. 2323
39. 7967

*7.3 Pseudoprimes (optional)

In Theorem 7.7, we found that if n is a prime, then $a^n \equiv a \pmod{n}$ for every integer a. As a result, if this congruence fails for some integer b, that is, if $b^n \not\equiv b \pmod{n}$ for some integer n, then n cannot be a prime. Thus, in order to show that n is composite, it suffices to produce an integer b such that $b^n \not\equiv b \pmod{n}$. This provides a test for compositeness, as the following example shows.

EXAMPLE 7.12 Verify that 33 is a composite number.

PROOF

If 33 were a prime, then $2^{33} \equiv 2 \pmod{33}$. But

$$\begin{aligned} 2^{33} &= (2^5)^6 \cdot 2^3 \equiv (-1)^6 \cdot 8 \equiv 8 \pmod{33} \\ &\not\equiv 2 \pmod{33} \end{aligned}$$

Therefore, 33 is not a prime, as expected. ■

We still have not answered whether the converse of Theorem 7.7 is true: If $a^n \equiv a \pmod{n}$, then n is a prime. Interestingly, the ancient Chinese mathematicians claimed that if $2^n \equiv 2 \pmod{n}$, then n must be a prime. In support of this conjecture, we find that the claim holds for all positive integers $n \le 340$. Unfortunately, however, it fails when $n = 341 = 11 \cdot 31$, a composite number discovered by Sarrus in 1819. The following example confirms this.

EXAMPLE 7.13 Show that $2^{341} \equiv 2 \pmod{341}$.

PROOF

By Fermat's little theorem, $2^{10} \equiv 1 \pmod{11}$, so $2^{341} = (2^{10})^{34} \cdot 2 \equiv 1^{34} \cdot 2 \equiv 2 \pmod{11}$. Also, $2^5 \equiv 1 \pmod{31}$, so $2^{341} = (2^5)^{68} \cdot 2 \equiv 1^{68} \cdot 2 \equiv 2 \pmod{31}$. Therefore, by Theorem 4.8, $2^{341} \equiv 2 \pmod{[11, 31]}$; that is, $2^{341} \equiv 2 \pmod{341}$, although 341 is a composite number. ■

This example voids the Chinese claim and thus disproves the converse of Fermat's little theorem.

It follows from the preceding discussion that $n = 341$ is the smallest composite number such that $2^n \equiv 2 \pmod{n}$. The next three are 561, 645, and 1105 (see Exercises 5–8). Such numbers, although not as important as primes, have been given a name of their own, so we make the following definition.

Pseudoprimes

A composite number n is called a **pseudoprime** if $2^n \equiv 2 \pmod{n}$. Thus, the first four pseudoprimes are 341, 561, 645, and 1105; they are all odd. The smallest even pseudoprime is 161,038, found in 1950.

Pseudoprimes appear to be rare and sparsely spaced. For example, there are 455,052,512 primes less than 10 billion, but only 14,884 pseudoprimes less than 10 billion. But this should not lead us to draw any false conclusions, since the next theorem establishes categorically the infinitude of pseudoprimes. First, we pave the way for its proof by introducing two lemmas.

LEMMA 7.3 Let m and n be positive integers such that $m|n$. Then $2^m - 1|2^n - 1$.

PROOF

Since $m|n$, $n = km$ for some positive integer k. Then

$$\begin{aligned} 2^n - 1 &= 2^{km} - 1 \\ &= (2^m - 1)\left[2^{(k-1)m} + 2^{(k-2)m} + \cdots + 2^m + 1\right] \end{aligned}$$

Therefore, $2^m - 1|2^n - 1$. ■

LEMMA 7.4 If n is an odd pseudoprime, then $N = 2^n - 1$ is also an odd pseudoprime.

PROOF

Let n be an odd pseudoprime. Then n is composite and $2^n \equiv 2 \pmod{n}$. But n is odd, so $2^{n-1} \equiv 1 \pmod{n}$.

Since n is composite, let $n = rs$, where $1 < r, s < n$. Since $r|n$, by Lemma 7.3, $2^r - 1|2^n - 1$; that is, $2^r - 1|N$, so N is a composite.

It remains to show that $2^N \equiv 2 \pmod{N}$. To this end, since $2^n \equiv 2 \pmod{n}$, $n|2^n - 2$, so $2^n - 2 = kn$ for some integer k; that is, $N - 1 = kn$. Therefore, $2^{N-1} - 1 = 2^{kn} - 1$.

Again, by Lemma 7.3, $N = 2^n - 1|2^{kn} - 1$, so $2^{N-1} - 1 \equiv 0 \pmod{N}$; that is, $2^{N-1} \equiv 1 \pmod{N}$. So $2^N \equiv 2 \pmod{N}$.

Thus, if n is an odd pseudoprime, then $2^n - 1$ is a larger odd pseudoprime. ■

This lemma gives us a recipe for constructing a larger odd pseudoprime from a given odd pseudoprime. For example, since 341 is an odd pseudoprime, $2^{341} - 1$ is a larger odd pseudoprime.

We can now establish the existence of infinitely many pseudoprimes.

THEOREM 7.9 There are infinitely many pseudoprimes.

***Robert Daniel Carmichael** (1879–1967) was born in Goodwater, Alabama. He received his B.A. from Lineville College in 1898 and Ph.D. from Princeton three years later under the guidance of George D. Birkhoff. His dissertation was a significant contribution to difference equations. He taught at Alabama Presbyterian College, Indiana University, and the University of Illinois, where he later served as the dean of the graduate school for 15 years. Carmichael was president of the Mathematical Association of America, editor-in-chief of* The American Mathematical Monthly, *and a significant contributor to difference equations, number theory, relativity theory, group theory, and mathematical philosophy.*

PROOF

By Lemma 7.4, we can construct an infinite number of odd pseudoprimes $n_{i+1} = 2^{n_i} - 1$ for $i = 0, 1, 2, \ldots$ from a given odd pseudoprime n_i. Since 341 is such a pseudoprime, by choosing $n_0 = 341$, we obtain the odd pseudoprimes $n_0 < n_1 < n_2 < \cdots$. Thus, there is an infinite number of pseudoprimes. ■

The following two questions about pseudoprimes remain unresolved:

- Are there infinitely many square pseudoprimes?
- Are there infinitely many primes p such that $2^{p-1} \equiv 1 \pmod{p^2}$?

The smallest pseudoprimes that are not square-free are $1{,}194{,}649 = 1093^2$; $12{,}327{,}121 = 3511^2$; and $3{,}914{,}864{,}773 = 29 \cdot 113 \cdot 1093^2$.

Carmichael Numbers

Besides base 2, there are other bases a and composite numbers n such that $a^{n-1} \equiv 1 \pmod{n}$. For example, $3^{90} \equiv 1 \pmod{91}$ and $4^{14} \equiv 1 \pmod{15}$. (See Exercises 10 and 11.)

In 1907, the American mathematician Robert D. Carmichael established the existence of composite numbers n such that $a^{n-1} \equiv 1 \pmod{n}$ for all positive integers a relatively prime to n. Such composite numbers are called **Carmichael numbers**. One such number is 561, as the following example shows.

EXAMPLE 7.14 Show that 561 is a Carmichael number.

PROOF

Since $561 = 3 \cdot 11 \cdot 17$, 561 is a composite number. So it remains to show that $a^{560} \equiv 1 \pmod{561}$ for all positive integers a relatively prime to 561.

By Fermat's little theorem, $a^2 \equiv 1 \pmod 3$, $a^{10} \equiv 1 \pmod{11}$, $a^{16} \equiv 1 \pmod{17}$. Therefore, $a^{560} = (a^2)^{280} \equiv 1 \pmod 3$, $a^{560} = (a^{10})^{56} \equiv 1 \pmod{11}$, and $a^{560} = (a^{16})^{35} \equiv 1 \pmod{17}$. So, by Theorem 4.8, $a^{560} \equiv 1 \pmod{[3, 11, 17]}$; that is, $a^{560} \equiv 1 \pmod{561}$, as desired. ■

In fact, 561 is the smallest Carmichael number. The next two are $1105 = 5 \cdot 13 \cdot 17$ and $1729 = 7 \cdot 13 \cdot 19$. There are four more numbers below 10,000, and 1547 less than 10 billion. In 1992, using high-powered computers, Richard G. E. Pinch (1954–) at Cambridge University found that there are 105,212 less than one quadrillion. And the search continues.

As we saw, Carmichael numbers are sparsely spaced and appear to be rare. In 1992, however, Andrew Granville, Carl Pomerance, and Red Alford of the University of Georgia established the existence of infinitely many Carmichael numbers.

Interestingly, it has been found that one Carmichael number can be a factor of another; for instance, 1729 and $63{,}973 = 7 \cdot 13 \cdot 19 \cdot 37$ are Carmichael numbers, and $1729|63{,}973$. It was also established in 1948 that the product of two Carmichael numbers can also be a Carmichael number; for example, 1729, $294{,}409 = 37 \cdot 73 \cdot 109$ and $509{,}033{,}161 = 1729 \cdot 294{,}409$ are Carmichael numbers.

In 1990, H. Dubner and H. Nelson discovered two Carmichael numbers that are products of three Carmichael numbers; one is 97 digits long and the other 124 digits long.

EXERCISES 7.3

Using Lemma 7.3, factor each.

1. $2^{10} - 1$
2. $2^{14} - 1$
3. $2^{15} - 1$
4. $2^{21} - 1$

Verify that $2^n \equiv 2 \pmod n$ for each value of n.

5. 561
6. 645
7. 1105
8. 161,038

Verify each.

9. $2^{340} \not\equiv 2 \pmod{340}$
10. $3^{90} \equiv 1 \pmod{91}$
11. $4^{14} \equiv 1 \pmod{15}$
12. $5^{123} \equiv 1 \pmod{124}$
13. $6^{34} \equiv 1 \pmod{35}$
14. $12^{64} \equiv 1 \pmod{65}$

Determine whether each is true.

15. $2^{90} \equiv 1 \pmod{91}$
16. $3^{340} \equiv 1 \pmod{341}$

Verify that each is a Carmichael number.

17. $1105 = 5 \cdot 13 \cdot 17$
18. $1729 = 7 \cdot 13 \cdot 19$
19. $2465 = 5 \cdot 17 \cdot 29$
20. $2821 = 7 \cdot 13 \cdot 31$

There can exist positive integers a and n, and a prime p with $p \nmid a$ such that $a^{p-1} \equiv 1 \pmod{p^n}$. Exercises 21–26 present six such instances. Verify each.

21. $7^4 \equiv 1 \pmod{5^2}$
22. $3^{10} \equiv 1 \pmod{11^2}$
23. $19^6 \equiv 1 \pmod{7^3}$
24. $19^{12} \equiv 1 \pmod{13^2}$
25. $239^{12} \equiv 1 \pmod{13^4}$
26. $38^{16} \equiv 1 \pmod{17^2}$

7.4 Euler's Theorem

Fermat's little theorem enables us to work with congruences involving only prime moduli. It tells us that there is a positive integer $f(p)$ such that $a^{f(p)} \equiv 1 \pmod{p}$, where $f(p) = p - 1$. So an obvious question arises: *Can we extend Fermat's little theorem to congruences with arbitrary moduli m?* In other words, is there an exponent $f(m)$ such that $a^{f(m)} \equiv 1 \pmod{m}$, where $(a, m) = 1$? Before we answer this, we will find it helpful to study the following example.

EXAMPLE 7.15 Determine if there exists a positive integer $f(m)$ such that $a^{f(m)} \equiv 1 \pmod{m}$ for $m = 4$, 9, and 12, where a is a positive integer $\leq m$ and relatively prime to it.

SOLUTION

1. With $m = 4$, there are two positive integers $a \leq m$ and relatively prime to it, namely, 1 and 3: $1^2 \equiv 1 \pmod{4}$ and $3^2 \equiv 1 \pmod{4}$. So when $m = 4$, $f(m) = 2$ works.
2. For $m = 9$, there are six residues ≤ 9 and relatively prime to it: 1, 2, 4, 5, 7, and 8. After computing their first sixth powers, we find that

$$1^6 \equiv 1 \pmod{9} \qquad 2^6 \equiv 1 \pmod{9} \qquad 4^6 \equiv 1 \pmod{9}$$
$$5^6 \equiv 1 \pmod{9} \qquad 7^6 \equiv 1 \pmod{9} \qquad 8^6 \equiv 1 \pmod{9}$$

(See Table 7.2.) Thus, when $m = 9, f(m) = 6$ does the job.

a	a^2	a^3	a^4	a^5	a^6
1	1	1	1	1	1
2	4	8	7	5	1
4	7	1	4	7	1
5	7	8	4	2	1
7	4	1	7	4	1
8	1	8	1	8	1
					↑ all ones

Table 7.2

a	a^2	a^3	a^4
1	1	1	1
5	1	5	1
7	1	7	1
11	1	11	1
	↑ all ones		↑ all ones

Table 7.3

3. There are four positive integers ≤ 12 and relatively prime to 12; namely, 1, 5, 7, and 11. Let us compute the first four powers of each modulo 12. It follows from Table 7.3 that $1^4 \equiv 1 \pmod{12}$, $5^4 \equiv 1 \pmod{12}$, $7^4 \equiv 1 \pmod{12}$, and $11^4 \equiv 1 \pmod{12}$. Once again, we have a candidate for $f(12)$,

Leonhard Euler *(1707–1783) was born in Basel, Switzerland. His father, a mathematician and a Calvinist pastor, wanted him also to become a pastor. Although Euler had different ideas, he followed his father's wishes, and studied Hebrew and theology at the University of Basel. His hard work at the university and remarkable ability brought him to the attention of the well-known mathematician Johann Bernoulli (1667–1748). Realizing the young Euler's talents, Bernoulli persuaded the boy's father to change his mind, and Euler pursued his studies in mathematics.*

At the age of 19, Euler brought out his first paper. His paper failed to win the Paris Academy Prize in 1727; however, he won it 72 times in later years.

Euler was the most prolific mathematician, making significant contributions to every branch of mathematics. With his phenomenal memory, he had every formula at his fingertips. A genius, he could work anywhere and under any conditions. Euler belongs to a class by himself.

namely, 4. [Notice that 2 also works: $1^2 \equiv 1 \pmod{12}$, $5^2 \equiv 1 \pmod{12}$, $7^2 \equiv 1 \pmod{12}$, and $11^2 \equiv 1 \pmod{12}$.] ■

From the three cases presented in this example, we can make an educated guess about the exponent $f(m)$: It is the number of positive integers $\leq m$ and relatively prime to m. Accordingly, we now turn to an important number-theoretic function, named after the great Swiss mathematician Leonhard Euler.

Euler's Phi Function

Let m be a positive integer. Then **Euler's phi function** $\varphi(m)$ denotes the number of positive integers $\leq m$ and relatively prime to m.

It follows by Example 7.15 that $\varphi(4) = 2$, $\varphi(9) = 6$, and $\varphi(12) = 4$. Since $1 \leq 1$ and relatively prime to 1, $\varphi(1) = 1$.

EXAMPLE 7.16 Compute $\varphi(11)$ and $\varphi(18)$.

SOLUTION

Since 11 is a prime, every positive integer < 11 is relatively prime to 11, so $\varphi(11) = 10$.

There are six positive integers ≤ 18 and relatively prime to it, namely, 1, 5, 7, 11, 13, and 17. Therefore, $\varphi(18) = 6$. (We can verify that each raised to the sixth power is congruent to 1 modulo 18.) ■

The following lemma, which we will use again later, shows the value of $\varphi(m)$ if m is a prime.

LEMMA 7.5 A positive integer p is a prime if and only if $\varphi(p) = p - 1$.

PROOF

Let p be a prime. Then there are $p-1$ positive integers $\leq p$ and relatively prime to p, so $\varphi(p) = p - 1$.

Conversely, let p be a positive integer such that $\varphi(p) = p - 1$. Let $d|p$, where $1 < d < p$. Since there are exactly $p - 1$ positive integers $< p$, d is one of them, and $(d, p) \neq 1$; so $\varphi(p) < p - 1$, a contradiction. Thus, p must be a prime. ■

We can now examine Euler's theorem, the next milestone in the development of number theory and one of its most celebrated results. Proved by Euler in 1760, it extends Fermat's little theorem to arbitrary moduli. But before we study the theorem, we need to lay some groundwork in the form of a lemma.

In Lemma 7.2, we found that the least residues of the integers $a, 2a, \ldots, (p-1)a$ modulo p are a rearrangement of the integers $1, 2, \ldots, (p-1)$, where a is any integer with $p \nmid a$. The following lemma extends this result to an arbitrary modulus. Its proof resembles quite closely that of Lemma 7.2, but first an example to shed some light on its proof.

EXAMPLE 7.17 Let $m = 12$ and $a = 35$, so $(a, m) = (35, 12) = 1$. By Example 7.15, the least residues modulo 12 that are relatively prime to 12 are 1, 5, 7, and 11. Multiply each by 35: $35 \cdot 1 \equiv 11 \pmod{12}$, $35 \cdot 5 \equiv 7 \pmod{12}$, $35 \cdot 7 \equiv 5 \pmod{12}$, and $35 \cdot 11 \equiv 1 \pmod{12}$. Thus, the least residues of $35 \cdot 1$, $35 \cdot 5$, $35 \cdot 7$, and $35 \cdot 11$ modulo 12 are a rearrangement of the least residues 1, 5, 7, and 11 modulo 12. ■

LEMMA 7.6 Let m be a positive integer and a any integer with $(a, m) = 1$. Let $r_1, r_2, \ldots, r_{\varphi(m)}$ be the positive integers $\leq m$ and relatively prime to m. Then the least residues of the integers $ar_1, ar_2, \ldots, ar_{\varphi(m)}$ modulo m are a permutation of the integers $r_1, r_2, \ldots, r_{\varphi(m)}$.

PROOF

Again, the proof consists of two parts. First, we will show that $(ar_i, m) = 1$ for every i. Then we will show that no two numbers ar_i and ar_j can be congruent modulo m if $i \neq j$, where $1 \leq i < j \leq \varphi(m)$.

To show that each ar_i is relatively prime to m:

Suppose $(ar_i, m) > 1$. Let p be a prime factor of (ar_i, m). Then $p|ar_i$ and $p|m$. Since $p|ar_i$, $p|a$ or $p|r_i$. If $p|r_i$, then $p|r_i$ and $p|m$, so $(r_i, m) \neq 1$, a contradiction. So

$p|a$. This coupled with $p|m$ implies $p|(a, m)$, again a contradiction. Thus, $(ar_i, m) = 1$; that is, the integers $ar_1, ar_2, \ldots, ar_{\varphi(m)}$ are relatively prime to m.

To show that no two of the integers ar_i can be congruent modulo m; that is, $ar_i \not\equiv ar_j$, where $1 \le i < j \le \varphi(m)$:

To this end, suppose $ar_i \equiv ar_j \pmod{m}$. Since $(a, m) = 1$, by Theorem 4.6, $r_i \equiv r_j \pmod{m}$. But r_i and r_j are least residues modulo m, so $r_i = r_j$. Thus, if $i \neq j$, then $ar_i \not\equiv ar_j \pmod{m}$.

Thus, the least residues of $ar_1, ar_2, \ldots, ar_{\varphi(m)}$ modulo m are distinct and are $\varphi(m)$ in number. So they are a permutation of the least residues $r_1, r_2, \ldots, r_{\varphi(m)}$ modulo m. ■

The following example demonstrates the technique used to prove Euler's theorem.

EXAMPLE 7.18 Let $m = 12$ and $a = 35$. The least residues modulo 12 that are relatively prime to 12 are 1, 5, 7, and 11, so $\varphi(12) = 4$. By Example 7.17, the least residues of $35 \cdot 1, 35 \cdot 5, 35 \cdot 7$, and $35 \cdot 11$ modulo 12 are a rearrangement of the least residues 1, 5, 7, and 11. Therefore,

$$(35 \cdot 1)(35 \cdot 5)(35 \cdot 7)(35 \cdot 11) \equiv 1 \cdot 5 \cdot 7 \cdot 11 \pmod{12}$$

That is,

$$35^4 \cdot (1 \cdot 5 \cdot 7 \cdot 11) \equiv 1 \cdot 5 \cdot 7 \cdot 11 \pmod{12}$$

But $(1 \cdot 5 \cdot 7 \cdot 11, 12) = 1$, so by Theorem 4.6, $35^4 \equiv 1 \pmod{12}$; that is, $35^{\varphi(12)} \equiv 1 \pmod{12}$. ■

We are now ready to present Euler's theorem. With Lemma 7.6 in hand, the proof is very short, but still elegant.

THEOREM 7.10 **(Euler's Theorem)** Let m be a positive integer and a any integer with $(a, m) = 1$. Then $a^{\varphi(m)} \equiv 1 \pmod{m}$.

PROOF

Let $r_1, r_2, \ldots, r_{\varphi(m)}$ be the least residues modulo m that are relatively prime to m. Then, by Lemma 7.6, the integers $ar_1, ar_2, \ldots, ar_{\varphi(m)}$ are congruent modulo m to $r_1, r_2, \ldots, r_{\varphi(m)}$ in some order. Consequently,

$$(ar_1)(ar_2)\cdots(ar_{\varphi(m)}) \equiv r_1 r_2 \cdots r_{\varphi(m)} \pmod{m}$$

That is,

$$a^{\varphi(m)} r_1 r_2 \cdots r_{\varphi(m)} \equiv r_1 r_2 \cdots r_{\varphi(m)} \pmod{m}$$

Since each r_i is relatively prime to m, $(r_1 r_2 \cdots r_{\varphi(m)}, m) = 1$; so, by Theorem 4.6, $a^{\varphi(m)} \equiv 1 \pmod{m}$. ■

The following example illustrates Euler's theorem.

EXAMPLE 7.19 Let $m = 24$ and a any integer relatively prime to 24. There are eight positive integers ≤ 24 and relatively prime to 24, namely, 1, 5, 7, 11, 13, 17, 19, and 23; so $\varphi(24) = 8$. Thus, by Euler's theorem, $a^8 \equiv 1 \pmod{24}$. In particular, let $a = 77$. Since $77 = 7 \cdot 11$, $(77, 24) = 1$, so $77^8 \equiv 1 \pmod{24}$.

We can confirm this using the fundamental properties of congruence:

$$77 \equiv 5 \pmod{24}$$

Therefore,

$$77^8 \equiv 5^8 \pmod{24}$$

Since $5^2 \equiv 1 \pmod{24}$, $77^8 \equiv 5^8 \equiv (5^2)^4 \equiv 1^4 \equiv 1 \pmod{24}$, as expected. ■

Euler's theorem is useful for finding remainders of numbers involving large exponents even if the divisor is composite, provided the divisor is relatively prime to the base. The following example illustrates this.

EXAMPLE 7.20 Find the remainder when 245^{1040} is divided by 18.

SOLUTION

Since $245 \equiv 11 \pmod{18}$, $245^{1040} \equiv 11^{1040} \pmod{18}$. Since $(11, 18) = 1$, by Euler's theorem, $11^{\varphi(18)} \equiv 11^6 \equiv 1 \pmod{18}$. Therefore, $11^{1040} = (11^6)^{173} \cdot 11^2 \equiv 1^{173} \cdot 13 \equiv 13 \pmod{18}$. Thus, the desired remainder is 13. ■

Using Lemma 7.5, the following corollary deduces Fermat's little theorem from Euler's theorem.

COROLLARY 7.3 **(Fermat's Little Theorem)** Let p be a prime and a any integer such that $p \nmid a$. Then $a^{p-1} \equiv 1 \pmod{p}$.

PROOF

By Euler's theorem, $a^{\varphi(p)} \equiv 1 \pmod{p}$. But $\varphi(p) = p - 1$, by Lemma 7.5, so $a^{p-1} \equiv 1 \pmod{p}$. ■

Using Euler's theorem, we can extend Theorems 7.5 and 7.6 to an arbitrary modulus m in an obvious way, as the following two theorems show. We leave their proofs as exercises.

THEOREM 7.11 Let m be a positive integer and a any integer with $(a, m) = 1$. Then $a^{\varphi(m)-1}$ is an inverse of a modulo m.

THEOREM 7.12 Let m be a positive integer and a any integer with $(a, m) = 1$. Then the solution of the linear congruence $ax \equiv b \pmod{m}$ is given by $x \equiv a^{\varphi(m)-1}b \pmod{m}$.

The following example uses this result.

EXAMPLE 7.21 Solve the linear congruence $35x \equiv 47 \pmod{24}$.

SOLUTION

The congruence can be simplified as $11x \equiv -1 \pmod{24}$. Since $(11, 24) = 1$, by Theorem 7.12,

$$
\begin{aligned}
x &\equiv 11^{\varphi(24)-1} \cdot (-1) \equiv 11^7 \cdot (-1) \pmod{24} \\
&\equiv (11^2)^3 \cdot 11 \cdot (-1) \equiv 1^3 \cdot (-11) \pmod{24} \\
&\equiv 13 \pmod{24}
\end{aligned}
$$ ■

In order to find an inverse of a modulo m in Theorem 7.11 and hence to solve the congruence $ax \equiv b \pmod{m}$, we need to compute $\varphi(m)$. Lemma 7.5 gives its value if m is a prime. Suppose m is composite. It is not practical to list all positive integers $\leq m$ and relatively prime to it, and then count them. In Section 8.1 we will return to this issue and derive a formula for $\varphi(m)$ for every positive integer m.

Additionally, Euler's theorem can be used to develop a formula to solve a system of linear congruences with pairwise relatively prime moduli. (See Exercises 62–67.)

Next, we give a generalization of Euler's theorem. Its proof is fairly straightforward, so we leave its proof as an exercise. See Exercises 59 and 60.

THEOREM 7.13 **(Koshy, 1996)** Let $m_1, m_2, \ldots, m_k$ be any positive integers and a any integer such that $(a, m_i) = 1$ for $1 \leq i \leq k$. Then

$$a^{[\varphi(m_1), \varphi(m_2), \ldots, \varphi(m_k)]} \equiv 1 \pmod{[m_1, m_2, \ldots, m_k]}$$ ■

The next result follows from Theorem 7.10.

COROLLARY 7.4 Let $m_1, m_2, \ldots, m_k$ be pairwise relatively prime integers and a any integer such that $(a, m_i) = 1$ for $1 \le i \le k$. Then

$$a^{[\varphi(m_1),\varphi(m_2),\ldots,\varphi(m_k)]} \equiv 1 \pmod{m_1 m_2 \cdots m_k}$$ ■

EXERCISES 7.4

Compute $\varphi(m)$ for each integer m.

1. 8 2. 15 3. 21 4. 28

Use the modulus $m = 15$ and $a = 28$ for Exercises 5–6.

5. List the positive integers $\le m$ and relatively prime to it.
6. Multiply each by a and find their least residues modulo m.

7–8. Redo Exercises 5 and 6 with $m = 28$ and $a = 15$.

9. Verify that $a^6 \equiv 1 \pmod{18}$ for $a = 1, 5, 7, 11, 13$, and 17.
10. Using the values of $\varphi(m)$ for $m \le 15$, make a conjecture on the evenness of $\varphi(m)$.

Let m be a positive integer and a any positive integer $\le m$ and relatively prime to it. Verify Euler's theorem for each modulus.

11. 6 12. 10 13. 15 14. 28

Find the remainder when the first integer is divided by the second.

15. 7^{1020}, 15 16. 25^{2550}, 18
17. 79^{1776}, 24 18. 199^{2020}, 28

Using Euler's theorem, find the ones digit in the decimal value of each.

19. 17^{6666} 20. 23^{7777}

Using Euler's theorem, find the ones digit in the hexadecimal value of each.

21. 7^{1030} 22. 13^{4444}

Solve each linear congruence.

23. $7x \equiv 8 \pmod{10}$ 24. $23x \equiv 17 \pmod{12}$
25. $25x \equiv 13 \pmod{18}$ 26. $17x \equiv 20 \pmod{24}$
27. $143x \equiv 47 \pmod{20}$ 28. $79x \equiv 17 \pmod{25}$

If m and n are relatively prime, then $\varphi(mn) = \varphi(m) \cdot \varphi(n)$. Using this fact, compute each.

29. $\varphi(15)$ 30. $\varphi(35)$
31. $\varphi(105)$ 32. $\varphi(462)$

Compute $\sum_{d|n} \varphi(d)$ for each n.

33. 7 34. 10 35. 12 36. 17

37. Conjecture a formula for $\sum_{d|n} \varphi(d)$ using Exercises 33–36.

38–41. Compute the value of $\sum_{d|n} (-1)^{n/d} \varphi(d)$ for each n in Exercises 33–36.

42. Conjecture a formula using Exercises 38–41.

Verify each.

43. $1 + 9 + 9^2 + \cdots + 9^{23} \equiv 0 \pmod{35}$.
44. $1 + 11 + 11^2 + \cdots + 11^{31} \equiv 0 \pmod{51}$.

Prove or disprove each.

45. $\varphi((a, b)) = (\varphi(a), \varphi(b))$
46. $\varphi([a, b]) = [\varphi(a), \varphi(b)]$

Prove each.

47. Let m be a positive integer and a any integer with $(a, m) = 1$. Then $a^{\varphi(m)-1}$ is an inverse of a modulo m.

48. Let m be a positive integer and a any integer with $(a, m) = 1$. Then the solution of the linear congruence $ax \equiv b \pmod{m}$ is given by $x \equiv a^{\varphi(m)-1} b \pmod{m}$.
49. If a and b are relatively prime, then $a^{\varphi(b)} + b^{\varphi(a)} \equiv 1 \pmod{ab}$. (M. Charosh, 1982)
50. If p and q are distinct primes, then $p^{q-1} + q^{p-1} \equiv 1 \pmod{pq}$.
51. Let a and m be positive integers such that $(a, m) = 1 = (a - 1, m)$. Then $1 + a + a^2 + \cdots + a^{\varphi(m)-1} \equiv 0 \pmod{m}$.
52. Every integer n with $(n, 10) = 1$ divides some integer N consisting of all 1s. For example, $3|111$.
(*Hint*: Use Euler's theorem.)

Compute $\varphi(p^n)$ for the given values of p and n.

53. $p = 2, n = 3$
54. $p = 2, n = 4$
55. $p = 3, n = 3$
56. $p = 5, n = 2$
57. Using Exercises 53–56, predict the value of $\varphi(p^n)$.

*58. Prove that $\varphi(p^n) = p^n - p^{n-1}$, where $n \geq 1$.

Prove Theorem 7.13 using

59. Theorem 4.8.
60. Induction.

*61. Show that the solutions of the linear system $x \equiv a_i \pmod{m_i}$, where $1 \leq i \leq k$, and $(m_i, m_j) = 1$ if $i \neq j$, are given by

$$x \equiv a_1 M_1^{\varphi(m_1)} + a_2 M_2^{\varphi(m_2)} + \cdots + a_k M_k^{\varphi(m_k)} \pmod{M}$$

where $M = m_1 m_2 \cdots m_k$ and $M_i = M/m_i$.

62–67. Using Exercise 61, solve the linear systems 5–10 in Exercises 6.1.

68. Let $a_n = 2^n + 1$ and $m = n + k\varphi(n)$, where k is a positive integer. Prove that $a_m | a_n$. (J. Linkovskiĭ-Condé, 1980)

CHAPTER SUMMARY

We have studied three celebrated results that have played a significant role in the development of number theory: Wilson's theorem, Fermat's little theorem, and Euler's theorem.

Wilson's Theorem

- A positive integer a is self-invertible modulo p if and only if $a \equiv \pm 1 \pmod{p}$. (p. 322)
- If p is prime, then $(p-1)! \equiv -1 \pmod{p}$. (**Wilson's theorem**) (p. 323)
- If n is a positive integer such that $(n-1)! \equiv -1 \pmod{n}$, then n is a prime. (p. 324)

Fermat's Little Theorem

- If a is an integer such that $p \nmid a$, then the integers $1a, 2a, 3a, \ldots, (p-1)a$ modulo p are a permutation of the integers $1, 2, 3, \ldots, (p-1)$. (p. 327)
- If a is an integer such that $p \nmid a$, then $a^{p-1} \equiv 1 \pmod{p}$. (**Fermat's little theorem**) (p. 328)
- If a is an integer such that $p \nmid a$, then a^{p-2} is an inverse of a modulo p. (p. 330)

- If a is an integer such that $p \nmid a$, then the solution of the congruence $ax \equiv b \pmod{p}$ is given by $x \equiv a^{p-2}b \pmod{p}$. (p. 331)
- If a is an integer such that $p \nmid a$, then $a^p \equiv a \pmod{p}$. (p. 333)
- Let $p_1, p_2, \ldots, p_k$ be any distinct primes, a any positive integer, and $\ell = [p_1 - 1, p_2 - 1, \ldots, p_k - 1]$. Then $a^{\ell+1} \equiv a \pmod{p_1 p_2 \cdots p_k}$. (J. E. Phythian, 1970) (p. 334)

The Pollard p − 1 Factoring Method

- Let $r = 2^{k!} \pmod{n}$. Then $(r - 1, n)$ is a nontrivial factor of n, provided $r \not\equiv 1 \pmod{n}$. (p. 335)

Pseudoprimes

- A composite number n such that $2^n \equiv 2 \pmod{n}$ is a pseudoprime. (p. 338)
- The smallest pseudoprime is 341. (p. 338)
- If m and n are positive integers such that $m|n$, then $2^m - 1 | 2^n - 1$. (p. 338)
- If n is an odd pseudoprime, then so is $2^n - 1$. (p. 338)
- There is an infinite number of odd pseudoprimes. (p. 338)
- A composite number n such that $a^{n-1} \equiv 1 \pmod{n}$ for all positive integers a relatively prime to n is a **Carmichael number**. (p. 339)
- The smallest Carmichael number is 561. (p. 340)
- There are infinitely many Carmichael numbers. (p. 340)

Euler's Theorem

- Euler's phi function $\varphi(m)$ denotes the number of positive integers $\leq m$ and relatively prime to m. (p. 342)
- $\varphi(p) = p - 1$. (p. 343)
- Let a be any integer with $(a, m) = 1$. Let $r_1, r_2, \ldots, r_{\varphi(m)}$ be the positive integers $\leq m$ and relatively prime to m. Then the integers $ar_1, ar_2, \ldots, ar_{\varphi(m)}$ modulo m are the same as $r_1, r_2, \ldots, r_{\varphi(m)}$ in some order. (p. 343)
- Let a be any integer with $(a, m) = 1$. Then $a^{\varphi(m)} \equiv 1 \pmod{m}$. (**Euler's theorem**) (p. 344)
- Let a be any integer with $(a, m) = 1$. Then $a^{\varphi(m)-1}$ is an inverse of $a \pmod{m}$. (p. 346)
- Let a be any integer with $(a, m) = 1$. Then the solution of the congruence $ax \equiv b \pmod{m}$ is given by $x \equiv a^{\varphi(m)-1}b \pmod{m}$. (p. 346)
- Let $m_1, m_2, \ldots, m_k$ be any positive integers and a any integer such that $(a, m_i) = 1$ for $1 \leq i \leq k$. Then $a^{[\varphi(m_1), \varphi(m_2), \ldots, \varphi(m_k)]} \equiv 1 \pmod{[m_1, m_2, \ldots, m_k]}$. (Koshy, 1996) (p. 346)

REVIEW EXERCISES

Verify Wilson's theorem for each prime p.

1. 19
2. 23

Verify each.

3. $(12-1)! \not\equiv -1 \pmod{12}$
4. $(15-1)! \not\equiv -1 \pmod{15}$

Verify Fermat's little theorem for each integer a and the corresponding prime p.

5. $a = 19, p = 23$
6. $a = 20, p = 31$

Compute the remainder when the first integer is divided by the second.

7. 18^{4567}, 13
8. 31^{1706}, 23
9. 55^{1876}, 12
10. 715^{1863}, 28
11. $13^{16} + 17^{12}$, 221
12. $23^{42} + 43^{22}$, 989
13. $11^{19} + 19^{11}$, 209
14. $23^{29} + 29^{23}$, 667
15. $18^{20} + 25^{6}$, 450
16. $35^{32} + 51^{24}$, 1785

Find the ones digit in the base-eleven representation of each integer.

17. 15^{1942}
18. 24^{1010}

Using the Pollard $p-1$ method, find a nontrivial factor of each.

19. 5899
20. 9353

Compute $\varphi(m)$ for each m.

21. 16
22. 17
23. 200
24. 3675

Determine the number of least residues that are invertible modulo m for each m.

25. 17
26. 20
27. 25
28. 28

Verify Euler's theorem for each m.

29. 16
30. 20

Find the last two digits in the decimal value of each.

31. 273^{1961}
32. 1309^{1732}

Solve each linear congruence.

33. $33x \equiv 23 \pmod{13}$
34. $94x \equiv 32 \pmod{19}$
35. $65x \equiv 27 \pmod{18}$
36. $255x \equiv 63 \pmod{28}$

Verify that each is a pseudoprime.

37. 2047

38. 18705

Verify each.

39. $1 + 7 + 7^2 + \cdots + 7^{17} \equiv 0 \pmod{19}$
40. $1 + 14 + 14^2 + \cdots + 14^{10} \equiv 0 \pmod{27}$
41. $12^{65} \equiv 12 \pmod{65}$
42. $15^{341} \equiv 15 \pmod{341}$
43. $28^{87} \equiv 28 \pmod{87}$
44. $35^{51} \equiv 35 \pmod{51}$
45. $38^{16} \equiv 1 \pmod{17^2}$
46. $11^{70} \equiv 1 \pmod{71^2}$

Compute $\sum_{d|n} \varphi(d)$ for each n.

47. 8
48. 11
49. 18
50. 28

51–54. Compute $\sum_{d|n} (-1)^{n/d}\varphi(d)$ for each n in Exercises 47–50.

Verify that each is a Carmichael number.

55. $8911 = 7 \cdot 19 \cdot 67$
56. $6601 = 7 \cdot 23 \cdot 41$

Prove each, where p is an odd prime.

57. Let a be any integer such that $p \nmid a$. Then $\sum_{i=1}^{p-1} ai \equiv 0 \pmod{p}$.
58. The least nonzero residues $1, 2, \ldots, (p-1)/2$ modulo p are congruent to $-(p-1)/2, \ldots, -2, -1$, in some order.
59. If $p \equiv 3 \pmod{4}$, then $((p-1)/2)! \equiv \pm 1 \pmod{p}$.

SUPPLEMENTARY EXERCISES

By Wilson's theorem, $(p-1)! \equiv -1 \pmod{p}$, so $W(p) = \dfrac{(p-1)!+1}{p}$ is an integer. If $W(p) \equiv 0 \pmod{p}$, then p is a **Wilson prime**.

1. Show that 5 and 13 are Wilson primes. (They were found in 1953 by Goldberg by an exhaustive computer search; the next larger Wilson prime is 563. In spite of continued searches, no other Wilson prime has been found less than 4 million. Also, nothing is known about the infinitude of Wilson primes.)
2. Prove that p is a Wilson prime if and only if $(p-1)! \equiv -1 \pmod{p^2}$.

In 1982, Albert Wilansky of Lehigh University reported that his brother-in-law Harold Smith had a telephone number 493-7775 with the property that the sum of its digits equals the sum of the digits of its prime factors. Since $4{,}937{,}775 = 3 \cdot 5 \cdot 5 \cdot 65{,}837$, $4 + 9 + 3 + 7 + 7 + 7 + 5 = 3 + 5 + 5 + 6 + 5 + 8 + 3 + 7$, so it has the said property. Such a number is a **Smith number**. The smallest Smith number is 4. In 1987, W. L. McDaniel showed that there are infinitely many palindromic Smith numbers.

3. Show that 202, 265, 666, and 1111 are Smith numbers.
4. There are six Smith numbers < 100. Find them.
5. In 1917, R. Ratat gave four solutions to the equation $\varphi(n) = \varphi(n+1)$, namely, 1, 3, 15, and 104. A year later, R. Goormaghtigh added four more to the list: 164, 194, 255, and 495. In 1974, T. E. Moore found a new solution, namely, 65535. Verify that they are indeed solutions of the equation.

Let $f(n) = \varphi(n) + \varphi^2(n) + \varphi^3(n) + \cdots + \varphi(1)$, where $\varphi^i(n) = \varphi(\varphi^{i-1}(n))$ and $\varphi^1(n) = \varphi(n)$. (D. L. Silverman, 1982)

6. Compute $f(5)$ and $f(8)$.
7. Prove that $f(2^k) = 2^k$.

Prove each, where p is any prime, and m and n are any positive integers.

8. Every positive integer n is a factor of some integer N consisting of 0s and 1s.
9. $\dfrac{(np-1)!}{(n-1)!p^{n-1}} \equiv (-1)^n \pmod{p}$. (H. Sazegar, 1993)
10. Let $m \geq 2$ such that $\dfrac{(nm-1)!}{(n-1)!m^{n-1}} \equiv (-1)^n \pmod{m}$. Then m is a prime. (*Hint*: Use contradiction to prove this converse of Sazegar's result.)

COMPUTER EXERCISES

Write a program to perform each task, where p is a prime ≤ 100.

1. Read in a prime p and list all least residues modulo p that are self-invertible.
2. Read in a prime p and verify Wilson's theorem.
3. Find all Wilson primes ≤ 1000.
4. Read in a positive integer $n \leq 100$ and list all primes of the form $n! + 1$.
5. Verify that $2^n \equiv 2 \pmod{n}$ for all primes ≤ 340.
6. Verify that $2^{341} \equiv 2 \pmod{341}$.
7. List all positive integers $n \leq 341$ such that $2^n \equiv 2 \pmod{n}$. Identify those that are composite numbers.

Verify each.

8. $a^{23} \equiv a \pmod{23}$ for $0 \leq a \leq 22$.
9. $a^{31} \equiv a \pmod{31}$ for $0 \leq a \leq 30$.

10. Find all primes $p \leq 1000$ such that $2^{p-1} \equiv 1 \pmod{p^2}$.

ENRICHMENT READINGS

1. A. H. Beiler, *Recreations in the Theory of Numbers*, Dover, New York, 1966, 39–53.
2. H. Dubner and H. Nelson, "Carmichael Numbers which Are the Product of Three Carmichael Numbers," *J. Recreational Mathematics*, 22 (1990), 2–6.
3. D. H. Lehmer, "On the Converse of Fermat's Theorem," *The American Mathematical Monthly*, 43 (1936), 347–348.
4. C. S. Ogilvy and J. T. Anderson, *Excursions in Number Theory*, Dover, New York, 1966.
5. S. M. Ruiz, "An Algebraic Identity Leading to Wilson's Theorem," *The Mathematical Gazette* 80 (Nov. 1996), 579–582.

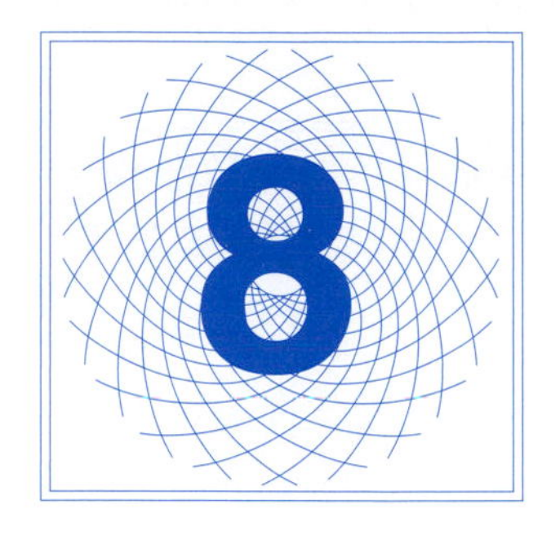

8 Multiplicative Functions

It is the man, not the method, that solves the problem.
—H. MASCHKE

In the preceding chapter, we explored Euler's phi function, one of the most important number-theoretic functions (also known as **arithmetic functions**). Arithmetic functions are defined for all positive integers. Euler's phi function belongs to a large class of arithmetic functions called **multiplicative functions**.

In addition to revisiting the phi function, we will learn three new multiplicative functions, τ (tau), σ (sigma), and μ (mu)† and study some of their fascinating properties. We will also investigate two classes of positive integers, perfect numbers and Mersenne primes, and see how they are related.

8.1 Euler's Phi Function Revisited

Recall from Section 7.4 that $\varphi(n)$ denotes the number of positive integers $\leq n$ and relatively prime to it. We found in Lemma 7.5 that if n is a prime, then $\varphi(n) = n - 1$. Suppose n is not a prime; is there a way to compute $\varphi(n)$?

We will now develop a formula to compute $\varphi(n)$ for any positive integer n using its prime-power decomposition. To this end, first we derive a formula for $\varphi(p^e)$ and then develop a mechanism for computing $\varphi(mn)$ when $(m, n) = 1$.

To achieve this goal, we first introduce multiplicative functions.

† τ, σ, and μ are lowercase Greek symbols.

Multiplicative Function

A number-theoretic function f is **multiplicative** if $f(mn) = f(m)f(n)$ whenever m and n are relatively prime.

EXAMPLE 8.1 The constant function $f(n) = 1$ is multiplicative, since $f(mn) = 1 = 1 \cdot 1 = f(m)f(n)$. So is the function $g(n) = n^k$, k being a fixed integer, since $g(mn) = (mn)^k = m^k n^k = g(m)g(n)$. ■

Notice that in both cases, we did not have to assume that $(m, n) = 1$; nevertheless they have the desired property. (We shall use a special case of g in Corollary 8.1.)

The following theorem, called the **fundamental theorem for multiplicative functions**, enables us to compute the value of a multiplicative function f for any positive integer, provided we know its values for prime powers in n. For example, suppose we know the values of $f(p^a)$ and $f(q^b)$, where p and q are distinct primes and a and b positive integers. Then $f(p^a q^b) = f(p^a)f(q^b)$, since f is multiplicative. More generally, we have the following result.

THEOREM 8.1 Let f be a multiplicative function and n a positive integer with canonical decomposition $n = p_1^{e1} p_2^{e2} \cdots p_k^{ek}$. Then $f(n) = f(p_1^{e1})f(p_2^{e2}) \cdots f(p_k^{ek})$.

PROOF (by induction on the number of distinct primes in n)

If $k = 1$, that is, if $n = p_1^{e1}$, then $f(n) = f(p_1^{e1})$, so the theorem is trivially true.

Assume it is true for any integer with canonical decomposition consisting of k distinct primes: $f(n) = f(p_1^{e1})f(p_2^{e2}) \cdots f(p_k^{ek})$.

Let n be any integer with $k + 1$ distinct primes in its canonical decomposition, say, $n = p_1^{e1} p_2^{e2} \cdots p_{k+1}^{ek+1}$. Since $(p_1^{e1} \cdots p_k^{ek}, p_{k+1}^{ek+1}) = 1$ and f is multiplicative, $f(p_1^{e1} \cdots p_k^{ek} p_{k+1}^{ek+1}) = f(p_1^{e1} \cdots p_{kk}^{e})f(p_{k+1}^{ek+1}) = f(p_1^{e1}) \cdots f(p_k^{ek})f(p_{k+1}^{ek+1})$, by the inductive hypothesis. Therefore, by induction, the result is true for any positive integer n. ■

This theorem is invaluable in our quest for finding a formula for $\varphi(n)$. If we know φ is multiplicative, and if we know the value of $\varphi(p^e)$, then we can use the theorem to derive the formula. We now take up these issues one by one in the next two theorems.

THEOREM 8.2 Let p be a prime and e any positive integer. Then $\varphi(p^e) = p^e - p^{e-1}$.

PROOF

$$\begin{aligned}
\varphi(p^e) &= \text{number of positive integers} \le p^e \text{ and relatively prime to it} \\
&= \begin{pmatrix}\text{number of positive} \\ \text{integers} \le p^e\end{pmatrix} - \begin{pmatrix}\text{number of positive integers} \le p^e \\ \text{and } \textit{not} \text{ relatively prime to it}\end{pmatrix}
\end{aligned}$$

The positive integers $\leq p^e$ and *not* relatively prime to it are the various multiples of p, namely, $p, 2p, 3p, \ldots, (p^{e-1})p$, and they are p^{e-1} in number. Thus, $\varphi(p^e) = p^e - p^{e-1}$. ■

The following example demonstrates this theorem.

EXAMPLE 8.2 Compute $\varphi(8)$, $\varphi(81)$, and $\varphi(15{,}625)$.

SOLUTION

$$\varphi(8) = \varphi(2^3) = 2^3 - 2^2 = 8 - 4 = 4$$
$$\varphi(81) = \varphi(3^4) = 3^4 - 3^3 = 54$$
$$\varphi(15{,}625) = \varphi(5^6) = 5^6 - 5^5 = 12{,}500$$

Thus, there are four positive integers ≤ 8 and relatively prime to it; they are 1, 3, 5, and 7. ■

Notice that the value of $\varphi(p^e)$ can also be written as $\varphi(p^e) = p^e\left(1 - \frac{1}{p}\right)$. You will find this version useful in Theorem 8.4.

The Monkey and Coconuts Puzzle Revisited Once Again

Next, we revisit the monkey and coconuts riddle, as an application of Theorems 7.10 and 8.2.

EXAMPLE 8.3 Using Theorems 7.10 and 8.2, solve the monkey and coconuts riddle.

SOLUTION

Let n denote the least possible number of coconuts in the original pile and z each sailor's share after the final division. Then, by Example 3.23,

$$1024n - 15625z = 11529$$

That is,

$$1024n \equiv 11529 \pmod{15625}$$
$$n \equiv 1024^{-1} \cdot 11529 \pmod{15625}$$
$$\equiv 1024^{\varphi(15625)-1} \cdot (-4096) \pmod{15625}$$

But $\varphi(15625) = 12500$, by Example 8.2. Therefore,

$$n \equiv 1024^{12499} \cdot (-4096) \pmod{15625} \tag{8.1}$$

We can evaluate this using a scientific calculator, but it is time consuming. The key steps are listed below, and the details can be filled in as an exercise:

$$1024^2 \equiv 1701 \pmod{15625} \qquad 1024^{16} \equiv -3899 \pmod{15625}$$
$$1024^{32} \equiv -924 \pmod{15625}$$

Therefore,

$$\begin{aligned} 1024^{12499} &= (1024^{32})^{390} \cdot 1024^{16} \cdot 1024^2 \cdot 1024 \\ &\equiv (-924)^{390} \cdot (-3899) \cdot 1701 \cdot 1024 \pmod{15625} \end{aligned} \tag{8.2}$$

$$924^6 \equiv 701 \pmod{15265} \qquad 924^{18} \equiv 3351 \pmod{15265}$$
$$924^{72} \equiv 4651 \pmod{15265} \qquad 924^{216} \equiv 6451 \pmod{15265}$$

Therefore,

$$\begin{aligned} 924^{390} &= 924^{216} \cdot (924^{72})^2 \cdot 924^{18} \cdot (924^6)^2 \\ &\equiv 6451 \cdot 4651^2 \cdot 3351 \cdot 701^2 \equiv 6451 \cdot 6801 \cdot 3351 \cdot 701^2 \pmod{15625} \\ &\equiv 6451 \cdot (6801 \cdot 701) \cdot (3351 \cdot 701) \equiv 6451 \cdot 1876 \cdot 5301 \pmod{15625} \\ &\equiv 6451 \cdot 7176 \pmod{15625} \end{aligned}$$

So, by congruence (8.2), $1024^{12499} \equiv 6451 \cdot 7176 \cdot 3224 \pmod{15625}$. Therefore, by congruence (8.1),

$$\begin{aligned} n &\equiv (6451 \cdot 3224) \cdot 7176 \cdot (-4096) \equiv (1149 \cdot 7176) \cdot (-4096) \pmod{15625} \\ &\equiv (-4776)(-4096) \equiv 4776 \cdot 4096 \pmod{15625} \\ &\equiv 15621 \pmod{15625} \end{aligned}$$

Thus, the minimum number of coconuts in the original pile is 15,621. ■

Before we tackle the second issue directly, we will study an example, which suggests a method for showing that φ is multiplicative. Suppose we would like to compute $\varphi(28)$. If we know that φ is multiplicative, then $\varphi(28) = \varphi(4 \cdot 7) = \varphi(4)\varphi(7) = 2 \cdot 6 = 12$. The following example confirms this result.

EXAMPLE 8.4 Let $m = 4$ and $n = 7$. Then $(m, n) = 1$ and $mn = 28$. To find $\varphi(mn) = \varphi(28)$, we list the positive integers ≤ 28 in four rows of 7 each and then ignore the ones that are *not* relatively prime to 28 (see Table 8.1):

1	5	9	13	17	21	25
2	6	10	14	18	22	26
3	7	11	15	19	23	27
4	8	12	16	20	24	28

Table 8.1

Clearly, the first element in the second and fourth rows is not relatively prime to m; in fact, no element in either row is relatively prime to m. So none of them is relatively prime to mn.

Consequently, the positive integers ≤ 28 and relatively prime to it must come from the $2 = \varphi(4)$ remaining rows:

1	5	9	13	17	21	25
3	7	11	15	19	23	27

Each of them is relatively prime to m. Each row contains $6 = \varphi(7)$ elements relatively prime to 7:

1	5	9	13	17	25
3	11	15	19	23	27

The resulting array contains 12 elements and they are indeed relatively prime to 28. Thus, $\varphi(28) = 12 = 2 \cdot 6 = \varphi(4)\varphi(7)$. ■

This example contains the essence of the proof of the next theorem, which confirms what you probably have been guessing all along. First, we need the following lemma.

LEMMA 8.1 Let m and n be relatively prime positive integers, and r any integer. Then the integers $r, m + r, 2m + r, \ldots, (n - 1)m + r$ are congruent modulo n to $0, 1, 2, \ldots, (n - 1)$ in some order.

PROOF

It suffices to show that no two elements in the list are congruent modulo n. To this end, suppose $km + r \equiv \ell m + r \pmod{n}$, where $0 \leq k, \ell < n$. Then $km \equiv \ell m \pmod{n}$. But $(m, n) = 1$, so $k \equiv \ell \pmod{n}$. Since k and ℓ are least residues modulo n, this implies $k = \ell$.

Thus, if $k \neq \ell$, then $km + r \not\equiv \ell m + r \pmod{n}$; that is, no two elements in the given list yield the same least residue. But it contains n elements, so their least residues modulo n are a rearrangement to the integers 0 through $n - 1$. ■

The following example illustrates this lemma.

EXAMPLE 8.5 Let $m = 4$ and $n = 7$, so $(m, n) = (4, 7) = 1$. Choose $r = 3$. Then the list $r, m + r, 2m + r, \ldots, (n-1)m + r$ becomes 3, 7, 11, 15, 19, 23, and 27. Their least residues modulo 7 are 3, 0, 4, 1, 5, 2, and 6, which are a permutation of 0, 1, 2, 3, 4, 5, and 6. ■

Now, we are prepared to present the theorem.

THEOREM 8.3 The function φ is multiplicative.

PROOF

(The argument mirrors Example 8.4, so look for parallels if or when the proof gets confusing.) Let m and n be positive integers such that $(m, n) = 1$. We would like to show that $\varphi(mn) = \varphi(m)\varphi(n)$.

Arrange the integers 1 through mn in m rows of n each:

$$
\begin{array}{cccccc}
1 & m+1 & 2m+1 & \ldots & (n-1)m+1 & \\
2 & m+2 & 2m+2 & \ldots & (n-1)m+2 & \\
3 & m+3 & 2m+3 & \ldots & (n-1)m+3 & \\
 & & & \vdots & & \\
r & m+r & 2m+r & \ldots & (n-1)m+r & \leftarrow r\text{th row} \\
 & & & \vdots & & \\
m & 2m & 3m & \ldots & nm &
\end{array}
$$

Let r be a positive integer $\leq m$ such that $(r, m) > 1$. We will show that no element of the rth row in the array is relatively prime to mn. Let $d = (r, m)$. Then $d|r$ and $d|m$, so $d|km + r$ for any integer k; that is, d is a factor of every element in the rth row.

Thus, no element in the rth row is relatively prime to m and hence to mn if $(r, m) > 1$; in other words, the elements in the array relatively prime to mn come from the rth row only if $(r, m) = 1$. By definition, there are $\varphi(m)$ such integers r and hence $\varphi(m)$ such rows.

Now let us concentrate on the rth row, where $(r, m) = 1$:

$$r, m + r, 2m + r, \ldots, (n-1)m + r$$

By Lemma 8.1, their least residues modulo n are a permutation of $0, 1, 2, \ldots, (n-1)$ of which $\varphi(n)$ are relatively prime to n. Therefore, exactly $\varphi(n)$ elements in the rth row are relatively prime to n and hence to mn.

Thus, there are $\varphi(m)$ rows containing positive integers relatively prime to mn, and each row contains $\varphi(n)$ elements relatively prime to it. So the array contains $\varphi(m)\varphi(n)$ positive integers $\leq mn$ and relatively prime to mn; that is, $\varphi(mn) = \varphi(m)\varphi(n)$. ■

Using Lemma 7.5 and Theorems 8.2 and 8.3, we can compute $\varphi(n)$ for any positive integer n, as the following example demonstrates.

EXAMPLE 8.6 Evaluate $\varphi(221)$ and $\varphi(6125)$.

SOLUTION

- $$\begin{aligned} \varphi(221) &= \varphi(13 \cdot 17) = \varphi(13) \cdot \varphi(17), \quad \text{by Theorem 8.3} \\ &= 12 \cdot 16, \quad \text{by Lemma 7.5} \\ &= 192 \end{aligned}$$

- $$\begin{aligned} \varphi(6125) &= \varphi(5^3 \cdot 7^2) = \varphi(5^3)\varphi(7^2), \quad \text{by Theorem 8.3} \\ &= (5^3 - 5^2)(7^2 - 7), \quad \text{by Theorem 8.2} \\ &= 4200 \end{aligned}$$

■

Theorems 8.2 and 8.3 can now be effectively applied to derive an explicit formula for $\varphi(n)$ using its canonical decomposition.

THEOREM 8.4 Let $n = p_1^{e_1} p_2^{e_2} \cdots p_k^{e_k}$ be the canonical decomposition of a positive integer n. Then

$$\varphi(n) = n\left(1 - \frac{1}{p_1}\right)\left(1 - \frac{1}{p_2}\right)\cdots\left(1 - \frac{1}{p_k}\right)$$

PROOF

Since φ is multiplicative, by Theorem 8.1,

$$\begin{aligned} \varphi(n) &= \varphi\left(p_1^{e_1}\right)\varphi\left(p_2^{e_2}\right)\cdots\varphi\left(p_k^{e_k}\right) \\ &= p_1^{e_1}\left(1 - \frac{1}{p_1}\right)p_2^{e_2}\left(1 - \frac{1}{p_2}\right)\cdots p_k^{e_k}\left(1 - \frac{1}{p_k}\right), \quad \text{by Theorem 8.2} \end{aligned}$$

$$= p_1^{e_1} p_2^{e_2} \cdots p_k^{e_k} \left(1 - \frac{1}{p_1}\right)\left(1 - \frac{1}{p_2}\right) \cdots \left(1 - \frac{1}{p_k}\right)$$

$$= n\left(1 - \frac{1}{p_1}\right)\left(1 - \frac{1}{p_2}\right) \cdots \left(1 - \frac{1}{p_k}\right)$$ ■

The following example illustrates this theorem.

EXAMPLE 8.7 Compute $\varphi(666)$ and $\varphi(1976)$.

SOLUTION

- $$666 = 2 \cdot 3^2 \cdot 37$$
$$\varphi(666) = 666\left(1 - \frac{1}{2}\right)\left(1 - \frac{1}{3}\right)\left(1 - \frac{1}{37}\right)$$
$$= 216$$

[*An interesting observation*: $\varphi(666) = 6 \cdot 6 \cdot 6$]

- $$1976 = 2^3 \cdot 13 \cdot 19$$
$$\varphi(1976) = 1976\left(1 - \frac{1}{2}\right)\left(1 - \frac{1}{13}\right)\left(1 - \frac{1}{19}\right)$$
$$= 864$$ ■

You must have observed in Exercise 10 in Section 7.4, and in Examples 8.6 and 8.7, that $\varphi(n)$ is nearly always even. In fact, it is odd only if $n = 1$ or 2. We leave the proof as an exercise.

Before pursuing another result involving φ, we investigate $\sum_{d|n} \varphi(d)n$ in an example.

EXAMPLE 8.8 Let $n = 28$ and $d|28$. Let C_d denote the class of those positive integers $m \le n$, where $(m, n) = d$. Since 28 has six positive factors 1, 2, 4, 7, 14, and 28, there are six such classes:

$$C_1 = \{1, 3, 5, 9, 11, 13, 15, 17, 19, 23, 25, 27\} \qquad C_2 = \{2, 6, 10, 18, 22, 26\}$$
$$C_4 = \{4, 8, 12, 16, 20, 24\} \qquad C_7 = \{7, 21\}$$
$$C_{14} = \{14\} \qquad C_{28} = \{28\}$$

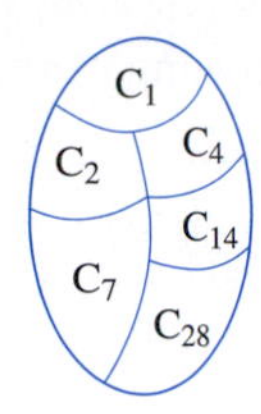

Figure 8.1

They contain $12 = \varphi(28) = \varphi(28/1)$, $6 = \varphi(14) = \varphi(28/2)$, $6 = \varphi(7) = \varphi(28/4)$, $2 = \varphi(4) = \varphi(28/7)$, $1 = \varphi(2) = \varphi(28/14)$, and $1 = \varphi(1) = \varphi(28/28)$ elements, respectively. Also, they form a partitioning of the set of positive integers ≤ 28, as Figure 8.1 shows.

Therefore, the sum of the numbers of elements in the various classes must equal 28; that is, $12 + 6 + 6 + 2 + 1 + 1 = 28$. In other words, $\varphi(28) + \varphi(14) + \varphi(7) + \varphi(4) + \varphi(2) + \varphi(1) = 28$; that is, $\sum_{d|28} \varphi(d) = 28$. ■

More generally, we have the following result. The technique behind its proof, illustrated in the preceding example, was originally conceived by Gauss.

THEOREM 8.5 Let n be a positive integer. Then $\sum_{d|n} \varphi(d) = n$.

PROOF

We partition the set of positive integers 1 through n into various classes C_d as follows, where $d|n$. Let m be a positive integer $\leq n$. Then m belongs to class C_d if and only if $(m, n) = d$; that is, if and only if $(m/d, n/d) = 1$. The number of elements in C_d equals the number of positive integers $\leq n/d$ and relatively prime to it, namely, $\varphi(n/d)$; thus, each class C_d contains $\varphi(n/d)$ elements.

Since there is a class corresponding to every factor d of n and every integer m belongs to exactly one class, the sum of the elements in the various classes must yield the total number of elements. That is,

$$\sum_{d|n} \varphi(n/d) = n$$

But as d runs over the divisors of n, so does n/d. Consequently,

$$\sum_{d|n} \varphi(n/d) = \sum_{d|n} \varphi(d), \quad \text{thus} \quad \sum_{d|n} \varphi(d) = n$$

■

The following example illustrates this theorem.

EXAMPLE 8.9 Verify that $\sum_{d|18} \varphi(d) = 18$.

PROOF

The positive divisors of 18 are 1, 2, 3, 6, 9, and 18. So

$$\sum_{d|18} \varphi(d) = \varphi(1) + \varphi(2) + \varphi(3) + \varphi(6) + \varphi(9) + \varphi(18)$$

$$= 1 + 1 + 2 + 2 + 6 + 6 = 18$$

■

EXERCISES 8.1

1. Let f be the number-theoretic function such that $f(n) = 0$ for every positive integer n. Show that f is multiplicative.

Compute $\varphi(n)$ for each n.

2. 56
3. 341
4. 561
5. 1105
6. 2047
7. 6860
8. 98,865
9. 183,920
10. Show that $\varphi(5186) = \varphi(5187) = \varphi(5188)$. (These are the only three known consecutive integers with this property.)

Compute $\varphi(p!)$ for each prime p.

11. 3
12. 5
13. 7
14. 11

Find the positive integers n such that

15. $\varphi(n) = n$
16. $\varphi(n) = 4$
17. $\varphi(n) = 6$
18. $\varphi(n) = 12$
19. Derive a formula for $\varphi(pq)$, where p and q are twin primes.

Find the twin primes p and q if

20. $\varphi(pq) = 120$
21. $\varphi(pq) = 288$
22. If p and q are twin primes with $p < q$, show that $\varphi(q) = \varphi(p) + 2$.
23. Can $\varphi(n) = 2n$ for any positive integer? If yes, find such an integer.

Prove each, where m, n, and e are arbitrary positive integers and p is any prime.

24. Let f be a multiplicative function. Then $f(1) = 1$.
25. If $n = 2^k$, then $\varphi(n) = n/2$.
26. Let f_n denote a Fermat prime. Then $\varphi(f_n) = f_n - 1$.
27. $\varphi(4n) = 2\varphi(n)$, where n is odd.
28. $\varphi(2n) = \begin{cases} \varphi(n) & \text{if } n \text{ is odd} \\ 2\varphi(n) & \text{if } n \text{ is even} \end{cases}$
29. If $n = 2^j$, where $j \geq 1$, then $n = 2\varphi(n)$.
30. If $n = 2\varphi(n)$, then $n = 2^j$, where $j \geq 1$.
31. If $n = 2^j 3^k$, where $j, k \geq 1$, $n = 3\varphi(n)$.
32. If $n = 3\varphi(n)$, then $n = 2^j 3^k$, where $j, k \geq 1$.
33. $\varphi(2^{2k+1})$ is a square.
34. If $\varphi(p^e)$ is a square, then $p - 1$ must be a square and e must be odd.
35. If $p \nmid n$, then $\varphi(pn) = (p - 1)\varphi(n)$.
36. If $\varphi(pn) = (p - 1)\varphi(n)$, then $p \nmid n$.
37. $\varphi(\varphi(p^e)) = p^{e-2}\varphi(p(p - 1))$, where $e \geq 2$.
38. If $m|n$, then $\varphi(m)|\varphi(n)$.
39. Let $(m, n) = p$. Then $\varphi(mn) = \dfrac{p}{p-1}\varphi(m)\varphi(n)$.
40. Deduce from Exercise 39 that $\varphi(p^2) = p(p - 1)$.
41. Let $(m, n) = d$. Prove that $\varphi(mn) = \dfrac{d}{\varphi(d)}\varphi(m)\varphi(n)$.

42. Deduce from Exercise 41 that φ is multiplicative.

Prove each, where n and e are positive integers.

43. $\varphi(n^2) = n\varphi(n)$
44. $\varphi(n^3) = n^2\varphi(n)$
45. $\varphi(n^e) = n^{e-1}\varphi(n)$
46. Using Exercise 29, compute $\varphi(256)$ and $\varphi(4096)$.

Evaluate each using Exercise 41.

47. $\varphi(48)$
48. $\varphi(90)$
49. $\varphi(375)$
50. $\varphi(1690)$

Evaluate each using Exercise 45.

51. $\varphi(16)$
52. $\varphi(81)$
53. $\varphi(2401)$
54. $\varphi(1728)$

Let $S(n)$ denote the sum of the positive integers $\leq n$ and relatively prime to it. Compute $S(n)$ for each value of n.

55. 6
56. 8
57. 9
58. 12
59. Using Exercises 55–58, predict a formula for $S(n)$.

Let m and n be positive integers and p a prime. Prove each.

60. Let $n \equiv 0 \pmod 4$. Then $\varphi(n/2) = \varphi(n)/2$.
61. If $n-1$ and $n+1$ are twin primes, then $\varphi(n/2) \leq n$. (H. Demir, 1960)
62. $\varphi\left(\dfrac{m}{n}\right) = \dfrac{\varphi(m)}{\varphi(n)}$ if and only if $m = nk$, where $(n,k) = 1$. (L. Marvin, 1975)
63. There are infinitely many positive integers n such that $\varphi(n) = n/3$, but none such that $\varphi(n) = n/4$.

8.2 The Tau and Sigma Functions

We now turn to two additional number-theoretic functions: τ (tau) and σ (sigma). We will show that both are multiplicative and will derive formulas for $\tau(n)$ and $\sigma(n)$ using the canonical decomposition of n. Both functions are employed in the study of perfect numbers, as you will see in the next section.

Let us begin with the definition of the tau function.

The Tau Function

Let n be a positive integer. Then $\tau(n)$ denotes the number of positive factors of n; that is,

$$\tau(n) = \sum_{d|n} 1$$

The following example illustrates this simple definition.

EXAMPLE 8.10 Evaluate $\tau(18)$ and $\tau(23)$.

SOLUTION

- The positive divisors of 18 are 1, 2, 3, 6, 9, and 18, so $\tau(18) = 6$.
- 23, being a prime, has exactly two positive divisors, so $\tau(23) = 2$. ■

It follows by definition that if n is a prime, then $\tau(n) = 2$; conversely, if $\tau(n) = 2$, then n is a prime.

Next we define the sigma function.

The Sigma Function

Let n be a positive integer. Then $\sigma(n)$ denotes the sum of the positive factors of n; that is,

$$\sigma(n) = \sum_{d|n} d$$

The following example illustrates this definition.

EXAMPLE 8.11 Evaluate $\sigma(12)$ and $\sigma(28)$.

SOLUTION

- The positive divisors of 12 are 1, 2, 3, 4, 6, and 12; so

$$\sigma(12) = 1 + 2 + 3 + 4 + 6 + 12 = 28$$

- The positive divisors of 28 are 1, 2, 4, 7, 14, and 28; so

$$\sigma(28) = 1 + 2 + 4 + 7 + 14 + 28 = 56$$

■

Again, if n is a prime, then it has exactly two positive factors, 1 and n, so $\sigma(n) = n + 1$; on the other hand, if $\sigma(n) = n + 1$, then n must be a prime.

To derive formulas for $\tau(n)$ and $\sigma(n)$, we need to show that both functions are multiplicative. To this end, we introduce a new function F.

Let f be a multiplicative function. Then F is defined by

$$F(n) = \sum_{d|n} f(d)$$

For example,

$$\begin{aligned} F(12) &= \sum_{d|12} f(d) \\ &= f(1) + f(2) + f(3) + f(4) + f(6) + f(12) \end{aligned}$$

What properties does F possess? For example, is it multiplicative? The following example shows we can compute the value of $F(28)$ provided we know the values of $F(4)$ and $F(7)$. Follow the steps carefully; we will need them soon.

EXAMPLE 8.12 Determine if $F(mn) = F(m)F(n)$, where $m = 4$ and $n = 7$.

SOLUTION

Clearly, $(m, n) = (4, 7) = 1$. Then

$$\begin{aligned}
F(4 \cdot 7) &= \sum_{d|28} f(d) \\
&= f(1) + f(2) + f(4) + f(7) + f(14) + f(28) \\
&= f(1 \cdot 1) + f(1 \cdot 2) + f(1 \cdot 4) + f(1 \cdot 7) + f(2 \cdot 7) + f(4 \cdot 7) \\
&= f(1)f(1) + f(1)f(2) + f(1)f(4) + f(1)f(7) \\
&\quad + f(2)f(7) + f(4)f(7), \quad \text{since } f \text{ is multiplicative} \\
&= [f(1) + f(2) + f(4)]f(1) + [f(1) + f(2) + f(4)]f(7) \\
&= [f(1) + f(2) + f(4)][f(1) + f(7)] \\
&= \sum_{d|4} f(d) \cdot \sum_{d|7} f(d) \\
&= F(4)F(7)
\end{aligned}$$

■

More generally, we have the following result, which is the cornerstone of this section; it shows that F is indeed multiplicative.

THEOREM 8.6 If f is a multiplicative function, then $F(n) = \sum_{d|n} f(d)$ is also multiplicative.

PROOF

Let m and n be relatively prime positive integers. We would like to show that $F(mn) = F(m)F(n)$.

By definition,

$$F(mn) = \sum_{d|mn} f(d)$$

Since $(m, n) = 1$, every positive divisor d of mn is the product of a unique pair of positive divisors d_1 of m and d_2 of n, where $(d_1, d_2) = 1$. Therefore,

$$F(mn) = \sum_{\substack{d_1|m \\ d_2|n}} f(d_1 d_2)$$

But since f is multiplicative, $f(d_1d_2) = f(d_1)f(d_2)$. So

$$F(mn) = \sum_{\substack{d_1|m \\ d_2|n}} f(d_1d_2) = \sum_{d_2|n}\Big[\sum_{d_1|m} f(d_1)\Big]f(d_2)$$

$$= \sum_{d_2|n} F(m)f(d_2), \quad \text{by the definition of } F$$

$$= F(m)\sum_{d_2|n} f(d_2)$$

$$= F(m)F(n)$$

Thus, F is multiplicative. ■

With this theorem at our disposal, we can easily conclude that both τ and σ are multiplicative, as the following corollary shows.

COROLLARY 8.1 The tau and sigma functions are multiplicative.

PROOF

In Example 8.1 we found that the constant function $f(n) = 1$ and the identity function $g(n) = n$ are multiplicative. Therefore, by Theorem 8.6, the functions

$$\sum_{d|n} f(d) = \sum_{d|n} 1 = \tau(n) \quad \text{and} \quad \sum_{d|n} g(d) = \sum_{d|n} d = \sigma(n)$$

are multiplicative; that is, if $(m, n) = 1$, then $\tau(mn) = \tau(m)\tau(n)$ and $\sigma(mn) = \sigma(m)\sigma(n)$. ■

The following example demonstrates these results.

EXAMPLE 8.13 Compute $\tau(36)$ and $\sigma(36)$.

SOLUTION

Because $36 = 4 \cdot 9$, where $(4, 9) = 1$, by Corollary 8.1,

$$\tau(36) = \tau(4) \cdot \tau(9) = 3 \cdot 3 = 9$$

and

$$\sigma(36) = \sigma(4) \cdot \sigma(9) = (1 + 2 + 4)(1 + 3 + 9) = 91$$ ■

Corollary 8.1 certainly takes us a step closer to the derivation of the formulas for $\tau(n)$ and $\sigma(n)$, but first we need to develop the formulas for $\tau(p^e)$ and $\sigma(p^e)$.

THEOREM 8.7 Let p be any prime and e any positive integer. Then $\tau(p^e) = e + 1$ and $\sigma(p^e) = \dfrac{p^{e+1} - 1}{p - 1}$.

PROOF

The positive factors of p^e are of the form p^i, where $0 \le i \le e$; there are $e + 1$ of them, so $\tau(p^e) = e + 1$. Also,

$$\sigma(p^e) = \sum_{i=0}^{e} p^i = \frac{p^{e+1} - 1}{p - 1}$$ ■

For example, $\tau(81) = \tau(3^4) = 5$; that is, 81 has five positive factors, namely, 1, 3, 9, 27, and 81; and $\sigma(81) = \sigma(3^4) = \dfrac{3^5 - 1}{3 - 1} = 121$.

The following theorem gives the two formulas promised earlier. They are a consequence of Corollary 8.1 and Theorem 8.7.

THEOREM 8.8 Let n be a positive integer with canonical decomposition $n = p_1^{e_1} p_2^{e_2} \cdots p_k^{e_k}$. Then $\tau(n) = (e_1 + 1)(e_2 + 1) \cdots (e_k + 1)$ and

$$\sigma(n) = \frac{p_1^{e_1+1} - 1}{p_1 - 1} \cdot \frac{p_2^{e_2+1} - 1}{p_2 - 1} \cdots \frac{p_k^{e_k+1} - 1}{p_k - 1}$$

PROOF

Since τ is multiplicative, by Corollary 8.1,

$$\begin{aligned} \tau(n) &= \tau\left(p_1^{e_1}\right) \cdot \tau\left(p_2^{e_2}\right) \cdots \tau\left(p_k^{e_k}\right) \\ &= (e_1 + 1)(e_2 + 1) \cdots (e_k + 1), \quad \text{by Theorem 8.7} \end{aligned}$$

Since σ is multiplicative,

$$\begin{aligned} \sigma(n) &= \sigma(p_1^{e_1}) \cdot \sigma(p_2^{e_2}) \cdots \sigma(p_k^{e_k}) \\ &= \frac{p_1^{e_1+1} - 1}{p_1 - 1} \cdot \frac{p_2^{e_2+1} - 1}{p_2 - 1} \cdots \frac{p_k^{e_k+1} - 1}{p_k - 1}, \quad \text{by Theorem 8.7} \end{aligned}$$ ■

Notice that the formulas for $\tau(n)$ and $\sigma(n)$ can be rewritten as follows:

$$\tau(n) = \prod_{i=1}^{k} (e_i + 1) \quad \text{and} \quad \sigma(n) = \prod_{i=1}^{k} \frac{p_i^{e_i+1} - 1}{p_i - 1}$$

The following example demonstrates this theorem.

EXAMPLE 8.14 Compute $\tau(6120)$ and $\sigma(6120)$.

SOLUTION

First, we find the canonical decomposition of 6120: $6120 = 2^3 \cdot 3^2 \cdot 5 \cdot 17$. Therefore,

$$\tau(6120) = (3+1)(2+1)(1+1)(1+1) = 48$$

and

$$\begin{aligned}\sigma(6120) &= \frac{2^{3+1}-1}{2-1} \cdot \frac{3^{2+1}-1}{3-1} \cdot \frac{5^{1+1}-1}{5-1} \cdot \frac{17^{1+1}-1}{17-1} \\ &= 15 \cdot 13 \cdot 6 \cdot 18 \\ &= 21{,}060\end{aligned}$$

■

A Brainteaser (optional)

We now turn to a fascinating brainteaser.

Marilyn vos Savant, who is listed in *The Guinness Book of World Records* as having the highest IQ ever recorded, writes a regular column on puzzles in *Parade Magazine*. The following is from the book *Ask Marilyn* (1992), a collection of puzzles from her column:

> There are 1000 tenants and 1000 apartments. The first tenant opens all the doors. The second tenant closes every other door. The third tenant goes to every third door, opens it if it is closed and closes it if it is open. The fourth tenant goes to every fourth door, closes it if it is open and opens it if it is closed. This continues with each tenant until the 1000th tenant closes the 1000th door. How many doors remain open?

The solution employs three simple properties from number theory:

- *Every non-square integer has an even number of positive factors.*

Assume N is *not* a square. Then, by the Fundamental Theorem of Arithmetic,

$$N = p_1^{e_1} p_2^{e_2} \cdots p_t^{e_t}$$

where $p_1, p_2, \ldots, p_t$ are distinct prime numbers and *not* all exponents e_i are even. (If all exponents are even, then N would be a square.) By Theorem 8.8, $\tau(N) = (e_1+1)(e_2+1)\cdots(e_t+1)$. Since at least one of the exponents e_i is odd, $e_t + 1$ is even, so the product $(e_1+1)(e_2+1)\cdots(e_t+1)$ is an even integer.

- *Every square has an odd number of positive factors.*

To establish this result, let M be a square. Then

$$M = \left(p_1^{f_1} p_2^{f_2} \cdots p_k^{f_k}\right)^2 = p_1^{2f_1} p_2^{2f_2} \cdots p_k^{2f_k}$$

Then $\tau(M) = (2f_1 + 1)(2f_2 + 1) \cdots (2f_k + 1)$. This, being the product of odd integers, is clearly an odd integer.

- *There are* $\lfloor \sqrt{n} \rfloor$ *squares* $\leq n$.

To see this, suppose there are k squares $\leq n$. Then k is the largest positive integer such that $k^2 \leq n < (k+1)^2$, so $k \leq \sqrt{n} < k+1$; thus, $k = \lfloor \sqrt{n} \rfloor$.

Before we apply these results to solve the puzzle, we study a miniversion with 10 tenants and 10 apartments. The first tenant opens all 10 doors; the second tenant closes the 2nd, 4th, 6th, 8th, and 10th doors; the third closes the 3rd door, opens the 6th door, and closes the 9th door; the fourth tenant opens the 4th and 8th doors. Continuing like this, the 10th tenant closes the 10th door. These data are summarized in Table 8.2, where O indicates the door is open and C indicates the door is closed.

Door / *Tenant*	1	2	3	4	5	6	7	8	9	10
1	O	O	O	O	O	O	O	O	O	O
2	.	C	.	C	.	C	.	C	.	C
3	.	.	C	.	.	O	.	.	C	.
4	.	.	.	O	.	.	.	O	.	.
5	.	.	.	.	C	.	.	.	.	O
6	.	.	.	.	.	C	.	.	.	.
7	.	.	.	.	.	.	C	.	.	.
8	.	.	.	.	.	.	.	C	.	.
9	.	.	.	.	.	.	.	.	O	.
10	.	.	.	.	.	.	.	.	.	C

Table 8.2

It follows from the table that doors 1, 4, and 9 remain open at the end, so the number of such doors is three. (Notice that $3 = \lfloor \sqrt{10} \rfloor$; can you predict the answer to the given problem? Construct tables like Table 8.2 for 13 tenants and 13 apartments, 18 tenants and 18 apartments, and 25 tenants and 25 apartments, and look for a pattern.)

Returning to the original problem, recall that the first tenant opens all doors. Consider the nth tenant, where $2 \leq n \leq 1000$.

case 1 Let n be a square, where $n^2 \leq 1000$. Since n has an odd number of positive factors, the last person to touch an open door will close it. Thus, every nth door will remain open if n is a square. The number of such doors equals the number of squares ≤ 1000, namely, $\lfloor\sqrt{1000}\rfloor = 31$.

case 2 Suppose n is *not* a square, where $n^2 \leq 1000$. Since n has an even number of positive factors, the last person to touch an open door will close it. In other words, every nth door will remain closed if n is not a square.

Thus, since the two cases are disjoint, $31 + 0 = 31$ doors will remain open. They are doors numbered $1, 4, 9, 16, 25, \ldots, 900$, and 961.

Can you generalize this puzzle? Suppose there are m tenants and m apartments, and the first tenant opens all doors. The jth tenant closes every jth door if it is open and opens it otherwise, where $2 \leq j \leq m$. How many doors will remain open at the end?

EXERCISES 8.2

Compute $\tau(n)$ for each n.

1. 43
2. 1560
3. 2187
4. 44,982

5–8. Compute $\sigma(n)$ for each n in Exercises 1–4.

List the positive factors of each, where p and q are distinct primes.

9. pq
10. pq^2
11. p^2q
12. p^2q^3

13–16. Find the sum of the positive divisors of each number in Exercises 9–12.

Let p and q be distinct primes in Exercises 17 and 18.

17. List the positive factors of p^iq^j.
18. Find the sum of the positive factors of p^iq^j.
19. Identify the positive integers with exactly two positive divisors.
20. Identify the positive integers with exactly three positive divisors.

21. Let $n = p_1p_2\cdots p_k$ be a product of k distinct primes. Find $\tau(n)$ and $\sigma(n)$.

Use $n = 2^{2^e}$ for Exercises 22 and 23.

22. Find $\tau(n)$ and $\sigma(n)$.
23. Find the product of the positive divisors of n.
24. Find the product of the positive divisors of p^e.
25. Find the product of the positive divisors of p^aq^b.
26. In 1638, the French mathematician and philosopher René Descartes (1596–1650) showed that $\sigma(p^e) - p^e = \dfrac{p^e - 1}{p - 1}$. Verify this.

Let $n = 2^{p-1}(2^p - 1)$, where p and $2^p - 1$ are primes. Find each.

27. $\tau(n)$
28. $\sigma(n)$

Compute $\sigma(n)$ for each n.

29. 6
30. 28
31. 496
32. 8128
33. Predict the pattern observed in Exercises 29–32.

Let n be the product of a pair of twin primes, p being the smaller of the two.

34. Find $\tau(n)$.
35. Show that $\sigma(p+2) = \sigma(p) + 2$.
36. Show that $\sigma(n) = (p+1)(p+3)$.
37. Find p for which $\sigma(p)$ is odd.

Verify each. (D. E. Iannucci, 2002)

38. $\varphi(\sigma(666)) = 2\varphi(666)$
39. $\sigma(668) = 2\sigma(\varphi(668))$
40. $\sigma(\varphi(667)) = 2\sigma(667)$
41. $\varphi(665) = 2\varphi(666)$

Prove each, where m, n, and e are positive integers and p is a prime.

42. $\sum_{d|n} \frac{1}{d} = \frac{\sigma(n)}{n}$
43. $\varphi(p) + \sigma(p) = 2p$
44. If $\varphi(p^e) + \sigma(p^e) = 2p^e$, then $e = 1$.
45. If $\tau(n)$ is odd, then n is a square.
46. If n is a square, then $\tau(n)$ is odd.
47. If $\tau(n)$ is a prime, then n is of the form p or p^{2e}.
48. If n is a power of 2, then $\sigma(n)$ is odd.
49. If p is odd, then $1 + p + p^2 + \cdots + p^k$ is odd if and only if k is even.
50. If n is a square, then $\sigma(n)$ is odd.
51. If $\sigma(n)$ is odd, then n is a square or twice a square.
52. Let n and $\sigma(n)$ be odd. Then n must be a square.
53. If $m|n$, then $\frac{\sigma(m)}{m} \leq \frac{\sigma(n)}{n}$.
54. $\frac{\sigma(p^e)}{p^e} < \frac{p}{p-1}$ (P. A. Weiner, 2000)
55. $\frac{\sigma(n)}{n} < \prod_{p|n} \frac{p}{p-1}$, where $n \geq 2$. (P. A. Weiner, 2000)

Let $\sigma_k(n)$ denote the sum of the kth powers of the positive factors of n; that is, $\sigma_k(n) = \sum_{\substack{d>0 \\ d|n}} d^k$. Clearly $\sigma_1(n) = \sigma(n)$. Compute each.

56. $\sigma_2(12)$
57. $\sigma_2(18)$
58. $\sigma_3(23)$
59. $\sigma_3(28)$

Find a formula for $\sigma_k(n)$ for each n.

60. p
61. p^e
62. $p^a q^b$
63. Prove that $\sigma_k(n)$ is multiplicative.
64. Derive a formula for $\sigma_k(p^a q^b)$. (*Hint*: Use Exercise 65.)
65. Derive a formula for $\sigma_k(n)$, where $n = \prod_{i=1}^{k} p_i^{e_i}$. (*Hint*: Use Exercises 65 and 66.)

Compute each using Exercise 66.

66. $\sigma_2(16)$
67. $\sigma_3(18)$
68. $\sigma_3(36)$
69. $\sigma_4(84)$

*70. Prove that the product of the positive divisors of a positive integer n is $n^{\tau(n)/2}$.

8.3 Perfect Numbers

We can use the sigma function to study a marvelous class of numbers, called **perfect numbers**. The term *perfect numbers* was coined by the Pythagoreans. The ancient Greeks thought these numbers had mystical powers and held them to be "good" numbers. They were also studied by the early Hebrews; Rabbi Josef ben Jehuda in the twelfth century recommended their study in his book, *Healing of Souls*.

Historically, some biblical scholars considered 6 a perfect number, because they believed God created the world in six days and God's work is perfect. St. Augustine, on the other hand, believed God's work to be perfect because 6 is a perfect number. He writes, "Six is a number perfect in itself, and not because God created all things in six days; rather the inverse is true; God created all things in six days because this number is perfect. And it would remain perfect even if the work of the six days did not exist."

The Pythagoreans regarded 6 as the symbol of "marriage and health and beauty on account of the integrity of its parts and the agreement existing in it."

What is mystical about 6? The Pythagoreans observed that 6 equals the sum of its proper factors: $6 = 1 + 2 + 3$. The next two perfect numbers are 28 and 496:

$$28 = 1 + 2 + 4 + 7 + 14$$

$$496 = 1 + 2 + 4 + 8 + 16 + 31 + 62 + 124 + 248$$

Their discovery is sometimes attributed to the Greek mathematician Nichomachus (ca. A.D. 100). Notice that the moon orbits the earth every 28 days, the second perfect number.

We can now formalize the definition of a perfect number.

Perfect Number

A positive integer n is a **perfect number** if the sum of its proper factors equals n. Thus, n is perfect if $\sigma(n) - n = n$, that is, if $\sigma(n) = 2n$.

The first eight perfect numbers are

$$\begin{aligned} 6 &= 2(2^2 - 1) \\ 28 &= 2^2(2^3 - 1) \\ 496 &= 2^4(2^5 - 1) \\ 8128 &= 2^6(2^7 - 1) \\ 33{,}550{,}336 &= 2^{12}(2^{13} - 1) \\ 8{,}589{,}869{,}056 &= 2^{16}(2^{17} - 1) \\ 137{,}438{,}691{,}328 &= 2^{18}(2^{19} - 1) \\ 2{,}305{,}843{,}008{,}139{,}952{,}128 &= 2^{30}(2^{31} - 1) \end{aligned}$$

The perfect eye chart

6
28
496
8128
33550336
8589869056
137438691328
2305843008139952128
2658455991569831744654692615953842176

of which only the first four were known to the ancient Greeks; they are listed in Nichomachus' *Introductio Arithmeticae*. The next perfect number was discovered by the Greek mathematician Hudalrichus Regius around 1536. The Italian mathematician Pietro Antonio Cataldi (1548–1626) discovered the next two in 1588. Euler discovered the eighth perfect number in 1750.

Interestingly, a medieval German nun, Hrotsvit, a Benedictine in the Abbey of Gandersheim in Saxony and the first known woman German poet, listed the first four perfect numbers in her tenth-century play, *Sapientia*.

Mathematicians of the Middle Ages, basing their assumptions on the first four perfect numbers, conjectured that

- There is a perfect number between any two consecutive powers of 10; that is, there is a perfect number of n digits long for every positive integer n; and
- Perfect numbers end alternately in 6 and 8.

Unfortunately, both conjectures are false. There are no perfect numbers that are five digits long. Even perfect numbers do end in 6 or 8, but *not* alternately; for instance, the fifth and sixth even perfect numbers end in 6; the next four end in 8.

Notice that every perfect number in the preceding list is even and is of the form $2^{p-1}(2^p - 1)$, where p and $2^p - 1$ are primes. We should be doubly impressed that Euclid proved that every such number is a perfect number, as the following theorem confirms.

THEOREM 8.9 **(Euclid)** If n is an integer ≥ 2 such that $2^n - 1$ is a prime, then $N = 2^{n-1}(2^n - 1)$ is a perfect number.

PROOF

Since $2^n - 1$ is a prime, $\sigma(2^n - 1) = 1 + (2^n - 1) = 2^n$. Because σ is multiplicative,

$$\begin{aligned}\sigma(N) &= \sigma(2^{n-1})\sigma(2^n - 1) = (2^n - 1)(2^n) \\ &= 2 \cdot 2^{n-1}(2^n - 1) = 2N\end{aligned}$$

Thus, N is a perfect number, as anticipated. ■

About 2000 years after Euclid's discovery, Euler proved that the converse of this theorem is also true: If $N = 2^{n-1}(2^n - 1)$ is an even perfect number, then $2^n - 1$ is a prime. Theorems 8.9 and 8.10 categorically characterize even perfect numbers.

THEOREM 8.10 **(Euler)** If $N = 2^{n-1}(2^n - 1)$ is an even perfect number, then $2^n - 1$ is a prime.

PROOF

Let N be of the form $2^e s$, where s is odd and $e \geq 1$. Since N is perfect,

$$\sigma(N) = 2N = 2^{e+1}s$$

Clearly, $(2^e, s) = 1$, so

$$\begin{aligned}\sigma(N) &= \sigma(2^e s) = \sigma(2^e)\sigma(s) \\ &= (2^{e+1} - 1)\sigma(s)\end{aligned}$$

Thus,

$$2^{e+1}s = (2^{e+1} - 1)\sigma(s) \tag{8.3}$$

Since $(2^{e+1}, 2^{e+1} - 1) = 1$, it follows by Corollary 3.4 that $2^{e+1}|\sigma(s)$, so $\sigma(s) = 2^{e+1}t$ for some positive integer t. Substituting for $\sigma(s)$ in equation (8.3),

$$2^{e+1}s = (2^{e+1} - 1)2^{e+1}t \tag{8.4}$$

$$s = (2^{e+1} - 1)t \tag{8.5}$$

This implies $t|s$ and $t < s$, since $t = s$ implies $e = 0$, a contradiction.

We will now show that $t = 1$. To this end, equation (8.5) can be rewritten as

$$s + t = 2^{e+1}t$$

$$s + t = \sigma(s) \tag{8.6}$$

This shows t is the sum of the proper factors of s, but, by equation (8.5), t is itself a proper factor of s. So, for the relationship (8.6) to hold, t must be 1.

Thus, $s + 1 = \sigma(s)$, so s has exactly two positive factors 1 and s. Consequently, $s = 2^{e+1} - 1$ must be a prime.

Thus, $N = 2^e(2^{e+1} - 1)$, where $2^{e+1} - 1$ is a prime. ■

? Although this theorem furnishes a remarkable formula for constructing even perfect numbers, it is not known whether there are infinitely many even perfect numbers; the answer has eluded number theorists all over the world in spite of their relentless pursuit.

We now present a fascinating problem proposed in 1990 by Peter L. Montgomery of the University of California at Los Angeles and John L. Selfridge of Northern Illinois University at DeKalb. It explicitly identifies a very special class of even perfect numbers. The solution[†] is a bit lengthy and needs to be followed carefully.

EXAMPLE 8.15 Find all perfect numbers of the form $n^n + 1$.

SOLUTION

Let $N = n^n + 1$.

[†] The solution presented here is based on the one by D. E. Iannucci and G. L. Cohen of Temple University. Selfridge raised the problem for odd N at the 1990 Western Number Theory Conference after solving it for N even. Montgomery solved it during the conference; this is why both are credited with proposing the problem.

case 1 Let n be odd. Since N is an even perfect number, so N must be of the form $N = 2^{m-1}(2^m - 1)$, where $2^m - 1$ is a prime.

Clearly, N can be factored as $N = n^n + 1 = (n+1)r$, where $r = n^{n-1} - n^{n-2} + \cdots - n + 1$. We now claim that $(n+1,\ r) = 1$. To show this, notice that since n is odd, r is odd and $n+1$ is even. Let $n + 1 = 2^s t$, where t is an odd integer ≥ 1. Then $N = 2^s tr$, where both t and r are odd. Since N is an even perfect number, this is possible only if $t = 1$; so $n + 1 = 2^s$ and hence $(n+1,\ r) = 1$. (Notice that if $r = 1$, then $N = n^n + 1 = n + 1$; so $n = 1$. Then $N = 2$, which is not a perfect number.)

Since $N = 2^{m-1}(2^m - 1) = (n+1)r = 2^s r$, where $2^m - 1$ is a prime and r is odd, $2^s = 2^{m-1} = n + 1$ and $r = 2^m - 1 = 2 \cdot 2^{m-1} - 1 = 2(n+1) - 1 = 2n + 1$. Therefore,

$$N = n^n + 1 = (n+1)(2n+1) = 2n^2 + 3n + 1$$

This yields

$$n^n = 2n^2 + 3n$$
$$n^{n-1} = 2n + 3$$

Since n is an integer, this equation has a unique solution 3. (See Figure 8.2. It can be verified algebraically also; see Exercise 9.) Then $N = 3^3 + 1 = 28$. Thus, 28 is the only even perfect number of the desired form.

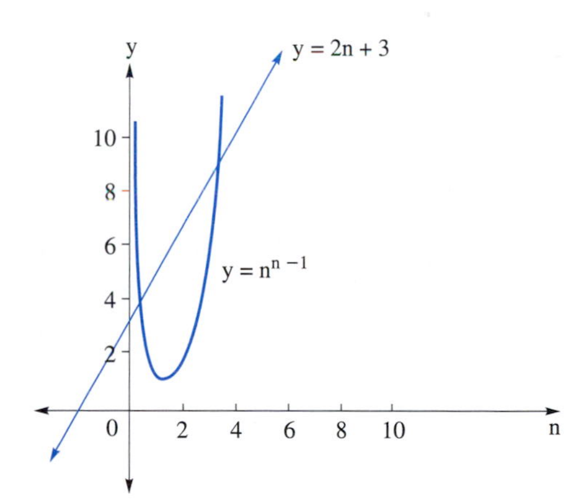

Figure 8.2

case 2 Let n be even, say, $n = 2k$. Then N is odd, n^n is a square, and $n^n \equiv -1 \pmod{N}$.

We claim that $3 \nmid N$, for suppose $3|N$. Then $n^n \equiv -1 \pmod 3$; that is,

$$\begin{aligned} (2k)^{2k} &\equiv -1 \pmod 3 \\ 4^k \cdot k^{2k} &\equiv 2 \pmod 3 \\ 1 \cdot k^{2k} &\equiv 2 \pmod 3 \\ k^{2k} &\equiv 2 \pmod 3 \end{aligned} \tag{8.7}$$

Clearly, $k \not\equiv 0$ or 1 modulo 3. If $k \equiv 2 \pmod 3$, then congruence (8.7) yields

$$\begin{aligned} 2^{2k} &\equiv 2 \pmod 3 \\ 4^k &\equiv 2 \pmod 3 \\ 1 &\equiv 2 \pmod 3, \quad \text{a contradiction} \end{aligned}$$

So k cannot be congruent to 0, 1, or 2, which is absurd. Thus, $3 \nmid N$.

By Touchard's theorem,[†] $N = 12m + 1$ or $36m + 9$ for some integer m. If $N = 36m + 9$, then $3|N$, a contradiction. So $N = 12m + 1$; that is, $n^n = 12m$. Since $3|12m$, $3|n^n$, so $3|n$. Thus, $2|n$ and $3|n$, so $6|n$.

Let $N = a^6 + 1$, where $a = n^{n/6} > 1$. Then N can be factored as

$$N = (a^2 + 1)(a^4 - a^2 + 1) \tag{8.8}$$

We will now see that these factors of N are relatively prime. To this end, let p be a common prime factor of the two factors $a^2 + 1$ and $a^4 - a^2 + 1$. Since

$$\begin{aligned} a^4 - a^2 + 1 &= (a^4 + 2a^2 + 1) - 3a^2 \\ &= (a^2 + 1)^2 - 3a^2 \\ &= (a^2 + 1)^2 - 3(a^2 + 1) + 3 \end{aligned}$$

$p|3$; that is, $p = 3$. This implies $3|N$, a contradiction; so the factors $a^2 + 1$ and $a^4 - a^2 + 1$ are relatively prime. Besides, since N is odd, both factors are also odd.

Since N is perfect and σ is multiplicative, equation (8.8)

$$\sigma(N) = \sigma(a^2 + 1) \cdot \sigma(a^4 - a^2 + 1)$$

That is,

$$2N = \sigma(a^2 + 1) \cdot \sigma(a^4 - a^2 + 1)$$

Since N is odd, one of the factors on the RHS must be odd. But, if m and $\sigma(m)$ are both odd, then m is a square. This implies that either $a^2 + 1$ or $a^4 - a^2 + 1$ is a square.

[†] Every odd perfect number is of the form $12m + 1$ or $36m + 9$.

But $a^2 < a^2 + 1 < (a+1)^2$ and $(a^2 - 1)^2 < a^4 - a^2 + 1 < (a^2)^2$, so neither can be a square, a contradiction.

Consequently, there are no odd perfect numbers of the form $n^n + 1$.

Thus, 28 is the only perfect number of the desired form. ■

Odd Perfect Numbers

The question remains unanswered as to whether there are any odd perfect numbers N. Although a host of conditions that N must satisfy have been established, no one has been successful in finding one, in spite of large computer searches with modern supercomputers. For example, in 1953, J. Touchard of France established that N must be of the form $12k + 1$ or $36k + 9$. Fifty years later, W. Chau of Soft Techies Corporation, E. Brunswick, New Jersey, showed that if N is of the form $36k + 9$, then it must be of the form $108k + 9$, $108k + 35$, or $324k + 81$; it must have at least eight different prime factors (E. Z. Chein, 1979; P. Hagis, 1980); if N has exactly eight distinct prime factors, then the smallest prime factor must be 3, 5, or 7; N must be of the form $p^{4a+1}n^2$, where p is a prime of the form $4m + 1$, $a \geq 0$, and $p \nmid n$; and in 1991 R. P. Brent, G. L. Cohen, and H. J. J. te Riele showed that it must be greater than 10^{300}. In 1998, G. L. Cohen of the University of Technology, Sydney, and P. Hagis, Jr., of Temple University proved that the largest prime factor of an odd perfect number exceeds 10^6; and three years earlier, D. E. Iannucci showed that the second prime factor exceeds 10^4 and the third prime factor exceeds 100. In 2000, Paul A. Weiner of St. Mary's University of Minnesota established that if $3\sigma(n) = 5n$ for some integer n, then $5n$ is an odd perfect number. There is, however, a strong belief in the mathematical community that there may not be any odd perfect numbers.

By Theorem 8.10, the search for even perfect numbers boils down to finding primes of the form $2^m - 1$, so we need to examine them closely. By Lemma 7.3, if m is composite, then $2^m - 1$ is also composite. Therefore, m must be a prime for $2^m - 1$ to be a prime. We will analyze such numbers in the following section.

EXERCISES 8.3

1. If $2p$ is a perfect number, show that $p = 3$. Assume p is a prime.

Let $n = 2^{p-1}(2^p - 1)$ be an even perfect number, where p is a prime. Show that

2. $\varphi(n) = 2^{p-1}(2^{p-1} - 1)$
3. $\varphi(n) = n - 2^{2p-2}$
4. Show that $n = 2^{10}(2^{11} - 1)$ is not a perfect number.
5. Show that every even perfect number is a triangular number. (J. Broscius, 1652)
6. Show that the sum of the first $2^p - 1$ positive integers is an even perfect number, where p and $2^p - 1$ are primes.

7. Show that pq is not a perfect number, where p and q are distinct primes.
8. Show that p^e is not a perfect number, where p is a prime.
9. Solve the equation $n^{n-1} = 2n + 3$, where n is a positive integer.

Prove each.

10. The sum of the cubes of the first n odd positive integers is $n^2(2n^2 - 1)$.
11. Every even perfect number $2^{p-1}(2^p - 1)$ is the sum of the cubes of the first $2^{(p-1)/2}$ odd positive integers.
12. The digital root of every even perfect number > 6 is one.
13. The product of two even perfect numbers cannot be a perfect number.
14. Let n be a perfect number. Then $\sum_{\substack{0<d<n \\ d|n}} (d/n) = 1$.
15. Let $n = 2^{p-1}(2^p - 1)$ be an even perfect number. Then $\prod_{d|n} d = n^p$. (P. A. Lindstrom, 2004)
16. Suppose $\dfrac{\sigma(n)}{n} = \dfrac{5}{3}$. Then $5n$ is an odd perfect number. (P. A. Weiner, 2000)

Every odd perfect number n, if it exists, is of the form $n = p^e m^2$, where p is an odd prime and m is odd. Using this fact, prove each.

17. $p \equiv e \equiv 1 \pmod 4$
18. $n \equiv p \pmod 8$
19. $n \equiv 1 \pmod 4$

A positive integer n is **deficient** if $\sigma(n) < 2n$ and **abundant** if $\sigma(n) > 2n$. Such numbers were also originally studied by the Pythagoreans. For example, $\sigma(9) = 1 + 3 + 9 = 13 < 2 \cdot 9$, so 9 is a deficient number, whereas $\sigma(12) = 1 + 2 + 3 + 4 + 6 + 12 = 28 > 2 \cdot 12$, so 12 is an abundant number. Determine if each number is deficient or abundant, where $M_p = 2^p - 1$.

20. 23 21. 88 22. 24
23. 315 24. $2^{10}M_{11}$ 25. $2^{22}M_{23}$
26. $2^p(2^p - 1)$, where p and $2^p - 1$ are primes.
27. $2^{k-1}(2^k - 1)$, where $2^k - 1$ is a composite.

A positive integer n is ***k*-perfect** if $\sigma(n) = kn$, where $k \geq 2$. (A perfect number is 2-perfect.) Verify the following.

28. 672 is 3-perfect. 29. 30,240 is 4-perfect.

Two positive integers m and n are **amicable** (or **friendly**) if $\sigma(m) - m = n$ and $\sigma(n) - n = m$, that is, if $\sigma(m) = m + n = \sigma(n)$. For example, $\sigma(220) - 220 = 1 + 2 + 4 + 5 + 10 + 11 + 20 + 22 + 44 + 55 + 110 = 284$ and $\sigma(284) - 284 = 1 + 2 + 4 + 71 + 142 = 220$, so 220 and 284 are an amicable pair, found by Pythagoras around 540 B.C. Interestingly enough, this pair is mentioned in the Bible in Genesis 32:14. The next smallest pair, 1184 and 1210, was discovered in 1866 by a 16-year-old Italian student, B. N. I. Paganini. It is not known if there are infinitely many amicable pairs. Verify that each is an amicable pair. (The pairs 2620 and 2924, and 6232 and 6368 were discovered during 1747–1750 by Euler.)

30. 1184, 1210 31. 2620, 2924 32. 6232, 6368

The Arabian mathematician Thabit ben Korrah developed an algorithm for constructing amicable numbers: If $a = 3 \cdot 2^n - 1$, $b = 3 \cdot 2^{n-1} - 1$, and $c = 9 \cdot 2^{2n-1} - 1$ are all primes, then $2^n ab$ and $2^n c$ are amicable, where $n \geq 2$.

33. Find two amicable pairs using Korrah's method.
34. Verify that $2^n ab$ and $2^n c$ are amicable numbers.
35. In 1978, Elvin J. Lee made an interesting observation about the numbers 220 and 284: The sum of the first 17 primes is $2 \cdot 220$ and that of their squares is $59 \cdot 284$. Verify this.
36. A positive integer n is **superperfect** if $\sigma(\sigma(n)) = 2n$. Verify that 16 is superperfect.

Let $\nu(n) = \prod_{\substack{d|n \\ d<n}} d$, the product of the proper divisors of n. Some positive integers have the property that $\nu(n) = n^k$ for some positive integer k. For example, $\nu(12) = 1 \cdot 2 \cdot 3 \cdot 4 \cdot 6 = 12^2$. Verify that $\nu(n) = n^k$ for each integer n.

37. 20 38. 45 39. 24 40. 48
41. Verify that $\nu(pq) = pq$, where p and q are distinct primes.
42. Find a formula for $\nu(p^a q^b)$.

Prove each, where p is a prime and n any positive integer.

43. Every prime is deficient.
44. The product of two distinct odd primes is deficient.
45. Any positive power of a prime is deficient.
46. If $2 \nmid p$ and $2p$ is deficient, then $p \geq 5$.
47. A number of the form $2^e p$, where $2 \nmid p$, is abundant if $e \geq \lfloor \lg(p+1) \rfloor - 1$.
48. If n is p-perfect and $p \nmid n$, then pn is $(p+1)$-perfect.
49. Every number of the form 2^k, where $2^{k+1} - 1$ is a prime, is superperfect.
50. Let $n = 2^{k-1}(2^k - 1)$ be an even perfect number. Then $\sigma(\sigma(n)) = 2^k(2^{k+1} - 1)$.
51. If m and n are amicable, then

$$\left(\sum_{d|m} \frac{1}{d}\right)^{-1} + \left(\sum_{d|n} \frac{1}{d}\right)^{-1} = 1$$

52. If n is 3-perfect, and $3 \nmid n$, then $3n$ is 4-perfect.
53. $\nu(p^e)$ is a power of p^e if and only if e is an odd integer ≥ 3.
54. $\nu(p^a q^b)$ is a power of $p^a q^b$ if and only if $(a+1)(b+1)$ is an even integer ≥ 4.
55. $\nu(n)$ is a power of n if and only if $\tau(n)$ is an even integer ≥ 4.
56. $\nu(n) = n^{\tau(n)/2-1}$

The **harmonic mean** m of the numbers $a_1, a_2, \ldots, a_n$ is the reciprocal of the arithmetic mean of their reciprocals; that is,

$$\frac{1}{m} = \frac{1}{n}\left(\sum_{i=1}^{n} \frac{1}{a_i}\right)$$

57. Show that the harmonic mean $h(n)$ of the positive factors of a positive integer n is given by $h(n) = n\tau(n)/\sigma(n)$.
58. Find the harmonic mean of the positive factors of a perfect number n.

8.4 Mersenne Primes

It was originally thought that if m is a prime, then $2^m - 1$ is also a prime. However, in 1536, Hudalrichus Regius found that it fails when $m = 11$: $2^{11} - 1 = 2047 = 23 \cdot 89$.

Mersenne Primes

Numbers of the form $2^m - 1$ were studied extensively by the French mathematician and Franciscan monk Marin Mersenne. Accordingly, they are called **Mersenne numbers**, a name given to them by W. W. Rouse Ball of Trinity College, Cambridge, England. Primes of the form $M_p = 2^p - 1$ are **Mersenne primes**.

In 1644, Mersenne wrote in his *Cogitata Physica-Mathematica* that M_p is a prime for $p = 2, 3, 5, 7, 13, 17, 19, 31, 67, 127$, and 257 and composite for other primes < 257. No one knew how he arrived at this claim. His statement contains some omissions and errors. In any case, it took over three centuries to settle his claim.

In 1814, Peter Barlow wrote in *A New Mathematical and Philosophical Dictionary*: "Euler ascertained that $2^{31} - 1 = 2{,}147{,}483{,}647$ is a prime number; and this

***Marin Mersenne** (1588–1648), "best known as the priest-scientist who facilitated the cross-fertilization of the most eminent minds of his time," was born in Soultière, France. He was baptized on the same day and christened as Marin since it was the feast of the Nativity of Mary. After attending the College de Mans and the Jesuit College at La Flêche, he went to Paris to study theology and became a Minim friar in 1611. Science began to dominate his religious thought, and in 1624 he accepted the Copernican theory that the sun, and not the earth, was the center of the universe. Mersenne corresponded with many scientists and philosophers, including René Descartes; his residence became a meeting place for such eminent thinkers as Fermat, Girard Desargues, and Fr. Pierre Gassendi. He even came to the defense of Descartes and Galileo when their works were attacked by the Church. He also made important contributions to music and acoustics.*

is the greatest at present known to be such, and, consequently, the last of the above perfect numbers, which depends upon this, is the greatest perfect number known at present, and probably the greatest that ever will be discovered; for, as they are merely curious without being useful, it is not likely that any person will attempt to find one beyond it."† As it turns out, Barlow underestimated human curiosity, and he could not have foreseen the power of computers.

In 1876, Lucas proved that M_{67} is composite, although he did not provide any factors; but in October 1903, the American mathematician Frank Nelson Cole provided a factorization:

$$2^{67} - 1 = 193,707,721 \times 761,838,257,287$$

It is said that Cole spent his Sunday afternoons for 20 years trying to find the two factors.

In 1883, I. M. Pervushin showed that $M_{61} = 2^{61} - 1$ is a prime, which Mersenne missed. R. E. Powers discovered that $2^{89} - 1$ and $2^{107} - 1$ are primes, in 1911 and 1914, respectively. In 1922, M. Kraitchik showed that $M_{257} = 2^{257} - 1$ is composite. Ironically, on March 27, 1936, the *New York Herald Tribune* erroneously reported that $2^{257} - 1$ was discovered to be a prime by Samuel I. Krieger of Chicago; it was shown in 1931, and then in 1947 using a desk calculator, and then reconfirmed in 1952 using computers, that $2^{257} - 1$ is actually a composite.

?

The question of whether there are infinitely many Mersenne primes is still unanswered. If there are, then there would be an infinitude of even perfect numbers and hence of perfect numbers.

† D. Shanks, *Solved and Unsolved Problems in Number Theory*, Vol. 1, Spartan Books (1962).

It is also not known if every M_p is square-free; it also remains unresolved.

Mersenne primes M_p appear to be scarce as p increases. In 1963, Donald B. Gillies of the University of Illinois conjectured that there are about two such primes p in the interval $[n, 2n]$. Interestingly enough, his conjecture is consistent with the observed frequency of primes p. It also agrees well with the Eberhart conjecture that for the ith Mersenne prime M_p, $p \approx 1.5^i$. For example, when $i = 23$, $p \approx 1.5^{23} \approx 11223$, which is not that far from the actual value of $p = 11213$.

Modern computers have become a powerful tool for finding larger Mersenne primes. For instance, the next five larger Mersenne primes corresponding to $p = 521, 607, 1279, 2203$, and 2281 were discovered in 1952; the next one corresponding to $p = 3217$ in 1957; the next two with $p = 4253$ and 4423 in 1961; the next three were found in 1963 at the University of Illinois at Urbana-Champaign (see the meter stamp in Figure 8.3). In 1971, a still larger prime was found by Bryant Tuckerman of International Business Machines (IBM); see Figure 8.4, which shows the top portion of IBM's office envelope spreading its news.

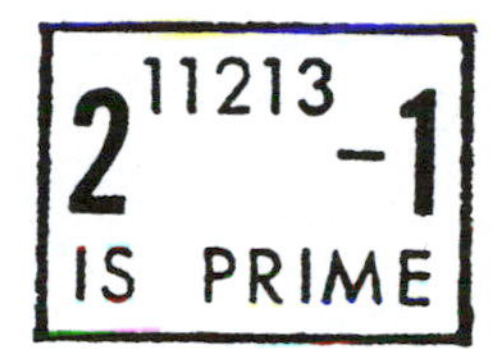

Figure 8.3

Figure 8.4

As of 1994, 33 Mersenne primes had been discovered; the 33rd largest known Mersenne prime, M_{859433}, was discovered in 1993 by David Slowinski of Harwell Laboratory, England; it took 7.2 hours on a Cray C90 supercomputer to determine its primality; its decimal value is 258,716 digits long (see Figure 8.5). The next two larger ones were also discovered by Slowinski.

The largest known prime by the year 1999, $M_{6972593}$, was discovered by N. Hajrawala, G. Woltman, and S. Kurowski; its decimal value is 2,098,960 digits long. So the largest known even perfect number in 1999 was $2^{6972592}(2^{6972593} - 1)$; its decimal value is 4,197,919 digits long.

Worldwide Number Search Inspires Prime Competition

Baltimore Sun

BALTIMORE – It just didn't add up.

Technicians at US West, the Denver-based telephone company, couldn't understand last May why directory-assistance computers were grinding away for minutes to find the numbers when they normally needed seconds. At one point, the slowdown threatened to shutter the company's Phoenix service center.

Alarmed that hackers were afoot, US West scrambled its Intrusion Response Team. The squad of computer specialists combed through the company's computer network and found a mysterious software program running on more than 2,500 machines.

The case has brought to light a mathematical treasure hunt taking place around the globe. It's called the Great Internet Mersenne Prime Search—GIMPS for short—and it has captured the imagination of everyone from PhD's to grade-schoolers. Their goal is to hunt down world-record prime numbers.

Most record-seekers now focus their attention on Mersenne primes. These special prime numbers are the Hope Diamonds of the mathematical world, as large as they are rare. Named after the 17th-century French mathematician Marin Mersenne, they are primes generated by the formula 2 to the nth power minus 1, where n is also a prime.

Just 37 have been found in all of human history. The most recent was unearthed in January by a 19-year-old student using PCs in the computer lab at California State University, Dominguez Hills, in Los Angeles County. It had a whopping 909,526 digits, making it the largest Mersenne prime yet found.

Until recently it took a supercomputer to flush out these elephantine numbers. But in 1996 a retired Orlando, Fla., computer programmer and a California engineer devised a way to use home computers to find them. The idea was to link PCs through the Internet, turning them into a single, massively parallel supercomputer. They wrote the software, and the Great Internet Mersenne Prime Search was born.

Today more than 4,000 number lovers around the world are using GIMPS software to hunt for Mersenne primes, each vying for 15 minutes of fame and a $1,100 cash prize. Collectively, the group churns through 280 billion calculations per second, a computing punch roughly equivalent to five of the world's most powerful supercomputers working full-steam.

The GIMPS software is designed to look for primes when its host PC isn't occupied with other tasks. On a Web site (www.mersenne.org), it gets an untested number and then grinds through a formula to determine whether it is prime. The calculations can take days or weeks to complete. If the number turns out to be a dud, the process is repeated. But if it turns out to be prime, it should mean fame and fortune.

Figure 8.5

The Mersenne prime $M_{25964951}$ was discovered in 2005 by Martin Nowak, an eye surgeon in Germany (see Figure 8.6). It is 7,816,230 digits long and took 50 days of computations on his 1 GHz Pentium 4 computer. Today, the largest known such prime is $2^{30402457} - 1$ with 9,152,052 digits; it was discovered on December 15, 2005, by C. Cooper and S. R. Boone of Central Missouri State University.

Table 8.3 lists the 43 known Mersenne primes M_p, the number of decimal digits in each, the number of decimal digits in the corresponding perfect numbers $2^{p-1}M_p$, the years of discovery of M_p, and their discoverers.

Rank	Prime, p	No. of Digits in M_p	No. of Digits in $2^{p-1}M_p$	Year of Discovery	Discoverer(s)
1	2	1	1	unknown	Pythagoreans
2	3	1	2	unknown	Pythagoreans
3	5	2	3	unknown	Pythagoreans
4	7	3	4	unknown	Pythagoreans
5	13	4	8	15th century	H. Regius
6	17	6	10	1588	P. A. Cataldi
7	19	6	12	1588	P. A. Cataldi
8	31	10	19	1772	L. Euler
9	61	19	37	1883	I. M. Pervushin
10	89	27	54	1911	R. E. Powers
11	107	33	65	1914	R. E. Powers & E. Fauquembergue
12	127	39	77	1876	E. Lucas
13	521	157	314	1952	D. H. Lehmer
14	607	183	366	1952	D. H. Lehmer
15	1279	386	770	1952	D. H. Lehmer
16	2203	664	1327	1952	D. H. Lehmer
17	2281	687	1373	1952	D. H. Lehmer
18	3217	969	1937	1957	H. Riesel
19	4253	1281	2561	1961	A. Hurwitz
20	4423	1332	2663	1961	A. Hurwitz
21	9689	2917	5834	1963	D. B. Gillies
22	9941	2993	5985	1963	D. B. Gillies
23	11,213	3376	6751	1963	D. B. Gillies
24	19,937	6002	12,003	1971	B. Tuckerman
25	21,701	6533	13,066	1978	L. Nickel & C. Noll
26	23,209	6987	13,973	1979	C. Noll
27	44,497	13,395	26,790	1979	D. Slowinski & H. Nelson
28	86,243	25,962	51,924	1983	D. Slowinski
29	110,503	33,265	66,530	1988	W. N. Colquitt & L. Welch, Jr.
30	132,049	39,751	79,502	1983	D. Slowinski
31	216,091	65,050	130,100	1985	D. Slowinski
32	756,839	227,832	455,663	1992	D. Slowinski & P. Gage
33	859,433	258,716	517,430	1993	D. Slowinski & P. Gage
34	1,257,787	378,632	757,263	1996	D. Slowinski & P. Gage
35	1,398,269	420,921	841,842	1996	J. Armengaud & G. Woltman
36	2,976,221	895,932	1,791,864	1997	G. Spence & G. Woltman
37	3,021,377	900,526	1,819,050	1998	R. Clarkson *et al.*
38	6,972,593	2,098,960	4,197,919	1999	N. Hajrawala *et al.*
39	13,466,917	4,053,946	8,107,892	2001	M. Cameron
40	20,996,011	6,320,430	12,640, 858	2003	M. Shafer
41	24,036,583	7,235,733	14,471,465	2004	J. Findley
42	25,964,951	7,816,230	15,632,458	2005	M. Novak
43	30,402,457	9,152,052	18,304,103	2005	C. Cooper & S. R. Boone
44	32,582,657	9,808,358	19,616,715	2006	C. Cooper & S. R. Boone

Table 8.3 *The 44 known Mersenne primes.*

Prime Number Is Largest Ever
Reuters
EAGAN, Minn. – Scientists announced yesterday that they have discovered the largest prime number found to date—a 258,716-digit behemoth that would take eight newspaper pages to print.

Prime numbers are those that can be divided only by themselves or 1. Simple examples are 2, 3, 5, 7 and 11. There are an infinite number of them, but they do not occur in a regular sequence, meaning that supercomputers are needed to hunt for them.

Cray Research Inc. said its supercomputer had chased down the new number—two multiplied by itself 859,433 times, minus 1.

The previous largest such number, tracked down in 1992, had 227,832 digits.

Figure 8.6

Their discoveries often generated considerable media publicity. In October 1978, for instance, the discovery of the 25th Mersenne prime, M_{21701}, was carried by every news agency in the United States and announced by Walter Cronkite on the CBS Evening News. The discovery of $M_{20,996,011}$ by 26-year-old Michigan State University chemical engineering student M. Shafer was announced by Peter Jennings on ABC World News Tonight on December 11, 2003. The discovery of $M_{25,964,951}$ made *The New York Times* on March 29, 2005 (see Figure 8.7).

The discovery of M_{21701} by L. Nickell and C. Noll was reported erroneously in *The Times of London* on November 17, 1978: "Two 18-year-old American students have discovered with the help of a computer at California State University the biggest known prime number, the number two to the 21,701st power." Fortunately, a correction was soon published by *The Times*.

The Great Internet Mersenne Prime Search

The Great Internet Mersenne Prime Search (GIMPS), based in Orlando, Florida, was formed in 1996 by George Woltman for discovering record Mersenne primes. Like the recent discoverers, with a powerful personal computer, you can join the global search by downloading the necessary software for free at www.Mersenne.org. You can share the thrill of discovering larger primes and make history in the process.

The *Electronic Frontier Foundation* has announced a $100,000 cash award for the discovery of the first 10-million-digit prime. The GIMPS participant who discovers it will receive $50,000; charity will receive $25,000; and the rest will fund new discoveries. So join the fun.

Now, Can You Find Its Square Root?

Kenneth Chang

An eye surgeon in Germany has discovered the world's largest known prime number—or at least his computer did.

The surgeon, Dr. Martin Nowak of Michelfeld, is among thousands of participants in the Great Internet Mersenne Prime Search, one of several big projects that tap idle computers worldwide.

Last month, Dr. Nowak's Pentium 4 computer concluded that a number it had been crunching on for more than 50 days was indeed prime with only two integer divisors, 1 and itself.

A different computer using different software verified the result.

The number, rendered in exponential shorthand, is $2^{25,964,951} - 1$. It has 7,816,230 digits, and if printed in its entirety, would fill 235 pages of this newspaper.

In addition, it falls in a rare category of primes known as Mersenne primes, which can be written as $2^n - 1$, where n is also prime.

The first few Mersenne primes are easily verifiable—inserting 2, 3 and 5 for n produces 3, 7, and 31, all prime—but the math quickly becomes overwhelming for larger values.

In 1644, Marin Mersenne a French monk, published a list of 11 prime numbers—the highest being 257—for which he asserted that $2^n - 1$ was also prime.

That list was not fully checked until 1947, three centuries later. Mersenne turned out to be wrong about two numbers on his list and had missed three others, but his name still remains attached to the concept. Even with computers to speed up the search, Dr. Nowak's number is still only the 42nd Mersenne prime to be found.

The announcement did not, however, cause much of a stir because what mathematicians really want to know is: Are there an infinite number of such numbers? "Finding an additional prime doesn't enlighten us very much," said Dr. Andrew M. Odlyzko, a mathematician at the University Minnesota.

The search nevertheless goes on, on about 75,000 computers. Begun in 1996 by George Woltman, a computer scientist, the project has discovered eight Mersenne primes.

For those who want to join, to the free software is available at www.mersenne.org. The Web site informs would-be volunteers, "Your chance of finding a new Mersenne prime is about 1 in 150,000."

Figure 8.7

A New Mersenne Conjecture

In 1989, P. T. Bateman and J. L. Selfridge of Northern Illinois University and S. S. Wagstaff, Jr., of Purdue University made an interesting conjecture about Mersenne primes:

If two of the following statements about an odd prime p are true, then the third one is also true:

- $p = 2^k \pm 1$ or $p = 4^k \pm 3$.
- M_p is prime.
- $(2^p + 1)/3$ is prime.

For example, let $p = 7$. Then $7 = 2^3 - 1$ and $(2^7 + 1)/3 = 43$ is prime. As we already know, M_7 is a prime. Thus, the conjecture holds when $p = 7$. In fact, it has been verified that the conjecture holds for all $p < 100{,}000$.

Number of Digits in M_p

We can easily pre-determine the number of digits in the Mersenne number M_p. First, recall that every odd prime p is of the form $4k + 1$ or $4k + 3$. If $p = 4k + 1$, then $2^p = 2^{4k+1} = (2^4)^k \cdot 2 \equiv 6^k \cdot 2 \equiv 6 \cdot 2 \equiv 2 \pmod{10}$; likewise, if $p = 4k + 3$, then $2^p \equiv 8 \pmod{10}$. Thus, $2^p = M_p + 1$ ends in 2 or 8. Consequently, M_p ends in 1 or 7 and hence has the same number of digits as 2^p.

To compute the number of digits in 2^p, notice that $\log 2^p = p \cdot \log 2$. Therefore,

$$\text{Number of digits in } 2^p = 1 + \text{characteristic of } p \log 2 = \lceil p \log 2 \rceil$$

For example, $M_{25964951}$ contains $\lceil 25964951 \log 2 \rceil = \lceil 0.301029995664 \times 25964951 \rceil = 7{,}816{,}230$ digits, as expected.

Interestingly, the largest known composite Mersenne number is M_p with $p = 39051 \times 2^{6001} - 1$, discovered by W. Keller in 1987.

It follows from Supplementary Exercise 14 in Chapter 3 that any two distinct Mersenne numbers are relatively prime; that is, if p and q are distinct primes, then $(M_p, M_q) = 1$.

Primality of Mersenne Numbers

A host of conditions exist for testing the primality of Mersenne numbers. The following theorem presents a possible prime factor of the Mersenne number M_p. Its proof involves quadratic residues, developed in Chapter 11, so we omit it. A simple proof is given in the classic book *An Introduction to the Theory of Numbers* by G. H. Hardy and E. M. Wright.

THEOREM 8.11 **(Euler)** Let $p = 4k + 3$ be a prime, where $k > 1$. Then $2p + 1$ is a prime if and only if $2^p \equiv 1 \pmod{2p + 1}$. ■

It follows by this theorem that if $p = 4k + 3$ and $2p + 1$ are primes, where $k > 1$, then $2p + 1 | M_p$ and M_p is composite.

The following example demonstrates an interesting application of this theorem. It was proposed as a problem in 1988 by David Grannis of Vancouver, British Columbia.

EXAMPLE 8.16 Find a factor of the Mersenne number $M_{1000151}$.

SOLUTION

Both $p = 1{,}000{,}151 = 4 \cdot 250{,}037 + 3$ and $2p+1 = 2{,}000{,}303$ are primes. Therefore, by Theorem 8.11, $2{,}000{,}303 | M_{1000151}$. (This simple solution was provided in 1991 by Warut Roonguthai of Bangkok, Thailand.) ■

Before presenting a primality test for Mersenne primes in Theorem 8.12, we prepare the way with Lemma 8.2. But first, let us look at an example.

Let a and n be relatively prime positive integers. Then, by Euler's theorem, $a^{\varphi(n)} \equiv 1 \pmod{n}$. Often, however, there can be exponents k smaller than $\varphi(n)$ such that $a^k \equiv 1 \pmod{n}$, as the following example illustrates.

EXAMPLE 8.17 Let $n = 12$; then $\varphi(n) = \varphi(12) = 4$. The least residues a modulo 12 that are relatively prime to 12 are 1, 5, 7, and 11; by Euler's theorem $a^{\varphi(n)} = a^4 \equiv 1 \pmod{12}$. But $1^2 \equiv 1 \pmod{12}$, $5^2 \equiv 1 \pmod{12}$, $7^2 \equiv 1 \pmod{12}$, and $11^2 \equiv 1 \pmod{12}$; so $k = 2$ is the least positive exponent such that $a^k \equiv 1 \pmod{12}$. [Notice that $k|\varphi(n)$.] ■

More generally, we have the following result.

LEMMA 8.2 Let a, m, and n be positive integers with $(a, n) = 1$, and k the smallest positive integer such that $a^k \equiv 1 \pmod{n}$. Then $a^m \equiv 1 \pmod{n}$ if and only if $k|m$.

PROOF

Suppose $a^m \equiv 1 \pmod{n}$. By the division algorithm, $m = kq + r$ for some integers q and r, where $0 \le r < k$. Then

$$a^m = a^{kq+r} = (a^k)^q \cdot a^r$$

Since $a^k \equiv 1 \pmod{n}$ and $a^m \equiv 1 \pmod{n}$, this yields

$$1 \equiv 1^q \cdot a^r \pmod{n}$$

$$1 \equiv a^r \pmod{n}$$

That is,

$$a^r \equiv 1 \pmod{n}, \qquad \text{where } 0 \le r < k$$

If $r > 0$, this would contradict the minimality of k. So $r = 0$ and hence $m = kq$. Thus, $k|m$.

Conversely, let $k|m$, so let $m = kq$ for some integer q. Then

$$a^m = a^{kq} = (a^k)^q$$
$$\equiv 1^q \equiv 1 \pmod{n}$$

Thus, $a^m \equiv 1 \pmod{n}$ if and only if $k|m$. ■

This lemma has an immediate corollary, and we will revisit it in Section 10.1.

COROLLARY 8.2 Let a and n be relatively prime positive integers, and k the smallest positive integer such that $a^k \equiv 1 \pmod{n}$. Then $k|\varphi(n)$. ■

We can now undertake a primality test for Mersenne primes.

THEOREM 8.12 **(Fermat, 1640)** If p is an odd prime, every prime factor of M_p is of the form $2kp + 1$, where k is a positive integer.

PROOF

Let q be a prime factor of M_p. (Clearly, q is odd.) Then $q|M_p$, so $2^p \equiv 1 \pmod{q}$. Let k be the smallest positive integer such that $2^k \equiv 1 \pmod{q}$. Then, by Lemma 8.2, $k|p$. But $k \neq 1$, for if $k = 1$, then $2^1 \equiv 1 \pmod{q}$; that is, $q = 1$, a contradiction. Therefore, $k = p$; that is, p is the smallest positive integer such that $2^p \equiv 1 \pmod{q}$.

By Fermat's little theorem, $2^{q-1} \equiv 1 \pmod{q}$, so by Lemma 8.2, $p|q - 1$. Let $q - 1 = pm$ for some positive integer m. Since $q - 1$ is even and p is odd, m must also be even, say, $m = 2k$ for some positive integer k. Then $q - 1 = 2pk$; that is, $q = 2kp + 1$.

Thus, if p is odd, every prime factor of M_p is of the form $2kp + 1$. ■

The following two examples illustrate this test.

EXAMPLE 8.18 Verify that M_{11} is a composite number.

PROOF

$M_{11} = 2^{11} - 1 = 2047$. By Theorem 8.12, every prime factor of M_{11} is of the form $22k + 1$. If M_{11} is composite, by Theorem 2.11, it must have a prime factor $\leq \lfloor\sqrt{M_{11}}\rfloor$, that is, ≤ 45. There is exactly one prime of the form $22k + 1$ and ≤ 45, namely, 23. Since $23|M_{11}$, M_{11} is composite. ■

EXAMPLE 8.19 Determine whether M_{19} is a prime.

PROOF

$M_{19} = 2^{19} - 1 = 524287$. If M_{19} is composite, it must have a prime factor $\leq \lfloor\sqrt{M_{19}}\rfloor$, that is, ≤ 724. By Theorem 8.12, every prime factor of M_{19} is of the form $38k + 1$ and ≤ 725; such primes are 191, 229, 419, 457, 571, and 647. None of them divides M_{19}, so M_{19} is a prime. ■

Lucas–Lehmer Test

Next, we turn to the Lucas–Lehmer test, an extremely efficient primality test for Mersenne primes, developed in 1877 by Lucas and then refined in 1930 by the American mathematician Derrick H. Lehmer. Lucas used his version to establish the primality of M_{127}, the largest Mersenne number ever to be checked without the help of a computing device.

The Lucas–Lehmer test, used since 1930 to prove the primality of Mersenne primes, is based on the number sequence $4, 14, 194, 37634, 1416317954, \ldots$; it is defined recursively as follows:

$$S_1 = 4$$
$$S_k = S_{k-1}^2 - 2, \quad k \geq 2$$

According to the test, M_p is a prime if and only if $S_{p-1} \equiv 0 \pmod{M_p}$, where p is an odd prime. Lehmer used this test to prove the primality of M_{521}, M_{607}, M_{1279}, M_{2203}, and M_{2281} using the National Bureau of Standard's Western Automatic Computer (SWAC). He also reconfirmed, using SWAC for 48 seconds, that M_{257} is composite, a task that had taken 700 work-hours 20 years earlier.

The test is formally presented in the following theorem.

THEOREM 8.13 **(Lucas–Lehmer Test)**† Let $p \geq 3$. Then the Mersenne number M_p is prime if and only if $S_{p-1} \equiv 0 \pmod{M_p}$, where S_k is the least residue modulo M_p defined recursively as

$$S_1 = 4$$
$$S_k \equiv S_{k-1}^2 - 2 \pmod{M_p}, \quad k \geq 2$$

■

The next two examples illustrate this test.

† A proof of the test can be found in W. Sierpinski, *Elementary Theory of Numbers*, 2nd edition, North-Holland, Amsterdam, 1988.

***Derrick Henry Lehmer** (1905–1991) was born in Berkeley, California. (His father, Derrick Norman Lehmer, was a professor at Berkeley.) After graduating from Berkeley in physics in 1927, he received his Ph.D. in mathematics from Brown three years later. During the Great Depression, he worked at the California Institute of Technology, the Institute of Advanced Study, and Lehigh; and in 1940 he joined the faculty at Berkeley and remained there until his retirement in 1972.*

Lehmer, known as the father of computational number theory, shared with his wife, Emma, a life-long fascination with number theory. A prolific writer, he published extensively on Lucas functions, primality testing, factoring, power residues, continued fractions, Bernoulli numbers and polynomials, Diophantine equations, cyclotomy, and combinatorics. The Lucas–Lehmer primality test for Mersenne numbers is the result of his investigations into what are now called Lehmer functions, which he discussed in his dissertation.

Lehmer was a founding father of the journal Mathematical Tables and Aids to Computation, *which became* Mathematics of Computation *in 1960.*

EXAMPLE 8.20 Using the Lucas–Lehmer test, verify that M_{13} is a prime.

PROOF

Here $p = 13$ and $M_{13} = 2^{13} - 1 = 8191$. Compute S_2 through S_{12} modulo M_{13}:

$$S_2 \equiv 4^2 - 2 \equiv 14 \pmod{M_{13}} \qquad S_3 \equiv 14^2 - 2 \equiv 194 \pmod{M_{13}}$$
$$S_4 \equiv 194^2 - 2 \equiv 4870 \pmod{M_{13}} \qquad S_5 \equiv 4870^2 - 2 \equiv 3953 \pmod{M_{13}}$$
$$S_6 \equiv 3953^2 - 2 \equiv -2221 \pmod{M_{13}} \qquad S_7 \equiv 2221^2 - 2 \equiv 1857 \pmod{M_{13}}$$
$$S_8 \equiv 1857^2 - 2 \equiv 36 \pmod{M_{13}} \qquad S_9 \equiv 36^2 - 2 \equiv 1294 \pmod{M_{13}}$$
$$S_{10} \equiv 1294^2 - 2 \equiv 3470 \pmod{M_{13}} \qquad S_{11} \equiv 3470^2 - 2 \equiv 128 \pmod{M_{13}}$$
$$S_{12} \equiv 128^2 - 2 \equiv 0 \pmod{M_{13}}$$

Since $S_{12} \equiv 0 \pmod{M_{13}}$, M_{13} is a prime as expected. ■

EXAMPLE 8.21 Using the Lucas–Lehmer test, verify that M_{11} is not a prime.

PROOF

As in the preceding example, we compute S_2 through S_{10} modulo M_{11}. You may verify each:

$$S_2 \equiv 14 \pmod{M_{11}} \qquad S_3 \equiv 194 \pmod{M_{11}} \qquad S_4 \equiv 788 \pmod{M_{11}}$$
$$S_5 \equiv 701 \pmod{M_{11}} \qquad S_6 \equiv 119 \pmod{M_{11}} \qquad S_7 \equiv -170 \pmod{M_{11}}$$
$$S_8 \equiv 240 \pmod{M_{11}} \qquad S_9 \equiv 282 \pmod{M_{11}} \qquad S_{10} \equiv 1736 \pmod{M_{11}}$$

Since $S_{10} \not\equiv 0 \pmod{M_{11}}$, M_{11} is not a prime, as we already knew. ■

To date, the largest known prime that is *not* a Mersenne prime is $27653 \cdot 2^{9167433} + 1$, discovered by Samuel Yates in 2005; it is 2,759,677 digits long.

The following problem was proposed in 1989 by Jeffrey Shallit of Dartmouth College, New Hampshire.

EXAMPLE 8.22 Prove that $\sigma(n)$ is a power of 2 if and only if n is the product of distinct Mersenne primes.

PROOF

Let $n = p_1^{e_1} p_2^{e_2} \cdots p_k^{e_k}$ be the canonical decomposition of n. Then $\sigma(n) = \prod_{i=1}^{k} \frac{p_i^{e_i+1} - 1}{p_i - 1}$. Assume it is a power of 2. Let p^e be an arbitrary prime-power in the canonical decomposition of n. Then $\sigma(p^e) = \frac{p^{e+1} - 1}{p - 1} = p^e + \cdots + p + 1$ must be a power of 2, so both p and e must be odd.

Let $e = 2s + 1$. Then

$$\begin{aligned} \sigma(p^e) &= p^{2s+1} + \cdots + p + 1 \\ &= (p+1)(p^{2s} + p^{2s-2} + \cdots + p^2 + 1) \end{aligned} \tag{8.9}$$

Since $(p+1) | \sigma(p^e)$ and $\sigma(p^e)$ is a power of 2, $p + 1$ must be a power of 2, so p is a Mersenne prime.

It remains to show that $e = 1$; that is, $s = 0$. Suppose $s > 0$. From equation (8.9), since $p^{2s} + p^{2s-2} + \cdots + p^2 + 1$ is a power of 2 and p is odd, s must be odd, so let $s = 2t + 1$. Then

$$\begin{aligned} p^{2s} + p^{2s-2} + \cdots + p^2 + 1 &= p^{4t+2} + p^{4t} + \cdots + p^2 + 1 \\ &= (p^2 + 1)(p^{4t} + p^{4t-4} + \cdots + p^4 + 1) \end{aligned}$$

is a power of 2, so $p^2 + 1$ must be a power of 2. Therefore, $4 | (p^2 + 1)$; that is, $p^2 \equiv -1 \pmod 4$, a contradiction. Thus, $s = 0$ and hence $e = 1$. Consequently, n is the product of distinct Mersenne primes.

Conversely, let $n = \prod_i p_i$ be the product of Mersenne primes $p_i = 2^{m_i} - 1$. Then $\sigma(n) = \prod_i \sigma(p_i) = \prod_i (p_i + 1) = \prod_i 2^{m_i} = 2^{\sum_i m_i}$ is a power of 2. ■

Pascal's Triangle and Mersenne Numbers

There is an intriguing relationship between Mersenne numbers and Pascal's triangle. The numbers in rows $1, 3, 7, 15, 31, 63, \dots$, which are Mersenne numbers, are all odd and each of the other rows contains at least one even entry. The following theorem shows this is always the case; that is, every entry in row n, where n is a Mersenne number, is odd; the proof given below is due to Rade M. Dacic of Belgrade, Serbia.

THEOREM 8.14 A positive integer n is a Mersenne number if and only if every binomial coefficient $\binom{n}{r}$ is odd, where $0 \le r \le n$.

PROOF

Let $n = 2^s - 1$, where $s \ge 0$. Then

$$\binom{n}{r} = \binom{2^s - 1}{r} = \frac{2^s - 1}{1} \cdot \frac{2^s - 2}{2} \cdots \frac{2^s - r}{r} \tag{8.10}$$

Let $1 \le i \le r$ and $i = 2^a b$, where $0 \le a \le s$ and b is odd. Then

$$\frac{2^s - i}{i} = \frac{2^s - 2^a b}{2^a b} = \frac{2^{s-a} - b}{b}$$

which is a quotient of odd integers. Thus, every factor on the RHS of equation (8.10) is a quotient of odd integers, so the product is an odd integer; that is, every entry $\binom{n}{r}$ in row n is odd.

Conversely, suppose every binomial coefficient $\binom{n}{r}$ in row n is odd. Let n be odd, but not a Mersenne number. Then $2^{m-1} < n < 2^m$ for some positive integer m. So $n = 2^{m-1} + 2k + 1$, where $0 \le k \le 2^{m-2} - 1$. Let $r = 2k + 2$. Then

$$\begin{aligned}\binom{n}{r} &= \binom{n}{r-1} \cdot \frac{n-r+1}{r} \\ &= s \cdot \frac{2^{m-1}}{2k+2} \\ &= s \cdot \frac{2^{m-2}}{k+1}\end{aligned}$$

where s is an integer. If $k+1 < 2^{m-2}$, not all 2s on the RHS can be canceled; so $\binom{n}{r}$ is even. If $k+1 = 2^{m-2}$, then $\binom{n}{r} = 0$, still an even integer. Both cases contradict the hypothesis, so n must be a Mersenne number. ■

Pascal's Triangle and Even Perfect Numbers

Suppose we replace each even number in Pascal's triangle by a white dot (0) and each odd number by a blue dot (1). Figure 8.8 shows the resulting delightful binary pattern, *Pascal's binary triangle*.

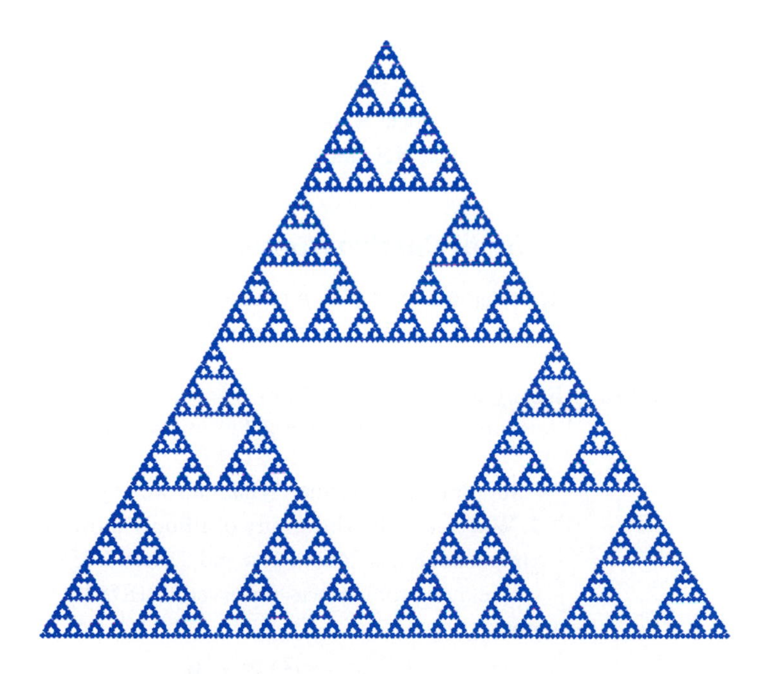

Figure 8.8 *Pascal's binary triangle.*

It follows by Theorem 8.14 that if n is Mersenne prime, then row n consists of blue dots only. There are exactly four such rows n, where $1 \leq n \leq 100$, namely, rows 1, 3, 7, and 31.

In addition to the aesthetic beauty of the binary triangle, it contains a fascinating treasure. To see it, consider the centrally located triangles ∇_n pointing downward and having their bases in row 2^n, where $n \geq 1$. Since the base of ∇_n contains $2^n - 1$ zeros, the number of zeros inside ∇_n equals

$$\begin{aligned} N &= \frac{(2^n - 1)(2^n - 1 + 1)}{2} \\ &= 2^{n-1}(2^n - 1) \end{aligned}$$

which is a perfect number if $2^n - 1$ is a Mersenne prime.

For example, the base of ∇_4 lies in row 32 and it contains $2^5 - 1 = 31$ zeros. So the number of zeros inside ∇_4 equals

$$\begin{aligned} 31 + 30 + \cdots + 2 + 1 &= \frac{31 \cdot 32}{2} \\ &= 496 \end{aligned}$$

which is the third perfect number.

More generally, every number $N = 2^{n-1}(2^n - 1)$ is represented by ∇_n. Consequently, every even perfect number $N = 2^{p-1}(2^p - 1)$ is represented by ∇_p, where $2^p - 1$ is a prime; in other words, even perfect numbers are represented geometrically by a subsequence of the sequence $\{\nabla_n\}$, as observed in 1956 by Alan L. Brown of South Orange, New Jersey.

Parity of Catalan Numbers

Mersenne numbers play a pivotal role in the parity of Catalan numbers, as the next theorem shows. We omit the proof for brevity; see Exercise 47.

THEOREM 8.15 **(Koshy and Salmassi, 2004)** The Catalan number C_n is odd if and only if n is a Mersenne number, where $n \geq 1$.

Suppose M_p is prime. What can we say about the primality of M_{M_p}? In 1954, D. J. Wheeler of the University of Illinois showed that $M_{M_{13}} = 2^{M_{13}} - 1 = 2^{8191} - 1$ is composite; it has 2466 digits and $338{,}193{,}759{,}479 | M_{M_{13}}$.

A related problem was observed in 1876 by Catalan. To this end, notice that the numbers

$$\begin{aligned} c_1 &= 2^2 - 1 = M_2 & c_2 &= 2^{c_1} - 1 = M_3 \\ c_3 &= 2^{c_2} - 1 = M_7 & c_4 &= 2^{c_3} - 1 = M_{127} \end{aligned}$$

are all primes. The primality of c_n remains unresolved, as is the infinitude of such primes.

?

EXERCISES 8.4

1. Find the binary representations of the first five Mersenne numbers M_n.
2. Find the binary representations of the reciprocals of the first five Mersenne numbers.

Compute the number of digits in each Mersenne number M_p for the given prime p.

3. 2281
4. 19,937
5. 110,503
6. 756,839

Compute the last digit in each Mersenne number M_p for the given prime p.

7. 127
8. 2281
9. 11,213
10. 132,049

Compute the last two digits in each Mersenne number M_p for the given prime p.

11. 127
12. 1279
13. 9941
14. 110,503

Compute the last three digits in each Mersenne number M_p for the given prime p.

15. 1279
16. 9941
17. 110,503
18. 756,839
19. Show that every Fermat number f_n ends in 7, where $n \geq 2$.
20. Find the number of digits in the Fermat number f_n.

Find the number of digits in the Fermat number f_n for the given value of n.

21. 13
22. 19
23. 23
24. 31

Find the binary expansion of each.

25. f_n
26. M_p
27. $2^{p-1}M_p$

Let a be an arbitrary least residue relatively prime to a modulus m. Find the least positive exponent k such that $a^k \equiv 1 \pmod{m}$ for each m.

28. 7
29. 10
30. 15
31. 18

Verify that each Mersenne number M_p is a prime for the given prime p.

32. 5
33. 7
34. 13
35. 17

Verify that each Mersenne number M_p is composite for the given prime p. Provide a factor in each case.

36. 23
37. 29
38. 37
39. 43
40. Add the next line to the following number pattern.

$$28 = 1^3 + 3^3$$
$$496 = 1^3 + 3^3 + 5^3 + 7^3$$
$$8128 = 1^3 + 3^3 + 5^3 + 7^3 + 9^3 + 11^3 + 13^3 + 15^3$$
$$\vdots$$

41. Beiler states in his book *Recreations in the Theory of Numbers* that every even perfect number $2^n(2^{n+1} - 1)$ is the sum of the cubes of the first $2^{n/2}$ odd positive integers, where n is an even integer > 2. Confirm this.
42. Is $\sum(2i-1)^3$ always an even perfect number, where $1 \leq i \leq 2^{n/2}$? If not, give a counterexample.
43. Prove that the sum $S = \sum_{i=1}^{m}(2i-1)^3$ ends in 6 or 28, where $m = 2^{n/2}$ and n is even.
44. Find the number of digits in the perfect number $N = 2^{p-1}M_p$.

Compute the digital root of each.

45. M_p
46. $2^{p-1}M_p$

*47. The Catalan number C_n is odd if and only if n is a Mersenne number, where $n \geq 1$.

***August Ferdinand Möbius** (1790–1868) was born in Schulfpforta near Hamburg, Germany. His father was a dance teacher and his mother a descendant of Martin Luther. He was home-taught until he was 13; by then he had shown an interest in mathematics. After receiving formal education in Schulfpforta, in 1809 he entered Leipzig University, where he intended to study law, but instead decided to pursue mathematics, physics, and astronomy. In 1813 he went to Göttingen to study with Gauss and then to Halle to study mathematics with Johann F. Pfaff. In the following year, he received his doctorate from Leipzig, where he became professor of astronomy and remained until his death.*

Möbius made contributions to astronomy, mechanics, affine and projective geometry, statics, optics, and number theory. He is well known for his discovery of the one-sided surface, the Möbius strip, formed by joining the ends of a rectangular strip of paper after giving it a half-twist.

*8.5 The Möbius Function (optional)

The Möbius function μ† is an important number-theoretic function discovered by the German mathematician August Ferdinand Möbius. It plays an important role in the study of the distribution of primes. The Möbius function is defined as follows.

The Möbius Function μ

Let n be a positive integer. Then

$$\mu(n) = \begin{cases} 1 & \text{if } n = 1 \\ 0 & \text{if } p^2 | n \text{ for some prime } p \\ (-1)^k & \text{if } n = p_1 p_2 \cdots p_k, \text{ where the } p_i\text{'s are distinct primes} \end{cases}$$

For example, $\mu(2) = -1$, $\mu(3) = -1$, $\mu(4) = 0$, $\mu(12) = 0$, $\mu(35) = \mu(5 \cdot 7) = (-1)^2 = 1$, and $\mu(672) = \mu(2^5 \cdot 3 \cdot 7) = 0$.

† μ is the Greek letter *mu*.

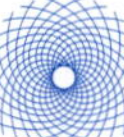

In words, μ assigns -1, 0, or 1 to each positive integer; $\mu(n) = 0$ if the canonical decomposition of n contains a square, that is, if n is not square-free; it is $(-1)^k$ if it consists of k distinct prime factors.

Before we determine whether μ is multiplicative, let us study the next example.

EXAMPLE 8.23 Determine whether $\mu(mn) = \mu(m)\mu(n)$, where $m = 15$ and $n = 28$.

SOLUTION

$m = 15 = 3 \cdot 5$, so $\mu(m) = 1$, by definition; $n = 28 = 2^2 \cdot 7$, so $\mu(n) = 0$, again by definition. Then $mn = 28 \cdot 15 = 2^2 \cdot 3 \cdot 5 \cdot 7$.

Since mn is not square-free, $\mu(mn) = 0$. Thus, $\mu(mn) = 0 = 1 \cdot 0 = \mu(m)\mu(n)$. ■

More generally, we have the following result, which confirms what we already suspected.

THEOREM 8.16 The function μ is multiplicative.

PROOF

Let m and n be relatively prime positive integers. If $m = 1$ or $n = 1$, then clearly $\mu(mn) = \mu(m)\mu(n)$. Suppose m or n (but not both) is divisible by p^2 for some prime p. Then $\mu(m)\mu(n) = 0$. Since $p^2|m$ or $p^2|n$, then $p^2|mn$, so $\mu(mn) = 0$. Thus, $\mu(mn) = \mu(m)\mu(n)$.

Finally, suppose both m and n are square-free, so let $m = p_1p_2\ldots p_r$ and $n = q_1q_2\ldots q_s$, where the p_is and the q_js are distinct primes, since $(m, n) = 1$. So $\mu(m) = (-1)^r$ and $\mu(n) = (-1)^s$.

Then $mn = p_1p_2\cdots p_rq_1q_2\cdots q_S$, a product of distinct primes. Therefore,

$$\mu(mn) = (-1)^{r+s} = (-1)^r \cdot (-1)^s = \mu(m)\mu(n)$$

Thus, in every case, $\mu(mn) = \mu(m)\mu(n)$, so μ is multiplicative. ■

Next, we develop a formula for $\sum_{d|n} \mu(d)$. When $n = 1$, $\sum_{d|1} \mu(d) = \mu(1) = 1$. If $n > 1$, we can compute the sum using the canonical decomposition of n and Theorem 8.6, provided we know the sum when n is a prime-power p^e. For this we need to introduce a new function, as the following lemma shows.

LEMMA 8.3 Let $F(n) = \sum_{d|n} \mu(d)$. Then $F(p^e) = 0$, where $e > 1$.

PROOF

$$
\begin{aligned}
F(p^e) &= \sum_{d|p^e} \mu(d) \\
&= \sum_{i=0}^{e} \mu(p^i) \\
&= \mu(1) + \mu(p) + \mu(p^2) + \cdots + \mu(p^e) \\
&= 1 + (-1) + 0 + \cdots + 0 = 0
\end{aligned}
$$

■

EXAMPLE 8.24 Illustrate Lemma 8.3 for $n = 81$.

SOLUTION

$$
\begin{aligned}
\sum_{d|81} \mu(d) &= \sum_{d|3^4} \mu(d) \\
&= \mu(1) + \mu(3) + \mu(3^2) + \mu(3^3) + \mu(3^4) \\
&= 1 + (-1) + 0 + 0 + 0 = 0
\end{aligned}
$$

■

We are now ready to display and prove the desired formula in the following theorem.

THEOREM 8.17 Let n be a positive integer. Then

$$
\sum_{d|n} \mu(d) = \begin{cases} 1 & \text{if } n = 1 \\ 0 & \text{otherwise} \end{cases}
$$

PROOF

If $n = 1$, $\sum_{d|1} \mu(d) = \mu(1) = 1$. So let $n > 1$ and let $n = p_1^{e_1} p_{2}^{e_2} \cdots p_k^{e_k}$ be the canonical decomposition of n. Let $F(n) = \sum_{d|n} \mu(d)$. Since μ is multiplicative, so is F by Theorem 8.6. Therefore,

$$
\begin{aligned}
F(n) &= \prod_{i=1}^{k} F(p^{e_i}) \\
&= \prod_{i=1}^{k} (0) = 0, \quad \text{by Lemma 8.3}
\end{aligned}
$$

■

The following example illustrates this theorem.

EXAMPLE 8.25 Compute $\sum_{d|18} \mu(d)$.

SOLUTION

$$\sum_{d|18} \mu(d) = \mu(1) + \mu(2) + \mu(3) + \mu(6) + \mu(9) + \mu(18)$$
$$= 1 + (-1) + (-1) + (-1)^2 + 0 + 0 = 0$$

which agrees with the theorem. ■

Theorem 8.17 plays a crucial role in the derivation of the Möbius inversion formula. Its derivation involves double summations over positive divisors, which can be confusing, so we use the following example to clarify it.

EXAMPLE 8.26 Let f be a number-theoretic function. Show that

$$\sum_{d|6} \sum_{d'|(6/d)} \mu(d)f(d') = \sum_{d'|6} \sum_{d|(6/d')} f(d')\mu(d)$$

PROOF

$$\begin{aligned}
&\sum_{d|6} \sum_{d'|(6/d)} \mu(d)f(d') \\
&\quad= \sum_{d'|6} \mu(1)f(d') + \sum_{d'|3} \mu(2)f(d') + \sum_{d'|2} \mu(3)f(d') + \sum_{d'|1} \mu(6)f(d') \\
&\quad= \mu(1)\sum_{d'|6} f(d') + \mu(2)\sum_{d'|3} f(d') + \mu(3)\sum_{d'|2} f(d') + \mu(6)\sum_{d'|1} f(d') \\
&\quad= \mu(1)[f(1) + f(2) + f(3) + f(6)] + \mu(2)[f(1) + f(3)] \\
&\qquad + \mu(3)[f(1) + f(2)] + \mu(6)f(1) \\
&\quad= f(1)[\mu(1) + \mu(2) + \mu(3) + \mu(6)] + f(2)[\mu(1) + \mu(3)] \\
&\qquad + f(3)[\mu(1) + \mu(2)] + f(6)[\mu(1)] \\
&\quad= f(1)\sum_{d|6} \mu(d) + f(2)\sum_{d|3} \mu(d) + f(3)\sum_{d|2} \mu(d) + f(6)\sum_{d|1} \mu(d) \\
&\quad= \sum_{d'|6} f(d') \sum_{d|(6/d')} \mu(d) \\
&\quad= \sum_{d'|6} \sum_{d|(6/d')} f(d')\mu(d)
\end{aligned}$$

(Notice that as d' runs over the positive divisors 1, 2, 3, and 6, $6/d'$ runs over them in the reverse order. We can also rewrite the double sum as $\sum_{dd'|6}\sum f(d')\mu(d)$.) ■

We can now turn to the next major result.

THEOREM 8.18 **(Möbius Inversion Formula)** Let f be a number-theoretic function and let $F(n) = \sum_{d|n} f(d)$. Then

$$f(n) = \sum_{d|n} \mu(d)F(n/d) \tag{8.11}$$

PROOF

$F(n) = \sum_{d|n} f(d)$. Then $F(n/d) = \sum_{d'|(n/d)} f(d')$. So

$$\mu(d)F(n/d) = \mu(d) \sum_{d'|(n/d)} f(d') = \sum_{d'|(n/d)} \mu(d)f(d')$$

$$\sum_{d|n} \mu(d)F(n/d) = \sum_{d|n} \sum_{d'|(n/d)} \mu(d)f(d')$$

As d runs over the positive divisors of n, so does d'; also $dd'|n$. Therefore, the sum on the RHS is the sum of all values of $\mu(d)f(d')$ as dd' runs over the positive factors of n; that is, the sum of all its values for all pairs d and d' such that $d'|n$ and $d|(n/d')$. That is,

$$\sum_{d|n} \sum_{d'|(n/d)} \mu(d)f(d') = \sum_{d'|n} \sum_{d|(n/d')} \mu(d)f(d')$$

Thus,

$$\sum_{d|n} \mu(d)F(n/d) = \sum_{d^{l}|n} f(d') \left[\sum_{d|(n/d')} \mu(d) \right]$$

But, by Theorem 8.17, $\sum_{d|(n/d')} \mu(d)$ equals 1 if $n/d' = 1$; that is, if $n = d'$, and 0 otherwise. Thus

$$\sum_{d|n} \mu(d)F(n/d) = f(d') \cdot 1, \quad \text{where } d' = n$$

$$= f(n)$$

In other words, $f(n) = \sum_{d|n} \mu(d)F(n/d)$. ■

As d runs over the positive factors of n, so does n/d. Therefore, the inversion formula (8.11) can also be written as

$$f(n) = \sum_{d|n} \mu(n/d)F(d)$$

Notice that the definition $F(n) = \sum_{d|n} f(d)$ expresses F in terms of f, whereas the inversion formula (8.11) expresses f in terms of F.

To illustrate the inversion formula, recall that

$$\tau(n) = \sum_{d|n} 1 \quad \text{and} \quad \sigma(n) = \sum_{d|n} d$$

Because both the constant function $f(n) = 1$ and the identity function $g(n) = n$ are multiplicative, it follows by Theorem 8.18 that

$$1 = \sum_{d|n} \mu(d)\tau(n/d) = \sum_{d|n} \mu(n/d)\tau(d) \tag{8.12}$$

and

$$n = \sum_{d|n} \mu(d)\sigma(n/d) = \sum_{d|n} \mu(n/d)\sigma(d) \tag{8.13}$$

The following example illustrates these results.

EXAMPLE 8.27 Verify formulas (8.12) and (8.13) for $n = 6$.

PROOF

- $$\begin{aligned}\textstyle\sum_{d|6} \mu(d)\tau(6/d) &= \mu(1)\tau(6) + \mu(2)\tau(3) + \mu(3)\tau(2) + \mu(6)\tau(1)\\ &= 1 \cdot 4 + (-1) \cdot 2 + (-1) \cdot 2 + (-1)^2 \cdot 1\\ &= 1\end{aligned}$$

- $$\begin{aligned}\textstyle\sum_{d|6} \mu(d)\sigma(6/d) &= \mu(1)\sigma(6) + \mu(2)\sigma(3) + \mu(3)\sigma(2) + \mu(6)\sigma(1)\\ &= 1 \cdot 12 + (-1) \cdot 4 + (-1) \cdot 3 + (-1)^2 \cdot 1\\ &= 6\end{aligned}$$

■

Using the inversion formula, the following theorem derives an explicit formula for $\varphi(n)$. We leave its proof as an exercise (see Exercise 43).

THEOREM 8.19 $\varphi(n) = n\sum_{d|n} \frac{\mu(d)}{d}$. ■

The following example illustrates this result.

EXAMPLE 8.28 Verify the formula in Theorem 8.19 for $n = 12$.

PROOF

By Theorem 8.4, $\varphi(12) = \varphi(2^2 \cdot 3) = 4$. Let us now compute the RHS:

$$\begin{aligned}
12\sum_{d|12} \frac{\mu(d)}{d} &= \sum_{d|12} (12/d)\mu(d) \\
&= 12\mu(1) + 6\mu(2) + 4\mu(3) + 3\mu(4) + 2\mu(6) + 1\mu(12) \\
&= 12 \cdot 1 + 6 \cdot (-1) + 4(-1) + 3 \cdot 0 + 2 \cdot (-1)^2 + 1 \cdot 0 \\
&= 4 = \varphi(12)
\end{aligned}$$

■

We conclude this section with the following theorem; it shows that the converse of Theorem 8.18 is also true. Once again, the proof contains double sums, so we need to proceed carefully.

THEOREM 8.20 Let F and f be number-theoretic functions such that $f(n) = \sum_{d|n} \mu(d)F(n/d)$. Then $F(n) = \sum_{d|n} f(d)$.

PROOF

By the definition of f,

$$f(d) = \sum_{d'|d} \mu(d')F(d/d')$$

$$\sum_{d|n} f(d) = \sum_{d|n} \sum_{d'|d} \mu(d')F(d/d')$$

Letting $d/d' = k$, this equation yields

$$\sum_{d|n} f(d) = \sum_{d|n} \sum_{kd'=d} \mu(d')F(k)$$

$$= \sum_{kd'|n} \mu(d')F(k)$$

$$= \sum_{k|n} F(k)\Big[\sum_{d'|(n/k)} \mu(d')\Big]$$

By Theorem 8.17, $\sum_{d'|(n/k)} \mu(d')$ equals 1 if $n = k$, and 0 otherwise. So the equation becomes

$$\sum_{d|n} f(d) = F(k) \cdot (1), \quad \text{where } n = k$$

$$= F(n)$$

That is, $F(n) = \sum_{d|n} f(d)$, the desired result. ■

EXERCISES 8.5

Compute $\mu(n)$ for each n, where p is an odd prime.

1. 101
2. 496
3. 2047
4. 11,319
5. p
6. p^{13}
7. $2^{p-1}(2^p - 1)$
8. $\varphi(\varphi(M_{11}))$

Verify formula (8.12) for each n.

9. 5
10. 6
11. 10
12. 13

13–16. Verify formula (8.13) for each n in Exercises 9–12.

Using Theorem 8.19, compute $\varphi(n)$ for each n.

17. 23
18. 28
19. 36
20. 1352

Using the definition of μ, verify Theorem 8.16 for each canonical factorization of n.

21. pq
22. pqr
23. p^2qr
24. $p_1p_2\cdots p_k$

25–28. Evaluate $\sum_{d|n} \mu(d)\tau(d)$ for each canonical decomposition of each n in Exercises 21–24.

Evaluate $\sum_{d|n} \mu(d)\sigma(d)$ for each canonical decomposition of the given integer n.

29. p
30. pq
31. p^2q
32. pqr
33. Using Exercises 29–32, predict the value of $\sum_{d|n} \mu(d)\sigma(d)$, where $n = \prod_{i=1}^{k} p_i$.

Another useful number-theoretic function that resembles the μ function is the **Liouville function** λ (lambda), introduced by the French mathematician Joseph Liouville (1809–1882). It is defined by

$$\lambda(1) = 1$$

$$\lambda(n) = (-1)^{e_1+e_2+\cdots+e_k}$$

where $n = p_1^{e_1}p_2^{e_2}\cdots p_k^{e_k}$. Compute $\lambda(n)$ for each n.

34. 17
35. 104
36. 990
37. 3024

Compute $\sum_{d|n} \lambda(d)$ for each n.

38. 9
39. 12
40. 16
41. 28

Prove each.

42. $\frac{\mu(n)}{n}$ is multiplicative.

43. $\varphi(n) = n\sum_{d|n} \frac{\mu(d)}{d}$

44. $\sum_{d|n} \frac{\mu(d)}{d} = 1 - \frac{1}{p}$, where $n = p^e$.

45. $\sum_{d|n} \frac{\mu(d)}{d} = \left(1 - \frac{1}{p}\right)\left(1 - \frac{1}{q}\right)$, where $n = p^a q^b$.

46. $\sum_{d|n} \frac{\mu(d)}{d} = \prod_{i=1}^{k}\left(1 - \frac{1}{p_i}\right)$, where $n = \prod_{i=1}^{k} p_i^{e_i}$.

47. Using Theorem 8.19, prove that $\varphi(p^e) = p^e - p^{e-1}$.

48. Using Theorem 8.19, prove that φ is multiplicative.

49. λ is multiplicative.

50. $\lambda(n) = 1$ if n is a square.

51. $\sum_{d|n} \lambda(d) = \begin{cases} 1 & \text{if } n \text{ is a perfect square} \\ 0 & \text{otherwise.} \end{cases}$

52. Using Exercise 46, derive a formula for $\varphi(n)$.

*53. Let F be a multiplicative function and f a number-theoretic function such that $F(n) = \sum_{d|n} f(d)$. Prove that f is also multiplicative.

CHAPTER SUMMARY

Five important multiplicative functions have played a significant role in the development of number theory: φ, τ, σ, μ, and λ. The first three play a pivotal role in the study of perfect numbers, Mersenne primes, amicable numbers, abundant numbers, and deficient numbers. They satisfy a variety of useful and beautifully appealing properties.

Multiplicative Functions

- A number-theoretic function f is **multiplicative** if $f(mn) = f(m)f(n)$ whenever $(m, n) = 1$. (p. 356)
- If f is multiplicative and $n = \prod_i p_i^{e_i}$, then $f(n) = \prod_i f(p_i^{e_i})$. (p. 356)

Euler's Phi Function φ

- $\varphi(n)$ = number of positive integers $\le n$ and relatively prime to it. (p. 342)
- $\varphi(p^e) = p^e - p^{e-1} = p^e(1 - 1/p)$ (p. 356)
- If $(m, n) = 1$, then the integers $r, m + r, 2m + r, \dots, (n - 1)m + r$ are congruent modulo n to $0, 1, 2, \dots, (n - 1)$ in some order. (p. 359)
- φ is multiplicative. (p. 360)
- If $n = \prod_i p_i^{e_i}$, then $\varphi(n) = n\prod_i(1 - 1/p_i)$. (p. 361)
- If $n \ge 3$, then $\varphi(n)$ is even. (p. 362)
- $\sum_{d|n} \varphi(d) = n$ (p. 363)

The Tau and Sigma Functions τ and σ

- $\tau(n) = \sum_{d|n} 1$ = number of positive factors of n. (p. 365)
- $\sigma(n) = \sum_{d|n} d$ = sum of positive factors of n. (p. 366)
- If f is multiplicative, so is $F(n) = \sum_{d|n} f(d)$. (p. 367)
- Both τ and σ are multiplicative. (p. 368)
- $\tau(p^e) = e + 1$ and $\sigma(p^e) = (p^{e+1} - 1)/(p - 1)$ (p. 369)
- If $n = \prod_i p_i^{e_i}$, then $\tau(n) = \prod_i (e_i + 1)$ and $\sigma(n) = \prod_i \dfrac{p_i^{e_i+1} - 1}{p_i - 1}$. (p. 369)

Perfect Numbers

- A positive integer n is **perfect** if $\sigma(n) = 2n$. (p. 374)
- (**Euclid's theorem**) Every integer $N = 2^{n-1}(2^n - 1)$, where $2^n - 1$ is a prime, is a perfect number. (p. 375)
- (**Euler's theorem**) Every even perfect number is of the form $2^{n-1}(2^n - 1)$, where $2^n - 1$ is a prime. (p. 375)
- Even perfect numbers end in 6 or 8. (p. 375)
- The infinitude of even perfect numbers is unresolved. (p. 376)
- The existence of odd perfect numbers is unsettled. (p. 379)

Mersenne Numbers

- Numbers of the form $2^m - 1$ are Mersenne numbers. Such numbers that are primes are **Mersenne primes** M_p. For $2^m - 1$ to be a prime, m must be a prime. (p. 381)
- (**Euler's theorem**) Let $p = 4k + 3$ be a prime, where $k > 1$. Then $2p + 1$ is a prime if and only if $2^p \equiv 1 \pmod{2p + 1}$. (p. 388)
- Let $(a, n) = 1$ and k the least positive integer such that $a^k \equiv 1 \pmod{n}$. Then $k|n$. In particular, $k|\varphi(n)$. (p. 389)
- Every prime factor of M_p is of the form $2kp + 1$, where p is an odd prime. (p. 390)
- (**Lucas–Lehmer test**) M_p is a prime if and only if $S_{p-1} \equiv 0 \pmod{M_p}$, where $S_1 = 4$ and $S_k \equiv S_{k-1}^2 - 2 \pmod{M_p}$. (p. 391)
- A positive integer n is a Mersenne number if and only if every binomial coefficient $\binom{n}{r}$ is odd. (p. 394)
- The Catalan number C_n is odd if and only if $n = 0$ or n is a Mersenne number, where $n \geq 1$. (p. 396)

The Möbius Function μ

- $\mu(n) = \begin{cases} 1 & \text{if } n = 1 \\ 0 & \text{if } p^2 | n \text{ for some prime } p \\ (-1)^k & \text{if } n \text{ is the product of } k \text{ distinct primes.} \end{cases}$ (p. 398)
- μ is multiplicative. (p. 399)
- $\sum_{d|n} \mu(d) = \begin{cases} 1 & \text{if } n = 1 \\ 0 & \text{otherwise.} \end{cases}$ (p. 400)
- (**Möbius Inversion Formula**) Let f be a number-theoretic function and let $F(n) = \sum_{d|n} f(d)$. Then $f(n) = \sum_{d|n} \mu(d) F(n/d)$. (p. 402)
- $\varphi(n) = n \sum_{d|n} \frac{\mu(d)}{d}$ (p. 404)
- Let F and f be number-theoretic functions such that $f(n) = \sum_{d|n} \mu(d) F(n/d)$. Then $F(n) = \sum_{d|n} f(d)$. (p. 404)

REVIEW EXERCISES

Evaluate each.

1. $\sum_{\substack{1 \le n \le 2020 \\ (n,2020)=1}} 1$
2. $\sum_{\substack{1 \le n \le 5850 \\ (n,5850)=1}} 1$
3. Until 1509, mathematicians believed that odd abundant numbers did not exist. Then Charles de Bouvelles (1470–1553) showed that $45{,}045 = 3^2 \cdot 5 \cdot 7 \cdot 11 \cdot 13$ and that its odd multiples are odd abundant numbers. Nearly 400 years later, in 1891, Lucas showed that $945 = 3^3 \cdot 5 \cdot 7$ is the smallest odd abundant number. Show that 945 and 45,045 are abundant numbers.
4. Verify that $17{,}296 = 2^4 \cdot 23 \cdot 47$ and $18{,}416 = 2^4 \cdot 1151$ are amicable numbers. (This pair, discovered by Fermat in 1636, was the second to be found; the original pair was found by the Pythagoreans in 540 B.C.)
5. Verify that $12{,}285 = 3^3 \cdot 5 \cdot 7 \cdot 13$ and $14{,}595 = 3 \cdot 5 \cdot 7 \cdot 139$ are amicable numbers. (This smallest odd amicable pair was discovered in 1939 by B. H. Brown.)

Verify that each Mersenne number M_p is composite for the indicated prime p. Furnish a factor in each case.

6. 47
7. 53
8. Verify that 64 is superperfect.

Evaluate each.

9. $\sum_{i=0}^{n} \sigma(2^i)$

10. $\sum_{i=0}^{n} \varphi(p^i)$

11. Show that the sum of two multiplicative functions need not be multiplicative.
12. Find the product of the positive factors of the even perfect number $n = 2^{p-1}(2^p - 1)$.

Prove each, where f and g are multiplicative functions and p is any prime.

13. fg is multiplicative.
14. f/g is multiplicative.
15. Every even perfect number is a hexagonal number.
16. No twin primes can be an amicable pair.
17. Let n be the product of distinct Mersenne primes. Then $\sigma(n)$ is a power of 2.
18. Let f_n be a Fermat prime. Then $\sigma(f_n)$ is even.
19. Let f_n be a Fermat prime. Then $\sigma(f_n) - \varphi(f_n) = 2$.
20. If $n \geq 4$, then $\sum_{k=1}^{n} \mu(k!) = -1$.

*21. Let $(a, p) = 1$. Then $a^{\varphi(p^e)} \equiv 1 \pmod{p^e}$.
(*Hint*: Use the binomial theorem.)

22. Prove Euler's theorem using Exercise 21.

*23. Every even perfect number ends in 6 or 8.
(*Hint*: Consider $p \pmod{10}$.)

*24. Every even perfect number ends in 6 or 28.

*25. The only 3-perfect numbers of the form $2^k \cdot 3 \cdot p$, where p is odd, are 120 and 672.

*26. Find all even perfect numbers that are superperfect.

SUPPLEMENTARY EXERCISES

1. Find two consecutive abundant numbers.
2. Find three consecutive abundant numbers. (S. Kravitz, 1994)

Korrah's formula for an amicable pair M, N can be developed as follows. Suppose $M = dab$ and $N = dc$, where $d = (M, N)$ and a, b, and c are distinct odd primes.

3. Using the conditions $\sigma(M) = \sigma(N) = M + N$, show that

$$(a+1)(b+1) = c+1 \quad (8.14)$$

and

$$\sigma(d)(a+1)(b+1) = d(ab+c) \quad (8.15)$$

4. Using equations (8.14) and (8.15), show that

$$\sigma(d)(a+1)(b+1) = d(2ab+a+b) \tag{8.16}$$

5. Let $d = 2^n$. Show that equation (8.16) can be written as

$$[a-(2^n-1)][b-(2^n-1)] = 2^{2n} \tag{8.17}$$

6. By equation (8.17), $a-(2^n-1) = 2^{n+m}$ and $b-(2^n-1) = 2^{n-m}$ for some integer m. Solve for a, b, and c using these equations.
7. Deduce Korrah's formula from the solutions in Exercise 4.
8. Using equation (8.16), show that $\sigma(d)/d = 2 - 1/g$, where $g = (p+1)(q+1)/[(p+1)+(q+1)]$.
9. Let $g = 9/2$ and $d = 3^2 \cdot 7 \cdot 13$. Show that

$$(2a-7)(2b-7) = 81 \tag{8.18}$$

(*Hint*: Use Exercise 8.)
10. Find the possible values of a, b, and c using equation (8.18).
11. Using the values of a, b, and c, find the corresponding amicable pair.
12. In 1951, the Dutch electrical engineer Balthazar van der Pol (1889–1959) established the recurrence relation

$$\frac{n^2(n-1)}{6}\sigma(n) = \sum_{k=1}^{n-1}\left(3n^2 - 10k^2\right)\sigma(k)\sigma(n-k)$$

Using this formula, find a recurrence relation for $\sigma(n)$ for $n = 2, 3, 4$, and 5.
13. Let $m = m_1 m_2 \cdots m_n$, where $(m_i, m_j) = 1$ for $i \neq j$. Prove that $\sum_{i=1}^{n} m_i^{\varphi(m)/\varphi(m_i)} \equiv n - 1 \pmod{m}$. (J. O. Silva, 1996)
14. Show that every even perfect number > 6 is one more than nine times a triangular number. (C. F. Eaton, 1995)
15. Using the formulas in Exercise 10, compute $\sigma(n)$ for $n = 2, 3, 4$, and 5.
16. Let $s^k(n) = s(s^{k-1}(n))$, where $k \geq 2$ and $s^1(n) = s(n) = \sigma(n) - n$. A number n such that $s^k(n) = n$ for some integer k is a **sociable number**. Show that 12496 is a sociable number. (The sociable numbers 12496, 14288, 15472, 14536, and 14264 were discovered in 1918 by P. Poulet.)

Let $n = \prod_{i=1}^{k} p_i^{e_i}$ be the canonical decomposition of n. Prove each.

17. $\sum_{d|n} \mu(d)\tau(d) = (-1)^k$

18. $\sum_{d|n} \mu(d)\sigma(d) = (-1)^k \prod_{i=1}^{k} p_i$
19. $\sum_{d|n} \mu(d)\varphi(d) = \prod_{i=1}^{k} (2 - p_i)$
20. Let f be a multiplicative function. Prove that $\sum_{d|n} \mu(d)f(d) = \prod_{i=1}^{k} [1 - f(p_i)]$.

Using the formula in Exercise 20, deduce the formula in

21. Exercise 17.
22. Exercise 18.
23. Exercise 19.

Using the formula in Exercise 20, deduce a formula for each.

24. $\sum_{d|n} d\mu(d)$
25. $\sum_{d|n} \frac{\mu(d)}{d}$
26. Derive a formula for $\sum_{d=1}^{n} \mu(d)\lambda(d)$.

A positive integer n is **near-perfect** if the sum of its proper factors is $n - 1$; that is, if $\sigma(n) = 2n - 1$. It is not known if odd near-perfect numbers exist.

27. Show that 16 is near-perfect.
28. Prove that every power of 2 is near-perfect.

COMPUTER EXERCISES

Write a program to perform each task.

1. Read in a positive integer $n \leq 1000$. Compute $\varphi(n)$ and list all positive integers $\leq n$ and relatively prime to it.
2. Solve the monkey and coconuts riddle in Example 8.3 by solving congruence (8.1) in Section 8.1.
3. Read in a positive integer $n \leq 1000$. Compute $\tau(n)$ and $\sigma(n)$.
4. Read in a Mersenne number $2^m - 1$ and determine whether it is a prime; if it is not, find a factor.
5. Using the Lucas–Lehmer test, determine whether a Mersenne number is a prime.
6. Read in an integer of the form $2^{n-1}(2^n - 1)$, and determine whether it is a perfect number.
7. Read in a positive integer and determine whether it is deficient, perfect, or abundant.
8. Read in an even perfect number of the form $2^{p-1}(2^p - 1)$ and compute its ones digit.

9. Read in two positive integers m and n, and determine if they are amicable.
10. Read in a positive integer n and determine if it is
    (a) k-perfect; if yes, find k.
    (b) Superperfect.
11. Construct Pascal's binary triangle through row 50.
12. Find all positive integers $3 \leq n \leq 10^4$ such that
    (a) $\sigma(n) = \sigma(n-1) + \sigma(n-2)$
    (b) $\tau(n) = \tau(n-1) + \tau(n-2)$
13. Read in a positive integer n, and determine $\mu(n)$ and $\lambda(n)$.
14. Read in a positive integer n and a least residue a modulo n, where $(a, n) = 1$. Find the least positive integer k such that $a^k \equiv 1 \pmod{n}$.

ENRICHMENT READINGS

1. W. W. R. Ball, *Mathematical Recreations and Essays*, Macmillan, New York, 1973, 65–73.
2. P. T. Bateman *et al.*, "The New Mersenne Conjecture," *The American Mathematical Monthly*, 96 (Feb. 1989), 125–128.
3. A. H. Beiler, *Recreations in the Theory of Numbers*, 2nd ed., Dover, New York, 1966, 11–30.
4. T. Koshy, "Digital Roots of Mersenne Primes and Even Perfect Numbers," *The Mathematical Gazette*, 89 (Nov. 2005), 464–466.
5. T. Koshy, "The Ends of a Mersenne Prime and an Even Perfect Number," *J. Recreational Mathematics*, 29 (1998), 196–202.
6. A. R. G. MacDivitt, "The Most Recently Discovered Prime Number," *The Mathematical Gazette*, 63 (1979), 268–270.
7. M. R. Schroeder, "Where Is the Next Mersenne Number Hiding?" *The Mathematical Intelligencer*, 5 (1983), 31–33.
8. M. T. Whalen and G. L. Miller, "Odd Abundant Numbers: Some Interesting Observations," *J. Recreational Mathematics*, 22 (1990), 257–261.

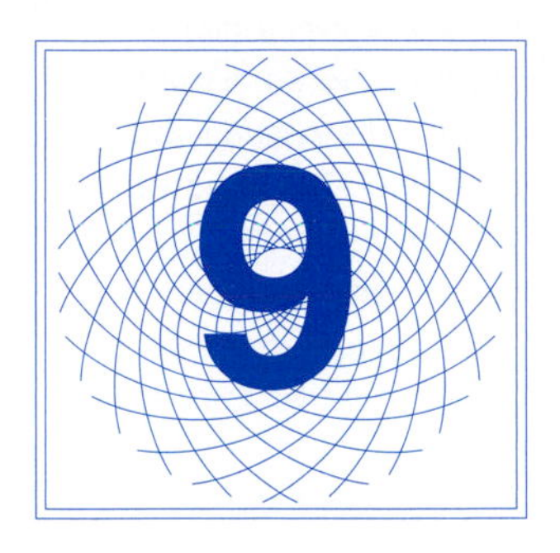

9 Cryptology

Mathematicians are like Frenchmen: whatever you say to them they translate into their own language and forthwith it is something entirely different.
—GOETHE

The great English number theorist Godfrey H. Hardy apparently believed that number theory had no practical applications. In his book *A Mathematician's Apology*, Hardy wrote that Theorem 2.10, which established the infinitude of primes, had only the slightest practical importance. Despite his opinion, ingenious mathematicians over the years, as we saw in Chapter 5, have discovered many practical and interesting applications of number theory.

We will now study several useful and charming applications that show that despite Hardy's conservative judgment, human creativity can turn virtually every aspect of mathematical knowledge to some practical use.

One exciting application of number theory is *cryptology*, the study of secrecy systems, which can be traced back to the early Egyptians. For centuries, a powerful tool in military and diplomatic circles, cryptology has become indispensable in commerce as well. Governments often want to keep policy decisions secret until an appropriate time; multinational corporations protect proprietary research and development, and marketing strategies.

In 1917, at the height of World War I, Germany cabled the Mexican government that it would commence submarine warfare and promised Arizona, New Mexico, and Texas to Mexico if it would join the Axis against the United States, in the event that the United States entered the war. The cable was intercepted, the code broken by British intelligence, the message passed on to President Woodrow Wilson, and the rest is history.

***Godfrey Harold Hardy** (1877–1947), an eminent English number theorist, was born in Cranleigh, England. Even as a child, he showed a precocious interest in mathematics. At the age of thirteen, he left Cranleigh School, where his father was a master, and moved to Winchester College. In 1896, he entered Trinity College, Cambridge, and was elected a fellow four years later. Ten years later, Hardy became a lecturer at Cambridge University, a position he held until 1919. He plunged into research, wrote many papers in analysis, and completed his well-known book,* A Course of Pure Mathematics *(1908). The text, designed for undergraduates, provided the first rigorous exposition of analysis, and transformed mathematics teaching forever.*

In 1919, Hardy left Cambridge to become Savilian professor of geometry at Oxford University, where he also was an active researcher. He was succeeded at Cambridge by John E. Littlewood (1885–1977). Eleven years later, Hardy returned to Cambridge, where he remained until his retirement in 1942. They had the most remarkable and productive partnership in the history of mathematics; they coauthored about 100 papers.

Hardy's most spectacular contribution to the mathematical community is generally considered to be his 1913 discovery of the unsophisticated Indian mathematical genius Srinivasa Ramanujan (1887–1920), whom Hardy brought to England in April 1914. Their relentless collaboration produced many spectacular discoveries.

Today, electronic banking and computer data banks commonly use encryption for secrecy and security. In 1984, R. Sedgewick of the University of Illinois noted that "a computer user wants to keep his computer files just as private as papers in his file cabinet, and a bank wants electronic funds transfer to be just as secure as funds transfer by armored car."

Recent developments in computer technology and sophisticated techniques in cryptology have revolutionized information security, protecting secret communications over insecure channels such as telephone lines and microwaves from being accessed by unauthorized users. See Figure 9.1.

Cryptography and Cryptanalysis

Cryptology consists of *cryptography* and *cryptanalysis*. The word *cryptography* is derived from the Greek words *kryptos*, meaning *hidden*, and *graphein*, meaning *to write*. **Cryptography** is the art and science of concealing the meaning of confidential communications from all except the intended recipients. **Cryptanalysis** deals with breaking secret messages. During World War II, 30,000 people were engaged in cryptographic work. The breaking of Japan's Purple machine code by U.S. cryptanalysts shortly before the attack on Pearl Harbor led to the Allied victory in the

A New Encryption System Would Protect a Coveted Digital Data Stream—Music on the Web

Sabra Chartrand

As the Internet continues to influence the evolution of intellectual property law and policy, an issue currently generating tremendous controversy is the free and anonymous swapping of digital music files.

Various companies have proposed terms of encryption as solutions to the problem. Now add another candidate: three mathematicians at Brown University have capped six years of research with a patent for an encryption code they say will make it impractical—if not impossible—to infringe copyrighted data like digital music.

The mathematicians, Jeffrey Hoffmein and Jill Pipher, both of Pawtucket, R.I., and Joseph Silverman of Needham, Mass., patented a system they said could quickly encode every second of a data stream with a different encryption key. That means that a typical three-minute song could be scrambled into 180 different codes; anyone taking the time to break a single code would be rewarded with only one second of music.

Like other encryption systems, the new invention grew out of advanced mathematical formulas. NTRU's technology differs from other encryption processes, Mr. Crenshaw said, because it relies on a mathematical system called a "convolution product" to make it faster and more efficient. With that kind of math, he said, encoding requires only one step, while decoding requires only two. Some other encryption systems need more than 1,000, he said.

The invention uses what is called "public key" encryption, which does not require the sender and receiver to privately exchange code keys to complete a transaction. Mr. Crenshaw said that when a person ordered music online, his computer or music player would provide the encoding key to the server computer of a Web site dispensing the music.

Figure 9.1

Pacific. Today the U.S. government and business employ tens of thousands of people and spend billions of dollars annually on cryptology.

Cryptography is by no means the exclusive domain of professionals. Franklin Delano Roosevelt, when he was 21, used a simple code in his diary. American poet Edgar Allan Poe, who was a skilled cryptanalyst, wrote that human ingenuity could invent no code that human ingenuity could not crack. Section 9.4, however, will prove otherwise.

Before we turn to some number-theoretic secrecy systems, we must define our terminology. **Plaintext** is the original message that is to be transmitted in secret form. **Ciphertext** is its secret version. A **cipher** is a method of translating plaintext to ciphertext. The **key** is an explicit formulation of the cipher, so the job of the cryptanalyst is to discover the key and then break the code. The process of converting plaintext to ciphertext is **enciphering** (or **encrypting**) and the converting device the encryptor. The reverse process by the intended recipient who knows the key is **deciphering** (or **decrypting**) and it is accomplished by a **decryptor**. The encryptor and decryptor may be algorithms executed by people or computers. Thus, the method used by an unintended receiver to recover the original message is **cryptanalysis**. A **cryptosystem** is a system for encrypting a plaintext to a ciphertext using a key.

This chapter presents five cryptosystems—affine, Hill, exponentiation, RSA, and knapsack—based on modular arithmetic. The first three are **conventional** and the last two are **public-key**. In a **conventional cryptosystem**, pictured in Figure 9.2, the encryption key, from which the decryption key can be found fairly quickly, is kept secret from unintended users of the system. In a public-key system, the enciphering key is made public while only the intended receiver knows the deciphering key.

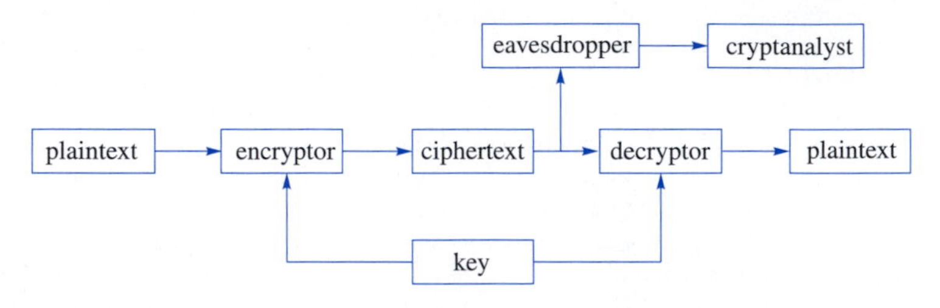

Figure 9.2

We now turn to our first cryptosystem.

9.1 Affine Ciphers

We will restrict our discussion to plaintext messages written in capital letters of the English alphabet and ignore blank spaces and punctuation marks. In all cryptosystems we first translate each letter to a number. A convenient way of doing this is by numbering the letters A through Z by their **ordinal numbers** 00 through 25, respectively, as Table 9.1 shows. Using this scheme, we translate the plaintext into a numeric message which is then enciphered into a numeric ciphertext. Each number is then replaced by a letter. The recipient of the ciphertext substitutes the ordinal number for each letter and uses the key to decipher the numeric message by substituting letters for the various numbers.

Letter	A	B	C	D	E	F	G	H	I	J	K	L	M	N	O	P	Q	R	S	T	U	V	W	X	Y	Z
Ordinal Number	00	01	01	03	04	05	06	07	08	09	10	11	12	13	14	15	16	17	18	19	20	21	22	23	24	25

Table 9.1

Substitution Ciphers

In a **substitution cipher**, we substitute a letter of the alphabet for each letter of the plaintext. It is, in fact, a **permutation cipher**, since each substitution is a permutation of the letters of the alphabet. Since there are 26! permutations of the letters, there is a total of 26! possible substitution ciphers; one of them is the trivial one, where each letter is substituted for itself.

Caesar Cipher

Around 50 B.C. the Roman emperor Julius Caesar (100–44 B.C.) sent encoded messages to his general, Marcus T. Cicero (106–43 B.C.), during the Gallic Wars, using a substitution cipher based on modular arithmetic. A Caesar cipher shifts each letter by three places to the right, with the last three letters X, Y, and Z shifted to A, B, and C respectively, in a cyclic fashion:

A	B	C	D	E	F	G	H	I	J	K	L	M	N	O	P	Q	R	S	T	U	V	W	X	Y	Z
↕	↕	↕	↕	↕	↕	↕	↕	↕	↕	↕	↕	↕	↕	↕	↕	↕	↕	↕	↕	↕	↕	↕	↕	↕	↕
D	E	F	G	H	I	J	K	L	M	N	O	P	Q	R	S	T	U	V	W	X	Y	Z	A	B	C

Let P denote the ordinal number of a plaintext letter and C that of the corresponding ciphertext letter. Then the Caesar cipher can be described by the congruence

$$C \equiv P + 3 \pmod{26} \tag{9.1}$$

where $0 \leq P, C \leq 25$.

Ciphertext is often grouped into blocks of five letters to prevent short words from being quickly recognized by cryptanalysts. The following example illustrates the Caesar enciphering algorithm.

EXAMPLE 9.1 Encipher the message HAVE A NICE DAY using the Caesar key.

SOLUTION

step 1 *Using Table 9.1 replace each letter by its ordinal number.*

07 00 21 04 00 13 08 02 04 03 00 24

step 2 *Apply the Caesar transformation* $C \equiv P + 3 \pmod{26}$.

The resulting numbers are 10 03 24 07 03 16 11 05 07 06 03 01. For example, when $P = 24$, $C \equiv 24 + 3 \equiv 01 \pmod{26}$.

step 3 *Substitute the letter corresponding to each ordinal number and group them in blocks of five.*

The resulting ciphertext message is KDYHD QLFHG DB. ■

To decipher such a ciphertext, the recipient simply reverses the steps. From the congruence $C \equiv P + 3 \pmod{26}$, we have the deciphering formula $P \equiv C - 3 \pmod{26}$, which enables us to recover the original plaintext, as the following example demonstrates.

EXAMPLE 9.2 Decipher the ciphertext KDYHD QLFHG DB in Example 9.1.

SOLUTION

step 1 *Using Table 9.1, replace each number with its ordinal number.*

10 03 24 07 03 16 11 05 07 06 03 01

step 2 *Using the deciphering formula* $P \equiv C - 3 \pmod{26}$, *retrieve the numeric plaintext.*
The resulting numeric string is

07 00 21 04 00 13 08 02 04 03 00 24

step 3 *Translate these numbers back to the alphabetic format.*
This yields HAVEA NICED AY.

step 4 *Regroup the letters to recover the original message:* HAVE A NICE DAY. ■

Shift Ciphers

Clearly there is nothing sacred about the choice of the **shift factor** 3 in the Caesar cipher. It is one possible choice out of all the shift ciphers $C \equiv P + k \pmod{26}$, where k is the shift factor and $0 \le k \le 25$. There are 26 possible shift ciphers, one of which is $C \equiv P \pmod{26}$; that is, $C = P$.

A shift cipher is a substitution cipher. By substituting one letter for another, a cryptanalyst can crack a code by using the universally available knowledge of the relative frequency distribution of letters in ordinary text. The most frequently occurring letters in the ciphertext correspond to those in the plaintext. For example, E is the most frequently occurring letter in an arbitrary text, occurring about 12.5% of the time; the next three letters are T, A, and O, occurring about 9%, 8%, and 8% of the time, respectively. Table 9.2 shows the relative frequencies of the various letters in the English alphabet.

Letter	A	B	C	D	E	F	G	H	I	J	K	L	M	N	O	P	Q	R	S	T	U	V	W	X	Y	Z
Relative Frequency in %	8	1.5	3	4	12.5	2	2	5.5	7	0.1	0.7	4	2.5	7	8	2	0.1	6	6.5	9	3	1	2	0.2	2	0.1

Table 9.2

The following example illustrates how this table can be used in cryptanalysis. However, for short and selective messages, the percentages might not be helpful. Consider, for instance, the following well-known passage from President John F. Kennedy's inaugural address in 1961: ASK NOT WHAT YOUR COUNTRY CAN DO FOR YOU, ASK WHAT YOU CAN DO FOR YOUR COUNTRY. This sentence does not contain a single E, and the most frequent letter in it is O.

EXAMPLE 9.3 Assuming that the following ciphertext was created by the shift cipher $C \equiv P + k \pmod{26}$, decipher it:

SLABZ ULCLY ULNVA PHALV BAVMM LHYIB

ASLAB ZULCL YMLHY AVULN VAPHA L

SOLUTION

The given ciphertext can be cracked if we can determine the value of k. To this end, first we construct a frequency table for the letters in the ciphertext, as in Table 9.3.

Letter	A	B	C	D	E	F	G	H	I	J	K	L	M	N	O	P	Q	R	S	T	U	V	W	X	Y	Z
Frequency	9	4	2	0	0	0	0	4	1	0	0	12	3	2	0	2	0	0	2	0	4	5	0	0	4	2

Table 9.3

The most frequently occurring letter in the ciphertext is L, so our best guess is that it must correspond to the plaintext letter E. Since their ordinal numbers are 11 and 4, this implies $11 \equiv 4 + k \pmod{26}$; that is, $k = 7$. Then $C \equiv P + 7 \pmod{26}$, so $P \equiv C - 7 \pmod{26}$. Using this congruence, we can now determine the ordinal number of each letter in the plaintext, as Table 9.4 shows. It follows from the table that the plaintext, after regrouping the blocks, is LET US NEVER NEGOTIATE OUT OF FEAR BUT LET US NEVER FEAR TO NEGOTIATE, another passage from President Kennedy's inaugural address.

Ciphertext Letter	A	B	C	D	E	F	G	H	I	J	K	L	M	N	O	P	Q	R	S	T	U	V	W	X	Y	Z
	00	01	02	03	04	05	06	07	08	09	10	11	12	13	14	15	16	17	18	19	20	21	22	23	24	25
Plaintext Letter	19	20	21	22	23	24	25	00	01	02	03	04	05	06	07	08	09	10	11	12	13	14	15	16	17	18
	T	U	V	W	X	Y	Z	A	B	C	D	E	F	G	H	I	J	K	L	M	N	O	P	Q	R	S

Table 9.4

■

In Example 9.3, our initial guess did in fact produce an intelligible message. On the other hand, if it had resulted in gobbledygook, then we would continue the preceding procedure with the next frequently occurring letters until we succeeded.

Affine Ciphers

Shift ciphers belong to a large family of affine ciphers defined by the formula

$$C \equiv aP + k \pmod{26} \tag{9.2}$$

where a is a positive integer ≤ 25 and $(a, 26) = 1$.

The condition that $(a, 26) = 1$ guarantees that as P runs through the least residues modulo 26, so does C; it ensures that congruence (9.2) has a unique solution for P, by Corollary 4.6:

$$P \equiv a^{-1}(C - k) \pmod{26} \tag{9.3}$$

Since $(a, 26) = 1$, there are $\varphi(26) = 12$ choices for a, so there are $12 \cdot 26 = 312$ affine ciphers. One of them is the identity transformation $C \equiv P \pmod{26}$, corresponding to $a = 1$ and $k = 0$.

When $a = 5$ and $k = 11$, $C \equiv 5P + 11 \pmod{26}$. If $P = 8$, then $C \equiv 5 \cdot 8 + 11 \equiv 25 \pmod{26}$, so under the affine cipher $C \equiv 5P + 11 \pmod{26}$, the letter I is transformed into Z and the letter Q into N. Table 9.5 shows the plaintext letters and the corresponding ciphertext letters created by this affine cipher, which shifts A to L, and in which each successive letter is paired with every fifth letter.

Plaintext Letter	A	B	C	D	E	F	G	H	I	J	K	L	M	N	O	P	Q	R	S	T	U	V	W	X	Y	Z
	00	01	02	03	04	05	06	07	08	09	10	11	12	13	14	15	16	17	18	19	20	21	22	23	24	25
Ciphertext Letter	11	16	21	00	05	10	15	20	25	04	09	14	19	24	03	08	13	18	23	02	07	12	17	22	01	06
	L	Q	V	A	F	K	P	U	Z	E	J	O	T	Y	D	I	N	S	X	C	H	M	R	W	B	G

Table 9.5

The following example illustrates the encrypting procedure for this affine cipher.

EXAMPLE 9.4 Using the affine cipher $C \equiv 5P + 11 \pmod{26}$, encipher the message THE MOON IS MADE OF CREAM CHEESE.

SOLUTION

Since most of the work has been done in Table 9.5, we group the letters into blocks of length five:

THEMO ONISM ADEOF CREAM CHEES E

Then replace each letter by the corresponding ciphertext letter in the table. The resulting encrypted message is CUFTD DYZXT LAFDK VSFLT VUFFX F. ■

The following example demonstrates how to decrypt a message generated by an affine cipher.

EXAMPLE 9.5 Decipher the ciphertext message OZKFZ XPDDA created by the affine cipher $C \equiv 5P + 11 \pmod{26}$.

SOLUTION

Since $C \equiv 5P + 11 \pmod{26}$, $P \equiv 5^{-1}(C - 11) \equiv 21(C - 11) \equiv 21C + 3 \pmod{26}$. For example, when $C = 14$, $P \equiv 21 \cdot 14 + 3 \equiv 11 \pmod{26}$. Thus, the ciphertext letter O is decrypted as L. The other letters can be deciphered in a similar fashion. (We could also use Table 9.5 in the reverse order.) This yields the message LIFEI SGOOD. Reassembling the blocks, we find that the original plaintext is LIFE IS GOOD. ■

If a cryptanalyst knows that the enciphered message was generated by an affine cipher, then he or she will be able to break the cipher using the frequency counts of letters in Table 9.2, as the following example shows.

EXAMPLE 9.6 Cryptanalyze the ciphertext BYTUH NCGKN DUBIH UVNYX HUTYP QNGYV IVROH GSU that was generated by an affine cipher.

SOLUTION

Assume the cipher we are searching for is $C \equiv aP + k \pmod{26}$. To make an educated guess as to which are the most frequently occurring letters in the plaintext, construct a frequency table of letters in the ciphertext, as Table 9.6 shows. According to the table, the most commonly occurring letter in the ciphertext is U, so it is reasonable to assume that it corresponds to the plaintext letter E; that is, $20 \equiv 4a + k \pmod{26}$. Now there are three choices for the next most commonly occurring letter, namely, H, N, and Y. If we assume H corresponds to T, then $7 \equiv 19a + k \pmod{26}$.

Ciphertext Letter	A	B	C	D	E	F	G	H	I	J	K	L	M	N	O	P	Q	R	S	T	U	V	W	X	Y	Z
Frequency	0	2	1	1	0	0	3	4	2	0	1	0	0	4	1	1	1	1	1	2	5	3	0	1	4	0

Table 9.6

Thus, we have

$$4a + k \equiv 20 \pmod{26}$$

$$19a + k \equiv 7 \pmod{26}$$

Solving this linear system, we get $a \equiv 13 \pmod{26}$ and $k \equiv 20 \pmod{26}$, so $C \equiv 13P + 20 \pmod{26}$. But $(13, 26) \neq 1$, so this is not a valid cipher. Thus, our guess that H corresponds to T was not a valid one.

So let us assume that N corresponds to T. This yields the linear system

$$4a + k \equiv 20 \pmod{26}$$

$$19a + k \equiv 13 \pmod{26}$$

Solving this system, $a \equiv 3 \pmod{26}$ and $k \equiv 8 \pmod{26}$. Since $(3, 26) = 1$, this yields a valid cipher $C \equiv 3P + 8 \pmod{26}$. Then $P \equiv 3^{-1}(C - 8) \equiv 9(C - 8) \equiv 9C + 6 \pmod{26}$.

Using this deciphering formula, next we construct Table 9.7, which displays the plaintext letters corresponding to the ciphertext ones. Using the table, we can translate the given encryptic message as POVER TYIST HEPAR ENTOF REVOL UTION ANDCR IME, that is, POVERTY IS THE PARENT OF REVOLUTION AND CRIME, a statement made by the Greek philosopher Aristotle. (It would be interesting to check if the third choice leads to an intelligent plaintext message.)

Ciphertext Letter	A	B	C	D	E	F	G	H	I	**J**	K	L	M	N	O	P	Q	R	S	T	U	V	**W**	X	Y	Z
	00	01	02	03	04	05	06	07	08	**09**	10	11	12	13	14	15	16	17	18	19	20	21	**22**	23	24	25
Plaintext Letter	06	15	24	07	16	25	08	17	00	**09**	18	01	10	19	02	11	20	03	12	21	04	13	**22**	05	14	23
	G	P	Y	H	Q	Z	I	R	A	**J**	S	B	K	T	C	L	U	D	M	V	E	N	**W**	F	O	X
										↑													↑			

Table 9.7 ■

An interesting bonus: It follows from Table 9.7 that the plaintext letters J and W are not affected by the transformation $C \equiv 3P + 8 \pmod{26}$. They are said to be left **fixed** by the cipher. See Exercises 15–18 also.

By and large, a ciphertext generated by an affine cipher does not provide adequate security. One way to make breaking complicated is by using a finite sequence of affine ciphers $C \equiv a_i P + k_i \pmod{26}$, as Figure 9.3 shows, where $1 \le i \le n$. Such a cipher is the **product** (or **composition**) of the n ciphers. Exercises 22–25 further explore such ciphers.

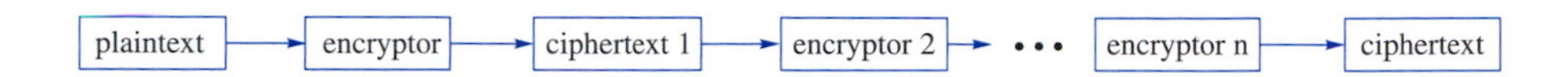

Figure 9.3

Vigenère Ciphers

Another option is to use the enciphering scheme developed by the French cryptographer B. de Vigenère in 1586. The Vigenère cryptosystem employs a **keyword** $w_1 w_2 \dots w_n$ of length n and n shift ciphers $C \equiv P_i + k_i \pmod{26}$ to each plaintext block of length n, where k_i is the ordinal number of the letter w_i and $1 \le i \le n$.

The following example illustrates Vigenère encrypting.

EXAMPLE 9.7 Using the keyword CIPHER and a Vigenère cipher, encrypt the message CRYPTOGRAPHY IS FUN.

SOLUTION

Since the ordinal numbers of the letters C, I, P, H, E, and R in the word CIPHER are 02, 08, 15, 07, 04, and 17, respectively, they serve as the shift factors for each shift cipher for every block. So the six shift ciphers are $C \equiv P + k \pmod{26}$, where $k = 2$, 8, 15, 7, 4, and 17.

Since the keyword is a six-letter word, first we group the letters of the plaintext into blocks of length six: CRYPTO GRAPHY ISFUN.

Now apply the ith cipher to the letter w_i in each block, where $1 \le i \le n$. For instance, consider the first block CRYPTO. Since the ordinary numbers of its letters are 02, 17, 24, 15, 19, and 14, respectively, add to them the key values 2, 8, 15, 7, 4, and 17 in that order modulo 26. The resulting numbers are 4, 25, 13, 22, 23, and 5, and the corresponding letters are E, Z, N, W, X, and F, respectively, so the first ciphertext block is EZNWXF. The other two blocks are similarly transformed to IZPWLP and KAUBR, as Table 9.8 shows. Thus, the resulting ciphertext is EZNWXF IZPWLP KAUBR.

It is important to remember that an affine cipher substitutes the very same letter C for each occurrence of the plaintext letter P, whereas a Vigenère cipher need not. A Vigenère cipher may substitute the same letter C for different plaintext letters. For instance, in the preceding example the plaintext letters A and Y are enciphered

Plaintext Block	C	R	Y	P	T	O	G	R	A	P	H	Y	I	S	F	U	N
	02	17	24	15	19	14	06	17	00	15	07	24	08	18	05	20	13
Ciphertext Block	04	25	13	22	23	05	08	25	15	22	11	15	10	00	20	01	17
	E	Z	N	W	X	F	I	Z	P	W	L	P	K	A	U	B	R

Table 9.8

into P. This makes both encrypting and decrypting in Vigenère more difficult. The two Rs are transformed into Z because they occupy the same spot in their respective blocks. ■

EXERCISES 9.1

Using the Caesar cipher, encipher each proverb.

1. ALL IS WELL THAT ENDS WELL.
2. ALL THAT GLITTERS IS NOT GOLD.

Decipher each ciphertext created by the Caesar cipher.

3. QHFHV VLWBL VWKHP RWKHU RILQY HQWLR Q
4. PDWKH PDWLF VLVWK HTXHH QRIWK HVFLH QFHV

Encipher each quotation using the shift cipher $C \equiv P + 11 \pmod{26}$.

5. NO LEGACY IS SO GREAT AS HONESTY. (W. Shakespeare)
6. THERE IS NO ROYAL ROAD TO GEOMETRY. (Euclid)

Decrypt each quotation below encrypted by the shift cipher $C \equiv P + k \pmod{26}$.

7. GVZRV FGURO RFGZR QVPVA R
8. NSOZX YNHJF SDBMJ WJNXF YMWJF YYTOZ XYNHJ JAJWD BMJWJ

Encipher each using the affine cipher $C \equiv 3P + 7 \pmod{26}$.

9. A THING OF BEAUTY IS A JOY FOR EVER. (John Keats)
10. A JOURNEY OF A THOUSAND MILES MUST BEGIN WITH A SINGLE STEP. (Lao-Tzu)

11–12. Encrypt the messages in Exercises 9 and 10 using the cipher $C \equiv 7P + 10 \pmod{26}$.

The enciphered messages in Exercises 13 and 14 were generated by the affine cipher $C \equiv 5P + 3 \pmod{26}$. Decipher each.

13. UMXIZ NBPUV APMXK X
14. XEXKT IVSTP IZPRQ XPPRP QVIVS TPIZP RQXPP

A plaintext letter is left **fixed** by a cipher if it remains the same in the ciphertext generated by the cipher. Find the letters left fixed by each affine cipher.

15. $C \equiv 5P + 11 \pmod{26}$
16. $C \equiv 7P + 13 \pmod{26}$
17. $C \equiv 5P + 14 \pmod{26}$
18. $C \equiv 9P + 18 \pmod{26}$

Cryptanalyze each ciphertext created by an affine cipher $C \equiv aP + k \pmod{26}$.

19. IRCCH EKKEV CLLFK EIOKL XKKLF ILIGM EKOIV EKKE
20. KARRH HRSLR VUXER FKSRH HDHKA RYREL RYKDV SKAFK QDEKN RDHRS VNXA

21. Find the total number of affine ciphers possible.

Encipher the message, SEND MORE MONEY, using the product of the given affine ciphers.

22. $C \equiv 3P + 7 \pmod{26}$, $C \equiv 5P + 8 \pmod{26}$
23. $C \equiv 5P + 7 \pmod{26}$, $C \equiv 7P + 5 \pmod{26}$

Cryptanalyze each ciphertext generated by the product of two affine ciphers.
(*Hint*: The product of two affine ciphers is also an affine cipher.)

24. GIPJU QDHQG PCUHG XKPGJ LJPOX RGPUL PXRLJ APRGC VLGXJ U
25. ZLFYL FCZFP TBLOO RSBYL FQPON CRELA JOSLE LYCRE RSB

Encrypt each message using the keyword CIPHER for a Vigenère cipher.

26. SEND MORE MONEY.
27. MATHEMATICS IS THE DOOR AND THE KEY TO THE SCIENCES.

Decrypt each ciphertext generated by a Vigenère cipher using the keyword MATH.

28. TETS FHBZ IETS FH
29. XIYL IOGA IABA

9.2 Hill Ciphers

The ciphers we just studied do not provide adequate protection from the cryptanalysts. For this we turn to a new class of ciphers called **block ciphers** (or **polygraphic ciphers**), developed by Lester S. Hill of Hunter College in 1929. In a block cryptosystem, we substitute for each block of plaintext letters of length n a ciphertext block of the same length n, where $n \geq 2$.

Block ciphers with $n = 2$ are called **digraphic ciphers**. In such a system, we group the letters of the plaintext into blocks of length two, adding a dummy letter X at the end, if necessary, to make all blocks of the same length, and then replace each letter with its ordinal number. Each plaintext block P_1P_2 is then replaced by a numeric ciphertext block C_1C_2, where C_1 and C_2 are different linear combinations of P_1 and P_2 modulo 26:

$$C_1 \equiv aP_1 + bP_2 \pmod{26}$$
$$C_2 \equiv cP_1 + dP_2 \pmod{26} \tag{9.4}$$

where $(ad - bc, 26) = 1$. (This condition is needed by Theorem 6.4 to uniquely solve the linear system for P_1 and P_2.) Then we translate each number into a ciphertext letter; the resulting text is the ciphertext.

The following example illustrates this algorithm.

EXAMPLE 9.8 Using the 2×2 linear system

$$C_1 \equiv 5P_1 + 13P_2 \pmod{26}$$
$$C_2 \equiv 3P_1 + 18P_2 \pmod{26}, \tag{9.5}$$

encipher the message SLOW AND STEADY WINS THE RACE. (Aesop, *The Hare and the Tortoise*)

SOLUTION

step 1 *Assemble the plaintext into blocks of length two*:

SL OW AN DS TE AD YW IN ST HE RA CE

step 2 *Replace each letter by its cardinal number*:

18 11	14 22	00 13	03 18	19 04	00 03
24 22	08 13	18 19	07 04	17 00	02 04

step 3 *Using the linear system* (9.5), *convert each block into a ciphertext numeric block*:
When $P_1 = 18$ and $P_2 = 11$, we have

$$C_1 \equiv 5 \cdot 18 + 13 \cdot 11 \equiv 25 \pmod{26}$$
$$C_2 \equiv 3 \cdot 18 + 18 \cdot 11 \equiv 18 \pmod{26}$$

so the first block 18 11 is converted into 25 18. Transforming the other blocks in a similar fashion yields the numeric string

25 18	18 22	13 00	15 21	17 25	13 02
16 00	01 24	25 06	09 15	07 25	10 00

step 4 *Translate the numbers into letters.*
The resulting ciphertext is ZS SW NA PV RZ NC QA BY ZG JP HZ KA. ■

Matrices are useful in the study of Hill cryptosystems. For example, that the linear system (9.5) can be written as

$$\begin{bmatrix} C_1 \\ C_2 \end{bmatrix} \equiv \begin{bmatrix} 5 & 13 \\ 3 & 18 \end{bmatrix} \begin{bmatrix} P_1 \\ P_2 \end{bmatrix} \pmod{26}$$

Since $\Delta = \begin{vmatrix} 5 & 13 \\ 3 & 18 \end{vmatrix} = 51$ and $(51, 26) = 1$, the matrix $\begin{bmatrix} 5 & 13 \\ 3 & 18 \end{bmatrix}$ is invertible modulo 26, with inverse $\begin{bmatrix} 8 & 13 \\ 3 & 21 \end{bmatrix}$ modulo 26. So the deciphering procedure can be effected using the congruence

$$\begin{bmatrix} P_1 \\ P_2 \end{bmatrix} \equiv \begin{bmatrix} 8 & 13 \\ 3 & 21 \end{bmatrix} \begin{bmatrix} C_1 \\ C_2 \end{bmatrix} \pmod{26} \tag{9.6}$$

as the following example demonstrates. (A scientific calculator, such as TI-86, can facilitate the computations.)

EXAMPLE 9.9 Using congruence (9.6), decipher the ciphertext

ZS SW NA PV RZ NC QA BY ZG JP HZ KA.

SOLUTION

Translating the ciphertext letters into numbers, we get

25 18 18 22 13 00 15 21 17 25 13 02
16 00 01 24 25 06 09 15 07 25 10 00

The plaintext numbers corresponding to the block 25 18 are given by

$$\begin{bmatrix} P_1 \\ P_2 \end{bmatrix} \equiv \begin{bmatrix} 8 & 13 \\ 3 & 21 \end{bmatrix} \begin{bmatrix} 25 \\ 18 \end{bmatrix} \equiv \begin{bmatrix} 18 \\ 11 \end{bmatrix} \pmod{26}$$

so $P_1 = 18$ and $P_2 = 11$. The other blocks can be converted similarly. The ensuing plaintext numeric string is

18 11 14 22 00 13 03 18 19 04 00 03
24 22 08 13 18 19 07 04 17 00 02 04

which yields the plaintext SL OW AN DS TE AD YW IN ST HE RA CE, that is, SLOW AND STEADY WINS THE RACE. ■

It is obvious from the preceding two examples that the size of a block can be any size $n \geq 2$, and that the enciphering and deciphering tasks can be accomplished by choosing an $n \times n$ enciphering matrix A modulo 26 such that $(|A|, 26) = 1$, where $|A|$ denotes the determinant of A. Let $P_1, P_2, \ldots, P_n$ be the ordinal numbers of an arbitrary plaintext block and $C_1, C_2, \ldots, C_n$ the corresponding ciphertext numbers.

Let

$$P = \begin{bmatrix} P_1 \\ P_2 \\ \vdots \\ P_n \end{bmatrix} \quad \text{and} \quad C = \begin{bmatrix} C_1 \\ C_2 \\ \vdots \\ C_n \end{bmatrix}$$

The congruence $C \equiv AP \pmod{26}$ provides the enciphering recipe, as the following example shows for $n = 3$. (Once again, a scientific calculator can speed up your computations and minimize the number of errors.)

EXAMPLE 9.10 Using the matrix

$$A = \begin{bmatrix} 3 & 2 & 6 \\ 5 & 7 & 11 \\ 13 & 4 & 1 \end{bmatrix}$$

encrypt the English proverb A PROVERB IS THE CHILD OF EXPERIENCE.

SOLUTION

First, notice that $|A| = -261 \equiv 25 \pmod{26}$, so $(|A|, 26) = 1$. Since A is a 3×3 matrix, split the plaintext into blocks of length three:

APR OVE RBI STH ECH ILD OFE XPE RIE NCE

The corresponding numeric string is

00 15 17 14 21 04 17 01 08 18 19 07 04 02 07
08 11 03 14 05 04 23 15 04 17 08 04 13 02 04

The first numeric ciphertext block is given by

$$\begin{bmatrix} C_1 \\ C_2 \\ C_3 \end{bmatrix} \equiv \begin{bmatrix} 3 & 2 & 6 \\ 5 & 7 & 11 \\ 13 & 4 & 1 \end{bmatrix} \begin{bmatrix} 00 \\ 15 \\ 17 \end{bmatrix} \equiv \begin{bmatrix} 2 \\ 6 \\ 25 \end{bmatrix} \pmod{26}$$

that is, 02 06 25. Continuing in this fashion, we get the numeric string

02 06 25 04 01 10 23 24 25 04 14 05 06 07 15
12 20 21 24 19 24 19 04 25 13 03 23 15 19 25

Convening this into the alphabetic form, we get the desired ciphertext CGZ EBK XYZ EOF GHP MUV YTY TEZ NDX PTZ. ■

To decipher such an encrypted message, we use the fact $P \equiv A^{-1}C \pmod{26}$, as the following example demonstrates.

EXAMPLE 9.11 Using the enciphering matrix A in the preceding example, decrypt the ciphertext CGZ EBK XYZ EOF GHP MUV YTY TEZ HVL PTZ.

SOLUTION

Since $(|A|, 26) = 1$, A^{-1} exists and

$$A^{-1} \equiv \begin{bmatrix} 11 & 4 & -6 \\ -8 & -3 & 3 \\ -7 & 12 & -11 \end{bmatrix} \pmod{26}$$

(We could use any method, such as Gaussian elimination or adjoints, to find A^{-1}, keeping in mind that we are using modular arithmetic. In any case, we can verify that $AA^{-1} \equiv I_3 \pmod{26}$, where I_3 is the 3×3 identity matrix.)

Substituting the numeric equivalents for the ciphertext letters yields the string

02 06 25	04 01 10	23 24 25	04 14 05	06 07 15
12 20 21	24 19 24	19 04 25	07 21 11	15 19 25

To decrypt each block, we employ the formula $P \equiv A^{-1}C \pmod{26}$; that is,

$$\begin{bmatrix} P_1 \\ P_2 \\ P_3 \end{bmatrix} \equiv \begin{bmatrix} 11 & 4 & -6 \\ -8 & -3 & 3 \\ -7 & 12 & -11 \end{bmatrix} \begin{bmatrix} 02 \\ 06 \\ 25 \end{bmatrix} \equiv \begin{bmatrix} 00 \\ 15 \\ 17 \end{bmatrix} \pmod{26}$$

so the first numeric plaintext block is 00 15 17. Continuing in this fashion, we get all blocks:

00 15 17	14 21 04	17 01 08	18 19 07	04 02 07
08 11 03	14 05 04	23 15 04	17 08 04	13 02 04

The corresponding plaintext is APR OVE RBI STH ECH ILD OFE XPE RIE NCE, that is, A PROVERB IS THE CHILD OF EXPERIENCE. ■

Because Hill ciphers deal with blocks, they are more difficult to break. A cryptanalyst could, however, employ the publicly known data about the relative frequency distribution of words of length n to crack the cipher, similar to the case of affine ciphers. When $n = 2$, for instance, there are $26 \cdot 26 = 676$ possible words of length two. So using their frequency counts in an arbitrary plaintext with those of two-letter words in the ciphertext, a cryptanalyst might be able to guess the enciphering matrix A and hence A^{-1}. Obviously, as n gets larger, this task becomes infeasible.

In the next three sections, we will study ciphers that are more difficult to break than Hill ciphers.

EXERCISES 9.2

Using the enciphering matrix A in Example 9.8, encrypt each message.

1. HAVE A NICE DAY.
2. ENJOY THE WEEKEND.

Using the deciphering matrix A^{-1} in Example 9.9, decrypt each.

3. MW AP IC HT IC NH DS
4. NF XT BY ZC SU AO NZ

Using the enciphering key $\begin{bmatrix} 1 & 25 & 25 \\ 25 & 1 & 24 \\ 2 & 9 & 5 \end{bmatrix}$, encipher each plaintext.

5. TIME AND TIDE WAIT FOR NO MAN. (Proverb)
6. THE PEN IS MIGHTIER THAN THE SWORD. (E. G. Bulwer-Lytton)

Using the deciphering key $\begin{bmatrix} 7 & 18 & 19 \\ 15 & 1 & 19 \\ 17 & 17 & 0 \end{bmatrix}$, decipher each ciphertext.

7. ZTH QLJ MOA NLG GPN EXA OCA QTY
8. IGR LDX LRR CIU DIH YVM DYF NBT

Find the blocks of letters left fixed by each enciphering matrix in a block cipher.

9. $\begin{bmatrix} 5 & 13 \\ 3 & 18 \end{bmatrix}$
10. $\begin{bmatrix} 3 & 2 & 6 \\ 5 & 7 & 11 \\ 13 & 4 & 1 \end{bmatrix}$

Let A be the $n \times n$ enciphering matrix of a Hill cipher and B that of another with the same size, so $C \equiv AP \pmod{26}$ and $C \equiv BP \pmod{26}$, where $(|A|, 26) = 1 = (|B|, 26)$. Then $C \equiv B(AP) \equiv (BA)P \pmod{26}$ is the enciphering formula for the product of the two ciphers. Encipher each message using the product cipher formed by the Hill cipher with $A = \begin{bmatrix} 2 & 11 \\ 5 & 13 \end{bmatrix}$ followed by that with $B = \begin{bmatrix} 7 & 15 \\ 3 & 4 \end{bmatrix}$.

11. VANITY FAIR
12. PRIDE AND PREJUDICE
13. Find the blocks of letters left fixed by the product of the block ciphers with the enciphering matrices $A = \begin{bmatrix} 2 & 11 \\ 5 & 13 \end{bmatrix}$ and $B = \begin{bmatrix} 7 & 15 \\ 3 & 4 \end{bmatrix}$.

The following messages were generated by the product of the block ciphers with the enciphering matrices in Exercise 13. Decrypt each.

14. RA XU PV CM EC IS SN XF
15. CZ MH UP GJ DU TC KN DC CX

9.3 Exponentiation Ciphers

The class of **exponentiation ciphers** was developed by Stephen C. Pohlig and Martin E. Hellman of Stanford University in 1978. Exponentiation ciphers provide an interesting confluence of the euclidean algorithm, modular exponentiation, and Fermat's little theorem.

Let p, the **exponentiation modulus**, be an odd prime and let e be a positive integer such that $(e, p-1) = 1$. Since we use the numbers 00 through 25 to represent the letters of the alphabet, clearly $p > 25$; thus, $p \geq 29$. (In fact, we will observe later that the security of the exponentiation cryptosystem is directly related to the size of p, so in practice we choose p to be extremely large.) As we will see shortly, e uniquely determines the ciphertext numeric string corresponding to a given plaintext numeric string, so e serves as the **enciphering exponent**.

To encrypt a plaintext, first translate it into a numeric string using the two-digit ordinal representations in Table 9.1. Then assemble the numbers into blocks of length $2m$ such that the numeric face value of *every* block is $< p$; in other words, choose m as the largest integer such that the number formed by the concatenation of m 25s is $< p$; that is, $\underbrace{2525\ldots25}_{m\text{ 25s}} < p$. For example, if $p = 3037$, then $m = 2$, since $2525 < 3037 < 252525$. This makes sense since p is the modulus.

Now, convert each plaintext numeric block P of length $2m$ into a ciphertext numeric block of the same length using the enciphering congruence

$$C \equiv P^e \pmod{p} \tag{9.7}$$

where $0 \leq P, C \leq p-1$.

The following example illustrates this encrypting procedure.

EXAMPLE 9.12 Using $p = 3037$ as the exponentiation modulus and $e = 31$ as the enciphering exponent, encrypt the message SILENCE IS GOLDEN.

SOLUTION

Using Table 9.1, the plaintext yields the numeric sequence

18 08 11 04 13 02 04 08 18 06 14 11 03 04 13

Since $2525 < p < 252525$, choose $m = 2$ and group the numbers in blocks of length four:

1808 1104 1302 0408 1806 1411 0304 1323

(The last block has been padded with a 23 for X at the end to make all blocks of the same length.)

Now translate each block into a ciphertext block using the enciphering congruence $C \equiv P^{31} \pmod{3037}$. For instance, when $P = 1808$, by modular exponentiation,

$$\begin{aligned} C &\equiv 1808^{16+8+4+2+1} \pmod{3037} \\ &\equiv 1151 \cdot 85 \cdot 1236 \cdot 1052 \cdot 1808 \equiv 1450 \pmod{3037} \end{aligned}$$

so $C = 1450$. The remaining blocks can be computed similarly. The resulting ciphertext is 1450 0186 1435 0523 1894 2531 2340 0990. ■

To decipher a ciphertext generated by the enciphering formula (9.7), notice that $(e, p-1) = 1$; so e has an inverse d modulo $p-1$; that is, $e \cdot d \equiv 1 \pmod{p-1}$. Then $ed = 1 + q(p-1)$ for some integer q. Thus, to recover the plaintext P from the ciphertext block C, we raise C to the power d and reduce it modulo p:

$$C^d \equiv (P^e)^d = P^{ed} = P^{1+q(p-1)} = P(P^{p-1})^q \equiv P \cdot 1^q \equiv P \pmod{p}$$

using Fermat's little theorem. Since the enciphering key uniquely determines d modulo $p-1$, the congruence $C^d \equiv P \pmod{p}$ provides the deciphering algorithm, as the following example demonstrates.

EXAMPLE 9.13 Using the exponentiation modulus $p = 3037$ and the enciphering key $e = 31$, decipher the ciphertext 1450 0186 1435 0523 1894 2531 2340 0990.

SOLUTION

First, notice that $(e, p-1) = (31, 3036) = 1$. We need to compute the inverse d of 31 modulo $p - 1 = 3036$. Using the euclidean algorithm, we can verify that $15 \cdot 3036 - 1469 \cdot 31 = 1$, so $(-1469) \cdot 31 \equiv 1567 \cdot 31 \equiv 1 \pmod{3036}$. Thus, the deciphering key is 1567 modulo 3036.

Since $P \equiv C^{1567} \pmod{3037}$, to decipher a block we raise it to the power 1567 and reduce it modulo 3037 using modular exponentiation. For instance,

$$\begin{aligned} 1450^{1567} &\equiv 1450^{1024+512+16+8+4+2+1} \pmod{3037} \\ &\equiv 2777 \cdot 2304 \cdot 633 \cdot 1947 \cdot 1048 \cdot 896 \cdot 1450 \pmod{3037} \\ &\equiv 1808 \pmod{3037} \end{aligned}$$

Thus, the corresponding plaintext numeric block is 1808. The other blocks can be found similarly. The resulting numeric string is 1808 1104 1302 0408 1806 1411 0304 1323. This yields the plaintext SILENCE IS GOLDEN. ■

As this example demonstrates, once the values of p and e are known, a cryptanalyst can break the cipher. First, he or she must find the deciphering key d and then apply modular exponentiation to each ciphertext block. This, however, is not an easy task for very large primes p.

However, fast algorithms do exist for finding d with only small factors for $p-1$. So to avoid this problem, we choose $p = 2q + 1$, where q is a large prime.

As another application of modular exponentiation, a common key k known only to two individuals can be established in such a way that it would be computationally infeasible for a cryptanalyst to crack it. This can be accomplished without exchanging their enciphering keys at all. To see this, let p be a large prime as before and let x be a positive integer such that $(x, p) = 1$, known to both people. Each person chooses his own key e_i, where $(e_i, p-1) = 1$ and $1 \leq i \leq 2$. The first individual sends the other the integer y_1, where $y_1 \equiv x^{e_1} \pmod{p}$, and the second individual then sends the first person integer y_2, where $y_2 \equiv x^{e_2} \pmod{p}$. The first person determines the common key e by computing $e \equiv y_2^{e_1} \equiv (x^{e_2})^{e_1} \equiv x^{e_1 e_2} \pmod{p}$ and the second by computing $e \equiv y_1^{e_2} \equiv (x^{e_1})^{e_2} \equiv x^{e_1 e_2} \pmod{p}$, where $0 < e < p$. Because e_1 and e_2 are known only to the two individuals, e is known only to them. It is computationally infeasible for an unauthorized individual to determine it in a reasonable amount of time.

This technique can obviously be extended to a network of n individuals with individual keys $e_1, e_2, \ldots, e_n$ who want to share a common key $e \equiv x^{e_1 e_2 \cdots e_n} \pmod{p}$ for secret communication.

EXERCISES 9.3

1. Find the number of letters grouped for an exponentiation cipher in a plaintext numeric block that is 12 digits long.
2. Find the smallest prime that can be used as the modulus in an exponentiation cryptosystem if the letters are grouped in blocks of two letters.
3. Show that the plaintext AB is left fixed by every exponentiation cipher.

With $p = 3037$ as the exponentiation modulus and $e = 31$ as the enciphering exponent, encipher each message.

4. ALL IS WELL.
5. HAVE A NICE DAY.

Using $p = 2549$ as the exponentiation modulus and $e = 11$ as the enciphering exponent, encrypt each message.

6. NO PAINS NO GAINS.
7. NOTHING TO EXCESS. (Solon)

Each ciphertext below was generated by an exponentiation cipher with $p = 3037$ and $e = 31$. Decipher each.

8. 0790 0778 1509 0499
9. 0624 1435 2669 0998

Each ciphertext below was created by an exponentiation cipher with $p = 2333$ and $e = 13$. Decrypt each.

10. 1194 1693 2202 1185 0008
11. 1560 1250 0522 0631 1505

Two persons would like to share secret messages by using a common key and an exponentiation cipher with $p = 131$. Using $x = 2$ as in the text, compute the common key e for the given pair of individual keys.

12. 11, 23
13. 7, 17

14–15. Determine the common deciphering key in Exercises 12 and 13.

9.4 The RSA Cryptosystem

In a conventional cipher system, the enciphering key is known only to the sender and the intended receiver. Since once the enciphering key is known, an unauthorized individual can discover the deciphering key in a short time. Consequently, before coded messages are sent, the key must be transmitted over a secure communication channel.

However, in 1976, Whitfield Diffie and Martin E. Hellman of Stanford University proposed a revolutionary cipher system, called a **public-key cryptosystem**, that makes it unnecessary to keep the key away from unauthorized users. In a public-key system, the enciphering algorithm E of every user of the system is made public as in a telephone directory, while the corresponding decrypting algorithm D is known only to the intended user. Although the encryption key E is public knowledge, it is computationally infeasible to employ it to discover the decryption key D, so it is virtually impossible for a cryptanalyst to crack the system.

Although Diffie and Hellman did not provide a practical implementation of a public-key cipher system, they developed three properties such a cryptosystem must have:

- Each user must have an encryption key E (which is made public) and a decryption key D (which is kept secret) such that $M = E(D(M)) = D(E(M))$ for every message M. Thus, the algorithms E and D are inverse operations.
- It is computationally easy for the user to compute the keys E and D.
- It is computationally infeasible for an unauthorized user to employ the encryption key E to develop the decryption key D, ensuring the security of the system.

How does such a cipher system work? Suppose there are n users of the system. Each person i has an encryption key E_i in the public directory, where $1 \leq i \leq n$. For him to send a message P to person j, he looks up j's encryption key E_j and then sends him the encrypted message $C = E_j(P)$. Then j applies his secret deciphering algorithm D_j to C to recover the original plaintext P, since $D_j(C) = D_j(E_j(P)) = P$. No other person k can crack the message C since $D_k(C) = D_k(E_j(P)) \neq P$, when $k \neq j$.

In 1978, Ronald L. Rivest, Adi Shamir, and Leonard Adelman of the Massachusetts Institute of Technology developed a practical way of implementing Diffie and Hellman's elegant concept. Popularly known as the **RSA cryptosystem**, this public-key system is an exponentiation cipher system based on modular exponentiation and Euler's theorem. (RSA is an acronym for **R**ivest, **S**hamir, and **A**delman.) See Figures 9.4 and 9.5.

Computer Science Prize to Honor 3 Forerunners of Internet Security

John Markoff

The Association of Computing Machinery plans to announce today that Ronald Le Rivest, Adi Shamir and Leonard M. Adleman will receive the 2002 A. M. Turing Award for their development work in public-key cryptography.

The award, which carries a $100,000 prize financed by the Intel Corporation, is gives annually to leading researchers in the field of computer science.

Working at the Massachusetts Institute of Technology in 1977, the three men developed the RSA algorithm, which is widely used today as a basic mechanism for secure Internet transactions, as well as in the banking and credit card industries.

The strength of this approach is that it provides highly secure communications over distances between parties that have never previously been in contact.

Dr. Rivest now teaches in the electrical engineering and computer science department at M.I.T.

Dr. Shamir is a professor in the applied mathematics department at the Weizmann Institute of Science in Israel.

Dr. Adleman is a professor of computer science and of molecular biology at the University of Southern California.

Figure 9.4

A Prime Argument in Patent Debate

Simson Garfinkel SPECIAL TO THE GLOBE

In a move that will likely inflame the debate over the government's patent application procedures, a California mathematician has received what is believed to be the first patent on a prime number.

But collecting royalties for its use might be difficult.

Actually, Roger Schlafly has patented two prime numbers, but only when they are used together. According to the US Patent and Trade office, the numbers are trade-marked under patent No. 5,373,560, a figure that doesn't nearly approach the size of the two patented numbers themselves—one is 150 digits long, the other 300 digits.

The patent, titled "partial modular reduction method," was awarded to Schlafly, an independent mathematician and specialist in the field of cryptography, in December but only recently came to public attention.

The patent claims a new technique for finding certain kinds of prime numbers, which can be used to rapidly perform the kinds of mathematical operations necessary for public key cryptography.

(A prime number is a number that cannot be evenly divided by any number other than 1 and itself. The numbers 2, 3, 13 and 29 are all prime and are not covered by any known patent. Public key cryptography is a technique, based on prime number theory, that allows two individuals to exchange secret messages by computer.)

"I'm sure if you just went to someone and said, 'Can you patent a prime number?' they would say 'No, that's ridiculous,'" said Schlafly, interviewed from his home in Soquel, near Santa Cruz, Calif. Schlafly said he developed the patented algorithm while working on a program called SECRET AGENT, which is used to encrypt electronic mail. He added the patent claims for the two prime numbers as an experiment. "I was kind of interested in pushing the system to see how far you could go with allowable claims."

(*continued*)

Figure 9.5

Nevertheless, Schlafly said, the two prime numbers satisfy the patent office's conditions for patentability: They are useful, have never been used before by anyone else, and their use for this particular technique is not obvious.

Others see the prime number patent as evidence that the patent office has lost its grip on the patenting process.

"That's outrageous," said Pamela Samuelson, a professor of law at the University of Pittsburgh and an expert on software patents and copyrights.

"It also seems inconsistent with some of the recent decisions issued by the Federal Circuit [Court of Appeals] ... Unless you claim some physical structure [that is used by] an algorithm or a data structure, you can't patent it."

Nearly two years ago, the patent office awarded a sweeping patent that covered the field of multimedia to Compton's New Media. At the time, an outraged computer industry argued that there was nothing new or novel in Compton's programs that deserved a patent. Eventually, the patent office reconsidered the Compton's patent, and threw it out.

Whether or not that will happen with Schlafly's patent remains to be seen. Under most circumstances, patents are invalid if the invention that they described is published before the patent application is filed.

"There are entire journals and conference proceedings devoted to the general subject of this application," says Gregory Aharonian, who published the Internet Patent News Services and maintains a database of several hundred thousand pieces of software art. But few software patents that have been awarded in recent years cite any prior art other than previous patents, Aharonian says.

But whereas the algorithm may be covered under the doctrine of prior art, says Aharonian, the prime numbers themselves are probably patentable. "The claiming of certain prime numbers as part of an encryption process doesn't seem to me to be unnatural," said Aharonian. "I can claim certain specific chemicals as part of a chemical engineering process, so why not a specific number as part of a math engineering process?"

The numbers claimed in the patent are 512 bits and 1,024 bits long, or roughly 150 and 300 decimal digits. While these numbers are quite large by everyday standards, they are typical of the size of numbers used for cryptographic processes. By design, the numbers are so large that it is exceedingly unlikely that a person could guess them or otherwise intentionally discover what they are.

The two principle techniques of public key cryptography were discovered and patented by scientists at Stanford University and at the Massachusetts Institute of Technology in the 1970s. In 1990, they were both licensed to Public Key Partners, a holding company based in California. Last year, Schlafly filed suit against PKP in federal court, claiming that the PKP patents are invalid.

Regarding his own patent, Schlafly said, its real value is the technique that it describes for finding the special prime numbers, rather than the two specific prime numbers that it describes. "I really don't anticipate somebody reading this patent and saying, 'look, here's a good prime number, let's use it!'" he said.

Nevertheless, the patent gives Schlafly the legal right to sue anybody in the United States for using his numbers without permission. "I suppose that you can tell people that if they want to license these prime numbers, they should just call me up."

Figure 9.5

The Enciphering Algorithm

In an RSA system, the enciphering key is a pair (e, n) of positive integers e and n, where the enciphering modulus n is the product of two very large and distinct primes p and q, each about 100 digits long, and $(e, \varphi(n)) = 1$. To encrypt a plaintext message, as in the exponentiation cryptosystem, we group the plaintext numeric equivalents into blocks of length $2m$, with padding at the end if necessary. Then we convert

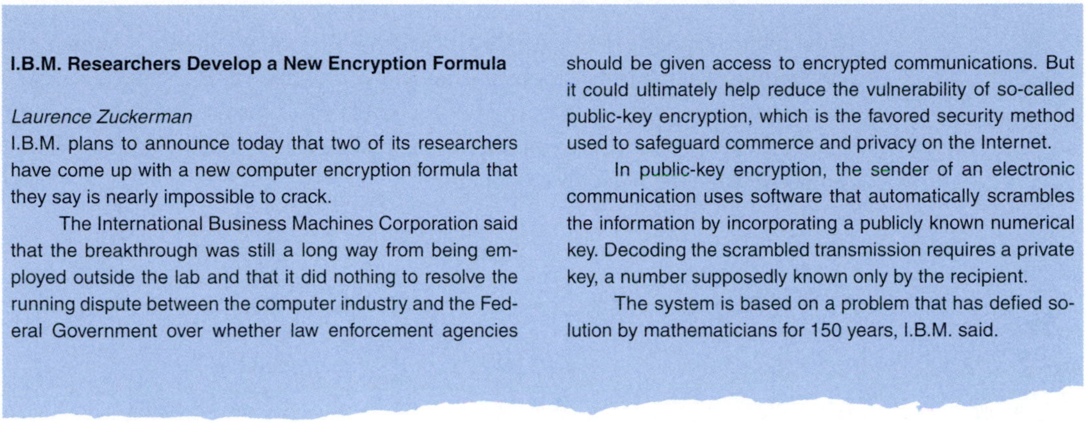

I.B.M. Researchers Develop a New Encryption Formula

Laurence Zuckerman

I.B.M. plans to announce today that two of its researchers have come up with a new computer encryption formula that they say is nearly impossible to crack.

The International Business Machines Corporation said that the breakthrough was still a long way from being employed outside the lab and that it did nothing to resolve the running dispute between the computer industry and the Federal Government over whether law enforcement agencies should be given access to encrypted communications. But it could ultimately help reduce the vulnerability of so-called public-key encryption, which is the favored security method used to safeguard commerce and privacy on the Internet.

In public-key encryption, the sender of an electronic communication uses software that automatically scrambles the information by incorporating a publicly known numerical key. Decoding the scrambled transmission requires a private key, a number supposedly known only by the recipient.

The system is based on a problem that has defied solution by mathematicians for 150 years, I.B.M. said.

Figure 9.6

each block P into a ciphertext block C using the encrypting congruence

$$C = E(P) \equiv P^e \pmod{n} \tag{9.8}$$

where $0 \leq C, P < n$. See Figure 9.6.

The following example illustrates this algorithm.

EXAMPLE 9.14 Using the RSA enciphering modulus $n = 2773$ and the enciphering key $e = 21$, encrypt the message SILENCE IS GOLDEN.

SOLUTION

As in Example 9.12, after the numeric translation and grouping into blocks, the plaintext yields

$$1808 \ 1104 \ 1302 \ 0408 \ 1806 \ 1411 \ 0304 \ 1323$$

Now, using modular exponentiation and formula (9.8), convert each block P into a ciphertext block C:

$$C \equiv P^e = P^{21} \pmod{2773}$$

For instance, when $P = 1808$,

$$C \equiv 1808^{21} \equiv 1808^{16+4+1} \equiv 1511 \cdot 666 \cdot 1808 \equiv 0010 \pmod{2773}$$

The other blocks can be found similarly. The ensuing ciphertext message is 0010 0325 2015 2693 2113 2398 2031 1857. ■

The Deciphering Algorithm

To decipher a ciphertext C generated by an RSA system, we need to compute the inverse d of the enciphering exponent e modulo $\varphi(n)$, which exists since $(e, \varphi(n)) = 1$. Then $de \equiv 1 \pmod{\varphi(n)}$; that is, $de = 1 + k\varphi(n)$ for some constant k. Knowing the deciphering exponent d, we can recover the plaintext P by raising both sides of congruence (9.8) to the power d modulo n:

$$\begin{aligned} C &= P^e \pmod{n} \\ C^d &\equiv (P^e)^d = P^{ed} = P^{1+k\varphi(n)} \pmod{n} \\ &= P \cdot [P^{\varphi(n)}]^k \equiv P \cdot 1^k = P \pmod{n} \end{aligned} \tag{9.9}$$

where, by Euler's theorem, $P^{\varphi(n)} \equiv 1 \pmod{n}$, if $(P, n) = 1$. The pair (d, n) is the **deciphering key**.

Even in the highly unlikely event that $(P, n) \neq 1$, the RSA algorithm works. To see this, let $n = pq$. Then $(P, n) = p$, q, or pq. Since $P < n$, $(P, n) \neq n$. When $(P, n) = p$, $(P, q) = 1$, so by Fermat's little theorem, $P^{q-1} \equiv 1 \pmod{q}$. Since $de \equiv 1$ $(\text{mod } (p-1)(q-1))$, $de = 1 + k(p-1)(q-1)$ for some integer k. Therefore,

$$P^{de} = P \cdot (P^{q-1})^{k(p-1)} \equiv P \cdot 1^{k(p-1)} \equiv P \pmod{q}$$

That is,

$$C^d \equiv P \pmod{q}$$

When $(P, n) = p$, $C^d \equiv P^{de} \equiv 0 \equiv P \pmod{p}$. Thus, $C^d \equiv P \pmod{p}$ and $C^d \equiv P$ $(\text{mod } q)$, so $C^d \equiv P \pmod{n}$. The case $(P, n) = q$ yields the same conclusion.

For instance, if p and q are 100-digit primes, the probability of such an occurrence of a plaintext block is extremely negligible, namely, less than $2 \cdot 10^{-99}$. See Supplementary Exercises 6 and 7.

The following example demonstrates the decrypting algorithm D.

EXAMPLE 9.15 Decrypt the ciphertext message 0010 0325 2015 2693 2113 2398 2031 1857 that was created using the RSA enciphering key $(e, n) = (21, 2773)$.

SOLUTION

Because $\varphi(n) = \varphi(2773) = \varphi(47 \cdot 59) = 46 \cdot 58 = 2668 = 127 \cdot 21 + 1$, $(-127) \cdot 21 \equiv 1 \pmod{2668}$; that is, $2541 \cdot 21 \equiv 1 \pmod{2668}$, so the deciphering exponent is $d = 2541$. Because $P \equiv C^d \pmod{n}$, raise each ciphertext C to the power 2541 modulo 2773. For instance, when $C = 0010$:

$$
\begin{aligned}
P &\equiv 0010^{2541} \equiv 10^{2541} \pmod{2773} \\
&\equiv 10^{2048+256+128+64+32+8+4+1} \pmod{2773} \\
&\equiv 1024 \cdot 2431 \cdot 2500 \cdot 1366 \cdot 2127 \cdot 74 \cdot 1681 \cdot 10 \pmod{2773} \\
&\equiv 1808 \pmod{2773}
\end{aligned}
$$

as expected. The other blocks can be decrypted similarly. ■

Digital Signatures

The property $E(D(M)) = M$, found in public-key cryptosystems, can be effectively used to transmit "digitally signed" messages. This is a practical and highly desirable feature, since such a cipher system ensures authentication and protects against forgeries. Such digital signatures are widely used in electronic banking.

Interestingly, in June 2000, President Bill Clinton signed into law a bill allowing businesses and consumers to enter into legally binding arrangements with electronic rather than handwritten signatures. *E-signing*, as the new process is called, is expected to spur new technologies, accelerate electronic transactions, and save billions of dollars in administrative costs. See Figure 9.7.

To see how signed messages work in public-key cipher systems and, in particular, RSA systems, suppose that person i wishes to send person j a signed message P. First, person i applies his secret deciphering algorithm D_i to P. This yields $D_i(P) \equiv P^{d_i} \pmod{n}$; he then applies j's enciphering algorithm E_j to it, since E_j is public knowledge. This produces the message $E_j(D_i(P)) \equiv P^{d_i e_j} \pmod{n}$. Person i now sends this convoluted message to j.

To decipher this message, allegedly sent by person i, first person j applies his deciphering key to it to yield

$$D_j(E_j(D_l(P))) \equiv (P^{d_i e_j})^{d_j} = (P^{e_j d_j})^{d_i} \equiv P^{d_i} \pmod{n} = D_i(P)$$

because D_j and E_j are inverse operations. He then applies person i's public encryption algorithm E_i to it to yield

$$E_i(D_i(P)) \equiv (P^{d_i})^{e_i} = P^{d_i e_i} \equiv P \pmod{n}$$

Once again, because E_i and D_i are inverse operations, this operation produces the original plaintext P. This ensures that the original message was in fact sent by person i and nobody else, since $E_i(D_k(P)) \neq P$ if $k \neq i$. Consequently, i can never claim that he did not send the plaintext P, since he is in sole possession of the secret key D_i.

As these two examples demonstrate, both encryption and decryption become tedious as n gets larger and larger; so for an RSA system to be realistically useful, n must be very large. Both processes require fast computers for implementation.

E-Signing Law Seen as a Boon to E-Business

Barnaby J. Feder

The law President Clinton signed last week allowing businesses and consumers to seal a variety of legally binding arrangements with electronic rather than hand-written signatures raised the speed limit on e-business development, analysts say.

They project that many enterprises awash in documents, especially financial services, real estate and the government itself, will accelerate efforts to use computer transactions to limit paperwork once the law takes effect in October. Such a transition is expected to save billions of dollars annually in administrative costs and cut some online transactions—like setting up a trading account or applying for a home loan—from days to minutes.

"E-signing," as the new process is called, is also expected to spur a variety of technologies that provide digital variations on penning one's name on paper, including the use of coded messages, penlike styluses or thumb prints on electronic pads, or camera shots of the signer's face or eye.

The new law came after 46 states and many foreign countries had adopted laws encouraging online deal-making, and many online businesses have already incorporated such capabilities. So it may be hard to calculate the immediate financial impact of the federal law on the major companies supplying such technology.

"It validates the market but it won't really add to their revenue in the near term," said Mark Fernandes of Merrill Lynch, who follows companies that provide the software and support services needed for electronic business.

All of them rely on public key encryption, a technology invented in the 1970's but not widely used until the e-commerce wave hit the Internet. Such systems use a combination of public and private keys, or snippets of numbers, to pass secure messages through a trusted third party, or certification authority. The system not only allows a recipient to be assured the message came from the party that claims to have sent it but also that it has not been tampered with.

The digital signature such systems produce looks nothing like a scrawled John Hancock—in fact, it is invisible. As a result, many entrepreneurs are betting that other systems will be used instead of public key encryption, or in addition to it, to complete e-commerce deals. President Clinton signed the bill into law on Friday with a smart card—a credit card-sized device programmed to work in combination with a password furnished by the user. Such systems are already widely used in Europe.

Figure 9.7

To choose n, first find two large primes p and q, about 100 digits long. Then $n = pq$ is about 200 digits long. That the value of n is public information does not imply that its prime factors are publicly known. The factoring of a 200-digit number is an extremely time-consuming proposition.

Once p and q have been selected, the enciphering exponent e must be chosen in such a way that $(e, \varphi(n)) = 1$. One way to do this is by choosing a prime greater than both p and q.

The exponent e must also be chosen so that $2^e > n$; this ensures that every plaintext block, except 0 and 1, will be subjected to reduction modulo n. Otherwise, since $C \equiv P^e \pmod{n}$, P can be recovered by taking the eth root of C.

The deciphering exponent d can easily be computed using the euclidean algorithm, where $de \equiv 1 \pmod{\varphi(n)}$ and $\varphi(n) = (p-1)(q-1)$.

Publishing the enciphering key (e, n) does not compromise security, because a cryptanalyst must know the value of $\varphi(n)$ to compute the deciphering expo-

Cracking Huge Numbers

Ivars Peterson

It's easy to multiply two large prime numbers to obtain a larger number as the answer. But the reverse process—factoring a large number to determine its components—presents a formidable challenge. The problem appears so hard that the difficulty of factoring underlies the so-called RSA method of encrypting digital information.

An international team of computer scientists, mathematicians, and other experts recently succeeded in finding the factors of a 129-digit number (see fig. 1) suggested seventeen years ago as a test of the security of the RSA cryptographic scheme.

114, 381, 625, 757, 888, 867, 669, 235, 779, 976, 146, 612, 010, 218, 296, 721, 242, 362, 562, 561, 842, 935, 706, 935, 245, 733, 897, 830, 597, 123, 563, 958, 705, 058, 989, 075, 147, 599, 290, 026, 879, 543, 541 $=$ 3, 490, 529, 510, 847, 650, 949, 147, 849, 619, 903, 898, 133, 417, 764, 638, 493, 387, 843, 990, 820, 577 $\times$ 32, 769, 132, 993, 266, 709, 549, 961, 988, 190, 834, 461, 413, 177, 642, 967, 992, 942, 539, 798, 288, 533

Fig. 1
The number and its two prime factors

This feat and other work now complicate encoding schemes used for national and commercial security.

The effort required the use of more than 600 computers scattered throughout the world. Partial results were sent electronically to graduate student Derek Atkins at the Massachusetts Institute of Technology, who assembled and passed the calculations on to Arjen K. Lenstra of Bell Communications Research in Morristown, New Jersey. In the final step, which by itself consumed forty-five hours of computer time, Lenstra used these data and a MasPar MP-1 computer with 16000 processors to compute the factors.

"It was a nice piece of work—a huge computation done over 8 months," says Burton S. Kaliski Jr. of RSA Data Security in Redwood City, California.

The magnitude of the effort required to factor a 129-digit number demonstrates the strength of the RSA cryptosystem, which typically involves numbers of 155 or more digits. However, steady improvements in factoring methods are likely to force the use of significantly larger numbers in the future to ensure security. More worrisome are the consequences of new research apparently proving that under certain circumstances, factoring may actually be easy.—From *Science News*, 7 May 1994.

Figure 9.8

nent d. Clearly, $\varphi(n)$ can be computed if p and q are known, since $\varphi(n) = \varphi(pq) = (p-1)(q-1)$. Since computing $\varphi(n)$ involves the factoring of n, it is an equally difficult task. Since p and q are 100 digits long and $n = pq$ is about 200 decimal digits long, the fastest known factorization algorithm will take about four billion years of computing time on the fastest available computer, as Table 9.9[†] shows. Although this could change with time and technology, the RSA system is virtually secure at present. If faster factorization techniques and faster computers become available, then the size of the factors can be increased accordingly to maintain the security of the system. See Figure 9.8.

Note that the primes p and q can be computed from $\varphi(n)$. See Exercises 9 and 10.

[†] Based on R. L. Rivest *et al.*, "A Method for Obtaining Digital Signatures and Public-Key Cryptosystems," *Communications of the ACM*, 21 (Feb. 1978), 120–126.

Number of digits	*Time*
50	3.9 hours
75	104 days
100	74 years
200	3.8×10^9 years
300	4.9×10^{15} years
500	4.2×10^{25} years

Table 9.9

Also, to prevent a cryptanalyst from resorting to special techniques to factor n, both p and q should be of about the same size, with $p-1$ and $q-1$ having large prime factors and $(p-1, q-1)$ small.

However, if d is known then $ed-1$, a multiple of $\varphi(n)$ can be computed; knowing a multiple of $\varphi(n)$, n can be factored fairly easily using an algorithm developed in 1976 by G. L. Miller.

EXERCISES 9.4

Using the RSA enciphering key $(e, n) = (11, 2867)$, encrypt each message.

1. SEAFOOD
2. OPEN DOOR

3–4. Redo Exercises 1 and 2 using the RSA enciphering key $(e, n) = (17, 2867)$.

Each ciphertext below was generated by the RSA enciphering key $(e, n) = (11, 2867)$. Decipher each.

5. 1420 0614 1301 1694
6. 1959 1384 1174 2050

Decrypt each ciphertext below that was created by the RSA enciphering key $(e, n) = (17, 2867)$.

7. 0579 0341 0827 1511
8. 0592 2131 2584 2188

Let $n = pq$, where p and q are primes with $p > q$. [Exercises 9–11 show that if n and $\varphi(n)$ are known, then the prime factors of n can be determined.]

9. Show that $p+q = n - \varphi(n) + 1$.
10. Show that $p-q = \sqrt{(p+q)^2 - 4n}$.
11. Express p and q in terms of n and $\varphi(n)$.
12. Using Exercises 9–11, determine the primes p and q if $n = pq = 3869$ and $\varphi(n) = 3744$.
13. Redo Exercise 12 if $n = 3953$ and $\varphi(n) = 3828$.

Anne and Betsey would like to send each other a signed message using an RSA cipher. Their encryption keys are $(13, 2747)$ and $(17, 2747)$, respectively. Find the signed cipher message sent by

14. Anne if the plaintext message is MARKET.
15. Betsey if the plaintext message is INPUT.

With the enciphering keys as before, find the plaintext sent by

16. Anne if her signed message to Betsey is 1148 0194 2715.
17. Betsey if her signed message to Anne is 1130 2414 2737.

9.5 Knapsack Ciphers

In 1978, Ralph C. Merkle and Martin E. Hellman, both electrical engineers at Stanford University, developed a public-key cryptosystem based on the **knapsack problem**, a celebrated problem in combinatorics. It can be stated as follows: *Given a knapsack of volume S and n items of various volumes $a_1, a_2, \ldots, a_n$, which of the items can fill the knapsack*? In other words, given the positive integers $a_1, a_2, \ldots, a_n$, called **weights**, and a positive integer S, solve the LDE

$$S = a_1x_1 + a_2x_2 + \cdots + a_nx_n \tag{9.10}$$

where $x_i = 0$ or 1. [Note that S is the dot product of the vectors $(a_1, a_2, \ldots, a_n)$ and $(x_1, x_2, \ldots, x_n)$.] The knapsack problem may have no solutions, one solution, or more than one solution.

For example, the knapsack problem $3x_1 + 5x_2 + 9x_3 + 19x_4 + 37x_5 = 45$ has one solution $(1, 1, 0, 0, 1)$, since $3 + 5 + 0 + 0 + 37 = 45$. On the other hand, the knapsack problem $3x_1 + 5x_2 + 8x_3 + 13x_4 + 21x_5 = 34$ has two solutions; they are $(0, 0, 0, 1, 1)$ and $(0, 1, 1, 0, 1)$, because $0 + 0 + 0 + 13 + 21 = 34 = 0 + 5 + 8 + 0 + 21$. But the problem $5x_1 + 14x_2 + 15x_3 + 27x_4 + 11x_5 = 23$ has no solutions.

Solving a knapsack problem is usually a very difficult task. An obvious, but certainly impractical, method is to check the various 2^n possibilities for a solution $(x_1, x_2, \ldots, x_n)$, where $x_i = 0$ or 1, until a solution emerges or all cases have been exhausted. Even the best-known method for solving the problem requires about $2^{n/2}$ computational operations, so for $n = 100$ a computer solution becomes computationally infeasible.

Nonetheless, problem (9.10) can be solved fairly easily if the weights have special properties. For instance, if $a_i = 2^{i-1}$, then $S = x_1 + 2x_2 + 2^2x_3 + \cdots + 2^{n-1}x_n$ has a solution $(x_1, x_2, \ldots, x_n)$ if $(x_n, x_{n-1}, \ldots, x_1)_{\text{two}} = S$.

It is also easy to solve it if $\sum_{i=1}^{j-1} a_i < a_j$, where $2 \le j \le n$. A sequence with this property is said to be **superincreasing**. For example, consider the sequence 3, 5, 9, 19, 37. Because $3 < 5$, $3 + 5 < 9$, $3 + 5 + 9 < 19$, and $3 + 5 + 9 + 19 < 37$, the sequence is superincreasing.

The following example shows how to solve a knapsack problem with superincreasing weights.

EXAMPLE 9.16 Solve the knapsack problem $3x_1 + 5x_2 + 9x_3 + 19x_4 + 37x_5 = 45$.

SOLUTION

Since the weights are superincreasing and since $3 + 5 + 9 + 19 < 37 < 45$, $x_5 = 1$. Then $3x_1 + 5x_2 + 9x_3 + 19x_4 = 8$. Since $19 > 8$ and $9 > 8$, $x_4 = 0 = x_3$. This yields $3x_1 + 5x_2 = 8$, so $x_1 = 1 = x_2$. Thus the solution to the given problem is $(1, 1, 0, 0, 1)$. ■

An Algorithm for Solving the Knapsack Problem with Superincreasing Weights

This solution can be generalized to derive a solution to problem (9.10) with superincreasing weights, if a solution exists. It is given by the following algorithm:

$$x_n = \begin{cases} 1 & \text{if } S \geq a_n \\ 0 & \text{otherwise} \end{cases}$$

Once x_n is determined, the remaining components $x_{n-1}, x_{n-2}, \ldots, x_1$ can be computed using the formula

$$x_j = \begin{cases} 1 & \text{if } S - \sum_{i=j+1}^{n} a_i x_i \geq a_j \\ 0 & \text{otherwise} \end{cases}$$

where $j = n - 1, n - 2, \ldots, 1$. As the preceding example demonstrates and these formulas indicate, we must work from right to left to find a solution.

To see why these formulas work, suppose $x_n = 0$ when $S \geq a_n$. Then $S = \sum_{i=1}^{n} a_i x_i \leq \sum_{i=1}^{n-1} a_i < a_n$, which is a contradiction. Therefore, $x_n = 1$ when $S \geq a_n$. It also implies that $x_n = 0$ if $S < a_n$.

Now let $1 \leq j \leq n - 1$. Assume $x_j = 0$ when $S - \sum_{i=j+1}^{n} a_i x_i \geq a_j$. Then $S - \sum_{i=j+1}^{n} a_i x_i = \sum_{i=1}^{j} a_i x_i = \sum_{i=1}^{j-1} a_i x_i \leq \sum_{i=1}^{j-1} a_i < a_j$, again a contradiction. Thus, both halves do hold.

The following example illustrates this algorithm.

EXAMPLE 9.17 Solve the knapsack problem $2x_1 + 3x_2 + 7x_3 + 13x_4 + 27x_5 = 39$.

SOLUTION

First, notice that the sequence of weights is superincreasing. Here $S = 39$ and $(a_1, a_2, a_3, a_4, a_5) = (2, 3, 7, 13, 27)$. Because $S \geq a_5$, $x_5 = 1$. Then $2x_1 + 3x_2 + 7x_3 + 13x_4 = 12 < 13(= a_4)$, so $x_4 = 0$. This yields, $2x_1 + 3x_2 + 7x_3 = 12 > 7(= a_3)$, so $x_3 = 1$. Then $2x_1 + 3x_2 = 5 > 3$, so $x_2 = 1$. This implies $x_1 = 1$. Thus, the solution is $(1, 1, 1, 0, 1)$; that is, $2 + 3 + 7 + 0 + 27 = 39$. ■

The Enciphering Algorithm

We can build a public-key system based on knapsack problems with superincreasing weights $a_1, a_2, \ldots, a_n$. To this end, choose a positive integer $m > 2a_n$ and a positive integer w relatively prime to m. Now form the sequence $b_1, b_2, \ldots, b_n$, where $b_i \equiv wa_i \pmod{m}$, $0 \le b_i < m$. This sequence need not be superincreasing.

A user of the knapsack cryptosystem makes the **enciphering sequence** $b_1, b_2, \ldots, b_n$ public in a directory, keeping secret the original sequence $a_1, a_2, \ldots, a_n$, the **enciphering modulus** m, and the **multiplier** w. Before encrypting a plaintext, convert it into a bit string using the five-digit binary equivalents in Table 9.10. The string is then partitioned into blocks P of length n, where n is the number of elements in the enciphering sequence. If the last block does not have n bits, then pad it with enough 1s, so all blocks will be of the same length n. Now transform each numeric plaintext block $x_1x_2\ldots x_n$ into the sum

$$S = b_1x_1 + b_2x_2 + \cdots + b_nx_n \tag{9.11}$$

The sums thus generated form the ciphertext message.

Letter	*Binary Equivalent*	*Letter*	*Binary Equivalent*
A	00000	N	01101
B	00001	O	01110
C	00010	P	01111
D	00011	Q	10000
E	00100	R	10001
F	00101	S	10010
G	00110	T	10011
H	00111	U	10100
I	01000	V	10101
J	01001	W	10110
K	01010	X	10111
L	01011	Y	11000
M	01100	Z	11001

Table 9.10

The following example illustrates this method.

EXAMPLE 9.18 Using the knapsack cipher based on the superincreasing weights 6, 8, 15, and 31, modulus $m = 65$, and multiplier $w = 12$, encipher the message ON SALE.

SOLUTION

First, notice that the sequence has $n = 4$ elements, $m > 2a_4$, and $(m, w) = (65, 12) = 1$.

step 1 *Multiply each element in the sequence by 12 and reduce each product modulo 65*:
$6 \cdot 12 \equiv 7 \pmod{65}$, $8 \cdot 12 \equiv 31 \pmod{65}$, $15 \cdot 12 \equiv 50 \pmod{65}$, and $31 \cdot 12 \equiv 47 \pmod{65}$. The resulting enciphering sequence is 7, 31, 50, 47.

step 2 *Using Table 9.10, translate the letters into binary and then group the bits into blocks of length 4. Pad the last block with 1s if necessary*:
01110 01101 10010 00000 01011 00100. This yields 0111 0011 0110 0100 0000 0101 1001 0011.

step 3 *Find the ciphertext message*:
To this end, convert each block into a sum by multiplying the bits by the elements 7, 31, 50, and 47 of the enciphering sequence and then by adding the products. For example, $0 \cdot 7 + 1 \cdot 31 + 1 \cdot 50 + 1 \cdot 47 = 128$. Similarly, the other sums are 97, 81, 31, 0, 78, 54, and 97. Thus, the ciphertext message is 128 97 81 31 0 78 54 97. ■

The Deciphering Algorithm

Deciphering a knapsack ciphertext is equally easy. Multiply equation (9.11) by w^{-1} modulo m, which exists since $(m, w) = 1$:

$$w^{-1}S \equiv \sum_{i=1}^{n} w^{-1} b_i x_i \pmod{m}$$

$$\equiv \sum_{i=1}^{n} (w^{-1} b_i) x_i \equiv \sum_{i=1}^{n} a_i x_i \pmod{m}$$

Because $m > 2a_n$ and $2a_n > \sum_{i=1}^{n} a_i x_i$, $m > \sum_{i=1}^{n} a_i x_i$. Let $S' \equiv w^{-1}S \pmod{m}$, where $0 \le S' < m$. Then $S' = \sum_{i=1}^{n} a_i x_i$. This knapsack problem can be solved because the original coefficients $a_1, a_2, \ldots, a_n$ are superincreasing. The unique solution $(x_1, x_2, \ldots, x_n)$ yields the block $x_1 x_2 \ldots x_n$. After finding all the blocks, all we need to do is regroup the bits into blocks of five bits and then substitute the letter corresponding to each block.

The following example illustrates this algorithm.

EXAMPLE 9.19 Decipher the knapsack ciphertext message 128 97 81 31 0 78 54 97 created with modulus $m = 65$, multiplier $w = 12$, and the enciphering sequence 7, 31, 50, 47.

SOLUTION

step 1 *Use the euclidean algorithm to find the inverse of w modulo m.*
Since $38 \cdot 12 \equiv 1 \pmod{65}$, $w^{-1} \equiv 38 \pmod{65}$.

step 2 *Construct a knapsack problem for each numeric ciphertext block and solve it.*

$$S = b_1x_1 + b_2x_2 + b_3x_3 + b_4x_4$$

$$128 = 7x_1 + 31x_2 + 50x_3 + 47x_4$$

Multiply both sides by $w^{-1} \equiv 38$ modulo 65:

$$38 \cdot 128 \equiv 38 \cdot 7x_1 + 38 \cdot 31x_2 + 38 \cdot 50x_3 + 38 \cdot 47x_4 \pmod{65}$$

This yields $54 = 6x_1 + 8x_2 + 15x_3 + 31x_4$. Solving, $x_1x_2x_3x_4 = 0111$. Similarly, we get the other blocks: 0011 0110 0100 0000 0101 1001 0011.

step 3 *Recover the plaintext by regrouping the bits into blocks of length five, and then replace each with the corresponding letter.*
This yields the original message ON SALE. (Verify this.) ■

A Drawback of the Knapsack Cryptosystem

The Merkle–Hellman knapsack cryptosystem does not possess the property $E(D(M)) = M$ as proposed for a public-key system by Diffie and Hellman. Consequently, it is not a candidate for a signature system.

Initially, the Merkle–Hellman system generated a great deal of interest since the encryption and decryption algorithms are easier and faster to implement. It seemed to be a major breakthrough, since it is based on a difficult problem. In April 1982, however, A. Shamir established otherwise; he developed an efficient algorithm for solving knapsack problems involving the encryption weights $b_1, b_2, \ldots, b_n$, where $b_i \equiv wa_i \pmod{m}$ and $a_1, a_2, \ldots, a_n$ is a superincreasing sequence. The flaw lies in the fact that multiplying a_i by w and then reducing it modulo m does not hide a_i well enough.

Since 1982, several knapsack ciphers have been proposed and broken. For instance, the scheme proposed by R. L. Graham and A. Shamir was broken by

L. M. Adelman in 1983. More recently, a new knapsack cipher based on finite fields in abstract algebra was proposed by B. Chor and R. L. Rivest. Referring to Poe's claim that any code could be cracked, they remarked, "At the moment we do not know of any attacks capable of breaking this system in a reasonable amount of time."

EXERCISES 9.5

Determine whether the given sequence is superincreasing.

1. 3, 5, 10, 19, 36
2. 3, 6, 12, 24, 48

Solve each knapsack problem with superincreasing weights.

3. $x_1 + 2x_2 + 4x_3 + 8x_4 + 16x_5 = 23$
4. $3x_1 + 6x_2 + 12x_3 + 24x_4 + 48x_5 = 57$
5. $4x_1 + 5x_2 + 11x_3 + 23x_4 + 45x_5 = 60$
6. $2x_1 + 3x_2 + 6x_3 + 12x_4 + 24x_5 + 48x_6 + 96x_7 = 65$
7. Using the superincreasing sequence 3, 6, 12, 24, construct the knapsack enciphering sequence with modulus $m = 53$ and multiplier $w = 23$.
8. Redo Exercise 7 with the superincreasing sequence 2, 3, 7, 13, 29, and with $m = 63$ and $w = 25$.

Encrypt each message using Exercise 7.

9. SELL ALL.
10. EUREKA.

11–12. Using Exercise 8, encipher the messages in Exercises 9 and 10.

Each knapsack ciphertext below was generated with modulus 65, multiplier 12, and the enciphering sequence 7, 31, 50, 47. Decrypt each.

13. 54 47 47 57 97 81 97 57 50 31
14. 104 47 47 81 104 47 104 54 57 31

Each knapsack ciphertext was created with modulus 53, multiplier 23, and the enciphering sequence 16, 32, 11, 22. Decipher each.

15. 65 33 48 16 70 00 49 38 48 27
16. 33 33 38 48 33 32 49 16 33

CHAPTER SUMMARY

This chapter discussed the art of secrecy systems, a widely used application of number theory in the form of cryptology. A cipher system can be conventional or public-key. In a conventional cryptosystem, the enciphering and deciphering keys are kept secret between the sender and the intended receiver; in a public-key system, the enciphering key is published in a public directory.

The conventional systems presented here are affine, Vigenère, Hill, and exponentiation.

Affine Ciphers

- $C \equiv aP + k \pmod{26}$, where a is a positive integer ≤ 25 and $(a, 26) = 1$. (p. 420)
- When $a = 1$ and $k = 3$, it yields the Caesar cipher. (p. 417)

Vigenère Ciphers

- Vigenère ciphers employ a keyword $w_1 w_2 \dots w_n$ of length n and n shift ciphers $C_i \equiv P_i + k_i \pmod{26}$. (p. 423)

Hill Ciphers

- Hill ciphers are block ciphers that convert plaintext blocks P of length n into ciphertext blocks of the same length using an $n \times n$ enciphering matrix A: $C \equiv AP \pmod{26}$. (p. 425)

Exponentiation Ciphers

- $C \equiv P^e \pmod{p}$, where $0 \leq P, C < p$, and $(e, p - 1) = 1$; e is the encryption exponent of the cryptosystem. Exponentiation ciphers employ the euclidean algorithm, modular exponentiation, and Fermat's little theorem. The multiplicative inverse d of e modulo p serves as the deciphering exponent: $P \equiv C^d \pmod{p}$. (p. 431)

The RSA Cryptosystems

- $C = E(P) \equiv P^e \pmod{n}$, where $0 \leq P, C < n$, $n = pq$, and $(e, \varphi(n)) = 1$. (p. 437)
- $P = D(C) \equiv C^d \pmod{n}$, where d is the multiplicative inverse of e modulo $\varphi(n)$. (p. 438)
- The RSA system uses the euclidean algorithm, modular exponentiation, and Euler's theorem. Since $E(D(M)) = M$ for any message M, the RSA system enables the transmission of digitally signed messages. (p. 439)

Knapsack Ciphers

- Based on the classic knapsack problem, $S = \sum a_i x_i$, where the weights a_i are superincreasing, $x_i = 1$ or 0, and $1 \leq i \leq n$, the knapsack system makes the sequence $b_1, b_2, \dots, b_n$ and $m > 2a_n$ public, where $b_i \equiv wa_i \pmod{m}$, $0 \leq b_i < m$, and $(w, m) = 1$. (p. 445)
- The decrypting strategy involves solving the knapsack problem $S' = \sum a_i x_i$, where S' is the least residue of $w^{-1}S$ modulo m. Unlike the RSA system, knapsack ciphers do not have the property $E(D(M)) = M$, so they are not a signature system. (p. 446)

REVIEW EXERCISES

Encipher each using the affine cipher $C \equiv 5P + 11 \pmod{26}$.

1. NO ROSE WITHOUT A THORN.
2. THE HIGHEST RESULT OF EDUCATION IS TOLERANCE. (Helen Keller)

Decipher each ciphertext created by the affine cipher $C \equiv 7P + 13 \pmod{26}$.

3. QHOJP BSPCS
4. JPNWH HIJNM NISSS

Cryptanalyze each ciphertext created by an affine cipher $C \equiv aP + k \pmod{26}$.

5. VDGVT VLONN
6. JAMWM KJWJW TBBBB

Using the keyword SECRET for a Vigenère cipher, encipher each message.

7. FOR SALE.
8. EXIT ONLY.

Use the matrix $A = \begin{bmatrix} 2 & 11 & 5 \\ 7 & 0 & 4 \\ 9 & 3 & 8 \end{bmatrix}$ for Exercises 9–13.

Using the Hill enciphering matrix A, encrypt each message.

9. GOODBYE.
10. VIOLETS ARE BLUE.

Each ciphertext below was generated by the Hill enciphering matrix A. Decrypt each.

11. ZXB UYW NUM
12. DAT SKO DOB UQR
13. Find the blocks left fixed by the Hill enciphering matrix A.
14. Find the blocks left fixed by the Hill encrypting matrix $\begin{bmatrix} 3 & 5 \\ 8 & 13 \end{bmatrix}$.

With $p = 2729$ as the exponentiation modulus and $e = 37$ as the enciphering exponent, encrypt each message.

15. LABOR DAY
16. MARATHON

Decipher each ciphertext created by an exponentiation cipher with $p = 2729$ and $e = 29$.

17. 2740 2652 0996
18. 0920 1279 0466 1146 1575

Using the RSA enciphering key $(e, n) = (23, 3599)$, encrypt each message.

19. CLOSED BOOK
20. TOP SECRET

Decipher each ciphertext generated by the RSA enciphering key $(e, n) = (23, 3599)$.

21. 0710 0854 0182 1587
22. 1549 1816 2376 0699

Ann and Bob would like to send each other a secret message using the RSA enciphering keys (17, 2537) and (13, 2537), respectively. Find the signed ciphertext sent by:

23. Ann if the plaintext is MAIL.
24. Bob if the plaintext is FINE.

With the RSA enciphering keys as defined, find the plaintext message sent by

25. Ann if her signed message to Bob is 1206 1821.
26. Bob if his signed message to Ann is 0386 1611.

Determine whether the given sequence is superincreasing.

27. 3, 5, 8, 13, 21
28. 2, 3, 6, 12, 24

Solve each knapsack problem.

29. $2x_1 + 3x_2 + 6x_3 + 12x_4 + 24x_5 = 17$
30. $3x_1 + 4x_2 + 9x_3 + 19x_4 + 43x_5 = 55$

Encipher each message using the knapsack cipher with modulus 65, multiplier 17, and superincreasing sequence 2, 3, 6, 12.

31. FOR SALE
32. TOP RANK

The knapsack ciphertext here was created with the enciphering sequence 36, 5, 22, 13, modulus 43, and multiplier 12. Decrypt each.

33. 27 00 22 22 58 22 05 54 41 76 22 35 40
34. 27 49 22 76 35 22 18 54 13 35

SUPPLEMENTARY EXERCISES

1. Define the product cipher resulting from the affine ciphers $C \equiv aP + b \pmod{26}$ and $C \equiv cP + d \pmod{26}$, where $(a, 26) = 1 = (c, 26)$.
2. Find a deciphering formula to decipher a ciphertext generated by the affine ciphers $C \equiv aP + b \pmod{26}$ and $C \equiv cP + d \pmod{26}$, where $(ac, 26) = 1$.
3. Find the number of blocks of letters left fixed by the enciphering matrix $\begin{bmatrix} a & b \\ c & d \end{bmatrix}$.
4. Prove that the product of two Hill ciphers is a Hill cipher.
5. Let A and B be two enciphering matrices of the same size for two different block ciphers. Does the product cipher formed by A followed by B yield the same cipher as the one formed by B followed by A? If not, why not?
6. Show that the probability that a plaintext block P selected at random is not relatively prime to the RSA enciphering modulus $n = pq$ is $\dfrac{1}{p} + \dfrac{1}{q} - \dfrac{1}{pq}$.

7. Suppose the primes p and q in the RSA enciphering modulus $n = pq$ are 100 digits long. Show that the probability that an arbitrarily selected block is not relatively prime to n is less than $2 \cdot 10^{-99}$.
8. Show that the sequence of positive integers $a_1, a_2, \ldots, a_n$ is superincreasing, where $a_i = 2^{i-1}$ and $1 \le i \le n$.
9. Show that the sequence of positive integers $a_1, a_2, \ldots, a_n$ is superincreasing if $a_{i+1} > 2a_i$, where $1 \le i \le n - 1$.

COMPUTER EXERCISES

Write a program to perform each task. Ignore all blank spaces and punctuation marks in all plaintext messages.

1. Read in a plaintext and encipher it using the Caesar cipher.
2. Read in a positive integer k and construct a table showing the alphabetic letters, the corresponding ciphertext letters created by the shift cipher $C \equiv P + k \pmod{26}$ and their ordinal numbers, as in Table 9.4.
3. Read in a plaintext, a shift factor k, and encipher it using the shift cipher $C \equiv P + k \pmod{26}$.
4. Read in a ciphertext encrypted by the Caesar cipher and decrypt it.
5. Read in a ciphertext encrypted by the shift cipher and the shift factor k, and decrypt it.
6. Read in two positive integers a and k, and a ciphertext enciphered using the affine cipher $C \equiv aP + k \pmod{26}$, where $(a, 26) = 1$. Decrypt it.
7. Read in two positive integers a and k with $(a, 26) = 1$. Find the letters of the plaintext left fixed by the affine cipher $C \equiv aP + k \pmod{26}$.
8. Read in two positive integers a and k with $(a, 26) = 1$ and a plaintext. Encipher it using the affine cipher $C \equiv aP + k \pmod{26}$.
9. Read in arbitrary text and construct a percent frequency distribution of letters in the text.
10. Read in a ciphertext enciphered using a shift cipher $C \equiv P + k \pmod{26}$. Using Table 9.2, cryptanalyze it.
11. Read in a ciphertext enciphered using an affine cipher $C \equiv aP + k \pmod{26}$. Using Table 9.2, cryptanalyze it.
12. Read in two positive integers a and k with $(a, 26) = 1$ and a plaintext. Construct a table that shows the alphabetic letters, the corresponding ciphertext letters generated by the cipher $C \equiv aP + k \pmod{26}$, and their ordinal numbers, as in Table 9.5.
13. Read in a plaintext and an $n \times n$ enciphering matrix A for Hill encipherment. Translate it into ciphertext.

14. Read in a ciphertext encrypted by a Hill enciphering matrix A and decrypt it.
15. Read in a plaintext, a prime modulus p for an exponentiation cipher, and an enciphering exponent e. Convert it into ciphertext using modular exponentiation.
16. Read in a numeric ciphertext generated by an exponentiation cipher with a prime modulus p and an enciphering exponent e. Translate it into plaintext.
17. Read in a plaintext and encrypt it using the RSA enciphering key (e, n).
18. Read in a numeric ciphertext generated using an RSA enciphering key (e, n). Translate it into plaintext.
19. Read in the enciphering keys (e_1, n) and (e_2, n) of Anne and Betsey. Send each a signed message by the other. Convert the received message by each to recover the original message.
20. Read in the weights of a knapsack problem and determine whether they are superincreasing.
21. Read in the superincreasing weights of a knapsack problem and solve it.
22. Read in a superincreasing sequence, a modulus m, and a multiplier w. Compute the corresponding knapsack enciphering sequence.
23. Read in a superincreasing sequence, a modulus m, a multiplier w, and a plaintext. Translate it into a knapsack ciphertext.
24. Read in a knapsack ciphertext, the enciphering sequence that generated it, the knapsack modulus m, and multiplier w. Convert it into plaintext.

ENRICHMENT READINGS

1. W. Diffie and M. E. Hellman, "New Directions in Cryptography," *IEEE Transactions on Information Theory*, 22 (Nov. 1976), 644–654.
2. H. Feistel, "Cryptography and Computer Privacy," *Scientific American*, 228 (May 1973), 15–23.
3. M. E. Hellman, "The Mathematics of Public-Key Cryptography," *Scientific American*, 241 (Aug. 1979), 146–157.
4. P. Hilton, "Cryptanalysis in World War II—and Mathematics Education," *Mathematics Teacher*, 77 (Oct. 1984), 548–552.
5. P. Lefton, "Number Theory and Public-Key Cryptography," *Mathematics Teacher* (Jan. 1991), 54–62.
6. D. Luciano and G. Prichett, "Cryptology: From Caesar Ciphers to Public-Key Cryptosystems," *The College Mathematics Journal*, 18 (Jan. 1987), 2–17.
7. R. C. Merkle and M. E. Hellman, "Hiding Information and Signatures in Trapdoor Knapsacks," *IEEE Transactions on Information Theory*, 24 (Sept. 1976), 525–530.

8. R. L. Rivest *et al.*, "A Method for Obtaining Digital Signatures and Public-Key Cryptosystems," *Communications of the ACM*, 21 (Feb. 1978), 120–126.
9. J. Smith, "Public-Key Cryptography," *Byte*, 8 (Jan. 1983), 198–218.

Primitive Roots and Indices

An expert problem solver must be endowed with two incompatible qualities—a restless imagination and a pertinent pertinacity.
—HOWARD W. EVES

In this chapter, we will continue to study the least residues modulo a positive integer m, this time using three important concepts: the order of an element a, the primitive root modulo m, and the index of a. We will then identify those positive integers that possess primitive roots and study some interesting applications. Anyone familiar with group theory in abstract algebra will find the first two sections familiar territory.

10.1 The Order of a Positive Integer

Let m be a positive integer, and a any positive integer such that $(a, m) = 1$. Then, by Euler's theorem, there is a positive exponent e such that $a^e \equiv 1 \pmod{m}$, namely, $e = \varphi(m)$. In general, $\varphi(m)$ need not be the smallest such exponent. By the well-ordering principle, there is always such a least positive exponent.

For example, let us compute the least residues of the first $6 = \varphi(7)$ powers of every positive least residue a modulo 7 and look for the smallest such exponent in each case. For convenience, they are summarized in Table 10.1. The smallest positive exponent e such that $a^e \equiv 1 \pmod 7$ for each positive residue a is circled in the table; they are 1, 3, 6, 3, 6, and 2 for $a = 1, 2, 3, 4, 5$, and 6, respectively. Such an exponent e is called the order of a modulo 7, a concept introduced by Gauss.

a	a^2	a^3	a^4	a^5	a^6
①	1	1	1	1	1
2	4	①	2	4	1
3	2	6	4	5	①
4	2	①	4	2	1
5	4	6	2	3	①
6	①	6	1	6	1

Table 10.1

The Order of a Positive Integer

Let m and a be any positive integers such that $(a, m) = 1$. Then the least positive exponent e such that $a^e \equiv 1 \pmod{m}$ is the **order** of a modulo m. It is denoted by $\text{ord}_m\, a$, or simply ord a, if omitting the modulus does not lead to confusion.

The term *order* is borrowed here from group theory. (If you have already studied group theory, you should find this definition and this section relatively familiar.)

It follows from Table 10.1 that $\text{ord}_7\, 1 = 1$, $\text{ord}_7\, 2 = \text{ord}_7\, 4 = 3$, $\text{ord}_7\, 3 = \text{ord}_7\, 5 = 6$, and $\text{ord}_7\, 6 = 2$. The following example illustrates the definition further.

EXAMPLE 10.1 Compute $\text{ord}_{13}\, 5$ and $\text{ord}_{13}\, 7$.

SOLUTION

First, notice that $(5, 13) = 1 = (7, 13)$. To evaluate each order, we compute the least residues of powers of 5 and 7 modulo 13 until we reach the residue 1. (Feel free to introduce negative residues when convenient.)

$$5^2 \equiv -1 \pmod{13}, \qquad 5^3 \equiv -5 \pmod{13}, \qquad 5^4 \equiv 1 \pmod{13}$$

Thus, 4 is the least positive exponent e such that $5^e \equiv 1 \pmod{13}$, so $\text{ord}_{13}\, 5 = 4$.

To evaluate $\text{ord}_{13}\, 7$, notice that

$$\begin{array}{lll} 7^2 \equiv -3 \pmod{13} & 7^3 \equiv 5 \pmod{13} & 7^4 \equiv -4 \pmod{13} \\ 7^5 \equiv -2 \pmod{13} & 7^6 \equiv -1 \pmod{13} & 7^7 \equiv 6 \pmod{13} \\ 7^8 \equiv 3 \pmod{13} & 7^9 \equiv -5 \pmod{13} & 7^{10} \equiv 4 \pmod{13} \\ 7^{11} \equiv 2 \pmod{13} & 7^{12} \equiv 1 \pmod{13} & \end{array}$$

Thus, $\text{ord}_{13}\, 7 = 12$. ■

It appears from this example that to compute $\text{ord}_m\, a$, we need to compute a^k modulo m for every positive integer $k \le \varphi(m)$. Fortunately, the following theorem helps us eliminate many of them as possible candidates for $\text{ord}_m\, a$.

THEOREM 10.1 Let a be a positive integer such that $(a, m) = 1$ and $\text{ord}_m a = e$. Then $a^n \equiv 1 \pmod{m}$ if and only if $e|n$.

PROOF

Suppose $a^n \equiv 1 \pmod{m}$. By the division algorithm, there are integers q and r such that $n = qe + r$, where $0 \le r < e$. Then

$$a^n = a^{qe+r} = (a^e)^q \cdot a^r$$
$$\equiv 1^q \cdot a^r \equiv a^r \pmod{m}$$

But $a^n \equiv 1 \pmod{m}$, so $a^r \equiv 1 \pmod{m}$, where $0 \le r < e$. Since e is the least positive integer such that $a^e \equiv 1 \pmod{m}$ and $r < e$, this forces $r = 0$. Thus, $n = qe$ and hence $e|n$.

Conversely, let $e|n$. Then $n = be$ for some positive integer b. Therefore,

$$a^n = a^{be} = (a^e)^b$$
$$\equiv 1^b \equiv 1 \pmod{m}$$

This completes the proof. ■

This theorem has a very useful corollary that provides a practical tool for computing $\text{ord}_m a$.

COROLLARY 10.1 Let a be a positive integer such that $(a, m) = 1$. Then $\text{ord}_m a|\varphi(m)$. In particular, if p is a prime and $p \nmid a$, then $\text{ord}_p a|p - 1$.

PROOF

By Euler's theorem, $a^{\varphi(m)} \equiv 1 \pmod{m}$. Therefore, by Theorem 10.1, $\text{ord}_m a|\varphi(m)$.

The special case follows since $\varphi(p) = p - 1$. ■

This result narrows down considerably the list of possible candidates for $\text{ord}_m a$ to the set of positive factors of $\varphi(m)$. Consequently, to compute $\text{ord}_m a$, we do not need to look at all positive powers of a that are $\le \varphi(m)$, but need only consider those positive powers d of a, where $d|\varphi(m)$. The following two examples illustrate this.

EXAMPLE 10.2 Compute $\text{ord}_{21} 5$.

SOLUTION

First, notice that $\varphi(21) = \varphi(3 \cdot 7) = \varphi(3)\varphi(7) = 2 \cdot 6 = 12$. The positive factors d of $\varphi(21) = 12$ are 1, 2, 3, 4, 6, and 12, so only these are the possible values of $\text{ord}_{21} 5$.

To find it, compute 5^d modulo 21 for each d until the residue becomes 1:

$$5^1 \equiv 5 \pmod{21} \qquad 5^2 \equiv 4 \pmod{21} \qquad 5^3 \equiv -1 \pmod{21}$$
$$5^4 \equiv -5 \pmod{21} \quad \text{but} \quad 5^6 \equiv 1 \pmod{21}$$

Thus, we conclude that $\text{ord}_{21} 5 = 6$. ■

Suppose $a^i \equiv a^j \pmod{m}$. Then a reasonable question to ask is: *How are i and j related*? This is answered by the following corollary.

COROLLARY 10.2 Let $\text{ord}_m a = e$. Then $a^i \equiv a^j \pmod{m}$ if and only if $i \equiv j \pmod{e}$.

PROOF

Suppose $a^i \equiv a^j \pmod{m}$ and $i \geq j$. Since $(a, m) = 1$, $(a^j, m) = 1$. So, by Corollary 4.6, a^{-j} exists modulo m. Therefore,

$$a^i \cdot a^{-j} \equiv a^j \cdot a^{-j} \pmod{m}$$

That is,

$$a^{i-j} \equiv 1 \pmod{m}$$

Thus, by Theorem 10.1, $e|i - j$; that is, $i \equiv j \pmod{e}$.

Conversely, let $i \equiv j \pmod{e}$, where $i \geq j$. Then $i = j + ke$ for some integer k. Therefore,

$$\begin{aligned} a^i = a^{j+ke} &= a^j \cdot (a^e)^k \\ &\equiv a^j \cdot 1^k \equiv a^j \pmod{m} \end{aligned}$$

which is the desired result. ■

The following example illustrates this result. It will be useful to us later.

EXAMPLE 10.3 Recall from Example 10.2 that $\text{ord}_{21} 5 = 6$. You may verify that $5^{14} \equiv 5^2 \pmod{21}$, where $14 \equiv 2 \pmod{6}$. But $5^{17} \not\equiv 5^3 \pmod{21}$, since $17 \not\equiv 3 \pmod{6}$. ■

Suppose we know that $\text{ord}_m a = e$. How then is $\text{ord}_m(a^k)$ related to e, where $k > 0$? This is answered by the following theorem.

THEOREM 10.2 Let $\text{ord}_m a = e$ and k any positive integer. Then $\text{ord}_m(a^k) = \dfrac{e}{(e, k)}$.

PROOF

Let $\text{ord}_m(a^k) = r$ and $d = (e, k)$. Then $e = sd$ and $k = td$, where s and t are positive integers such that $(s, t) = 1$. Since

$$(a^k)^s = (a^{td})^s = (a^{sd})^t = (a^e)^t$$
$$\equiv 1^t \equiv 1 \pmod{m}$$

by Theorem 10.1, $r|s$.

Since $\text{ord}_m(a^k) = r$, $(a^k)^r = a^{kr} \equiv 1 \pmod{m}$, so $e|kr$. Thus, $sd|kr$ and hence, $sd|tdr$. So $s|tr$. But $(s, t) = 1$, so $s|r$.

Thus, $r|s$ and $s|r$. Therefore, $s = r$; that is,

$$\text{ord}_m(a^k) = r = s = \frac{e}{d} = \frac{e}{(e, k)}$$ ■

The following example illustrates this theorem.

EXAMPLE 10.4 In Example 10.2, we found that $\text{ord}_{21} 5 = 6$. Therefore, by Theorem 10.2, $\text{ord}_{21}(5^9) = \frac{6}{(6, 9)} = \frac{6}{3} = 2$.

To confirm this, notice that

$$5^2 \equiv 4 \pmod{21} \qquad 5^4 \equiv 16 \pmod{21} \qquad 5^8 \equiv 4 \pmod{21}$$
$$5^9 \equiv -1 \pmod{21} \qquad 5^{18} \equiv 1 \pmod{21}$$

So $\text{ord}_{21}(5^9) = 2$, as expected. ■

Theorem 10.2 leads us to the following result.

COROLLARY 10.3 Let $\text{ord}_m a = e$ and k any positive integer. Then $\text{ord}_m(a^k) = e$ if and only if $(e, k) = 1$.

PROOF

By Theorem 10.2, $\text{ord}_m(a^k) = \frac{e}{(e, k)}$. This equals e if and only if $(e, k) = 1$. ■

For instance, by Example 10.2, $\text{ord}_{21} 5 = 6$. Therefore, $\text{ord}_{21}(5^{11}) = 6$, since $(11, 6) = 1$. We can confirm this by direct computation.

Once again, let a be a positive integer such that $(a, m) = 1$. Then, by Corollary 10.1, $\text{ord}_m a|\varphi(m)$; so the maximum possible value of $\text{ord}_m a$ is $\varphi(m)$. Such least residues do exist. For example, in Example 10.1 we found that $\text{ord}_{13} 7 = 12 = \varphi(13)$. Such least residues possess remarkable properties and consequently deserve special attention, so we make the following definition.

Primitive Roots

Let α be a positive integer such that $(\alpha, m) = 1$. Then α is a **primitive root** modulo m if $\text{ord}_m \alpha = \varphi(m)$.

The following two examples illustrate this definition.

EXAMPLE 10.5 It follows by Table 10.1 that $\text{ord}_7 3 = 6 = \varphi(7) = \text{ord}_7 5$; so both 3 and 5 are primitive roots modulo 7.

In Example 10.1, we found that $\text{ord}_{13} 7 = 12 = \varphi(13)$, so 7 is a primitive root modulo 13. ■

EXAMPLE 10.6 Verify that 2 is a primitive root modulo 9.

SOLUTION

Since $\varphi(9) = 6$, it suffices to show that $2^6 \equiv 1 \pmod 9$ and $2^k \not\equiv 1 \pmod 9$ if $0 < k < 6$. Since $\text{ord}_9 2 = 1, 2, 3,$ or 6, we compute 2^1, 2^2, 2^3, and 2^6 modulo 9:

$$2^1 \equiv 2 \pmod 9 \qquad 2^2 \equiv 4 \pmod 9 \qquad 2^3 \equiv -1 \pmod 9 \quad \text{and} \quad 2^6 \equiv 1 \pmod 9$$

Thus, $\text{ord}_9 2 = 6 = \varphi(9)$, and hence 2 is a primitive root modulo 9. ■

Examples 10.5 and 10.6 might give the impression that every positive integer m has a primitive root. However, this is not always the case. For example, there are no primitive roots modulo 12. Note that there are $\varphi(12) = 4$ positive integers less than 12 and relatively prime to it, namely, 1, 5, 7, and 11. But $\text{ord}_{12} 1 = 1$ and $\text{ord}_{12} 5 = \text{ord}_{12} 7 = \text{ord}_{12} 11 = 2$; so none of them are primitive roots.

Primitive Roots Modulo Fermat Primes f_n

Next, we take a look at the primitive roots modulo Fermat primes f_n, where $n \ge 0$. Clearly, 2 is a primitive root modulo the Fermat primes $f_0 = 3$ and $f_1 = 5$. The following example demonstrates that these are the only Fermat primes for which 2 is a primitive root.

EXAMPLE 10.7 Show that 2 is *not* a primitive root modulo any Fermat prime f_n, where $n \ge 2$.

PROOF

We have

$$2^{2^n} + 1 = f_n \equiv 0 \pmod{f_n}$$

so

$$2^{2^n} \equiv -1 \pmod{f_n}$$

Then

$$\begin{aligned} 2^{2^{n+1}} &\equiv 1 \pmod{f_n} \\ \text{ord}_{f_n} 2 &\leq 2^{n+1} \\ &< 2^{2^n}, \quad \text{because } 2^n > n+1 \text{ for } n \geq 2 \\ &= \varphi(f_n) \end{aligned}$$

Thus, 2 is not a primitive root modulo f_n, where $n \geq 2$. ■

In Section *10.4, we shall conclusively identify those positive integers that possess primitive roots. The following theorem plays an important role in our search.

THEOREM 10.3 If α is a primitive root modulo m, then the least residues of $\alpha, \alpha^2, \ldots, \alpha^{\varphi(m)}$ modulo m are a permutation of the $\varphi(m)$ positive integers $\leq m$ and relatively prime to m.

PROOF

It suffices to show that $\alpha, \alpha^2, \ldots, \alpha^{\varphi(m)}$ are relatively prime to m and no two of them are congruent modulo m.

- Since $(\alpha, m) = 1$, by Corollary 3.2, $(\alpha^k, m) = 1$ for every positive integer k.
- To show that no two of the first $\varphi(m)$ powers of α are congruent modulo m, assume that $\alpha^i \equiv \alpha^j \pmod{m}$, where $1 \leq i, j \leq \varphi(m)$. Assume further that $i \leq j$. Then, by Corollary 10.2, $i \equiv j \pmod{\varphi(m)}$. But $i, j \leq \varphi(m)$, so $i = j$. Thus, no two of the powers of α are congruent modulo m.

Thus, the least residues of $\alpha, \alpha^2, \ldots, \alpha^{\varphi(m)}$ modulo m are a rearrangement of the $\varphi(m)$ positive integers $\leq m$ and relatively prime to m. ■

The following example illustrates this theorem.

EXAMPLE 10.8 Let $m = 18$. There are $\varphi(18) = 6$ positive integers ≤ 18 and relatively prime to 18. They are 1, 5, 7, 11, 13, and 17. You may verify that $\alpha = 5$ is a primitive root modulo 18. The first $\varphi(18) = 6$ powers of 5 are 5, 5^2, 5^3, 5^4, 5^5, and 5^6. Their least residues modulo 18 are 5, 7, 17, 13, 11, and 1, respectively; they are a rearrangement of the residues 1, 5, 7, 11, 13, and 17, as expected. ■

Theorem 10.3 has a powerful corollary. It gives us the exact number of primitive roots modulo m, if they exist.

COROLLARY 10.4 If m has a primitive root, then it has $\varphi(\varphi(m))$ primitive roots. In particular, if m is a prime p, then it has $\varphi(p-1)$ primitive roots.

PROOF

Let α be a primitive root modulo m. Then, by Theorem 10.3, the least residues of $\alpha, \alpha^2, \ldots, \alpha^{\varphi(m)}$ modulo m are distinct and relatively prime to m. By Corollary 10.3, $\text{ord}_m(\alpha^k) = \varphi(m)$ if and only if $(k, \varphi(m)) = 1$; that is, α^k is a primitive root modulo m if and only if $(k, \varphi(m)) = 1$. But there are $\varphi(\varphi(m))$ positive integers $\leq \varphi(m)$ and relatively prime to $\varphi(m)$. Thus, m has $\varphi(\varphi(m))$ primitive roots.

The special case follows trivially since $\varphi(p) = p - 1$. ■

This proof provides a constructive method for finding all $\varphi(\varphi(m))$ primitive roots modulo m from a given primitive root α modulo m. They are given by α^k, where $(k, \varphi(m)) = 1$, as the following example demonstrates.

EXAMPLE 10.9 Using the fact that 5 is a primitive root modulo 54, find the remaining incongruent primitive roots.

SOLUTION

By Corollary 10.4, 54 has $\varphi(\varphi(54)) = \varphi(18) = 6$ primitive roots. They are given by 5^k, where $(k, 18) = 1$. The positive integers ≤ 18 and relatively prime to it are 1, 5, 7, 11, 13, and 17, so the corresponding primitive roots are given by $5^1, 5^5, 5^7, 5^{11}, 5^{13}$, and 5^{17} modulo 54, that is, 5, 47, 41, 29, 23, and 11, respectively. Thus, the remaining primitive roots modulo 54, in increasing order, are 11, 23, 29, 41, and 47. ■

The following example employs the special case in Corollary 10.4.

EXAMPLE 10.10 Find the incongruent primitive roots modulo 19.

SOLUTION

By trial and error, we find that 2 is a primitive root modulo 19. Therefore, by Corollary 10.5, 19 has $\varphi(18) = 6$ primitive roots 2^k, where $(k, 18) = 1$. Thus they are 2^1, 2^5, 2^7, 2^{11}, 2^{13}, and 2^{17} modulo 19, that is, 2, 3, 10, 13, 14, and 15 in ascending order. ■

EXERCISES 10.1

Evaluate each.

1. $\text{ord}_7 3$
2. $\text{ord}_8 5$
3. $\text{ord}_{11} 5$
4. $\text{ord}_{13} 8$
5. Given that $\text{ord}_{23} 9 = 11$, find $\text{ord}_{23} 14$.
6. Given that $\text{ord}_{19} 11 = 3$, find $\text{ord}_{19} 8$.
7. Using the congruence $5^8 \equiv 1 \pmod{13}$, compute $\text{ord}_{13} 5$.
8. Using the congruence $3^9 \equiv -1 \pmod{19}$, compute $\text{ord}_{19} 3$.

Using the given order of the least residue a of a prime p, find the order of the given element b modulo p.

9. $\text{ord}_{13} 4 = 6$, $b = 4^5$
10. $\text{ord}_{17} 8 = 16$, $b = 8^9$
11. $\text{ord}_{11} 7 = 10$, $b = 7^4$
12. $\text{ord}_{17} 2 = 8$, $b = 2^6$
13. Show, by an example, that $\text{ord}_p(-a) \neq \text{ord}_p a$.

Verify each.

14. $\text{ord}_{11} 3 = 5$
15. $3^4 + 3^3 + 3^2 + 3 + 1 \equiv 0 \pmod{11}$
16. $\text{ord}_{13} 4 = 6$
17. $4^5 + 4^4 + \cdots + 4 + 1 \equiv 0 \pmod{13}$

Find the least positive integer m such that

18. $\text{ord}_m 9 = 16$
19. $\text{ord}_m 5 = 22$
20. $\text{ord}_m 5 = 6$
21. $\text{ord}_m 7 = 10$
22. Make a conjecture using Exercises 18–21.

Give a counterexample to disprove each statement, where p is a prime.

23. $\text{ord}_p(p - a) = \text{ord}_p a$.
24. If $d | \varphi(m)$, then there is a least residue a modulo m with order d.
25. Every prime p has an even number of primitive roots.
26. Without direct computation, show that $\text{ord}_8 a \leq 2$, where $(a, 8) = 1$.
27. Using the fact that $\text{ord}_{13} 5 = 4$, compute the remainder when 5^{1001} is divided by 13.
28. Using the fact that 6 is a primitive root modulo 41, compute the remainder when 6^{2020} is divided by 41.

Using the orders of the given least residues a and b modulo m, compute $\text{ord}_m(ab)$.

29. $\text{ord}_9 7 = 3$, $\text{ord}_9 8 = 2$
30. $\text{ord}_{13} 3 = 3$, $\text{ord}_{13} 5 = 4$

Assuming each prime p has a primitive root, find the number of primitive roots modulo p.

31. 11
32. 17
33. 29
34. 101

Find the incongruent primitive roots modulo the given prime. Assume each has a primitive root.

35. 7
36. 11
37. 13
38. 17
39. Using Exercises 35–38, make a conjecture about the product of the incongruent primitive roots modulo an odd prime p.
40. Show that 8 has no primitive roots.
41. Show that 12 has no primitive roots.
42. Let p be an odd prime. Then every prime factor of M_p is of the form $2kp + 1$. Using this fact, find the smallest prime factor of M_{11}. (See Exercise 54 for a proof of this fact.)
43. Let p and q be primes > 3 and $q | R_p$, where R_p is the repunit with p ones. Then q must be of the form $2kp + 1$. Using this fact, find the least prime factors of R_5, R_7, R_{13}, and R_{41}. (See Exercise 55 for a proof of this fact.)

Prove each, where p is a prime, and a and m are positive integers.

44. If a prime $p > 3$ has a primitive root, then it has an even number of primitive roots.
45. Let a^{-1} be a multiplicative inverse of a modulo m. Then $\text{ord}_m(a^{-1}) = \text{ord}_m a$.
46. A least residue a is a primitive root modulo m if and only if a^{-1} is a primitive root modulo m.
47. Let $(a, m) = 1$ such that $\text{ord}_m a = m - 1$. Then m is a prime.
48. Let p be an odd prime and $\text{ord}_p r = e$. Then $r^{e-1} + r^{e-2} + \cdots + r + 1 \equiv 0 \pmod{p}$.
49. Let $p \equiv 1 \pmod{4}$ be a prime and a a positive integer such that $p \nmid a$. Then a is a primitive root modulo p if and only if $p - a$ is also a primitive root.
50. Let α be a primitive root modulo a positive integer $m \geq 3$. Then $\alpha^{\varphi(m)/2} \equiv -1 \pmod{m}$.

51. Let $\text{ord}_m a = hk$. Then $\text{ord}_m(a^h) = k$.
52. Let $\text{ord}_m a = h$ and $\text{ord}_m b = k$, where $(h, k) = 1$. Then $\text{ord}_m ab = hk$.
53. Let p and q be odd primes such that $q|a^p - 1$. Then either $q|a - 1$ or $q = 2kp + 1$ for some integer k.
54. Let p be an odd prime. Then every prime factor of M_p is of the form $2kp + 1$. (*Hint*: Use Exercise 53.)
55. Let p and q be primes > 3 and $q|R_p$, where R_p is the repunit with p ones. Then q must be of the form $2kp + 1$.
56. The odd prime factors of the integers $n^2 + 1$ are of the form $4k + 1$.
57. The odd prime factors of the integers $n^4 + 1$ are of the form $8k + 1$.
58. There are infinitely primes of the form $4k + 1$. [*Hint*: Assume there is only a finite number of such primes, $p_1, p_2, \ldots, p_r$. Then consider $N = (2p_1p_2 \cdots p_r)^2 + 1$ and use Exercise 56.]
59. There are infinitely many primes of the form $8k + 1$. [*Hint*: Assume there is only a finite number of such primes, $p_1, p_2, \ldots, p_r$. Then consider $N = (2p_1p_2 \cdots p_r)^4 + 1$ and use Exercise 57.]

10.2 Primality Tests

We can use the concept of the order of an integer to develop primality tests. Lucas' theorem, discovered in 1876, provides one such test; it is based on the fact that a positive integer n is prime if and only if $\varphi(n) = n - 1$.

THEOREM 10.4 **(Lucas' Theorem)** Let n be a positive integer. If there is a positive integer x such that $x^{n-1} \equiv 1 \pmod{n}$ and $x^{(n-1)/q} \not\equiv 1 \pmod{n}$ for all prime factors q of $n - 1$, then n is prime.

PROOF

Let $\text{ord}_n x = e$. Since $x^{n-1} \equiv 1 \pmod{n}$, by Theorem 10.1, $e|n - 1$. We would like to show that $e = n - 1$, so assume that $e \neq n - 1$. Since $e|n - 1$, $n - 1 = ke$ for some integer $k > 1$. Let q be a prime factor of k. Then:

$$x^{(n-1)/q} = x^{ke/q} = (x^e)^{(k/q)}$$
$$\equiv 1 \pmod{n},$$

which is a contradiction. So $e = n - 1$; that is, $\text{ord}_n x = n - 1 = \varphi(n)$, because $n - 1 = \text{ord}_n x | \varphi(n) \leq n - 1$. Thus, n is a prime. ■

The following example illustrates this test.

EXAMPLE 10.11 Using Lucas' theorem, show that $n = 1117$ is a prime.

SOLUTION

We shall choose $x = 2$ to show that n satisfies the conditions of the test.

First, notice that

$$\begin{aligned} 2^{1116} &= (2^{100})^{11} \cdot 2^{16} \\ &\equiv 293^{11} \cdot 750 \equiv 70 \cdot 750 \equiv 1 \pmod{1117} \end{aligned}$$

Since $1116 = 2^2 \cdot 3^2 \cdot 31$, the prime factors of $n - 1 = 1116$ are 2, 3, and 31.

When $q = 2$,

$$\begin{aligned} 2^{(n-1)/q} &= 2^{558} = (2^{50})^{11} \cdot 2^8 \\ &\equiv 69^{11} \cdot 256 \equiv 1069 \cdot 256 \equiv -1 \pmod{1117}; \end{aligned}$$

when $q = 3$,

$$\begin{aligned} 2^{(n-1)/q} &= 2^{372} = (2^{50})^7 \cdot 2^{22} \\ &\equiv 69^7 \cdot 1086 \equiv 112 \cdot 1086 \equiv 996 \pmod{1117}; \end{aligned}$$

when $q = 31$,

$$\begin{aligned} 2^{(n-1)/q} &= 2^{36} = (2^{10})^3 \cdot 2^6 \\ &\equiv (-93)^3 \cdot 64 \equiv 1000 \cdot 64 \equiv 331 \pmod{1117} \end{aligned}$$

Thus, $2^{1116/q} \not\equiv 1 \pmod{1117}$ for all prime factors q of 1116. Therefore, by Lucas' theorem, 1117 is a prime. ■

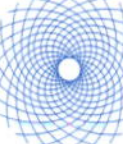

Note: As this example indicates, a scientific calculator, such as TI-86, with a built-in *mod* operator will speed up your computation.

We can refine Lucas' theorem to yield the following more efficient primality test.

COROLLARY 10.5 Let n be an odd positive integer. If there is a positive integer x such that $x^{(n-1)/2} \equiv -1 \pmod{n}$ and $x^{(n-1)/q} \not\equiv 1 \pmod{n}$ for all odd prime factors q of $n-1$, then n is prime.

PROOF

Since $x^{(n-1)/2} \equiv -1 \pmod{n}$, $x^{n-1} = (x^{(n-1)/2})^2 \equiv (-1)^2 \equiv 1 \pmod{n}$. Furthermore, $x^{(n-1)/q} \not\equiv 1 \pmod{n}$, when $q = 2$ or q is any odd prime factor of $n-1$. Thus, both conditions in Lucas' theorem are satisfied, so n is prime. ■

The following example illustrates this refined test.

EXAMPLE 10.12 Using Corollary 10.5, verify that $n = 1213$ is a prime.

SOLUTION

We shall use $x = 5$ here. Since $n - 1 = 1212 = 2^2 \cdot 3 \cdot 101$, the odd prime factors of $n-1$ are 3 and 101.

Notice that

$$\begin{aligned} 5^{(n-1)/2} &= 5^{606} = (5^{100})^6 \cdot 5^6 \\ &\equiv (-252)^6 \cdot 1069 \equiv 497 \cdot 1069 \equiv -1 \pmod{1213} \end{aligned}$$

When $q = 3$,

$$\begin{aligned} 5^{(n-1)/q} &= 5^{404} = (5^{100})^4 \cdot 5^4 \\ &\equiv (-252)^4 \cdot 625 \equiv 21 \cdot 625 \equiv 995 \pmod{1213} \end{aligned}$$

When $q = 101$,

$$\begin{aligned} 5^{(n-1)/q} &= 5^{12} = 5^{10} \cdot 5^2 \\ &\equiv (-238) \cdot 25 \equiv 115 \pmod{1213} \end{aligned}$$

Thus, in both cases, $5^{(n-1)/q} \not\equiv 1 \pmod{1213}$, so 1213 is a prime. ■

EXERCISES 10.2

Verify that each number is a prime, using Lucas' theorem and the given value of x.

1. 101, $x = 2$
2. 257, $x = 3$
3. 773, $x = 3$
4. 823, $x = 3$

Verify that each number is a prime, using Corollary 10.5 and the given value of x.

5. 127, $x = 3$
6. 241, $x = 7$
7. 577, $x = 5$
8. 797, $x = 2$
9. Let f_n denote the nth Fermat number. Suppose there exists a positive integer x such that $x^{2^{2^n}} \equiv 1 \pmod{f_n}$ and $x^{2^{2^n-1}} \not\equiv 1 \pmod{f_n}$. Prove that f_n is a prime.
10. Using Exercise 9, show that both f_2 and f_3 are primes.

10.3 Primitive Roots for Primes

In Corollary 10.4, we found that if a positive integer m has a primitive root, then it has $\varphi(\varphi(m))$ primitive roots. The corollary, however, does not assure us that every positive integer m has a primitive root. For example, 8 does not have a primitive root. To see this, notice that $\varphi(8) = 4$. For a positive integer a to be a primitive root modulo 8, $(a, 8) = 1$ and hence a must be odd. So $a \equiv \pm 1$ or $\pm 3 \pmod 8$. Then $a^2 \equiv 1 \pmod 8$. Thus $\text{ord}_8\, a \leq 2$. Consequently, $\text{ord}_8\, a \neq \varphi(8)$, so a cannot be a primitive root.

The obvious question is: *What kind of positive integers m have primitive roots*? First, we need to show that every prime has a primitive root. To this end, we need to lay some groundwork by using polynomial congruences.

Let $f(x)$ be a polynomial with integral coefficients. An integer α is a **solution** of $f(x) \equiv 0 \pmod m$ if $f(\alpha) \equiv 0 \pmod m$. Clearly, if $\beta \equiv \alpha \pmod m$, then β is also a solution modulo m.

EXAMPLE 10.13 Consider the polynomial congruence

$$f(x) = x^2 - x + 1 \equiv 0 \pmod{13}.$$

It has two incongruent solutions modulo 13, namely 4 and 10:

$$f(4) \equiv 16 - 4 + 1 \equiv 0 \pmod{13}$$

$$f(10) \equiv 100 - 10 + 1 \equiv 0 \pmod{13}$$

But the congruence $2x^2 + 3x + 4 \equiv 0 \pmod 5$ has no solutions. (Verify this.) ■

The following theorem about the number of solutions of polynomial congruences $f(x) \equiv 0 \pmod p$ plays a pivotal role in the existence proof of primitive roots for primes.

THEOREM 10.5 **(Lagrange's Theorem)** Let $f(x) = \sum_{i=0}^{n} a_i x^i$ be a polynomial of degree $n \geq 1$ with integral coefficients, where $p \nmid a_n$. Then the congruence $f(x) \equiv 0 \pmod p$ has at most n incongruent solutions modulo p.

PROOF **(by induction on *n*)**
When $n = 1$, $f(x) = a_1 x + a_0$, where $p \nmid a_1$. Since $(p, a_1) = 1$, the congruence $a_1 x + a_0 \equiv 0 \pmod p$ has a unique solution, by Corollary 4.6. So when $n = 1$, $f(x) \equiv 0 \pmod p$ has at most one solution. Thus the theorem is true when $n = 1$.

Now assume it is true for polynomials of degree $k-1$. Let $f(x) = \sum_{i=0}^{k} a_i x^i$ be a polynomial of degree k, where $p \nmid a_k$. If $f(x) \equiv 0 \pmod{p}$ has no solutions, then the result follows.

So assume that it has at least one solution α, where $0 \leq \alpha < p$. Let $q(x)$ be the quotient and r (an integer) be the remainder when $f(x)$ is divided by $x - \alpha$, where $q(x)$ is a polynomial of degree $k-1$ with integral coefficients. (This follows by the remainder theorem.) Then

$$f(x) = (x-\alpha)q(x) + r$$

Then

$$f(\alpha) = (\alpha-\alpha)q(\alpha) + r$$
$$0 \equiv 0 + r \pmod{p}$$
$$r \equiv 0 \pmod{p}$$

Therefore,

$$f(x) \equiv (x-\alpha)q(x) \pmod{p}$$

where degree $q(x) \leq k-1$. Let β be any other incongruent solution of $f(x) \equiv 0 \pmod{p}$, where $0 \leq \beta < p$. Then

$$f(\beta) \equiv (\beta-\alpha)q(\beta) \pmod{p}$$
$$0 \equiv (\beta-\alpha)q(\beta) \pmod{p}$$

Since $\beta \not\equiv \alpha \pmod{p}$, this implies $q(\beta) \equiv 0 \pmod{p}$. Thus, every solution of $f(x) \equiv 0 \pmod{p}$, different from α, is a solution of $q(x) \equiv 0 \pmod{p}$. Clearly, every solution of $q(x) \equiv 0 \pmod{p}$ is also a solution of $f(x) \equiv 0 \pmod{p}$. Since $\deg q(x) \leq k-1$, by the inductive hypothesis, $q(x) \equiv 0 \pmod{p}$ has at most $k-1$ solutions, so $f(x) \equiv 0 \pmod{p}$ has at most $1 + (k-1) = k$ solutions.

Thus, by induction, the theorem is true for every polynomial of degree $n \geq 1$. ■

For example, the polynomial $f(x) = x^2 - x + 1$ in Example 10.13 has degree two and the congruence $f(x) \equiv 0 \pmod{13}$ has at most two solutions modulo 13. The polynomial $g(x) = 2x^2 + 3x + 4$ also has degree two, but the congruence $g(x) \equiv 0 \pmod{5}$ has no solutions modulo 5; in any case, it has at most two solutions.

The following result is a very important consequence of this theorem. It plays a crucial role in establishing the existence of primitive roots for primes.

COROLLARY 10.6 If p is a prime and $d|p-1$, then the congruence $x^d - 1 \equiv 0 \pmod{p}$ has exactly d incongruent solutions modulo p.

PROOF

By Fermat's little theorem, the congruence $x^{p-1} - 1 \equiv 0 \pmod{p}$ has exactly $p - 1$ solutions modulo p, namely, 1 through $p - 1$. Since $d|p - 1$,

$$\begin{aligned} x^{p-1} - 1 &= (x^d - 1)(x^{p-1-d} + x^{p-1-2d} + \cdots + x^d + 1) \\ &= (x^d - 1)g(x) \end{aligned}$$

where $g(x) = x^{p-1-d} + x^{p-1-2d} + \cdots + x^d + 1$ is a polynomial of degree $p - 1 - d$. By Lagrange's theorem, $g(x) \equiv 0 \pmod{p}$ has at most $p - 1 - d$ incongruent solutions. Therefore, $x^d - 1 \equiv 0 \pmod{p}$ has at least $(p - 1) - (p - 1 - d) = d$ incongruent solutions. But, again, by Lagrange's theorem, $x^d - 1 \equiv 0 \pmod{p}$ has at most d incongruent solutions. Thus, it has exactly d incongruent solutions modulo p. ■

The following example illustrates this result.

EXAMPLE 10.14 Find the incongruent solutions of the congruence $x^3 - 1 \equiv 0 \pmod{13}$.

SOLUTION

Since $x^3 - 1 = (x - 1)(x^2 + x + 1)$, the congruence $x^3 - 1 \equiv 0 \pmod{13}$ implies $x - 1 \equiv 0 \pmod{13}$ or $x^2 + x + 1 \equiv 0 \pmod{13}$. The congruence $x - 1 \equiv 0 \pmod{13}$ yields $x = 1$. Because $x^2 + x + 1 \equiv x^2 + x - 12 \equiv (x + 4)(x - 3) \equiv 0 \pmod{13}$, $x \equiv 3 \pmod{13}$ or $x \equiv -4 \equiv 9 \pmod{13}$. (Verify both.) It has no other incongruent solutions. Thus, the given congruence has exactly three incongruent solutions: 1, 3, and 9. ■

Wilson's theorem, presented in Section 7.1, can be derived elegantly from Corollary 10.6, as shown below. But first, we should note that Lagrange's theorem can be restated as follows: Let $f(x) = \sum_{i=0}^{n} a_i x^i$ be a polynomial of degree n with integral coefficients. If the congruence $f(x) \equiv 0 \pmod{p}$ has more than n incongruent solutions, then $a_i \equiv 0 \pmod{p}$ for every i.

COROLLARY 10.7 **(Wilson's Theorem)** If p is a prime, then $(p - 1)! \equiv -1 \pmod{p}$.

PROOF

[The essence of the proof lies in cleverly selecting a suitable polynomial $f(x)$.] Let $f(x) = (x-1)(x-2)\cdots(x-p+1) - x^{p-1} + 1$. Clearly, $f(x)$ is a polynomial of degree $p-2$ with integral coefficients. By Fermat's little theorem, $x^{p-1} - 1 \equiv 0 \pmod{p}$ has $p-1$ incongruent solutions. Each is also a solution of $(x-1)(x-2)\cdots(x-p+1) \equiv 0 \pmod{p}$. Therefore, $f(x) \equiv 0 \pmod{p}$ has $p - 1$ incongruent solutions, one more than the degree of $f(x)$. Therefore, every coefficient of $f(x)$ must be congruent to 0 modulo p. In particular, the constant term $f(0)$ must be congruent to 0 modulo p. But

$$\begin{aligned} f(0) &= (-1)(-2)\cdots[-(p-1)] - 0 + 1 \\ &= (-1)^{p-1}(p-1)! + 1 \end{aligned}$$

Therefore, $(-1)^{p-1}(p-1)! + 1 \equiv 0 \pmod{p}$; that is, $(p-1)! \equiv (-1)^p \pmod{p}$.

If $p = 2$, then $(-1)^p \equiv 1 \equiv -1 \pmod{p}$; if p is odd, then $(-1)^p = -1$. Thus, in both cases, $(p-1)! \equiv -1 \pmod{p}$. ■

Next, we turn to a major result on the number of incongruent residues of order d modulo p. However, before we do, let us study an example that will illuminate the proof of the theorem.

EXAMPLE 10.15 Let $p = 19$ and $d|p-1$. Let $\psi(d)$[†] denote the number of incongruent residues of order d modulo p. Compute $\psi(d)$ and $\varphi(d)$ for each d, and $\sum_{d|p-1} \psi(d)$.

SOLUTION

(The details are left for you to fill in.) Because $d|18$, $d = 1, 2, 3, 6, 9$, or 18. The number $\psi(d)$ of incongruent residues of order d, the incongruent residues of order d, and $\varphi(d)$ are listed in Table 10.2 for various values of d. (Verify them.)

d	1	2	3	6	9	18
Incongruent residues of order d	1	18	7, 11	8, 12	4, 5, 6, 9 16, 17	2, 3, 10, 13 14, 15
$\psi(d)$	1	1	2	2	6	6
$\varphi(d)$	1	1	2	2	6	6

Table 10.2

It follows from the table that

$$\begin{aligned} \sum_{d|p-1} \psi(d) &= \sum_{d|18} \psi(d) \\ &= 1 + 1 + 2 + 2 + 6 + 6 = 18 \\ &= p - 1 \end{aligned}$$

■

[†] ψ is the lowercase Greek letter *psi*.

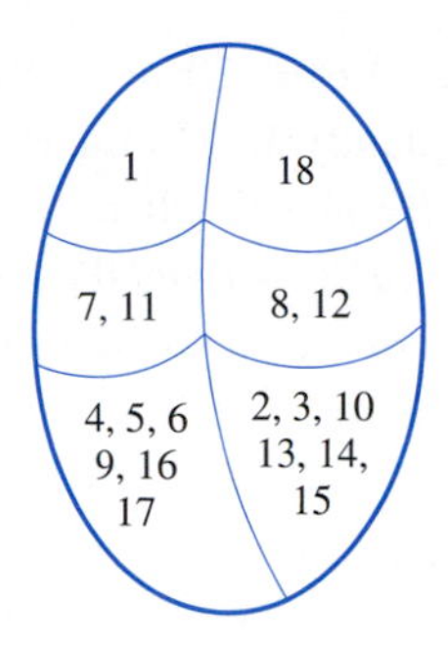

Figure 10.1 *A partitioning of the set of positive residues modulo 19.*

Let's pursue this example a bit further. Notice that the incongruent residues of order d modulo 19 form a partitioning of the set of positive residues modulo 19, as Figure 10.1 shows; $\psi(d)$ denotes the number of elements in each class. Interestingly enough, $\psi(d) = \varphi(d)$ for each d. (See Exercises 12–15.)

We now arrive at the main result, proved by the French mathematician Adrien-Marie Legendre in 1785.

THEOREM 10.6 Let p be a prime and d a positive factor of $p - 1$. Then there are exactly $\varphi(d)$ incongruent integers of order d modulo p.

PROOF

For every positive factor d of $p - 1$, let $\psi(d)$ denote the number of positive residues modulo p that have order d. Because there are $p - 1$ positive residues and each has a unique order d, the positive residues of order d form a partitioning of the set of positive residues. Therefore,

$$\sum_{d|p-1} \psi(d) = p - 1$$

But, by Theorem 8.6,

$$\sum_{d|p-1} \varphi(d) = p - 1$$

Therefore,

$$\sum_{d|p-1} \psi(d) = \sum_{d|p-1} \varphi(d) \tag{10.1}$$

Next, we need to show that $\psi(d) = \varphi(d)$ for every d. To this end, we consider two cases.

case 1 Let $\psi(d) = 0$. Then, clearly, $\psi(d) < \varphi(d)$, so $\psi(d) \le \varphi(d)$.

case 2 Let $\psi(d) \neq 0$. Then there must be an integer a of order d modulo p. Consequently, by Corollary 10.3, the d integers $a, a^2, \ldots, a^d$ are incongruent modulo p. Besides, each is a solution of the congruence $x^d - 1 \equiv 0 \pmod{p}$, since $(a^k)^d = (a^d)^k \equiv 1 \pmod{p}$, where $1 \leq k \leq d$. Therefore, by Corollary 10.6, they are the d incongruent solutions of the congruence $x^d - 1 \equiv 0 \pmod{p}$ and $\text{ord}_p(a^k)|d$ by Theorem 10.1.

But, by Corollary 10.3, $\text{ord}_p(a^k) = \text{ord}_p\, a = d$ if and only if $(k, d) = 1$. Since there are $\varphi(d)$ positive integers $\leq d$ and relatively prime to d, there are exactly $\varphi(d)$ residues of a^k modulo p that have order d. Therefore, $\psi(d) = \varphi(d)$.

Thus, in both cases, $\psi(d) \leq \varphi(d)$. So, for equality (10.1) to hold, we must have $\psi(d) = \varphi(d)$ for all d. In other words, there are exactly $\varphi(d)$ incongruent integers (or residues) of order d modulo p. ■

The following example illustrates this theorem.

EXAMPLE 10.16 Find the number of incongruent integers of order d modulo 13, where $d|12$.

SOLUTION

Since $d|12$, $d = 1, 2, 3, 4, 6,$ or 12. Let $\psi(d)$ denote the number of incongruent residues of order d modulo 13. Then

$$\psi(1) = \varphi(1) = 1 \qquad \psi(2) = \varphi(2) = 1 \qquad \psi(3) = \varphi(3) = 2$$
$$\psi(4) = \varphi(4) = 2 \qquad \psi(6) = \varphi(6) = 2 \qquad \psi(12) = \varphi(12) = 4$$

It will be useful to find the least positive residues modulo 13 of order d (see Exercise 12). Since $\psi(12) = 4$, it follows that there are four primitive roots modulo 13. ■

In the following corollary, Theorem 10.6 yields a class of positive integers that have primitive roots.

COROLLARY 10.8 Every prime p has $\varphi(p-1)$ incongruent primitive roots.

PROOF

Since $p-1|p-1$, by Theorem 10.6, there are $\varphi(p-1)$ incongruent integers of order $p-1$ modulo p. Each of them, by definition, is a primitive root. Therefore, there are $\varphi(p-1)$ primitive roots modulo p. ■

For instance, $p = 19$ has $\varphi(18) = 6$ incongruent primitive roots. Try to find them.

The fact that every prime has a primitive root was established by Euler in 1773. He even constructed a list of primitive roots modulo primes ≤ 37.

Notice that the proof of Corollary 10.8 is nonconstructive; that is, it does not tell us how to find the primitive roots modulo primes. They need to be found by direct computation. In 1839, the German mathematician Karl Gustave Jacob Jacobi published an extensive table of primitive roots modulo primes < 1000. For convenience, Table T.3 at the end of this book lists the least positive primitive roots modulo primes $p \leq 100$.

EXERCISES 10.3

Find the incongruent roots of the congruence $f(x) \equiv 0 \pmod{7}$ for each polynomial $f(x)$.

1. $x^2 + 3$
2. $x^3 + 1$
3. $x^2 + x + 1$
4. $x^3 + 2x^2 + 3x + 4$

Find the incongruent roots of the congruence $f(x) \equiv 0 \pmod{11}$ for each polynomial $f(x)$.

5. $x^2 + 3$
6. $x^2 - x - 1$
7. $2x^2 + 3x + 1$
8. $2x^3 + x^2 - 3x + 1$
9. Find the incongruent roots modulo 7 of the polynomial $x^d - 1$ for every factor d of 6.
10. Find the incongruent roots modulo 13 of the polynomial $x^d - 1$ for every factor d of 12.
11. Verify that 5 is a factor of every coefficient of the polynomial

$$f(x) = (x-1)(x-2)(x-3)(x-4) - x^4 + 1$$

Find the number of incongruent integers of order d modulo the given prime p.

12. $d = 4, p = 13$
13. $d = 6, p = 19$
14. $d = 11, p = 23$
15. $d = 48, p = 97$
16. Let α be a primitive root modulo an odd prime p. What can you say about α^k if $(k, p-1) = 1$?
17. Let α be a primitive root modulo an odd prime p. Can the least residue of $\alpha^{(p-1)/2}$ modulo p be 1?
18. Let α be a primitive root modulo an odd prime p. Find the least residue of $\alpha^{(p-1)/2}$ modulo p.

Both 3 and 5 are primitive roots modulo 7. Verify each.

19. $3^5 + 3^4 + \cdots + 3 + 1 \equiv 0 \pmod{7}$
20. $5^5 + 5^4 + \cdots + 5 + 1 \equiv 0 \pmod{7}$

Disprove each statement in Exercises 21 and 22.

21. The product of two primitive roots modulo an odd prime p is a primitive root modulo p.
22. The sum of two primitive roots modulo an odd prime p is a primitive root modulo p.
23. Show, by an example, that the sum of two primitive roots modulo an odd prime p can also be a primitive root.
24. The prime $p = 7$ has two primitive roots α. Find $\mathrm{ord}_p(-\alpha)$ in each case.
25. The prime $p = 11$ has four primitive roots α. Find $\mathrm{ord}_p(-\alpha)$ in each case.
26. Let α be a primitive root modulo a prime $p \equiv 3 \pmod{4}$. Using Exercises 24 and 25, make a conjecture about $\mathrm{ord}_p(-\alpha)$.

Using the fact that α is a primitive root modulo an odd prime, determine whether $-\alpha$ is also a primitive root.

27. $\alpha = 2, p = 5$
28. $\alpha = 2, p = 13$
29. $\alpha = 5, p = 17$
30. $\alpha = 2, p = 29$
31. Using Exercises 27–30, make a conjecture about the prime p for which both α and $-\alpha$ are primitive roots.
32. Find an odd prime p such that α is a primitive root modulo p, but $-\alpha$ is not.
33. Using Exercises 35–38 in Exercises 10.1, make a conjecture about the product of incongruent primitive roots modulo a prime p.

Let $p = 13$ and d a positive factor of $p - 1$. Let $\psi(d)$ denote the number of positive residues of order d modulo p.

34. Find $\psi(d)$ for each d.
35. Verify that $\sum_{d|p-1} \psi(d) = p - 1$.
36. Verify that $\psi(d) = \varphi(d)$ for each d.

37–39. Redo Exercises 34–36 with $p = 17$.

Find the least residues of order d modulo p for each positive factor d of $p-1$, where

40. $p = 13$
41. $p = 23$

Using the given primitive root α modulo a prime p, find the remaining least incongruent primitive roots modulo p.

42. $\alpha = 2, p = 13$
43. $\alpha = 3, p = 17$
44. $\alpha = 5, p = 23$
45. $\alpha = 3, p = 31$

Prove each.

46. Let α be a solution of the congruence $f(x) \equiv 0 \pmod{m}$ and $\beta \equiv \alpha \pmod{m}$, where $f(x)$ is a polynomial with integral coefficients. Then β is also a solution of the congruence.
47. Let α be a primitive root modulo an odd prime p. Then $\alpha^{p-2} + \alpha^{p-3} + \cdots + \alpha + 1 \equiv 0 \pmod{p}$.
48. The product of the incongruent primitive roots modulo a prime p is congruent to 1 modulo p.
49. If α is a primitive root modulo a prime $p \equiv 1 \pmod 4$, then $\alpha^{(p-1)/4}$ satisfies the congruence $x^2 + 1 \equiv 0 \pmod{p}$.
50. If α is a primitive root modulo a prime $p \equiv 3 \pmod 4$, then $\alpha^{(p-3)/4}$ cannot be a solution of the congruence $x^2 + 1 \equiv 0 \pmod{p}$.
51. If α is a primitive root modulo a prime $p \equiv 3 \pmod 4$, then $\text{ord}_p(-\alpha) = \dfrac{p-1}{2}$.
52. Let α be a primitive root modulo an odd prime p. Then $-\alpha$ is a primitive root modulo p if and only if $p \equiv 1 \pmod 4$.
53. Let α be a primitive root modulo p^j, where p is an odd prime, α is odd, and $j \geq 1$. Then α is also a primitive root modulo $2p^j$.
54. Let α be a primitive root modulo p^j, where p is an odd prime, α is even, and $j \geq 1$. Then $\alpha + p^j$ is a primitive root modulo $2p^j$.

*10.4 Composites with Primitive Roots (optional)

In the preceding section, we established that every prime p has a primitive root; in fact, it has $\varphi(p-1)$ primitive roots. We will now identify the class of positive integers that possess primitive roots; it consists of $1, 2, 4, p^k$, and $2p^k$, where p is an odd prime and k any positive integer. In Example 10.9, for instance, we found that $54 = 2 \cdot 3^3$ has (six) incongruent primitive roots. The development of this major result is a bit complicated. We begin by showing that p^2 has a primitive root, but first, we need to study an example.

EXAMPLE 10.17 Notice that $\alpha = 2$ is the only primitive root modulo $p = 3$. It is also a primitive root modulo $p^2 = 9$: $2^{\varphi(9)} = 2^6 \equiv 1 \pmod 9$, and $2^k \not\equiv 1 \pmod 9$ for $k < 6$. Thus, α is a primitive root modulo both p and p^2.

Likewise, 3 is a primitive root modulo both 5 and 5^2. (Verify this.) ■

Before moving on to our first result, we add a lemma to shorten its proof.

LEMMA 10.1 Let α be a primitive root modulo an odd prime p. Then $\text{ord}_{p^2}(\alpha + p) \neq p - 1$.

PROOF (by contradiction)
Let $\beta = \alpha + p$. Assume that $\text{ord}_{p^2} \beta = p - 1$. Then $\beta^{p-1} \equiv 1 \pmod{p^2}$. We have

$$\begin{aligned}
\beta^{p-1} &= (\alpha + p)^{p-1} \\
&= \alpha^{p-1} + (p-1)\alpha^{p-2}p + \binom{p-1}{2}\alpha^{p-3}p^2 + \cdots + p^{p-1} \\
1 &\equiv \alpha^{p-1} + p(p-1)\alpha^{p-2} \pmod{p^2} \\
1 &\equiv 1 - p\alpha^{p-2} \pmod{p^2}
\end{aligned}$$

This yields $p\alpha^{p-2} \equiv 0 \pmod{p^2}$; that is, $\alpha^{p-2} \equiv 0 \pmod{p}$. This is a contradiction, since α is a primitive root. Thus,

$$\text{ord}_{p^2} \beta = \text{ord}_{p^2}(\alpha + p) \neq p - 1$$ ■

The following example illustrates this result.

EXAMPLE 10.18 Notice that $\alpha = 5$ is a primitive root modulo 7. Verify that

$$\text{ord}_{p^2}(\alpha + p) = \text{ord}_{49} 12 \neq \text{ord}_7 5 = \text{ord}_p \alpha$$

PROOF
Notice that $12^2 \equiv -3 \pmod{49}$, $12^3 \equiv 13 \pmod{49}$, $12^6 \equiv 22 \pmod{49}$, $12^7 \equiv 19 \pmod{49}$, and $12^{21} \equiv -1 \pmod{49}$. Therefore, $\text{ord}_{49} 12 = 42 \neq 6 = \text{ord}_7 5$. ■

The following theorem shows that p^2 has a primitive root for every odd prime p.

THEOREM 10.7 If α is a primitive root modulo an odd prime p, then either α or $\alpha + p$ is a primitive root modulo p^2.

PROOF
Since α is a primitive root modulo p, $\text{ord}_p \alpha = p - 1$. Let $\text{ord}_{p^2} \alpha = e$. Then $\alpha^e \equiv 1 \pmod{p^2}$, so $e|\varphi(p^2)$, but $\varphi(p^2) = p(p-1)$. Therefore, $e|p(p-1)$.

Since $\alpha^e \equiv 1 \pmod{p^2}$, $\alpha^e \equiv 1 \pmod{p}$, so $p - 1|e$. Then $e = k(p-1)$ for some integer k. Therefore, $k(p-1)|p(p-1)$, so $k|p$. Thus, either $k = 1$ or $k = p$; that is, either $e = p - 1$ or $e = p(p-1)$.

case 1 Let $e = p(p-1)$. Then $e = \varphi(p^2)$, so α is a primitive root modulo p^2.

case 2 Let $e = p-1$. We shall show that $\beta = \alpha + p$ is a primitive root modulo p^2. Since $\beta \equiv \alpha \pmod{p}$, β is also a primitive root modulo p. Therefore, by the preceding discussion, $\text{ord}_{p^2}\,\beta = p-1$ or $p(p-1)$. But, by Lemma 10.1, $\text{ord}_{p^2}\,\beta \neq p-1$. So $\text{ord}_{p^2}\,\beta = p(p-1) = \varphi(p^2)$. Thus, $\beta = \alpha + p$ is a primitive root modulo p^2. ■

This theorem shows that the square of every odd prime has a primitive root. It also provides a mechanism for finding such a primitive root, as the following example illustrates.

EXAMPLE 10.19 Recall from Example 10.17 that $\alpha = 3$ is a primitive root modulo both 5 and 5^2. In Example 10.18, we found that $\alpha = 5$ is a primitive root modulo 7. Although it is not a primitive root modulo 49, $\alpha + p = 5 + 7 = 12$ is a primitive root modulo 49. (Verify this.) ■

Now we can show that every power p^k of an odd prime p has a primitive root. We know that it is true for $k = 1$ and 2. So it suffices to show that it is true for $k \geq 3$. Again, we split the proof into smaller units, for the sake of clarity.

LEMMA 10.2 Let α be a primitive root modulo an odd prime p such that $\alpha^{p-1} \not\equiv 1 \pmod{p^2}$. Then $\alpha^{p^{k-2}(p-1)} \not\equiv 1 \pmod{p^k}$ for every integer $k \geq 2$.

PROOF **(by induction on *k*)**
When $k = 2$,

$$\alpha^{p^{k-2}(p-1)} = \alpha^{p-1} \not\equiv 1 \pmod{p^2}$$

by the hypothesis. Thus, the statement is true when $k = 2$.

Assume it is true for an arbitrary integer $t \geq 2$:

$$\alpha^{p^{t-2}(p-1)} \not\equiv 1 \pmod{p^t}$$

Since $(\alpha, p) = 1$, $(\alpha, p^{t-1}) = 1$. So, by Euler's theorem,

$$\alpha^{\varphi(p^{t-1})} \equiv 1 \pmod{p^{t-1}}$$

That is,

$$\alpha^{p^{t-2}(p-1)} \equiv 1 \pmod{p^{t-1}}$$

Then

$$\alpha^{p^{t-2}(p-1)} = 1 + qp^{t-1} \tag{10.2}$$

for some integer q. By the inductive hypothesis, this implies $p \nmid q$.

Now we take the pth power of both sides of equation (10.2) and expand the RHS using the binomial theorem:

$$\begin{aligned}\alpha^{p^{t-1}(p-1)} &= (1 + qp^{t-1})^p \\ &= 1 + qp^t + \binom{p}{2} q^2 p^{2(t-1)} + \cdots + q^p p^{p(t-1)} \\ &\equiv 1 + qp^t \pmod{p^{t+1}}\end{aligned}$$

Since $p \nmid q$, this implies

$$\alpha^{p^{t-1}(p-1)} \not\equiv 1 \pmod{p^{t+1}}$$

Thus, by induction, the given statement is true for every integer $k \geq 2$. ■

This lemma enables us to complete the next segment of our proof.

THEOREM 10.8 Every power p^k of an odd prime p has a primitive root, where $k \geq 2$.

PROOF

Let α be a primitive root modulo p. If α is also a primitive root modulo p^2, then $\alpha^{p-1} \not\equiv 1 \pmod{p^2}$. On the other hand, if α is not a primitive root modulo p^2, then, by Theorem 10.7, $\beta = \alpha + p$ is a primitive root modulo p^2,where $\beta \equiv \alpha \pmod{p}$ and $\beta^{p-1} \not\equiv 1 \pmod{p^2}$. Thus, in both cases, p^2 has a primitive root γ such that $\gamma^{p-1} \not\equiv 1 \pmod{p^2}$. (*Note*: $\gamma = \alpha$ if α is a primitive root modulo p^2; otherwise, $\gamma = \beta$.) Therefore, by Lemma 10.2,

$$\gamma^{p^{k-2}(p-1)} \not\equiv 1 \pmod{p^k}$$

for every integer $k \geq 2$.

Next we need to show that γ is in fact a primitive root modulo p^k; that is, $\text{ord}_{p^k} \gamma = \varphi(p^k)$. To this end, assume that $\text{ord}_{p^k} \gamma = e$. Then $e|\varphi(p^k)$, where $\varphi(p^k) = p^{k-1}(p-1)$. Since $\gamma^e \equiv 1 \pmod{p^k}$, $\gamma^e \equiv 1 \pmod{p}$. This implies $p - 1|e$.

Let $e = (p-1)t$ for some integer t. Then $(p-1)t|p^{k-1}(p-1)$; that is, $t|p^{k-1}$. So $t = p^i$, where $0 \leq i \leq k-1$ and hence $e = p^i(p-1)$. If $i \leq k-2$, then

$$\gamma^{p^i(p-1)} \equiv 1 \pmod{p^k}$$

which is a contradiction. Therefore, $i = k-1$ and hence $\text{ord}_{p^k} \gamma = p^{k-1}(p-1) = \varphi(p^k)$. Thus, γ is a primitive root modulo p^k for every $k \geq 2$. ■

Two interesting observations: (1) The proof of this theorem gives us a bonus. It shows every primitive root modulo p^2 is also a primitive root modulo p^k, where $k \geq 2$. (2) So, by Theorems 10.7 and 10.8, a primitive root α modulo an odd prime p can be employed to find a primitive root γ modulo p^k. If α is a primitive root modulo p^2, then $\gamma = \alpha$; otherwise, $\gamma = \alpha + p$.

The following example illustrates both cases.

EXAMPLE 10.20 In Example 10.19, we found that 3 is a primitive root modulo both 5 and 5^2. So, by Theorem 10.8, 3 is a primitive root modulo every power of 5. For instance, it is a primitive root modulo $5^6 = 15{,}625$. To verify this, we shall just show that $3^{\varphi(5^6)/2} = 3^{2\cdot 5^5} = 3^{6250} \equiv -1 \pmod{5^6}$, leaving out the details.

Since $3^{125} \equiv 11693 \pmod{5^6}$ and $3^{2000} \equiv 2501 \pmod{5^6}$ (verify both),

$$\begin{aligned} 3^{6250} &= (3^{2000})^3 \cdot (3^{125})^2 \\ &\equiv 2501^3 \cdot 11693^2 \pmod{5^6} \\ &\equiv 15624 \equiv -1 \pmod{5^6} \end{aligned}$$

Remember that in Example 10.19, we found that $\alpha = 5$ is a primitive root modulo 7, but not of 49. However, $\alpha + p = 5 + 7 = 12$ is a primitive root modulo 7^2. So, by Theorem 10.8, 12 is a primitive root modulo 7^k, where $k \geq 2$. For example, 12 is a primitive root modulo $7^5 = 16,807$. To verify this, we can show that $12^{\varphi(7^5)/2} = 12^{3\cdot 7^4} = 12^{7203} \equiv -1 \pmod{7^5}$, leaving out the details.

We have $12^3 \equiv 1728 \pmod{7^5}$, $12^{200} \equiv -4336 \pmod{7^5}$ and $12^{7000} \equiv -4184 \pmod{7^5}$. (Verify them.) Therefore,

$$\begin{aligned} 12^{7203} &= 12^{7000} \cdot 12^{200} \cdot 12^3 \\ &\equiv (-4184)(-4336)(1728) \equiv -1 \pmod{7^5} \end{aligned}$$

■

Next, we can determine whether the integer 2^k has primitive roots. Clearly, 1, 2, and 4 have primitive roots, namely, 1, 1, and 3, respectively. At this point, the natural question to ask is: Does 2^k have primitive roots if $k \geq 3$? Before we can answer, we introduce two lemmas that enable us to shorten the proof of the next theorem, Theorem 10.9.

LEMMA 10.3 The square of every odd integer is congruent to 1 modulo 8.

PROOF

Let a be an odd integer, say, $a = 2i + 1$ for some integer i. Then $a^2 = 4i^2 + 4i + 1 = 4i(i+1) + 1$. Since $2|i(i+1)$, $8|4i(i+1)$, so $a^2 \equiv 1 \pmod 8$. ■

LEMMA 10.4 Let a be an odd integer and $t \geq 3$. Then $a^{2^{t-2}} \equiv 1 \pmod{2^t}$.

PROOF (by induction on t)

The given conclusion is clearly true when $t = 3$, by Lemma 10.3. So assume it is true for an arbitrary integer $k \geq 3$:

$$a^{2^{k-2}} \equiv 1 \pmod{2^k}$$

$$a^{2^{k-2}} = 1 + q \cdot 2^k \quad \text{for some integer } q$$

Then

$$\begin{aligned} a^{2^{k-1}} &= (a^{2^{k-2}})^2 = (1 + q \cdot 2^k)^2 \\ &= 1 + q \cdot 2^{k+1} + q^2 2^{2k} \\ &\equiv 1 \pmod{2^{k+1}} \end{aligned}$$

Thus, if the statement holds for $t = k$, it also holds for $t = k + 1$. Therefore, by induction, the conclusion is true for every integer $t \geq 3$. ■

Using this lemma, we can conclude that 2^k has no primitive roots if $k \geq 3$, as the following theorem shows.

THEOREM 10.9 The integer 2^k has no primitive roots if $k \geq 3$.

PROOF (by contradiction)

Suppose 2^k has a primitive root α. Then $\text{ord}_{2^k} \alpha = \varphi(2^k) = 2^{k-1}$. But, since $(\alpha, 2) = 1$, α is odd. Then, by Lemma 10.4, $\alpha^{2^{k-2}} \equiv 1 \pmod{2^k}$. Consequently, $\text{ord}_{2^k} \alpha \leq 2^{k-2}$, which is a contradiction. Thus, 2^k has no primitive roots for $k \geq 3$. ■

By virtue of this theorem, the integers 8, 16, 32, 64, and so on, possess no primitive roots.

Next we can prove that a positive integer cannot have a primitive root, if it is divisible by two distinct odd primes, or if it has the form $2^i p^j$, where $i \geq 2$ and p is an odd prime. To this end, we need the following lemma.

LEMMA 10.5 The integer ab possesses no primitive roots if $a, b > 2$ and $(a, b) = 1$.

PROOF (by contradiction)

Suppose ab has a primitive root α. Then $(\alpha, ab) = 1$ and $\alpha^{\varphi(ab)} \equiv 1 \pmod{ab}$. Since $(\alpha, ab) = 1$, $(\alpha, a) = 1 = (\alpha, b)$.

Let $d = (\varphi(a), \varphi(b))$. Since a, $b > 2$, both $\varphi(a)$ and $\varphi(b)$ are even by Theorem 8.5, so $d \geq 2$. Besides, since $d|\varphi(a)$ and $d|\varphi(b)$, $\dfrac{\varphi(a)\varphi(b)}{d} = \dfrac{\varphi(ab)}{d}$ is an integer. But $d > 1$, so $\dfrac{\varphi(ab)}{d} < \varphi(ab)$.

Since $(\alpha, a) = 1$ and $\alpha^{\varphi(a)} \equiv 1 \pmod a$,

$$\begin{aligned}\alpha^{\varphi(ab)/d} &= \alpha^{\varphi(a)\varphi(b)/d} = [\alpha^{\varphi(a)}]^{\varphi(b)/d} \quad [\textit{Note: } d|\varphi(b).]\\ &\equiv 1^{\varphi(b)/d} \equiv 1 \pmod a\end{aligned}$$

Similarly, $\alpha^{\varphi(ab)/d} \equiv 1 \pmod b$. Therefore, $\alpha^{\varphi(ab)/d} \equiv 1 \pmod{ab}$, which is a contradiction, since α is a primitive root modulo ab and $\varphi(ab)/d < \varphi(ab)$. Thus, ab has no primitive roots. ■

For example, 20 has no primitive roots, since $20 = 4 \cdot 5$, where $4, 5 > 2$ and $(4, 5) = 1$. Likewise, $150 = 6 \cdot 25$ has no primitive roots.

THEOREM 10.10 A positive integer has no primitive roots if it has two distinct odd prime factors, or if it is of the form $2^i p^j$, where p is an odd prime and $i \geq 2$.

PROOF

Suppose a positive integer n has two distinct odd prime factors p and q. Then, by Lemma 10.5, $p^i q^j$ and hence n has no primitive roots.

On the other hand, let $n = 2^i p^j$, where $i \geq 2$ and p is an odd prime. Again, by Lemma 10.5 with $a = 2^i$ and $b = p^j$, $n = ab = 2^i p^j$ does not have a primitive root. This concludes the proof. ■

EXAMPLE 10.21 The integer $1125 = 3^2 \cdot 5^3$ has no primitive roots, since it is divisible by two distinct odd primes. Likewise, $3780 = 2^2 \cdot 3^3 \cdot 5 \cdot 7$ also has no primitive roots. The integer $19{,}208 = 2^3 \cdot 7^4$ also has none since it has the form $2^i p^j$, where $i \geq 2$. ■

Theorem 10.10 brings us a giant step forward in our search for positive integers with primitive roots. It narrows the list considerably to a list of integers of the form $n = 2^i p^j$, where p is an odd prime. If $i = 0$ and $j = 0$, then $n = 1$ has a primitive root. On the other hand, if $j > 0$, then by Corollary 10.8 and Theorems 10.7 and 10.8, $n = p^j$ has a primitive root. If $i \geq 2$, then, by Theorem 10.10, n has no primitive roots.

We will now take up the remaining case $n = 2p^j$ in the following theorem.

THEOREM 10.11 The integer $n = 2p^j$, where p is an odd prime, has a primitive root.

PROOF

Let α be a primitive root modulo p^j. (Such an integer exists by Corollary 10.10 and Theorem 10.8.) So

$$\alpha^{\varphi(p^j)} \equiv 1 \pmod{p^j}$$

case 1 Suppose α is odd. (We shall show that α is a primitive root modulo n.) Since $\varphi(n) = \varphi(2p^j) = \varphi(2)\varphi(p^j) = \varphi(p^j)$,

$$\alpha^{\varphi(n)} = \alpha^{\varphi(p^j)} \equiv 1 \pmod{p^j} \tag{10.3}$$

Since α is odd, $\alpha \equiv 1 \pmod 2$, so

$$\alpha^{\varphi(p^j)} \equiv 1 \pmod 2 \tag{10.4}$$

Therefore, by congruences (10.3) and (10.4),

$$\alpha^{\varphi(n)} \equiv 1 \pmod{2p^j}; \quad \text{that is,} \quad \alpha^{\varphi(n)} \equiv 1 \pmod n$$

Suppose $\text{ord}_n\,\alpha = e < \varphi(n) = \varphi(2p^j)$. Then $\alpha^e \equiv 1 \pmod{p^j}$. Thus $\varphi(p^j) \le e < \varphi(p^j)$, which is clearly a contradiction. Therefore, $\text{ord}_n\,\alpha = \varphi(n)$ and α is a primitive root modulo n.

case 2 Suppose α is even. Then $\beta = \alpha + p^j$ is odd, so

$$\beta^{\varphi(p^j)} \equiv 1 \pmod 2$$

Besides, since $\beta \equiv \alpha \pmod{p^j}$, $\beta^{\varphi(n)} \equiv \alpha^{\varphi(p^j)} \equiv 1 \pmod{2p^j}$; that is, $\beta^{\varphi(n)} \equiv 1 \pmod n$. As in case 1, it follows that β is a primitive root modulo n.

Thus, in both cases, $n = 2p^j$ has a primitive root. ■

The following example illustrates this theorem.

EXAMPLE 10.22 Let $n = 38 = 2 \cdot 19$. By Example 10.10, 3 is a primitive root modulo 19. So, by Theorem 10.11, 3 is also a primitive root modulo 38. (Verify this.)

On the other hand, 10 is also a primitive root modulo 19. Since 10 is even, $10 + 19 = 29$ is a primitive root modulo 38, by Theorem 10.11. To verify this, we shall just show that $29^{\varphi(38)/2} = 29^9 \equiv -1 \pmod{38}$, leaving out the details. Since $29 \equiv -9 \pmod{38}$, $29^3 \equiv -7 \pmod{38}$. Thus $29^9 = (29^3)^3 \equiv (-7)^3 \equiv -1 \pmod{38}$. ■

In conclusion, we can now combine the results in Corollary 10.8 and Theorems 10.7, 10.8, 10.9, and 10.11 into Theorem 10.12, which conclusively identifies the integers with primitive roots. It was published by Gauss in 1801.

THEOREM 10.12 The only positive integers that possess primitive roots are 1, 2, 4, p^k, and $2p^k$, where p is an odd prime and k a positive integer. ■

Accordingly, the first 12 positive integers that have primitive roots are 1, 2, 3, 4, 5, 6, 7, 9, 10, 11, 13, and 14; the integers 8, 12, and 15 do not have primitive roots.

EXERCISES 10.4

Verify that 3 is a primitive root modulo each.

1. 5^2
2. 5^3
3. 5^4
4. 5^5
5. Verify that 3 is a primitive root modulo 7^2.
6. Verify that 12 is a primitive root modulo 7^4.

Find a primitive root modulo p^2 for each odd prime p.

7. 11
8. 13
9. 17
10. 19

Find a primitive root modulo p^k for each odd prime p and $k \geq 2$.

11. 3, $k = 4$
12. 5, $k = 3$
13. 7, $k = 3$
14. 23, $k = 2$

Two is a primitive root modulo 5. Determine whether each is a primitive root modulo 5^2.

15. 2
16. $2 + 5$

Five is a primitive root modulo 7. Determine whether each is a primitive root modulo 7^2.

17. 5
18. $5 + 7$

Using the given primitive root α modulo the odd prime p, find a primitive root modulo n.

19. $\alpha = 3, p = 5, n = 10$
20. $\alpha = 5, p = 23, n = 1058$
21. $\alpha = 2, p = 3, n = 486$
22. $\alpha = 6, p = 13, n = 4394$

Determine whether each integer has a primitive root.

23. 46
24. 486
25. 1024
26. 1029
27. 2187
28. 5324
29. 11,466
30. 742,586

Find the incongruent primitive roots modulo each.

31. 22
32. 26
33. 3^3
34. 3^4

Prove each.

35. If p is an odd prime, both p^k and $2p^k$ have the same number of primitive roots.
36. If a positive integer n (>7) has no primitive roots, then n^m has no primitive roots for any integer $m \geq 1$.

10.5 The Algebra of Indices

The concept of an index, which is analogous to a logarithm, was introduced by Gauss in his *Disquisitiones Arithmeticae*. As we will see shortly, the concept of index is very useful for solving certain congruences and for computing remainders.

Let α be a primitive root modulo a positive integer m. (Recall from Theorem 10.12 that $m = 1, 2, 4, p^k$, or $2p^k$, where p is an odd prime.) Then, by Theorem 10.3, the least residues of $\alpha, \alpha^2, \ldots, \alpha^{\varphi(m)}$ modulo m are a permutation of the $\varphi(m)$ positive integers $\leq m$ and relatively prime to it. For instance, in Example 10.10 we found that $\alpha = 5$ is a primitive root modulo 18 and the least residues of 5, 5^2, 5^3, 5^4, 5^5, and 5^6 are a rearrangement of the $\varphi(18) = 6$ positive integers 1, 5, 7, 11, 13, and 17 that are ≤ 18 and relatively prime to it.

Let a be a positive integer ≤ 18 and relatively prime to it. Then $a \equiv 5^k \pmod{18}$ for some positive integer k, where $1 \leq k \leq 6$. For instance, let $a = 13$; then $k = 4$ since $13 \equiv 5^4 \pmod{18}$. Accordingly, we say that 4 is the index of 13 to the base 5 modulo 18 and make the following definition.

Index

Let m be a positive integer with a primitive root α, and a a positive integer such that $(a, m) = 1$. Then the least positive integer k such that $\alpha^k \equiv a \pmod{m}$ is called the **index of a to the base α modulo m**. It is denoted by $\text{ind}_\alpha\, a$ or simply ind a when no confusion arises. Note that $1 \leq k \leq \varphi(m)$.

The following example illustrates this definition.

EXAMPLE 10.23 The integer 5 is a primitive root modulo 18. Notice that

$$5^1 \equiv 5 \pmod{18} \qquad 5^2 \equiv 7 \pmod{18} \qquad 5^3 \equiv 17 \pmod{18}$$
$$5^4 \equiv 13 \pmod{18} \qquad 5^5 \equiv 11 \pmod{18} \qquad 5^6 \equiv 1 \pmod{18}$$

Consequently,

$$\text{ind}_5\, 5 = 1 \qquad \text{ind}_5\, 7 = 2 \qquad \text{ind}_5\, 17 = 3$$
$$\text{ind}_5\, 13 = 4 \qquad \text{ind}_5\, 11 = 5 \qquad \text{ind}_5\, 1 = 6$$

Suppose we choose a different primitive root modulo 18, say, 11. Then,

$$11^1 \equiv 11 \pmod{18} \qquad 11^2 \equiv 13 \pmod{18} \qquad 11^3 \equiv 17 \pmod{18}$$
$$11^4 \equiv 7 \pmod{18} \qquad 11^5 \equiv 5 \pmod{18} \qquad 11^6 \equiv 1 \pmod{18}$$

Consequently,

$$\text{ind}_{11}\, 5 = 5 \qquad \text{ind}_{11}\, 7 = 4 \qquad \text{ind}_{11}\, 17 = 3$$
$$\text{ind}_{11}\, 13 = 2 \qquad \text{ind}_{11}\, 11 = 1 \qquad \text{ind}_{11}\, 1 = 6$$

Notice that, in general, $\text{ind}_5\, a \neq \text{ind}_{11}\, a$. For instance, $2 = \text{ind}_5\, 7 \neq \text{ind}_{11}\, 7 = 4$. Consequently, the value of $\text{ind}_\alpha\, a$ depends on the primitive root α (and the modulus m). ■

It follows from the definition that, as in the case of logarithms, $\text{ind}_\alpha a$ is a positive exponent. Notice that $\alpha^{\text{ind}_\alpha a} \equiv a \pmod{m}$ and that $\text{ind}_\alpha a$ is the least such positive exponent, where $1 \le \text{ind}_\alpha a \le \varphi(m)$.

Suppose $a \equiv b \pmod{m}$. To see how ind a and ind b are related, let us assume that α is a primitive root modulo m. Then $\alpha^{\text{ind}_\alpha a} \equiv a \pmod{m}$ and $\alpha^{\text{ind}_\alpha b} \equiv b \pmod{m}$. Because $a \equiv b \pmod{m}$, $\alpha^{\text{ind}_\alpha a} \equiv \alpha^{\text{ind}_\alpha b} \pmod{m}$. Then, by Corollary 10.2, $\text{ind}_\alpha a = \text{ind}_\alpha b$. Thus, $a \equiv b \pmod{m}$ if and only if $\text{ind}_\alpha a = \text{ind}_\alpha b$.

For example, $67 \equiv 13 \pmod{18}$. Recall from Example 10.23 that $\text{ind}_5 13 = 4$. Since $5^4 \equiv 67 \pmod{18}$, $\text{ind}_5 67 = 4$. Thus, $\text{ind}_5 13 = \text{ind}_5 67$.

The property $\alpha^{\text{ind}_\alpha a} \equiv a \pmod{m}$ reminds us of the logarithmic property, $b^{\log_b a} = a$ for any legal base b and any positive real number a. Likewise, the property $\text{ind}_\alpha a = \text{ind}_\alpha b$ if and only if $a \equiv b \pmod{m}$ reminds us of another logarithmic property: $\log_b x = \log_b y$ if and only if $x = y$.

Indices obey three additional properties, analogous to the following logarithmic properties:

- $\log_b 1 = 0$
- $\log_b(xy) = \log_b x + \log_b y$
- $\log_b(x^n) = n \log_b x$

They are presented in the following theorem.

THEOREM 10.13 Let m be a positive integer with a primitive root α, and a and b be positive integers relatively prime to m. Then:

- $\text{ind}_\alpha 1 \equiv 0 \pmod{\varphi(m)}$
- $\text{ind}_\alpha(ab) \equiv \text{ind}_\alpha a + \text{ind}_\alpha b \pmod{\varphi(m)}$
- $\text{ind}_\alpha(a^n) \equiv n \cdot \text{ind}_\alpha a \pmod{\varphi(m)}$

PROOF

(1) Since α is a primitive root modulo m, $\varphi(m)$ is the least positive integer such that $\alpha^{\varphi(m)} \equiv 1 \pmod{m}$. Consequently, $\text{ind}_\alpha 1 = \varphi(m) \equiv 0 \pmod{\varphi(m)}$.

(2) By definition, $\alpha^{\text{ind}_\alpha a} \equiv a \pmod{m}$ and $\alpha^{\text{ind}_\alpha b} \equiv b \pmod{m}$. Therefore,

$$ab \equiv \alpha^{\text{ind}_\alpha a} \cdot \alpha^{\text{ind}_\alpha b} \equiv \alpha^{\text{ind}_\alpha a + \text{ind}_\alpha b} \pmod{m}$$

Again, by definition, $ab \equiv \alpha^{\text{ind}_\alpha(ab)} \pmod{m}$. Thus,

$$\alpha^{\text{ind}_\alpha(ab)} \equiv \alpha^{\text{ind}_\alpha a + \text{ind}_\alpha b} \pmod{m}$$

Therefore, by Corollary 10.2, $\text{ind}_\alpha(ab) \equiv \text{ind}_\alpha a + \text{ind}_\alpha b \pmod{\varphi(m)}$.

(3) By definition, $\alpha^{\text{ind}_\alpha(a^n)} \equiv a^n \pmod{m}$. But

$$\alpha^{n \cdot \text{ind}_\alpha a} = (\alpha^{\text{ind}_\alpha a})^n \equiv a^n \pmod{m}$$

Thus,

$$\alpha^{\text{ind}_\alpha(a^n)} \equiv \alpha^{n\cdot\text{ind}_\alpha a} \pmod{m}$$

$$\text{ind}_\alpha(a^n) \equiv n \cdot \text{ind}_\alpha a \pmod{(m)}$$ ■

EXAMPLE 10.24 Verify properties (2) and (3) of Theorem 10.13 with $\alpha = 5$, $m = 18$, $a = 11$, $b = 13$, and $n = 7$.

SOLUTION

From Example 10.23, $\text{ind}_5\, 11 = 5$ and $\text{ind}_5\, 13 = 4$.

(1) $\text{ind}_5\, 11 + \text{ind}_5\, 13 = 5 + 4 \equiv 3 \pmod{6}$. [*Note*: $\varphi(18) = 6$.]
By direct computation,

$$\begin{aligned}\text{ind}_5(11 \cdot 13) &= \text{ind}_5\, 17 = 3\\ &\equiv \text{ind}_5\, 11 + \text{ind}_5\, 13 \pmod{6}\end{aligned}$$

(2) $7 \cdot \text{ind}_5\, 11 = 7 \cdot 5 \equiv 5 \pmod{6}$
By direct computation, $\text{ind}_5(11^7) = \text{ind}_5\, 11 \equiv 5 \pmod{6}$

Therefore, $_5(11^7) \equiv 7 \cdot \text{ind}_5\, 11 \pmod{6}$. ■

Just as we can use logarithms to convert multiplication problems to addition problems, we can use Theorem 10.13 to do the same. Accordingly, indices are useful in solving congruences of the form $ax^b \equiv c \pmod{m}$ and $a^{bx} \equiv c \pmod{m}$, where $(a, m) = 1$. The following three examples illustrate this technique.

EXAMPLE 10.25 Solve the congruence $11x \equiv 7 \pmod{18}$.

SOLUTION

Since 5 is a primitive root modulo 18 by Example 10.8, we take ind_5 of both sides of the given congruence:

$$\text{ind}_5(11x) \equiv \text{ind}_5\, 7 \pmod{\varphi(18)}$$

By Theorem 10.13, this yields,

$$\text{ind}_5\, 11 + \text{ind}_5\, x \equiv \text{ind}_5\, 7 \pmod{6}$$

But, by Example 10.23, $\text{ind}_5\, 11 = 5$ and $\text{ind}_5\, 7 = 2$. Therefore:

$$5 + \text{ind}_5 x \equiv 2 \pmod{6}$$

$$\text{ind}_5 x \equiv 3 \pmod{6}$$

$$x \equiv 5^3 \equiv 17 \pmod{18}$$

We can verify this by direct substitution. (*Note*: This method requires the availability of indices of positive integers ≤ 18 and relatively prime to it.) ■

The following two examples involve a knowledge of the indices to the base 2 modulo 13. (Notice that 2 is a primitive root modulo 13.) So, for convenience, we construct a necessary table, as Table 10.3 shows.

a	1	2	3	4	5	6	7	8	9	10	11	12
$\text{ind}_2\, a$	12	1	4	2	9	5	11	3	8	10	7	6

Table 10.3

EXAMPLE 10.26 Solve the congruence $8x^5 \equiv 3 \pmod{13}$.

SOLUTION

We have $8x^5 \equiv 3 \pmod{13}$. Take ind_2 of both sides:

$$\text{ind}_2(8x^5) \equiv \text{ind}_2 3 \pmod{12}$$

Applying Theorem 10.13 twice, this yields

$$\text{ind}_2 8 + 5\,\text{ind}_2 x \equiv \text{ind}_2 3 \pmod{12}$$

Using Table 10.3, this becomes

$$3 + 5\,\text{ind}_2 x \equiv 4 \pmod{12}$$

$$5\,\text{ind}_2 x \equiv 1 \pmod{12}$$

$$\text{ind}_2 x \equiv 5 \pmod{12}$$

$$x \equiv 6 \pmod{13}, \quad \text{by Table 10.3}$$

Again, we can verify this by direct computation. ■

Note: In this example, we have used an index table to make solving this relatively difficult problem surprisingly easy. In fact, there is nothing sacred about the choice of 2 as the base. We can use any primitive root modulo 13 as the base. See Exercises 17 and 19.

Indices are useful for solving congruences with variable exponents, as the next example illustrates.

EXAMPLE 10.27 Solve the congruence $11^{3x} \equiv 5 \pmod{13}$.

SOLUTION

Take ind_2 of both sides of the congruence:

$$\text{ind}_2(11^{3x}) \equiv \text{ind}_2 5 \pmod{12}$$

Using Theorem 10.13, this yields

$$3x \cdot \text{ind}_2 11 \equiv \text{ind}_2 5 \pmod{12}$$

Now use Table 10.3:

$$(3x) \cdot 7 \equiv 9 \pmod{12}$$
$$7x \equiv 3 \pmod{4}$$
$$x \equiv 1 \pmod{4}$$
$$x \equiv 1, 5, \text{ or } 9 \pmod{13}$$

Thus, the given congruence has three incongruent solutions. (See Exercises 18 and 20 also.) ■

EXERCISES 10.5

1. Let α be a primitive root modulo an odd prime p and $(\alpha, p) = 1$. Evaluate $\text{ind}_\alpha \alpha$.
2. Let α be a primitive root modulo a positive integer m. Find $\text{ind}_\alpha 1$.

Let α be a primitive root modulo an odd prime p and $(a, p) = 1$. Then $\text{ind}_\alpha(p - a) \equiv \text{ind}_\alpha a + \frac{p-1}{2} \pmod{p-1}$. This formula enables us to compute $\text{ind}_\alpha(p - a)$ using $\text{ind}_\alpha a$. In Exercises 3–6, use the given data to compute the corresponding index.

3. $p = 13$, $\text{ind}_2 5 = 9$, $\text{ind}_2 8$
4. $p = 13$, $\text{ind}_2 9 = 8$, $\text{ind}_2 4$
5. $p = 17$, $\text{ind}_3 11 = 7$, $\text{ind}_3 6$
6. $p = 19$, $\text{ind}_2 13 = 5$, $\text{ind}_2 6$

Using the preceding formula for $\text{ind}_\alpha(p - a)$, complete each table.

7. $p = 7$,

a	1	2	3	4	5	6
$\text{ind}_3 a$	6	2	1	.	.	.

8. $p = 11$,

a	1	2	3	4	5	6	7	8	9	10
$\text{ind}_7\, a$	10	3	4	6	2	.	.	.	.	.

Let m be a positive integer with a primitive root α and a a positive integer relatively prime to m. Let b be a multiplicative inverse of a modulo m. Then $\text{ind}_\alpha\, b = \varphi(m) - \text{ind}_\alpha\, a$. Using this fact, compute $\text{ind}_\alpha\, b$ for the given values of m, α, and a.

9. $m = 13, \alpha = 2, a = 5$
10. $m = 17, \alpha = 3, a = 7$
11. $m = 19, \alpha = 2, a = 8$
12. $m = 18, \alpha = 5, a = 13$

Let m be a positive integer with a primitive root α. Let a and b be positive integers such that $(a, m) = 1 = (b, m)$ and $\text{ind}_\alpha\, b = \varphi(m) - \text{ind}_\alpha\, a$. Then a and b are multiplicative inverses of each other modulo m. Using this fact, determine if the given integers a and b are multiplicative inverses of each other for the given value of m.

13. $m = 13, \alpha = 2, a = 6, b = 11$
14. $m = 17, \alpha = 3, a = 12, b = 10$
15. $m = 19, \alpha = 2, a = 5, b = 13$
16. $m = 18, \alpha = 11, a = 7, b = 13$

Solve each congruence using indices to the base 6.

17. $8x^5 \equiv 3 \pmod{13}$
18. $11^{3x} \equiv 5 \pmod{13}$

19–20. Solve the congruences in Exercises 17 and 18 using indices to the base 11.

Solve each congruence using indices.

21. $7x \equiv 13 \pmod{18}$
22. $5x \equiv 8 \pmod{17}$
23. $2x^4 \equiv 5 \pmod{13}$
24. $3x^2 \equiv 10 \pmod{13}$
25. $4x^3 \equiv 5 \pmod{17}$
26. $8^{5x} \equiv 5 \pmod{13}$
27. $7^{5x-1} \equiv 5 \pmod{13}$
28. $3^{4x+1} \equiv 10 \pmod{19}$

Using indices, determine the remainder when the first integer is divided by the second.

29. 23^{1001}, 13
30. 41^{1776}, 19
31. $5^{17} \cdot 7^{19}$, 13
32. $23^{111} + 111^{23}$, 17

Prove each.

33. Let α be a primitive root modulo a positive integer $m > 2$. Then $\text{ind}_\alpha (m-1) = \varphi(m)/2$.
34. Let α be a primitive root modulo an odd prime p. Then $\text{ind}_\alpha (p-1) = (p-1)/2$. (*Hint*: Use Exercise 33.)
35. Let α be a primitive root modulo a positive integer $m > 2$ and $(a, m) = 1$. Then $\text{ind}_\alpha (m-a) = \text{ind}_\alpha\, a + \varphi(m)/2$.
36. Let m be a positive integer with a primitive root α. Let a and b be positive integers such that $(a, m) = 1 = (b, m)$. Then a and b are multiplicative inverses of each other modulo m if and only if $\text{ind}_\alpha\, b = \varphi(m) - \text{ind}_\alpha\, a$.
37. Let p be an odd prime. Let b be a multiplicative inverse of an integer a modulo p, where $p \nmid a$. Then $\text{ind}\, b = p - 1 - \text{ind}\, a$.
38. Let p be an odd prime. Then the congruence $x^2 \equiv -1 \pmod{p}$ is solvable if and only if p is of the form $4k + 1$.
39. There are infinitely many primes of the form $4k + 1$. [*Hint*: Assume there is only a finite number of primes $p_1, p_2, \ldots, p_n$ of the form $4k + 1$. Using Exercise 38, show that $q = (p_1 p_2 \cdots p_n)^2 + 1$ has a prime factor of the form $4k + 1$, but different from $p_1, p_2, \ldots,$ and p_n.]
40. Let p be an odd prime. Then the congruence $x^4 \equiv -1 \pmod{p}$ is solvable if and only if p is of the form $8k + 1$.
41. There are infinitely many primes of the form $8k + 1$. [*Hint*: Assume there is only a finite number of primes $p_1, p_2, \ldots, p_n$ of the form $8k + 1$. Using Exercise 40, show that $q = (p_1 p_2 \cdots p_n)^4 + 1$ has a prime factor of the form $8k + 1$, but different from $p_1, p_2, \ldots,$ and p_n.]
*42. Let m be a positive integer with a primitive root and a a positive integer such that $(a, m) = 1$. Then the congruence $x^k \equiv a \pmod{m}$ is solvable if and only if $a^{\varphi(m)/d} \equiv 1 \pmod{m}$, where $d = (k, \varphi(m))$.
43. Let p be a prime and a a positive integer such that $p \nmid a$. Then the congruence $x^k \equiv a \pmod{p}$ is solvable if and only if $a^{(p-1)/d} \equiv 1 \pmod{p}$, where $d = (k, p-1)$. (*Hint*: Use Exercise 42.)

CHAPTER SUMMARY

In this chapter we studied three important concepts: the order of a positive integer modulo m, primitive root, and the index of a least residue modulo m. We learned their fundamental properties and a few applications, including the existence of infinitely many primes of certain types; two primality tests; and methods of finding remainders and solving special congruences.

The Order of a Positive Integer

- The order of a positive integer a, where $(a, m) = 1$, is the least positive exponent e such that $a^e \equiv 1 \pmod{m}$. It is denoted by $\text{ord}_m a$ or simply $\text{ord}\, a$. (p. 456)
- Let $\text{ord}_m a = e$. Then $a^n \equiv 1 \pmod{m}$ if and only if $e|n$. (p. 457)
- In particular, $\text{ord}_m a|\varphi(m)$. (p. 457)
- If p is a prime, then $\text{ord}_m a|p-1$. (p. 457)
- Let $\text{ord}_m a = e$. Then $a^i \equiv a^j \pmod{m}$ if and only if $i \equiv j \pmod{e}$. (p. 458)
- Let $\text{ord}_m a = e$ and k any positive integer. Then $\text{ord}_m(a^k) = e/(e, k)$. (p. 458)
- $\text{ord}_m(a^k) = e$ if and only if $(e, k) = 1$. (p. 459)

Primitive Root

- A positive integer α such that $(\alpha, m) = 1$ is a primitive root modulo m if $\text{ord}_m \alpha = \varphi(m)$. (p. 460)
- If α is a primitive root modulo m, then the least residues of $\alpha, \alpha^2, \ldots, \alpha^{\varphi(m)}$ modulo m are a permutation of the $\varphi(m)$ positive integers $\leq m$ and relatively prime to it. (p. 461)
- If m has a primitive root, then it has $\varphi(\varphi(m))$ primitive roots. (p. 462)
- If a prime p has a primitive root, then it has $\varphi(p-1)$ primitive roots. (p. 462)

Primality Tests

- (**Lucas' theorem**) Let n be a positive integer. If there is a positive integer x such that $x^{n-1} \equiv 1 \pmod{n}$ and $x^{(n-1)/q} \not\equiv 1 \pmod{n}$ for all prime factors q of $n-1$, then n is prime. (p. 464)
- Let n be an odd positive integer. If there is a positive integer x such that $x^{(n-1)/2} \equiv -1 \pmod{n}$ and $x^{(n-1)/q} \not\equiv 1 \pmod{n}$ for all odd prime factors q of $n-1$, then n is prime. (p. 465)

Primitive Roots for Primes

- (**Lagrange's theorem**) Let $f(x) = \sum_{i=0}^{n} a_i x^i$ be a polynomial of degree $n \geq 1$ with integral coefficients, where $p \nmid a_n$. Then the congruence $f(x) \equiv 0 \pmod{p}$ has at most n incongruent solutions modulo p. (p. 467)

- If p is a prime and $d|p-1$, then the congruence $x^d - 1 \equiv 0 \pmod{p}$ has exactly d incongruent solutions modulo p. (p. 469)
- (**Wilson's theorem**) If p is a prime, then $(p-1)! \equiv -1 \pmod{p}$. (p. 469)
- Let p be a prime and $d|p-1$. Then there are exactly $\varphi(d)$ incongruent integers of order d modulo p. (p. 471)
- Every prime p has $\varphi(p-1)$ incongruent primitive roots. (p. 472)

Composites with Primitive Roots

- If α is a primitive root modulo an odd prime p, then $\text{ord}_{p^2}(\alpha + p) \neq p - 1$. (p. 475)
- If α is a primitive root modulo an odd prime p, then either α or $\alpha + p$ is a primitive root modulo p^2. (p. 475)
- Let α be a primitive root modulo an odd prime p such that $\alpha^{p-1} \not\equiv 1 \pmod{p^2}$. Then $\alpha^{p^{k-2}(p-1)} \not\equiv 1 \pmod{p^k}$ for every integer $k \geq 2$. (p. 476)
- Every power p^k of an odd prime p has a primitive root, where $k \geq 2$. (p. 477)
- The square of every odd integer is congruent to 1 modulo 8. (p. 478)
- Let a be an odd integer and $t \geq 3$. Then $a^{2^{t-2}} - \equiv 1 \pmod{2^t}$. (p. 479)
- The integer 2^k has no primitive roots if $k \geq 3$. (p. 479)
- The integer ab possesses no primitive roots if $a, b > 2$ and $(a, b) = 1$. (p. 479)
- A positive integer has no primitive roots if it has two distinct odd prime factors or if it is of the form $2^i p^j$, where p is an odd prime and $i \geq 2$. (p. 480)
- The integer $n = 2p^j$, where p is an odd prime, has a primitive root. (p. 480)
- The only positive integers that possess primitive roots are 1, 2, 4, p^k, and $2p^k$, where p is an odd prime and k a positive integer. (p. 482)

Index of an Integer

- Let α be a primitive root modulo m and $(a, m) = 1$. The least positive integer k such that $\alpha^k \equiv a \pmod{m}$ is the **index** of a to the base α modulo m, denoted by $\text{ind}_\alpha a$ or simply ind a. For a given modulus m, it depends on the choice of α. (p. 483)
- Let m be a positive integer with a primitive root α, and a and b be positive integers relatively prime to m. Then:
  - $\text{ind}_\alpha 1 \equiv 0 \pmod{\varphi(m)}$
  - $\text{ind}_\alpha(ab) \equiv \text{ind}_\alpha a + \text{ind}_\alpha b \pmod{\varphi(m)}$
  - $\text{ind}_\alpha(a^n) \equiv n \cdot \text{ind}_\alpha a \pmod{\varphi(m)}$ (p. 484)

REVIEW EXERCISES

Evaluate each.

1. $\text{ord}_7\, 4$ 2. $\text{ord}_9\, 4$ 3. $\text{ord}_{11}\, 4$ 4. $\text{ord}_{13}\, 4$

Using the given order of the least residue a of a prime p, compute the order of the element b modulo p.

5. $\text{ord}_{13}\, 5 = 4$, $b = 5^7$ 6. $\text{ord}_{17}\, 8 = 8$, $b = 8^6$

Using the given orders of the least residues a and b, compute $\text{ord}_m(ab)$.

7. $\text{ord}_{13}\, 5 = 4$, $\text{ord}_{13}\, 9 = 3$ 8. $\text{ord}_{18}\, 7 = 3$, $\text{ord}_{18}\, 17 = 2$

Find the number of primitive roots modulo each.

9. 24 10. 38 11. 1024 12. 33,614

Determine whether each integer has a primitive root.

13. 1723 14. 2116 15. 48,778 16. 167,042

Using the given primitive root α modulo m, find the remaining primitive roots.

17. $\alpha = 3$, $m = 50$ 18. $\alpha = 2$, $m = 81$
19. Let α be a primitive root modulo p^j, where p is an odd prime and $j \geq 1$. Find a primitive root β modulo $2p^j$.

Using the given primitive root α modulo each prime p, find a primitive root modulo n.

20. $\alpha = 3$, $p = 7$, $n = 2p$ 21. $\alpha = 5$, $p = 23$, $n = p^2$
22. $\alpha = 2$, $p = 29$, $n = p^2$ 23. $\alpha = 2$, $p = 5$, $n = 2p^3$

Find the incongruent primitive roots modulo each.

24. 10 25. 50 26. 54 27. 81
28. 98 29. 121 30. 125 31. 162
32. Using the fact that $\text{ord}_{15}\, 7 = 4$, compute the remainder when 37^{2002} is divided by 15.
33. Using the fact that 6 is a primitive root modulo 109, compute the remainder when 424^{2076} is divided by 109.
34. Find $\text{ord}_p(p - a)$ if $\text{ord}_p\, a = q$, where p and q are odd primes.

Compute each, where 3 and 5 are primitive roots modulo 14.

35. $\text{ind}_3\, 11$ 36. $\text{ind}_3\, 3$ 37. $\text{ind}_5\, 13$ 38. $\text{ind}_3\, 9$

Using the modulus 50 and the fact that $\text{ind}_3\, 13 = 17$ and $\text{ind}_3\, 47 = 11$, compute each.

39. $\text{ind}_3(13 \cdot 47)$ 40. $\text{ind}_3(47^4)$

Let α be a primitive root modulo a positive integer $m > 2$ and $(a, m) = 1$. Then $\text{ind}_\alpha(m-a) \equiv \text{ind}_\alpha a + \varphi(m)/2 \pmod{\varphi(m)}$. Using this fact, $\alpha = 3$, $m = 14$, and the given $\text{ind}_\alpha a$, compute $\text{ind}_\alpha(m-a)$.

41. $\text{ind}_3 5 = 5$

42. $\text{ind}_3 13 = 3$

Solve each congruence using indices.

43. $3x^7 \equiv 4 \pmod{11}$

44. $5x^3 \equiv 8 \pmod{13}$

45. $5^{4x-1} \equiv 11 \pmod{17}$

46. $13^{3x-4} \equiv 16 \pmod{19}$

Using indices, find the remainder when the first integer is divided by the second.

47. 50^{1976}, 13

48. 1030^{1030}, 17

Verify that each integer n is a prime, using Lucas' theorem and the given value of x.

49. $n = 137, x = 3$

50. $n = 1193, x = 3$

Verify that each integer n is a prime, using Corollary 10.5 and the given value of x.

51. $n = 137, x = 3$

52. $n = 709, x = 2$

SUPPLEMENTARY EXERCISES

Let f_n denote the nth Fermat number, where $n \geq 0$.

1. Prove that $\text{ord}_{f_n} 2 | 2^{n+1}$.
2. Find $\text{ord}_{f_n} 2$, where f_n is a prime.
3. Prove that $\text{ord}_p 2 = 2^{n+1}$, where p is a prime factor of f_n.
4. Prove that every prime factor of f_n is of the form $2^{n+1}k + 1$. (This was shown by Euler in 1747; eight years earlier, he had shown that every such factor must be of the form $2^{t+1}k + 1$. (*Hint*: Use Exercise 3.)
5. Using the fact $\text{ord}_{19} 7 = 3$, find $\text{ord}_{19} 8$. (*Hint*: $8 = 7 + 1$.)
6. Using the fact $\text{ord}_{31} 26 = 6$, find $\text{ord}_{31} 25$. (*Hint*: $25 = 26 - 1$.)
7. Let $\text{ord}_p a = 3$, where p is an odd prime. Prove that $\text{ord}_p(a+1) = 6$.
8. Let $\text{ord}_p(a+1) = 6$, where p is an odd prime. Prove that $\text{ord}_p a = 3$.

Let k, m, and n be any positive integers. Prove or disprove each.

9. $\text{ord}_{mn} 10 = [\text{ord}_m 10, \text{ord}_n 10]$, where $(m, 10) = (n, 10) = (m, n) = 1$. (C. Cooper and R. E. Kennedy, 1995)
10. If p is a prime > 3, then $\text{ord}_{p^k} 10 = p^{k-1} \cdot \text{ord}_p 10$. (C. Cooper and R. E. Kennedy, 1995)

*11. Find an odd prime p and a primitive root g modulo p such that $1 < g < p$ and g is not a primitive root modulo p^2. (S. W. Golomb, 1993)

COMPUTER EXERCISES

Write a program to perform each task.

1. Read in a prime p and find the order of each least residue modulo p.
2. Read in a positive integer m. Find the order of each least residue modulo m, if it exists.
3. Read in a positive integer m and list all its primitive roots, if they exist.
4. Find the smallest prime p with a primitive root α such that it is not a primitive root modulo p^2.
5. Make a list of primes $p \leq 100$ and the smallest primitive root modulo each.
6. Read in a primitive root α modulo a prime p and print the remaining incongruent primitive roots modulo p.
7. Read in a primitive root α modulo p^j, where p is an odd prime and $j \geq 1$. Using α, find a primitive root modulo $2p^j$.
8. Read in an odd prime p. Find a primitive root α modulo p and use it to find a primitive root modulo p^2 and $2p^j$, where $j \geq 1$.
9. Read in the first 10 odd primes p. Find a primitive root α modulo p and use it to find a primitive root modulo p^2 and $2p^j$ for each p, where $j \geq 1$. Print the output in tabular form.
10. Read in an odd prime p. Find a primitive root α modulo p and use it to construct a table of indices of every least positive residue a modulo p. Use the table to pair the least residues that are multiplicative inverses of each other.

ENRICHMENT READINGS

1. J. D. Dixon, "Factorization and Primality Tests," *The American Mathematical Monthly*, 91 (1984), 333–353.
2. N. Robbins, "Calculating a Primitive Root (mod p^e)," *The Mathematical Gazette*, 59 (1975), 195.

Quadratic Congruences

... mathematical proofs, like diamonds, are hard as well as clear, and will be touched by nothing but strict reasoning.

—JOHN LOCKE

We studied the solvability of linear congruences in Section 4.2 and discussed primitive roots in Chapter 10. Now we turn to quadratic congruences. This includes the concept of a quadratic residue; a test for an integer to be a quadratic residue; two powerful notations—the Legendre symbol and the Jacobi symbol; the fascinating law of quadratic reciprocity, which is one of the jewels of number theory; and a primality test for Fermat numbers.

11.1 Quadratic Residues

We begin by considering the quadratic congruence

$$Ax^2 + Bx + C \equiv 0 \pmod{p} \tag{11.1}$$

where p is an odd prime and $p \nmid A$. (If $p|A$, then it reduces to a linear congruence.) Since p is odd and $p \nmid A$, $p \nmid 4A$. So we multiply both sides of congruence (11.1) by $4A$ to yield a perfect square on the LHS:

$$4A(Ax^2 + Bx + C) \equiv 0 \pmod{p} \tag{11.2}$$

But

$$\begin{aligned} 4A(Ax^2 + Bx + C) &= 4A^2x^2 + 4ABx + 4AC \\ &= (2Ax + B)^2 - \big(B^2 - 4AC\big) \end{aligned}$$

Therefore, congruence (11.2) can be rewritten as

$$(2Ax+B)^2 \equiv \left(B^2 - 4AC\right) \pmod{p} \tag{11.3}$$

which is of the form

$$y^2 \equiv a \pmod{p} \tag{11.4}$$

where $y = 2Ax + B$ and $a = B^2 - 4AC$.

Since these steps are reversible, this discussion shows that congruence (11.1) is solvable if and only if congruence (11.4) is solvable.

The following numeric example demonstrates this.

EXAMPLE 11.1 Solve the quadratic congruence $3x^2 - 4x + 7 \equiv 0 \pmod{13}$.

SOLUTION

$$3x^2 - 4x + 7 \equiv 0 \pmod{13}$$

Multiply both sides by $4 \cdot 3 = 12$:

$$36x^2 - 48x + 84 \equiv 0 \pmod{13}$$

That is,

$$(6x-4)^2 \equiv (16-84) \pmod{13}$$
$$(6x-4)^2 \equiv 10 \pmod{13}$$

Let $y = 6x - 4$. Then $y^2 \equiv 10 \pmod{13}$. This congruence has exactly two solutions, $y \equiv 6, 7 \pmod{13}$. (Verify this.)

Therefore, the solutions of the congruence are given by those of the linear congruences $6x - 4 \equiv 6 \pmod{13}$ and $6x - 4 \equiv 7 \pmod{13}$, namely, $x \equiv 6, 4 \pmod{13}$. Verify this, too. ■

Notice that the quadratic congruence in this example has exactly two solutions. But the next example shows that not every quadratic congruence has a solution.

EXAMPLE 11.2 Solve, if possible, the quadratic congruence $3x^2 + 7x + 5 \equiv 0 \pmod{13}$.

SOLUTION

The congruence $3x^2 + 7x + 5 \equiv 0 \pmod{13}$ yields $(6x+7)^2 \equiv 2 \pmod{13}$ (verify this). But the square of none of the least residues modulo 13 yields 2. So this congruence, and hence the given one, is not solvable. ■

Since congruences (11.1) and (11.4) are equivalent, meaning they have exactly the same solutions when solvable, we restrict our study to congruences of the form

$$x^2 \equiv a \pmod{p} \tag{11.5}$$

Since $x^2 \equiv 10 \pmod{13}$ has exactly two solutions, but $x^2 \equiv 2 \pmod{13}$ has none, we are tempted to ask: *When is congruence (11.5) solvable? When solvable, how many incongruent solutions does it have modulo p?*

To answer the second question first, suppose that $p|a$. Then $x^2 \equiv 0 \pmod{p}$, so $x \equiv 0 \pmod{p}$ is the only solution. Now assume $p \nmid a$. Then congruence (11.5) has exactly two incongruent solutions.

To see this, let α be a solution of (11.5): $\alpha^2 \equiv a \pmod{p}$. Let $\beta = p - \alpha$. Then $\beta^2 = (p-\alpha)^2 \equiv (-\alpha)^2 \equiv \alpha^2 \equiv a \pmod{p}$. So β is also a solution of the congruence. Besides, α and β are incongruent, since if $\beta \equiv \alpha \pmod{p}$, then $p - \alpha \equiv \alpha \pmod{p}$; that is, $-\alpha \equiv \alpha \pmod{p}$, so $2\alpha \equiv 0 \pmod{p}$. But $(2, p) = 1$; therefore, $\alpha \equiv 0 \pmod{p}$, which is a contradiction. Thus, α and $p - \alpha$ are two incongruent solutions of congruence (11.5).

Suppose congruence (11.5) has a third solution γ. Then $\gamma^2 \equiv \alpha^2 \pmod{p}$, so $p|\gamma^2 - \alpha^2$. Then either $\gamma \equiv \alpha \pmod{p}$ or $\gamma \equiv -\alpha \equiv \beta \pmod{p}$. Consequently, congruence (11.5) has no more than two solutions.

We have thus established the following result.

LEMMA 11.1 Let p be an odd prime and a an integer such that $p \nmid a$. Then the congruence $x^2 \equiv a \pmod{p}$ has either no solutions or exactly two incongruent solutions. ■

This discussion shows that if we can find one solution α, we can find the other by simply taking its additive inverse $-\alpha$. For instance, in Example 11.1 we found that 6 is a solution of $x^2 \equiv 10 \pmod{13}$; so the other solution is $-6 \equiv 7 \pmod{13}$, as expected.

Before we answer the question concerning when congruence (11.5) is solvable, we need to make the following definition.

Quadratic Residue

Let m be a positive integer and a any integer such that $(a, m) = 1$. Then a is a **quadratic residue** of m if the congruence $x^2 \equiv a \pmod{m}$ is solvable; otherwise, it is a **quadratic nonresidue** of m.

Notice that if $b \equiv a \pmod{m}$, and if a is a quadratic residue of m, then b is also a quadratic residue of m. Accordingly, we confine our discussion of quadratic residues to the least residues modulo m.

The following example illustrates the definition.

EXAMPLE 11.3 Find the quadratic residues and nonresidues of $p = 13$.

SOLUTION

Notice that

$$\begin{array}{ll} 1^2 \equiv 1 \equiv 12^2 \pmod{13} & 2^2 \equiv 4 \equiv 11^2 \pmod{13} \\ 3^2 \equiv 9 \equiv 10^2 \pmod{13} & 4^2 \equiv 3 \equiv 9^2 \pmod{13} \\ 5^2 \equiv 12 \equiv 8^2 \pmod{13} & 6^2 \equiv 10 \equiv 7^2 \pmod{13} \end{array}$$

Accordingly, 13 has exactly six quadratic residues, namely, 1, 3, 4, 9, 10, and 12; and it has six quadratic nonresidues also, namely, 2, 5, 6, 7, 8, and 11. (In 1973, R. H. Hudson of the University of South Carolina proved that 13 is the only prime p that has more than $\sqrt{p}$ consecutive quadratic nonresidues.) ■

This example provides us with two interesting bonuses:

- The prime 13 has the same number of quadratic residues and nonresidues, namely, 6; and
- They form a partitioning of the set of positive residues of 13 (see Figure 11.1).

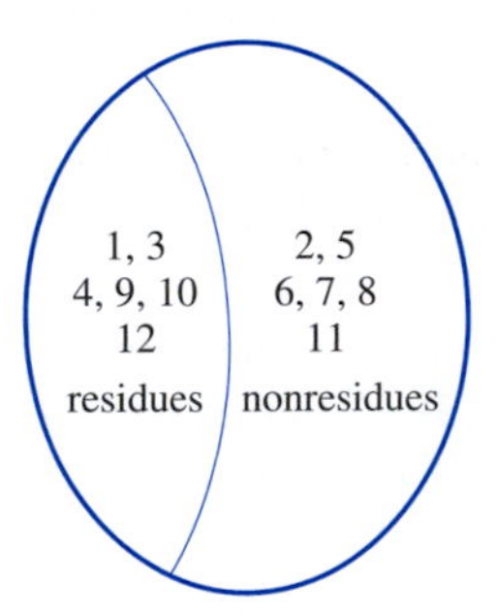

Figure 11.1 *The set of positive residues of 13.*

The following theorem shows that it is not a coincidence that 13 has the same number of quadratic residues and nonresidues.

THEOREM 11.1 Every odd prime p has exactly $(p-1)/2$ quadratic residues and $(p-1)/2$ quadratic nonresidues.

PROOF

Suppose p has k (incongruent) quadratic residues. By Lemma 11.1, each yields two incongruent solutions, so the total number of solutions is $2k$. But there are $p-1$

squares of the least positive residues, 1 through $p-1$. So $2k = p-1$; that is, $k = (p-1)/2$. Thus, there are $(p-1)/2$ quadratic residues and $(p-1)/2$ quadratic nonresidues. ■

We still have not answered the question we asked earlier: When is congruence (11.5) solvable? This is answered by the following theorem, developed by Euler.

THEOREM 11.2 **(Euler's Criterion)** Let p be an odd prime. Then a positive integer a with $p \nmid a$ is a quadratic residue of p if and only if $a^{(p-1)/2} \equiv 1 \pmod{p}$.

PROOF

Suppose that a is a quadratic residue of p. Then the congruence $x^2 \equiv a \pmod{p}$ has a solution α, where $(p, \alpha) = 1$. Consequently, by Fermat's little theorem, $\alpha^{p-1} \equiv 1 \pmod{p}$. Then $a^{(p-1)/2} \equiv (\alpha^2)^{(p-1)/2} = \alpha^{p-1} \equiv 1 \pmod{p}$.

Conversely, suppose that $a^{(p-1)/2} \equiv 1 \pmod{p}$. By Corollary 10.8, p has a primitive root β. Then $a \equiv \beta^k \pmod{p}$ for some positive integer k, where $1 \le k \le p-1$. Then $\beta^{k(p-1)/2} \equiv a^{(p-1)/2} \equiv 1 \pmod{p}$. Since β is a primitive root modulo p, $\mathrm{ord}_p\,\beta = p-1 | k(p-1)/2$; that is, k must be an even integer, say, $k = 2i$. Then $a \equiv \beta^{2i} \equiv (\beta^i)^2 \pmod{p}$, so a is a quadratic residue of p.

This completes the proof. ■

The following example demonstrates this test.

EXAMPLE 11.4 Determine whether 10 and 7 are quadratic residues of 13.

SOLUTION

- Notice that $10^{(13-1)/2} = 10^6 \equiv (-3)^6 \equiv 1 \pmod{13}$, so, by Euler's criterion, 10 is a quadratic residue of 13, as found in Example 11.3. (Consequently, the congruence $x^2 \equiv 10 \pmod{13}$ is solvable.)
- This time, we compute $7^{(13-1)/2} \pmod{13}$: $7^{(13-1)/2} \equiv 7^6 \equiv (7^3)^2 \equiv 5^2 \equiv -1 \pmod{13}$. Since $7^6 \not\equiv 1 \pmod{13}$, by Euler's criterion, 7 is a quadratic nonresidue of 13, as found in Example 11.3. ■

In Theorem 11.2, suppose $a^{(p-1)/2} \not\equiv 1 \pmod{p}$. Then a is a quadratic nonresidue. We can then tell exactly what the least residue of $a^{(p-1)/2}$ modulo p is. To this end, notice that, by Fermat's little theorem, $a^{p-1} \equiv 1 \pmod{p}$. Since p is odd and $a^{p-1} - 1 = [a^{(p-1)/2} + 1][a^{(p-1)/2} - 1]$, this implies either $a^{(p-1)/2} \equiv 1 \pmod{p}$ or $a^{(p-1)/2} \equiv -1 \pmod{p}$. But $a^{(p-1)/2} \not\equiv 1 \pmod{p}$, so $a^{(p-1)/2} \equiv -1 \pmod{p}$. Thus, if a is a quadratic nonresidue, then $a^{(p-1)/2} \equiv -1 \pmod{p}$.

Conversely, let a be an integer such that $p \nmid a$ and $a^{(p-1)/2} \equiv -1 \pmod{p}$. Then a cannot be a quadratic residue, since if it is, then, by Euler's criterion, $a^{(p-1)/2} \equiv$

$1 \pmod{p}$. This implies $-1 \equiv 1 \pmod{p}$; that is, $p = 2$, which is a contradiction. Thus, if $a^{(p-1)/2} \equiv -1 \pmod{p}$, then a must be a quadratic nonresidue.

Accordingly, we have the following result.

COROLLARY 11.1 Let p be an odd prime. Then a positive integer a, where $p \nmid a$, is a quadratic nonresidue if and only if $a^{(p-1)/2} \equiv -1 \pmod{p}$. ■

For instance, in Example 10.12 we found that $5^{(1213-1)/2} \equiv 5^{606} \equiv -1 \pmod{1213}$. So, by this corollary, 5 is a quadratic nonresidue of 1213.

It follows, by Euler's criterion, that congruence (11.5) is solvable if and only if $a^{(p-1)/2} \equiv 1 \pmod{p}$. Although Theorem 11.2 provides a test for determining the solvability of the congruence, it is not practical when p is fairly large. We will explore this further in the next section.

EXERCISES 11.1

Solve each quadratic congruence, if possible.

1. $x^2 \equiv 1 \pmod{6}$
2. $x^2 \equiv 1 \pmod{8}$
3. $x^2 \equiv 1 \pmod{12}$
4. $x^2 \equiv 3 \pmod{6}$
5. $x^2 \equiv 5 \pmod{6}$
6. $7x^2 \equiv 1 \pmod{18}$
7. $3x^2 \equiv 5 \pmod{7}$
8. $7x^2 \equiv 6 \pmod{13}$
9. $4x^2 \equiv 7 \pmod{11}$
10. $4x^2 + 4x - 3 \equiv 0 \pmod{5}$
11. $2x^2 + 3x + 1 \equiv 0 \pmod{7}$
12. $2x^2 + x + 1 \equiv 0 \pmod{11}$
13. $25x^2 + 70x + 37 \equiv 0 \pmod{13}$

Find the number of quadratic residues of each integer.

14. 17
15. 12
16. 19
17. 101

Find the quadratic residues of each integer.

18. 8
19. 18
20. 17
21. 23

Determine whether the given integer a is a quadratic residue of the corresponding prime p.

22. $a = 5, p = 23$
23. $a = 7, p = 29$
24. $a = 2, p = 37$
25. $a = 3, p = 47$

Verify that the congruences $x^2 \equiv a \pmod{p}$ and $x^2 \equiv b \pmod{p}$ are not solvable, but $x^2 \equiv ab \pmod{p}$ is solvable for the given values of a, b, and p. (See Exercise 34.)

26. $a = 3, b = 5, p = 7$
27. $a = 7, b = 10, p = 11$
28. Find the number of quadratic residues of the Fermat prime f_n.

Prove each.

29. Every primitive root modulo an odd prime p is a quadratic nonresidue.
30. The integer $p - 1$ is a quadratic residue of an odd prime p if and only if $p \equiv 1 \pmod{4}$.
31. Let a be a quadratic residue of an odd prime p. Then $p - a$ is a quadratic residue if and only if $p \equiv 1 \pmod{4}$.
32. Let a be a quadratic residue of an odd prime p. Then $p - a$ is a quadratic nonresidue if and only if $p \equiv 3 \pmod{4}$.
33. The product of two quadratic residues of an odd prime p is a quadratic residue.
34. The product of two quadratic nonresidues of an odd prime p is a quadratic residue.
35. The square of a quadratic nonresidue of an odd prime p is a quadratic residue.
36. The product of a quadratic residue and a quadratic nonresidue of an odd prime p is a quadratic nonresidue.

37. The multiplicative inverse of a quadratic residue of an odd prime is also a quadratic residue.
(*Hint*: Use Exercise 36.)
38. The integer $p-1$ is a quadratic nonresidue of every Mersenne prime M_p.
39. An integer a is a quadratic residue of a Mersenne prime M_p if and only if $p-a$ is a quadratic nonresidue of M_p.
40. If $p = 2^n + 1$ is a prime, then every quadratic nonresidue of p is a primitive root of p.
41. Let p be an odd prime such that $(a, p) = 1 = (b, p)$. Then either all three quadratic congruences $x^2 \equiv a \pmod{p}$, and $x^2 \equiv b \pmod{p}$, $x^2 \equiv ab \pmod{p}$ are solvable or exactly one of them is solvable.
(*Hint*: Use Exercises 33 and 34.)

11.2 The Legendre Symbol

Although Euler's criterion provides us with a beautiful test for determining the solvability of the congruence $x^2 \equiv a \pmod{p}$, computing $a^{(p-1)/2} \pmod{p}$ is tedious when p and a are large. For example, it is not easy to apply the test for determining the solvability of $x^2 \equiv 3797 \pmod{7297}$. (See Example 11.18.) So we now present the Legendre symbol, a powerful notation introduced by the French mathematician Adrien-Marie Legendre in his 1798 book, *Essai sur la Theorie de Nombres*. Legendre's *Theories des Nombres* and Gauss' *Disquisitiones Arithmeticae* were the standard works on number theory for many years.

The Legendre symbol, as we will see shortly, greatly simplifies our task of determining the solvability of congruence (11.5).

The Legendre Symbol

Let p be an odd prime and a any integer such that $p \nmid a$. The **Legendre symbol** (a/p) is defined by

$$(a/p) = \begin{cases} 1 & \text{if } a \text{ is a quadratic residue of } p \\ -1 & \text{otherwise} \end{cases}$$

Notice that the Legendre symbol (a/p) is *not* defined if $p|a$. The next example illustrates this definition.

EXAMPLE 11.5 In Example 11.3, we found that the residues 1, 3, 4, 9, 10, and 12 are quadratic residues of 13, whereas the residues 2, 5, 6, 7, 8, and 11 are not. Therefore, $(1/13) = (3/13) = (4/13) = (9/13) = (10/13) = (12/13) = 1$, whereas $(2/13) = (5/13) = (6/13) = (7/13) = (8/13) = (11/13) = -1$. ■

***Adrien-Marie Legendre** (1752–1833) was born into a well-to-do Parisian family and studied at the Collège Mazarin. His first published articles appeared in 1774 in a treatise on mechanics by his mathematics professor, although he was not given credit in the book. He was a professor of mathematics at the École Militaire in Paris from 1775 to 1780 and afterward appointed professor at then École Normale. In 1782 Legendre won the Berlin Academy prize for his essay on ballistics. His memoir,* Recherches d'Analyse Indeterminee, *published in 1785, contains a nonrigorous account of the law of quadratic reciprocity, as well as many applications, a discussion of the decomposition of positive integers as the sum of three squares, and a Statement of Dirichlet's theorem on the existence of infinitely many primes in arithmetic progressions (Theorem 3.16). In 1823 he provided a beautiful demonstration of Fermat's last theorem for the case n = 5.*

Although best known for his work on elliptic integrals, Legendre made significant contributions to number theory, calculus of variations, spherical harmonics, and geometry. His Éléments de Géométrie, *a pedagogical improvement of Euclid's* Elements, *was so popular that it went through numerous editions and translations, and was used as a text for over a century, the first English translation appearing in 1819.*

Legendre had a second edition of his number theory book published in 1808, a third edition in 1830 in two volumes under the title Théories des Nombres, *and a fourth edition in 1900.*

A disciple of Euler and Lagrange, a marvelous calculator, and a skillful analyst, Legendre raised "numerous questions that were fruitful subjects of investigation for mathematicians of the 19th century."

Using the Legendre symbol, it follows that $(a/p) \equiv a^{(p-1)/2} \pmod{p}$. Accordingly, Euler's criterion can be rewritten as follows.

THEOREM 11.3 **(Euler's Criterion)** Let p be an odd prime. Then a positive integer a with $p \nmid a$ is a quadratic residue of p if and only if $(a/p) = 1$. ■

In other words, $x^2 \equiv a \pmod{p}$ is solvable if and only if $(a/p) = 1$. For instance, since $(10/13) = 1$, $x^2 \equiv 10 \pmod{13}$ is solvable. (See Example 11.5.) But $(7/13) = -1$, so $x^2 \equiv 7 \pmod{13}$ is not solvable.

Thus, to determine the solvability of the congruence $x^2 \equiv 3797 \pmod{7297}$, we need to compute the symbol $(3797/7297)$. But how do we evaluate it? We do not have enough tools to work with the symbol, so we present three fundamental properties of the symbol in the following theorem.

THEOREM 11.4 Let p be an odd prime, and a and b be any integers with $p \nmid ab$. Then

(1) If $a \equiv b \pmod{p}$, then $(a/p) = (b/p)$.
(2) $(a/p)(b/p) = (ab/p)$
(3) $(a^2/p) = 1$

PROOF

(1) Suppose $a \equiv b \pmod{p}$. Then the congruence $x^2 \equiv a \pmod{p}$ is solvable if and only if $x^2 \equiv b \pmod{p}$ is solvable. Therefore, $(a/p) = (b/p)$.

(2) By Euler's criterion, $(ab/p) \equiv (ab)^{(p-1)/2} \equiv a^{(p-1)/2}b^{(p-1)/2} \equiv (a/p)(b/p) \pmod{p}$. Again, since p is odd and the value of a Legendre symbol is 1 or -1, this is so if and only if equality holds.

(3) By part (2), $(a^2/p) = (a/p)(a/p)$. But $(a/p) = \pm 1$. So $(a^2/p) = 1$ in both cases. This completes the proof. ■

Property (1) in the theorem can also be proved as follows. Suppose $a \equiv b \pmod{p}$ and the congruence $x^2 \equiv a \pmod{p}$ is solvable; that is, $(a/p) = 1$. Since $a \equiv b \pmod{p}$, $x^2 \equiv b \pmod{p}$ is also solvable. Therefore, $(b/p) = 1 = (a/p)$. On the other hand, suppose $x^2 \equiv a \pmod{p}$ is not solvable; that is, $(a/p) = -1$. Since $a \equiv b \pmod{p}$, $x^2 \equiv b \pmod{p}$ is also not solvable. So $(b/p) = -1 = (a/p)$. Thus, in both cases, $(a/p) = (b/p)$.

Property (3) can also be proven using congruence. Since $x^2 \equiv a^2 \pmod{p}$ is always solvable, $(a^2/p) = 1$.

The preceding properties have beautiful consequences (see Exercises 33–36 in Section 11.1):

- If $a \equiv b \pmod{p}$, then either both are quadratic residues or both are quadratic nonresidues.
- The product of two quadratic residues is a quadratic residue.
- The product of a quadratic residue and a quadratic nonresidue is a quadratic nonresidue.
- The product of two quadratic nonresidues is a quadratic residue.
- The square of every integer relatively prime to p is a quadratic residue.

Properties (2) and (3) can be employed to evaluate the Legendre symbol (a^2b/p), where $p \nmid ab$, provided we know the value of (b/p). To see this, notice that

$$\begin{aligned}(a^2b/p) &= (a^2/p)(b/p), \quad \text{by property (2)}\\ &= (b/p), \quad \text{by property (3)}\end{aligned}$$

For example, suppose we know that $(7/31) = 1$. Then $(28/31) = (4/31)(7/31) = 1 \cdot (7/31) = (7/31) = 1$. (We will see later how to compute $(7/31)$ without using Euler's criterion.)

Using Euler's criterion, we can now identify the primes for which -1 is a quadratic residue.

COROLLARY 11.2 If p is an odd prime, then $(-1/p) = (-1)^{(p-1)/2}$. That is,

$$(-1/p) = \begin{cases} 1 & \text{if } p \equiv 1 \pmod{4} \\ -1 & \text{if } p \equiv -1 \pmod{4} \end{cases}$$

PROOF

By Euler's criterion,

$$
\begin{aligned}
(-1/p) &\equiv (-1)^{(p-1)/2} \pmod{p} \\
&= (-1)^{(p-1)/2} \quad \text{since } (-1)^{(p-1)/2} = \pm 1 \\
&= \begin{cases} 1 & \text{if } p \text{ is of the form } 4k+1 \\ -1 & \text{if } p \text{ is of the form } 4k+3 \end{cases} \\
&= \begin{cases} 1 & \text{if } p \equiv 1 \pmod{4} \\ -1 & \text{if } p \equiv -1 \pmod{4} \end{cases}
\end{aligned}
$$

■

According to this corollary, -1 is a quadratic residue of p if and only if $p \equiv 1 \pmod{4}$; that is, $x^2 \equiv p-1 \pmod{p}$ is solvable if and only if $p \equiv 1 \pmod{4}$. For example, $x^2 \equiv 12 \pmod{13}$ is solvable, but $x^2 \equiv 22 \pmod{23}$ is not.

Corollary 11.2 can now be used to evaluate Legendre symbols of the form $(-a^2/p)$, as the following example shows.

EXAMPLE 11.6 Evaluate $(-4/41)$ and $(-9/83)$.

SOLUTION

- $$
\begin{aligned}
(-4/41) &= (4/41)(-1/41), \quad \text{by property (2)} \\
&= (-1/41) \quad \text{by property (3)} \\
&= 1, \quad \text{by Corollary 11.2}
\end{aligned}
$$
- $$
\begin{aligned}
(-9/83) &= (9/83)(-1/83) \\
&= (-1/83) \\
&= -1
\end{aligned}
$$

■

Another interesting application of Theorem 11.4 and Corollary 11.2 is that they can be used to establish the existence of infinitely many primes of the form $4n+1$, as the following example shows (see Section 3.4).

EXAMPLE 11.7 Prove that there are infinitely many primes of the form $4n+1$.

PROOF (by contradiction)

Assume that there is only a finite number of such primes, say, $p_1, p_2, \ldots, p_k$. Let $N = (2p_1p_2 \cdots p_k)^2 + 1$. Since N is odd, it must have an odd prime factor p. Then $N \equiv 0 \pmod{p}$, so $(2p_1p_2 \cdots p_k)^2 \equiv -1 \pmod{p}$.

By property (3) in Theorem 11.4, $((2p_1p_2\cdots p_k)^2/p) = 1$ and so, by property (1), $(-1/p) = 1$. Then, by Corollary 11.2, p must be of the form $4n+1$. So $p = p_i$ for some i, where $1 \le i \le k$. This implies, $N \equiv 1 \pmod{p}$, which is a contradiction. Thus, there are infinitely many primes of the desired form. ■

Property (2) in Theorem 11.4 can be applied to evaluate Legendre symbols of the form (q^i/p), where $p \nmid q$, as the following corollary shows.

COROLLARY 11.3 Let p be an odd prime, q a prime such that $p \nmid q$, and i a positive integer. Then $(q^i/p) = (q/p)^i$. ■

The following example illustrates this.

EXAMPLE 11.8 Using the fact that $(5/17) = -1$, compute $(125/17)$ and $(15625/17)$.

SOLUTION

- $$\begin{aligned}(125/17) &= (5^3/17)\\ &= (5/17)^3, \quad \text{by Corollary 11.3}\\ &= (-1)^3 = -1\end{aligned}$$
- $$\begin{aligned}(15625/17) &= (5^6/17)\\ &= (5/17)^6 = (-1)^6 = 1\end{aligned}$$ ■

Returning to Theorem 11.4, we find that property (2) can obviously be extended to any finite number of primes not divisible by p. Accordingly, we have the following result, which follows by induction.

COROLLARY 11.4 Let p be an odd prime and let $\prod_{i=1}^{n} p_i^{e_i}$ be the canonical decomposition of a, where $(a,p) = 1$. Then $(a/p) = \prod_{i=1}^{n} (p_i/p)^{e_i}$.

PROOF

Since $(a, p) = 1$, $(p_i, p) = 1$ for every i. So $(p_i^{e_i}/p) = (p_i/p)^{e_i}$, by Corollary 11.3. Thus,

$$(a/p) = \prod_{i=1}^{n} (p_i^{e_i}/p) = \prod_{i=1}^{n} (p_i/p)^{e_i}$$ ■

This result can be employed to evaluate (a/p), provided we know the value of (p_i/p) for every prime factor p_i of a, as the following example illustrates.

EXAMPLE 11.9 Using the fact that $(2/23) = 1$ and $(5/23) = -1$, compute $(5000/23)$.

SOLUTION

Notice that $5000 = 2^3 5^4$. So, by Corollary 11.4,

$$\begin{aligned}(5000/23) &= (2/23)^3(5/23)^4 \\ &= 1^3 \cdot (-1)^4 = 1\end{aligned}$$ ■

How did we know that $(2/23) = 1$ and $(5/23) = -1$? We could certainly use Euler's criterion to evaluate each, but we would like to avoid that tedious undertaking. Instead, we can derive additional properties of the Legendre symbol in the rest of this section and in the next, which will enable us to compute (a/p). To this end, we now prove an elegant criterion due to Gauss, although this also is theoretical in nature. Its proof is a bit long and complicated, so we will first study two examples to clarify the proof.

EXAMPLE 11.10 Let $p = 23$ and $a = 5$. Let ν† denote the number of least positive residues of the $11 = (p-1)/2$ integers $1 \cdot 5, 2 \cdot 5, 3 \cdot 5, \ldots, 11 \cdot 5$ modulo p that exceed $p/2$. Find ν and determine whether $(5/23) = (-1)^\nu$.

SOLUTION

Notice that the least positive residues of the integers $1 \cdot 5, 2 \cdot 5, 3 \cdot 5, 4 \cdot 5, 5 \cdot 5, 6 \cdot 5, 7 \cdot 5, 8 \cdot 5, 9 \cdot 5, 10 \cdot 5$, and $11 \cdot 5$ modulo 23 are 5, 10, 15, 20, 2, 7, 12, 17, 22, 4, and 9, respectively. Clearly, five of them exceed $p/2 = 11.5$, so $\nu = 5$.

To evaluate $(5/23)$, we apply Euler's criterion:

$$5^{(p-1)/2} = 5^{11} = \left(5^5\right)^2 \cdot 5 \equiv (-3)^2 \cdot 5 \equiv -1 \pmod{23}$$

Thus, $(5/23) = -1$ and hence $(5/23) = (-1)^\nu$. ■

This example shows that 5 is a quadratic nonresidue of 23; furthermore, the quadratic nature of 5 modulo 23 is determined by the value of ν. This fact is not a coincidence and is the essence of the next theorem, but first, let us examine one more example.

† ν is the lower case Greek letter *nu*.

EXAMPLE 11.11 **(Example 11.10 continued)** There are $v = 5$ least positive residues $> p/2$, namely, 12, 15, 17, 20, and 22. Call them s_1 through s_5. Then the integers $p - s_1$ through $p - s_5$ are $23 - 12$, $23 - 15$, $23 - 17$, $23 - 20$, and $23 - 22$ respectively, namely, 11, 8, 6, 3, and 1, respectively; no two of them are congruent modulo 23.

There are $k = 11 - v = 11 - 5 = 6$ residues, r_1 through r_6, that are $< p/2$, namely, 2, 4, 5, 7, 9, and 10; no two of them are congruent modulo 23 either.

Furthermore, none of them is congruent to 11, 8, 6, 3, or 1 modulo 23. Thus the residues 2, 4, 5, 7, 9, 10, 11, 8, 6, 3, and 1 are positive and $\leq (p-1)/2$. (Amazingly enough, they are a permutation of the residues 1 through $(p-1)/2$ modulo p.) ■

We are now ready for the next milestone in our journey, discovered by Gauss in 1808. The proof is a bit long, so follow it patiently.

THEOREM 11.5 **(Gauss' Lemma)** Let p be an odd prime and a an integer such that $p \nmid a$. Let ν denote the number of least positive residues of the integers $a, 2a, 3a, \ldots, [(p-1)/2]a$ that exceed $p/2$. Then $(a/p) = (-1)^{\nu}$.

PROOF

Let $r_1, r_2, \ldots, r_k$ be the least positive residues of the integers $a, 2a, 3a, \ldots, [(p-1)/2]a$ modulo p that are $\leq p/2$, and $s_1, s_2, \ldots, s_\nu$ those that exceed $p/2$. Then $k + \nu = (p-1)/2$.

Now, consider the integers $r_1, r_2, \ldots, r_k, p - s_1, p - s_2, \ldots, p - s_\nu$. Each is positive and less than $p/2$. We would like to show that no two of them are congruent modulo p.

First, notice that no two r_is are congruent, since if $r_i \equiv r_j \pmod{p}$, then $t_i a \equiv t_j a \pmod{p}$ for some t_i and t_j, where $i < j$ and $1 \leq t_i,\ t_j \leq (p-1)/2$. But $p \nmid a$, so $t_i \equiv t_j \pmod{p}$, which is impossible. Thus, no two r_is are congruent. Likewise, no two s_is and hence no two $p - s_i$s are congruent modulo p.

Next we would like to show that no r_i is congruent to any $p - s_j$. If $r_i \equiv p - s_j \pmod{p}$, then $r_i \equiv -s_j \pmod{p}$, so $r_i + s_j \equiv 0 \pmod{p}$. This is impossible, since both r_i and s_j are less than $p/2$ and hence $r_i + s_j < p$. Thus, no r_i is congruent to $p - s_j$ modulo p.

Consequently, the positive integers $r_1, r_2, \ldots, r_k, p - s_1, p - s_2, \ldots, p - s_\nu$ are all $< p/2$ and are incongruent modulo p. Since there are $k + \nu = (p-1)/2$ of them, they must be the same as the least residues $1, 2, \ldots, (p-1)/2$. Therefore,

$$r_1 r_2 \cdots r_k (p - s_1)(p - s_2) \cdots (p - s_\nu) \equiv 1 \cdot 2 \cdots \left(\frac{p-1}{2}\right) \pmod{p}$$

That is,

$$(-1)^{\nu} r_1 r_2 \cdots r_k s_1 s_2 \cdots s_\nu \equiv \left(\frac{p-1}{2}\right)! \pmod{p} \tag{11.6}$$

But $r_1, r_2, \ldots, r_k$, $s_1, s_2, \ldots, s_\nu$ are the least positive residues of $a, 2a, \ldots, \left(\frac{p-1}{2}\right)a$. Therefore,

$$r_1 r_2 \cdots r_k s_1 s_2 \cdots s_\nu \equiv a(2a)(3a) \cdots \left(\frac{p-1}{2}\right)a \pmod{p}$$

Thus,

$$(-1)^\nu a(2a)(3a) \cdots \left(\frac{p-1}{2}\right)a \equiv \left(\frac{p-1}{2}\right)! \pmod{p}, \quad \text{by equation (11.6).}$$

Thus,

$$(-1)^\nu a^{(p-1)/2} \left(\frac{p-1}{2}\right)! \equiv \left(\frac{p-1}{2}\right)! \pmod{p}$$

But $p \nmid ((p-1)/2)!$, so

$$(-1)^\nu a^{(p-1)/2} \equiv 1 \pmod{p}$$

That is,

$$a^{(p-1)/2} \equiv (-1)^\nu \pmod{p}$$

But, by Euler's criterion,

$$(a/p) \equiv a^{(p-1)/2} \pmod{p}$$

Thus, $(a/p) = (-1)^\nu$, because $(a/p) = \pm 1$ and p is an odd prime. ■

EXAMPLE 11.12 Evaluate (10/13) and (7/13) using Gauss' lemma.

SOLUTION

- We have $p = 13$, $a = 10$, and $(p-1)/2 = 6$. The least positive residues of the integers $1 \cdot 10, 2 \cdot 10, 3 \cdot 10, 4 \cdot 10, 5 \cdot 10$, and $6 \cdot 10$ modulo 13 are 10, 7, 4, 1, 11, and 8, respectively. Exactly $\nu = 4$ of them are greater than $p/2 = 6.5$. Therefore, by Gauss' lemma, $(10/13) = (-1)^4 = 1$.
- With $a = 7$, the least positive residues of the integers $1 \cdot 7, 2 \cdot 7, 3 \cdot 7, 4 \cdot 7, 5 \cdot 7$, and $6 \cdot 7$ modulo 13 are 7, 1, 8, 2, 9, and 3, respectively. Since $\nu = 3$ of them are greater than $p/2 = 6.5$, $(7/13) = (-1)^3 = -1$.

  (Notice that these values agree with the ones found in Example 11.5.) ■

The following is an immediate consequence of Gauss' lemma.

COROLLARY 11.5 Let p be an odd prime with $p \nmid a$. Let ν denote the number of least positive residues of the integers $a, 2a, \ldots, [(p-1)/2]a$ that exceed $p/2$. Then $(a/p) = 1$ if and only if ν is even. ■

An elegant application of Gauss' lemma is that it can be used to determine the quadratic nature of 2 modulo an odd prime p, as the next theorem shows. First, we will study an example.

EXAMPLE 11.13 Evaluate $(2/13)$ using Gauss' lemma.

SOLUTION

Here $p = 13$. By Gauss' lemma, $(2/13) = (-1)^{\nu}$, where ν denotes the number of least positive residues of the integers $1 \cdot 2, 2 \cdot 2, 3 \cdot 2, 4 \cdot 2, 5 \cdot 2$, and $6 \cdot 2$ modulo 13 that exceed $p/2 = 6.5$. Notice that they are all less than p. This time, we shall find ν in a different way.

$$\begin{aligned}
\nu &= \text{number of the residues } 2r \text{ that exceed } p/2 \\
&= (p-1)/2 - (\text{number of positive integers } 2r < p/2) \\
&= 6 - (\text{number of positive integers} r < p/4) \\
&= 6 - \lfloor p/4 \rfloor \\
&= 6 - \lfloor 13/4 \rfloor = 6 - 3 = 3
\end{aligned}$$

Therefore, $(2/13) = (-1)^3 = -1$. (See Example 11.5 also.) ■

This example paves the way for the following important result.

THEOREM 11.6 Let p be an odd prime. Then

$$(2/p) = \begin{cases} 1 & \text{if } p \equiv \pm 1 \pmod 8 \\ -1 & \text{if } p \equiv \pm 3 \pmod 8 \end{cases}$$

PROOF

By Gauss' lemma, $(2/p) = (-1)^{\nu}$, where ν denotes the number of least positive residues of the integers $1 \cdot 2, 2 \cdot 2, 3 \cdot 2, \ldots, \left(\frac{p-1}{2}\right) \cdot 2$ modulo p that are greater than $p/2$. Each of them is positive and less than p, so they are $(p-1)/2$ least residues modulo p. Thus,

$$\begin{aligned}
\nu &= \text{number of the residues } 2r \text{ that exceed } p/2 \\
&= (p-1)/2 - (\text{number of positive integers } 2r < p/2) \\
&= (p-1)/2 - (\text{number of positive integers } r < p/4) \\
&= (p-1)/2 - \lfloor p/4 \rfloor
\end{aligned} \tag{11.7}$$

case 1 Let $p \equiv 1 \pmod{8}$. Then $p = 8k + 1$ for some integer k. So,

$$\nu = (p-1)/2 - \lfloor p/4 \rfloor = 4k - 2k = 2k$$

case 2 Let $p \equiv -1 \pmod{8}$. Then $p = 8k - 1$ for some integer k. So,

$$\nu = (p-1)/2 - \lfloor p/4 \rfloor = (4k-1) - (2k-1) = 2k$$

case 3 Let $p \equiv 3 \pmod{8}$. Then $p = 8k + 3$ for some integer k. So,

$$\nu = (p-1)/2 - \lfloor p/4 \rfloor = (4k+1) - 2k = 2k + 1$$

case 4 Let $p \equiv -3 \pmod{8}$. Then $p = 8k - 3$ for some integer k. So,

$$\nu = (p-1)/2 - \lfloor p/4 \rfloor = (4k-2) - (2k-1) = 2k - 1$$

Thus, if $p \equiv \pm 1 \pmod{8}$, then ν is even, so $(2/p) = 1$; if $p \equiv \pm 3$, ν is odd and hence $(2/p) = -1$. ■

It follows from this theorem that 2 is a quadratic residue of an odd prime p if and only if $p \equiv \pm 1 \pmod{8}$; that is, $x^2 \equiv 2 \pmod{p}$ is solvable if and only if $p \equiv \pm 1 \pmod{8}$.

Using this result, we can now compute Legendre symbols of the form $(\pm 2a^2/p)$, where p is an odd prime and $p \nmid a$, as the following example illustrates.

EXAMPLE 11.14 Compute (8/19) and (22/31).

SOLUTION

- $$\begin{aligned}(8/19) &= (4 \cdot 2/19) = (4/19)(2/19), && \text{by Theorem 11.4}\\ &= (2/19), && \text{by Theorem 11.4}\\ &= -1, && \text{by Theorem 11.6}\end{aligned}$$
- $$\begin{aligned}(13/31) &= (-18/31), && \text{by Theorem 11.4}\\ &= (9/31)(2/31)(-1/31), && \text{by Theorem 11.4}\\ &= (-1/31), && \text{by Theorems 11.4 and 11.6}\\ &= -1, && \text{by Corollary 11.2}\end{aligned}$$ ■

The following example is an interesting application of Theorem 11.6 and Euler's criterion. It illustrates the combined power of the two results.

EXAMPLE 11.15 Verify that $9973|(2^{4986}+1)$.

SOLUTION

Notice that 9973 is a prime and is $\equiv 5 \pmod 8$. So, by Theorem 11.6, $(2/9973) = -1$. Thus, by Euler's criterion, $(2/9973) \equiv 2^{4986} \equiv -1 \pmod{9973}$, so $9973|(2^{4986}+1)$. ■

In this example, we could use properties of congruence and arrive at the same conclusion by showing that $2^{4986} \equiv -1 \pmod{9973}$, but this would involve tedious and time-consuming computations. The dual power of Theorem 11.6 and Euler's criterion makes our job a lot easier.

Returning to Theorem 11.6, we can restate the formula for $(2/p)$ in a compact way, as the following result shows.

COROLLARY 11.6 Let p be an odd prime. Then $(2/p) = (-1)^{(p^2-1)/8}$.

PROOF

We consider the four cases as in the preceding proof. If $p \equiv 1 \pmod 8$, then $\dfrac{p^2-1}{8} = \left(\dfrac{p-1}{8}\right)(p+1)$ is an even integer; if $p \equiv -1 \pmod 8$, then $\dfrac{p^2-1}{8} = \left(\dfrac{p+1}{8}\right)(p-1)$ is also an even integer. Thus, if $p \equiv \pm 1 \pmod 8$, then $(-1)^{(p^2-1)/8} = 1 = (2/p)$, by Theorem 11.6.

On the other hand, let $p \equiv \pm 3 \pmod 8$. Then $p = 8k \pm 3$ for some integer k. Therefore,

$$\frac{p^2-1}{8} = \frac{64k^2 \pm 48k + 8}{8} = 8k^2 \pm 6k + 1$$

which is clearly an odd integer. Consequently, $(-1)^{(p^2-1)/8} = -1 = (2/p)$, again, by Theorem 11.6. ■

For example, $(2/13) = (-1)^{(169-1)/8} = (-1)^{21} = -1$. (See Example 11.13 also.)

The following example, proposed by O. N. Dalton of Texas in 1982, is an interesting application of this corollary and Theorem 11.6. (See Example 11.27 also.)

EXAMPLE 11.16 Let p be a prime of the form $4n \pm 1$. Compute the value of n^n modulo p.

SOLUTION

Let $p = 4n + r$, where $r = \pm 1$. Then $4n \equiv -r \pmod{p}$. So $(4n)^n \equiv (-r)^n \pmod{p}$; that is,

$$2^{2n} n^n \equiv (-r)^n \pmod{p} \tag{11.8}$$

case 1 Let $r = 1$. Then $2n = (p-1)/2$, so

$$\begin{aligned} 2^{2n} &= 2^{(p-1)/2} \equiv (2/p) = (-1)^{(p^2-1)/8} = (-1)^{n(2n+1)} \\ &= \left[(-1)^{2n+1}\right]^n \equiv (-1)^n \pmod{p} \end{aligned}$$

So congruence (11.8) yields

$$(-1)^n n^n \equiv (-1)^n \pmod{p}; \quad \text{thus,} \quad n^n \equiv 1 \pmod{p}$$

case 2 Let $r = -1$. Then $2^{2n} n^n \equiv 1 \equiv p + 1 \pmod{p}$. Since $(2/p) \equiv 2^{(p-1)/2} = 2^{2n-1} \pmod{p}$, this implies $(2/p)n^n \equiv (p+1)/2 \pmod{p}$; that is, $n^n \equiv (2/p)(p+1)/2 \pmod{p}$. Since $p \equiv -1 \pmod{4}$, $p \equiv 3$ or 7 modulo 8.

If $p \equiv 3 \pmod{8}$, $(2/p) = -1$, so $n^n \equiv -(p+1)/2 \equiv (p-1)/2 = 2n-1 \pmod{p}$. On the other hand, if $p \equiv 7 \pmod{8}$, $(2/p) = 1$, so $n^n \equiv (p+1)/2 = 2n \pmod{p}$.

Thus,

$$n^n = \begin{cases} 1 \pmod{p} & \text{if } p \equiv 1 \text{ or } 5 \pmod{8} \\ 2n \pmod{p} & \text{if } p \equiv 7 \pmod{8} \\ 2n - 1 \pmod{p} & \text{if } p \equiv 3 \pmod{8} \end{cases}$$ ■

If $p = 4n \pm 1$ is a prime, it is easy to show that n is a quadratic residue of p. See Exercise 44.

Theorem 11.6 also has a fine application to primitive roots. Although we established in Chapter 10 that the integers 1, 2, 4, p^k, and $2p^k$ have primitive roots, we did not provide a constructive mechanism for finding them. We shall now see that $2 \cdot (-1)^{(p-1)/2}$ is a primitive root modulo primes of the form $2p + 1$, where p is an odd prime.

THEOREM 11.7 If p and $2p+1$ are odd primes, then $2 \cdot (-1)^{(p-1)/2}$ is a primitive root modulo $2p+1$.

PROOF

Let $q = 2p + 1$ and $\alpha = 2 \cdot (-1)^{(p-1)/2}$. Clearly, $(\alpha, p) = 1$ and $\varphi(q) = 2p$. It suffices to show that $\text{ord}_q \alpha = 2p$.

case 1 Let $p \equiv 1 \pmod{4}$. Then $q \equiv 3 \pmod{8}$, so $(2/q) = -1$, by Theorem 11.6. But, by Euler's criterion, $(2/q) \equiv 2^{(q-1)/2} \pmod{q}$, so $2^p = 2^{(q-1)/2} \equiv (2/q) \equiv$

$-1 \pmod{q}$. Then $\alpha^{2p} = [2 \cdot (-1)^{(p-1)/2}]^{2p} = 2^{2p} \cdot (-1)^{p(p-1)} = 2^{2p} \cdot 1 = 2^{2p} \equiv (-1)^2 \equiv 1 \pmod{q}$. Thus, $\text{ord}_q\,\alpha | 2p$, so $\text{ord}_q\,\alpha = 1, 2, p$, or $2p$.

Clearly, $\text{ord}_q\,\alpha \neq 1$. If $\text{ord}_q\,\alpha = 2$, then $\alpha^2 \equiv 1 \pmod{q}$. Since $\alpha^2 = 4$, this means $4 \equiv 1 \pmod{q}$; so $q = 3$; this implies $p = 1$, which is a contradiction.

Now suppose that $\text{ord}_q\,\alpha = p$. Then $\alpha^p \equiv 1 \pmod{q}$. This implies, $[2 \cdot (-1)^{(p-1)/2}]^p = (2 \cdot 1)^p = 2^p \equiv 1 \pmod{q}$, since $(p-1)/2$ is even. This is also a contradiction, since $2^p \equiv -1 \pmod{q}$ from the first paragraph. Therefore, $\text{ord}_q\,\alpha \neq p$.

Thus, $\text{ord}_q\,\alpha = 2p$ and α is a primitive root modulo q.

case 2 Let $p \equiv -1 \pmod 4$. Then $q \equiv -1 \pmod 8$, so $(2/q) = 1$, by Theorem 11.6. Then $2^p \equiv 1 \pmod{q}$ and $\alpha^{2p} \equiv 1 \pmod{q}$. Consequently, $\text{ord}_q\,\alpha = 1, 2, p$, or $2p$.

Proceeding, as in case 1, we can show that $\text{ord}_q\,\alpha \neq 1$ or 2. So assume $\text{ord}_q\,\alpha = p$. Then $\alpha^p \equiv 1 \pmod{q}$. Since p and $(p-1)/2$ are odd, this implies $[2 \cdot (-1)^{(p-1)/2}]^p = [2 \cdot (-1)]^p = -2^p \equiv 1 \pmod{q}$; that is, $2^p \equiv -1 \pmod{q}$. This is a contradiction, so $\text{ord}_q\,\alpha \neq p$. Consequently, $\text{ord}_q\,\alpha = 2p$ and α is a primitive root modulo q.

Thus, in both cases, α is a primitive root. ■

It follows by this theorem that 2 is a primitive root modulo the primes 11, 59, 83, and 107, and -2 is a primitive root modulo 7, 23, and 47. (Verify these.)

Interestingly, there is a similar result that shows 2 is a primitive root modulo yet another class of primes: If p and $4p+1$ are primes, then 2 is a primitive root modulo $4p+1$. (See Exercise 62.)

EXERCISES 11.2

Evaluate each Legendre symbol, using Euler's criterion.

1. $(5/7)$
2. $(3/11)$
3. $(7/11)$
4. $(11/17)$

Evaluate each, using Corollary 11.2.

5. $(16/17)$
6. $(18/19)$
7. $(-1/29)$
8. $(-1/47)$

Using the fact that $(2/23) = 1 = (3/23)$ and $(5/23) = -1$, evaluate each.

9. $(128/23)$
10. $(125/23)$
11. $(600/23)$
12. $(1250/23)$

Using the fact that $(3/19) = -1 = (7/19)$, compute each.

13. $(27/19)$
14. $(63/19)$
15. $(147/19)$
16. $(9261/19)$

Let p be an odd prime with $p \nmid a$. Let v denote the number of least positive residues of the integers $a, 2a, 3a, \ldots, [(p-1)/2]a$ that exceed $p/2$. Find the value of v for the given values of p and a.

17. $p = 13,\ a = 3$
18. $p = 13,\ a = 5$
19. $p = 17,\ a = 4$
20. $p = 19,\ a = 7$

Using Gauss' lemma, evaluate each.

21. $(5/13)$
22. $(4/17)$
23. $(7/19)$
24. $(13/31)$

Compute each.

25. $(2/19)$
26. $(2/23)$
27. $(2/41)$
28. $(2/43)$
29. $(13/31)$
30. $(-50/29)$
31. $(41/43)$
32. $(110/59)$

33. If p and $4p + 1$ are primes, then 2 is a primitive root modulo $4p + 1$. Using this fact, find five primes for which 2 is a primitive root. (See Exercise 62.)

Using the fact that $(3/p) = \begin{cases} 1 & \text{if } p \equiv \pm 1 \pmod{12} \\ -1 & \text{if } p \equiv \pm 5 \pmod{12} \end{cases}$

evaluate each. (See Exercise 32 in Section 11.3.)

34. $(3/17)$
35. $(12/19)$
36. $(-3/31)$
37. $(35/47)$
38. Let $p \equiv 3 \pmod 4$ and $q = 2p + 1$ be primes. Then $q|M_p$. Using this fact, verify that $23|M_{11}$ and $47|M_{23}$. (See Exercise 52.)

Let p be an odd prime with a primitive root α and a a positive integer $\leq p - 1$. Then there exists an integer k such that $\alpha^k \equiv a \pmod p$, where $1 \leq k \leq p - 1$. Then a is a quadratic residue of p if and only if k is even. Using this fact and the given primitive root α, find the quadratic residues of the corresponding prime.

39. $p = 13, \alpha = 2$
40. $p = 17, \alpha = 3$

Verify each. (*Hint*: Use Euler's criterion.)

41. $1913|(2^{956} - 1)$
42. $2029|(2^{1014} - 1)$

Prove each, where p is an odd prime and $(a, p) = 1 = (b, p)$.

43. If $a \equiv b \pmod p$, then either both a and b are quadratic residues or both are quadratic nonresidues.
44. If $p = 4n \pm 1$ is a prime, then n is a quadratic residue of p. (O. N. Dalton, 1981)
45. Let $p \equiv 1 \pmod 4$ and a be a quadratic residue of p. Then $p - a$ is a quadratic residue of p.
46. Let $p \equiv 3 \pmod 4$ and a be a quadratic residue of p. Then $p - a$ is a quadratic nonresidue of p.
47. Let $p \equiv \pm 1 \pmod 8$ and a be a quadratic residue of p. Then $2a$ is a quadratic residue of p.
48. Let $p \equiv 3 \pmod 8$ and a be a quadratic residue of p. Then $p - 2a$ is a quadratic residue of p.
49. $(1/p) + (2/p) + \cdots + ((p-1)/p) = 0$.
(*Hint*: Use Theorem 11.1.)
50. $(-2/p) = \begin{cases} 1 & \text{if } p \equiv 1 \text{ or } 3 \pmod 8 \\ -1 & \text{if } p \equiv -1 \text{ or } -3 \pmod 8 \end{cases}$
51. Let q be a prime factor of M_p. Then 2 is a quadratic residue of q if and only if q is of the form $2kp + 1$.
52. Let $p \equiv 3 \pmod 4$ and $q = 2p + 1$ be primes. Then $q|M_p$. (Euler)

Let a and b be positive integers such that $ab \equiv 1 \pmod p$. Then:

53. $(a/p) = (b/p)$
54. $(a(a+1)/p) = ((b+1)/p)$
55. Let $p > 3$. Then p divides the sum of its quadratic residues.
56. Every primitive root modulo a Fermat prime f_n is a quadratic nonresidue.

Let p be an odd prime with a primitive root α and a a positive integer $\leq p - 1$. Then there exists an integer k such that $\alpha^k \equiv a \pmod p$, where $1 \leq k \leq p - 1$. Using this fact, prove each.

57. The integer a is a quadratic residue of p if and only if k is even.
58. There are exactly $(p - 1)/2$ quadratic residues of p.
59. Using PMI, prove Corollary 11.4.

Prove each.

*60. There are infinitely many primes of the form $8n - 1$. [*Hint*: Assume there is only a finite number of such primes, $p_1, p_2, \ldots, p_k$. Consider the integer $N = (4p_1p_2\cdots p_k)^2 - 2$ and apply Theorem 11.6.]

*61. $\sum_{a=1}^{p-2} (a(a+1)/p) = -1$
(*Hint*: The integer a has a multiplicative inverse b. Then use Exercises 54 and 58.)

*62. If p and $4p + 1$ are primes, then 2 is a primitive root modulo $4p + 1$.

*63. Every quadratic nonresidue of a Fermat prime f_n is a primitive root modulo f_n.

*64. Using the technique used in the proof of Theorem 11.6, derive a formula for $(3/p)$, where p is a prime > 3.

11.3 Quadratic Reciprocity

Let p be an odd prime and $a = \prod_i p_i^{e_i}$ be the canonical decomposition of a, where $(p, p_i) = 1$ for all p_i. Then, as we found in Corollary 11.4, the Legendre symbol (a/p) can be evaluated, provided we know the value of (p_i/p), where p and p_i are distinct primes.

To evaluate the Legendre symbol, suppose that p and q are distinct odd primes. Suppose we know the value of (p/q). Can we then compute (q/p)? In other words, if p is a square modulo q, is q a square modulo p? Astonishingly, such a remarkable relationship exists between them. Based on numerical evidence, it was conjectured by Euler in 1783 and by Lagrange in 1785. Two years later, Legendre restated the relationship, the *law of quadratic reciprocity*, in its present elegant form, and provided a long but incomplete proof in the *Mémoires* of the French Academy. He tried another proof in his 1798 *Essai*, but that also was imperfect.

Gauss began his work on quadratic residues before he entered the University of Göttingen. After a year of intense study and perhaps unaware of the earlier work by Euler, Lagrange, and Legendre, Gauss gave the first complete, rigorous proof of the law; he was only 18 years old then. He called it *the fundamental theorem*, "the gem of higher arithmetic." "For a whole year," he later wrote, "this theorem tormented me and absorbed my greatest efforts until, at last, I obtained the proof explained in the fourth section of the *Disquisitiones Arithmeticae*." Gauss took credit for the law, claiming that a theorem belongs to the one who provides the first demonstration of it. An irate Legendre complained: "This excessive impudence is unbelievable in a man who has sufficient personal merit not to have the need of appropriating the discoveries of others." In any case, in 1808 Legendre adopted the proof by his young critic. Gauss was so intrigued by the law that he went on to publish seven more independent proofs, one in 1796, 1801, and 1805, two in 1808, and two in 1818; he also searched for an analogue in the theory of cubic and biquadratic residues.

Since that time, mathematicians have attempted to construct newer proofs of the law. In 1830, Jacobi supplied a proof that Legendre called superior to Gauss'. Since 1796, more than 190 proofs have been published, the most recent in 2004 by S. Y. Kim of McMaster University, Ontario, Canada, in *The American Mathematical Monthly*.

The following lemma, which appears complicated, paves the way for the law of quadratic reciprocity.

LEMMA 11.2 Let p and q be distinct odd primes. Then

$$\sum_{k=1}^{(p-1)/2} \left\lfloor \frac{kq}{p} \right\rfloor + \sum_{k=1}^{(q-1)/2} \left\lfloor \frac{kp}{q} \right\rfloor = \frac{(p-1)}{2} \cdot \frac{(q-1)}{2}$$

Before we prove the lemma, let us study the following example for a better understanding of its proof.

EXAMPLE 11.17 Verify Lemma 11.2 with $p = 7$ and $q = 11$.

SOLUTION

$$\sum_{k=1}^{(p-1)/2} \left\lfloor \frac{kq}{p} \right\rfloor = \sum_{k=1}^{3} \left\lfloor \frac{11k}{7} \right\rfloor$$

$$= \left\lfloor \frac{11 \cdot 1}{7} \right\rfloor + \left\lfloor \frac{11 \cdot 2}{7} \right\rfloor + \left\lfloor \frac{11 \cdot 3}{7} \right\rfloor = 1 + 3 + 4 = 8$$

$$\sum_{k=1}^{(q-1)/2} \left\lfloor \frac{kp}{q} \right\rfloor = \sum_{k=1}^{5} \left\lfloor \frac{7k}{11} \right\rfloor$$

$$= \left\lfloor \frac{7 \cdot 1}{11} \right\rfloor + \left\lfloor \frac{7 \cdot 2}{11} \right\rfloor + \left\lfloor \frac{7 \cdot 3}{11} \right\rfloor + \left\lfloor \frac{7 \cdot 4}{11} \right\rfloor + \left\lfloor \frac{7 \cdot 5}{11} \right\rfloor$$

$$= 0 + 1 + 1 + 2 + 3 = 7$$

Therefore, LHS $= 8 + 7 = 15 = 3 \cdot 5 =$ RHS. ■

Unfortunately, although this example illustrates the lemma, the algebraic approach does not shed any light on its proof. Therefore, we demonstrate it geometrically.

Notice that $\left\lfloor \frac{11k}{7} \right\rfloor$ is the number of positive integers $\le \frac{11k}{7}$; that is, $\left\lfloor \frac{11k}{7} \right\rfloor$ equals the number of **lattice points** (which are points with integral coordinates on the cartesian plane) that lie on the line $x = k$, above the x-axis, and below the line $y = \frac{11}{7}x$. Therefore, $\sum_{i=1}^{3} \left\lfloor \frac{11k}{7} \right\rfloor$ is the number of lattice points inside or on the polygon ABCD in Figure 11.2. It equals $1 + 3 + 4 = 8$.

On the other hand, $\left\lfloor \frac{7k}{11} \right\rfloor$ equals the number of lattice points that lie on the line $y = k$, but to the right of the y-axis, and above line $x = \frac{7}{11}y$; that is, $y = \frac{11}{7}x$. Thus, $\sum_{k=1}^{5} \left\lfloor \frac{7k}{11} \right\rfloor$ is the number of lattice points that lie on or inside the polygon ADEF in the figure; it equals $0 + 1 + 1 + 2 + 3 = 7$. Thus the total number of lattice points $= 8 + 7 = 15 = 3 \cdot 5 =$ number of lattice points inside or on the rectangle BCEF.

Ferdinand Gotthold Eisenstein *(1823–1852), born in Berlin, entered the University of Berlin at the late age of 20 due to ill health. But by then he had mastered the work of Gauss, Dirichlet, and Jacobi. In 1844, he published two proofs of the law of quadratic reciprocity, and the analogous laws of cubic reciprocity and biquadratic reciprocity. Four years later in Berlin he was imprisoned briefly by the Prussian army for his revolutionary activities in Berlin. Elected to the Berlin Academy of Sciences as Jacobi's successor in 1852, he made significant contributions to number theory and algebra.*

Eisenstein died of tuberculosis at the age of 29.

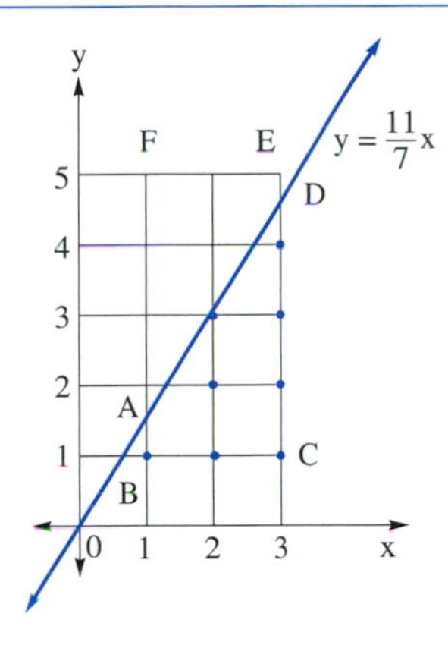

Figure 11.2

We now employ this geometric approach to establish the lemma. It is due to the German mathematician Ferdinand Eisenstein, a student of Gauss at Berlin.

PROOF **(of the lemma)**

Let $S(p,q) = \sum_{k=1}^{(p-1)/2} \left\lfloor \frac{kq}{p} \right\rfloor$ and $S(q,p) = \sum_{k=1}^{(q-1)/2} \left\lfloor \frac{kp}{q} \right\rfloor$. Since p and q are distinct odd primes, kq/p and kp/q are never integers. Since $\left\lfloor \frac{kq}{p} \right\rfloor$ is the number of positive integers $\le \frac{kq}{p}$, $\left\lfloor \frac{kq}{p} \right\rfloor$ gives the number of lattice points on the vertical line $x = k$, above the x-axis, and below the line $y = \frac{q}{p}x$. (Notice that no points on the line $y = \frac{q}{p}x$ are lattice points when $x < p$.) Therefore, $S(p,q)$ denotes the number of lattice points above the x-axis, below the line $y = \frac{q}{p}x$, and on the vertical lines $x = k$, where $1 \le k \le \frac{p-1}{2}$. Referring to Figure 11.3, $S(p,q)$ equals the number of lattice points on or inside the polygon ABCDE.

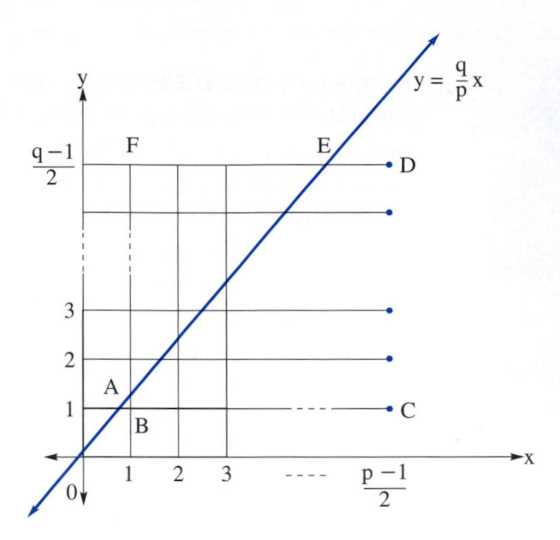

Figure 11.3

A similar argument shows that $S(q, p)$ denotes the total number of lattice points lying to the right of the y-axis, above the line $x = \frac{p}{q}y$; that is, $y = \frac{q}{p}x$, and on the lines $y = k$, where $1 \le k \le \frac{q-1}{2}$. In other words, $S(q, p)$ denotes the number of lattice points inside or on ΔAEF.

Thus, $S(p, q) + S(q, p)$ denotes the total number of lattice points inside or on the rectangle BCDF, namely, $\left(\frac{p-1}{2}\right) \cdot \left(\frac{q-1}{2}\right)$. This yields the desired result. ■

At last we arrive at the law of quadratic reciprocity, which is quite aesthetically appealing. The proof given is Gauss' third proof of the law, and hinges on his earlier lemma and Lemma 11.2, just proved. The proof begins identically to that of Gauss' lemma, so you may review its proof.

THEOREM 11.8 **(Law of Quadratic Reciprocity)** Let p and q be distinct odd primes. Then

$$(p/q)(q/p) = (-1)^{\frac{p-1}{2} \cdot \frac{q-1}{2}}$$

PROOF

Let $r_1, r_2, \ldots, r_k$ denote the least positive residues of the integers $q, 2q, \ldots, \left(\frac{p-1}{2}\right)q$ modulo p that are $\le p/2$ and $s_1, s_2, \ldots, s_v$ those that exceed $p/2$. Clearly, $k + v = \frac{p-1}{2}$ and $(q/p) = (-1)^v$.

In the proof of Gauss' lemma, we established that the $\frac{p-1}{2}$ integers $r_1, r_2, \ldots, r_k, p-s_1, p-s_2, \ldots, p-s_\nu$ are a permutation of the integers $1, 2, \ldots, (p-1)/2$. Therefore,

$$\sum_{i=1}^{k} r_i + \sum_{j=1}^{\nu}(p - s_j) = \sum_{k=1}^{(p-1)/2} k$$

$$= \frac{\left(\frac{p-1}{2}\right)\left(\frac{p+1}{2}\right)}{2}$$

Thus,

$$\sum_{i=1}^{k} r_i + \nu p - \sum_{j=1}^{\nu} s_j = \frac{p^2-1}{8}$$

Let $R = \sum_{i=1}^{k} r_i$ and $S = \sum_{j=1}^{\nu} s_j$. Then, this yields,

$$\frac{p^2-1}{8} = R + \nu p - S \tag{11.9}$$

Returning to the preceding integers kq, where $1 \leq k \leq \frac{p-1}{2}$, $\left\lfloor \frac{kq}{p} \right\rfloor$ denotes the quotient when kq is divided by p. Let t_k denote the remainder when kq is divided by p, where $0 \leq t_k < p$. Thus t_k is the least residue of kq modulo p. Then, by the division algorithm,

$$kq = \left\lfloor \frac{kq}{p} \right\rfloor \cdot p + t_k \quad \text{where } 0 \leq t_k \leq p-1$$

Therefore,

$$\sum_{k=1}^{(p-1)/2} kq = \sum_{k=1}^{(p-1)/2} \left\lfloor \frac{kq}{p} \right\rfloor \cdot p + \sum_{k=1}^{(p-1)/2} t_k$$

That is,

$$q \sum_{k=1}^{(p-1)/2} k = pS(p,q) + R + S$$

$$q \cdot \left(\frac{p^2-1}{8}\right) = pS(p,q) + R + S \tag{11.10}$$

Subtracting equation (11.9) from equation (11.10), we get

$$(q-1) \cdot \left(\frac{p^2-1}{8}\right) = p[S(p,q) - \nu] + 2S$$

Since the LHS and $2S$ are even, it follows that $S(p, q) - v$ is even. Therefore, $(-1)^{S(p,q)-v} = 1$; that is, $(-1)^{S(p,q)} = (-1)^v$.

But, by Gauss' lemma, $(q/p) = (-1)^v$. Therefore, $(q/p) = (-1)^{S(p,q)}$. Similarly, $(p/q) = (-1)^{S(q,p)}$. Therefore,

$$\begin{aligned}(p/q)(q/p) &= (-1)^{S(q,p)} \cdot (-1)^{S(p,q)} \\ &= (-1)^{S(q,p)+S(p,q)} \\ &= (-1)^{\frac{p-1}{2} \cdot \frac{q-1}{2}}, \quad \text{by Lemma 11.2}\end{aligned}$$

This concludes the proof. ■

The law of quadratic reciprocity can be restated in a more practical form, as the following corollary shows.

COROLLARY 11.7 Let p and q be distinct odd primes. Then

$$(q/p) = \begin{cases} (p/q) & \text{if } p \equiv 1 \pmod 4 \text{ or } q \equiv 1 \pmod 4 \\ -(p/q) & \text{if } p \equiv q \equiv 3 \pmod 4 \end{cases}$$

PROOF

If $p \equiv 1 \pmod 4$, then $(p-1)/2$ is even; so $(p-1)(q-1)/4$ is even. Therefore, by the law of quadratic reciprocity, $(p/q)(q/p) = 1$. But $(p/q) = \pm 1 = (q/p)$, so $(q/p) = (p/q)$. Similarly, if $q \equiv 1 \pmod 4$, then $(q/p) = (p/q)$.

On the other hand, assume that $p \equiv q \equiv 3 \pmod 4$. Then $(p-1)/2$, $(q-1)/2$, and hence $(p-1)/2 \cdot (q-1)/2$ are odd. Therefore, again by the law of quadratic reciprocity, $(p/q)(q/p) = -1$. Thus, $(q/p) = -(p/q)$. ■

For example, since $17 \equiv 1 \pmod 4$, $(17/29) = (29/17)$; and since $23 \equiv 3 \pmod 4$ and $47 \equiv 3 \pmod 4$, $(23/47) = -(47/23)$.

Corollary 11.7, together with Theorems 11.4 and 11.6, and Corollaries 11.2 and 11.4, can be applied to evaluate any Legendre symbol (a/p), where p is an odd prime and $p \nmid a$, as the following example illustrates. It demonstrates the power of the law of quadratic reciprocity.

EXAMPLE 11.18 Compute $(152/43)$ and $(3797/7297)$.

SOLUTION

- Notice that $152 \equiv 23 \pmod{43}$, so $(152/43) = (23/43)$. Since $23 \equiv 3 \pmod 4$ and $43 \equiv 3 \pmod 4$, by Corollary 11.7, $(23/43) = -(43/23) = -(20/23) = -(4/23)(5/23) = -(5/23) = 1$. Therefore, $(152/43) = (23/43) = 1$. [Consequently, the congruence $x^2 \equiv 152 \pmod{43}$ is solvable.]

- First, notice that both 3797 and 7297 are primes; also, $3797 \equiv 1 \pmod{4}$ and $7297 \equiv 1 \pmod{4}$. Therefore, by the law of quadratic reciprocity,

$$\begin{aligned}(3797/7297) &= (7297/3797)\\ &= (3500/3797) = (2^2 \cdot 5^3 \cdot 7/3797)\\ &= (2^2/3797)(5^3/3797)(7/3797)\\ &= (2^2/3797)(5/3797)^3(7/3797) \qquad (11.11)\end{aligned}$$

By Theorem 11.4, $(2^2/3797) = 1$; furthermore, $(5/3797) = (3797/5) = (2/5) = -1$, by Theorem 11.6; $(7/3797) = (3797/7) = (3/7) = -1$; therefore, by equation (11.11), $(3797/7297) = 1 \cdot (-1)^3(-1) = 1$. [Thus the congruence $x^2 \equiv 3797 \pmod{7297}$ is also solvable.] ■

The following example is a simple but interesting application of the law of quadratic reciprocity and Theorem 11.6.

EXAMPLE 11.19 Show that $1! + 2! + 3! + \cdots + n!$ is never a square, where $n > 3$.

***PROOF* (by contradiction)**

Let $N = 1! + 2! + 3! + \cdots + n!$. Assume that $N = x^2$ for some positive integer x. Since $n > 3$, $N > 5$. Then $(N/5) = (x^2/5) = 1$.

Since $N \equiv 1! + 2! + 3! + 4! \equiv 3 \pmod{5}$, $(N/5) = (3/5)$. But $(3/5) = (5/3) = (2/3) = -1$, so $(N/5) = -1$, which is a contradiction. Thus, N cannot be a perfect square, where $n > 3$. (Notice that N is a square if $n = 0$, 1, or 3.) ■

The following example, a bit complicated, is also a fine application of the law of quadratic reciprocity. In addition to Theorem 11.4, it employs two additional facts:

- The product of two integers of the form $5k + 1$ is also of the same form.
- If p is an odd prime $\neq 5$, then $(5/p) = 1$ if and only if $p \equiv \pm 1 \pmod{5}$.

Try to confirm both. See Exercises 21 and 29.

EXAMPLE 11.20 Prove that there are infinitely many primes of the form $10k - 1$.

PROOF

Let $N = 5(n!)^2 - 1$, where $n > 1$. Clearly, N is of the form $10k - 1$. Let p be a prime factor of N. Clearly, p must be odd. Since $p|N$, $5(n!)^2 \equiv 1 \pmod{p}$, so $(5(n!)^2/p) = (1/p) = 1$; that is, $(5/p)((n!)^2/p) = (5/p) = 1$. Thus, $p \equiv \pm 1 \pmod{5}$, so p is of the form $5k \pm 1$. Since N is of the form $5k-1$, not all prime factors of N can be of the form $5k+1$. Therefore, N must have a prime factor q of the form $5k-1$.

If k is an odd integer $2j+1$, then $q = 5k-1 = 5(2j+1)-1 = 10j+4$ is not a prime. Therefore, k must be even and hence q must be of the form $10k-1$.

Notice that if $q \le n$, then $q|n!$. But $q|N$, so $q|-1$, which is a contradiction. Thus, $q > n$.

Thus, for every positive integer $n > 1$, there is a prime factor q greater than n and of the form $10k-1$. In other words, there are infinitely many such primes; they all end in the digit 9. (See Exercise 42 for an alternate proof.) ■

The next example is a bonus. It is a fine confluence of the binomial theorem, Fermat's little theorem, Euler's criterion, and the law of quadratic reciprocity.

EXAMPLE 11.21 Let F_n denote the nth Fibonacci number and p an odd prime $\ne 5$. Then

$$F_p \equiv \begin{cases} 1 \pmod{p} & \text{if } p \equiv \pm 1 \pmod{5} \\ -1 \pmod{p} & \text{if } p \equiv \pm 2 \pmod{5} \end{cases}$$

SOLUTION

Recall from Section 2.6 that

$$F_n = \frac{\alpha^n - \beta^n}{\alpha - \beta} = \frac{\alpha^n - \beta^n}{\sqrt{5}} \qquad \leftarrow \text{Binet's formula}$$

where $\alpha = \dfrac{1+\sqrt{5}}{2}$ and $\beta = \dfrac{-1}{\alpha} = \dfrac{1-\sqrt{5}}{2}$.

Thus,

$$\sqrt{5}F_p = \left(\frac{1+\sqrt{5}}{2}\right)^p - \left(\frac{1-\sqrt{5}}{2}\right)^p$$

$$\sqrt{5}2^p F_p = (1+\sqrt{5})^p - (1-\sqrt{5})^p$$

$$= \sum_{r=0}^{p} \binom{p}{r} (\sqrt{5})^r - \sum_{r=0}^{p} \binom{p}{r} (-\sqrt{5})^r$$

$$= 2\left[\binom{p}{1}(\sqrt{5}) + \binom{p}{3}(\sqrt{5})^3 + \binom{p}{5}(\sqrt{5})^5 + \cdots + \binom{p}{p}(\sqrt{5})^p\right]$$

That is,

$$2^{p-1}F_p = \binom{p}{1} + \binom{p}{3}5 + \binom{p}{5}5^2 + \cdots + \binom{p}{p}5^{(p-1)/2}$$

Since $p|\binom{p}{k}$ for $1 \le k \le p-1$ and $\binom{p}{p} = 1$, this implies $2^{p-1}F_p \equiv 5^{(p-1)/2} \pmod{p}$. By Fermat's little theorem, $2^{p-1} \equiv 1 \pmod{p}$ and by Euler's criterion, $5^{(p-1)/2} \equiv (5/p) \pmod{p}$. Thus $F_p \equiv (5/p) \pmod{p}$. But

$$(5/p) = \begin{cases} 1 & \text{if } p \equiv \pm 1 \pmod{5} \\ -1 & \text{if } p \equiv \pm 2 \pmod{5} \end{cases}$$

Therefore,

$$F_p \equiv \begin{cases} 1 \pmod{p} & \text{if } p \equiv \pm 1 \pmod{5} \\ -1 \pmod{p} & \text{if } p \equiv \pm 2 \pmod{5} \end{cases}$$

■

For example, $41 \equiv 1 \pmod{5}$ and $F_{41} = 165{,}580{,}141 \equiv 1 \pmod{41}$, whereas $43 \equiv 3 \pmod{5}$ and $F_{43} = 433{,}494{,}437 \equiv 42 \pmod{43}$, as expected.

In 1877, T. Pepin of France employed the law of quadratic reciprocity to develop an elegant test for determining the primality of the Fermat number f_n. His test is the essence of the following theorem.

THEOREM 11.9 **(Pepin's Test, 1877)** The Fermat number f_n is a prime if and only if $3^{(f_n-1)/2} \equiv -1 \pmod{f_n}$, where $n \ge 1$.

PROOF

Assume that $f_n = 2^{2^n} + 1$ is a prime. Since $f_n \equiv 1 \pmod{4}$, by the law of quadratic reciprocity, $(3/f_n) = (f_n/3)$. But $f_n \equiv (-1)^{2^n} + 1 \equiv 2 \pmod{3}$, so $(f_n/3) = (2/3) = -1$. Thus $(3/f_n) = -1$.

But, by Euler's criterion, $(3/f_n) \equiv 3^{(f_n-1)/2} \pmod{f_n}$. Therefore,

$$3^{(f_n-1)/2} \equiv -1 \pmod{f_n}$$

Conversely, assume that $3^{(f_n-1)/2} \equiv -1 \pmod{f_n}$; so $3^{(f_n-1)/2} \equiv -1 \pmod{p}$ for a prime factor p of f_n. Then $3^{f_n-1} \equiv 1 \pmod{p}$, so $\text{ord}_p 3 | f_n - 1$; that is, $\text{ord}_p 3 | 2^{2^n}$. Consequently, $\text{ord}_p 3 = 2^k$ for some positive integer k.

We would like to show that $k = 2^n$. Suppose that $k < 2^n$; then $2^n - k - 1 \ge 0$. Since $3^{2^k} \equiv 1 \pmod{p}$, $(3^{2^k})^{2^{2^n-k-1}} \equiv 1 \pmod{p}$; that is, $3^{2^{2^n-1}} \equiv 3^{(f_n-1)/2} \equiv 1 \pmod{p}$. This yields $1 \equiv -1 \pmod{p}$; that is, $p = 2$, which is a contradiction. Therefore, $k = 2^n$ and hence $\text{ord}_p 3 = f_n - 1$. By Fermat's little theorem, $\text{ord}_p 3 \le p-1$. Therefore, $f_n - 1 = \text{ord}_p 3 \le p-1$, where $p|f_n$. This implies that $f_n = p$, a prime. ■

The following example illustrates this test.

EXAMPLE 11.22 Show that $f_4 = 2^{2^4} + 1 = 65,537$ is a prime.

PROOF

By Pepin's test, it suffices to show that $3^{(f_4-1)/2} = 3^{2^{15}} = 3^{32768} \equiv -1 \pmod{f_4}$. We have $3^8 \equiv 6561 \pmod{f_4}$, $3^{20} \equiv 19390 \pmod{f_4}$, $3^{40} \equiv -13669 \pmod{f_4}$, $3^{60} \equiv -10282 \pmod{f_4}$, $3^{200} \equiv -28787 \pmod{f_4}$, $3^{500} \equiv 26868 \pmod{f_4}$, and $3^{32000} \equiv 27748 \pmod{f_4}$. Therefore:

$$\begin{aligned} 3^{32768} &= 3^{32000} \cdot 3^{500} \cdot 3^{200} \cdot 3^{60} \cdot 3^{8} \\ &\equiv (27748)(26868)(-28787)(-10282)(6561) \pmod{f_4} \\ &\equiv -1 \pmod{f_4} \end{aligned}$$

Thus, f_4 is a prime. ■

In 1905, J. C. Moorehead and A. E. Western, independently applied Pepin's test and established the compositeness of the 39-digit Fermat number f_7, although no factors were provided. The actual prime factorization of f_7 was provided 65 years later, in 1970, by J. Brillhart and M. A. Morrison:

$$f_7 = (2^9 \times 116,503,103,764,643 + 1)(2^9 \times 11,141,971,095,088,142,685 + 1)$$

In 1909, Moorehead and Western, this time working together, proved the composite nature of f_8, again using Pepin's test. However, the smallest prime factor of f_8 was not discovered until 1980, when R. P. Brent and J. M. Pollard found it to be 1,238,926,361,552,897. The other prime factor, found shortly thereafter by H. C. Williams, is 62 digits long.

The search for Fermat primes did not end there. In 1960, G. A. Paxson used the test to determine the composite nature of f_{13}. Two years later, J. L. Selfridge and A. Hurwitz, again using Pepin's test, demonstrated the compositeness of f_{14}, without finding any of its prime factors. In 1980, W. Keller determined that f_{9448} is composite, with $19 \times 2^{9450} + 1$ as a factor. In the same year, G. Gostin discovered that 31,065,037,602,817 is a prime factor of f_{17}.

In 1988, Brent successfully factored $f_{11} = 2^{2048} + 1$; f_{11} is 617 digits long and 319,489 is the smallest prime factor of f_{11}. In the same year, J. Young and D. Buell applied Pepin's test using a supercomputer, to determine that f_{20}, which is 315,653 digits long, is composite. Two years later, Pollard factored the 155-digit f_9 into three primes, the smallest of which is 2,424,833; the other two are 49 and 99 digits long. It took Pollard two months to factor f_9 using techniques in algebraic number fields and a worldwide network of 1000 computers.

n	*Status of f_n*	*Year*	*Discoverer(s)*
0–4	primes	1640	P. Fermat
5	composite	1732	L. Euler
6	composite	1880	F. Landry (at age 82)
7	composite	1905	J. C. Moorehead & A. E. Western
8	composite	1909	J. C. Moorehead & A. E. Western
9	composite	1903	A. E. Western
10	composite	1952	R. M. Robinson
11	composite	1899	A. J. C. Cunningham
12	composite	1877	I. M. Pervouchine & E. Lucas
13	composite	1960	G. A. Paxson
14	composite	1961	A. Hurwitz & J. L. Selfridge
15	composite	1925	M. Kraitchik
16	composite	1953	J. L. Selfridge
17	composite	1980	G. B. Gostin
18	composite	1903	A. E. Western
19	composite	1962	H. Riesel
20	composite	1988	J. Young & D. A. Buell
21	composite	1963	C. P. Wrathall
22	composite	1993	R. E. Crandall *et al.*
23	composite	1878	I. M. Pervouchine
24	composite	1999	E. Mayer *et al.*
25	composite	1963	C. P. Wrathall
26	composite	1963	C. P. Wrathall
27	composite	1963	C. P. Wrathall
28	composite	1997	T. Taura
29	composite	1980	G. B. Gostin & P. B. Mclaughlin
30	composite	1963	C. P. Wrathall

Table 11.1

Although Fermat numbers f_{10} through f_{30} are known to be composite, not all their prime factors have been discovered; f_{14}, f_{20}, f_{22}, and f_{24} are known to be composite, but no prime factors are known. Thus, $f_4 = 65{,}537$ remains the largest known Fermat prime.

Table 11.1 summarizes the primality status of Fermat numbers f_n, where $0 \leq n \leq 30$, their discoverers, and the years of discovery. As of January 1, 2006, 225 Fermat numbers are known to be composite, 258 prime factors are known, and f_5 through f_{11} are completely factored. Besides, the largest known composite Fermat number is $f_{2478782}$, discovered in 2003 by Cosgrave, Jobling, Woltman, and Gallot: $3 \cdot 2^{2478785} + 1$ is a factor.

In 1878, Pepin's test was generalized by François Proth,† although he never gave a proof.

† François Proth (1852–1879) was a self-taught farmer, who lived near Verdun, France.

THEOREM 11.10 **(Proth's Theorem, 1878)** Let $N = k \cdot 2^n + 1$, where $0 < k < 2^n$. Suppose there is a positive integer a such that $(a/N) = -1$. Then N is a prime if and only if $a^{(N-1)/2} \equiv -1 \pmod N$. ■

Since the proof of Proth's theorem involves results not yet discussed, we omit the proof.†

EXERCISES 11.3

Verify Lemma 11.2 for the given values of p and q.

1. $p = 5, q = 11$
2. $p = 13, q = 11$

Compute each Legendre symbol.

3. $(261/47)$
4. $(-267/61)$
5. $(176/241)$
6. $(1188/379)$
7. $(-1776/1013)$
8. $(-2020/3593)$
9. $(1428/2411)$
10. $(3533/4133)$

Let p and q be odd primes such that $p = 4a + q$. Then $(a/p) = (a/q)$ (see Exercise 31). Using this fact, evaluate each.

11. $(3/17)$
12. $(10/43)$
13. $(43/191)$
14. $(114/479)$

Let p be a prime $\equiv 3 \pmod 4$. Then the solutions of the congruence $x^2 \equiv a \pmod p$ are $x \equiv \pm a^{(p+1)/4} \pmod p$ (see Exercise 30). Using this fact, solve each quadratic congruence.

15. $x^2 \equiv 13 \pmod{23}$
16. $x^2 \equiv 17 \pmod{43}$

Using Pepin's test, if possible, verify that each Fermat number is a prime.

17. f_0
18. f_1
19. f_2
20. f_3
21. Derive a formula for $(5/p)$, where p is an odd prime $\neq 5$.

Compute the least residue of the Fibonacci number F_p modulo p corresponding to each prime p. (*Hint*: Use Example 11.21.)

22. 53
23. 79
24. 89
25. 97

Solve each quadratic congruence with composite modulus. Each has four incongruent solutions. [*Hint*: To solve $x^2 \equiv a \pmod{pq}$, solve $x^2 \equiv a \pmod p$ and $x^2 \equiv a \pmod q$, and then use the CRT.]

26. $x^2 \equiv 13 \pmod{391}$
27. $x^2 \equiv 17 \pmod{817}$
28. In 1891, Lucas proved that the prime factors of $2^{4q} + 1$ are of the form $16hq + 1$. Using this fact, find a prime factor of the Fermat numbers f_5 and f_6.

Prove each.

29. The product of two integers of the form $5k + 1$ is also of the same form.
30. Let p be a prime $\equiv 3 \pmod 4$. Then $x \equiv \pm a^{(p+1)/4} \pmod p$ are solutions of the congruence $x^2 \equiv a \pmod p$, where a is a quadratic residue of p.
31. Let p and q be odd primes such that $p = 4a + q$. Then $(a/p) = (a/q)$.
32. Let p be a prime > 3. Then
$$(3/p) = \begin{cases} 1 & \text{if } p \equiv \pm 1 \pmod{12} \\ -1 & \text{if } p \equiv \pm 5 \pmod{12} \end{cases}$$
33. Let p be a prime > 3. Then
$$(-3/p) = \begin{cases} 1 & \text{if } p \equiv 1 \pmod 6 \\ -1 & \text{if } p \equiv 5 \pmod 6 \end{cases}$$

† For a proof, see K. H. Rosen, *Elementary Number Theory and Its Applications*, Addison-Wesley, Boston, 2005.

34. Let p be an odd prime $\neq 5$. Then $(5^{-1}/p) = (p^{-1}/5)$.
35. Let f_n denote the nth Fermat number. Then $f_n \equiv 1 \pmod 4$, where $n \geq 1$. (*Hint*: Use induction.)
36. Let f_m and f_n denote distinct Fermat primes. Then $(f_m/f_n) = (f_n/f_m)$. (*Hint*: Use Exercise 35.)
37. Let f_n be a Fermat prime and M_p a Mersenne prime, where $n > 0$ and $p > 2$. Then $(f_n/M_p) = (M_p/f_n)$.
38. Three is a primitive root modulo every Fermat prime f_n, where $n \geq 1$. (*Hint*: Use Pepin's test.)
39. Let M_p and M_q be distinct Mersenne primes > 3. Then $(M_p/M_q) = -(M_q/M_p)$.
40. Let p and q be distinct odd primes. The congruence $x^2 \equiv a \pmod{pq}$ is solvable if both $x^2 \equiv a \pmod p$ and $x^2 \equiv a \pmod q$ are solvable.
41. There are infinitely many primes of the form $6n+1$. [*Hint*: Assume that there are only finitely many such primes, $p_1, p_2, \ldots, p_k$. Then consider $N = (2p_1p_2\cdots p_k)^2 + 3$ and use Exercise 33.]

*42. Prove by contradiction that there are infinitely many primes of the form $10k-1$.

*43. Derive a formula for $(7/p)$, where p is an odd prime $\neq 7$.

*44. Let p be an odd prime $\neq 5$, and F_n the nth Fibonacci number. Then

$$F_{p+1} \equiv \begin{cases} 1 \pmod p & \text{if } p \equiv \pm 1 \pmod 5 \\ 0 \pmod p & \text{if } p \equiv \pm 2 \pmod 5 \end{cases}$$

11.4 The Jacobi Symbol

Our discussion of the solvability of the quadratic congruence $x^2 \equiv a \pmod m$ led us to the definition of the Legendre symbol (a/p) and a detailed investigation of its properties in Sections 11.1–11.3. We will now generalize the Legendre symbol to the Jacobi symbol, which was introduced in 1846 by the German mathematician Karl G. J. Jacobi. In the Jacobi symbol (a/m), the modulus m need not be a prime, but must be odd and relatively prime to a. We now define it in terms of the Legendre symbol.

The Jacobi Symbol

Let m be an odd positive integer with the canonical decomposition $m = \prod_{i=1}^{k} p_i^{e_i}$, and a any integer with $(a, m) = 1$. Then the **Jacobi symbol** (a/m) is defined by

$$(a/m) = \left(a \Big/ \prod_{i=1}^{k} p_i^{e_i}\right) = \prod_{i=1}^{k} (a/p_i)^{e_i}$$

where (a/p_i) denotes the familiar Legendre symbol.

Although we are using the same notation for both symbols, it should be clear from the context whether the symbol is Legendre or Jacobi. Bear in mind that the symbol (a/m) is the Legendre symbol if and only if m is a prime.

The following example clarifies the definition.

Karl Gustav Jacob Jacobi *(1804–1851) was born into a wealthy family in Potsdam, Germany. After graduating in 1821 from the local Gymnasium, where he excelled in Greek, Latin, history, and mathematics, he pursued his mathematical interest at the University of Berlin. There he mastered the works of Euler, Lagrange, and other leading mathematicians, and received his Ph.D. in 1825. Although his professional career began at the age of 20 at Berlin, there was little prospect for promotion, so in 1826 he joined the faculty at the University of Königsberg. Becoming a full professor there in 1832, he remained there until he became a royal prisoner in 1842.*

An early founder of the theory of determinants, he developed (independently of Abel) the theory of elliptic functions, and invented the functional determinant, Jacobian. *He made important contributions to number theory, calculus of variations, analytical mechanics, and dynamics.*

A prolific writer, Jacobi died of smallpox in Berlin.

EXAMPLE 11.23 Evaluate the Jacobi symbols $(55/273)$ and $(364/935)$.

SOLUTION

- Notice that $273 = 3 \cdot 7 \cdot 13$. So, by the definition of the Jacobi symbol,

$$\begin{aligned}(55/273) &= (55/3)(55/7)(55/13)\\ &= (1/3)(-1/7)(3/13) = 1 \cdot (-1)(13/3)\\ &= -(1/3) = -1\end{aligned}$$

- $935 = 5 \cdot 11 \cdot 17$, so

$$\begin{aligned}(364/935) &= (364/5)(364/11)(364/17)\\ &= (4/5)(1/11)(7/17) = 1 \cdot 1 \cdot (7/17)\\ &= (7/17) = (17/7) = (3/7)\\ &= -(7/3) = -(1/3) = -1\end{aligned}$$

■

We can now ask if knowing the value of (a/m) helps us to determine the solvability of the congruence $x^2 \equiv a \pmod{m}$. From Sections 11.1 and 11.2, we know the answer if m is a prime p: It is solvable if and only if $(a/p) = 1$.

First, suppose that $x^2 \equiv a \pmod{m}$ is solvable, where m is composite and p_i is a prime factor of m. Then $x^2 \equiv a \pmod{p_i}$ is solvable, so $(a/p_i) = 1$. Therefore,

$$(a/m) = \prod_{i=1}^{k} (a/p_i)^{e_i} = \prod_{i=1}^{k} 1^{e_i} = 1$$

Thus, if $x^2 \equiv a \pmod{m}$ is solvable, then the Jacobi symbol $(a/m) = 1$.

On the other hand, assume that $(a/m) = 1$. Unfortunately, this does not imply that the congruence $x^2 \equiv a \pmod{m}$ is solvable. For example, notice that $(2/33) = (2/3 \cdot 11) = (2/3)(2/11) = (-1)(-1) = 1$, but $x^2 \equiv 2 \pmod{33}$ has no solutions. (Verify this. See Exercise 1.)

Interestingly enough, the Jacobi symbol and the Legendre symbol share several analogous properties, as the following two theorems show.

THEOREM 11.11 Let m be an odd positive integer, and a and b be any integers with $(a, m) = 1 = (b, m)$. Then

(1) If $a \equiv b \pmod{m}$, then $(a/m) = (b/m)$.
(2) $(ab/m) = (a/m)(b/m)$
(3) $(a^2/m) = 1$

PROOF

Let $m = \prod_{i=1}^{k} p_i^{e_i}$ be the canonical decomposition of m.

(1) Since $a \equiv b \pmod{m}$, $a \equiv b \pmod{p_i}$, so $(a/p_i) = (b/p_i)$, by Theorem 11.4. Therefore,

$$(a/p_i)^{e_i} = (b/p_i)^{e_i}$$

Thus,

$$(a/m) = \prod_{i=1}^{k} (a/p_i)^{e_i} = \prod_{i=1}^{k} (b/p_i)^{e_i} = (b/m)$$

(2) By Theorem 11.4, $(ab/p_i) = (a/p_i)(b/p_i)$. Therefore,

$$(ab/p_i)^{e_i} = (a/p_i)^{e_i}(b/p_i)^{e_i}$$

Thus,

$$(ab/m) = \prod_{i=1}^{k} (ab/p_i)^{e_i} = \prod_{i=1}^{k} (a/p_i)^{e_i}(b/p_i)^{e_i}$$

$$= \prod_{i=1}^{k} (a/p_i)^{e_i} \prod_{i=1}^{k} (b/p_i)^{e_i} = (a/m)(b/m)$$

(3) Since $(a^2/m) = (a/m)^2$, by property (2), the desired result follows. ■

The following example demonstrates the first two properties.

EXAMPLE 11.24 Let $m = 231 = 3 \cdot 7 \cdot 11$.

- First, notice that $211 \equiv -20 \pmod{231}$. Then

$$\begin{aligned}(211/231) &= (211/3 \cdot 7 \cdot 11) = (211/3)(211/7)(211/11) \\ &= (1/3)(1/7)(2/11) \\ (-20/231) &= (-20/3 \cdot 7 \cdot 11) = (-20/3)(-20/7)(-20/11) \\ &= (1/3)(1/7)(2/11)\end{aligned}$$

  Thus, $(211/231) = (-20/231)$. [You may verify that $(211/231) = -1$.]

- $$\begin{aligned}(4 \cdot 50/231) &= (4 \cdot 50/3 \cdot 7 \cdot 11) = (4 \cdot 50/3)(4 \cdot 50/7)(4 \cdot 50/11) \\ &= (4/3)(50/3)(4/7)(50/7)(4/11)(50/11) \\ &= [(4/3)(4/7)(4/11)][(50/3)(50/7)(50/11)] \\ &= (4/3 \cdot 7 \cdot 11)(50/3 \cdot 7 \cdot 11) = (4/231)(50/231)\end{aligned}$$

  (You may verify that $(4 \cdot 50/231) = 1$.) ■

The next theorem generalizes Corollary 11.2 and Theorem 11.6, but we add a lemma to make its proof simpler and shorter.

LEMMA 11.3 Let m be an odd positive integer with canonical decomposition $\prod_{i=1}^{k} p_i^{e_i}$. Then

(1) $\sum_{i=1}^{k} e_i(p_i - 1)/2 \equiv (m-1)/2 \pmod{2}$

(2) $\sum_{i=1}^{k} e_i(p_i^2 - 1)/8 \equiv (m^2 - 1)/8 \pmod{2}$

PROOF

First, notice that $p_i - 1 \equiv 0 \pmod{2}$ and $p_i^2 - 1 \equiv 0 \pmod{8}$.

(1) We write p_i as $p_i = 1 + (p_i - 1)$, so

$$p_i^{e_i} = \left[1 + (p_i - 1)\right]^{e_i}$$

Using the binomial theorem, since $p_i - 1$ is even, this yields

$$p_i^{e_i} \equiv 1 + e_i(p_i - 1) \pmod{4}$$

Therefore,

$$m = \prod_{i=1}^{k} p_i^{e_i} \equiv \prod_{i=1}^{k} \big[1 + e_i(p_i - 1)\big] \pmod 4 \tag{11.12}$$

But

$$\big[1 + e_i(p_i - 1)\big]\big[1 + e_j(p_j - 1)\big] \equiv 1 + e_i(p_i - 1) + e_j(p_j - 1) \pmod 4$$

So, by congruence (11.12),

$$m \equiv 1 + \sum_{i=1}^{k} e_i(p_i - 1) \pmod 4$$

Thus, $\sum_{i=1}^{k} e_i(p_i - 1)/2 \equiv (m-1)/2 \pmod 2$.

(2) Since $p_i^2 - 1 \equiv 0 \pmod 8$, by the binomial theorem,

$$\begin{aligned} p_i^{2e_i} &= (p_i^2)^{e_i} = \big[1 + (p_i^2 - 1)\big]^{e_i} \\ &\equiv 1 + e_i(p_i^2 - 1) \pmod{64} \end{aligned}$$

But

$$\big[1 + e_i(p_i^2 - 1)\big]\big[1 + e_j(p_j^2 - 1)\big] \equiv 1 + e_i(p_i^2 - 1) + e_j(p_j^2 - 1) \pmod{64}$$

Therefore,

$$\begin{aligned} m^2 = \prod_{i=1}^{k} p_i^{2e_i} &\equiv \prod_{i=1}^{k} \big[1 + e_i\big(p_i^2 - 1\big)\big] \pmod{64} \\ &\equiv 1 + \sum_{i=1}^{k} e_i(p_i^2 - 1) \pmod{64} \end{aligned}$$

Thus, $\sum_{i=1}^{k} e_i(p_i^2 - 1)/8 \equiv (m^2 - 1)/8 \pmod 2$. This concludes the proof. ■

We are now ready for the next major result.

THEOREM 11.12 Let m be an odd positive integer. Then

(1) $(-1/m) = (-1)^{(m-1)/2}$

(2) $(2/m) = (-1)^{(m^2-1)/8}$

PROOF

Let $m = \prod_{i=1}^{k} p_i^{e_i}$ be the canonical decomposition of m. Then

$$(1)\qquad (-1/m) = \prod_{i=1}^{k}(-1/p_i)^{e_i} = \prod_{i=1}^{k}\left[(-1)^{(p_i-1)/2}\right]^{e_i}, \quad \text{by Corollary 11.2}$$
$$= (-1)^{\sum_{i=1}^{k} e_i(p_i-1)/2}$$
$$= (-1)^{(m-1)/2}, \quad \text{by Lemma 11.3}$$

$$(2)\qquad (2/m) = \prod_{i=1}^{k}(2/p_i)^{e_i}$$
$$= \prod_{i=1}^{k}(-1)^{e_i(p_i^2-1)/8}, \quad \text{by Corollary 11.6}$$
$$= (-1)^{\sum_{i=1}^{k} e_i(p_i^2-1)/8}$$
$$= (-1)^{(m^2-1)/8}, \quad \text{by Lemma 11.3} \qquad \blacksquare$$

For example,

$$(-1/39) = (-1)^{(39-1)/2} = (-1)^{19} = -1$$
$$(2/819) = (-1)^{(819^2-1)/8} = (-1)^{83845} = -1$$

We now present the analogous law of quadratic reciprocity for Jacobi symbol. It employs both the earlier version and Lemma 11.3.

THEOREM 11.13 **(The Generalized Law of Quadratic Reciprocity)** Let m and n be relatively prime odd positive integers. Then

$$(m/n)(n/m) = (-1)^{\frac{m-1}{2}\cdot\frac{n-1}{2}}$$

PROOF

Let $m = \prod_{i=1}^{r} p_i^{a_i}$ and $n = \prod_{i=1}^{s} q_j^{b_j}$ be the canonical decompositions of m and n, respectively. Then, by definition,

$$(m/n) = \prod_{j=1}^{s}(m/q_j)^{b_j} = \prod_{j=1}^{s}\left[\prod_{i=1}^{r}(p_i/q_j)^{a_i b_j}\right]$$
$$= \prod_{j=1}^{s}\prod_{i=1}^{r}(p_i/q_j)^{a_i b_j}$$

and

$$(n/m) = \prod_{i=1}^{r}(n/p_i)^{a_i} = \prod_{i=1}^{r}\left[\prod_{j=1}^{s}(q_j/p_i)^{b_j a_i}\right]$$
$$= \prod_{i=1}^{r}\prod_{j=1}^{s}(q_j/p_i)^{b_j a_i}$$

Therefore,

$$(m/n)(n/m) = \prod_{i=1}^{r}\prod_{j=1}^{s}[(p_i/q_j)(q_j/p_i)]^{a_i b_j}$$
$$= \prod_{i=1}^{r}\prod_{j=1}^{s}\left[(-1)^{\frac{p_i-1}{2}\cdot\frac{q_j-1}{2}}\right]^{a_i b_j}$$
$$= (-1)^{\sum_i\sum_j a_i b_j(\frac{p_i-1}{2})(\frac{q_j-1}{2})}$$
$$= (-1)^{[\sum_i a_i(\frac{p_i-1}{2})][\sum_j b_j(\frac{q_j-1}{2})]}$$
$$= (-1)^{\frac{m-1}{2}\cdot\frac{n-1}{2}}, \quad \text{by Lemma 11.3}$$

This concludes the proof. ■

The following two examples demonstrate the power of this generalized version.

EXAMPLE 11.25 Using the generalized law of quadratic reciprocity, compute the Jacobi symbol (221/399).

SOLUTION

By the generalized law of quadratic reciprocity,

$$(221/399) = (-1)^{\frac{221-1}{2}\cdot\frac{399-1}{2}}(399/221) = (399/221)$$
$$= (178/221) = (2/221)(89/221) \tag{11.13}$$

By Theorem 11.12, $(2/221) = (-1)^{(221^2-1)/8} = -1$.

By the generalized law of quadratic reciprocity,

$$(89/221) = (-1)^{\frac{89-1}{2}\cdot\frac{221-1}{2}}(221/89)$$
$$= (221/89) = (43/89)$$
$$= (-1)^{\frac{43-1}{2}\cdot\frac{89-1}{2}}(89/43)$$
$$= (89/43) = (3/43)$$
$$= (-1)^{\frac{3-1}{2}\cdot\frac{43-1}{2}}(43/3)$$

$$= -(43/3) = -(1/3)$$
$$= -1$$

Therefore, by equation (11.13), $(221/399) = (-1)(-1) = 1$. ■

The generalized law of quadratic reciprocity is extremely useful for evaluating the Legendre symbol (a/p), where a is an odd composite number and p an odd prime with $p \nmid a$, as the following example illustrates.

EXAMPLE 11.26 Using the generalized law of quadratic reciprocity, evaluate $(391/439)$.

SOLUTION

By the generalized law of quadratic reciprocity,

$$\begin{aligned}(391/439) &= (-1)^{\frac{391-1}{2}\cdot\frac{439-1}{2}}(439/391) = -(439/391)\\ &= -(48/391) = -(16/391)(3/391)\\ &= -(3/391) = -(-1)^{\frac{3-1}{2}\cdot\frac{391-1}{2}}(391/3)\\ &= (391/3) = (1/3)\\ &= 1\end{aligned}$$

■

The following example, a continuation of Example 11.16, is a fine application of the generalized law of quadratic reciprocity.

EXAMPLE 11.27 Let p be a prime of the form $4n \pm 1$. Prove that every positive factor d of n is a quadratic residue of p.

PROOF

Let $p = 4n + r$, where $r = \pm 1$. Let $d = 2^s t$, where $s \geq 0$ and t is odd.

If $s \geq 1$, then $p \equiv r \pmod 8$, so $(2/p) = 1$, by Theorem 11.6. Thus, for $s \geq 0$, $(d/p) = (2^s t/p) = (2/p)^s (t/p) = (t/p)$. So $(d/p) = 1$ if $t = 1$. If $t \neq 1$, then, by the generalized law of quadratic reciprocity,

$$(t/p) = (-1)^{\frac{t-1}{2}\cdot\frac{p-1}{2}}(p/t)$$

If $r = 1$, then $(p-1)/2$ is even and $(p/t) = (1/t) = 1$, so $(t/p) = 1$. If $r = -1$, then $(p-1)/2$ is odd, so $(t/p) = (-1)^{(t-1)/2}(-1/t) = (-1/t)^2 = 1$. Thus, in both cases, $(d/p) = (t/p) = 1$, so d is a quadratic residue of p. ■

EXERCISES 11.4

1. Find the quadratic residues of 33.

Evaluate each Jacobi symbol.

2. $(2/21)$
3. $(3/35)$
4. $(12/25)$
5. $(23/65)$
6. $(52/129)$
7. $(442/385)$
8. $(-68/665)$
9. $(-198/2873)$

Evaluate each Jacobi symbol.

10. $(2/15)$
11. $(17/33)$

Verify that each congruence is not solvable.

12. $x^2 \equiv 2 \pmod{15}$
13. $x^2 \equiv 17 \pmod{33}$

Compute each.

14. $(3/7^2)$
15. $(3/5^3 \cdot 7^5)$
16. $(3/5^7 \cdot 7^5 \cdot 11^3)$
17. $(3/5 \cdot 7^3 \cdot 13^6)$
18. Let m be an odd positive integer with prime-power factorization $p^a q^b r^c$, where $p \equiv q \equiv r \equiv \pm 5 \pmod{12}$. Under what conditions will $(3/m) = 1$? (*Hint*: Study Exercises 14–17.)
19. Let m be an odd positive integer with prime-power factorization $p^a q^b r^c s^d$, where $p \equiv q \equiv r \equiv \pm 5 \pmod{12}$ and $s \equiv \pm 1 \pmod{12}$. Under what conditions will $(3/m) = 1$? (*Hint*: Study Exercises 14–18.)

Prove each.

20. Let m be an odd integer such that $(a/m) = -1$. Then the congruence $x^2 \equiv a \pmod{m}$ is not solvable.
21. Let m be an odd positive integer. Then

$$(-1/m) = \begin{cases} 1 & \text{if } m \equiv 1 \pmod{4} \\ -1 & \text{if } m \equiv -1 \pmod{4} \end{cases}$$

22. Let m be an odd positive integer. Then

$$(2/m) = \begin{cases} 1 & \text{if } m \equiv \pm 1 \pmod{8} \\ -1 & \text{if } m \equiv \pm 3 \pmod{8} \end{cases}$$

23. Let m and n be relatively prime odd positive integers. Then

$$(n/m) = \begin{cases} (m/n) & \text{if } m \equiv 1 \text{ or } n \equiv l \pmod{4} \\ -(m/n) & \text{if } m \equiv n \equiv 3 \pmod{4} \end{cases}$$

*24. Let m be an odd positive integer such that $3 \nmid m$. Then $(3/m) = 1$ if and only if the sum of the exponents of the prime factors $\equiv \pm 5 \pmod{12}$ of m is even.

*11.5 Quadratic Congruences with Composite Moduli (optional)

Thus far, we have focused on solving quadratic congruences $x^2 \equiv a \pmod{p}$, where p is an odd prime and $p \nmid a$. In fact, we have enough tools to solve quadratic congruences even if the modulus m is the product of a finite number of distinct odd primes. For instance, let p and q be distinct odd primes such that $(a, pq) = 1$. Since $(a/pq) = (a/p)(a/q)$, $x^2 \equiv a \pmod{pq}$ is solvable if both $x^2 \equiv a \pmod{p}$ and $x^2 \equiv a \pmod{q}$ are solvable. Consequently, if we know their solutions, we can then apply the CRT to generate the solutions of $x^2 \equiv a \pmod{pq}$, as the following two examples illustrate.

EXAMPLE 11.28 Solve the quadratic congruence $x^2 \equiv 15 \pmod{187}$.

SOLUTION

First, notice that $187 = 11 \cdot 17$, and $(15/11) = 1 = (15/17)$, so the congruences $x^2 \equiv 15 \pmod{11}$ and $x^2 \equiv 15 \pmod{17}$ are solvable. Thus, $x^2 \equiv 15 \pmod{187}$ is also solvable.

You may verify that the two incongruent solutions of $x^2 \equiv 15 \pmod{11}$ are $x \equiv \pm 2 \pmod{11}$ and those of $x^2 \equiv 15 \pmod{17}$ are $x \equiv \pm 7 \pmod{17}$. Therefore, by the CRT, the given congruence has four incongruent solutions: $x \equiv \pm 24, \pm 75 \pmod{187}$; that is, $x \equiv 24, 75, 112, 163$ modulo 187. ■

If $p \equiv 3 \pmod 4$ and $x^2 \equiv a \pmod p$ is solvable, then its solutions are known explicitly. (See Exercise 30 in Section 11.3.) Consequently, in such cases, $x^2 \equiv a \pmod{pq}$ can be solved fairly easily, as the following example shows.

EXAMPLE 11.29 Solve the congruence $x^2 \equiv 6 \pmod{437}$.

SOLUTION

First, notice that $437 = 19 \cdot 23$ and $(6/19) = 1 = (6/23)$, so the congruences $x^2 \equiv 6 \pmod{19}$ and $x^2 \equiv 6 \pmod{23}$ are solvable. When $p \equiv 3 \pmod 4$, the solutions of $x^2 \equiv a \pmod p$ are $x \equiv \pm a^{(p+1)/4} \pmod p$. Since $19 \equiv 3 \equiv 23 \pmod 4$, the solutions of $x^2 \equiv 6 \pmod{19}$ are $x \equiv \pm 6^{(19+1)/4} \equiv \pm 6^5 \equiv \pm 5 \pmod{19}$, and those of $x^2 \equiv 6 \pmod{23}$ are $x \equiv \pm 6^{(23+1)/4} \equiv \pm 6^6 \equiv \pm 12 \pmod{23}$. Thus, by the CRT, the solutions of $x^2 \equiv 6 \pmod{437}$ are $x \equiv \pm 81, \pm 195 \pmod{437}$; that is, $x \equiv 81, 195, 242, 356$ modulo 437. ■

It follows from these two examples that if the modulus m is the product of a finite number of distinct primes and $(a/p) = 1$ for every prime factor p of m, then the congruence can be solved. See Exercises 1–4.

We now turn our attention to the case where m is a prime-power p^n, p being odd. Two questions we can reasonably ask are:

- When is the congruence $x^2 \equiv a \pmod{p^n}$ solvable?
- When it is solvable, how do we find the solutions?

Before answering, let us study the following example, since it should clarify the proof of the next theorem.

EXAMPLE 11.30 Find a solution of the congruence $x^2 \equiv 23 \pmod{7^3}$.

SOLUTION

We shall illustrate the strategy step by step.

step 1 *Solve the congruence* $x^2 = 23 \pmod 7$.
Since $23 \equiv 2 \pmod 7$ and $(2/7) = 1$, $x^2 \equiv 23 \pmod 7$ is solvable; its solutions are $x \equiv 3, 4 \pmod 7$.

step 2 *Construct a solution of* $x^2 \equiv 23 \pmod{7^2}$.
Since 3 is a solution of $x^2 \equiv 23 \pmod 7$, $3^2 = 23 + 7i$ for some integer i, namely, $i = -2$; thus, $9 = 23 + (-2) \cdot 7$. We now ingeniously look for a solution of the form $3 + 7j$, so square it:

$$\begin{aligned}(3 + 7j)^2 &= 9 + 42j + 49j^2\\ &\equiv 9 + 42j \pmod{7^2}\\ &\equiv [23 + (-2) \cdot 7] + 42j \pmod{7^2}\\ &\equiv 23 + 7(-2 + 6j) \pmod{7^2}\end{aligned}$$

Now choose j such that $-2 + 6j \equiv 0 \pmod 7$; that is, $6j \equiv 2 \pmod 7$; thus, choose $j \equiv 5 \pmod 7$ or $j = 5$. Then $3 + 7j \equiv 3 + 7 \cdot 5 \equiv 38 \pmod{7^2}$ and $38^2 \equiv 23 \pmod{7^2}$. Thus, 38 is a solution of $x^2 \equiv 23 \pmod{7^2}$.

step 3 *Now use 38 to generate a solution of* $x^2 \equiv 23 \pmod{7^3}$.
Since $38^2 \equiv 23 \pmod{7^2}$, $38^2 = 23 + k \cdot 7^2$ for some integer k, namely, 29: $38^2 = 23 + 29 \cdot 7^2$. Now look for a solution of the form $38 + 7^2\ell$. Since

$$\begin{aligned}(38 + 7^2\ell)^2 &= 38^2 + 76 \cdot 7^2\ell + 7^4\ell^2\\ &\equiv 38^2 + 76 \cdot 7^2\ell \pmod{7^3}\\ &\equiv (23 + 29 \cdot 7^2) + 76 \cdot 7^2\ell \pmod{7^3}\\ &\equiv 23 + 7^2(29 + 76\ell) \pmod{7^3}\end{aligned}$$

Choose ℓ such that $29 + 76\ell \equiv 0 \pmod 7$; that is, such that $1 + 6\ell \equiv 0 \pmod 7$; thus choose $\ell = 1$. Then $38 + 7^2\ell = 38 + 7^2 \cdot 1 \equiv 87 \pmod{7^3}$. Thus, $87^2 \equiv 23 \pmod{7^3}$, so 87 is a solution of the given congruence. [Now use the solution 4 of $x^2 \equiv 23 \pmod 7$, and steps 2 and 3 to find the other solution of $x^2 \equiv 23 \pmod{7^3}$; it is 256.] ■

This example shows that if the congruence $x^2 \equiv a \pmod p$ is solvable, then $x^2 \equiv a \pmod{p^n}$ is also solvable. Further, its solutions can be used step by step to generate the solutions of $x^2 \equiv a \pmod{p^n}$. This is the essence of the following theorem.

THEOREM 11.14 Let p be an odd prime, a any integer such that $p \nmid a$, and n any positive integer. Then the congruence $x^2 \equiv a \pmod{p^n}$ is solvable if and only if $(a/p) = 1$.

PROOF

Suppose that $x^2 \equiv a \pmod{p^n}$ is solvable. Then $x^2 \equiv a \pmod{p}$ is also solvable, so $(a/p) = 1$.

Conversely, let $(a/p) = 1$; that is, assume that $x^2 \equiv a \pmod{p}$ is solvable. We shall now prove by induction that $x^2 \equiv a \pmod{p^n}$ is solvable for every positive integer n. Clearly it is true when $n = 1$. So assume it is true for an arbitrary integer $k \geq 1$: $x^2 \equiv a \pmod{p^k}$ is solvable. We shall now show that $x^2 \equiv a \pmod{p^{k+1}}$ is also solvable by constructing a solution.

Let α be a solution of $x^2 \equiv a \pmod{p^k}$. Then $\alpha^2 \equiv a \pmod{p^k}$; that is, $\alpha^2 = a + ip^k$ for some integer i (see Step 2 in Example 11.30). We now generate a solution of the form $\alpha + jp^k$ of $x^2 \equiv a \pmod{p^{k+1}}$. Then

$$
\begin{aligned}
(\alpha + jp^k)^2 &= \alpha^2 + 2\alpha jp^k + j^2p^{2k} \\
&\equiv \alpha^2 + 2\alpha jp^k \pmod{p^{k+1}}, \quad \text{since } 2k \geq k+1 \\
&\equiv (a + ip^k) + 2\alpha jp^k \pmod{p^{k+1}} \\
&\equiv a + (i + 2\alpha j)p^k \pmod{p^{k+1}}
\end{aligned}
$$

Now choose j such that $i + 2\alpha j \equiv 0 \pmod{p}$. Such a j exists by Theorem 4.9, since $(2\alpha, p) = 1$. With such a j, $(\alpha + jp^k)^2 \equiv a \pmod{p^{k+1}}$. Thus, $\alpha + jp^k$ is a solution of $x^2 \equiv a \pmod{p^{k+1}}$.

Thus, by induction, $x^2 \equiv a \pmod{p^n}$ is solvable for every positive integer n. ■

This theorem provides a test to determine the solvability of $x^2 \equiv a \pmod{p^n}$, and an algorithm to construct a solution of the congruence from that of $x^2 \equiv a \pmod{p^{n-1}}$. Thus, knowing the solutions of $x^2 \equiv a \pmod{p}$, we can step by step build up to those of $x^2 \equiv a \pmod{p^n}$, as the following example illustrates.

EXAMPLE 11.31 In Example 11.30, we found that 87 is a solution of $x^2 \equiv 23 \pmod{7^3}$. Using Theorem 11.14, find the remaining solution.

SOLUTION

Recall that $x^2 \equiv 23 \pmod{7}$ has two solutions, 3 and 4 modulo 7, and we used 3 to arrive at the solution 87. To find the remaining solution we proceed as follows, where $a = 23$ and $p = 7$.

step 1 *Initialize α and k.* (k is the current exponent of the modulus 7.)
$\alpha = 4$ and $k = 1$.

step 2 *Express α^2 in the form $a + ip$ and solve for i.*
$\alpha^2 = a + ip$ yields $16 = 23 + 7i$, thus $i = -1$.

step 3 *Solve the linear congruence $i + 2\alpha j \equiv 0 \pmod{p}$ for j.*
Then $-1 + 2 \cdot 4j \equiv 0 \pmod{7}$; that is, $j \equiv 1 \pmod{7}$, so choose $j = 1$.

step 4 *Extract a solution of $x^2 \equiv a \pmod{p^2}$.*
$\alpha + jp = 4 + 1 \cdot 7 = 11$ is a solution of $x^2 \equiv 23 \pmod{7^2}$. (Verify this.)

step 5 *Update α and k.*
$\alpha = 11$ and $k = 2$.
With the new values of α and k, repeat steps 2–4 to find a solution of $x^2 \equiv a \pmod{p^3}$.

step 6 *Express α^2 in the form $a + ip^2$.*
$\alpha^2 = a + ip^2$ yields $121 = 23 + i \cdot 7^2$; thus $i = 2$.

step 7 *Solve the linear congruence $i + 2\alpha j \equiv 0 \pmod{p}$ for j.*
$2 + 2 \cdot 11j \equiv 0 \pmod{7}$ yields $j = 5$.

step 8 *Generate a solution of $x^2 \equiv a \pmod{p^3}$.*
$\alpha + jp^2 = 11 + 5 \cdot 7^2 = 256$ is a solution of $x^2 \equiv 23 \pmod{7^3}$, as expected. ■

Using the preceding two examples, we can solve $x^2 \equiv 23 \pmod{7^4}$ and $x^2 \equiv 23 \pmod{7^5}$. See Exercises 7 and 8.

Next we examine congruences of the form $x^2 \equiv a \pmod{2^n}$. Suppose a is even and $a = 2^b c \not\equiv 0 \pmod{2^n}$, where c is odd. Clearly, $b < n$. If b is odd, the congruence has no solution. (See Exercise 52.) If b is even, say, $b = 2i$, then $x^2 \equiv 2^{2i}c \pmod{2^n}$; that is, $(x/2^i)^2 \equiv c \pmod{2^{n-b}}$. This is of the form $y^2 \equiv c \pmod{2^k}$, where c is odd. Consequently, we restrict our investigation to the case where a is odd. Then, depending on the value of k, the congruence can have exactly one, two, or four solutions. This is the essence of the following theorem. A portion of its proof runs along the same lines as the proof of Theorem 11.14.

THEOREM 11.15 Let a be an odd integer and n any integer ≥ 3. Then the congruence

- $x^2 \equiv a \pmod{2}$ is solvable.
- $x^2 \equiv a \pmod{4}$ is solvable if and only if $a \equiv 1 \pmod{4}$.
- $x^2 \equiv a \pmod{2^n}$ is solvable if and only if $a \equiv 1 \pmod{8}$.

PROOF

- Because $a \equiv 1 \pmod{2}$, $x^2 \equiv 1 \pmod{2}$ has exactly one solution, namely, 1.
- Suppose $x^2 \equiv a \pmod{4}$ is solvable. Since a is odd, so is x^2. Then x must be odd and hence $x^2 \equiv 1 \pmod{8}$. (See Exercise 46.) Consequently, $x^2 \equiv 1 \pmod{4}$ and hence $a \equiv 1 \pmod{4}$.

Conversely, assume that $a \equiv 1 \pmod 4$. Then $x^2 \equiv 1 \pmod 4$ is solvable since it has (exactly) two incongruent solutions, namely, 1 and 3.

- Suppose $x^2 \equiv a \pmod{2^n}$ is solvable. Then, $x^2 \equiv 1 \pmod 8$, so $a \equiv 1 \pmod 8$. Conversely, suppose $a \equiv 1 \pmod 8$. We will then prove by induction that $x^2 \equiv a \pmod{2^n}$ is solvable for every $n \geq 3$. First, notice that $x^2 \equiv 1 \pmod 8$ is solvable with exactly four incongruent solutions, namely, 1, 3, 5, and 7.
  Now assume that $x^2 \equiv a \pmod{2^k}$ has a solution α for an arbitrary integer $k \geq 3$. Then $\alpha^2 \equiv a \pmod{2^k}$; that is, $\alpha^2 = a + i2^k$ for some integer i. We now generate a solution of $x^2 \equiv a \pmod{2^{k+1}}$ in the form $\alpha + j2^{k-1}$ (similar to that in Theorem 11.14). Then:

$$
\begin{aligned}
(\alpha + j2^{k-1})^2 &= \alpha^2 + \alpha j 2^k + j^2 2^{2k-2} \\
&\equiv \alpha^2 + \alpha j 2^k \pmod{2^{k+1}}, \quad \text{since } k \geq 3 \\
&\equiv (a + i2^k) + \alpha j 2^k \pmod{2^{k+1}} \\
&\equiv a + (i + \alpha j)2^k \pmod{2^{k+1}}
\end{aligned}
$$

  Now choose j such that $i + \alpha j \equiv 0 \pmod 2$. Since α is odd, such a j exists by Corollary 4.7. Then $(\alpha + j2^{k-1})^2 \equiv a \pmod{2^{k+1}}$. Consequently, $x^2 \equiv a \pmod{2^{k+1}}$ is solvable with $\alpha + j2^{k-1}$ as a solution. Thus, by induction, $x^2 \equiv a \pmod{2^n}$ is solvable for every integer $n \geq 3$. ■

We can take another example: Since $37 \not\equiv 1 \pmod 8$, it follows by the theorem that $x^2 \equiv 37 \pmod{64}$ is not solvable; likewise, $x^2 \equiv 5 \pmod 8$ is also not solvable.

Since every solution of $x^2 \equiv a \pmod{2^k}$ yields a solution of $x^2 \equiv a \pmod{2^{k+1}}$, where $k \geq 3$, and $x^2 \equiv a \pmod 8$ has four solutions, it follows that $x^2 \equiv a \pmod{2^n}$ has at least four solutions, where $n \geq 3$. In fact, if α is a solution, then it can be shown that $2^n - \alpha$ and $2^{n-1} \pm \alpha$ are also solutions. (See Exercise 53.) It can also be shown that it has no other incongruent solutions. Accordingly, we have the following result.

COROLLARY 11.8 If $a \equiv 1 \pmod 8$ and $n \geq 3$, then the congruence $x^2 \equiv a \pmod{2^n}$ has exactly four incongruent solutions. ■

The following example illustrates the theorem.

EXAMPLE 11.32 Solve the congruence $x^2 \equiv 17 \pmod{32}$.

SOLUTION

Since $17 \equiv 1 \pmod 8$, the congruence is solvable.

step 1 *Find a solution of* $x^2 \equiv a \pmod{2^3}$.
Since $1^2 \equiv 17 \pmod{2^3}$, $\alpha = 1$ is a solution of $x^2 \equiv 17 \pmod{2^3}$. Then $1^2 = 17 + 8i$, where $i = -2$.

step 2 *Find a solution of* $x^2 \equiv a \pmod{2^4}$ *with* $k = 3$.
(See the proof of Theorem 11.15.) Choose j such that $i + \alpha j = -2 + 1 \cdot j \equiv 0 \pmod 2$. Thus, we choose $j = 0$. Then $\alpha + j2^{k-1} = 1 + 0 \cdot 4 = 1$ is a solution of $x^2 \equiv 17 \pmod{2^4}$, which is obviously true.

step 3 *Update the values of* α, k, *and* i.
Clearly, $\alpha = 1$ and $k = 4$. Since $1^2 \equiv 17 \pmod{16}$, $1^2 = 17 + 16i$; so choose $i = -1$.

step 4 *Find a solution of* $x^2 \equiv a \pmod{2^5}$, *where* $k = 4$.
Choose j such that $i + \alpha j = -1 + 1 \cdot j \equiv 0 \pmod 2$, so choose $j = 1$. Then $\alpha + j2^{k-1} = 1 + 1 \cdot 2^3 = 9$ is a solution of $x^2 \equiv 17 \pmod{2^5}$. (You may verify this.)

step 5 *Find the remaining solutions of* $x^2 \equiv a \pmod{2^5}$.
They are given by $-9 = 23$ and $2^{n-1} \pm 9 = 2^4 \pm 9$ modulo 32.

Thus, the four solutions of the given congruence are 7, 9, 23, and 25 modulo 32. ■

As the theorem shows and this example illustrates, finding a solution of $x^2 \equiv a \pmod 8$ is a good starting place to construct a solution of $x^2 \equiv a \pmod{2^n}$, where $n \geq 4$. The exercises to follow provide ample opportunities for such a pursuit.

We now tie all pieces together. Let $m = 2^{e_0} \prod_i p_i^{e_i}$ be the canonical decomposition of m and $(a, m) = 1$. Theorems 11.14 and 11.15 enable us to solve the congruences $x^2 \equiv a \pmod{p^j}$ and $x^2 \equiv a \pmod{2^k}$. Consequently, $x^2 \equiv a \pmod m$ is solvable if and only if they are solvable. Accordingly, we have the following result.

THEOREM 11.16 Let m be a positive integer with canonical decomposition $2^{e_0} \prod_i p_i^{e_i}$ and a any integer with $(a, m) = 1$. Then $x^2 \equiv a \pmod m$ is solvable if and only if $x^2 \equiv a \pmod{2^{e_0}}$ and $x^2 \equiv a \pmod{p_i^{e_i}}$ are solvable. ■

Knowing the solutions of the congruences $x^2 \equiv a \pmod{2^{e_0}}$ and $x^2 \equiv a \pmod{p_i^{e_i}}$, we can solve the congruence $x^2 \equiv a \pmod m$ using the CRT, as the next example illustrates. Obviously, the task will be long and complicated if m contains several distinct prime factors.

EXAMPLE 11.33 Solve the quadratic congruence $x^2 \equiv 97 \pmod{7688}$.

SOLUTION

(We leave the details for you to fill in.) Since $7688 = 2^3 \cdot 31^2$, we first solve the congruences $x^2 \equiv 97 \pmod 8$ and $x^2 \equiv 97 \pmod{31^2}$, and then use their solutions to solve the given congruence.

step 1 *Solve* $x^2 \equiv 97 \pmod 8$.
Since $x^2 \equiv 1 \pmod 8$, $x \equiv 1, 3, 5, 7 \pmod 8$.

step 2 *Solve* $x^2 \equiv 97 \pmod{31}$.
Then $x^2 \equiv 4 \pmod{31}$, so $x \equiv 2, 29 \pmod{31}$.

step 3 *Solve* $x^2 \equiv 97 \pmod{31^2}$.
Using Theorem 11.14, the two solutions are $x \equiv 215, 746 \pmod{31^2}$.

step 4 *Use the solutions in steps 1 and 3 to solve* $x^2 \equiv 97 \pmod{7688}$.
By the CRT, it has eight incongruent solutions, namely, $x \equiv 215, 1707, 2137, 3629, 4059, 5551, 5981, 7473 \pmod{7688}$. ■

EXERCISES 11.5

Solve each quadratic congruence.

1. $x^2 \equiv 4 \pmod{35}$
2. $x^2 \equiv 23 \pmod{77}$
3. $x^2 \equiv 43 \pmod{221}$
4. $x^2 \equiv 69 \pmod{2431}$

Using the given solution of the congruence, solve the corresponding congruence.

5. 108 is a solution of $x^2 \equiv 3 \pmod{13^2}$; $x^2 \equiv 3 \pmod{13^3}$
6. 211 is a solution of $x^2 \equiv 15 \pmod{17^2}$; $x^2 \equiv 15 \pmod{17^3}$

Solve each congruence.
(*Hint*: Use Examples 11.30 and 11.31.)

7. $x^2 \equiv 23 \pmod{7^4}$
8. $x^2 \equiv 23 \pmod{7^5}$

Solve each congruence.

9. $x^2 \equiv 10 \pmod{13^2}$
10. $x^2 \equiv 10 \pmod{13^3}$
11. $x^2 \equiv 5 \pmod{11^2}$
12. $x^2 \equiv 5 \pmod{11^3}$
13. $x^2 \equiv 13 \pmod{17^2}$
14. $x^2 \equiv 17 \pmod{19^2}$
15. $x^2 \equiv 17 \pmod{64}$
16. $x^2 \equiv 17 \pmod{256}$
17. $x^2 \equiv 25 \pmod{32}$
18. $x^2 \equiv 25 \pmod{128}$
19. $x^2 \equiv 33 \pmod{64}$
20. $x^2 \equiv 33 \pmod{128}$
21. $x^2 \equiv 41 \pmod{32}$
22. $x^2 \equiv 41 \pmod{256}$
23. $x^2 \equiv 41 \pmod{1024}$

Solve each. (See Example 11.1 and Exercise 10 in Section 11.1.)

24. $3x^2 - 4x + 7 \equiv 0 \pmod{13^2}$
25. $4x^2 + 4x - 3 \equiv 0 \pmod{5^2}$

Solve each congruence.
(*Hint*: Find a least residue b such that $ab \equiv 1 \pmod p$.)

26. $2x^2 + 1 \equiv 0 \pmod{11}$
27. $2x^2 + 1 \equiv 0 \pmod{17}$
28. $3x^2 + 1 \equiv 0 \pmod{13}$
29. $3x^2 + 1 \equiv 0 \pmod{19}$

Solve each congruence, if possible.
(*Hint*: Use Exercises 1–4.)

30. $x^2 \equiv 4 \pmod{140}$
31. $x^2 \equiv 23 \pmod{308}$
32. $x^2 \equiv 43 \pmod{1768}$
33. $x^2 \equiv 13 \pmod{1156}$
34. $x^2 \equiv 5 \pmod{5324}$
35. $x^2 \equiv 17 \pmod{2888}$

Assuming that the congruence $x^2 \equiv a \pmod{m}$ is solvable, find the number of solutions for the indicated value of m, where p, q, and r are distinct odd primes, and a is an integer with $(a, pqr) = 1$.

36. $m = pq$
37. $m = p^2$
38. $m = 4pq$
39. $m = pqr$
40. $m = p^2q^3r^4$
41. $m = 8p^2q^3r^4$

Let $m = 2^{e_0}\prod_{i=1}^{k} p_i^{e_i}$ be the canonical decomposition of m and $(a, m) = 1$. Assuming that $x^2 \equiv a \pmod{m}$ is solvable, find its number of solutions in each case.

42. $e_0 = 0$
43. $e_0 = 1$
44. $e_0 = 2$
45. $e_0 \geq 3$

Prove each.

46. The square of every odd integer is congruent to 1 modulo 8.
47. Let a and b be two positive integers and p an odd prime such that $ab \equiv 1 \pmod{p}$. Then $(a/p) = (b/p)$.
48. Let p be an odd prime. Then the congruence $2x^2 + 1 \equiv 0 \pmod{p}$ is solvable if and only if $p \equiv 1$ or $p \equiv 3 \pmod{8}$. (*Hint*: Use Exercise 47.)
49. Let p be a prime > 3. Then $3x^2 + 1 \equiv 0 \pmod{p}$ if and only if $p \equiv 1 \pmod{6}$.
50. Let α be a solution of $x^2 \equiv a \pmod{p^n}$, where p is an odd prime and $p \nmid a$. Show that $p^n - \alpha$ is also a solution.
51. Let p be an odd prime and a an integer such that $p \nmid a$. If the congruence $x^2 \equiv a \pmod{p^n}$ is solvable, it has exactly two solutions, where $n \geq 2$. (*Hint*: Use Exercise 50.)
52. Let n be a positive integer and $a = 2^b c \not\equiv 0 \pmod{2^n}$, where b and c are odd integers. Then the congruence $x^2 \equiv a \pmod{2^n}$ is not solvable.
53. Let α be a solution of $x^2 \equiv a \pmod{2^n}$, where $n \geq 3$ and $a \equiv 1 \pmod{8}$. Then $2^n - \alpha$ and $2^{n-1} \pm \alpha$ are also solutions.

*54. The congruence $x^2 \equiv a \pmod{2^n}$, where $n \geq 3$ and $a \equiv 1 \pmod{8}$ has exactly four incongruent solutions. (*Hint*: Use Exercise 53.)

CHAPTER SUMMARY

In this chapter, we explored the theory of quadratic congruences via the concept of a quadratic residue, and established several criteria for determining the solvability of the congruence $x^2 \equiv a \pmod{p}$, where p is an odd prime and $p \nmid a$. The congruence has either two or no incongruent solutions.

Quadratic Residue

- An integer a is a **quadratic residue** of a positive integer m, where $(a, m) = 1$, if $x^2 \equiv a \pmod{m}$ has a solution; otherwise, it is a **quadratic nonresidue**. (p. 497)
- Every odd prime p has exactly $(p-1)/2$ quadratic residues and exactly $(p-1)/2$ quadratic nonresidues. (p. 498)

- (**Euler's criterion**) Let p be an odd prime. Then a positive integer a with $p \nmid a$ is a quadratic residue of p if and only if $a^{(p-1)/2} \equiv 1 \pmod{p}$. (p. 499)
- Let p be an odd prime. Then a positive integer a, where $p \nmid a$, is a quadratic nonresidue if and only if $a^{(p-1)/2} \equiv -1 \pmod{p}$. (p. 500)

The Legendre Symbol

- Let p be an odd prime and a any integer such that $p \nmid a$. The **Legendre symbol** (a/p) is defined by

$$(a/p) = \begin{cases} 1 & \text{if } a \text{ is a quadratic residue of } p \\ -1 & \text{otherwise} \quad \text{(p. 501)} \end{cases}$$

- **Euler's criterion** Let p be an odd prime. Then a positive integer a with $p \nmid a$ is a quadratic residue of p if and only if $(a/p) = 1$. (p. 502)
- Let p be an odd prime, and a and b be integers with $p \nmid ab$. Then:
  - If $a \equiv b \pmod{p}$, then $(a/p) = (b/p)$.
  - $(a/p)(b/p) = (ab/p)$.
  - $(a^2/p) = 1$. (p. 502)
- If p is an odd prime, then

$$(-1/p) = \begin{cases} 1 & \text{if } p \equiv 1 \pmod{4} \\ -1 & \text{if } p \equiv -1 \pmod{4} \quad \text{(p. 503)} \end{cases}$$

- Let p be an odd prime, q a prime such that $p \nmid q$, and i a positive integer. Then $(q^i/p) = (q/p)^i$. (p. 505)
- Let p be an odd prime and let $a = \prod_{i=1}^{n} p_i^{e_i}$ be the canonical decomposition of a, where $(a, p) = 1$. Then $(a/p) = \prod_{i=1}^{n} (p_i/p)^{e_i}$. (p. 505)
- **Gauss' Lemma** Let p be an odd prime and a an integer such that $p \nmid a$. Let ν denote the number of least positive residues of the integers $a, 2a, 3a, \ldots, [(p-1)/2]a$ that exceed $p/2$. Then $(a/p) = (-1)^{\nu}$. (p. 507)
- Let p be an odd prime with $p \nmid a$. Let v denote the number of least positive residues of the integers $a, 2a, \ldots, [(p-1)/2]a$ that exceed $p/2$. Then $(a/p) = 1$ if and only if v is even. (p. 508).
- Let p be an odd prime. Then

$$(2/p) = \begin{cases} 1 & \text{if } p \equiv \pm 1 \pmod{8} \\ -1 & \text{if } p \equiv \pm 3 \pmod{8} \quad \text{(p. 509)} \end{cases}$$

- Let p be an odd prime. Then $(2/p) = (-1)^{(p^2-1)/8}$. (p. 511)

- If p and $2p+1$ are odd primes, then $2(-1)^{(p-1)/2}$ is a primitive root modulo $2p+1$. (p. 512)
- Let p and q be distinct odd primes. Then

$$\sum_{k=1}^{(p-1)/2} \left\lfloor \frac{kq}{p} \right\rfloor + \sum_{k=1}^{(q-1)/2} \left\lfloor \frac{kp}{q} \right\rfloor = \frac{(p-1)}{2} \cdot \frac{(q-1)}{2} \quad \text{(p. 515)}$$

- **Law of Quadratic Reciprocity** Let p and q be distinct odd primes. Then

$$(p/q)(q/p) = (-1)^{\frac{(p-1)}{2} \cdot \frac{(q-1)}{2}} \quad \text{(p. 518)}$$

- Let p and q be distinct odd primes. Then

$$(q/p) = \begin{cases} (p/q) & \text{if } p \equiv 1 \pmod 4 \text{ or } q \equiv 1 \pmod 4 \\ -(p/q) & \text{if } p \equiv q \equiv 3 \pmod 4 \quad \text{(p. 520)} \end{cases}$$

- **Pepin's Test** The Fermat number f_n is a prime if and only if $3^{(f_n-1)/2} \equiv -1 \pmod{f_n}$. (p. 523)

The Jacobi Symbol

- Let m be an odd positive integer with the canonical decomposition $\prod_{i=1}^{k} p_i^{e_i}$, where $(a, m) = 1$. Then the **Jacobi symbol** (a/m) is defined by

$$(a/m) = \left(a \Big/ \prod_{i=1}^{k} p_i^{e_i} \right) = \prod_{i=1}^{k} (a/p_i)^{e_i}$$

where (a/p_i) denotes the Legendre symbol. (p. 527)

- Let m be an odd positive integer, and a and b be any integers with $(a, m) = 1 = (b, m)$. Then
  - If $a \equiv b \pmod m$, then $(a/m) = (b/m)$.
  - $(ab/m) = (a/m)(b/m)$
  - $(a^2/m) = 1$ (p. 529)
- Let m be an odd positive integer with prime-power decomposition $\prod_{i=1}^{k} p_i^{e_i}$. Then
  - $\sum_{i=1}^{k} e_i(p_i - 1)/2 \equiv (m-1)/2 \pmod 2$
  - $\sum_{i=1}^{k} e_i\big(p_i^2 - 1\big)/8 \equiv (m^2-1)/8 \pmod 2$. (p. 530)

- Let m be an odd positive integer. Then
  - $(-1/m) = (-1)^{(m-1)/2}$
  - $(2/m) = (-1)^{(m^2-1)/8}$. (p. 531)
- **The Generalized Quadratic Reciprocity Law** Let m and n be relatively prime odd positive integers. Then

$$(m/n)(n/m) = (-1)^{\frac{m-1}{2}\frac{n-1}{2}} \quad \text{(p. 532)}$$

- Let p be an odd prime, and a and n any positive integers such that $p \nmid a$. Then the congruence $x^2 \equiv a \pmod{p^n}$ is solvable if and only if $(a/p) = 1$. (p. 538)
- Let a be an odd integer and n any integer ≥ 3. Then the congruence
  - $x^2 \equiv a \pmod 2$ is solvable.
  - $x^2 \equiv a \pmod 4$ is solvable if and only if $a \equiv 1 \pmod 4$.
  - $x^2 \equiv a \pmod{2^n}$ is solvable if and only if $a \equiv 1 \pmod 8$. (p. 539)
- If $a \equiv 1 \pmod 8$ and $n \geq 3$, then $x^2 \equiv a \pmod{2^n}$ has exactly four incongruent solutions. (p. 540)
- Let m be a positive integer with canonical decomposition $2^{e_0} \prod_i p_i^{e_i}$ and a any integer with $(a, m) = 1$. Then the congruence $x^2 \equiv a \pmod m$ is solvable if and only if $x^2 \equiv a \pmod{2^{e_0}}$ and $x^2 \equiv a \pmod{p_i^{e_i}}$ are solvable. (p. 541)

REVIEW EXERCISES

Solve each quadratic congruence.

1. $x^2 \equiv 13 \pmod{17}$
2. $x^2 \equiv 31 \pmod{33}$

Evaluate each, where M_p is a Mersenne number, f_n a Fermat number, and p an odd prime, and $n > 0$.

3. $(116/73)$
4. $(1033/1999)$
5. $(1739/3749)$
6. $(2327/4367)$
7. $(1/M_p)$
8. $(1/f_n)$
9. $(-1/M_p)$
10. $(-1/f_n)$
11. $(3/M_p)$
12. $(-3/M_p)$
13. $(3/f_n)$
14. $(-3/f_n)$
15. $(5/p)$, where $p = n! + 1$ is a prime and $n \geq 5$.
16. $(5/p)$, where $p = n! - 1$ is a prime and $n \geq 5$.
17. $(3/p)$, where $p = 2^n + 1$ is a prime and n is even.
18. $(p/3)$, where $p = 2^n + 1$ is a prime and n is even.
19. $(5/p)$, where $p = 2^{4n} + 1$ is a prime.

20. $(p/5)$, where $p = 2^{4n} + 1$ is a prime.
21. Both 13 and 29 are quadratic nonresidues of 47. Using this fact, find a quadratic residue of 47.

Verify each. (*Hint*: Use Euler's criterion.)

22. $2999|(2^{1499} - 1)$
23. $3989|(2^{1994} + 1)$
24. $3347|(3^{1673} - 1)$
25. $4793|(3^{2396} + 1)$

Given that α is a solution of the congruence $x^2 \equiv a \pmod{p}$, where p is an odd prime. Find a solution of $x^2 \equiv 4a \pmod{p}$ for the given values of a, p, and α.

26. $a = 10$, $p = 13$, and $\alpha = 6$
27. $a = 9$, $p = 17$, and $\alpha = 14$

If p is a prime $\equiv \pm 1 \pmod{8}$, then $p|(2^{(p-1)/2} - 1)$. (See Exercise 51.) Using this fact, find a prime factor of each Mersenne number.

28. $2^{23} - 1$
29. $2^{83} - 1$
30. $2^{89} - 1$
31. $2^{1013} - 1$
32. Let p be an odd prime such that $q = 2p + 1$ is also a prime. If $p \equiv 1 \pmod{4}$, then 2 is a primitive root modulo q. Otherwise, -2 is a primitive root modulo q. (See Exercise 53.) Using these facts, find four primes for which 2 is a primitive root and four primes for which -2 is a primitive root.
33. Characterize the prime factors p of the integer $n^2 + 1$, where $n > 1$.

Solve each quadratic congruence. (See Exercises 11 and 12 in Section 1.)

34. $2x^2 + 3x + 1 \equiv 0 \pmod{7^2}$
35. $2x^2 + x + 1 \equiv 0 \pmod{11^2}$

Solve each congruence, if possible.

36. $x^2 \equiv 27 \pmod{253}$
37. $x^2 \equiv 53 \pmod{2431}$
38. $x^2 \equiv 5 \pmod{968}$
39. $x^2 \equiv 169 \pmod{9724}$
40. $x^2 \equiv 47 \pmod{17^3}$
41. $x^2 \equiv 226 \pmod{19^3}$
42. Let α be a solution of the congruence $x^2 \equiv a \pmod{p}$, where p is an odd prime and $p \nmid 4a$. Find a solution of $x^2 \equiv 4a \pmod{p}$.
43. Find a factor of the Mersenne number $2^{1000151} - 1$.

Prove each.

44. Let p be an odd prime, e an even positive integer, and a an integer such that $p \nmid a$. Then $(2^e a/p) = (a/p)$.
45. Let a be a positive integer and p an odd prime such that $p \nmid 4a$. Then the congruence $x^2 \equiv a \pmod{p}$ is solvable if and only if $x^2 \equiv 4a \pmod{p}$ is solvable.
46. Let p be a prime factor of a positive integer n and $q = n! + 1$ be a prime. Then $(p/q) = 1$.
47. Let p be a prime factor of a positive integer n and $q = n! - 1$ be a prime, where $p \equiv 1 \pmod{4}$. Then $(p/q) = 1$.

48. Let p and q be distinct odd primes. Then the congruence $x^2 \equiv q \pmod{p}$ is solvable if and only if $x^2 \equiv p \pmod{q}$ is solvable, unless $p \equiv q \equiv 3 \pmod{4}$.
49. If p is a prime $\equiv \pm 1 \pmod{8}$, then $p|[2^{(p-1)/2} - 1]$.
50. Let p be an odd prime such that $p|(a^{2n} + b^{2n})$, where $p \nmid ab$ and $n \geq 1$. Then p is of the form $4k + 1$.
51. Every quadratic nonresidue of a Fermat prime f_n is a primitive root.
52. Every primitive root modulo a Fermat prime is a quadratic nonresidue.
53. Let p be an odd prime such that $q = 2p + 1$ is also a prime. If $p \equiv 1 \pmod{4}$, then 2 is a primitive root modulo q. Otherwise, -2 is a primitive root modulo q.
54. There is an infinite number of primes of the form $8n - 1$.
    (*Hint*: Assume there are only finitely many such primes $p_1, p_2, \ldots, p_k$. Consider $N = (p_1 p_2 \cdots p_k)^2 - 2$.)
55. There is an infinite number of primes of the form $8n + 3$.
    (*Hint*: Assume there are only finitely many such primes $p_1, p_2, \ldots, p_k$. Consider $N = (p_1 p_2 \cdots p_k)^2 + 2$.)
56. There is an infinite number of primes of the form $6n + 1$.
    (*Hint*: Assume there are only finitely many such primes $p_1, p_2, \ldots, p_k$. Consider $N = (p_1 p_2 \cdots p_k)^2 + 3$.)
57. Three is a primitive root modulo every prime $p = 2^n + 1$, where $n > 1$.
58. Let p and q be distinct odd primes and $p^\star = p(-1)^{(p-1)/2}$. Then $(p^\star/q) = (q/p)$.
58. Let m and n be relatively prime odd integers and $m^\star = m(-1)^{(m-1)/2}$. Then $(m^\star/n) = (n/m)$.

SUPPLEMENTARY EXERCISES

Let $Z_p^\star$ denote the set of least positive residues modulo an odd prime p. Let $f : Z_p^\star \to \{\pm 1\}$ defined by $f(a) = (a/p)$. (The function f is a fine example of a **homomorphism** in group theory.)

1. Prove that $f(ab) = f(a) \cdot f(b)$, where $p \nmid ab$.
2. Prove that f is a surjection.
3. Identify the set K of least positive residues a modulo p such that $f(a) = 1$. (K is the **kernel** of f.)
4. Find K when $p = 13$.

Let p be an odd prime and a any integer such that $p \nmid a$. Then a is a **cubic residue** if $x^3 \equiv a \pmod{p}$ is solvable; it is a **biquadratic residue** if $x^4 \equiv a \pmod{p}$ is solvable.

5. Find the cubic residues of 5 and 7.
6. Find the biquadratic residues of 7 and 11.
7. Prove that every biquadratic residue of p is also a quadratic residue.

8. Show that a quadratic residue of p need not be a biquadratic residue.
9. Gauss proved that -1 is a biquadratic residue of p if and only if $p \equiv 1 \pmod 8$. Verify that -1 is a biquadratic residue of 17, but a biquadratic nonresidue of 13.

Let a be a nonsquare positive integer $\equiv 0$ or $1 \pmod 4$, p an odd prime with $p \nmid a$, and n a positive integer with prime factorization $\prod_{i=1}^{k} p_i^{e_i}$, where $(a, n) = 1$. Then the **Kronecker symbol** (a/n), named for the German mathematician Leopold Kronecker (1823–1891), is defined as follows:

$$(a/2) = \begin{cases} 1 & \text{if } a \equiv 1 \pmod 8 \\ -1 & \text{if } a \equiv 5 \pmod 8 \end{cases}$$

$$(a/p) = \text{Legendre symbol } (a/p) = \begin{cases} 1 & \text{if } a \text{ is a quadratic residue modulo } p \\ -1 & \text{otherwise} \end{cases}$$

$$(a/n) = \text{Jacobi symbol } (a/n) = \prod_{i=1}^{k} (a/p_i)^{e_i}$$

In Exercises 10–15, (a/n) denotes the Kronecker symbol. Evaluate each.

10. $(108/239)$
11. $(85/2)$
12. $(28/153)$
13. $(85/3861)$

Prove each.

14. $(a/2) = (2/a)$, if $2 \nmid a$.
15. Let m and n be positive integers such that $(a, mn) = 1$. Then $(a/mn) = (a/m)(a/n)$.

COMPUTER EXERCISES

Write a program to perform each task, where p is an odd prime.

1. Read in p and a positive integer a relatively prime to p. Find the incongruent solutions of the quadratic congruence $x^2 \equiv a \pmod p$.
2. Read in p and list the quadratic residues of p.
3. Read in p and two positive integers a and b relatively prime to p. Determine if they are quadratic residues of p. If neither of them is a quadratic residue, use them to find one.
4. Read in p and an integer a, where $p \nmid a$. Evaluate (a/p) using
   (a) Euler's criterion.
   (b) The law of quadratic reciprocity.

5. Read in p and a positive integer such that $p \nmid a$.
   (a) Compute (a/p).
   (b) Let ν denote the number of least residues of the integers $a, 2a, 3a, \ldots, [(p-1)/2]a$ that exceed $p/2$. Find ν.
   (c) Is $(a/p) = (-1)^{\nu}$?
6. Read in two distinct odd primes p and q, and verify the law of quadratic reciprocity.
7. Read in a prime $p \equiv \pm 1 \pmod 8$.
   (a) Verify that $2^{(p-1)/2} \equiv 1 \pmod p$.
   (b) Find a prime factor of $2^{(p-1)/2} - 1$.
8. Read in two distinct odd primes p and q, and a positive integer a such that $(a, pq) = 1$. Using the solutions of the congruences $x^2 \equiv a \pmod p$ and $x^2 \equiv a \pmod q$, solve $x^2 \equiv a \pmod{pq}$.
9. Read in an odd positive integer m and an integer a such that $(a, m) = 1$. Evaluate the Jacobi symbol (a/m).
10. Read in two relatively prime odd integers, and verify Theorems 11.12 and 11.13.
11. Read in a positive integer n and an integer a such that $p \nmid a$. Determine if the congruence $x^2 \equiv a \pmod{p^n}$ is solvable.
12. Read in a positive integer n and an integer a such that $p \nmid a$. Solve the congruence $x^2 \equiv a \pmod{p^n}$.
13. Read in an odd integer a and an integer $n \geq 3$. Determine if the congruence $x^2 \equiv a \pmod{2^n}$ is solvable; if so, find its solutions.
14. Read in an integer $m \geq 2$ and an integer a such that $(a, m) = 1$. Solve the congruence $x^2 \equiv a \pmod m$, if possible.

ENRICHMENT READINGS

1. H. M. Edwards, "Euler and Quadratic Reciprocity," *Mathematics Magazine*, 56 (Nov. 1983), 285–291.
2. M. Gerstenhaber, "The 152nd Proof of Quadratic Reciprocity," *The American Mathematical Monthly*, 70 (1963), 397–398.
3. D. E. Rowe, "Gauss, Dirichlet, and the Law of Biquadratic Reciprocity," *The Mathematical Intelligencer*, 10 (1988), 13–25.
4. W. Watkins, "The Quadratic Residues −1 and −3," *The American Mathematical Monthly*, 107 (Dec. 2000), 934.

Continued Fractions

If I have seen farther than other men, it is because I have stood on the shoulders of giants.
—ISAAC NEWTON

This chapter explores fractions of a special nature that we do not encounter in everyday life, fractions such as

$$\frac{113}{77} = 1 + \cfrac{1}{2 + \cfrac{1}{3 + \cfrac{1}{4 + \cfrac{1}{5 + \cdots}}}}$$

Such a multi-layered fraction is a **continued fraction**, a term coined by the English mathematician John Wallis (1616–1703). His book, *Opera Mathematica* (1695) contains some basic work on continued fractions. Aryabhata used them to solve specific LDEs. Italian mathematician Rafael Bombelli (1526–1573) is often credited with laying the foundation for the theory of continued fractions, since he attempted to approximate $\sqrt{13}$ by such fractions in his *L'Algebra Opera* (1572). In 1613, Italian mathematician Pietro Antonio Cataldi (1548–1626) pursued approximating $\sqrt{18}$ by continued fractions. The Dutch physicist and mathematician Christiaan Huygens (1629–1695) investigated such fractions for the design of a mathematical model for the planets in his *Descriptio Automati Planetari* (1703).

Although these mathematicians made contributions to the development of continued fractions, the modern theory of such fractions did not flourish until Euler, Johan Heinrich Lambert (1728–1777), and Lagrange embraced the topic. Euler studied them around 1730 and his *De Fractionlous Continious* (1737) contains much of his work. In 1759, he employed them to solve equations of the form $x^2 - Ny^2 = 1$, called *Pell's equation* (see Section 13.1). Seven years later, Lagrange developed the fundamental properties of periodic continued fractions.

In 1931, D. H. Lehmer and R. E. Powers developed a factoring method based on continued fractions. M. A. Morrison and J. Brillhart demonstrated the power of this method by factoring f_7 in 1974.

We now study a brief introduction to continued fractions.

12.1 Finite Continued Fractions

A **finite continued fraction** is an expression of the form

$$x = a_0 + \cfrac{1}{a_1 + \cfrac{1}{a_2 + \cfrac{1}{\ddots + \cfrac{1}{a_{n-1} + \cfrac{1}{a_n}}}}} \tag{12.1}$$

where each a_i is a real number, $a_0 \geq 0$, $a_{i+1} > 0$ and $i \geq 0$. The numbers $a_1, a_2, \ldots, a_n$ are the **partial quotients** of the finite continued fraction. The fraction is *simple* if each a_i is an integer.

Since this notation is a bit cumbersome to manage, the fraction is often written as

$$[a_0; a_1, a_2, a_3, \ldots, a_n]$$

where $a_0 = \lfloor x \rfloor$ and the semicolon separates the fractional part from the integral part.

For example,

$$[1; 2, 3, 4, 5, 6] = 1 + \cfrac{1}{2 + \cfrac{1}{3 + \cfrac{1}{4 + \cfrac{1}{5 + \cfrac{1}{6}}}}}$$

$$= \frac{1393}{972}$$

Although it follows from the definition that every simple finite continued fraction represents a rational number, we shall now formally prove it, using induction.

THEOREM 12.1 Every finite simple continued fraction represents a rational number.

PROOF

(We shall apply induction on the number of partial quotients.) Let $[a_0; a_1, a_2, \dots, a_n]$ be a finite simple continued fraction. When $n = 1$,

$$[a_0; a_1] = a_0 + \frac{1}{a_1} = \frac{a_0 a_1 + 1}{a_1}$$

is a rational number.

Now assume that every finite simple continued fraction with k partial quotients is a rational number, where $k \geq 1$. Then

$$[a_0; a_1, a_2, \dots, a_k, a_{k+1}] = a_0 + \frac{1}{[a_1; a_2, \dots, a_k, a_{k+1}]}$$

Since $[a_1; a_2, \dots, a_{k+1}]$ contains k partial quotients, it is a rational number r/s, where $s \neq 0$. Then

$$[a_0; a_1, a_2, \dots, a_k, a_{k+1}] = a_0 + \frac{1}{r/s} = a_0 + \frac{s}{r} = \frac{a_0 r + s}{r}$$

is a rational number.

Thus, by induction, $[a_0; a_1, a_2, \dots, a_n]$ is a rational number for every positive integer n. ■

The following theorem shows that the converse is also true: Every rational number can be represented by a finite simple continued fraction. This was discovered by Euler. The proof invokes the euclidean algorithm from Section 3.2.

THEOREM 12.2 Every rational number can be represented by a finite simple continued fraction.

PROOF

Let $x = a/b$ be a rational number, where $b > 0$. For convenience, we let $r_0 = a$ and $r_1 = b$. By the euclidean algorithm, we have

$$\begin{aligned}
r_0 &= r_1 q_1 + r_2, & 0 &< r_2 < r_1 \\
r_1 &= r_2 q_2 + r_3, & 0 &< r_3 < r_2 \\
r_2 &= r_3 q_3 + r_4, & 0 &< r_4 < r_3 \\
&\ \vdots \\
r_{n-2} &= r_{n-1} q_{n-1} + r_n, & 0 &< r_n < r_{n-1} \\
r_{n-1} &= r_n q_n
\end{aligned}$$

where the quotients $q_2, q_3, \ldots, q_n$ and the remainders $r_2, r_3, \ldots, r_n$ are positive.

It follows from these equations that

$$\begin{aligned}
\frac{a}{b} = \frac{r_0}{r_1} &= q_1 + \frac{r_2}{r_1} &&= q_1 + \frac{1}{r_1/r_2} \\
\frac{r_1}{r_2} &= q_2 + \frac{r_3}{r_2} &&= q_2 + \frac{1}{r_2/r_3} \\
\frac{r_2}{r_3} &= q_3 + \frac{r_4}{r_3} &&= q_3 + \frac{1}{r_3/r_4} \\
&\vdots \\
\frac{r_{n-2}}{r_{n-1}} &= q_{n-1} + \frac{r_n}{r_{n-1}} &&= q_{n-1} + \frac{1}{r_{n-1}/r_n} \\
\frac{r_{n-1}}{r_n} &= q_n
\end{aligned}$$

Substituting for r_1/r_2 in the first equation yields

$$\frac{a}{b} = q_1 + \frac{1}{q_2 + r_3/r_2}$$

Now substitute for r_2/r_3:

$$\frac{a}{b} = q_1 + \cfrac{1}{q_2 + \cfrac{1}{q_3 + \cfrac{1}{r_3/r_4}}}$$

Continuing like this, we get

$$\begin{aligned}
\frac{a}{b} &= q_1 + \cfrac{1}{q_2 + \cfrac{1}{q_3 + \cfrac{1}{\ddots + \cfrac{1}{q_{n-1} + \cfrac{1}{q_n}}}}} \\
&= [q_1; q_2, q_3, \ldots, q_{n-1}, q_n]
\end{aligned}$$

Thus, every rational number can be represented by a finite simple continued fraction. ■

The following example illustrates this algorithm.

EXAMPLE 12.1 Express $\dfrac{225}{157}$ as a finite simple continued fraction.

SOLUTION

By the euclidean algorithm, we have

$$
\begin{aligned}
225 &= 1 \cdot 157 + 68\\
157 &= 2 \cdot \ \ 68 + 21\\
68 &= 3 \cdot \ \ 21 + 5\\
21 &= 4 \cdot \ \ \ \ 5 + 1\\
5 &= 5 \cdot \ \ \ \ 1
\end{aligned}
$$

Thus,

$$
\begin{aligned}
\frac{225}{157} &= 1 + \frac{68}{157} \qquad\qquad = 1 + \frac{1}{157/68}\\
&= 1 + \cfrac{1}{2 + \cfrac{21}{68}} \qquad = 1 + \cfrac{1}{2 + \cfrac{1}{68/21}}\\
&= 1 + \cfrac{1}{2 + \cfrac{1}{3 + \cfrac{5}{21}}} = 1 + \cfrac{1}{2 + \cfrac{1}{3 + \cfrac{1}{21/5}}}\\
&= 1 + \cfrac{1}{2 + \cfrac{1}{3 + \cfrac{1}{4 + \cfrac{1}{5}}}}\\
&= [1; 2, 3, 4, 5]
\end{aligned}
$$

The Jigsaw Puzzle Revisited

In Section 3.2, we saw that the euclidean algorithm yields a jigsaw puzzle. We will now show how the numbers of different-size squares in the puzzle yield an interesting dividend. To this end, suppose we would like to convert $\dfrac{23}{13}$ into a simple continued fraction. By the euclidean algorithm, we have

$$\begin{aligned} 23 &= 1 \cdot 13 + 10 \\ 13 &= 1 \cdot 10 + 3 \\ 10 &= 3 \cdot \ 3 + 1 \\ 3 &= 3 \cdot \ 1 \end{aligned}$$

So $\frac{23}{13} = [1; 1, 3, 3]$.

It follows from Figure 12.1 that the 23×13 rectangle can be cut up into one 13×13 square, one 10×10 square, three 3×3 squares, and three 1×1 squares. Notice that the numbers of squares of the various sizes are the partial quotients in the continued fraction.

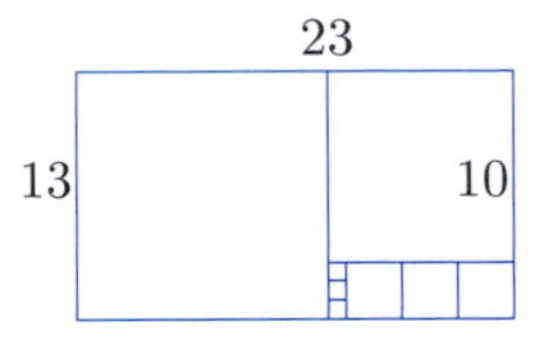

Figure 12.1

Suppose $a_n > 1$ in the finite simple continued fraction $[a_0; a_1, \ldots, a_n]$. Since $a_n = (a_n - 1) + \frac{1}{1}$, it follows that $[a_0; a_1, \ldots, a_n] = [a_0; a_1, \ldots, a_n - 1, 1]$. For example, $[1; 2, 3, 4, 5] = [1; 2, 3, 4, 1]$.

On the other hand, let $a_n = 1$. Then

$$[a_0; a_1, \ldots, a_n] = [a_0; a_1, \ldots, a_{n-1}, 1] = [a_0, a_1, \ldots, a_{n-1} + 1]$$

For example, $[1; 2, 3, 4, 1] = [1; 2, 3, 5]$.

Thus, every rational number can be written as a finite simple continued fraction in two different ways. In other words, the continued fraction representation of a rational number is not unique.

Next, we discuss approximations of continued fractions.

Convergents of a Continued Fraction

By truncating the continued fraction for $x = [a_0; a_1, \ldots, a_n]$ at the various plus signs [see equation (12.1)], we can generate a sequence $\{c_k\}$ of approximations of x, where $0 \le k \le n$; thus, $c_k = [a_0, a_1, \ldots, a_k]$; c_k is called the kth *convergent* of x, a concept introduced by Wallis in his *Opera Mathematica*.

For example, you may verify that

$$\frac{F_8}{F_7} = \frac{21}{13} = [1; 1, 1, 1, 1, 1, 1]$$

The various convergents are

$$\begin{aligned}
c_0 &= [1] &&= \frac{1}{1} = 1 \\
c_1 &= [1; 1] &&= \frac{2}{1} = 2 \\
c_2 &= [1; 1, 1] &&= \frac{3}{2} = 1.5 \\
c_3 &= [1; 1, 1, 1] &&= \frac{5}{3} \approx 1.6666666667 \\
c_4 &= [1; 1, 1, 1, 1] &&= \frac{8}{5} = 1.6 \\
c_5 &= [1; 1, 1, 1, 1, 1] &&= \frac{13}{8} = 1.625 \\
c_6 &= [1; 1, 1, 1, 1, 1, 1] &&= \frac{21}{13} \approx 1.6153846154
\end{aligned}$$

Some interesting observations:

- These convergents c_k approach the actual value $\frac{21}{13}$ as k increases, where $0 \le k \le 6$.
- The convergents c_{2k} approach it from below and the convergents c_{2k+1} from above; so the convergents are alternately less than and greater than $\frac{21}{13}$, except the last convergent; that is, $c_0 < c_2 < c_4 < \frac{21}{13} < c_5 < c_3 < c_1$; see Figure 12.2.

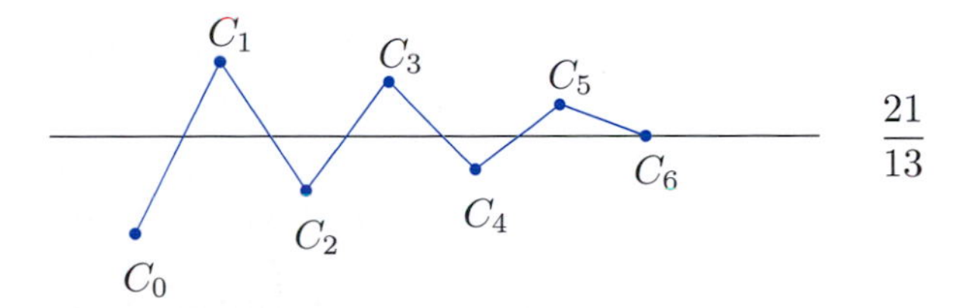

Figure 12.2

- The convergents display a remarkable pattern: $c_k = \frac{F_{k+2}}{F_{k+1}}$, $0 \le k \le 6$.

We shall return to these observations later.

We can facilitate the computation of the convergents $c_k = \frac{p_k}{q_k}$ by developing formulas for its numerator and denominator, as the next theorem shows.

THEOREM 12.3 The kth convergent of the finite simple continued fraction $[a_0; a_1, \ldots, a_n]$ is

$$c_k = \frac{p_k}{q_k}$$

where $2 \le k \le n$, and the sequences $\{p_k\}$ and $\{q_k\}$ are defined recursively as follows:

$$\begin{aligned} p_0 &= a_0 & q_0 &= 1 \\ p_1 &= a_0 a_1 + 1 & q_1 &= a_1 \\ p_k &= a_k p_{k-1} + p_{k-2} & q_k &= a_k q_{k-1} + q_{k-2} \end{aligned}$$

PROOF

We shall prove by induction that $c_k = \frac{p_k}{q_k}$ yields the kth convergent of the continued fraction for each value of k, where $0 \le k \le n$.

When $k = 0$,

$$c_0 = [a_0] = \frac{a_0}{1} = \frac{p_0}{q_0}$$

and when $k = 1$,

$$c_1 = [a_0; a_1] = a_0 + \frac{1}{a_1} = \frac{a_0 a_1 + 1}{a_1} = \frac{p_1}{q_1}$$

Thus, the theorem is true when $k = 0$ and $k = 1$.

Now assume that the formula for c_k works for an arbitrary integer m, where $2 \le m < n$. That is,

$$c_m = \frac{p_m}{q_m} = \frac{a_m p_{m-1} + p_{m-2}}{a_m q_{m-1} + q_{m-2}} \tag{12.2}$$

Then

$$\begin{aligned} c_{m+1} &= [a_0; a_1, \ldots, a_m, a_{m+1}] \\ &= \left[a_0; a_1, \ldots, a_{m-1}, a_m + \frac{1}{a_{m+1}}\right] \end{aligned}$$

Notice that the integers p_{m-1}, p_{m-2}, q_{m-1}, and q_{m-2} depend only on the partial quotients $a_0, a_1, \ldots, a_{m-1}$, and not on a_m. So the convergent c_{m+1} can be computed from formula (12.2) by replacing a_m with $a_m + \frac{1}{a_{m+1}}$:

$$c_{m+1} = \frac{\left(a_m + \dfrac{1}{a_{m+1}}\right)p_{m-1} + p_{m-2}}{\left(a_m + \dfrac{1}{a_{m+1}}\right)q_{m-1} + q_{m-2}}$$

$$= \frac{a_{m+1}(a_m p_{m-1} + p_{m-2}) + p_{m-1}}{a_{m+1}(a_m q_{m-1} + q_{m-2}) + q_{m-1}}$$

$$= \frac{a_{m+1}p_m + p_{m-1}}{a_{m+1}q_m + q_{m-1}}, \quad \text{by formula (12.2)}$$

Thus, by induction, the formula works for every value of k, where $0 \le k \le n$. ■

The following example illustrates this theorem.

EXAMPLE 12.2 Using Theorem 12.3, compute the convergents of the continued fraction $[2; 3, 1, 5] = \frac{52}{23}$.

SOLUTION

We have $a_0 = 2$, $a_1 = 3$, $a_2 = 1$, and $a_3 = 5$. First, we compute p_k and q_k for each k, where $0 \le k \le 3$:

$$p_0 = a_0 = 2 \qquad q_0 = 1$$
$$p_1 = a_0 a_1 + 1 = 2 \cdot 3 + 1 = 7 \qquad q_1 = a_1 = 3$$
$$p_2 = a_2 p_1 + p_0 = 1 \cdot 7 + 2 = 9 \qquad q_2 = a_2 q_1 + q_0 = 1 \cdot 3 + 1 = 4$$
$$p_3 = a_3 p_2 + p_1 = 5 \cdot 9 + 7 = 52 \qquad q_3 = a_3 q_2 + q_1 = 5 \cdot 4 + 3 = 23$$

Thus, the various convergents are

$$c_0 = \frac{p_0}{q_0} = \frac{2}{1} \qquad c_1 = \frac{p_1}{q_1} = \frac{7}{3}$$
$$c_2 = \frac{p_2}{q_2} = \frac{9}{4} \qquad c_3 = \frac{p_3}{q_3} = \frac{52}{23}$$

■

A table such as Table 12.1 can be used effectively to compute p_k and q_k, from which c_k can be computed. For example, $p_2 = 9 = 1 \cdot 7 + 2$ and $q_3 = 23 = 5 \cdot 4 + 3$.

k	**0**	**1**	**2**	**3**
a_k	2	3	1	5
p_k	2	7	9	52
q_k	1	3	4	23

Table 12.1

The next theorem reveals a Cassini-like relationship among the numerators and the denominators of two successive convergents of a finite simple continued fraction.

THEOREM 12.4 Let $c_k = \dfrac{p_k}{q_k}$ be the kth convergent of the simple continued fraction $[a_0; a_1, \ldots, a_n]$, where $1 \le k \le n$. Then $p_k q_{k-1} - q_k p_{k-1} = (-1)^{k-1}$.

PROOF **(by induction)**
Using the definitions of the sequences $\{p_k\}$ and $\{q_k\}$ in Theorem 12.3,

$$p_1 q_0 - q_1 p_0 = (a_0 a_1 + 1) \cdot 1 - a_1 a_0 = 1 = (-1)^{1-1}$$

So the formula works when $k = 1$.

Now assume that it is true for an arbitrary positive integer $k < n$:

$$p_k q_{k-1} - q_k p_{k-1} = (-1)^{k-1}$$

Then, by the recursive definition of p_k and q_k,

$$\begin{aligned} p_{k+1} q_k - q_{k+1} p_k &= (a_{k+1} p_k + p_{k-1}) q_k - (a_{k+1} q_k + q_{k-1}) p_k \\ &= -(p_k q_{k-1} - q_k p_{k-1}) \\ &= -(-1)^{k-1}, \quad \text{by the inductive hypothesis} \\ &= (-1)^k \end{aligned}$$

So the formula works for $k + 1$ also. Thus, by induction, the theorem is true for every positive integer $\le n$. ■

The following example illustrates this theorem.

EXAMPLE 12.3 Verify Theorem 12.4 using the convergents of the continued fraction $[2; 3, 1, 5]$.

SOLUTION
Using Example 12.2, we have

$$\begin{aligned} p_1 q_0 - q_1 p_0 &= 7 \cdot 1 - 3 \cdot 2 = 1 = (-1)^{1-1} \\ p_2 q_1 - q_2 p_1 &= 9 \cdot 3 - 4 \cdot 7 = -1 = (-1)^{2-1} \\ p_3 q_2 - q_3 p_2 &= 52 \cdot 4 - 23 \cdot 9 = 1 = (-1)^{3-1} \end{aligned}$$

Thus, $p_k q_{k-1} - q_k p_{k-1} = (-1)^{k-1}$ for every value of k, where $1 \le k \le 3$. ■

As in the case of Cassini's formula, Theorem 12.4 has an interesting byproduct about p_k and q_k, as the next corollary shows.

COROLLARY 12.1 Let $c_k = \dfrac{p_k}{q_k}$ be the kth convergent of the simple continued fraction $[a_0; a_1, \ldots, a_n]$. Then $(p_k, q_k) = 1$, where $1 \le k \le n$.

PROOF
Let $d = (p_k, q_k)$. Since $p_k q_{k-1} - q_k p_{k-1} = (-1)^{k-1}$, it follows by Theorem 2.4 that $d | (-1)^{k-1}$. But $d > 0$, so $d = 1$. Thus, $(p_k, q_k) = 1$, as desired. ■

For example, consider the convergents of the continued fraction $[2; 3, 1, 5]$ in Example 12.2. Notice that $(p_1, q_1) = (7, 3) = 1$; $(p_2, q_2) = (9, 4) = 1$; and $(p_3, q_3) = (52, 23) = 1$.

The following corollary shows another interesting consequence of Theorem 12.4.

COROLLARY 12.2 Let $c_k = \dfrac{p_k}{q_k}$ be the kth convergent of the simple continued fraction $[a_0; a_1, \ldots, a_n]$, where $1 \le k \le n$. Then

$$c_k - c_{k-1} = \frac{(-1)^{k-1}}{q_k q_{k-1}}$$

PROOF
By Theorem 12.3,

$$\begin{aligned} c_k - c_{k-1} &= \frac{p_k}{q_k} - \frac{p_{k-1}}{q_{k-1}} \\ &= \frac{p_k q_{k-1} - q_k p_{k-1}}{q_k q_{k-1}} \\ &= \frac{(-1)^{k-1}}{q_k q_{k-1}}, \quad \text{by Theorem 12.4} \end{aligned}$$ ■

This leads us to following result.

COROLLARY 12.3 Let $c_k = \dfrac{p_k}{q_k}$ be the kth convergent of the simple continued fraction $[a_0; a_1, \ldots, a_n]$. Then

$$c_k - c_{k-2} = \frac{a_k(-1)^k}{q_k q_{k-2}}$$

where $2 \le k \le n$.

PROOF
By Corollary 12.2,

$$c_k - c_{k-1} = \frac{(-1)^{k-1}}{q_k q_{k-1}}$$

and

$$c_{k-1} - c_{k-2} = \frac{(-1)^{k-2}}{q_{k-1}q_{k-2}}$$

Adding these two equations,

$$\begin{aligned} c_k - c_{k-2} &= \frac{(-1)^{k-1}}{q_k q_{k-1}} + \frac{(-1)^{k-2}}{q_{k-1}q_{k-2}} \\ &= \frac{(-1)^{k-2}(q_k - q_{k-2})}{q_k q_{k-1} q_{k-2}} \\ &= \frac{(-1)^{k-2}(a_k q_{k-1})}{q_k q_{k-1} q_{k-2}}, \quad \text{by Theorem 12.3} \\ &= \frac{(-1)^k a_k}{q_k q_{k-2}} \end{aligned}$$

■

We are now ready to confirm the observation made earlier (in Figure 12.2).

THEOREM 12.5 Let $c_k = \dfrac{p_k}{q_k}$ be the kth convergent of the simple continued fraction $[a_0; a_1, \dots, a_n]$. Then $c_{2i} < c_{2i+2}$, $c_{2i+3} < c_{2i+1}$, and $c_{2i} < c_{2j+1}$, where $i, j \geq 0$.

PROOF

By Corollary 12.3,

$$c_k - c_{k-2} = \frac{a_k(-1)^k}{q_k q_{k-2}}$$

where $2 \leq k \leq n$. Let $k = 2i + 2$, where $i \geq 0$. This equation yields

$$c_{2i+2} - c_{2i} = \frac{a_{2i+2}(-1)^{2i+2}}{q_{2i+2}q_{2i}}$$

Since $a_{2i+2}, q_{2i+2}, q_{2i} > 0$, the RHS is positive; so $c_{2i} < c_{2i+2}$; thus $c_0 < c_2 < c_4 < \cdots$.

Likewise, by letting $k = 2i + 3$ in Corollary 12.3, we get $c_{2i+3} < c_{2i+1}$; thus $c_1 > c_3 > c_5 > \cdots$.

Finally, by Corollary 12.2, $c_{2s} < c_{2s-1}$; so $c_{2i+2j} < c_{2i+2j-1}$. But $c_{2i+2j-1} < c_{2j-1}$. Thus $c_{2i} < c_{2i+2j} < c_{2i+2j-1} < c_{2j-1}$, so $c_{2i} < c_{2j-1}$ for every $i, j \geq 0$. In words, every even-numbered convergent is less than every odd-numbered convergent. ■

It follows from the theorem that

$$c_0 < c_2 < c_4 < \cdots < c_5 < c_3 < c_1$$

For example, returning to the convergents of the continued fraction $\frac{21}{13} = [1; 1, 1, 1, 1, 1, 1]$, notice that $c_0 < c_2 < c_4 < c_6, c_1 < c_3 < c_5$, and $c_0 < c_2 < c_4 < c_6 < c_5 < c_3 < c_1$.

Next we present an interesting application of finite simple continued fractions to LDEs.

Continued Fractions and LDEs

Recall from Theorem 4.1 that the LDE $ax + by = c$ is solvable if and only if $d \mid c$, where $d = (a, b)$. If x_0, y_0 is a particular solution, then it has infinitely many solutions $x = x_0 + (b/d)t, y = y_0 - (a/d)t$.

Continued fractions can be employed to solve LDEs. To see this, first consider the LDE $ax + by = 1$, where $b > 0$ and $(a, b) = 1$. Since a/b is a rational number, by Theorem 12.2, it can be represented by a continued fraction $[a_0; a_1, \ldots, a_n]$. Then

$$c_n = \frac{p_n}{q_n} = \frac{a}{b}$$

Since $(p_n, q_n) = 1 = (a, b)$, it follows that $a = p_n$ and $b = q_n$.

By Theorem 12.4, $p_n q_{n-1} - q_n p_{n-1} = (-1)^{n-1}$; so $aq_{n-1} - bp_{n-1} = (-1)^{n-1}$. When n is odd, it becomes $aq_{n-1} + b(-p_{n-1}) = 1$; so $x_0 = q_{n-1}, y_0 = -p_{n-1}$ is a solution of the LDE $ax + by = 1$. On the other hand, when n is even, it becomes $a(-q_{n-1}) + bp_{n-1} = 1$; so $x_0 = -q_{n-1}, y_0 = p_{n-1}$ is a solution.

When x_0, y_0 is a solution of the LDE $ax + by = 1$, $ax_0 + by_0 = 1$; so $a(cx_0) + b(cy_0) = c$; thus, cx_0, cy_0 is a particular solution of the LDE $ax + by = c$.

The following example illustrates this technique.

EXAMPLE 12.4 Using continued fractions, solve Mahavira's puzzle in Example 3.17.

SOLUTION

By Example 3.17, we have $63x - 23y = -7$; that is, $(-63)x + 23y = 7$. First, we find a particular solution of the LDE $(-63)x + 23y = 1$, where $(-63, 23) = 1$. To this end, we express $\frac{-63}{23}$ as a continued fraction, using the euclidean algorithm:

$$\begin{aligned} -63 &= (-3) \cdot 23 + 6 \\ 23 &= 3 \cdot 6 + 5 \\ 6 &= 1 \cdot 5 + 1 \\ 5 &= 5 \cdot 1 \end{aligned}$$

So $\frac{-63}{23} = [-3; 3, 1, 5]$. Then $c_2 = \frac{p_2}{q_2} = \frac{-11}{4}$ and $c_3 = \frac{p_3}{q_3} = \frac{-63}{23}$; so $p_2 = -11, q_2 = 4, p_3 = -63$, and $q_3 = 23$.

By Theorem 12.4, $p_3q_2 - q_3p_2 = (-1)^{3-1}$; that is, $(-63)\cdot 4 + 23\cdot 11 = 1$. Consequently, $x_0 = 4$, $y_0 = 11$ is a particular solution of the LDE $(-63)x + 23y = 1$. Therefore, $7x_0 = 28$, $7y_0 = 77$ is a particular solution of the LDE $(-63)x + 23y = 7$. So, by Theorem 3.20, its general solution is $x = 7x_0 + bt = 28 + 23t$, $y = 7y_0 - at = 77 + 63t$. (Notice that this is consistent with the solution in Example 3.17.) ■

EXERCISES 12.1

Rewrite each as a finite simple continued fraction.

1. $\frac{57}{23}$
2. $\frac{1199}{199}$
3. $\frac{-43}{17}$
4. $\frac{89}{55}$

Represent each continued fraction as a rational number.

5. $[1; 2, 3, 4, 5]$
6. $[1; 1, 1, 1, 1, 1]$
7. $[-3; 5, 4, 3, 2]$
8. $[5; 4, 3, 2, 1]$

Use Figure 12.3 to answer Exercises 9 and 10, where each smallest square is a 1×1 square. (R. Knott)

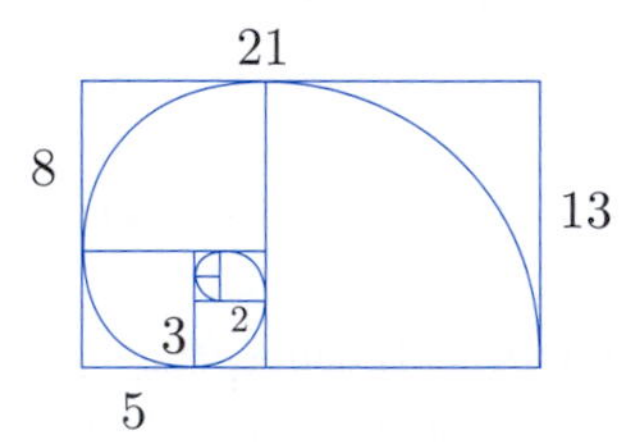

Figure 12.3

9. Find the continued fraction corresponding to the splitting up of the rectangle into squares.
10. Find the rational number (in lowest terms) represented by the continued fraction.

Using Theorem 12.3, compute the convergents of each continued fraction.

11. $[1; 1, 1, 1, 1, 1, 1]$
12. $[0; 2, 7, 7, 6]$
13. $[3; 1, 4, 2, 7]$
14. $[-2; 5, 4, 7, 1]$

The second and third convergents of the continued fraction $[1; 2, 3, 4, 5, 6, 7, 8]$ are $10/7$ and $43/30$, respectively. Using them, find each convergent:

15. c_4
16. c_6

The fourth and fifth convergents of the continued fraction $[1; 1, 1, 1, 1, 1, 1, 3]$ are $8/5$ and $13/8$, respectively. Using them, compute each convergent.

17. c_6
18. c_7

19. Let c_n denote the nth convergent of the finite continued fraction $[1; 1, 1, 1, \dots, 1]$. Prove that $c_n = \frac{F_{n+2}}{F_{n+1}}$, where $n \geq 1$.
20. Using the finite continued fraction $[1; 1, 1, 1, \dots, 1]$, prove that $F_{n+2}F_n - F_{n+1}^2 = (-1)^{n+1}$, where $n \geq 1$.

Let $c_k = \frac{p_k}{q_k}$ be the kth convergent of the simple continued fraction $[1; 2, 3, 4, 5, 6]$, where $1 \leq k \leq n$. Express each as a simple continued fraction.

21. $\frac{p_4}{p_3}$
22. $\frac{p_5}{p_4}$
23. $\frac{q_4}{q_3}$
24. $\frac{q_5}{q_4}$

25. Using Exercises 21 and 22, make a conjecture about $\frac{p_k}{p_{k-1}}$, where $\frac{p_k}{q_k}$ denotes the kth convergent of the simple continued fraction $[a_0; a_1, \dots, a_n]$, where $2 \leq k \leq n$.
26. Using Exercises 23 and 24, make a conjecture about $\frac{q_k}{q_{k-1}}$, where $\frac{p_k}{q_k}$ denotes the kth convergent of the simple continued fraction $[a_0; a_1, \dots, a_n]$, where $2 \leq k \leq n$.

Using the continued fraction of the given rational number r, find the continued fraction of $1/r$.

27. $57/23$
28. $1199/199$

29. Let r be a rational number with simple continued fraction $[a_0; a_1, \dots, a_n]$. Using Exercises 25 and 26, conjecture the continued fraction for $1/r$.
30. Establish the conjecture in Exercise 29.

Solve each LDE, using continued fractions.

31. $12x + 13y = 14$
32. $28x + 91y = 119$
33. $1776x + 1976y = 4152$
34. $1076x + 2076y = 3076$

12.2 Infinite Continued Fractions

Suppose there are infinitely many terms in the expression $[a_0; a_1, \ldots, a_n, \ldots]$, where $a_0 \geq 0$ and $a_i > 0$ for $i \geq 1$. Such a fraction is an *infinite continued fraction*. In particular, if each a_j is an integer, then it is an *infinite simple continued fraction*.

More generally, an infinite simple continued fraction is of the form

$$a_0 + \cfrac{b_1}{a_1 + \cfrac{b_2}{a_2 + \cfrac{b_3}{a_3 + \cdots}}}$$

where $a_0 \geq 0$, and a_i and b_{i+1} are integers for each i.

An interesting example of such a continued fraction is the identity for $\frac{4}{\pi}$, discovered in 1655 by Lord William V. Brouncker (1620–1684), the first president of the Royal Society. He discovered it by converting Wallis' celebrated infinite product

$$\frac{4}{\pi} = \frac{3 \cdot 3 \cdot 5 \cdot 5 \cdot 7 \cdot 7 \cdots}{2 \cdot 4 \cdot 4 \cdot 6 \cdot 6 \cdots}$$

into a continued fraction:

$$\frac{4}{\pi} = 1 + \cfrac{1^2}{2 + \cfrac{3^2}{2 + \cfrac{5^2}{2 + \cfrac{7^2}{2 + \cdots}}}}$$

This is the first recorded infinite continued fraction, but Brouncker did not provide a proof; it was given by Euler in 1775.

An infinite continued fraction for $\frac{\pi}{4}$ is

$$\frac{\pi}{4} = 1 + \cfrac{1^2}{3 + \cfrac{2^2}{5 + \cfrac{3^2}{7 + \cfrac{4^2}{9 + \cdots}}}}$$

In 1999, L. J. Lange of the University of Missouri developed an equally fascinating continued fraction for π:

$$\pi = 3 + \cfrac{1^2}{6 + \cfrac{3^2}{6 + \cfrac{5^2}{6 + \cfrac{7^2}{6 + \cdots}}}}$$

For convenience, we restrict our discussion to infinite simple continued fractions, where $b_i = 1$. The simplest such continued fraction is $[1; 1, 1, 1, \ldots]$.

One of the most astounding continued fractions was developed by the Indian mathematical genius Srinivasa Ramanujan, who studied them in 1908:

$$\left(\sqrt{\sqrt{5}\alpha - \alpha}\right)e^{2\pi/5} = \left[0; e^{-2\pi}, e^{-4\pi}, e^{-6\pi}, \ldots\right]$$

where α denotes the golden ratio. When Ramanujan communicated this marvelous result to Hardy in his first letter to him in 1913, Hardy was stunned by the discovery and could not derive it himself. Equally intriguing is its reciprocal:

$$\frac{e^{-2\pi/5}}{\sqrt{\sqrt{5}\alpha - \alpha}} = \left[1; e^{-2\pi}, e^{-4\pi}, e^{-6\pi}, \ldots\right]$$

Ramanujan discovered about 200 such infinite continued fractions.

How do we evaluate infinite simple continued fractions? We will answer this gradually. First, notice that although the continued fraction $[a_0; a_1, a_2, \ldots]$ is infinite, the convergents

$$c_n = [a_0; a_1, \ldots, a_n]$$

are finite, and hence represent rational numbers, so the properties of convergents from the previous section can be applied to these convergents also. Since

$$c_0 < c_2 < c_4 < \cdots < c_5 < c_3 < c_1$$

***Srinivasa Aiyangar Ramanujan** (1887–1920), the greatest Indian mathematician, was born in Erode, near Madras, the son of a bookkeeper at a cloth store in Kumbakonam. After two years of elementary school, he transferred to the high school at age seven. At ten, he placed first in the district primary examination. In 1903, his passion for mathematics was sparked when he borrowed a copy of George Schoobridge Carr's* A Synopsis of Elementary Results in Pure and Applied Mathematics *from a university student. Without any formal training or outside help, Ramanujan established the 6000 theorems in the book, stated without proofs or any explanation, and kept their proofs in a notebook.*

Graduating from high school in 1904, he entered the University of Madras on a scholarship. However, his neglect of all subjects except mathematics caused Ramanujan to lose the scholarship after a year, and he dropped out of college. He returned to the University after traveling through the countryside, but never graduated. During this period, he pursued his passion, rediscovering previously known results and discovering new ones in hypergeometric series and elliptic functions.

His marriage in 1909 compelled him to earn a living. Three years later, he secured a low-paying clerk's job with the Madras Port Trust. He published his first article in 1911 on Bernoulli numbers in the Journal of the Indian Mathematical Society *and two more the following year.*

In 1913, Ramanujan began corresponding with the eminent English mathematician Godfrey H. Hardy of Cambridge University. His first letter included more than 100 theorems, some without proofs. After examining them carefully, Hardy concluded that "they could only be written down by a mathematician of the highest class; they must be true because if they were not true, no one would have the imagination to invent them."

Ramanujan arrived in Cambridge in 1914 with the help of a scholarship arranged by Hardy. During his five-year stay, he and Hardy collaborated on a number of articles in the theory of partitions, analytic number theory, continued fractions, infinite series, and elliptic functions.

In 1917, Ramanujan became seriously ill. He was incorrectly diagnosed with tuberculosis; however, it is now believed that he suffered from a vitamin deficiency caused by his strict vegetarianism.

When Ramanujan was sick in a nursing home, Hardy visited him. Hardy told him that the number of the cab he came in, 1729, was a "rather dull number" and he hoped that it wasn't a bad omen. "No, sir," Ramanujan responded. "It is a very interesting number. It is the smallest number expressible as the sum of two cubes in two different ways."

In 1918, Ramanujan became one of the youngest members of the Fellow of the Royal Society and a fellow of Trinity College.

Ramanujan returned to India the following year. He pursued his mathematical passion even on his deathbed. His short but extremely productive life ended when he was only 32.

by Theorem 12.5, the sequence $\{c_{2n}\}$ is an increasing sequence that is bounded above by c_1, and the sequence $\{c_{2n+1}\}$ is a decreasing sequence that is bounded below by c_0. Consequently, both sequences have limits; that is, as n approaches infinity, sequence $\{c_{2n}\}$ approaches a limit ℓ; and the sequence $\{c_{2n+1}\}$ approaches a limit ℓ'; thus,

$$\lim_{n\to\infty} c_{2n} = \ell \quad \text{and} \quad \lim_{n\to\infty} c_{2n+1} = \ell'$$

The next theorem shows that $\ell = \ell'$.

THEOREM 12.6 Let $c_k = [a_0; a_1, \dots, a_k]$ denote the kth convergent of the simple continued fraction $[a_0; a_1, a_2, \dots]$. Then

$$\lim_{n\to\infty} c_{2n} = \lim_{n\to\infty} c_{2n+1}$$

PROOF

By Corollary 12.2,

$$\begin{aligned} c_{2n+1} - c_{2n} &= \frac{(-1)^{2n}}{q_{2n+1}q_{2n}} = \frac{1}{q_{2n+1}q_{2n}} \\ &< \frac{1}{q_{2n}^2}, \quad \text{since } q_{2n+1} > q_{2n} \end{aligned}$$

As n gets larger and larger, q_n and hence q_n^2 get larger and larger; then $\frac{1}{q_n^2}$ gets smaller and smaller, but never negative. So $\lim_{n\to\infty}(c_{2n+1} - c_{2n}) = 0$.

Thus,

$$\begin{aligned} \lim_{n\to\infty} c_{2n+1} - \lim_{n\to\infty} c_{2n} &= \lim_{n\to\infty}(c_{2n+1} - c_{2n}) \\ &= 0 \end{aligned}$$

So the two limits are equal. ■

It follows from this theorem that the sequences $\{c_{2n}\}$ and $\{c_{2n+1}\}$ of convergents of the continued fraction $[a_0; a_1, a_2, \dots]$ approaches a unique limit ℓ. This common limit is the *value* of the continued fraction:

$$\ell = \lim_{n\to\infty} c_n = [a_0; a_1, a_2, \dots]$$

For example, let c_n denote the nth convergent of the continued fraction $[1; 1, 1, 1, \dots]$, where $n \ge 0$. Then $c_n = \frac{F_{n+2}}{F_{n+1}}$; this can be established using induction (see Exercise 13). Thus,

$$c_n = \frac{p_n}{q_n} = \frac{F_{n+2}}{F_{n+1}}$$

[This relationship was first observed in 1753 by the English mathematician Robert Simson (1687–1768).] Since

$$\lim_{n\to\infty} c_n = \lim_{n\to\infty} \frac{F_{n+2}}{F_{n+1}}$$

(see Exercise 14), the sequence $\{c_n\}$ converges to the golden ratio α; that is, $[1; 1, 1, 1, \ldots] = \alpha$.

We can establish this fact by using an alternate route, without employing convergents. To this end, let $x = [1; 1, 1, 1, \ldots]$. Then $[1; 1, 1, 1, \ldots] = [1; [1; 1, 1, 1, \ldots]]$, so $x = [1; x] = 1 + \frac{1}{x}$. Then $x^2 - x - 1 = 0$. Solving it, we get $x = \frac{1 \pm \sqrt{5}}{2}$, but $x > 0$; so $x = \frac{1+\sqrt{5}}{2} = \alpha$; see Figure 12.4.

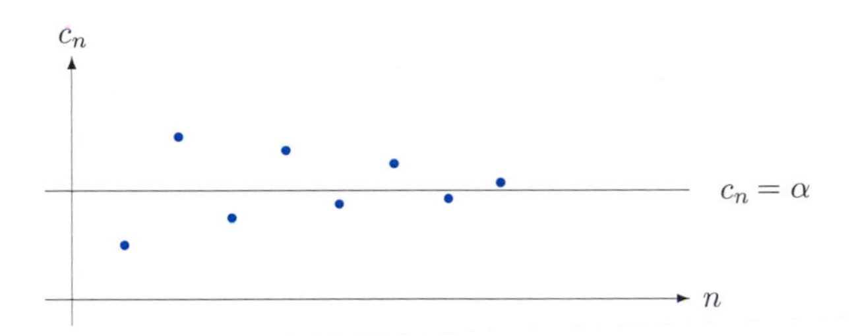

Figure 12.4

In Theorem 12.1, we proved that every finite simple continued fraction represents a rational number. We now show that every infinite simple continued fraction represents an irrational number.

THEOREM 12.7 The infinite simple continued fraction $[a_0; a_1, a_2, \ldots]$ represents an irrational number.

PROOF **(by contradiction)**

Let $x = [a_0; a_1, a_2, \ldots]$. Then x is the limit of the sequence $\{c_n\}$ of convergents $c_n = \frac{p_n}{q_n} = [a_0; , a_1, \ldots, a_n]$. Since

$$c_0 < c_2 < \cdots < c_{2n} < \underset{\substack{\uparrow \\ x}}{\cdots} < c_{2n+1} < \cdots < c_3 < c_1$$

it follows that $c_{2n} < x < c_{2n+1}$; so

$$0 < x - c_{2n} < c_{2n+1} - c_{2n}$$

By Corollary 12.2, this implies that

$$0 < x - \frac{p_{2n}}{q_{2n}} < \frac{1}{q_{2n+1}q_{2n}}$$

Suppose x is a rational number a/b, where $b > 0$. Then

$$0 < \frac{a}{b} - \frac{p_{2n}}{q_{2n}} < \frac{1}{q_{2n+1}q_{2n}}$$

That is,

$$0 < aq_{2n} - bp_{2n} < \frac{b}{q_{2n+1}}$$

Thus, $aq_{2n} - bp_{2n}$ is a positive integer $< \dfrac{b}{q_{2n+1}}$. But as n gets larger and larger, so does q_{2n+1}. Hence, there is an integer k such that $q_{2k+1} > b$, that is, $\dfrac{b}{q_{2k+1}} < 1$. Then $0 < aq_{2k} - bp_{2k} < 1$. This implies that $aq_{2k} - bp_{2k}$ is a positive integer < 1. Since this is impossible, x cannot be a rational number; in other words, x is an irrational number. ■

The next theorem shows that every irrational number x can be represented by an infinite simple continued fraction. Therefore, every irrational number can be approximated by a rational number. It provides an algorithm for constructing such a continued fraction and uses the floor function from Section 1.1. The proof is a bit long, so follow it carefully.

THEOREM 12.8 Let $x = x_0$ be an irrational number. Define the sequence $\{a_k\}_{k=0}^{\infty}$ of integers a_k recursively as follows:

$$a_k = \lfloor x_k \rfloor, \qquad x_{k+1} = \frac{1}{x_k - a_k}$$

where $k \geq 0$. Then $x = [a_0; a_1, a_2, \ldots]$.

PROOF

It follows from the recursive definition that a_k is an integer for every k.

We now establish by induction that x_k is an irrational number for every $k \geq 0$. To this end, first notice that x_0 is irrational and $a_0 = \lfloor x_0 \rfloor \neq x_0$. Besides, since x_0 is irrational, $x_0 - a_0$ is irrational; so $x_1 = \dfrac{1}{x_0 - a_0}$ is an irrational number.

Now assume that x_k is irrational for an arbitrary integer $k \geq 0$. Then $x_k - a_k$ and hence $\dfrac{1}{x_k - a_k}$ are irrational; that is, x_{k+1} is irrational. Thus, by induction, every x_k is an irrational number.

Next we will show that $a_k \geq 1$ for every $k \geq 1$. Since a_k is an integer and x_k is an irrational number, $a_k \neq x_k$; so $x_k - a_k > 0$. But $x_k - a_k = x_k - \lfloor x_k \rfloor < 1$; so $0 < x_k - a_k < 1$. Hence

$$x_{k+1} = \frac{1}{x_k - a_k} > 1$$

Consequently, $a_{k+1} = \lfloor x_{k+1} \rfloor \geq 1$ for every $k \geq 0$. That is, the integers $a_1, a_2, \ldots$ are all positive.

We will now show that $x = [a_0; a_1, a_2, \ldots]$. From the recursive formula

$$x_{k+1} = \frac{1}{x_k - a_k}$$

it follows that

$$x_k = a_k + \frac{1}{x_{k+1}}$$

where $k \geq 0$. Successively substituting for $x_1, x_2, x_3, \ldots$, this yields

$$\begin{aligned}
x_0 &= a_0 + \frac{1}{x_1} && = [a_0; x_1] \\
&= a_0 + \cfrac{1}{a_1 + \cfrac{1}{x_2}} && = [a_0; a_1, x_2] \\
&= a_0 + \cfrac{1}{a_1 + \cfrac{1}{a_2 + \cfrac{1}{x_3}}} && = [a_0; a_1, a_2, x_3] \\
&\vdots \\
&= a_0 + \cfrac{1}{a_1 + \cfrac{1}{a_2 + \cfrac{1}{\ddots + \cfrac{1}{a_n + \cfrac{1}{x_{n+1}}}}}} && = [a_0; a_1, a_2, \ldots, a_n, x_{n+1}]
\end{aligned}$$

where $n \geq 0$.

Finally, we must show that $x_0 = \lim_{n\to\infty} [a_0; a_1, \ldots, a_n, x_{n+1}]$. To this end, let $c_k = \frac{p_k}{q_k}$ denote the kth convergent of the continued fraction $[a_0; a_1, a_2, \ldots]$. Then, by Theorem 12.3,

$$\begin{aligned}
x_0 &= [a_0; a_1, \ldots, a_n, x_{n+1}] \\
&= \frac{x_{n+1}p_n + p_{n-1}}{x_{n+1}q_n + q_{n-1}}
\end{aligned}$$

Then

$$\begin{aligned} x_0 - c_n &= \frac{x_{n+1}p_n + p_{n-1}}{x_{n+1}q_n + q_{n-1}} - \frac{p_n}{q_n} \\ &= \frac{-(p_n q_{n-1} - q_n p_{n-1})}{(x_{n+1}q_n + q_{n-1})q_n} \\ &= \frac{(-1)^n}{(x_{n+1}q_n + q_{n-1})q_n}, \quad \text{by Theorem 12.4} \end{aligned}$$

Since $x_{n+1} > a_{n+1}$, this yields

$$\begin{aligned} |x_0 - c_n| &= \frac{1}{(x_{n+1}q_n + q_{n-1})q_n} \\ &< \frac{1}{(a_{n+1}q_n + q_{n-1})q_n} \\ &= \frac{1}{q_{n+1}q_n} \end{aligned}$$

As n gets larger and larger, so does q_n. Therefore, as n approaches infinity, $\dfrac{1}{q_{n+1}q_n}$ approaches zero; that is, $c_n \to x_0$ as $n \to \infty$. Thus,

$$x = x_0 = \lim_{n\to\infty} c_n = [a_0; a_1, a_2, \ldots]$$

as desired. ■

The next theorem shows that the infinite continued fraction representation of an irrational number is unique. We omit its proof in the interest of brevity; see Supplementary Exercise 4.

THEOREM 12.9 If $[a_0; a_1, a_2, \ldots]$ and $[b_0; b_1, b_2, \ldots]$ represent the same irrational number, then $a_k = b_k$ for every $k \geq 0$.

The following example illustrates the constructive algorithm in Theorem 12.8.

EXAMPLE 12.5 Express $\sqrt{13}$ as an infinite simple continued fraction.

SOLUTION

Let $x = x_0 = \sqrt{13}$. Then $a_0 = \lfloor\sqrt{13}\rfloor = 3$. By the recursive definition in Theorem 12.8, we have

$$x_1 = \frac{1}{x_0 - a_0} = \frac{1}{\sqrt{13}-3} = \frac{\sqrt{13}+3}{4} \qquad a_1 = \lfloor x_1 \rfloor = 1$$
$$x_2 = \frac{1}{x_1 - a_1} = \frac{\sqrt{13}-3}{4-\sqrt{13}} = \frac{\sqrt{13}+1}{3} \qquad a_2 = \lfloor x_2 \rfloor = 1$$
$$x_3 = \frac{1}{x_2 - a_2} = \frac{3}{\sqrt{13}-2} = \frac{\sqrt{13}+2}{3} \qquad a_3 = \lfloor x_3 \rfloor = 1$$
$$x_4 = \frac{1}{x_3 - a_3} = \frac{3}{\sqrt{13}-1} = \frac{\sqrt{13}+1}{4} \qquad a_4 = \lfloor x_4 \rfloor = 1$$
$$x_5 = \frac{1}{x_4 - a_4} = \frac{4}{\sqrt{13}-3} = \sqrt{13}+3 \qquad a_5 = \lfloor x_5 \rfloor = 6$$
$$x_6 = \frac{1}{x_5 - a_5} = \frac{1}{\sqrt{13}-3} = \frac{\sqrt{13}+3}{4} \qquad a_6 = \lfloor x_6 \rfloor = 1$$

Clearly, the pattern continues. Thus,

$$\sqrt{13} = [3;\, 1, 1, 1, 1, 6, 1, 1, 1, 1, 6, \ldots]$$

This is often written as

$$\sqrt{13} = \left[3;\, \overline{1, 1, 1, 1, 6}\right]$$

to indicate the periodic behavior. ■

An Infinite Continued Fraction for $\frac{1}{\alpha}$

In 1951, F. C. Ogg of Bowling Green State University discovered a sophisticated way of converting $\sqrt{5}-1$ into an infinite simple continued fraction:

$$\sqrt{5}-1 = 1+\sqrt{5}-2 = 1+\frac{1}{\sqrt{5}+2} = 1+\frac{1}{4+\sqrt{5}-2}$$
$$= 1+\cfrac{1}{4+\cfrac{1}{\sqrt{5}+2}}$$
$$= 1+\cfrac{1}{4+\cfrac{1}{4+\sqrt{5}-2}}$$

$$= 1 + \cfrac{1}{4 + \cfrac{1}{4 + \cfrac{1}{4 + \cdots}}}$$

$$= [1; 4, 4, 4, \ldots]$$

$$= [1; \overline{4}]$$

The various convergents of this continued fraction are $1, \frac{5}{4}, \frac{21}{17}, \frac{89}{72}, \frac{377}{305}, \ldots$. Now divide each by 2. The resulting numbers are $\frac{1}{2}, \frac{5}{8}, \frac{21}{34}, \frac{89}{144}, \frac{377}{610}, \ldots$; so the nth convergent of the continued fraction for $\frac{\sqrt{5}-1}{2}$ is $\frac{F_{3n+2}}{F_{3n+3}}$, where $n \geq 0$. Since

$$\lim_{n\to\infty} \frac{F_{3n+2}}{F_{3n+3}} = \lim_{n\to\infty} \frac{1}{F_{3n+3}/F_{3n+2}} = \frac{1}{\alpha} = \frac{\sqrt{5}-1}{2}$$

So the continued fraction for $\frac{\sqrt{5}-1}{2}$ is that of $\frac{1}{\alpha}$. Since $\alpha = [1; 1, 1, 1, \ldots]$, it follows that $\frac{1}{\alpha} = [0; 1, 1, 1, \ldots]$.

Using Theorem 12.8, the next example develops the infinite continued fraction for $e = 2.718281828\ldots$, the base of the natural logarithm. It was discovered by Euler in 1737.

EXAMPLE 12.6 Find the continued fraction expansion for e.

SOLUTION

We have $x_0 = e = 2.718281828\ldots$, so $a_0 = 2$. Using the algorithm in Theorem 12.8, we have

$$x_1 = \frac{1}{x_0 - a_0} = \frac{1}{0.7182818280\ldots} = 1.3922111920\ldots \qquad a_1 = 1$$

$$x_2 = \frac{1}{x_1 - a_1} = \frac{1}{0.3922111920\ldots} = 2.5496467725\ldots \qquad a_2 = 2$$

$$x_3 = \frac{1}{x_2 - a_2} = \frac{1}{0.5496467725\ldots} = 1.8193502627\ldots \qquad a_3 = 1$$

$$x_4 = \frac{1}{x_3 - a_3} = \frac{1}{0.8193502627\ldots} = 1.2204792571\ldots \qquad a_4 = 1$$

$$x_5 = \frac{1}{x_4 - a_4} = \frac{1}{0.2204792571\ldots} = 4.5355740627\ldots \qquad a_5 = 4$$

Continuing like this, we get

$$e = [2; 1, 2, 1, 1, 4, 1, 1, 6, 1, 1, 8, \ldots]$$

where the even partial quotients are separated by two 1s. ■

Euler also showed that

$$\frac{e-1}{e+1} = [0; 2, 6, 10, 14, 18, \ldots]$$

and

$$\frac{e^2-1}{e^2+1} = [0; 1, 3, 5, 7, 9, \ldots]$$

EXERCISES 12.2

Convert each into an infinite simple continued fraction.

1. $\sqrt{2}$
2. $\sqrt{3}$
3. $\sqrt{5}$
4. $\sqrt{12}$
5. π
6. $\sqrt{\pi}$

Compute the first five convergents of the continued fraction for each.

7. $\sqrt{2}$
8. $\sqrt{3}$
9. π
10. $\sqrt{\pi}$
11. Using the continued fraction $\pi = [3; 7, 15, 1, 292, 1, \ldots]$, compute the value of π correct to eight decimal places.
12. Using the continued fraction $e = [2; 1, 2, 1, 1, 4, 1, 1, 6, 1, 1, 8, \ldots]$, compute the value of e correct to six decimal places.
13. Let c_n denote the nth convergent of the continued fraction $[1; 1, 1, 1, 1, \ldots]$, where $n \geq 1$. Prove that $c_n = \dfrac{F_{n+2}}{F_{n+1}}$.
14. Prove that $\lim_{n\to\infty} \dfrac{F_{n+1}}{F_n} = \alpha$, the golden ratio.

Evaluate each simple infinite continued fraction.

15. $[F_n; F_n, F_n, F_n, \ldots]$
16. $[L_n; L_n, L_n, L_n, \ldots]$

CHAPTER SUMMARY

In this chapter, we presented a brief introduction to the theory of continued fractions. A continued fraction is simple if each partial quotient is an integer. We learned how to identify rational and irrational numbers, using their continued fraction representations.

Finite Simple Continued Fractions

- Compact notation $[a_0; a_1, \ldots, a_n]$ (p. 552)
- Every rational number can be represented by a finite simple continued fraction. (p. 553)
- $c_k = \frac{p_k}{q_k} = [a_0; a_1, \ldots, a_k]$ gives the kth convergent of the continued fraction. (p. 556)
- c_k can be defined recursively (p. 558):

$$\begin{aligned} p_0 &= a_0 & q_0 &= 1 \\ p_1 &= a_0 a_1 + 1 & q_1 &= a_1 \\ p_k &= a_k p_{k-1} + p_{k-2} & q_k &= a_k q_{k-1} + q_{k-2} \end{aligned}$$

- $p_k q_{k-1} - q_k p_{k-1} = (-1)^{k-1}$ (p. 560)
- $(p_k, q_k) = 1$ (p. 561)
- $c_k - c_{k-1} = \frac{(-1)^{k-1}}{q_k q_{k-1}}$ (p. 561)
- $c_k - c_{k-2} = \frac{a_k(-1)^k}{q_k q_{k-2}}$ (p. 561)
- $c_0 < c_2 < c_4 < \cdots < c_5 < c_3 < c_1$ (p. 562)
- Finite simple continued fractions can be used to solve LDEs. (p. 563)

Infinite Simple Continued Fractions

- Notation $[a_0; a_1, \ldots, a_n, \ldots]$ (p. 565)
- The sequence of convergents $\{c_n\}$ approaches a unique limit. (p. 568)
- An infinite simple continued fraction represents an irrational number. (p. 569)
- The continued fraction representation $[a_0; a_1, \ldots]$ of an irrational number $x = x_0$ can be found recursively:

$$a_k = \lfloor x_k \rfloor, \qquad x_{k+1} = \frac{1}{x_k - a_k} \quad \text{(p. 570)}$$

- Every irrational number has a unique continued fraction representation. (p. 572)

REVIEW EXERCISES

Rewrite each rational number as a continued fraction.

1. $\frac{47}{19}$
2. $\frac{-1023}{43}$

Represent each continued fraction as a rational number.

3. $[5; 4, 3, 2, 1]$
4. $[3; 1, 2, 1, 2, 1]$

The third and fourth convergents of the continued fraction $[0; 1, 3, 5, 7, 9, 11]$ are $16/21$ and $115/151$, respectively. Using them, compute each convergent.

5. c_5
6. c_6

Using the continued fraction for the given rational number r, find the continued fraction for $\frac{1}{r}$.

7. $\frac{25}{18}$
8. $\frac{464}{675}$

Using continued fractions, solve each LDE, if possible.

9. $43x + 23y = 33$
10. $33x + 55y = 93$
11. $76x + 176y = 276$
12. $365x + 185y = 135$

Using continued fractions, solve each Fibonacci LDE.

13. $144x + 89y = 23$
14. $233x - 144y = 19$

Rewrite each as an infinite simple continued fraction.

15. $\sqrt{7}$
16. $\sqrt{10}$
17. $\sqrt{e}$
18. $\lg 2$

19. Using the continued fraction

$$\frac{e-1}{e+1} = [0; 2, 6, 10, 14, 18, \ldots] \quad \text{(Euler, 1737)}$$

compute the value of $\frac{e-1}{e+1}$ correct to eight decimal places.

20. Using the continued fraction

$$\frac{e^2-1}{e^2+1} = [0; 1, 3, 5, 7, 9, \ldots] \quad \text{(Euler, 1737)}$$

compute the value of $\frac{e^2-1}{e^2+1}$ correct to eight decimal places.

21. Let r be a rational number < 1 with finite simple continued fraction $[a_0; a_1, \ldots, a_n]$. Prove that $\frac{1}{r} = [a_1; a_2, \ldots, a_n]$.
22. Using Cassini's formula, prove that $\lim_{n\to\infty}(c_n - c_{n-1}) = 0$, where c_n denotes the nth convergent of the continued fraction $[1; 1, 1, 1, 1, \ldots]$.

SUPPLEMENTARY EXERCISES

Let $\frac{p_k}{q_k}$ denote the kth convergent of the simple continued fraction $[a_0; a_1, \dots, a_n]$, where $a_0 > 0$ and $1 \le k \le n$. Prove each. (See Exercises 21–24 in Section 12.1.)

1. $\frac{p_k}{p_{k-1}} = [a_k; a_{k-1}, \dots, a_1, a_0]$
2. $\frac{q_k}{q_{k-1}} = [a_k; a_{k-1}, \dots, a_2, a_1]$
3. $q_k \ge 2^{k/2}$, where $k \ge 2$.
4. Theorem 12.9.

COMPUTER EXERCISES

Write a program to perform each task:

1. Express the rational numbers $\frac{3191}{2191}$ and $\frac{9587}{9439}$ as finite simple continued fractions.
2. Using continued fractions, compute the golden ratio correct to 50 decimal places.
3. Using continued fractions, solve the LDEs $5717x + 4799y = 3076$ and $9767x + 3919y = 6677$, if possible.
4. Using the continued fraction for e, compute e correct to 50 decimal places.
5. Using the continued fraction for π, compute it correct to 50 decimal places.
6. Using continued fractions, compute $\sqrt{1001}$ and $\sqrt{10001}$ correct to 50 decimal places.

ENRICHMENT READINGS

1. G. H. Hardy and E. M. Wright, *An Introduction to the Theory of Numbers*, 5th edition, Oxford, New York, 1995, 129–153.
2. I. Niven et al., *An Introduction to the Theory of Numbers*, 5th edition, Wiley, New York, 1991, 325–351.
3. C. S. Ogilvy and J. T. Anderson, *Excursions in Number Theory*, Dover, New York, 1988, 115–131.

Miscellaneous Nonlinear Diophantine Equations

And perhaps posterity will thank me for having shown it that the ancients did not know everything.
—PIERRE DE FERMAT

In this chapter we will deal with some important nonlinear diophantine equations and discover how to solve them, when possible. The most common nonlinear diophantine equations are $x^2+y^2=z^2$, $x^n+y^n=z^n$, and $x^2-Ny^2=\pm 1$. We will see how the solutions of $x^2+y^2=z^2$ and $x^2-Ny^2=\pm 1$ are related to Pythagorean triangles. We will also explore the celebrated Fermat's last theorem, as well as Beal's conjecture, which is related to it. We will also see that while some positive integers can be expressed as the sum of two squares, all can be written as the sum of four squares.

13.1 Pythagorean Triangles

The Pythagorean theorem is one of the most elegant and remarkable results in elementary mathematics. It states that the sum of the squares of the lengths of the legs of a right triangle equals the square of the length of its hypotenuse. Its converse is also true: If the sum of the squares of the lengths of two sides of a triangle equals the square of the length of its third side, then the triangle is a right triangle. Right triangles whose sides have integral lengths are called **Pythagorean triangles**.

This relationship was known even before Pythagoras. The Babylonian clay tablet (see Figure 13.1) in the G. A. Plympton Collection at Columbia University reveals that the Babylonians knew of the theorem more than 3500 years ago.

Figure 13.1 *Babylonian tablet.*

They described Pythagorean triangles with sides of the following lengths:

$$60, 45, 75; 72, 65, 97; 120, 119, 169; 360, 319, 481;$$

$$2700, 2291, 3541; 4800, 4601, 6649; 6480, 4961, 8161$$

The ancient Egyptians used right angles for surveying and resurveying their lands, the boundaries of which were often destroyed by Nile floods. According to S. J. Kolpas of Glendale Community College, around 2000 B.C., they discovered the simplest and universally known 3-4-5 Pythagorean triangle.

Around the same time, the ancient Indians also employed right angles. They used the 3-4-5 triangle and also found the 12-16-20, 5-12-13, and 8-15-17 triangles.

The Cairo Mathematical Papyrus (ca. 300 B.C.) is an Egyptian document that when it was examined in 1962 was found to contain 40 problems, 9 of which deal with the Pythagorean relation.

The ancient Greeks learned the technique of constructing the 3-4-5 triangle from the Egyptians. The Pythagoreans explored it and generalized it to all right triangles, resulting in the Pythagorean theorem.

It seems clear that the seed for the Pythagorean theorem was planted centuries before Pythagoras, although he is credited with its independent discovery and its first proof about 2500 years ago.

Interestingly, in modern times, six stamps featuring Pythagoras or the Pythagorean theorem have been issued; one by Nicaragua, one by Surinam, and four by Greece. The Greek stamp in Figure 13.2 provides a geometric illustration of the 3-4-5 Pythagorean triangle.

Numerous proofs of the Pythagorean theorem exist in mathematical literature, including one by James A. Garfield (1831–1881), who constructed a proof before he became the twentieth president of the United States. *The Pythagorean Proposition*, by E. S. Loomis, contains 230 different proofs of the theorem.

Figure 13.2 *A Greek stamp illustrating the Pythagorean theorem.*

Pythagorean Triples

Let x and y denote the lengths of the legs of a right triangle and z the length of its hypotenuse. Then, by the Pythagorean theorem, x, y, and z satisfy the diophantine equation

$$x^2 + y^2 = z^2 \tag{13.1}$$

The positive integral triplet x-y-z is called a **Pythagorean triple**. Thus, the task of finding all Pythagorean triangles is the same as that of finding all Pythagorean triples.

Clearly, 3-4-5 is a Pythagorean triple. Because $(3n)^2 + (4n)^2 = (5n)^2$, it follows that $3n$-$4n$-$5n$ is also a Pythagorean triple for every positive integer n. Thus, there are infinitely many Pythagorean triples.

For the curious-minded, there is a Pythagorean triplet that contains the beastly number: $216^2 + 630^2 = 666^2$; it can also be written as $(6 \cdot 6 \cdot 6)^2 + (666 - 6 \cdot 6)^2 = 666^2$, as observed by M. Keith in 2002.

Pythagorean Triples and Fibonacci Numbers

We can digress briefly to discuss how Fibonacci numbers can be used to construct Pythagorean triples. To this end, consider four consecutive Fibonacci numbers F_n, F_{n+1}, F_{n+2}, and F_{n+3}. Let $x = F_nF_{n+3}$, $y = 2F_{n+1}F_{n+2}$, and $z = F_{n+1}^2 + F_{n+2}^2$. Then $x^2 + y^2 = z^2$ (see Exercise 4), so x-y-z is a Pythagorean triple.

For example, let $n = 4$. Then $x = 3 \cdot 13 = 39$, $y = 2 \cdot 5 \cdot 8 = 80$, and $z = 5^2 + 8^2 = 89$. Since $39^2 + 80^2 = 89^2$, 39-80-89 is a Pythagorean triple.

Two Ancient Methods

Since $m^2 + [(m^2 - 1)/2]^2 = [(m^2 + 1)/2]^2$, where m is odd, this formula yields Pythagorean triples. The Pythagoreans are credited with its discovery.

A similar formula was discovered by the Greek philosopher Plato (ca. 427–347 B.C.):

$$(2m)^2 + (m^2 - 1)^2 = (m^2 + 1)^2$$

where m is any integer. This formula also yields infinitely many Pythagorean triples.

Unfortunately, neither of these two formulas yields all Pythagorean triples, so our task is to find them all, just as Euclid did in his *Elements*. To this end, we begin with a definition.

Primitive Pythagorean Triples

A Pythagorean triple x-y-z is **primitive** if $(x, y, z) = 1$.

For example, the Pythagorean triples 3-4-5 and 120-119-169 are primitive, whereas 6-8-10 and 60-45-75 are not.

In 1934, M. Willey of Mississippi and E. C. Kennedy of the University of Texas developed a delightful scheme for constructing mechanically any number of primitive Pythagorean triples, which manifest an intriguing pattern. The scheme appeared as a solution to a problem proposed in 1933 by Kennedy. Table 13.1 shows the scheme. Study the pattern, and add a few more lines to the pattern. See Exercises 5 and 6.

x	y	z
21	220	221
201	20200	20201
2001	2002000	2002001
20001	200020000	200020001
200001	2000020000	20000200001

Table 13.1

In our search for all Pythagorean triples, we first make an important observation. Let x-y-z be an arbitrary Pythagorean triple, where $(x, y, z) = d$. Then $x = du$, $y = dv$, and $z = dw$, where $(u, v, w) = 1$. Since $u^2 + v^2 = w^2$, u-v-w is also a Pythagorean triple. Thus, every Pythagorean triple is a multiple of a primitive Pythagorean triple.

In order to develop a formula that produces all primitive Pythagorean triples, we need to develop a series of lemmas.

LEMMA 13.1 If x-y-z is a primitive Pythagorean triple, then $(x, y) = (y, z) = (z, x) = 1$.

PROOF

Let x-y-z be a primitive Pythagorean triple and $(x, y) = d > 1$. Let p be a prime factor of d. Then $p|x$ and $p|y$. Since $x^2 + y^2 = z^2$, this implies $p|z^2$ and hence $p|z$.

Consequently, x-y-z is not a primitive Pythagorean triple, which is a contradiction. Thus, $(x, y) = 1$. Similarly, $(y, z) = 1 = (z, x)$. ■

The following lemma establishes that in a primitive Pythagorean triple x-y-z, exactly one of the numbers x and y is even; in other words, x and y have **opposite parity**.

LEMMA 13.2 If x-y-z is a primitive Pythagorean triple, then x and y have different parity.

PROOF

Suppose both x and y are even. Then $(x, y) \geq 2$. This violates Lemma 13.1, so both cannot be even.

Suppose both x and y are odd. Then $x^2 \equiv 1 \equiv y^2 \pmod 4$, so $z^2 = x^2 + y^2 \equiv 2 \pmod 4$. Since this is also impossible (this can be shown using the technique in Example 4.7), both cannot be odd, either.

Thus, exactly one of the integers x and y is even. ■

For example, 5-12-13 is a primitive Pythagorean triple, where 5 is odd and 12 is even; and 120-119-169 is a primitive Pythagorean triple, where 120 is even and 119 is odd.

This lemma has an immediate byproduct. Its proof can be completed as an exercise.

COROLLARY 13.1 If x-y-z is a primitive Pythagorean triple, where x is even, then y and z are odd. ■

We need one other lemma in order to find a formula for primitive Pythagorean triples. The following lemma, an application of the Fundamental Theorem of Arithmetic, proves that if the product of two relatively prime integers is a square, then both integers must be squares.

LEMMA 13.3 Let r and s be relatively prime integers such that rs is a square, then both r and s are also squares.

PROOF

Let $r = p_1^{e_1} p_2^{e_2} \cdots p_k^{e_k}$ and $s = q_1^{f_1} q_2^{f_2} \cdots q_l^{f_l}$ be the canonical decompositions of r and s, respectively. Since $(r, s) = 1$, the decompositions have no common prime factors; that is, $p_i \neq q_j$ for every i and j. Then, by the Fundamental Theorem of Arithmetic,

$$p_1^{e_1} p_2^{e_2} \cdots p_k^{e_k} q_1^{f_1} q_2^{f_2} \cdots q_l^{f_l}$$

is the prime-power decomposition of rs. Since rs is a square, it follows that each e_i and f_j must be even. Thus, both r and s are squares. ■

For example, $15{,}876 = 2^2 \cdot 3^4 \cdot 7^2$ and $75{,}625 = 5^4 \cdot 11^2$ are relatively prime; their product $15{,}876 \cdot 75{,}625 = (2 \cdot 3^2 \cdot 7 \cdot 5^2 \cdot 11)^2$ is a square; and so are $15{,}876 = (2 \cdot 3^2 \cdot 7)^2$ and $75{,}625 = (5^2 \cdot 11)^2$.

We are now ready to establish our main result, but we split its proof into two lemmas for the sake of brevity. Lemma 13.4 delineates the conditions all primitive Pythagorean triples must satisfy.

LEMMA 13.4 Let x-y-z be a primitive Pythagorean triple. Then there are relatively prime integers m and n with different parity such that $x = 2mn$, $y = m^2 - n^2$, and $z = m^2 + n^2$, where $m > n$.

PROOF

By Lemma 13.2, exactly one of the integers x and y is even. Without loss of generality, we assume that x is even; so both y and z are odd, by Corollary 13.1.

Then $z+y$ and $z-y$ are even, so $z+y = 2u$ and $z-y = 2v$ for some integers u and v. Therefore, $x^2 = z^2 - y^2 = (z+y)(z-y) = (2u)(2v) = 4uv$ and hence $(x/2)^2 = uv$.

We now claim that $(u, v) = 1$. To see this, let $d = (u, v) > 1$. Then $d|u$ and $d|v$; that is, $d|(z+y)$ and $d|(z-y)$. So $d|y$ and $d|z$, by Theorem 2.4. But this contradicts Lemma 13.1. Therefore, $(u, v) = 1$.

Since uv is a square, by Lemma 13.3, both u and v are squares. Thus, there are positive integers m and n such that $u = m^2$ and $v = n^2$. Then $x^2 = 4uv = 4m^2n^2$, so $x = 2mn$; $y = u - v = m^2 - n^2$; and $z = u + v = m^2 + n^2$. Since $y > 0$, $m > n$.

Next we show that $(m, n) = 1$. To this end, let $d = (m, n)$. Then $d|m$ and $d|n$, so $d|u$ and $d|v$, and hence $d|(u, v) = 1$. Thus, $d = (m, n) = 1$.

It remains to show that m and n have different parity. If both m and n are even, then both u and v would be even. This is impossible, since $(u, v) = 1$. If both m and n are odd, then $y = m^2 - n^2$ and $z = m^2 + n^2$ would be even. This is again a contradiction, since $(y, z) = 1$.

Thus, every primitive Pythagorean triple must satisfy the given conditions. ■

The integers m and n in this lemma are called the **generators** of the primitive Pythagorean triple x-y-z.

For example, consider the primitive Pythagorean triple 120-119-169. In this case $x = 120 = 2 \cdot 12 \cdot 5$, $y = 119 = 12^2 - 5^2$, and $z = 12^2 + 5^2$, where $12 > 5$; $(12, 5) = 1$; and 12 and 5 have opposite parity. The integers 12 and 5 are the generators of this primitive Pythagorean triple.

The following lemma establishes the converse of this lemma.

LEMMA 13.5 Let $x = 2mn$, $y = m^2 - n^2$, and $z = m^2 + n^2$, where m and n are relatively prime with different parity, and $m > n$. Then x-y-z is a primitive Pythagorean triple.

PROOF

Since

$$\begin{aligned} x^2 + y^2 &= (2mn)^2 + (m^2 - n^2)^2 \\ &= 4m^2n^2 + (m^4 + n^4 - 2m^2n^2) \\ &= (m^4 + n^4 + 2m^2n^2) = (m^2 + n^2)^2 \\ &= z^2 \end{aligned}$$

x-y-z is a Pythagorean triple.

So it remains to demonstrate that x-y-z is primitive. To confirm this, suppose that $(x, y, z) > 1$. Let p be a prime factor of (x, y, z). Then $p|y$ and $p|z$. Since m and n have different parity, both y and z are odd. So $p \neq 2$. Since $p|y$ and $p|z$, it follows that $p|2m^2$ and $p|2n^2$; but $p \neq 2$, so $p|m^2$ and $p|n^2$. Hence $p|m$ and $p|n$, so $(m, n) > 1$. Since this is a contradiction, it follows that $(x, y, z) = 1$. Thus, x-y-z is a primitive Pythagorean triple. ■

This lemma provides an algorithm for constructing a primitive Pythagorean triple, as the following example shows.

EXAMPLE 13.1 Let $m = 9$ and $n = 4$. Clearly, $m > n$; $(m, n) = 1$; and m and n have opposite parity. Therefore, by Lemma 13.5, x-y-z is a primitive Pythagorean triple, where $x = 2 \cdot 9 \cdot 4 = 72$, $y = 9^2 - 4^2 = 65$, and $z = 9^2 + 4^2 = 97$. ■

Combining Lemmas 13.4 and 13.5, we get the following elegant characterization of primitive Pythagorean triples.

THEOREM 13.1 Let x, y, and z be positive integers, where x is even. Then x-y-z is a primitive Pythagorean triple if and only if there are relatively prime integers m and n with different parity such that $x = 2mn$, $y = m^2 - n^2$, and $z = m^2 + n^2$, where $m > n$. ■

Table 13.2 lists all primitive Pythagorean triples with $m \leq 10$.

This table reveals some interesting patterns among the primitive Pythagorean triples x-y-z:

- Either x or y is divisible by 3.
- Exactly one of the numbers x, y, and z is divisible by 5.
- The number x is divisible by 4.
- The product of the lengths of the legs of a Pythagorean triangle is divisible by 12.
- The product of the lengths of the sides of a Pythagorean triangle is divisible by 60.

Generators		*Pythagorean Triples*		
m	n	$x = 2mn$	$y = m^2 - n^2$	$z = m^2 + n^2$
2	1	(4)	3	5
3	2	12	5	13
4	1	8	15	17
4	3	24	7	(25)
5	2	20	21	29
5	4	40	(9)	41
6	1	12	35	37
6	5	60	11	61
7	2	28	45	53
7	4	56	33	65
7	6	84	13	85
8	1	(16)	63	65
8	3	48	55	73
8	5	80	39	89
8	7	112	15	113
9	2	(36)	77	85
9	4	72	65	97
9	8	(144)	17	145
10	1	20	99	101
10	3	60	91	109
10	7	140	51	149
10	9	180	19	181

Table 13.2 *Pythagorean triples with* $m \leq 10$.

We can establish these results with little or no difficulty. See Exercises 15–19.

Next, we present certain Pythagorean triangles with special properties.

Intriguing Pythagorean Triangles

A close examination of Table 13.2 shows that the lengths of the legs of a **primitive Pythagorean triangle** can be consecutive integers. The triangles 3-4-5 and 20-21-29 are two such triangles. Such primitive Pythagorean triples x-y-z can be employed to construct a family of infinitely many primitive Pythagorean triples. See Exercise 30.

The table also shows that the lengths of the sides of a Pythagorean triangle can be squares. See the circled numbers in the table. Table 13.3 lists the lengths x, y, and z of the sides of four Pythagorean triangles, where z is a square, and Table 13.4 lists the lengths of the sides of four Pythagorean triangles, where x or y is a square.

Generators		Sides of a Pythagorean Triangle		
m	n	x	y	z
4	3	24	7	25
12	5	120	119	169
24	7	336	527	625
40	9	720	1519	1681

Table 13.3 *Pythagorean triangles with z a square.*

Generators		Sides of a Pythagorean Triangle		
m	n	x	y	z
2	1	4	3	5
5	4	40	9	41
8	1	16	63	65
13	12	312	25	313

Table 13.4 *Pythagorean triangles with x or y a square.*

Pythagorean Triangles with the Same Perimeter

Pythagorean triangles with the same perimeter do exist. They are rare and not easy to find, if primitive Pythagorean triples are used. Three such triangles are 7080-119-7081, 5032-3255-5993, and 168-7055-7057, generated by $m = 60$ and $n = 59$; $m = 68$ and $n = 37$; and $m = 84$ and $n = 1$, respectively. Their common perimeter is 14,280. Notice that the Pythagorean triangles 48-20-52, 24-45-51, and 40-30-50 share a smaller common perimeter, namely, 120.

Pythagorean Triangles with the Same Area

Do Pythagorean triangles with the same area exist? A close investigation of Table 13.2 reveals a pleasant surprise. It contains two such triangles, 20-21-29 and 12-35-37, with the same area, 210.

Pythagorean triangles with equal areas were studied by Diophantus, Fermat, and the English logician Lewis Carroll. A note in Carroll's diary, dated December 19, 1897, reads as follows: "Sat up last night till 4 A.M. [sic] over a tempting problem sent me from New York: to find three equal rational sided right triangles. I found two whose sides are 20, 21, 29 and 12, 35, 37 but could not find *three*."

The smallest area common to three primitive Pythagorean triangles is 13, 123, 110. Their sides are generated by 77, 38; 138, 5; and 78, 55. See Exercise 28.

***Lewis Carroll** (1832–1898), the son of a clergyman, was born in Daresbury, England. He graduated from Christ College, Oxford University, in 1854. He began teaching mathematics at his alma mater in 1855, where he spent most of his life. He became a deacon in the Church of England in 1861.*

Carroll's famous books, Alice in Wonderland *and its sequel,* Through the Looking-Glass and What Alice Found There, *have provided a lot of pleasure to both children and adults all over the world.* Alice in Wonderland *is available in more than 30 languages, including Arabic and Chinese, and also in Braille. The character is named for Alice Liddell, a daughter of the dean of Christ Church.*

Pythagorean Triangles with the Same Numerical Area and Perimeter

Interestingly, Pythagorean triangles with the same numerical areas and perimeters do exist. For example, let x-y-$z = 12$-5-13. Then $x + y + z = 30 = \frac{12 \cdot 5}{2} = \frac{1}{2}xy$.

To check if there are any others, we have $x^2 + y^2 = z^2$ and $x + y + z = \frac{xy}{2}$. Eliminating z between the two equations, we get

$$x(y-4) = 8 + 4(y-4)$$
$$x = 4 + \frac{8}{y-4}$$

This implies that $(y-4)|8$; so $y - 4 = 1, 2, 4,$ or 8; that is, $y = 5, 6, 8, 12$. Correspondingly, $x = 12, 8, 6, 5$. They yield two distinct Pythagorean triangles: 12-5-13 and 8-6-10.

Palindromic Pythagorean Triples

The Pythagorean triple 3-4-5 has the interesting property that each component is a palindrome. Such a triple is a **palindromic Pythagorean triple**. For example, 33-44-55 and 303-404-505 are both palindromic Pythagorean triples; the latter was discovered in 1997 by Patrick DeGeest of Belgium.

There is a systematic way of constructing an infinite number of such triples from the 3-4-5 triple. To see this, let s be a finite **binary word** beginning with a 1. Let s^R denote the word obtained by reversing order of the digits in s. Then their **concatenation** $t = ss^R$ is palindromic, and so are the numbers $3t$, $4t$, and $5t$. Consequently, $3t$-$4t$-$5t$ is also a palindromic Pythagorean triple.

EXERCISES 13.1

1. Rewrite symbolically the statement that the integers m and n have different parity.
2. Let x-y-z be a primitive Pythagorean triple, where x is even. Prove that y and z are odd.
3. Show that 3-4-5 is the only primitive Pythagorean triple consisting of consecutive integers.
4. Let F_n denote the nth Fibonacci number, where $n \geq 1$. Show that

$$(F_nF_{n+3})^2 + (2F_{n+1}F_{n+2})^2 = (F_{n+1}^2 + F_{n+2}^2)^2$$

5. Add the next two lines to Table 13.1.
6. Study the following primitive Pythagorean triples pattern. Add the next two lines.

$$\begin{array}{rcrcr} 41^2 & + & 840^2 & = & 841^2 \\ 401^2 & + & 80400^2 & = & 80401^2 \\ 4001^2 & + & 8004000^2 & = & 8004001^2 \\ 40001^2 & + & 800040000^2 & = & 800040001^2 \\ & & \vdots & & \end{array}$$

Find all primitive Pythagorean triples with the given value of m.

7. 11
8. 12

Study the following pattern of Pythagorean triples:

$$\begin{gathered} 3^2 + 4^2 = 5^2 \\ 5^2 + 12^2 = 13^2 \\ 7^2 + 24^2 = 25^2 \\ 9^2 + 40^2 = 41^2 \\ 11^2 + 60^2 = 61^2 \\ \vdots \end{gathered}$$

9. Add the next two lines.
10. Predict a formula for the nth line, where $n \geq 1$.
11. Establish the formula in Exercise 9.

12–14. Redo Exercises 9–11 with the following pattern of Pythagorean triples.

$$\begin{gathered} 8^2 + 15^2 = 17^2 \\ 12^2 + 35^2 = 37^2 \\ 16^2 + 63^2 = 65^2 \\ 20^2 + 99^2 = 101^2 \\ \vdots \end{gathered}$$

Let x-y-z be a primitive Pythagorean triple. Prove each.

15. Either x or y is divisible by 3.
16. Exactly one of the numbers x, y, or z is divisible by 5.
17. At least one of the numbers x, y, or z is divisible by 4.
18. The product of the lengths of the legs of a Pythagorean triangle is divisible by 12.
19. The product of the lengths of the sides of a Pythagorean triangle is divisible by 60.
20. Let x-y-z be a primitive Pythagorean triple such that $z = x + 1$. Prove that $x = 2n(n+1)$, $y = 2n + 1$, and $z = 2n(n+1) + 1$, where $n \geq 1$. (This formula characterizes the primitive Pythagorean triples with $z = x + 1$.)
    (*Hint*: Consider $z - x$ in the proof of Lemma 13.4.)
21. Let x-y-z be a primitive Pythagorean triple touch that $z = x + 2$. Prove that $x = 2m$, $y = m^2 - 1$, and $z = m^2 + 1$, where $m \geq 2$.

Consider a Pythagorean triangle with sides $x = 2mn$, $y = m^2 - n^2$, and $z = m^2 + n^2$, where x-y-z is a primitive Pythagorean triple.

22. Compute its perimeter.
23. Compute its area.
24. Compute its area if the hypotenuse is longer than the even leg by one.
25. Is it possible for the triangle to have its hypotenuse longer than its odd side by one? Justify your answer.
26. Compute the perimeter of an arbitrary Pythagorean triangle.
27. Compute the area of an arbitrary Pythagorean triangle.
28. Verify that the primitive Pythagorean triangles generated by 77, 38; 138, 5; and 78, 55 enclose the same area. (C. L. Shedd, 1945)
29. In 1943, W. P. Whitlock, Jr., studied the areas of over 1300 primitive Pythagorean triangles and made an interesting observation: The areas of only two of them

could be expressed using a single digit; they are the 3-4-5 and 1924-693-2045 triangles. Compute the area of the primitive Pythagorean triangle 1924-693-2045. (Watch for the beastly number.)

30. Let a_k be a positive integer defined by $a_k = 2a_{k-1} + a_{k-2}$, where $a_1 = 1$, $a_2 = 2$, and $k \geq 3$. Prove that $m = a_k$, $n = a_{k-1}$ generate a primitive Pythagorean triangle whose legs differ by unity. (W. P. Whitlock, Jr., 1943)

31. Let x_n-y_n-z_n be a primitive Pythagorean triple, where $y_n = x_n + 1$. Prove that x_{n+1}-y_{n+1}-z_{n+1} is also a primitive Pythagorean triple, where

$$\begin{bmatrix} x_{n+1} \\ y_{n+1} \\ z_{n+1} \end{bmatrix} = \begin{bmatrix} 3 & 2 & 1 \\ 3 & 2 & 2 \\ 4 & 3 & 2 \end{bmatrix} \begin{bmatrix} x_n \\ z_n \\ 1 \end{bmatrix}, \quad n \geq 1$$

(This yields a recursive algorithm for computing an infinite family of primitive Pythagorean triples.)

32. Show that $x = (m^2 - pn^2)/p$, $y = mn$, and $z = (m^2 + pn^2)/p$ is a solution of the diophantine equation $x^2 + py^2 = z^2$.

Consider the diophantine equation $1/x^2 + 1/y^2 = 1/z^2$.

*33. Find the smallest solution. (A. Dunn, 1980)

*34. Find the general solution. (A. Dunn, 1980)

13.2 Fermat's Last Theorem

In the preceding section, we established that there are infinitely many solutions of the diophantine equation $x^2 + y^2 = z^2$, where x, y, and z are positive integers. Now we can ask if the equation $x^3 + y^3 = z^3$ is solvable with positive integers. Or is $x^4 + y^4 = z^4$? More generally, is the diophantine equation

$$x^n + y^n = z^n \tag{13.2}$$

solvable with positive integers only, where $n \geq 3$?

Fermat's Conjecture

Fermat, around the year 1637, conjectured that Fermat's equation (13.2) has no positive integral solutions when $n \geq 3$. Unfortunately, he did not confirm the claim with a proof, which he had done on several occasions. In fact, in one of his many marginal notes in his copy of Claude Bachet de Méziriac's (1581–1638) Latin translation of Diophantus' *Arithmetica*, Fermat comments that he has "discovered a truly wonderful proof of this, but the margin is too small to contain it." Whether he indeed had a proof, or whether he, realizing its complexity, wrote the comment to challenge future mathematicians, we may never know. "Fermat's reputation for veracity should be strong evidence for believing he had a proof. Only once he has been found incorrect," writes Beiler in his delightful book, *Recreations in the Theory of Numbers*. He adds, "Posterity has wished many times that the margin of Bachet's *Diophantus* had been wider or Fermat less secretive" about his techniques.

Although no proof of Fermat's conjecture existed for over three centuries, the conjecture came to be known as *Fermat's Last Theorem* for two reasons: First, the name distinguishes it from Fermat's Little Theorem; and second, this was the last of his conjectures that was neither proved nor disproved.

In 1823 and then in 1850, the Academy of Science in Paris offered a prize for a correct proof. Unfortunately, this produced a wave of thousands of mathematical misadventures. A third prize was offered in 1883 by the Academy of Brussels.

When Gauss was told of the Paris Prize, he claimed that "Fermat's theorem as an isolated proposition has very little interest for me, because I could easily lay down a multitude of such proportions, which could neither prove nor dispose of." When the German mathematician David Hilbert (1862–1943) was asked, he said, "Before beginning I should have to put in three years of intensive study, and I haven't that much time to squander on a probable failure."

In 1908, the German physician and amateur mathematician F. Paul Wolfskehl bequeathed 100,000 marks to the Göttingen Academy of Sciences to be offered as a prize for a complete proof of Fermat's Last Theorem. As a result, from 1908–1911, a flood of over 1000 incorrect proofs were presented. According to mathematical historian Howard Eves, "Fermat's Last Theorem has the peculiar distinction of being the mathematical problem for which the greatest number of incorrect proofs have been published."

In 1770, Euler provided the first proof of Fermat's Last Theorem for the case $n = 3$, but his proof contained a few gaps. It was later perfected by Legendre. Fermat himself gave a proof for the case $n = 4$, employing the *method of infinite descent*, which we shall demonstrate shortly. Around 1825, Dirichlet and Legendre, capitalizing on Fermat's technique of infinite descent, independently confirmed the conjecture for $n = 5$. About fourteen years later, Lamé established the conjecture for $n = 7$.

Since the proof for $n = 3$ is complicated, we will omit it. We can instead go directly to the case $n = 4$ and establish the validity of the theorem as a corollary to the following stronger theorem. Its proof, employing Theorem 13.1, illustrates Fermat's technique of infinite descent, which is really rooted in the well-ordering principle. The essence of this method lies in constructing a solution "smaller" than a given positive integral solution.

THEOREM 13.2 **(Fermat)** The diophantine equation $x^4 + y^4 = z^2$ has no positive integral solutions.

PROOF

Let a-b-c be a solution of the equation, so $a^4 + b^4 = c^2$. Let $(a, b) = d$. Then $a = d\alpha$ and $b = d\beta$ for some positive integers α and β, where $(\alpha, \beta) = 1$. This implies $(d\alpha)^4 + (d\beta)^4 = c^2$, so $d^4|c^2$ and hence $d^2|c$. Therefore, $c = d^2\gamma$ for some integer γ.

Thus, $d^4(\alpha^4 + \beta^4) = d^4\gamma^2$, that is, $\alpha^4 + \beta^4 = \gamma^2$. In other words, α-β-γ is also a solution, where $(\alpha, \beta) = 1$. Thus, we can assume that $(a, b) = 1$. Since z is a positive integer, we also assume that a-b-c is a solution such that the value of c is the least among such solutions x-y-z of the given diophantine equation.

The equation $a^4 + b^4 = c^2$ can be rewritten as $(a^2)^2 + (b^2)^2 = c^2$, so a^2-b^2-c is a Pythagorean triple. Since $(a, b) = 1$, it follows by Exercise 72 in Section 3.1 that $(a^2, b^2) = 1$. Thus, a^2-b^2-c is in fact a primitive Pythagorean triple.

Therefore, by Theorem 13.1, there are positive integers m and n such that $a^2 = 2mn$, $b^2 = m^2 - n^2$, and $c = m^2 + n^2$, where $(m, n) = 1$, $m \not\equiv n \pmod 2$, and $m > n$ and b is odd.

Because $m \not\equiv n \pmod 2$, exactly one of them is even. To identify it, suppose n is odd, so m is even. Then $1 \equiv b^2 \equiv 0 - 1 \equiv 3 \pmod 4$, which is a contradiction. Therefore, n is even (and hence m is odd).

Let $n = 2q$. Then $a^2 = 4mq$, so $(a/2)^2 = mq$. (Remember, a is even.) Since $(m, n) = 1$, it follows that $(m, q) = 1$. Therefore, by Lemma 13.3, both m and q are squares. Let $m = t^2$ and $q = u^2$.

Because $n^2 + b^2 = m^2$, n-b-m is a Pythagorean triple. Furthermore, $(n, b) = 1$, so n-b-m is a primitive Pythagorean triple. Therefore, since n is even, again by Theorem 13.1, there are positive integers v and w such that $n = 2vw$, $b = v^2 - w^2$, and $m = v^2 + w^2$, where $v > w$, $(v, w) = 1$, and $v \not\equiv w \pmod 2$.

Then $vw = n/2 = q = u^2$. Because $(v, w) = 1$, it follows, again by Lemma 13.3, that $v = r^2$ and $w = s^2$ for some positive integers r and s.

Substituting for v, w, and m in the equation $v^2 + w^2 = m$, we get $r^4 + s^4 = t^2$. This shows that r-s-t is also a solution of the equation $x^4 + y^4 = z^2$, where

$$0 < t \le t^2 = m \le m^2 < m^2 + n^2 = c$$

Thus, we have systematically constructed a solution r-s-t of the given equation, where $0 < t < c$. This is a contradiction, since we assumed that the solution a-b-c has the least value of z. Hence, our assumption that the equation $x^4 + y^4 = z^2$ has positive integral solutions is invalid. This concludes the proof. ■

As a byproduct, this theorem establishes Fermat's conjecture for $n = 4$, as the following corollary shows.

COROLLARY 13.2 The diophantine equation $x^4 + y^4 = z^4$ has no positive integral solutions.

PROOF

Let a-b-c be a solution of the equation. Then $a^4 + b^4 = (c^2)^2$, showing that a-b-c^2 is a solution of the equation $x^4 + y^4 = z^2$. Since this contradicts Theorem 13.2, the result follows. ■

This corollary yields the following intriguing result.

COROLLARY 13.3 The lengths of the sides of a Pythagorean triangle cannot all be squares.

PROOF

Let x and y denote the lengths of the legs of a Pythagorean triangle and z the length of its hypotenuse. Suppose x, y, and z are squares, say, $x = u^2$, $y = v^2$, and $z = w^2$. Then $u^4 + v^4 = w^4$, which is impossible by Corollary 13.2. Thus, the lengths cannot all be squares. ■

The Rest of the Story

Corollary 13.2 plays an important role in establishing Fermat's Last Theorem for any exponent $n \geq 3$. It tells us that we need only concentrate on exponents that are odd primes.

We can see this as follows: By the Fundamental Theorem of Arithmetic, n is either a power of 2 or is divisible by an odd prime. If n is a power of 2, then $n = 4m$, where $m \geq 1$. Then the equation $x^n + y^n = z^n$ becomes $(x^m)^4 + (y^m)^4 = (z^m)^4$. This implies x^m-y^m-z^m is a solution of $x^4 + y^4 = z^4$, which is a contradiction. Thus, if n is a power of 2, equation (13.2) has no positive integral solutions.

On the other hand, let $n = mp$. Then equation (13.2) becomes $(x^m)^p + (y^m)^p = (z^m)^p$. So if we can show that the equation $x^p + y^p = z^p$ is not solvable, it will imply that equation (13.2) is not solvable when $n = mp$.

Thus, Fermat's Last Theorem can be established if we can show that the equation $x^p + y^p = z^p$ is not solvable for any odd prime, as Euler, Dirichlet and Legendre, and Lamé showed for $p = 3, 5$, and 7, respectively. For over 350 years, numerous tenacious mathematicians, in addition to these great ones, worked diligently to demolish what Beiler, in his book, calls "the stone wall" of Fermat's Last Theorem, "chipping off a piece of granite here and another there."

The German mathematician Ernst E. Kummer played a pivotal role in the development of a proof. In 1843, he submitted a purported proof to Dirichlet, who immediately found a flaw in his reasoning. Kummer returned to his search with added determination, developing a new class of numbers called **algebraic numbers**, and a new branch of modern algebra called the **theory of ideals**. Kummer succeeded in proving Fermat's Last Theorem for a large family of primes. In fact, all subsequent pursuits of a valid proof were based on Kummer's work.

In 1983, the German-born mathematician Gerd Faltings of Princeton University proved that the number of solutions of Fermat's equation is finite for $n \geq 3$, a result conjectured many years earlier by Louis Mordell. In 1988, the Japanese mathematician Yoichi Miyaoka claimed that he had a proof. Unfortunately, the stone wall remained insurmountable.

***Ernst Eduard Kummer** (1810–1893) was born at Sorau, Germany (now Zary, Poland). After his early education at the Gymnasium in Sorau, he entered the University of Halle in 1828 to study theology, but soon gave it up to pursue mathematics. After receiving his doctorate in 1831, Kummer taught at the Gymnasium in Sorau for a year and then at the Gymnasium at Liegnitz (now Legnica, Poland) for 10 years.*

In 1842, Kummer was appointed professor of mathematics at the University of Breslau (now Wroclaw, Poland), where he remained until 1855. When Dirichlet left the University of Berlin in 1855 to succeed Gauss at Göttingen, Kummer was appointed professor at Berlin.

In his quest for a proof of Fermat's Last Theorem, Kummer created the so-called algebraic numbers. His proof failed, since he assumed the fundamental theorem of arithmetic for such numbers, which he later restored by developing the theory of ideals. For this, Kummer was awarded the grand prize of the Paris Academy of Sciences in 1857, although he had not competed. He made significant contributions to the study of hypergeometric series and geometry.

A creative pioneer of nineteenth-century mathematics, Kummer died at Berlin after a productive career and quiet retirement.

The Stone Wall Crumbles and Wiles Meets Fermat

Finally, in June 1993, the stone wall started to crumble. The English mathematician Andrew Wiles of Princeton University announced at a number theory conference at Cambridge University, England, that he had solved Fermat's Last Theorem using elliptic functions and modular forms. The news was taken so seriously that it made the front page of *The New York News* (see Figure 13.3) and was covered by *Time* and *Newsweek*, and the *NBC Nightly News*. The discovery caused Wiles to be named one of "the 25 most important people of the year" in *People* magazine.

Five months later, a flaw was detected (see Figure 13.4), which was corrected in October 1994. The corrected version has withstood intense scrutiny by experts. Although the German prize offered in 1908 had lost most of its cash value due to inflation, Wiles still collected $50,000 in 1997 for his singular achievement.

Wiles' discovery was so outstanding that in 1997 the British Broadcasting Company (BBC) produced *The Proof*, an inspiring and delightful television documentary shown also in the United States on PBS. In the film, Wiles describes his seven years of relentless, solitary pursuit in his attic for a proof of Fermat's Last Theorem as follows:

"Perhaps I can best describe my experience of doing mathematics in terms of a journey through a dark unexplored mansion. You enter the first room of the man-

Andrew J. Wiles *(1953–) was born in Cambridge, England. From early childhood he had a fascination for mathematical problems; he loved to do them at school and home, and even made up new ones on his own. At age 10, he was browsing through mathematics books at a local public library. He was struck by one particular problem—Fermat's Last Theorem—in one of them. The fact that it looked simple and that it eluded the brilliance of mathematicians around the world for over three centuries really fascinated him. Solving the problem became his obsession.*

In 1971, Wiles entered Merton College, Oxford, and graduated in three years. He then went to Clare College, Cambridge, and received his Ph.D. in 1980 under John Coates in the theory of elliptic curves. This laid the foundation for his famous discovery of a proof of Fermat's Last Theorem in June 1993.

During the years 1977–1980, Wiles was a Junior Research Fellow at Clare College and a Benjamin Pierce Assistant Professor at Harvard University. In 1981, he became visiting professor at the Sonderforschungsbereich Theoretische Mathematik in Bonn, and later joined the Institute for Advanced Study in Princeton. The following year, he became a professor at Princeton University and a visiting professor at the University of Paris. As a Guggenheim Fellow, he then spent a year as a visiting professor at the Institut des Hautes Études Scientifiques and at the École Normale Supérieure. During 1988–1990, Wiles was a Royal Society Research Professor at Oxford. Since 1994, he has been Eugene Higgins Professor of Mathematics at Princeton.

In 1989, Wiles was elected a fellow of the Royal Society, London. He has received numerous honors, including the Schock Prize in Mathematics from the Royal Academy of Sciences (1995), Prix Fermat from the Universite Paul Sabatier and Matra Marconi Space (1995), the Wolf Prize in mathematics (1996), and the National Academy of Sciences Award in Mathematics (1996).

At Last, Shout of 'Eureka!' An Age-Old Math Mystery

Gina Kolata

More than 350 years ago, a French mathematician wrote a deceptively simple theorem in the margins of a book, adding that he had discovered a marvelous proof of it but lacked space to include it in the margin. He died without ever offering his proof, and mathematicians have been trying ever since to supply it.

Now, after thousands of claims of success that proved untrue, mathematicians say the daunting challenge, perhaps the most famous of unsolved mathematical problems, has at last been surmounted.

The problem is Fermat's last theorem, and its apparent conqueror is Dr. Andrew Wiles, a 40-year-old English mathematician who works at Princeton University. Dr. Wiles announced the result yesterday at the last of three lectures given over three days at Cambridge University in England.

Within a few minutes of the conclusion of his final lecture, computer mail messages were winging around the world as mathematicians alerted each other to the startling and almost wholly unexpected result.

Dr. Leonard Adelman of the University of Southern California said he received a message about an hour after Dr. Wiles's announcement. The frenzy is justified, he said.

(*continued*)

Figure 13.3

"It's the most exciting thing that's happened in—geez—maybe ever, in mathematics."

Impossible Is Possible

Mathematicians present at the lecture said they felt "an elation," said Dr. Kenneth Ribet of the University of California at Berkeley, in a telephone interview from Cambridge.

The theorem, an overarching statement about what solutions are possible for certain simple equations, was stated in 1637 by Pierre de Fermat, a 17th-century French mathematician and physicist. Many of the brightest-minds in mathematics have struggled to find the proof ever since, and many have concluded that Fermat, contrary to his tantalizing claim, had probably failed to develop one despite his considerable mathematical ability.

With Dr. Wiles' result, Dr. Ribet said, "the mathematical landscape has changed." He explained: "You discover that things that seemed completely impossible are more of a reality. This changes the way you approach problems, what you think is possible."

Dr. Barry Mazur, a Harvard University mathematician, also reached by telephone in Cambridge, said: "A lot more is proved than Fermat's last theorem. One could envision a proof of a problem, no matter how celebrated, that had no implications. But this is just the reverse. This is the emergence of a technique that is visibly powerful. It's going to prove a lot more."

Figure 13.3

How a Gap in the Fermat Proof Was Bridged

Gina Kolata

Fermat's last theorem, which has tantalized mathematicians for more than 350 years, has at last been solved, say those who have read the revised but not yet published proof. But the endgame of this furious chase has proved as full of last-minute surprises as a murder mystery.

For Dr. Andrew Wiles of Princeton University, the chief author of the proof, triumph had to be snatched from the jaws of disaster. His first proof, which aroused world-wide attention when announced two years ago, turned out to contain a gap, which Dr. Wiles found he was unable to cross alone.

The enormous intellectual trophy of having conquered the world's most famous mathematical problem seemed about to slip from his grasp. If he invited a well-known mathematician to help him bridge the gap, he would risk having to share the credit. What was needed, and what he pulled off, was a miraculous save with just the right collaborator.

On a mathematical level, Fermat's last theorem turns out to have extraordinarily deep roots, despite its apparent simplicity. The theorem is a special case of an overarching mathematical idea known as the Taniyama conjecture, which is itself a giant step toward the goal of the Langlands program, a grand unified theory of mathematics.

THE GAP

A 'Minor' Problem Turns Into a Crisis

Several minor faults were found and Dr. Wiles fixed them. But then, in the fall of 1993, a reviewer asked him to justify an assertion, in the midst of his proof, that a certain estimate was correct.

The gap at first seemed to be a minor one. But though the estimate seemed intuitively to be correct, proving it was a different matter.

Before Dr. Wiles' was willing to announce that he and Dr. Taylor had filled the gap, he asked a few leading experts to check his argument. One was Dr. Faltings in Germany, who said he read it in a week and was convinced it was correct. Now, Dr. Faltings has improved the proof, making it sleeker and easier to follow. Still, he said, most mathematicians who are expert enough in the field to read the proof will probably require a month to go through it.

Figure 13.4

sion and it's completely dark. You stumble around bumping into the furniture, but gradually you learn where each piece of furniture is. Finally, after six months or so, you find the light switch, you turn it on, and suddenly it's all illuminated. You can see exactly where you were. Then you move into the next room and spend another six months in the dark. So each of these breakthroughs, while sometimes they're momentary, sometimes over a period of a day or two, they are the culmination of—and couldn't exist without—the many months of stumbling around in the dark that proceed them."

Having solved his childhood passion and achieved world fame, he concludes at the end of the film that the journey is finally over: "There is a certain sense of sadness, but at the same time there is this tremendous sense of achievement. There's also a sense of freedom. I was so obsessed by this problem that I was thinking about it all the time—when I woke up in the morning, when I went to sleep at night—and that went on for eight years. That's a long time to think about one thing. That particular odyssey is now over. My mind is now at rest."

Wiles' monumental accomplishment has inspired several books and videos. More importantly, as R. K. Guy writes in his *Unsolved Problems in Number Theory*, "Fermat's Last Theorem ... has generated a great deal of good mathematics, whether goodness is fathomed by beauty, depth, or applicability."

Fermat's Last Theorem Goes to Broadway, Almost

Interestingly, Fermat's last theorem even has made it to off-Broadway. A musical called *Fermat's Last Tango*, based on the BBC film and Simon Singh's book *Fermat's Enigma*, premiered off-Broadway at the York Theater Company in New York City. It ran for about five weeks at the end of 2000; see Figure 13.5. The musical is available on videotape or DVD from the Clay Mathematics Institute, Cambridge, Massachusetts; visit the website www.claymath.org/events/fermtslasttango.htm.

The Diophantine Equation $x^4 - y^4 = z^2$

We now move on to the diophantine equation $x^4 - y^4 = z^2$, which is closely related to $x^4 + y^4 = z^2$. The proof by Fermat that this new equation also has no positive integral solutions runs similar to the proof of Theorem 13.2, so we omit it. See Exercise 17.

THEOREM 13.3 **(Fermat)** The diophantine equation $x^4 - y^4 = z^2$ has no positive integral solutions. ■

This theorem has a fascinating byproduct, which Fermat also jotted down in a margin of his copy of Diophantus' *Arithmetica*.

COROLLARY 13.4 **(Fermat)** The area of a Pythagorean triangle cannot be an integral square.

Fermat's Last Theorem: The Musical

Mathematics will show up on stage once again in November, this in a musical called *Fermat's Last Tango*, which will premiere Off Broadway at the York Theater Company in New York City. The musical, written by the husband-and-wife team of Joanne Sydney Lessner and Joshua Rosenblum, is inspired by the story of Andrew Wiles' proof of Fermat's Last Theorem, as made public by the PBS film "The Proof" and the book *Fermat's Enigma*, by Simon Singh.

The musical tells the story of Professor Daniel Keane, who comes up with a proof that Fermat couldn't possibly understand, finds a flaw in the proof, and then fixes the flaw under the watchful eye of Fermat and other dead mathematicians (who now all reside in the "After-Math," of course). Fermat, Pythagoras, Euclid, Newton, and Gauss all feature in the show. The tone is lighthearted and whimsical.

Towards the end of the story Fermat is shown running a nightmarish game show called "Prove My Theorem!" in which Professor Keane finds himself a contestant. The other great mathematicians end up getting tired of Fermat's antics, and decide to help the professor fix his proof. The music combines operetta, blues, pop, and, of course, tango. The performance is almost completely sung through, with very little spoken dialogue.

Both of the play's authors were music majors at Yale. Joshua Rosenblum says he briefly considered a mathematics minor, but in the end didn't get much beyond calculus, though he continues to nurture a recreational interest in mathematics. Both became fascinated by the story of the proof of Fermat's Last Theorem, and decided to write a musical on the subject. The show contains little mathematics, but what it does include it tries to get right. The authors have also included some references they hope will resonate with mathematicians. For example, one of the obstacles Fermat puts before Professor Keane in his game show is the fact that mathematics is a "young man's game."

Performances of *Fermat's Last Tango* will begin on November 21 and run through December 31, Tuesday through Saturday. The York Theatre Company is located at the Theatre at St. Peter's, Citicorp Center, 619 Lexington Avenue (at 54th Street), New York, NY 10022. For tickets, call 212-239-6200.

Figure 13.5

PROOF

Let x and y be the lengths of the legs, and z the length of the hypotenuse of the Pythagorean triangle. Then $x^2 + y^2 = z^2$ and the area of the triangle equals $xy/2$. Suppose the area is a square, say, $s^2 : xy/2 = s^2$. Then $(x+y)^2 = z^2 + 4s^2$ and $(x-y)^2 = z^2 - 4s^2$. This yields

$$(x+y)^2(x-y)^2 = (z^2+4s^2)(z^2-4s^2)$$
$$(x^2-y^2)^2 = z^4 - (2s)^4$$

This contradicts Theorem 13.3, so the area cannot be an integral square. ■

The following corollary also follows from Theorem 13.3. We leave its proof as a routine exercise.

COROLLARY 13.5 Let x and y be positive integers. Then $x^2 + y^2$ and $x^2 - y^2$ cannot both be squares. ■

Now we turn to a generalization of Fermat's Last Theorem.

Beal's Conjecture

In 1994, Texas millionaire and amateur mathematician Andrew Beal, the founder, owner, and chairman of the Beal Bank in Dallas, read in a local newspaper about Wiles' proof of Fermat's Last Theorem. He then conjectured that the equation $x^m + y^n = z^r$ has no positive integral solutions, where $(x, y, z) = 1$ and the exponents m, n, and r are integers ≥ 3. If Beal's conjecture can be established, then Fermat's last theorem would follow from it as an intriguing corollary. See Figure 13.6.

Figure 13.6 *Beal's problem,* Math Horizons, *February 1998.*

Beal, in the tradition of Paul Wolfskehl in 1908, has offered a cash prize of $50,000 to the first person to solve his problem; the prize is intended to motivate the mathematics community and to elevate Beal's conjecture to the level of the original Fermat's conjecture. Beal established a committee of three distinguished mathematicians, headed by R. Daniel Mauldin of the University of North Texas, to screen proofs of his conjecture submitted by potential prize-winners.

Since then, the prize has been increased to $100,000 for either a proof or a counterexample of Beal's conjecture. The prize will be awarded by *The American Mathematical Society*, which holds the prize money.

Andrew Beal

Beal's conjecture creates a new opportunity and a more difficult stone wall to challenge the mathematics community. If the past is any indication, then surmounting it could be a long and difficult task.

Catalan's Conjecture

As in the case with Fermat's equation, if one of the exponents in **Beal's equation** can be 1 or 2, then it does have solutions. Two such solutions are given by $1^1 + 2^3 = 3^2$ and $2^5 + 7^2 = 3^4$. The first of these two equations is the **Catalan relation**, named after Catalan, who conjectured that $8 = 2^3$ and $9 = 3^2$ are the only powers differing by unity; that is, the diophantine equation $x^m - y^n = 1$, where $m, n \geq 2$, has exactly one solution: $x = 3$, $y = 2$ and $m = 2$, $n = 3$.

Interestingly, more than 500 years before Catalan made the conjecture, Levi ben Gerson (1288–1344) of France showed that 3^2 and 2^3 are the only powers of 2 and 3 that differ by 1. In 1976, Robert Tijdeman of the University of Leiden, Netherlands, made significant progress when he proved that Catalan's diophantine equation has only a finite number of solutions. In 2000, Preda Mihailescu of the University of Paderborn, Germany, established Catalan's celebrated conjecture; he presented his proof at a number theory conference in Montreal in 2000.

A fact somewhat related to Catalan's solution: In 1992, H. Gauchman and I. Rosenholtz of East Illinois University found that 41 is the smallest prime that is *not* the difference (in either order) of a power of 2 and a power of 3: $41 \neq 3^m - 2^n$ or $2^n - 3^m$; see Exercise 18. In the following year, D. W. Kostwer, noticing that $41 = 2^5 + 3^2$, showed that 53 is the smallest prime that is neither a sum nor a difference (in either order) of a power of 2 and a power of 3.

Fermat–Catalan Conjecture

Combining Fermat's Last Theorem and Catalan's conjecture, a new conjecture has recently been formulated: The diophantine equation $x^r + y^s = z^t$ has only finitely many solutions, where x, y, and z are pairwise relatively prime, and $\frac{1}{r} + \frac{1}{s} + \frac{1}{t} < 1$.

So far, ten solutions are known, discovered by Mauldin in 1997:

$$
\begin{aligned}
1^r + 2^3 &= 3^2 & 2^5 + 7^2 &= 3^4\\
7^3 + 13^2 &= 2^9 & 2^7 + 17^3 &= 71^2\\
3^5 + 11^4 &= 122^2 & 17^7 + 76271^3 &= 21063928^2\\
1414^3 + 2213459^2 &= 65^7 & 9262^3 + 15312283^2 &= 113^7\\
43^8 + 96222^3 &= 30042907^2 & 33^8 + 1549034^2 &= 15613^3
\end{aligned}
$$

EXERCISES 13.2

1. Let x and y be positive integers. Prove that $x^2 + y^2$ and $x^2 - y^2$ cannot both be squares.
2. Let x and y be positive integers. Then both $x^2 + y^2$ and $x^2 - y^2$ cannot be k times a square, where k is a positive integer.
3. Prove that the area of a Pythagorean triangle cannot be twice a square.
4. Show that the diophantine equation $1/x^4 + 1/y^4 = 1/z^2$ has no positive integral solutions.

Prove each, where p is a prime.

5. If $x^{p-1} + y^{p-1} = z^{p-1}$, then $p|xyz$.
6. If $x^p + y^p = z^p$, then $p|(x+y-z)$.
7. Find a solution of the diophantine equation $x^4 + y^4 = 2z^2$.
8. Show that the diophantine equation $x^4 - y^4 = 2z^2$ has no positive integral solutions.
9. Show that no two of the numbers x, y, and z in a Pythagorean triple x-y-z can be squares; that is, the lengths of no two sides of a Pythagorean triangle can be squares.

Find a solution of each diophantine equation.

10. $x^2 + 2 = y^3$
11. $x^2 + 4 = y^3$

Use the diophantine equation $w^3 + x^3 + y^3 = z^3$ to answer Exercises 12–16.

12. Verify that 3-4-5-6 is a solution.
13. Prove that the equation has infinitely many solutions.
14. Show that 3-4-5-6 is the only solution of the equation consisting of consecutive integers.
15. In his book, Beiler gives a fancy, but complicated formula that can be used to find a special class of positive integral solutions of the equation:

$$a^3(a^3+b^3)^3 = b^3(a^3+b^3)^3 + a^3(a^3-2b^3)^3 + b^3(2a^3-b^3)^3$$

Verify that 9-12-15-18 and 28-53-75-84 are two solutions.

16. Beiler also gives another formula to find a similar class of solutions:

$$a^3(a^3+2b^3)^3 = a^3(a^3-b^3)^3 + b^3(a^3-b^3)^3 + b^3(2a^3+b^3)^3$$

Verify that 7-14-7-20 and 26-55-78-87 are both solutions.

*17. Prove that the diophantine equation $x^4 - y^4 = z^2$ has no positive integral solutions. (Fermat)

*18. Show that 41 is the smallest prime that is not the difference (in either order) of a power of 2 and a power of 3. (H. Gauchman and I. Rosenholtz, 1992)

13.3 Sums of Squares

Recall that if x-y-z is a Pythagorean triple, then z^2 can be written as the sum of two squares: $z^2 = x^2 + y^2$. There are several obviously related problems: *Can every positive integer be expressed as the sum of two squares of nonnegative integers? Three squares? Four squares? Is there a least positive integer n such that every positive integer can be written as the sum of n squares of nonnegative integers?* The problem of expressing positive integers as the sum of squares has been studied for centuries by mathematicians, including Diophantus, Fermat, Euler, and Lagrange.

Not all positive integers can be represented as the sum of two squares. For example, 3 cannot. So we will characterize those that are representable as the sum of two squares. Not all positive integers can be written as the sum of three squares, either. Interestingly, we will find later on that every positive integer can be represented as the sum of four squares. Now we will explore these problems step by step.

Sums of Two Squares

Of the first 100 positive integers, exactly 43 can be expressed as the sum of two squares. Those are the circled numbers in Table 13.5. Do you see a pattern among them? Do you observe a pattern among the uncircled ones?

1	**2**	3	**4**	**5**	6	7	**8**	**9**	**10**
11	12	**13**	14	15	**16**	**17**	**18**	19	**20**
21	22	23	24	**25**	**26**	27	28	**29**	30
31	**32**	33	**34**	35	**36**	**37**	38	39	**40**
41	42	43	44	**45**	46	47	48	**49**	**50**
51	**52**	**53**	54	55	56	57	**58**	59	60
61	62	63	**64**	**65**	66	67	**68**	69	70
71	**72**	**73**	**74**	75	76	77	78	79	**80**
81	**82**	83	84	**85**	86	87	88	**89**	**90**
91	92	93	94	95	96	**97**	**98**	99	**100**

Table 13.5

If the answer is no, look at some of the uncircled numbers, such as 3, 7, 11, 15, 19, and 23. Anything special about them? Yes, they are all congruent to 3 modulo 4. Such numbers are not representable as the sum of two squares, as the following lemma confirms.

LEMMA 13.6 If $n \equiv 3 \pmod 4$, then n is not representable as the sum of two squares.

PROOF

Suppose $n = x^2 + y^2$ for some integers x and y. Since $w^2 \equiv 0$ or $1 \pmod 4$ for any integer w, it follows that $n = x^2 + y^2 \equiv 0$, 1, or $2 \pmod 4$. Since $n \equiv 3 \pmod 4$, the result follows. ■

Suppose an integer n can be written as the sum of two squares. We can use such a representation to express infinitely many related integers also as such sums, as the next lemma shows. Its proof is a one-liner, so we leave it as an exercise.

LEMMA 13.7 If a positive integer n can be expressed as the sum of two squares, then so can k^2n for every positive integer k. ■

This lemma can be reworded as follows: A square times a representable number is also representable as the sum of two squares.

Lemma 13.7 explains why numbers such as $18 = 3^2 \cdot 2$ and $45 = 3^2 \cdot 5$ are circled in Table 13.5: $18 = 3^2 + 3^2$ and $45 = 6^2 + 3^2$.

Now, how did 20 and 26 make the list? Notice that $10 = 2 \cdot 5 = (1^2 + 1^2)(2^2 + 1^2) = 3^2 + 1^2$ and $26 = 2 \cdot 13 = (1^2 + 1^2)(3^2 + 2^2) = 5^2 + 1^2$. More generally, the product of two representable integers is also representable as the sum of two squares, as the following lemma confirms.

LEMMA 13.8 **(Diophantus)** The product of two sums of two squares is representable as the sum of two squares.

PROOF

Let $m = a^2 + b^2$ and $n = c^2 + d^2$ for some integers m and n. Then

$$mn = (a^2 + b^2)(c^2 + d^2) = (ac + bd)^2 + (ad - bc)^2$$

This identity can be verified by expanding both sides. ■

For example, let $m = 5 = 2^2 + 1^2$ and $n = 13 = 3^2 + 2^2$. Then

$$mn = 65 = (2 \cdot 3 + 1 \cdot 2)^2 + (2 \cdot 2 - 1 \cdot 3)^2 = 8^2 + 1^2$$

Thus, 65 is also a sum of two squares.

An easy way to establish the identity in Lemma 13.8 is by using complex numbers. See Exercise 6.

To see why a number like 99 is not representable in the desired form, let us study its canonical decomposition: $99 = 3^2 \cdot 11$. The factorization contains a prime factor p with an odd exponent, where $p \equiv 3 \pmod 4$, namely, $p = 11$. Such numbers are not representable; we shall confirm this in Theorem 13.4. But we can do so only after laying groundwork in the form of two more lemmas.

LEMMA 13.9 Let p be a prime. If $p \equiv 1 \pmod 4$, then there are positive integers x and y such that $x^2 + y^2 = kp$ for some positive integer k, where $k < p$.

PROOF

Since $p \equiv 1 \pmod 4$, by Corollary 11.2, -1 is a quadratic residue of p. In other words, there is a positive integer $a < p$ such that $a^2 \equiv -1 \pmod p$; that is, $a^2 + 1 = kp$ for some positive integer k. Thus, there are integers $x = a$ and $y = 1$ such that $x^2 + y^2 = kp$.

It remains to show that $k < p$. Since $a \le p - 1$ and $kp = a^2 + 1 < (p-1)^2 + 1$, $kp < p^2 - 2(p-1)$; so $kp < p^2$ and hence $k < p$, as desired. ■

For example, $p = 13 \equiv 1 \pmod 4$. Then $5^2 \equiv -1 \pmod{13}$, so $a = 5$ and $k = 2$. Notice that $a = 8$ also would work, since $8^2 \equiv -1 \pmod{13}$ in which case $k = 5$. In both cases, $k < 13$. Furthermore, $3^2 + 2^2 = 13$, so $x = 3$, $y = 2$, and $k = 1$.

The following lemma, employing the well-ordering principle and Lemmas 13.8 and 13.9, shows that if $p \equiv 1 \pmod 4$, then there exist positive integers x and y such that $p = x^2 + y^2$. It is the first major result for sums of two squares.

LEMMA 13.10 Every prime $p \equiv 1 \pmod 4$ can be written as the sum of two squares.

PROOF

Lemma 13.9 coupled with the well-ordering principle implies that there is a least positive integer m such that $mp = x^2 + y^2$ for some suitable positive integers x and y. We shall show that $m = 1$ by contradiction.

Suppose $m > 1$. Define integers r and s by

$$r \equiv x \pmod m \quad \text{and} \quad s \equiv y \pmod m \tag{13.3}$$

where

$$-m/2 < r,\ s \le m/2 \tag{13.4}$$

Then $r^2 + s^2 \equiv x^2 + y^2 = mp \equiv 0 \pmod m$, so $r^2 + s^2 = mn$ for some positive integer n. Therefore, $(r^2 + s^2)(x^2 + y^2) = (mn)(mp) = m^2np$. But, by Lemma 13.8, $(r^2 + s^2)(x^2 + y^2) = (rx + sy)^2 + (ry - sx)^2$. Thus,

$$(rx + sy)^2 + (ry - sx)^2 = m^2np \tag{13.5}$$

It follows by congruences (13.3) that

$$rx + sy \equiv x^2 + y^2 \equiv \pmod m \quad \text{and} \quad ry - sx \equiv xy - yx \equiv 0 \pmod m$$

These two congruences imply that both $(rx + sy)/m$ and $(ry - sx)/m$ are integers.

It now follows from equation (13.5) that

$$np = \left(\frac{rx + sy}{m}\right)^2 + \left(\frac{ry - sx}{m}\right)^2$$

which is a sum of two squares.

The inequalities (13.4) imply that $r^2 + s^2 \le (m/2)^2 + (m/2)^2 = m^2/2$. Thus, $mn \le m^2/2$; that is, $n \le m/2$. Therefore, $n < m$.

Now $n \ne 0$, since if $n = 0$, then $r^2 + s^2 = 0$; so $r = 0 = s$. Then $x \equiv 0 \equiv y \pmod{m}$, so $m|x$ and $m|y$. Therefore, $m^2|(x^2 + y^2)$: that is, $m^2|mp$. This implies $m|p$. Since $m < p$ by Lemma 13.9, m must be 1. This negates our hypothesis that $m > 1$. Therefore, $n \ge 1$.

Thus, n is a positive integer $< m$ such that np is a sum of two squares. This is a contradiction, since we assumed that m is the least such positive integer. Consequently, $m = 1$; that is, $p = x^2 + y^2$, as desired. ■

The proof of this lemma is an existence proof. It does not tell us how to find the integers x and y in the lemma.

THEOREM 13.4 If $p = 2$ or $p \equiv 1 \pmod{4}$, then p is expressible as the sum of two squares.

PROOF

Since $2 = 1^2 + 1^2$, the theorem follows by Lemma 13.10. ■

For example, $p = 61 \equiv 1 \pmod{4}$; $61 = 6^2 + 5^2$ and $p = 197 \equiv 1 \pmod{4}$; $197 = 14^2 + 1^2$.

We are now ready to identify those positive integers that can be written as the sum of two squares, as the following theorem shows.

THEOREM 13.5 A positive integer n is expressible as the sum of two squares if and only if the exponent of each of its prime factors congruent to 3 modulo 4 in the canonical decomposition of n is even.

PROOF

Suppose the canonical decomposition of n contains a prime factor p with odd exponent $2i + 1$, where $p \equiv 3 \pmod{4}$. Suppose also that $n = x^2 + y^2$ for some integers x and y.

Let $d = (x, y)$, $r = x/d$, $s = y/d$, and $m = n/d^2$. Then $(r, s) = 1$ and $r^2 + s^2 = m$.

Let p^j be the largest power of p that divides d. Since $m = n/d^2$, $p^{2i-2j+1}|m$, where $2i - 2j + 1 \ge 1$. Hence $p|m$.

Suppose $p|r$. Then since $s^2 = m - r^2$ and $p|m$, this implies that $p|s$. This is a contradiction, since $(r, s) = 1$. Therefore, $p \nmid r$.

Therefore, by Corollary 4.6, there is an integer t such that $rt \equiv s \pmod{p}$. Then

$$0 \equiv m = r^2 + s^2 \equiv r^2 + (rt)^2 = r^2(1 + t^2) \pmod{p}$$

Because $(p, r) = 1$, this implies $1 + t^2 \equiv 0 \pmod{p}$; that is, $t^2 \equiv -1 \pmod{p}$. So -1 is a quadratic residue of p. This is a contradiction, by Corollary 11.2. Thus, n cannot be the sum of two squares.

Conversely, suppose that the exponent of each prime factor of n congruent to 3 modulo 4 in the canonical decomposition of n is even. Then n can be written as $n = a^2b$, where b is a product of distinct primes $\not\equiv 3 \pmod{4}$. By Theorem 13.4 and Lemma 13.8, b is representable as the sum of two squares. Therefore, by Lemma 13.7, $a^2b = n$ can be expressed as the sum of two squares.

This completes the proof. ■

This theorem enables us to determine if a positive integer can be written as the sum of two squares.

EXAMPLE 13.2 Notice that $5733 = 3^2 \cdot 7^2 \cdot 13$. Although the prime factors 3 and 7 in the factorization are both congruent to 3 modulo 4, their exponents are even. The prime factor 13 is congruent to 1 modulo 4. Since $13 = 3^2 + 2^2$, by Theorem 13.5, 5733 can be written as the sum of two squares:

$$5733 = (3 \cdot 7)^2(3^2 + 2^2) = (3 \cdot 7 \cdot 3)^2 + (3 \cdot 7 \cdot 2)^2 = 63^2 + 42^2$$ ■

Sums of Three Squares

We have seen that some positive integers cannot be expressed as the sum of two squares. So we are tempted to ask if every positive integer can be represented as the sum of three squares. With an extra square to spare, we might be tempted to think so.

For example, $3 = 1^2 + 1^2 + 1^2, 6 = 2^2 + 1^2 + 1^2, 14 = 3^2 + 2^2 + 1^2$, and $26 = 4^2 + 3^2 + 1^2$. But, unfortunately, 7 has no such representation. So adding three squares does not work.

Sums of Four Squares

Will four squares suffice? The previous culprit, 7, does have such a representation: $7 = 2^2 + 1^2 + 1^2 + 1^2$. In fact, we will see later that adding four squares will work for every positive integer.

Although Diophantus seems to have conjectured that every positive integer can be represented as the sum of four squares, the French mathematician Bachet was the first to state it explicitly. In 1621, he verified it for positive integers ≤ 325, but could not provide a proof. Fifteen years later, Fermat claimed that he had discovered a proof using his method of infinite descent, but once again, he did not supply any details of his proof.

In 1730, Euler took up the challenge. Thirteen years later, he extended Lemma 13.8 to include products of sums of four squares, paving the way to an eventual proof. In 1751, he proved a fundamental result, marking another milestone in our journey: The diophantine equation $1 + x^2 + y^2 \equiv 0 \pmod{p}$ is solvable for every prime p. Still, Euler was unable to develop a proof. In 1770, Lagrange, employing Euler's techniques, constructed the first published proof. Three years later, Euler also succeeded, demonstrating a simpler proof.

We will now develop a proof of the four-squares problem, in several steps.

LEMMA 13.11 **(Euler, 1743)** The product of the sums of four squares can be expressed as the sum of four squares.

PROOF

The proof of this lemma follows from the algebraic identity

$$\begin{aligned}(a^2+b^2+c^2+d^2)(e^2+f^2+g^2+h^2)\\ &= (ae+bf+cg+dh)^2+(af-be+ch-dg)^2\\ &\quad+(ag-bh-ce+df)^2+(ah+bg-cf-de)^2\end{aligned}$$

which can be verified by expanding the two sides separately. ■

Notice that Lemma 13.8 follows from this lemma. See Exercise 23.

The following example illustrates this lemma.

EXAMPLE 13.3 We have $7 = 2^2 + 1^2 + 1^2 + 1^2$ and $11 = 3^2 + 1^2 + 1^2 + 0^2$. Therefore, by Lemma 13.11,

$$\begin{aligned}77 = 7 \cdot 11 &= (2^2+1^2+1^2+1^2)(3^2+1^2+1^2+0^2)\\ &= (2\cdot 3+1\cdot 1+1\cdot 1+1\cdot 0)^2+(2\cdot 1-1\cdot 3+1\cdot 0-1\cdot 1)^2\\ &\quad+(2\cdot 1-1\cdot 0-1\cdot 3+1\cdot 1)^2+(2\cdot 0+1\cdot 1-1\cdot 1-1\cdot 3)^2\\ &= 8^2+(-2)^2+0^2+(-3)^2 = 8^2+3^2+2^2+0^2\end{aligned}$$

■

The next lemma brings us a step closer.

LEMMA 13.12 **(Euler, 1751)** If p is an odd prime, then there are integers x and y such that $1 + x^2 + y^2 \equiv 0 \pmod{p}$, where $0 \le x, y < p/2$.

PROOF

Consider the set $A = \{0^2, 1^2, \ldots, [(p-1)/2]^2\}$. Let r^2 and s^2 be any two distinct elements of A such that $r^2 \equiv s^2 \pmod{p}$. Then $r \equiv \pm s \pmod{p}$. Since $r \neq s$ and $r, s < p$, $r \not\equiv s \pmod{p}$. If $r \equiv -s \pmod{p}$, then $p|(r+s)$. This is impossible since $0 < r + s < p$. Thus, no two distinct elements of A are congruent modulo p.

Likewise, no two elements of the set $B = \{-1 - 0^2, -1 - 1^2, \ldots, -1 - [(p-1)/2]^2\}$ are congruent modulo p.

Then $A \cup B$ contains $(p+1)/2 + (p+1)/2 = p + 1$ elements. Therefore, by the pigeonhole principle, two elements of $A \cup B$ must be congruent modulo p. This implies that some element in A must be congruent to some element in B modulo p; that is, $x^2 \equiv -1 - y^2 \pmod{p}$ for some integers x and y, where $0 \le x, y < p/2$. Thus, $1 + x^2 + y^2 \equiv 0 \pmod{p}$. ■

The following example illustrates this lemma.

EXAMPLE 13.4 Let $p = 13$. Then $A = \{0^2, 1^2, 2^2, 3^2, 4^2, 5^2, 6^2\} = \{0, 1, 4, 9, 16, 25, 36\}$ and $B = \{-1 - 0^2, -1 - 1^2, -1 - 2^2, -1 - 3^2, -1 - 4^2, -1 - 5^2, -1 - 6^2\} = \{-1, -2, -5, -10, -17, -26, -37\}$. By Lemma 13.12, some element x^2 in A is congruent to some element $-1 - y^2$ in B. Since $5^2 \equiv -1 \pmod{13}$, $1 + 5^2 + 0^2 \equiv 0 \pmod{13}$; so $x = 5$ and $y = 0$. Similarly, $1 + 4^2 + 3^2 \equiv 0 \pmod{13}$, so $x = 4$ and $y = 3$. (These solutions can easily be found by observing that $A = \{0, 1, 4, 9, 3, 12, 10\}$ and $B = \{12, 11, 8, 3, 9, 0, 2\}$ modulo 13.) ■

Lemma 13.12 yields a corollary, which is analogous to Lemma 13.9.

COROLLARY 13.6 If p is an odd prime, there exists a positive integer $k < p$ such that kp can be expressed as the sum of four squares.

PROOF

By Lemma 13.12, there are integers x and y such that $1 + x^2 + y^2 \equiv 0 \pmod{p}$ where $0 \le x, y < p/2$. So $x^2 + y^2 + 1^2 + 0^2 = kp$ for some positive integer k. Since $0 \le x, y < p/2$, $x^2 + y^2 + 1 < (p/2)^2 + (p/2)^2 + 1 = p^2/2 + 1 < p^2$. Hence, $kp < p^2$, so $k < p$. Thus, $kp = x^2 + y^2 + 1^2 + 0^2$, where $k < p$. ■

For instance, it follows by Example 13.4 that $4^2 + 3^2 + 1 \equiv 0 \pmod{13}$; so $4^2 + 3^2 + 1^2 + 0^2 = 13k$, where $k = 2 < 13 = p$.

The next lemma, which is analogous to Lemma 13.10, shows that every prime can be represented as the sum of four squares.

LEMMA 13.13 Every prime can be written as the sum of four squares.

PROOF

Let p be a prime. Since $2 = 1^2 + 1^2 + 0^2 + 0^2$, the result follows when $p = 2$. So assume that p is odd.

By the well-ordering principle and Corollary 13.6, there exists a smallest positive integer m such that

$$mp = w^2 + x^2 + y^2 + z^2 \tag{13.6}$$

for some integers w, x, y, and z, where $1 \le m < p$.

Suppose m is even, so mp is also even. Then w, x, y, and z must have the same parity, or exactly two of them must have odd parity. Rearranging the terms when necessary, let us assume that $w \equiv x \pmod 2$ and $y \equiv z \pmod 2$. Then $(w+x)/2$, $(w-x)/2$, $(y+z)/2$, and $(y-z)/2$ are integers and

$$\left(\frac{w+x}{2}\right)^2 + \left(\frac{w-x}{2}\right)^2 + \left(\frac{y+z}{2}\right)^2 + \left(\frac{y-z}{2}\right)^2 = \frac{w^2+x^2+y^2+z^2}{2} = \left(\frac{m}{2}\right)p$$

Thus, $(m/2)p$ can be expressed as the sum of four squares, where $m/2 < m$. This violates the minimality of m, so m is odd.

We shall now show that $m = 1$. To this end, assume $m > 1$. Let a, b, c, and d be nonnegative integers such that $w \equiv a \pmod m$, $x \equiv b \pmod m$, $y \equiv c \pmod m$, and $z \equiv d \pmod m$, where $-m/2 < a, b, c, d < m/2$. Then

$$a^2 + b^2 + c^2 + d^2 \equiv w^2 + x^2 + y^2 + z^2 \equiv 0 \pmod m$$

so

$$a^2 + b^2 + c^2 + d^2 = mn \tag{13.7}$$

where n is some nonnegative integer and $0 \le a^2 + b^2 + c^2 + d^2 < 4(m/2)^2 = m^2$. Hence, $0 \le mn < m^2$, so $0 \le n < m$.

If $n = 0$, then $a = b = c = d = 0$; so $w \equiv x \equiv y \equiv z \equiv 0 \pmod m$. Then $m^2|(w^2 + x^2 + y^2 + z^2)$; this implies $m^2|mp$, so $m|p$. This is a contradiction since $1 < m < p$. Thus, $n \ge 1$ and hence $1 \le n < m$.

Using equations (13.6) and (13.7), Lemma 13.11 yields

$$\begin{aligned}
&(w^2 + x^2 + y^2 + z^2)(a^2 + b^2 + c^2 + d^2) \\
&\quad = (wa + xb + yc + zd)^2 + (wb - xa + yd - zc)^2 \\
&\qquad + (wc - xd - ya + zb)^2 + (wd + xc - yb - za)^2
\end{aligned}$$

That is,

$$m^2 np = (mp)(mn) = r^2 + s^2 + t^2 + u^2 \tag{13.8}$$

where

$$r = wa + xb + yc + zd$$

$$s = wb - xa + yd - zc$$

$$t = wc - xd - ya + zb$$

$$u = wd + xc - yb - za$$

Notice that $r = wa + xb + yc + zd \equiv w^2 + x^2 + y^2 + z^2 \equiv 0 \pmod{m}$. Similarly, $s \equiv t \equiv u \equiv 0 \pmod{m}$. Thus r, s, t, and u are all divisible by m.

Thus equation (13.8) yields

$$np = (r/m)^2 + (s/m)^2 + (t/m)^2 + (u/m)^2$$

where $n < m$. This violates the choice of m, so $m = 1$. Hence, $p = w^2 + x^2 + y^2 + z^2$, a sum of four squares. This concludes the proof. ■

We are now ready to combine the various facts we have just learned, so we can establish that every positive integer can be expressed as the sum of four squares. Although the English mathematician Edward Waring (1734–1798) stated this in his 1770 edition of *Meditationes Algebraicae*, Lagrange proved it in the same year.

THEOREM 13.6 **(Lagrange, 1770)** Every positive integer can be written as the sum of four squares.

PROOF

Let n be a positive integer. Since $1 = 1^2 + 0^2 + 0^2 + 0^2$, the result is true when $n = 1$.

Now assume that $n \geq 1$. Let $n = \prod_i p_i^{e_i}$ be the canonical decomposition of n. By Lemma 13.13, each p_i can be written as the sum of four squares.

Therefore, by Lemma 13.11, $p_i^{e_i}$ and hence $\prod_i p_i^{e_i} = n$, can be written as the sum of four squares. ■

The following example demonstrates this result.

EXAMPLE 13.5 Write 15,795 as the sum of four squares.

SOLUTION

Notice that $15{,}795 = 3^5 \cdot 5 \cdot 13$. Also $3 = 1^2 + 1^2 + 1^2 + 0^2$, $5 = 2^2 + 1^2 + 0^2 + 0^2$, and $13 = 3^2 + 2^2 + 0^2 + 0^2$. Then:

$$\begin{aligned}
3^5 &= 3^4(1^2+1^2+1^2+0^2) = 9^2(1^2+1^2+1^2+0^2)\\
&= 9^2+9^2+9^2+0^2 \quad \text{and}\\
5\cdot 13 &= (2^2+1^2+0^2+0^2)(3^2+2^2+0^2+0^2)\\
&= (2\cdot 3+1\cdot 2+0\cdot 0+0\cdot 0)^2+(2\cdot 2-1\cdot 3+0\cdot 0-0\cdot 0)^2\\
&\quad +(2\cdot 0-1\cdot 0-0\cdot 3+0\cdot 2)^2+(2\cdot 0+1\cdot 0-0\cdot 2-0\cdot 3)^2\\
&= 8^2+1^2+0^2+0^2
\end{aligned}$$

Therefore,

$$\begin{aligned}
15795 &= 3^5\cdot 5\cdot 13 = (9^2+9^2+9^2+0^2)(8^2+1^2+0^2+0^2)\\
&= (9\cdot 8+9\cdot 1+9\cdot 0+0\cdot 0)^2+(9\cdot 1-9\cdot 8+9\cdot 0-0\cdot 0)^2\\
&\quad +(9\cdot 0-9\cdot 0-9\cdot 8+0\cdot 1)^2+(9\cdot 0+9\cdot 0-9\cdot 1-0\cdot 8)^2\\
&= 81^2+63^2+72^2+9^2 = 81^2+72^2+63^2+9^2
\end{aligned}$$

■

Next, we turn to a conjecture related to expressing positive integers as sums of powers.

In his *Meditationes Algebraicae*, Waring also stated without proof that every positive integer can be written as the sum of 9 cubes, 19 fourth powers, and so on. His conjecture can be rephrased formally as follows.

Waring's Conjecture

Every positive integer can be written as the sum of $g(k)$ kth powers of nonnegative integers.

It follows by Lagrange's theorem that $g(2) = 4$. In 1909, Hilbert established the existence of the number $g(k)$ for each positive integer k, but no formula for $g(k)$ was given.

In 1851, Jacobi observed that both 23 and 239 require nine cubes: $23 = 2\cdot 2^3 + 7\cdot 1^3$ and $239 = 2\cdot 4^3 + 4\cdot 3^3 + 3\cdot 1^3$; he also observed that 454 is the largest integer $\le 12{,}000$ that requires eight cubes and 8042 is the largest integer $\le 12{,}000$ that requires seven cubes: $454 = 7^3 + 4\cdot 3^3 + 3\cdot 1^3$ and $8042 = 16^3 + 12^3 + 2\cdot 10^3 + 6^3 + 2\cdot 1^3$; interestingly, 239, 454, and 8042 remain the largest known integers requiring 9, 8, and 7 cubes, respectively. In 1912, A. Wieferich and A. J. Kempner showed that $g(3) = 9$; Twenty-seven years later, L. E. Dickson of the University of Chicago established that 23 and 239 are the only integers that require nine cubes, and every integer greater than 239 can be written as a sum of at most eight cubes.

In 1859, Liouville showed that $g(4) \leq 53$ and in 1974 H. E. Thomas showed that $g(4) \leq 22$. In 1986, R. Balasubramanian, J. Deshouillers, and F. Dress showed that $g(4) = 19$.

In 1964, Jing-Run Chen of the Institute of Mathematics at Beijing, showed that $g(5) = 37$; and in 1940, S. S. Pillai established that $g(6) = 73$. It is also known that $g(7) = 143$, $g(8) = 279$, and $g(9) = 548$.

?

Interestingly enough, an explicit formula for $g(k)$ has been conjectured for $k \geq 6$:

$$g(k) = \lfloor 3/2 \rfloor^k + 2^k - 2$$

EXERCISES 13.3

Express each integer as the sum of two squares.

1. 13 2. 29 3. 41 4. 97
5. Let $n = x^2 + y^2$ for some positive integers x and y, and k an arbitrary positive integer. Show that $k^2 n$ can be written as the sum of two squares.
6. Using complex numbers, prove that $(a^2 + b^2)(c^2 + d^2) = (ac + bd)^2 + (ad - bc)^2$.

Using Lemma 13.8, express each product as the sum of two squares.

7. $13 \cdot 17$ 8. $41 \cdot 53$

Determine whether each integer can be written as the sum of two squares.

9. 101 10. 137 11. 233 12. 585

Express each integer as the sum of two squares.

13. 149 14. 193 15. 200 16. 833

Let p, q, and r be distinct primes such that $p \equiv 1 \pmod 4$ and $q \equiv 3 \equiv r \pmod 4$. Determine whether each number is expressible as the sum of two squares.

17. $2^3 p^3 q^4 r^5$ 18. $2p^5 q^4 r^6$
19. Show that one more than twice the product of two consecutive positive integers can be written as the sum of two squares.
20. Show that no positive integer $n \equiv 7 \pmod 8$ can be written as the sum of three squares.
21. Let x, y, and z be positive integers such that $x^2 + y^2 + z^2 \equiv 0 \pmod 4$. Show that $x \equiv y \equiv z \equiv 0 \pmod 2$.
22. Show that no integer N of the form $4^e(8n + 7)$ can be represented as the sum of three squares. (*Hint*: Use Exercise 20.)
23. Deduce Lemma 13.8 from Lemma 13.11.

Using the facts that $7 = 2^2 + 1^2 + 1^2 + 1^2$ and $31 = 5^2 + 2^2 + 1^2 + 1^2$, represent each as the sum of four squares.

24. 49 25. 217 26. 343 27. 29,791

Express each prime as the sum of four squares.

28. 43 29. 89 30. 197 31. 349

Write each integer as the sum of four squares.

32. 81 33. 840 34. 1,275 35. 64,125
36. Show that every integer $n > 169$ can be written as the sum of five squares. (*Hint*: Consider $m = n - 169$, and use the fact that $169 = 13^2 = 12^2 + 5^2 = 12^2 + 4^2 + 3^2 = 10^2 + 8^2 + 2^2 + 1^2$.)
37. Write 23 as the sum of cubes.

John Pell *(1611–1685) was born in Sussex, England. After studying at Steyning School, he entered Trinity College, Cambridge, at the age of thirteen. He received his B.A. in 1629 and M.A. in the following year. He then became an assistant schoolmaster, although his father, a clergyman, had wished him to follow his profession.*

Pell's scholarship in mathematics and languages earned him a faculty position at the University of Amsterdam in 1643. Three years later, at the invitation of the Prince of Orange, he moved to a new academy at Breda, where he remained until 1652.

In 1654, Pell entered the English diplomatic service, serving as Oliver Cromwell's diplomat to Switzerland for the next four years. After returning to England, he joined the clergy, and in 1673 he became a chaplain to the Bishop of London. Pell died in London in utter poverty.

13.4 Pell's Equation

The integer 10 has an intriguing property: One less than the number is a square: $10 - 1 = 3^2$; and one less than one-half of the number is also a square: $10/2 - 1 = 2^2$. The next number that has the same property is 290: $290 - 1 = 17^2$ and $290/2 - 1 = 12^2$.

Are there are other such numbers? If yes, how do we find them?

To discover this, let r be such a number. Then $r - 1 = x^2$ and $r/2 - 1 = y^2$ for some positive integers x and y. Eliminating r between the equations yields the diophantine equation $x^2 - 2y^2 = 1$. Clearly $x = 3, y = 2$ is a solution of the equation: $3^2 - 2 \cdot 2^2 = 1$. Likewise, $x = 17, y = 12$ is also a solution: $17^2 - 2 \cdot 12^2 = 1$.

More generally, we would like to solve the nonlinear diophantine equation $x^2 - Ny^2 = 1$. This equation is commonly called **Pell's equation**, after John Pell. However, the name, mistakenly given by Euler, is a misnomer because Pell did not make any significant contributions to the study of the equation.

As early as 800 A.D., Indian mathematicians knew how to solve Pell's equation. About 650 A.D., Brahmagupta claimed that "a person who can within a year solve the equation $x^2 - 92y^2 = 1$ is a mathematician." His claim makes sense when we realize that $x = 1151, y = 120$ is the least solution: $1151^2 - 92 \cdot 120^2 = 1$. The Indian mathematicians Acharya Jayadeva (ca. 1000) and Bhaskara developed a method for solving Pell's equation.

Fermat, writing in 1657 to Wallis and Brouncker, proposed the problem of showing that Pell's equation has infinitely many solutions, when $N > 0$ and square-free. Bernard Frenicle de Bessey had computed the least solutions of the equation for

***Brahmagupta** (ca. 598–ca. 670), the most prominent Indian astronomer and mathematician of the seventh century, was born in Bhillamala, Gujarat, India. He lived and worked at the astronomical observatory in Ujjain in Rajasthan, where he had access to the writings of Hero of Alexandria, Ptolemy, Diophantus, and Aryabhata. At the age of thirty, he wrote his masterwork,* Brahma-sphuta-siddhanta *(the revised astronomical system of Brahma). Two of its 21 chapters are devoted to mathematics. Translated into Arabic in 775, the book contains the rules of operation with zero: $a+0=a=a-0$, and $a\cdot 0=0$. (He claimed incorrectly that $0\div 0=1$.) In 665, he wrote his second book,* Khandakhadyaka, *on the astronomical system of Aryabhata. The mathematical contents of both works appeared in English in 1817.*

Brahmagupta considered himself primarily an astronomer, but pursued mathematics for its applicability to astronomy. Besides his work on indeterminate equations, he contributed to solving Pell's equation $x^2-Ny^2=1$. He showed that the related diophantine equation $x^2-Ny^2=-1$ could not be solved unless N is the sum of two squares. He developed many algebraic and geometric formulas, including the extension of Hero's formula to cyclic quadrilaterals and a proof of the Pythagorean theorem. Brahmagupta fairly accurately computed the circumference of the earth and the length of the calendar year.

Besides astronomy and mathematics, Brahmagupta had a keen interest in physics also. His claim that "bodies fall towards the earth as it is in the nature of the earth to attract bodies, just as it is in the nature of water to flow," indicates that he had a strong sense of the gravitational force.

Brahmagupta was considered the gem of all mathematicians by the 12th century Indian mathematician Bhaskara.

***Bhaskara** (1114–ca. 1185), an astrologer, astronomer, and mathematician of exceptional computational ability, was born in Bijapur, Karnataka, India. He became head of the famous astronomical observatory at Ujjain, Rajasthan. In 1150, he discovered several approximations of π: $3927/1250=3.1416$ for accurate work, $22/7$ for a rough estimate, and $\sqrt{10}$ for everyday use.*

Bhaskara's most important work, Siddhanta Siromani, *written in 1150, consists of four parts: The first two parts,* Leelavati *(which means* the beautiful *and happens to be his daughter's name) and* Beejaganita *(seed arithmetic), deal with arithmetic and algebra, respectively, and the other two with astronomy and geometry.*

Leelavati *delineates the rules of addition, subtraction, multiplication, division, squaring, cubing, square and cube roots, fractions, operations with zero: $a\pm 0=a$, $0^2=0=\sqrt{0}$, and $a\div 0=\infty$, linear and quadratic equations, and arithmetic and geometric sequences. In addition, it includes a symbol for zero and is the earliest known systematic exposition of the decimal system.* Leelavati *was translated into Persian in 1587 at the request of Emperor Akbar, and into English in 1816 and again in 1817.*

Bhaskara developed a formula for the surface area and volume of a sphere, the number of permutations with repetitions, $\sin(A+B)$, and $\sin 2A$. He also solved Pell's equation $x^2-Ny^2=1$ for $N=8, 11, 32, 61$, and 67.

$N\le 150$. Soon after Fermat's proposal, Wallis and Brouncker described a method for solving the equation, with no formal justification. In 1767, Euler made significant contributions toward a formal proof. The following year, 111 years after Fermat's

Bernard Frenicle de Bessy *(ca. 1605–1675), a gifted amateur French mathematician and counselor at the Cour des Monnaies, was born in Paris. In 1666, Louis XIV appointed him to the Academy of Sciences. Frenicle kept extensive correspondences with the most important mathematicians of his time, including René Descartes, Fermat, Mersenne, and John Wallis.*

Frenicle proved that the area of a Pythagorean triangle cannot be a square, a problem proposed to him by Fermat in 1640. He proved also that the legs of a Pythagorean triangle cannot both be squares and hence the area of a Pythagorean triangle cannot be twice a square.

In 1657, Fermat proposed to him a diophantine problem: Find an integer x such that $Nx^2 + 1$ is a square, where N is a non-square. Frenicle solved it for $N \leq 150$.

initial proposal, Lagrange provided a complete proof, using the theory of continued fractions.† In 1842, Dirichlet constructed an improved version of Lagrange's proof.

We now turn to solving Pell's equation.

Trivial Solutions

When $N = 0$, we halve $x = \pm 1$, but y is arbitrary. They are **trivial solutions**. If $N \leq -2$, $x^2 - Ny^2 \geq 0$; so the only solutions are the trivial ones with $y = 0$, namely, $x = \pm 1, y = 0$. If $N = -1$, then $x^2 + y^2 = 1$; again, the only solutions are the trivial ones, namely, $(\pm 1, 0)$, $(0, \pm 1)$.

Nontrivial Solutions

So we assume that $N > 0$. If N is a square m^2, then Pell's equation yields

$$1 = x^2 - m^2y^2 = (x - my)(x + my)$$

Then $x - my = 1 = x + my$ or $x - my = -1 = x + my$. Solving these two linear systems, we get a solution in each case.

Thus, when $N < 0$ or N is a square, we know the exact solutions, and they are finite in number. So we turn to the case when N is positive and square-free.

The simplest such Pell equation is when $N = 2$: $x^2 - 2y^2 = 1$. Because $3^2 - 2 \cdot 2^2 = 1, x = 3, y = 2$ is a solution; so is $x = 17, y = 12$. Thus, $x^2 - 2y^2 = 1$ has at least two solutions. In fact, we shall see later that it has infinitely many solutions.

† See I. Niven *et al.*, *An Introduction to the Theory of Numbers*, 5th ed., Wiley, New York, 1991, 351–359.

Next, we make thee important observations about the solutions:

- If x, y is a solution, then so are $x, -y$; and $-x, \pm y$. Thus, the nontrivial solutions occur in quadruples. So we can always assume that x and y are positive.
- Because $1+Ny^2 = x^2$, nontrivial solutions can be found by computing $1+Ny^2$ and then determining if it is a square. For example, $1 + 92 \cdot 120^2 = 1151^2$; so $x = 1151$, $y = 120$ is a solution of $x^2 - 92y^2 = 1$, as noted earlier. This method, although it sounds simple in theory, is not practical.
- Let a, b be a solution. Then $a^2 - Nb^2 = (a + b\sqrt{N})(a - b\sqrt{N})$, so numbers of the form $a + b\sqrt{N}$ play an important role in finding the solutions. For convenience, we then say that the irrational number $a + b\sqrt{N}$ **yields** a solution. For instance, $1151 + 120\sqrt{92}$ yields a solution of $x^2 - 92y^2 = 1$.

The next lemma enables us to construct new solutions from known ones.

LEMMA 13.14 If r and s yield solutions of Pell's equation $x^2 - Ny^2 = 1$, where $N > 0$ and is square-free, then so does rs.

PROOF

Let $r = a + b\sqrt{N}$ and $s = c + d\sqrt{N}$. Since

$$rs = \left(a + b\sqrt{N}\right)\left(c + d\sqrt{N}\right) = (ac + bdN) + (ad + bc)\sqrt{N}$$

(see Exercise 11), it follows that rs also yields a solution. ■

The following example illustrates this lemma.

EXAMPLE 13.6 $r = 3 + 2\sqrt{2}$ and $s = 17 + 12\sqrt{2}$ yield two solutions of $x^2 - 2y^2 = 1$. Find two additional solutions.

SOLUTION

By Lemma 13.14, $rs = (3 \cdot 17 + 2 \cdot 12 \cdot 2) + (3 \cdot 12 + 2 \cdot 17)\sqrt{2} = 99 + 70\sqrt{2}$ yields a solution: $99^2 - 2 \cdot 70^2 = 1$. Likewise, $r^2 = (3 \cdot 3 + 2 \cdot 2 \cdot 2) + (3 \cdot 2 + 2 \cdot 3)\sqrt{2} = 17 + 12\sqrt{2}$ also yields a solution. ■

As we noticed in this example, it follows by Lemma 13.14 that every positive power of r also yields a solution.

The following lemma helps us determine the least solution of $x^2 - Ny^2 = 1$, where $N > 0$ and is square-free.

LEMMA 13.15 Let $N > 0$ and square-free. Suppose $r = a + b\sqrt{N}$ and $s = c + d\sqrt{N}$ yield solutions of $x^2 - Ny^2 = 1$, where $a, b, c, d > 0$. Then $r < s$ if and only if $a < c$.

PROOF

Suppose $r < s$. If $a \geq c$, then $a^2 \geq c^2$. Since $a^2 = 1 + Nb^2$ and $c^2 = 1 + Nd^2$, this implies $b^2 \geq d^2$. So $b \geq d$. Then $a + b\sqrt{N} \geq c + d\sqrt{N}$; that is, $r \geq s$. This contradicts our hypothesis, so $a < c$.

Conversely, let $a < c$. Then $a^2 < c^2$, so $1 + Nb^2 < 1 + Nd^2$. This implies $b^2 < d^2$ and hence $b < d$. So $r = a + b\sqrt{N} < c + d\sqrt{N} = s$. Thus, if $a < c$, then $r < s$. This concludes the proof. ■

Using the well-ordering principle, it follows by this lemma that Pell's equation has a **least solution** α, β. For example, $\alpha = 3$, $\beta = 2$ is the least solution of $x^2 - 2y^2 = 1$, and $\alpha = 2$, $\beta = 1$ is the least solution of $x^2 - 3y^2 = 1$.

The least solution α, β of $x^2 - Ny^2 = 1$ can be used to compute infinitely many solutions. To see this, first notice that $\alpha^2 - N\beta^2 = 1$, so $(\alpha^2 - N\beta^2)^n = 1$ for every integer n. Then

$$x^2 - Ny^2 = (\alpha^2 - N\beta^2)^n$$
$$\left(x + y\sqrt{N}\right)\left(x - y\sqrt{N}\right) = \left(\alpha + \beta\sqrt{N}\right)^n \left(\alpha - \beta\sqrt{N}\right)^n$$

Equating factors with the same sign, we get

$$x + y\sqrt{N} = \left(\alpha + \beta\sqrt{N}\right)^n \quad \text{and} \quad x - y\sqrt{N} = \left(\alpha - \beta\sqrt{N}\right)^n$$

Solving this linear system yields

$$x = \left[\left(\alpha + \beta\sqrt{N}\right)^n + \left(\alpha - \beta\sqrt{N}\right)^n\right]/2$$

and

$$y = \left[\left(\alpha + \beta\sqrt{N}\right)^n - \left(\alpha - \beta\sqrt{N}\right)^n\right]/\left(2\sqrt{N}\right)$$

Thus, we have the following result.

THEOREM 13.7 Suppose $\alpha + \beta\sqrt{N}$ yields the least solution of $x^2 - Ny^2 = 1$, where $N > 0$ and is square-free. Then the equation has infinitely many solutions, given by

$$x = \left[\left(\alpha + \beta\sqrt{N}\right)^n + \left(\alpha - \beta\sqrt{N}\right)^n\right]/2$$
$$y = \left[\left(\alpha + \beta\sqrt{N}\right)^n - \left(\alpha - \beta\sqrt{N}\right)^n\right]/\left(2\sqrt{N}\right)$$ ■

The next example demonstrates this theorem.

EXAMPLE 13.7 Using the fact that $3 + 2\sqrt{2}$ yields the least solution of $x^2 - 2y^2 = 1$, find two new solutions.

SOLUTION

By Theorem 13.7,

$$x = [(3+2\sqrt{2})^n + (3-2\sqrt{2})^n]/2$$
$$y = [(3+2\sqrt{2})^n - (3-2\sqrt{2})^n]/(2\sqrt{2})$$

is a solution for every integer n. Notice that $n = 1$ yields the least solution.

When $n = 2$,

$$x = [(3+2\sqrt{2})^2 + (3-2\sqrt{2})^2]/2 = 17$$

and

$$y = [(3+2\sqrt{2})^2 - (3-2\sqrt{2})^2]/(2\sqrt{2}) = 12$$

when $n = 3$,

$$x = [(3+2\sqrt{2})^3 + (3-2\sqrt{2})^3]/2 = 99$$

and

$$y = [(3+2\sqrt{2})^3 - (3-2\sqrt{2})^3]/(2\sqrt{2}) = 70$$

Thus, $17 + 12\sqrt{2}$ and $99 + 70\sqrt{2}$ yield two new solutions. ■

Next, we pursue a related Pell's equation and show how its solutions are related to Pythagorean triples.

A Related Pell's Equation

Like the number 10, 48 enjoys an intriguing property. One more than the number is a square: $48 + 1 = 7^2$; one more than one-half of it is also a square: $48/2 + 1 = 5^2$. In fact, 48 is the smallest such positive integer. The next number that works is 1680: $1680 + 1 = 41^2$ and $1680/2 + 1 = 29^2$.

To see if there are other positive integers with the same property, let s be such an integer. Then $s + 1 = x^2$ and $s/2 + 1 = y^2$ for some positive integers x and y. These two equations yield the **related Pell's equation** $x^2 - 2y^2 = -1$.

Clearly, $x = 1 = y$ is a solution (It yields $s = 0$); in fact, it is the least solution. A second solution is $x = 7, y = 5 : 7^2 - 2 \cdot 5^2 = -1$; then $s = x^2 - 1 = 49 - 1 = 48$.

More generally, the infinitely many solutions of the nonlinear diophantine equation $x^2 - Ny^2 = -1$, when solvable, are given by the following theorem. Its proof follows along the same lines as that of Theorem 13.7, so in the interest of brevity, we leave it as an exercise.

THEOREM 13.8 Suppose $\alpha + \beta\sqrt{N}$ yields the least solution of $x^2 - Ny^2 = -1$, where $N > 0$ and is square-free. Then the equation has infinitely many solutions, given by

$$x = \left[\left(\alpha + \beta\sqrt{N}\right)^{2n-1} + \left(\alpha - \beta\sqrt{N}\right)^{2n-1}\right]/2$$
$$y = \left[\left(\alpha + \beta\sqrt{N}\right)^{2n-1} - \left(\alpha - \beta\sqrt{N}\right)^{2n-1}\right]/\left(2\sqrt{N}\right)$$ ■

The following example illustrates this theorem.

EXAMPLE 13.8 Using the fact that $1 + \sqrt{2}$ is the least solution of $x^2 - 2y^2 = -1$, find a new solution.

SOLUTION

We already know that 7, 5 is also a solution. So to find a third solution, we use Theorem 13.8 and the binomial theorem with $n = 3$. Then

$$x = \left[\left(1 + \sqrt{2}\right)^5 + \left(1 - \sqrt{2}\right)^5\right]/2 = 41$$

and

$$y = \left[\left(1 + \sqrt{2}\right)^5 - \left(1 - \sqrt{2}\right)^5\right]/\left(2\sqrt{2}\right) = 29$$

Thus, $41 + 29\sqrt{2}$ yields a new solution: $41^2 - 2 \cdot 29^2 = -1$. ■

Next, we present an interesting relationship between Pell's equation and its relative, and Pythagorean triples.

Primitive Pythagorean Triples Revisited

Recall from Section 13.1 that there are primitive Pythagorean triples x-y-z such that x and y are consecutive integers; 3-4-5, 21-20-29, and 119-120-169 are three such triples. The legs of the corresponding primitive Pythagorean triangles are consecutive integers.

For x and y to be consecutive integers, we must have $(m^2 - n^2) - 2mn = \pm 1$ for some integers m and n. This can be written in terms of Pell's equation as $(m - n)^2 - 2n^2 = \pm 1$. Consequently, such triples can be obtained alternately by solving the diophantine equations $u^2 - 2v^2 = \pm 1$.

For example, the least solution of $u^2 - 2v^2 = -1$ is $u = 1 = v$. This implies $m - n = 1 = n$, so $m = 2$ and $n = 1$. Correspondingly, $x = m^2 - n^2 = 3$ and $y = 2mn = 4$ are consecutive lengths of the legs of the 3-4-5 Pythagorean triangle.

The least solution of $u^2 - 2v^2 = 1$ is $u = 3, v = 2$. This implies $m - n = 3$ and $n = 2$; so $m = 5$ and $n = 2$. Then $x = m^2 - n^2 = 21$ and $y = 2mn = 20$, once again consecutive legs.

Fibonacci and Lucas Numbers Revisited

In their 1967 study of simple continued fractions, C. T. Long and J. H. Jordan of Washington State University discovered a close relationship among the golden ratio, F_n, L_n and the nonlinear diophantine equation $x^2 - Ny^2 = n$. The following results, discovered by them, "provide unusual characterization of both Fibonacci and Lucas numbers." We omit their proofs in the interest of brevity:

- The diophantine equation $x^2 - 5y^2 = -4$ is solvable in positive integers if and only if $x = L_{2n-1}$ and $y = F_{2n-1}$, where $n \geq 1$.
- The diophantine equation $x^2 - 5y^2 = 4$ is solvable in positive integers if and only if $x = L_{2n}$ and $y = F_{2n}$, where $n \geq 1$.

EXERCISES 13.4

Find two nontrivial solutions of Pell's equation with the given value of N.

1. 5 2. 7 3. 10 4. 11

Consider the Pell's equation $x^2 - 15y^2 = 1$.

5. Find a nontrivial solution.
6. Using the solution in Exercise 5, find two additional solutions.

7–8. Redo Exercises 5–6 with $x^2 - 14y^2 = 1$.

9. Using the fact that $x = 17, y = 12$ is a solution of $x^2 - 2y^2 = 1$, find a solution of $u^2 - 8v^2 = 1$.
10. Using the fact that $x = 97$, $y = 56$ is a solution of $x^2 - 3y^2 = 1$, find a solution of $u^2 - 12v^2 = 1$.
11. Let a, b, c, and d be any integers. Prove that $(a^2 - Nb^2)(c^2 - Nd^2) = (ac + Nbd)^2 - N(ad + bc)^2$
12. Using the fact that $2 + \sqrt{3}$ yields the least solution of $x^2 - 3y^2 = 1$, find two new solutions.
13. Using the fact that $5 + 2\sqrt{6}$ yields the least solution of $x^2 - 6y^2 = 1$, find two new solutions.

Suppose that $r = a + b\sqrt{N}$ yields a solution of $x^2 - Ny^2 = 1$. Prove each.

14. $1/r$ yields a solution.
15. r^n yields a solution for every $n \geq 1$.
16. r^{-n} yields a solution for every $n \geq 1$.
17. Suppose that $x_1 + y_1\sqrt{N} = \alpha + \beta\sqrt{N}$ yields the least solution of $x^2 - Ny^2 = 1$ and that x_n, y_n denotes an arbitrary solution, where $n \geq 1$. Show that
$$x_n = \alpha x_{n-1} + \beta N y_{n-1}, \qquad y_n = \beta x_{n-1} + \alpha y_{n-1}$$
is also a solution. (Koshy, 2000)
18. Deduce the recursive definition of solutions if $N = 2$. (Koshy, 2000)

In Exercises 19 and 20, a Pell's equation and its least solution are given. Using the recursive formula in Exercise 17, find two new solutions in each case.

19. $x^2 - 2y^2 = 1$; 3, 2 20. $x^2 - 23y^2 = 1$; 24, 5
21. Prove that the diophantine equation $x^2 - 3y^2 = -1$ has no solutions.
22. Prove that the diophantine equation $x^2 - Ny^2 = -1$ has no solutions if $N \equiv 3 \pmod 4$.
23. Establish Theorem 13.8.

Let x_n, y_n be an arbitrary solution of $x^2 - Ny^2 = -1$, where $x_1 = \alpha$, $y_1 = \beta$ is its least solution.

24. Define x_n, y_n recursively.
25. Deduce the recursive formula if $N = 2$.
26. Using Exercise 25 and the fact that $\alpha = 1 = \beta$ is the least solution of $x^2 - 2y^2 = -1$, compute x_2, y_2 and x_3, y_3.

27. Find a formula for the combined solutions of the diophantine equations $x^2 - 2y^2 = \pm 1$.
(*Hint*: Use Theorems 13.7 and 13.8.)

28. Let X-Y-Z be a primitive Pythagorean triangle with consecutive legs. Find a formula for X and Y.
(*Hint*: Use Exercise 27.)

CHAPTER SUMMARY

This chapter presented several nonlinear diophantine equations. Five of them are

Pythagorean equation	$x^2 + y^2 = z^2$
Fermat's equation	$x^n + y^n = z^n, n \geq 3$
Beal's equation	$x^m + y^n = z^r, m, n, r \geq 3$
Pell's equation	$x^2 - Ny^2 = 1$
Its relative	$x^2 - Ny^2 = -1$

Pythagorean Triple

- A triplet x-y-z such that $x^2 + y^2 = z^2$ is a Pythagorean triple. (p. 581)
- It is primitive if $(x, y, z) = 1$. (p. 582)
- If x-y-z is a primitive Pythagorean triple, then $(x, y) = (y, z) = (z, x) = 1$. (p. 582)
- If x-y-z is a primitive Pythagorean triple, then $x \not\equiv y \pmod 2$. (p. 583)
- If x-y-z is a primitive Pythagorean triple and x is even, then $y \equiv z \equiv 1 \pmod 2$. (p. 583)
- If $(r, s) = 1$ and rs is a square, then both r and s are also squares. (p. 583)
- Let x-y-z be a primitive Pythagorean triple. Then there are relatively prime integers m and n with different parity such that $x = 2mn$, $y = m^2 - n^2$, and $z = m^2 + n^2$, where $m > n$. (p. 584)
- Let $x = 2mn$, $y = m^2 - n^2$, and $z = m^2 + n^2$, where m and n are relatively prime integers with different parity and $m > n$. Then x-y-z is a primitive Pythagorean triple. (p. 584)
- Let x, y, z be positive integers, where x is even. Then x-y-z is a primitive Pythagorean triple if and only if there are relatively prime integers m and n with different parity such that $x = 2mn$, $y = m^2 - n^2$, and $z = m^2 + n^2$, where $m > n$. (p. 585)

Fermat's Last Theorem

- (Fermat's conjecture) The diophantine equation $x^n + y^n = z^n$ has no positive integral solutions if $n \geq 3$. (p. 590)

- The diophantine equation $x^4 + y^4 = z^2$ has no positive integral solutions. (p. 591)
- The diophantine equation $x^4 + y^4 = z^4$ has no positive integral solutions. (p. 592)
- Not all lengths of the sides of a Pythagorean triangle can be squares. (p. 593)
- In 1994, Wiles established Fermat's conjecture. (p. 594)
- The diophantine equation $x^4 - y^4 = z^2$ has no positive integral solutions. (p. 597)
- The area of a Pythagorean triangle cannot be an integral square. (p. 597)
- Let x and y be positive integers. Then $x^2 + y^2$ and $x^2 - y^2$ cannot both be squares. (p. 598)

Beal's Conjecture

- The diophantine equation $x^m + y^n = z^r$ has no positive integral solutions, where $(x, y, z) = 1$ and $m, n, r \geq 3$. (p. 599)

Sum of Two Squares

- If $n \equiv 3 \pmod 4$, then n cannot be written as the sum of two squares. (p. 603)
- If n can be written as the sum of two squares, so can $k^2 n$. (p. 603)
- The product of two sums of two squares can be expressed as the sum of two squares. (p. 603)
- If $p \equiv 1 \pmod 4$, there are positive integers x and y such that $x^2 + y^2 = kp$ for some positive integer $k < p$. (p. 604)
- Every prime $p \equiv 1 \pmod 4$ can be written as the sum of two squares. (p. 604)
- If $p = 2$ or $p \equiv 1 \pmod 4$, then p is expressible as the sum of two squares. (p. 605)
- A positive integer is expressible as the sum of two squares if and only if the exponent of each of its prime factors congruent to 3 (mod 4) in the canonical decomposition of n is even. (p. 605)

Sum of Four Squares

- The product of the sums of four squares can be written as the sum of four squares. (p. 607)
- If p is an odd prime, there exist integers x and y such that $1 + x^2 + y^2 \equiv 0 \pmod p$, where $0 \leq x, y < p/2$. (p. 607)
- If p is an odd prime, there is a positive integer $k < p$ such that kp can be written as the sum of four squares. (p. 608)

- Every prime can be written as the sum of four squares. (p. 608)
- Every positive integer can be written as the sum of four squares. (p. 610)

Waring's Conjecture

- Every positive integer can be written as the sum of $g(k)$ kth powers of nonnegative integers. (p. 611)

Pell's Equation $x^2 - Ny^2 = 1$

- Let $N > 0$ and square-free. Suppose that $r = a + b\sqrt{N}$ and $s = c + d\sqrt{N}$ yield solutions, where $a, b, c, d > 0$. Then $r < s$ if and only if $a < c$. (p. 616)
- If r and s yield solutions, then so does rs. (p. 616)
- Suppose $\alpha + \beta\sqrt{N}$ yields the least solution. Then the infinitely many solutions are given by

$$x = \left[(\alpha + \beta\sqrt{N})^n + (\alpha - \beta\sqrt{N})^n\right]/2$$
$$y = \left[(\alpha + \beta\sqrt{N})^n - (\alpha - \beta\sqrt{N})^n\right]/(2\sqrt{N}) \quad \text{(p. 617)}$$

- Suppose $\alpha + \beta\sqrt{N}$ yields the least solution of $x^2 - Ny^2 = -1$. Its infinitely many solutions are given by

$$x = \left[(\alpha + \beta\sqrt{N})^{2n-1} + (\alpha - \beta\sqrt{N})^{2n-1}\right]/2$$
$$y = \left[(\alpha + \beta\sqrt{N})^{2n-1} - (\alpha - \beta\sqrt{N})^{2n-1}\right]/(2\sqrt{N}) \quad \text{(p. 619)}$$

REVIEW EXERCISES

Study the following primitive Pythagorean triples pattern:

$$\begin{aligned} 69^2 &+ 260^2 &= 269^2 \\ 609^2 &+ 20600^2 &= 20609^2 \\ 6009^2 &+ 2006000^2 &= 2006009^2 \\ 60009^2 &+ 200060000^2 &= 200060009^2 \\ &\vdots \end{aligned}$$

1. Find the next two lines in the pattern.
2. Predict a formula for the nth equation.
3. Justify the formula in Exercise 2.

Use Table 13.6 to answer Exercises 4 and 5.

Generators		*Sides of a Pythagorean Triangle*		
M	N	X	Y	Z
3	1	6	8	10
6	3	36	27	45
10	6	120	64	136
15	10	300	125	325

Table 13.6

4. Predict the next two lines in the table.
5. Make a conjecture about the generators and a leg of a Pythagorean triangle.
6. Verify that the primitive Pythagorean triangles with generators 60, 59; 68, 37; and 84, 1 have the same perimeter. (A. S. Anema, 1949)
7. Show that every Pythagorean triple x-y-z, where x, y, and z are in arithmetic progression, is of the form $3n$-$4n$-$5n$, where $n \geq 1$.
8. Suppose that the generators of a Pythagorean triangle are consecutive triangular numbers. Show that the length of a leg of the triangle is a cube.
9. Verify that the Pythagorean triangles with generators 133, 88; 152, 35; 153, 133; and 133, 65 enclose the same area. Find the common area.
10. Let a, b, and c be positive integers such that $a^2 + ab + b^2 = c^2$. Prove that the Pythagorean triangles with generators c, a; c, b; and $a + b$, c enclose the same area. (Diophantus)
11. Let $a = m^2 + mn$, $b = 3n^2 - mn$, $c = m^2 - mn$, and $d = 3n^2 + mn$. Prove that the Pythagorean triangles with generators a, b, and c, d enclose the same area. (W. P. Whitlock, Jr., 1943)
12. Let $A = a^2 + 3b^2 = C = E$, $B = 3b^2 + 2ab - a^2$, $D = 3b^2 - 2ab - a^2$, and $F = 4ab$, where a and b are positive integers. Prove that the Pythagorean triangles generated by A, B; C, D; and E, F enclose the same area. (Euler)

Use the array in Table 13.7, studied in 1950 by H. L. Umansky of Emerson High School, Union City, New Jersey, to answer Exercises 13 and 14.

m \ n	1	2	3	4	5	...
1	·					
2	5					
3	10	13				
4	17	20	25			
5	26	29	34	41		
6	37	40	45	52	61	

Table 13.7

13. Add the next two lines.
14. Define the element $A(m, n)$ in row m and column n of the array, where $1 \leq m < n$.

Let m and n be the generators of a primitive Pythagorean triangle with a square perimeter P. (W. P. Whitlock, Jr., 1948)

15. Find P.
16. Show that 8, 1; 18, 7; and 78, 71 generate primitive Pythagorean triangles with square perimeters.
17. Prove that m must be even and n odd.
18. Prove that $M = 2m^2$ and $N = n^2 + 2mn - m^2$ generate a primitive Pythagorean triangle with perimeter P^2.

Determine whether each number is representable as the sum of two squares.

19. 131 20. 157 21. 341 22. 873

Express each number as the sum of two squares.

23. 289 24. 349 25. 449 26. 697

Using Lemma 13.8, express each product as the sum of two squares.

27. $37 \cdot 41$ 28. $74 \cdot 73$

Write each prime as the sum of four squares.

29. 179 30. 293 31. 337 32. 739

Write each integer as the sum of four squares.

33. $13 \cdot 17$ 34. $23 \cdot 43$ 35. 1360 36. 2394
37. Show that the diophantine equation $w^3 + x^3 = y^3 + z^3$ has infinitely many positive integral solutions.

Consider the diophantine equation $w^4 + x^4 = y^4 + z^4$.

38. Show that 59-158-133-134 is a solution.
39. Show that it has infinitely many integral solutions.

Find a nontrivial solution of each Pell's equation.

40. $x^2 - 18y^2 = 1$ 41. $x^2 - 20y^2 = 1$

Using the given least solution of the corresponding Pell's equation, find another solution.

42. 33, 8; $x^2 - 17y^2 = 1$ 43. 23, 4; $x^2 - 33y^2 = 1$

Using the given least solution of the corresponding Pell's equation and recursion, construct a new solution.

44. 25, 4; $x^2 - 39y^2 = 1$

45. 48, 7; $x^2 - 47y^2 = 1$

Use the diophantine equation $x^2 - 5y^2 = -1$ for Exercises 46 and 47.

46. Find the least solution.
47. Using the explicit formula in Theorem 13.7, find two new solutions.
48. Using recursion and the least solution in Exercise 46, find two new solutions.

SUPPLEMENTARY EXERCISES

1. In 1953, F. L. Miksa of Illinois found four primitive Pythagorean triangles whose areas contain all ten digits, each exactly once. They are generated by 320, 49; 122, 293; 322, 83; and 298, 179. Confirm Miksa's observation.

Verify each. (V. Thébault, 1946)

2. Each of the consecutive integers 370 and 371 equals the sum of the cubes of its digits.
3. Each of the consecutive integers 336,700 and 336,701 equals the sum of the cubes of the numbers formed by the digits grouped in twos.
4. Each of the consecutive integers 33,366,700 and 33,366,701 equals the sum of the cubes of the numbers formed by the digits grouped in threes.
5. Add the next line to the number pattern formed by Exercises 2–4.
6. A product of k different primes $\equiv 1$ or 3 (mod 8) can be written in the form $x^2 + 2y^2$ in 2^{k-1} ways. Express 561 in the desired form. (W. P. Whitlock, Jr., 1946)

The sum of $n + 1$ consecutive squares can sometimes be written as the sum of the squares of the next n consecutive integers. For example, $3^2 + 4^2 = 5^2$ and $10^2 + 11^2 + 12^2 = 13^2 + 14^2$. Express each as a sum of consecutive squares.

7. $21^2 + 22^2 + 23^2 + 24^2$

8. $36^2 + 37^2 + 38^2 + 39^2 + 40^2$

9. Predict the next equation in this number pattern.
10. Prove that the sum of $n + 1$ consecutive squares, beginning with $[n(2n + 1)]^2$ equals the sum of the squares of the next n consecutive squares.

The sum of consecutive squares can be a single square. For example, $1^2 + 2^2 + 3^2 + 4^2 + \cdots + 24^2 = 70^2$. Express each as a single square.

11. $18^2 + 19^2 + 20^2 + \cdots + 28^2$

12. $38^2 + 39^2 + 40^2 + \cdots + 48^2$

13. Jekuthiel Ginsburg of Yeshiva College (now Yeshiva University) observed that the positive factors 1, 2, 3, and 6 of 6 have 1, 2, 2, and 4 positive factors, respectively, and $(1 + 2 + 2 + 4)^2 = 1^3 + 2^3 + 2^3 + 4^3$. Verify that 30 has a similar property.

Study the following number pattern:

$$3(1^3) = 1^3 + 2(1^4)$$
$$3(1+3)^3 = 1^3 + 3^3 + 2(1^4 + 3^4)$$
$$3(1+3+6)^3 = 1^3 + 3^3 + 6^3 + 2(1^4 + 3^4 + 6^4)$$
$$\vdots$$

14. Add the next two lines.
15. Conjecture a formula for the nth line.
16. Find three consecutive positive integers x, y, and z such that $x^3 + y^3 + z^3$ is a cube.
17. Let w, x, y, and z be four consecutive positive integers such that $w^3 + x^3 + y^3 = z^3$. Show that $w = 3$. (In 1911, L. Aubry proved that 3, 4, and 5 are the only three consecutive integers, where the sum of each integer cubed is also a cube.)
18. Jacobi found that $2\left(\sum_{i=1}^{n} i\right)^4 = \sum_{i=1}^{n} i^5 + \sum_{i=1}^{n} i^7$. Verify the formula for $n = 4$.
19. Verify that each of the sums $14^2 + 36^2 + 58^2 + 90^2$ and $16^2 + 38^2 + 50^2 + 94^2$ equals the sum of the squares of the same bases with their digits reversed. (D. R. Kaprekar and R. V. Iyer, 1950)

In 1937, R. Goormaghtigh of Belgium discovered a delightful method for expressing sums of squares of integers as sums of squares with the digits reversed. It is based on the following fact:

$$\text{If } \sum_{i=1}^{n} x_i^2 = \sum_{i=1}^{n} y_i^2, \quad \text{then } \sum_{i=1}^{n} (10x_i + y_i)^2 = \sum_{i=0}^{n-1} (10y_{n-i} + x_{n-i})^2 \tag{13.9}$$

20. Establish formula (13.9).

Using formula (13.9), express each as a sum of three squares. (*Hint*: $4^2 + 5^2 + 6^2 = 8^2 + 3^2 + 2^2$.)

21. $48^2 + 53^2 + 62^2$
22. $43^2 + 52^2 + 68^2$

Using formula (13.9), express each as a sum of four squares. (*Hint*: $8^2 + 5^2 + 3^2 + 2^2 = 7^2 + 6^2 + 4^2 + 1^2$.)

23. $87^2 + 56^2 + 34^2 + 21^2$
24. $86^2 + 54^2 + 31^2 + 27^2$
25. Find four consecutive positive integers w, x, y, and z such that $w^3 + x^3 + y^3 + z^3$ is also a cube.
26. Find a solution of the diophantine equation $w^n + x^{n+1} + y^{n+2} + z^{n+3} = u^{n^2+3n+1}$. (N. Schaumberger, 1993)

27. Express the beastly number 666 as the sum of squares of consecutive primes. (M. Zerger, 1993)

COMPUTER EXERCISES

Write a program to perform each task.

1. Read in a positive integer $n \leq 25$ and print all primitive Pythagorean triples x-y-z, where $x \leq n$.
2. Read in a positive integer $n \leq 25$. Print the Fibonacci numbers F_n through F_{n+3}. Use them to print a Pythagorean triple x-y-z. Determine if it is a primitive Pythagorean triple.
3. Read in a positive integer n and a primitive Pythagorean triple x-y-z, where $y = x + 1$. Print n primitive Pythagorean triples.
4. Find four primitive Pythagorean triangles whose areas contain all ten digits. (F. L. Miksa, 1953)
5. Read in a positive integer n. Determine if it can be written as the sum of two squares. If yes, express it as such a sum.
6. Read in a positive integer n. Express it as the sum of four squares.
7. Read in a square-free positive integer N and the least solution of Pell's equation $x^2 - Ny^2 = 1$. Use recursion to print five new solutions.
8. Redo Exercise 6 with $x^2 - Ny^2 = -1$.

ENRICHMENT READINGS

1. A. D. Aczel, *Fermat's Last Theorem*, Four Walls Eight Windows, New York, 1996.
2. A. H. Beiler, *Recreations in the Theory of Numbers*, Dover, New York, 1966, 104–134, 248–268, 276–293.
3. N. Boston, "A Taylor-Made Plug for Wiles' Proof," *The College Mathematics Journal*, 26 (March 1995), 100–105.
4. K. Devlin, *Mathematics: The Science of Patterns*, W. H. Freeman, New York, 1996, 28–35, 200–208.
5. K. Devlin, "Move Over Fermat, Now It's Time for Beal's Problem," *Math Horizons* (Feb. 1998), 8–10.
6. M. Graham, "President Garfield and the Pythagorean Theorem," *Mathematics Teacher*, 69 (Dec. 1976), 686–687.
7. R. K. Guy, "What's Left," *Math Horizons* (April 1998), 5–7.

8. T. Jackson, "A Short Proof That Every Prime $p \equiv 3 \pmod 8$ Is of the Form $x^2 + 2y^2$," *The American Mathematical Monthly*, 107 (May 2000), 447.
9. R. D. Mauldin, "A Generalization of Fermat's Last Theorem: The Beal Conjecture and Prize Problem," *Notices of the American Mathematical Society*, 44 (Dec. 1997), 1436–1437.
10. S. Singh, *Fermat's Enigma*, Anchor Books, New York, 1997.
11. D. Veljan "The 2500-Year-Old Pythagorean Theorem," *Mathematics Magazine*, 73 (Oct. 2000), 259–272.

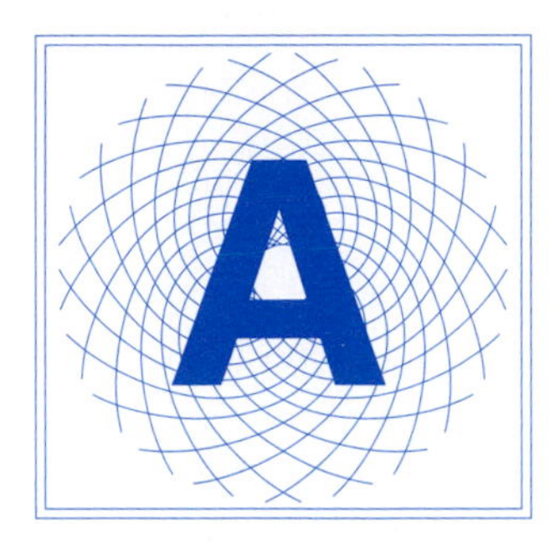

Appendix

A.1 Proof Methods

Proofs, no matter how simple or complicated, are the heart and soul of mathematics. They play a central role in its development and guarantee the correctness of mathematical results. No results are accepted as true unless they can be proved using logical reasoning.

Before we discuss the various proof techniques, we need several definitions and some facts from symbolic logic.

Proposition

A **proposition** (or **statement**) is a declarative sentence that is either true or false, but *not* both. It is often denoted by the letter p, q, r, s, or t. The truthfulness or falsity of a proposition is its **truth value**, denoted by T and F, respectively.

Compound Proposition

A **compound proposition** is formed by combining two or more simple propositions, using logic operators: $\wedge$ (and), $\vee$ (or), $\sim$ (not), $\rightarrow$ (implies), and $\leftrightarrow$ (if and only if). The **conjunction** $p \wedge q$ of two propositions p and q is true if and only if both components are true; their **disjunction** $p \vee q$ is true if at least one component is true. The negation $\sim p$ has the opposite truth value of p. An **implication** $p \rightarrow q$ (read *p implies q* or *if p, then q*) is false only if the **premise** p is true and the **conclusion** q is false. A **biconditional** $p \leftrightarrow q$ (read *p if and only if q*) is true if and only if both components have the same truth value.

p	q	$p \wedge q$	$p \vee q$	$\sim p$	$p \to q$	$p \leftrightarrow q$
T	T	T	T	F	T	T
T	F	F	T	F	F	F
F	T	F	T	T	T	F
F	F	F	F	T	T	T

Table A.1

Table A.1 shows the truth tables for the various compound propositions corresponding to the various possible pairs of truth values of p and q.

Three new implications can be constructed from a given implication $p \to q$. They are the **converse** $q \to p$, **inverse** $\sim p \to \sim q$, and the **contrapositive** $\sim q \to \sim p$.

A **tautology** is a compound proposition that is always true. A **contradiction** is a compound proposition that is always false. A **contingency** is a compound proposition that is neither a tautology nor a contradiction.

Logical Equivalence

Two compound propositions p and q are **logically equivalent** if they have identical truth values in every case, symbolized by $p \equiv q$. For example, $p \to q \equiv \sim p \vee q$ and $p \to q \equiv \sim q \to \sim p$.

Proof Methods

A **theorem** in mathematics is a true proposition. Theorems are often stated as implications H $\to$ C, where H denotes the hypothesis of the theorem and C its conclusion. **Proving** such a theorem means verifying that the proposition H $\to$ C is a tautology. This section presents six standard methods for proving theorems: **vacuous proof**, **trivial proof**, **direct proof**, **indirect proof**, **proof by cases**, and **existence proof** (see Figure A.1). Vacuous and trivial proofs are, in general, parts of larger and more complicated proofs.

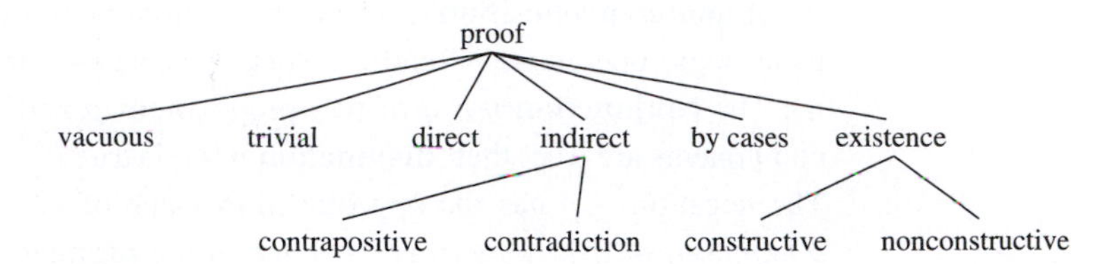

Figure A.1

Vacuous Proof

Suppose the hypothesis H of the implication $H \to C$ is false. Then the implication is true regardless of whether the conclusion C is true or false. Thus, if the hypothesis can be shown to be false, the theorem $H \to C$ is true by default; such a proof is a **vacuous proof**. Vacuous proofs, though rare, are necessary to handle special cases.

EXAMPLE A.1 Consider the statement, *If* $1 = 2$, *then* $3 = 4$. Since the hypothesis is false, the proposition is vacuously true. ■

Trivial Proof

Suppose the conclusion of the implication $H \to C$ is true. Again, the implication is true irrespective of the truth value of H. Consequently, if C can be shown to be true, such a proof is a **trivial proof**.

EXAMPLE A.2 Consider the statement, *If* $1 = 2$, *then Paris is in France*. This statement is trivially true since the conclusion is true. ■

Direct Proof

In a **direct proof** of the theorem $H \to C$, we assume that the given hypothesis H is true. Using the laws of logic or previously known facts, we then establish the desired conclusion C as the final step of a chain of implications: $H \to C_1, C_1 \to C_2, \ldots,$ and $C_n \to C$. Then $H \to C$.

The next example illustrates this method.

Often, theorems are stated in the form of sentences, so we first rewrite them symbolically and then work with the symbols, as the next example shows.

EXAMPLE A.3 Prove directly that the product of any two odd integers is an odd integer.

PROOF

Let x and y be any two odd integers. Then there exist integers m and n such that $x = 2m + 1$ and $y = 2n + 1$. Thus,

$$\begin{aligned} x \cdot y &= (2m + 1) \cdot (2n + 1) \\ &= 4mn + 2m + 2n + 1 \\ &= 2(2mn + m + n) + 1 \\ &= 2k + 1 \end{aligned}$$

where $k = 2mn + m + n$ is an integer. Thus, xy is also an odd integer. ■

Indirect Proof

There are two kinds of an **indirect proof** of the theorem $H \to C$: **proof of the contrapositive** and **proof by contradiction**. The first method is based on the law of the contrapositive, $H \to C \equiv \sim C \to \sim H$. In this method, we assume the desired conclusion C is false; then, using the laws of logic, we establish that the hypothesis H is also false. Once this is done, the theorem is proved.

The following example illustrates this method.

EXAMPLE A.4 Prove indirectly: If the square of an integer is odd, then the integer is odd.

PROOF OF THE CONTRAPOSITIVE

Let x be any integer such that x^2 is odd. We would like to prove that x must be an odd integer. In the indirect method, we assume the conclusion is false; that is, x is *not* odd, in other words, assume x is an even integer. Let $x = 2k$ for some integer k. Then $x^2 = (2k)^2 = 4k^2 = 2(2k^2)$, which is an even integer. This makes our hypothesis that x^2 is an odd integer false. Therefore, by the law of the contrapositive, our assumption must be wrong; so x must be an odd integer. Thus, if x^2 is an odd integer, then x is also an odd integer. ■

Proof by contradiction is based on the law of *reductio ad absurdum*: $H \to C \equiv [H \wedge (\sim C)] \to F$, where F denotes a contradiction. In this method, we assume the given hypothesis H is true, but the conclusion C is false. We then argue logically and reach a contradiction F.

The following example illustrates this method.

EXAMPLE A.5 Prove by contradiction: There is no largest prime number.

PROOF BY CONTRADICTION

(*Notice that the theorem has no explicit hypothesis.*) Suppose the given conclusion is false; that is, there is a largest prime number p. So the prime numbers we have are $2, 3, 5, \ldots, p$.

Let x denote the product of all these prime numbers plus one: $x = (2 \cdot 3 \cdot 5 \cdots p) + 1$. Clearly, $x > p$. When x is divided by each of the primes $2, 3, 5, \ldots, p$, we get 1 as the remainder. Therefore, x is *not* divisible by any of the primes and hence x must be a prime. Thus, x is a prime number greater than p.

But this contradicts the assumption that p is the largest prime, so our assumption is false. In other words, there is no largest prime number. ■

We now turn to the next proof technique.

Proof by Cases

Suppose we want to prove a theorem of the form $H_1 \vee H_2 \vee \cdots \vee H_n \to C$. Since $H_1 \vee H_2 \vee \cdots \vee H_n \to C \equiv (H_1 \to C) \wedge (H_2 \to C) \wedge \cdots \wedge (H_n \to C)$, $H_1 \vee H_2 \vee \cdots \vee H_n \to C$ is true if and only if each implication $H_i \to C$ is true. Consequently, we need only prove that each implication is true. Such a proof is a **proof by cases**, as the following example illustrates.

EXAMPLE A.6 There are two kinds of inhabitants, knights and knaves, on an island. Knights always tell the truth, whereas knaves always lie. Every inhabitant is either a knight or a knave. Two inhabitants are of the *same* type if they are both knights or both knaves. Let A, B, and C be three inhabitants of the island. Suppose A says, "B is a knave," and B says, "A and C are of the same type." Prove that C is a knave.[†]

PROOF BY CASES

Although this theorem is not explicitly of the form $(H_1 \vee H_2 \vee \cdots \vee H_n) \to C$, we artificially create two cases, namely, A is a knight and A is a knave.

case 1 Suppose A is a knight. Since knights always tell the truth, his statement that B is a knave is true. So B is a knave and hence his statement is false. Therefore, A and C are of different types, thus C is a knave.

case 2 Suppose A is a knave. Then his statement is false, so B is a knight. Since knights always tell the truth, B's statement is true. So A and C are of the same type, thus C is a knave.

Thus, in both cases, C is a knave. ■

We now turn to the existence proof method.

Existence Proof

Finally, theorems of the form $(\exists x)P(x)$ also occur in mathematics, where $\exists$ denotes the **existential quantifier**.[‡] To prove such a theorem, we must establish the existence of an object a for which $P(a)$ is true. Accordingly, such a proof is an **existence proof**.

[†] Based on R. M. Smullyan, *What Is the Name of This Book*?, Prentice-Hall, Englewood Cliffs, NJ, 1978.

[‡] The existential quantifier $\exists$ stands for *some, there exists* $a(n)$, or *there is a at least one*.

There are two kinds of an existence proof: **constructive existence proof** and **nonconstructive existence proof**. If we are able to find an object b such that $P(b)$ is true, such an existence proof is a **constructive proof**. The following example illustrates this method.

EXAMPLE A.7 Prove that there is a positive integer that can be expressed as the sum of two cubes in two different ways.

CONSTRUCTIVE PROOF

By the preceding discussion, all we need is to produce a positive integer b that has the required properties. Choose $b = 1729$. Since $1729 = 1^3 + 12^3 = 9^3 + 10^3$, 1729 is such an integer. This concludes the proof. ■

A **nonconstructive existence proof** of the theorem $(\exists x)P(x)$ does not provide us with an element a for which $P(a)$ is true, but rather establishes its existence by an indirect method, usually contradiction, as the following example illustrates.

EXAMPLE A.8 Prove that there is a prime number > 3.

NONCONSTRUCTIVE PROOF

Suppose there are no primes > 3. Then 2 and 3 are the only primes. Since every integer ≥ 2 can be expressed as a product of powers of primes, 25 must be expressible as a product of powers of 2 and 3; that is, $25 = 2^i 3^j$ for some integers i and j. But neither 2 nor 3 is a factor of 25, so 25 cannot be written in the form $2^i 3^j$, a contradiction. Consequently, there must be a prime > 3. ■

We conclude this section with a brief discussion of counterexamples.

Counterexample

Consider the statement, *Every girl is a brunette*. Is it true or false? Since we can find at least one girl who is not a brunette, the statement is false.

More generally, suppose you would like to show that the proposition $(\forall x)P(x)$ is false, where $\forall$ denotes the **universal quantifier**.† Since $\sim[(\forall x)P(x)] \equiv (\exists x)[\sim P(x)]$, the statement $(\forall x)P(x)$ is false if there exists an item for which $P(x)$ is false. Such an object x is a **counterexample**. Thus, to disprove the proposition $(\forall x)P(x)$, all we need is to produce a counterexample c for which $P(c)$ is false, as the next two examples illustrate.

† The universal quantifier $\forall$ stands for the word *all*, *each*, or *every*.

EXAMPLE A.9 Number theorists dream of finding formulas that generate prime numbers. One such formula was found by Euler (see Chapter 2), namely, $E(n) = n^2 - n + 41$. It yields a prime for $n = 1, 2, \ldots, 40$. Suppose we claim that the formula generates a prime for every positive integer n. Since $E(41) = 41^2 - 41 + 41 = 41^2$ is not a prime, 41 is a counterexample, thus disproving the claim. ■

EXAMPLE A.10 Around 1640, Fermat conjectured that numbers of the form $f(n) = 2^{2^n} + 1$ are prime numbers for all nonnegative integers n. For instance, $f(0) = 3, f(1) = 5, f(2) = 17, f(3) = 257$, and $f(4) = 65537$ are all primes. In 1732, however, Euler established the falsity of Fermat's conjecture by producing a counterexample. He showed that $f(5) = 2^{2^5} + 1 = 641 \times 6700417$, a composite number. ■

EXERCISES A.1

Determine whether each implication is vacuously true for the indicated value of n.

1. If $n \geq 1$, then $2^n \geq n$; $n = 0$
2. If $n \geq 4$, then $2^n \geq n^2$; $n = 0, 1, 2, 3$

Determine whether each implication is trivially true.

3. If n is a prime number, then $n^2 + n$ is an even integer.
4. If $n \geq 41$, then $n^3 - n$ is divisible by 3.

Prove each using the direct method.

5. The sum of any two even integers is even.
6. The sum of any two odd integers is even.
7. The square of an even integer is even.
8. The product of any two even integers is even.
9. The square of an odd integer is odd.
10. The product of any two odd integers is odd.
11. The product of any even integer and any odd integer is even.
12. The square of every integer of the form $3k + 1$ is also of the same form, where k is some integer.
13. The square of every integer of the form $4k + 1$ is also of the same form, where k is some integer.
14. The **arithmetic mean** $\frac{a+b}{2}$ of any two nonnegative real numbers a and b is greater than or equal to their **geometric mean** $\sqrt{ab}$.
    (*Hint*: consider $(\sqrt{a} - \sqrt{b})^2 \geq 0$.)

Prove each, using the law of the contrapositive.

15. If the square of an integer is even, then the integer is even.
16. If the square of an integer is odd, then the integer is odd.
17. If the product of two integers is even, then at least one of them must be an even integer.
18. If the product of two integers is odd, then both must be odd integers.

Prove the following by contradiction, where p is a prime number.

19. $\sqrt{2}$ is an irrational number.
20. $\sqrt{5}$ is an irrational number.
21. $\sqrt{p}$ is an irrational number.
22. *$\log_{10} 2$ is an irrational number.

Prove the following by cases, where n is an arbitrary integer.

23. $n^2 + n$ is an even integer.
24. $2n^3 + 3n^2 + n$ is an even integer.
25. $n^3 - n$ is divisible by 3.
    (*Hint*: Assume that every integer is of the form $3k$, $3k + 1$, or $3k + 2$.)

Prove each by the existence method, where $|x|$ denotes the absolute value of x.

26. There are integers x such that $x^2 = x$.
27. There are integers x such that $|x| = x$.

28. There are infinitely many integers that can be expressed as the sum of two cubes in two different ways.
29. The equation $x^2 + y^2 = z^2$ has infinitely many integer solutions.

Give a counterexample to disprove each statement.

30. The absolute value of every real number is positive.
31. The square of every real number is positive.
32. Every prime number is odd.
33. Every month has exactly 30 days.

Let a, b, and c be any real numbers. Then $a < b$ if and only if there is a positive real number x such that $a + x = b$. Using this fact, prove each.

34. If $a < b$ and $b < c$, then $a < c$.
35. If $a < b$, then $a + c < b + c$.
36. If $a + c < b + c$, then $a < b$.
37. Let a and b be any two real numbers such that $a \cdot b = 0$. Then either $a = 0$ or $b = 0$. (*Hint*: $p \to (q \vee r) \equiv (p \wedge \sim q) \to r$.)

*38. The formula $f(n) = n^2 - 79n + 1601$ yields a prime for $0 \le n \le 10$. Give a counterexample to disprove the claim that the formula yields a prime for every nonnegative integer n.

*39. Let $f(n) = 2^n - 1$, where n is a positive integer. For example, $f(2) = 3, f(3) = 7$, and $f(5) = 31$ are all primes. Give a counterexample to disprove the claim that if n is a prime, then $2^n - 1$ is a prime.

A.2 Web Sites

The following web sites provide valuable information for further exploration and enrichment. Since the status of a web site could change with time, not all of them may exist when you try; if they do not, use a search engine to locate sites that provide similar information.

1. **Number Theory Web** (`www.math.uga.edu/~ntheory/web.html`)
   Lists information of interest to number theorists.
2. **List of Mathematicians**
   (`http:/aleph0.clarku.edu/~djoyce/mathhist/chronology.html`)
   Gives a chronological list of mathematical works and mathematicians.
3. **Fibonacci and Lucas Numbers**
   (`http://jwilson/coe.uga.edu/EMT668/EMAT668...trypro/fib%26luc%26phi/AKEwrite-up12.html`)
   Shows how Fibonacci numbers, Lucas numbers, and the Golden Ratio are related.
4. **Euclidean Algorithm**
   There are over 20 web sites for euclidean algorithm. Five of them are:
   `www.math.umn.edu/~garrett/js/gcd.html`
   `www.unc.edu/~rowlet/Math81/notes/Euclidal.html`
   `www.csclub.uwaterloo.ca/~mpslager/compsci/GCD.html`
   `www.math.sc.edu/~sumner/numbertheory/euclidean/euclidean.html`
   `www.math.umass.edu/~dhayes/math597a/ugcd2/`

5. **Fermat Numbers** (www.vamri.xray.ufl.edu/proths/fermat.html)
   Lists the prime factors $k \cdot 2^n + 1$ of Fermat numbers f_m and complete factoring status.
6. **The Largest Known Primes** (www.utm.edu/research/primes/largest.html)
   Contains a brief introduction to prime numbers, and lists the 10 known largest primes, 10 largest known twin primes, 10 largest known Mersenne primes, 10 largest known factorial primes, and 10 largest known Sophie Germain primes.
7. **The Great Internet Mersenne Prime Search** (www.Mersenne.org)
   A very comprehensive web site for Mersenne primes and provides links to sites for other classes of primes, such as twin primes, Sophie Germain primes, Kelly primes, Cunningham chains, and Fermat primes. By signing up with GIMPS, you can join a worldwide network of mathematicians and computer scientists in the search for larger primes and get your name into the record book.
8. **Mersenne Primes: History, Theorems, and Lists**
   (www.utm.edu/research/primes/mersenne.shtml)
   Gives a brief history of Mersenne primes, perfect numbers, a list of known Mersenne primes, a brief history of the Lucas–Lehmer test, and some conjectures about Mersenne primes.
9. **Large Primes Found by SGU/Cray**
   (http//reality.sgi.com/csp/ioccc/noll/prime/prime_press.html)
   Lists all large primes discovered by computer scientists at Silicon Graphics' Cray Research Unit.
10. **The "Proth" Program and the Search for Primes**
    (http://perso.wanadoo.fr/yves.gallot/Dublin)
    Gives a history of the Proth program, primality tests, and the search for primes.
11. **The Law of Quadratic Reciprocity**
    There are over 20 web sites on quadratic reciprocity. Five of them are:
    www.izuser.uni-heidelberg.del/~hb3/rchrono.html
    www.seanet.com/~ksbrown/kmath075.htm
    www.math.umn.edu/~garrett/m/number_theory/quad_rec.shtml
    www.math.nmsu.edu/~history/schauspiel/schauspiel.html
    www.math.swt.edu/~haz/prob_sets/notes/node32.html
12. **Pythagorean triples**
    There are over 20 web sites for Pythagorean triples. Five of them are:
    www.math.clemson.edu/~rsimms/neat/math/pyth/
    www.maths.uts.edu.au/numericon/triples.html
    www.math.uic.edu/~fields/puzzle/triples.html

www.faust.fr.bw.schule.de/mhb/pythagen.htm
www.math.utah.edu/~alfeld/teaching/pt.html

13. **Palindromic Pythagorean Triples**
    (www.ping.be/~pingo6758/pythago.htm)
    Gives an algorithm for constructing palindromic Pythagorean triples from a known palindromic Pythagorean triple and lists several near-misses.
14. **Nova: The Proof** (www.pbs.org/wgbh/nova/proof)
    This web site is related to the delightful Nova program on Andrew Wiles' proof of Fermat's Last Theorem. It gives a transcript of an interview with Wiles, who spent seven years of his life in his attic cracking the problem.
15. **Fermat's Last Theorem**
    There are a number of sites on Fermat's Last Theorem, including one in Hebrew. Three of them are:
    www.cs.unb.ca/~alopez-o/math-faq/node22.html
    This answers a few basic questions about the theorem.
    www.best.com/~cgd/home/flt/flt01.htm
    This site provides a thorough overview of the proof of the theorem.
    www.-groups.dcs.st-and.ac.uk/~history/HistTopics/Fermat's_last_theorem.html
    This site gives a some biographical information on Fermat and a brief history of the problem.
16. **Waring's Problem**
    The following sites discuss the proofs of two special cases of Waring's conjecture; $g(3) = 9$ (Wieferich and Kemner, 1912) and $g(5) = 37$ (Chen, 1964). The web sites by valuable references for the problem:
    www.seanet.com/~ksbrown/kmath316.htm
    www.mathsoft.com/asolve/pwrs32/pwrs32.html
    www.math.niu.edu/~rusin/known-math/96/warings
17. **Generalized Fermat–Wiles Equation**
    (www.mathsoft.com/asolve/fermat/fermat.html)
    Examines a generalized version of Fermat's equation $x^n + y^n = z^n$.

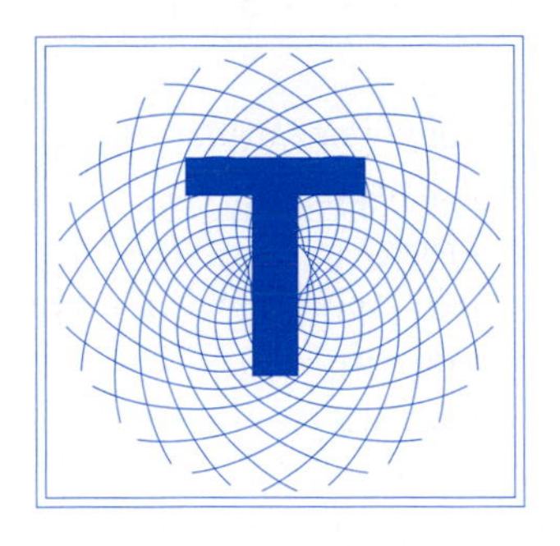

Tables

Table T.1 on pp. 642–648 lists the least prime factor of each odd positive integer $< 10{,}000$ and *not* divisible by 5. The beginning digit(s) is (are) listed on the left side and the last digit on the top of the column. For example, 3 is the smallest prime factor of 2001.

A dash indicates that the corresponding number is a prime. For instance, 2347 is a prime.

Table T.2 on pp. 649–651 gives the values of the arithmetic functions φ, τ, and σ for each positive integer ≤ 100. For example, $\varphi(28) = 12$, $\tau(28) = 6$, and $\sigma(28) = 56$.

Table T.3 on p. 652 lists the least primitive root r modulo each prime $p < 1000$. For example, 3 is the least primitive root modulo 17 and 5 the least primitive root modulo 73.

Table T.4 on pp. 653–656 gives the index of each positive integer ≤ 96 and to the base that is the least primitive root modulo a prime p. For example, 3 is the least primitive root modulo 17; $\text{ind}_3\, 12 = 13$, since $3^{13} \equiv 12 \pmod{17}$.

	1	3	7	9		1	3	7	9		1	3	7	9		1	3	7	9
0	–	–	–	3	40	–	13	11	–	80	3	11	3	–	120	–	3	17	3
1	–	–	–	–	41	3	7	3	–	81	–	3	19	3	121	7	–	–	23
2	3	–	3	–	42	–	3	7	3	82	–	–	–	–	122	3	–	3	–
3	–	3	–	3	43	–	–	19	–	83	3	7	3	–	123	–	3	–	3
4	–	–	–	7	44	3	–	3	–	84	29	3	7	3	124	17	11	29	–
5	3	–	3	–	45	11	3	–	3	85	23	–	–	–	125	3	7	3	–
6	–	3	–	3	46	–	–	–	7	86	3	–	3	11	126	13	3	7	3
7	–	–	7	–	47	3	11	3	–	87	13	3	–	3	127	31	19	–	–
8	3	–	3	–	48	13	3	–	3	88	–	–	–	7	128	3	–	3	–
9	7	3	–	3	49	–	17	7	–	89	3	19	3	29	129	–	3	–	3
10	–	–	–	–	50	3	–	3	–	90	17	3	–	3	130	–	–	–	7
11	3	–	3	7	51	7	3	11	3	91	–	11	7	–	131	3	13	3	–
12	11	3	–	3	52	–	–	17	23	92	3	13	3	–	132	–	3	–	3
13	–	7	–	–	53	3	13	3	7	93	7	3	–	3	133	11	31	7	13
14	3	11	3	–	54	–	3	–	3	94	–	23	–	13	134	3	17	3	19
15	–	3	–	3	55	19	7	–	13	95	3	8	3	7	135	7	3	23	3
16	7	–	–	13	56	3	–	3	–	96	31	3	–	3	136	–	29	–	37
17	3	–	3	–	57	–	3	–	3	97	–	7	–	11	137	3	–	3	7
18	–	3	11	3	58	7	11	–	19	98	3	–	3	23	138	–	3	19	3
19	–	–	–	–	59	3	–	3	–	99	–	3	–	3	139	13	7	11	–
20	3	7	3	11	60	–	3	–	3	100	7	17	19	–	140	3	23	3	–
21	–	3	7	3	61	13	–	–	–	101	3	–	3	–	141	17	3	13	3
22	13	–	–	–	62	3	7	3	17	102	–	3	13	3	142	7	–	–	–
23	3	–	3	–	63	–	3	7	3	103	–	–	17	–	143	3	–	3	–
24	–	3	13	3	64	–	–	–	11	104	3	7	3	–	144	11	3	–	3
25	–	11	–	7	65	3	–	3	–	105	–	3	7	3	145	–	–	31	–
26	3	–	3	–	66	–	3	23	3	106	–	–	11	–	146	3	7	3	13
27	–	3	–	3	67	11	–	–	7	107	3	29	3	13	147	–	3	7	3
28	–	–	7	17	68	3	–	3	13	108	23	3	–	3	148	–	–	–	–
29	3	–	3	13	69	–	3	17	3	109	–	–	–	7	149	3	–	3	–
30	7	3	–	3	70	–	19	7	–	110	3	–	3	–	150	19	3	11	3
31	–	–	–	11	71	3	23	3	–	111	11	3	–	3	151	–	17	37	7
32	3	17	3	7	72	7	3	–	3	112	19	–	7	–	152	3	–	3	11
33	–	3	–	3	73	17	–	11	–	113	3	11	3	17	153	–	3	29	3
34	11	7	–	–	74	3	–	3	7	114	7	3	31	3	154	23	–	7	–
35	3	–	3	–	75	–	3	–	3	115	–	–	13	19	155	3	–	3	–
36	19	3	–	3	76	–	7	13	–	116	3	–	3	7	156	7	3	–	3
37	7	–	13	–	77	3	–	3	19	117	–	3	11	3	157	–	11	19	–
38	7	–	3	–	78	11	3	–	3	118	–	7	–	29	158	3	–	3	7
39	17	3	–	3	79	7	13	–	17	119	3	–	3	11	159	37	3	–	3

Table T.1 *Factor table.*

	1	3	7	9		1	3	7	9		1	3	7	9		1	3	7	9
160	–	7	–	–	200	3	–	3	7	240	7	3	29	3	280	–	–	7	53
161	3	–	3	–	201	–	3	–	3	241	–	19	–	41	281	3	29	3	–
162	–	3	–	3	202	43	7	–	–	242	3	–	3	7	282	7	3	11	3
163	7	23	–	11	203	3	19	3	–	243	11	3	–	3	283	19	–	–	17
164	3	31	3	17	204	13	3	23	3	244	–	7	–	31	284	3	–	3	7
165	13	3	–	3	205	7	–	11	29	245	3	11	3	–	285	–	3	–	3
166	11	–	–	–	206	3	–	3	–	246	23	3	–	3	286	–	7	47	19
167	3	7	3	23	207	19	3	31	3	247	7	–	–	37	287	3	13	3	–
168	41	3	7	3	208	–	–	–	–	248	3	13	3	19	288	43	3	–	3
169	19	–	–	–	209	3	7	3	–	249	47	3	11	3	289	7	11	–	13
170	3	13	3	–	210	11	3	7	3	250	41	–	23	13	290	3	–	3	–
171	29	3	17	3	211	–	–	29	13	251	3	7	3	11	291	41	3	–	3
172	–	–	11	7	212	3	11	3	–	252	–	3	7	3	292	23	37	–	29
173	3	–	3	37	213	–	3	–	3	253	–	17	43	–	293	3	7	3	–
174	–	3	–	3	214	–	–	19	7	254	3	–	3	–	294	17	3	7	3
175	17	–	7	–	215	3	–	3	17	255	–	3	–	3	295	13	–	–	11
176	3	41	3	29	216	–	3	11	3	256	13	11	17	7	296	3	–	3	–
177	7	3	–	3	217	13	41	7	–	257	3	31	3	–	297	–	3	13	3
178	13	–	–	–	218	3	37	3	11	258	29	3	13	3	298	11	19	29	7
179	3	11	3	7	219	7	3	13	3	259	–	–	7	23	299	3	41	3	–
180	–	3	13	3	220	31	–	–	47	260	3	19	3	–	300	–	3	31	3
181	–	7	23	17	221	3	–	3	7	261	7	3	–	3	301	–	23	7	–
182	3	–	3	31	222	–	3	17	3	262	–	43	37	11	302	3	–	3	13
183	–	3	11	3	223	23	7	–	–	263	3	–	3	7	303	7	3	–	3
184	7	19	–	43	224	3	–	3	13	264	19	3	–	3	304	–	17	11	–
185	3	17	3	11	225	–	3	37	3	265	11	7	–	–	305	3	43	3	7
186	–	3	–	3	226	7	31	–	–	266	3	–	3	17	306	–	3	–	3
187	–	–	–	–	227	3	–	3	43	267	–	3	–	3	307	37	7	17	–
188	3	7	3	–	228	–	3	–	3	268	7	–	–	–	308	3	–	3	–
189	31	3	7	3	229	29	–	–	11	269	3	–	3	–	309	11	3	19	3
190	–	11	–	23	230	3	7	3	–	270	37	3	–	3	310	7	29	13	–
191	3	–	3	19	231	–	3	7	3	271	–	–	11	–	311	3	11	3	–
192	17	3	41	3	232	11	23	13	17	272	3	7	3	–	312	–	3	53	3
193	–	–	13	7	233	3	–	3	–	273	–	3	7	3	313	31	13	–	43
194	3	29	3	–	234	–	3	–	3	274	–	13	41	–	314	3	7	3	47
195	–	3	19	3	235	–	13	–	7	275	3	–	3	31	315	23	3	7	3
196	37	13	7	11	236	3	17	3	23	276	11	3	–	3	316	29	–	–	–
197	3	–	3	–	237	–	3	–	3	277	17	47	–	7	317	3	19	3	11
198	7	3	–	3	238	–	–	7	–	278	3	11	3	–	318	–	3	–	3
199	11	–	–	–	239	3	–	3	–	279	–	3	–	3	319	–	31	23	7

Table T.1 *(Continued)*

	1	3	7	9		1	3	7	9		1	3	7	9		1	3	7	9
320	3	–	3	–	360	13	3	–	3	400	–	–	–	19	440	3	7	3	–
321	13	3	–	3	361	23	–	–	7	401	3	–	3	–	441	11	3	7	3
322	–	11	7	–	362	3	–	3	19	402	–	3	–	3	442	–	–	19	43
323	3	53	3	41	363	–	3	–	3	403	29	37	11	7	443	3	11	3	23
324	7	3	17	3	364	11	–	7	41	404	3	13	3	–	444	–	3	–	3
325	–	–	–	–	365	3	13	3	–	405	–	3	–	3	445	–	61	–	7
326	3	13	3	7	366	7	3	19	3	406	31	17	7	13	446	3	–	3	41
327	–	3	29	3	367	–	–	–	13	407	3	–	3	–	447	17	3	11	3
328	17	7	19	11	368	3	29	3	7	408	7	3	61	3	448	–	–	7	67
329	3	37	3	–	369	–	3	–	3	409	–	–	17	–	449	3	–	3	11
330	–	3	–	3	370	–	7	11	–	410	3	11	3	7	450	7	3	–	3
331	7	–	31	–	371	3	47	3	–	411	–	3	23	3	451	13	–	–	–
332	3	–	3	–	372	61	3	–	3	412	13	7	–	–	452	3	–	3	7
333	–	3	47	3	373	7	–	37	–	413	3	–	3	–	453	23	3	13	3
334	13	–	–	17	374	3	19	3	23	414	41	3	11	3	454	19	7	–	–
335	3	7	3	–	375	11	3	13	3	415	7	–	–	–	455	3	29	3	47
336	–	3	7	3	376	–	53	–	–	416	3	23	3	11	456	–	3	–	3
337	–	–	11	31	377	3	7	3	–	417	43	3	–	3	457	7	17	23	19
338	3	17	3	–	378	19	3	7	3	418	37	47	53	59	458	3	–	3	13
339	–	3	43	3	379	17	–	–	29	419	3	7	3	13	459	–	3	–	3
340	19	41	–	7	380	3	–	3	31	420	–	3	7	3	460	43	–	17	11
341	3	–	3	13	381	37	3	11	3	421	–	11	–	–	461	3	7	3	31
342	11	3	23	3	382	–	–	43	7	422	3	41	3	–	462	–	3	7	3
343	47	–	7	19	383	3	–	3	11	423	–	3	19	3	463	11	41	–	–
344	3	11	3	–	384	23	3	–	3	424	–	–	31	7	464	3	–	3	–
345	7	3	–	3	385	–	–	7	17	425	3	–	3	–	465	–	3	–	3
346	–	–	–	–	386	3	–	3	53	426	–	3	17	3	466	59	–	13	7
347	3	23	3	7	387	7	3	–	3	427	–	–	7	11	467	3	–	3	–
348	59	3	11	3	388	–	11	13	–	428	3	–	3	–	468	31	3	43	3
349	–	7	13	–	389	3	17	3	7	429	7	3	–	3	469	–	13	7	37
350	3	31	3	11	390	47	3	–	3	430	11	13	59	31	470	3	–	3	17
351	–	3	–	3	391	–	7	–	–	431	3	19	3	7	471	7	3	53	3
352	7	13	–	–	392	3	–	3	–	432	29	3	–	3	472	–	–	29	–
353	3	–	3	–	393	–	3	31	3	433	61	7	–	–	473	3	–	3	7
354	–	3	–	3	394	7	–	–	11	434	3	43	3	–	474	11	3	47	3
355	53	11	–	–	395	3	59	3	37	435	19	3	–	3	475	–	7	67	–
356	3	7	3	43	396	17	3	–	3	436	7	–	11	17	476	3	11	3	19
357	–	3	7	3	397	11	29	41	23	437	3	–	3	29	477	13	3	17	3
358	–	–	17	37	398	3	7	3	–	438	13	3	41	3	478	7	–	–	–
359	3	–	3	59	399	13	3	7	3	439	–	23	–	53	479	3	–	3	–

Table T.1 *(Continued)*

	1	3	7	9		1	3	7	9		1	3	7	9		1	3	7	9
480	–	3	11	3	520	7	11	41	–	560	3	13	3	71	600	17	3	–3	
481	17	–	–	61	521	3	13	3	17	561	31	3	41	3	601	–	7	11	13
482	3	7	3	11	522	23	3	–	3	562	7	–	17	13	602	3	19	3	–
483	–	3	7	3	523	–	–	–	13	563	3	43	3	–	603	37	3	–	3
484	47	29	37	13	524	3	7	3	29	564	–	3	–	3	604	7	–	–	23
485	3	23	3	43	525	59	3	7	3	565	–	–	–	–	605	3	–	3	73
486	–	3	31	3	526	–	19	23	11	566	3	7	3	–	606	11	3	–	3
487	–	11	–	7	527	3	–	3	–	567	53	3	7	3	607	13	–	59	–
488	3	19	3	–	528	–	3	17	3	568	13	–	11	–	608	3	7	3	–
489	67	3	59	3	529	11	67	–	7	569	3	–	3	41	609	–	3	7	3
490	13	–	7	–	530	3	–	3	–	570	–	3	13	3	610	–	17	31	41
491	3	17	3	–	531	47	3	13	3	571	–	29	–	7	611	3	–	311	29
492	7	3	13	3	532	17	–	7	73	572	3	59	3	17	612	–	3	11	3
493	–	–	–	11	533	3	–	3	19	573	11	3	–	3	613	–	–	17	7
494	3	–	3	7	534	7	3	–	3	574	–	–	7	–	614	3	–	3	11
495	–	3	–	3	535	–	53	11	23	575	3	11	3	13	615	–	3	47	3
496	11	7	–	–	536	3	31	3	7	576	7	3	73	3	616	61	–	7	31
497	3	–	3	13	537	41	3	19	3	577	29	23	53	–	617	3	–	3	37
498	17	3	–	3	538	–	7	–	17	578	3	–	3	7	618	7	3	23	3
499	7	–	19	–	539	3	–	3	–	579	–	3	11	3	619	41	11	–	–
500	3	–	3	–	540	11	3	–	3	580	–	7	–	37	620	3	–	3	7
501	–	3	29	3	541	7	–	–	–	581	3	–	3	11	621	–	3	–	3
502	–	–	11	47	542	3	11	3	61	582	–	3	–	3	622	–	7	13	–
503	3	7	3	–	543	–	3	–	3	583	7	19	13	–	623	3	23	3	17
504	71	3	7	3	544	–	–	13	–	584	3	–	3	–	624	79	3	–	3
505	–	31	13	–	545	3	7	3	53	585	–	3	–	3	625	7	13	–	11
506	3	61	3	37	546	43	3	7	3	586	–	11	–	–	626	3	–	3	–
507	11	3	–	3	547	–	13	–	–	587	3	7	3	–	627	–	3	–	3
508	–	13	–	7	548	3	–	3	11	588	–	3	7	3	628	11	61	–	19
509	3	11	3	–	549	17	3	23	3	589	43	71	–	17	629	3	7	3	–
510	–	3	–	3	550	–	–	–	7	590	3	–	3	19	630	–	3	7	3
511	19	–	7	–	551	3	37	3	–	591	23	3	61	3	631	–	59	–	71
512	3	47	3	23	552	–	3	–	3	592	31	–	–	7	632	3	–	3	–
513	7	3	11	3	553	–	11	7	29	593	3	17	3	–	633	13	3	–	3
514	53	37	–	19	554	3	23	3	31	594	13	3	19	3	634	17	–	11	7
515	3	–	3	7	555	7	3	–	3	595	11	–	7	59	635	3	–	3	–
516	13	3	–	3	556	67	–	19	–	596	3	67	3	47	636	–	3	–	3
517	–	7	31	–	557	3	–	3	7	597	7	3	43	3	637	23	–	7	–
518	3	71	3	–	558	–	3	37	3	598	–	31	–	53	638	3	13	3	–
519	29	3	–	3	559	–	7	29	11	599	3	13	3	7	639	7	3	–	3

Table T.1 *(Continued)*

	1	3	7	9		1	3	7	9		1	3	7	9		1	3	7	9
640	37	19	43	13	680	3	–	3	11	720	19	3	–	3	760	11	–	–	7
641	3	11	3	7	681	7	3	17	3	721	–	–	7	–	761	3	23	3	19
642	–	3	–	3	682	19	–	–	–	722	3	31	3	–	762	–	3	29	3
643	59	7	41	47	683	3	–	3	7	723	7	3	–	3	763	13	17	7	–
644	3	17	3	–	684	–	3	41	3	724	13	–	–	11	764	3	–	3	–
645	–	3	11	3	685	13	7	–	19	725	3	–	3	7	765	7	3	13	3
646	7	23	29	–	686	3	–	3	–	726	53	3	13	3	766	47	79	11	–
647	3	–	3	11	687	–	3	13	3	727	11	7	19	29	767	3	–	3	7
648	–	3	13	3	688	7	–	71	83	728	3	–	3	37	768	–	3	–	3
649	–	43	73	67	689	3	61	3	–	729	23	3	–	3	769	–	7	43	–
650	3	7	3	23	690	67	3	–	3	730	7	67	–	–	770	3	–	3	13
651	17	3	7	3	691	–	31	–	11	731	3	71	3	13	771	11	3	–	3
652	–	11	61	–	692	3	7	3	13	732	–	3	17	3	772	7	–	–	59
653	3	47	3	13	693	29	3	7	3	733	–	–	11	41	773	3	11	3	71
654	31	3	–	3	694	11	53	–	–	734	3	7	3	–	774	–	3	61	3
655	–	–	79	7	695	3	17	3	–	735	–	3	7	3	775	23	–	–	–
656	3	–	3	–	696	–	3	–	3	736	17	37	53	–	776	3	7	3	17
657	–	3	–	3	697	–	19	–	7	737	3	73	3	47	777	19	3	7	3
658	–	29	7	11	698	3	–	3	29	738	11	3	83	3	778	31	43	13	–
659	3	19	3	–	699	–	3	–	3	739	19	–	13	7	779	3	–	3	11
660	7	3	–	3	700	–	47	7	43	740	3	11	3	31	780	29	3	37	3
661	11	17	13	–	701	3	–	3	–	741	–	3	–	3	781	73	13	–	7
662	3	37	3	7	702	7	3	–	3	742	41	13	7	17	782	3	–	3	–
663	19	3	–	3	703	79	13	31	–	743	3	–	3	43	783	41	3	17	3
664	29	7	17	61	704	3	–	3	7	744	7	3	11	3	784	–	11	7	47
665	3	–	3	–	705	11	3	–	3	745	–	29	–	–	785	3	–	3	29
666	–	3	59	3	706	23	7	37	–	746	3	17	3	7	786	7	3	–	3
667	7	–	11	–	707	3	11	3	–	747	31	3	–	3	787	17	–	–	–
668	3	41	3	–	708	73	3	19	3	748	–	7	–	–	788	3	–	3	7
669	–	3	37	3	709	7	41	47	31	749	3	59	3	–	789	13	3	53	3
670	–	–	19	–	710	3	–	3	–	750	13	3	–	3	790	–	7	–	11
671	3	7	3	–	711	13	3	11	3	751	7	11	–	73	791	3	41	3	–
672	11	3	7	3	712	–	17	–	–	752	3	–	3	–	792	89	3	–	3
673	53	–	–	23	713	3	7	3	11	753	17	3	–	3	793	7	–	–	17
674	3	11	3	17	714	37	3	7	3	754	–	19	–	–	794	3	13	3	–
675	43	3	29	3	715	–	23	17	–	755	3	7	3	–	795	–	3	73	3
676	–	–	67	7	716	3	13	3	67	756	–	3	7	3	796	19	–	31	13
677	3	13	3	–	717	71	3	–	3	757	67	–	–	11	797	3	7	3	79
678	–	3	11	3	718	43	11	–	7	758	3	–	3	–	798	23	3	7	3
679	–	–	7	13	719	3	–	3	23	759	–	3	71	3	799	61	–	11	19

Table T.1 *(Continued)*

	1	3	7	9		1	3	7	9		1	3	7	9		1	3	7	9
800	3	53	3	–	840	31	3	7	3	880	13	–	–	23	920	3	–	3	–
801	–	3	–	3	841	13	47	19	–	881	3	7	3	–	921	61	3	13	3
802	13	71	23	7	842	3	–	3	–	882	–	3	7	3	922	–	23	–	11
803	3	29	3	–	843	–	3	11	3	883	–	11	–	–	923	3	7	3	–
804	11	3	13	3	844	23	–	–	7	884	3	37	3	–	924	–	3	7	3
805	83	–	7	–	845	3	79	3	11	885	53	3	17	3	925	11	19	–	47
806	3	11	3	–	846	–	3	–	3	886	–	–	–	7	926	3	59	3	13
807	7	3	41	3	847	43	37	7	61	887	3	19	3	13	927	73	3	–	3
808	–	59	–	–	848	3	17	3	13	888	83	3	–	3	928	–	–	37	7
809	3	–	3	7	849	7	3	29	3	889	17	–	7	11	929	3	–	3	17
810	–	3	11	3	850	–	11	47	67	890	3	29	3	59	930	71	3	41	3
811	–	7	–	23	851	3	–	3	7	891	7	3	37	3	931	–	67	7	–
812	3	–	3	11	852	–	3	–	3	892	11	–	79	–	932	3	–	3	19
813	47	3	79	3	853	19	7	–	–	893	3	–	3	7	933	7	3	–	3
814	7	17	–	29	854	3	–	3	83	894	–	3	23	3	934	–	–	13	–
815	3	31	3	41	855	17	3	43	3	895	–	7	13	17	935	3	47	3	7
816	–	3	–	3	856	7	–	13	11	896	3	–	3	–	936	11	3	17	3
817	–	11	13	–	857	3	–	3	23	897	–	3	47	3	937	–	7	–	83
818	3	7	3	19	858	–	3	31	3	898	7	13	11	89	938	3	11	3	41
819	–	3	7	3	859	11	13	–	–	899	3	17	3	–	939	–	3	–	3
820	59	13	29	–	860	3	7	3	–	900	–	3	–	3	940	7	–	23	97
821	3	43	3	–	861	79	3	7	3	901	–	–	71	29	941	3	–	3	–
822	–	3	19	3	862	37	–	–	–	902	3	7	3	–	942	–	3	11	3
823	–	–	–	7	863	3	89	3	53	903	11	3	7	3	943	–	–	–	–
824	3	–	3	73	864	–	3	–	3	904	–	–	83	–	944	3	7	3	11
825	37	3	23	3	865	41	17	11	7	905	3	11	3	–	945	13	3	7	3
826	11	–	7	–	866	3	–	3	–	906	13	3	–	3	946	–	–	–	17
827	3	–	3	17	867	13	3	–	3	907	47	43	29	7	947	3	–	3	–
828	7	3	–	3	868	–	19	7	–	908	3	31	3	61	948	19	3	53	3
829	–	–	–	43	869	3	–	3	–	909	–	3	11	3	949	–	11	–	7
830	3	19	3	7	870	7	3	–	3	910	19	–	7	–	950	3	13	3	37
831	–	3	–	3	871	31	–	23	–	911	3	31	3	11	951	–	3	31	3
832	53	7	11	–	872	3	11	3	7	912	7	3	–	3	952	–	89	7	13
833	3	13	3	31	873	–	3	–	3	913	23	–	–	13	953	3	–	3	–
834	19	3	17	3	874	–	7	–	13	914	3	41	3	7	954	7	3	–	3
835	7	–	61	13	875	3	–	3	19	915	–	3	–	3	955	–	41	19	11
836	3	–	3	–	876	–	3	11	3	916	–	7	89	53	956	3	73	3	7
837	11	3	–	3	877	7	31	67	–	917	3	–	3	67	957	17	3	61	3
838	17	83	–	–	878	3	–	3	11	918	–	3	–	3	958	11	7	–	43
839	3	7	3	37	879	59	3	19	3	919	7	29	17	–	959	3	53	3	29

Table T.1 *(Continued)*

	1	3	7	9		1	3	7	9		1	3	7	9		1	3	7	9
960	–	3	13	3	970	89	31	18	7	980	3	–	3	17	990	–	3	–	3
961	7	–	59	–	971	3	11	3	–	981	–	3	–	3	991	11	23	47	7
962	3	–	3	–	972	–	3	71	3	982	7	11	31	–	992	3	–	3	–
963	–	3	23	3	973	37	–	7	–	983	3	–	3	–	993	–	3	19	3
964	31	–	11	–	974	3	–	3	–	984	13	3	43	3	994	–	61	7	–
965	3	7	3	13	975	7	3	11	3	985	–	59	–	–	995	3	37	3	23
966	–	3	7	3	976	43	13	–	–	986	3	7	3	71	996	7	3	–	3
967	19	17	–	–	977	3	29	3	7	987	–	3	7	3	997	13	–	11	17
968	3	23	3	–	978	–	3	–	3	988	41	–	–	11	998	3	67	3	7
969	11	3	–	3	979	–	7	97	41	989	3	13	3	19	999	97	3	13	3

Table T.1

n	$\phi(n)$	$\tau(n)$	$\sigma(n)$
1	1	1	1
2	1	2	3
3	2	2	4
4	2	3	7
5	4	2	6
6	2	4	12
7	6	2	8
8	4	4	15
9	6	3	13
10	4	4	18
11	10	2	12
12	4	6	28
13	12	2	14
14	6	4	24
15	8	4	24
16	8	5	31
17	16	2	18
18	6	6	39
19	18	2	20
20	8	6	42
21	12	4	32
22	10	4	36
23	22	2	24
24	8	8	60
25	20	3	31
26	12	4	42
27	18	4	40
28	12	6	56
29	28	2	30
30	8	8	72
31	30	2	32
32	16	6	63
33	20	4	48
34	16	4	54
35	24	4	48
36	12	9	91
37	36	2	38
38	18	4	60
39	24	4	56
40	16	8	90

Table T.2 *Values of some arithmetic functions.*

n	$\phi(n)$	$\tau(n)$	$\sigma(n)$
41	40	2	42
42	12	8	96
43	42	2	44
44	20	6	84
45	24	6	78
46	22	4	72
47	46	2	48
48	16	10	124
49	42	3	57
50	20	6	93
51	32	4	72
52	24	6	98
53	52	2	54
54	18	8	120
55	40	4	72
56	24	8	120
57	36	4	80
58	28	4	90
59	58	2	60
60	16	12	168
61	60	2	62
62	30	4	96
63	36	6	104
64	32	7	127
65	48	4	84
66	20	8	144
67	66	2	68
68	32	6	126
69	44	4	96
70	24	8	144
71	70	2	72
72	24	12	195
73	72	2	74
74	36	4	114
75	40	6	124
76	36	6	140
77	60	4	96
78	24	8	168
79	78	2	80
80	32	10	186

Table T.2 *(Continued)*

n	$\phi(n)$	$\tau(n)$	$\sigma(n)$
81	54	5	121
82	40	4	126
83	82	2	84
84	24	12	224
85	64	4	108
86	42	4	132
87	56	4	120
88	40	8	180
89	88	2	90
90	24	12	234
91	72	4	112
92	44	6	168
93	60	4	128
94	46	4	144
95	72	4	120
96	32	12	252
97	96	2	98
98	42	6	171
99	60	6	156
100	40	9	217

Table T.2

p	r	p	r	p	r	p	r
2	1	191	19	439	15	709	2
3	2	193	5	443	2	719	11
5	2	197	2	449	3	727	5
7	3	199	3	457	13	733	6
11	2	211	2	461	2	739	3
13	2	223	3	463	3	743	5
17	3	227	2	467	2	751	3
19	2	229	6	479	13	757	2
23	5	233	3	487	3	761	6
29	2	239	7	491	2	769	11
31	3	241	7	499	7	773	2
37	2	251	6	503	5	787	2
41	6	257	3	509	2	797	2
43	3	263	5	521	3	809	3
47	5	269	2	523	2	811	3
53	2	271	6	541	2	821	2
59	2	277	5	547	2	823	3
61	2	281	3	557	2	827	2
67	2	283	3	563	2	829	2
71	7	293	2	569	3	839	11
73	5	307	5	571	3	853	2
79	3	311	17	577	5	857	3
83	2	313	10	587	2	859	2
89	3	317	2	593	3	863	5
97	5	331	3	599	7	877	2
101	2	337	10	601	7	881	3
103	5	347	2	607	3	883	2
107	2	349	2	613	2	887	5
109	6	353	3	617	3	907	2
113	3	359	7	619	2	911	17
127	3	367	6	631	3	919	7
131	2	373	2	641	3	929	3
137	3	379	2	643	11	937	5
139	2	383	5	647	5	941	2
149	2	389	2	653	2	947	2
151	6	397	5	659	2	953	3
157	5	401	3	661	2	967	5
163	2	409	21	673	5	971	6
167	5	419	2	677	2	977	3
173	2	421	2	683	5	983	5
179	2	431	7	691	3	991	6
181	2	433	5	701	2	997	7

Table T.3 *Least primitive roots r modulo primes p.*

p	*Numbers*															
	1	**2**	**3**	**4**	**5**	**6**	**7**	**8**	**9**	**10**	**11**	**12**	**13**	**14**	**15**	**16**
3	2	1														
5	4	1	3	2										Indices		
7	6	2	1	4	5	3										
11	10	1	8	2	4	9	7	3	6	5						
13	12	1	4	2	9	5	11	3	8	10	7	6				
17	16	14	1	12	5	15	11	10	2	3	7	13	4	9	6	8
19	18	1	13	2	16	14	6	3	8	17	12	15	5	7	11	4
23	22	2	16	4	1	18	19	6	10	3	9	20	14	21	17	8
29	28	1	5	2	22	6	12	3	10	23	25	7	18	13	27	4
31	30	24	1	18	20	25	28	12	2	14	23	19	11	22	21	0
37	36	1	26	2	23	27	32	3	16	24	30	28	11	33	13	4
41	40	26	15	12	22	1	39	38	30	8	3	27	31	25	37	24
43	42	27	1	12	25	28	35	39	2	10	30	13	32	20	26	24
47	46	18	20	36	1	38	32	8	40	19	7	10	11	4	21	26
53	52	1	17	2	47	18	14	3	34	48	6	19	24	15	12	4
59	58	1	50	2	6	51	18	3	42	7	25	52	45	19	56	4
61	60	1	6	2	22	7	49	3	12	23	15	8	40	50	28	4
67	66	1	39	2	15	40	23	3	12	16	59	41	19	24	54	4
71	70	6	26	12	28	32	1	18	52	34	31	38	39	7	54	24
73	72	8	6	16	1	14	33	24	12	9	55	22	59	41	7	32
79	78	4	1	8	62	5	53	12	2	66	68	9	34	57	63	16
83	82	1	72	2	27	73	8	3	62	28	24	74	77	9	17	4
89	88	16	1	32	70	17	81	48	2	86	84	33	23	9	71	64
97	96	34	70	68	1	8	31	6	44	35	86	42	25	65	71	40

p	*Numbers*																
	17	**18**	**19**	**20**	**21**	**22**	**23**	**24**	**25**	**26**	**27**	**28**	**29**	**30**	**31**	**32**	**33**
19	10	9															
23	7	12	15	5	13	11								Indices			
29	21	11	9	24	17	26	20	8	16	19	15	14					
31	7	26	4	8	29	17	27	13	10	5	3	16	9	15			
37	7	17	35	25	22	31	15	29	10	12	6	34	21	14	9	5	20
41	33	16	9	34	14	29	36	13	4	17	5	11	7	23	28	10	18
43	38	29	19	37	36	15	16	40	8	17	3	5	41	11	34	9	31
47	16	12	45	37	6	25	5	28	2	29	14	22	35	39	3	44	27
53	10	35	37	49	31	7	39	20	42	25	51	16	46	13	33	5	23
59	40	43	38	8	10	26	15	53	12	46	34	20	28	57	49	5	17
61	47	13	26	24	55	16	57	9	44	41	18	51	35	29	59	5	21
67	64	13	10	17	62	60	28	42	30	20	51	25	44	55	47	5	32
71	49	58	16	40	27	37	15	44	56	45	8	13	68	60	11	30	57
73	21	20	62	17	39	63	46	30	2	67	18	49	35	15	11	40	61
79	21	6	32	70	54	72	26	13	46	38	3	61	11	67	56	20	69
83	56	63	47	29	80	25	60	75	56	78	52	10	12	18	38	5	14
89	6	18	35	14	82	12	57	49	52	39	3	25	59	87	31	80	85
97	89	78	81	69	5	24	77	76	2	59	18	3	13	9	46	74	60

Table T.4 *Indices.*

	Numbers															
p	**34**	**35**	**36**	**37**	**38**	**39**	**40**	**41**	**42**	**43**	**44**	**45**	**46**	**47**	**48**	**49**
37	8	19	18													
41	19	21	2	32	35	6	20					Indices				
43	23	18	14	7	4	33	22	6	21							
47	34	33	30	42	17	31	9	15	24	13	43	41	23			
53	11	9	36	30	38	41	50	45	32	22	8	29	40	44	21	23
59	41	24	44	55	39	37	9	14	11	33	27	48	16	23	54	36
61	48	11	14	39	27	46	25	54	56	43	17	34	58	20	10	38
67	65	38	14	22	11	58	18	53	63	9	61	27	29	50	43	46
71	55	29	64	20	22	65	46	25	33	48	43	10	21	9	50	2
73	29	34	28	64	70	65	25	4	47	51	71	13	54	31	38	66
79	25	37	10	19	36	35	74	75	58	49	76	64	30	59	17	28
83	57	35	64	20	48	67	30	40	81	71	26	7	61	23	76	16
89	22	63	34	11	51	24	30	21	10	29	28	72	73	54	65	74
97	27	32	16	91	19	95	7	85	39	4	58	45	15	84	14	62

	Numbers															
p	**50**	**51**	**52**	**53**	**54**	**55**	**56**	**57**	**58**	**59**	**60**	**61**	**62**	**63**	**64**	**65**
53	43	27	26													
59	13	32	47	22	35	31	21	30	29				Indices			
61	45	53	42	33	19	37	52	32	36	31	30					
67	31	37	21	57	52	8	26	49	45	36	56	7	48	35	6	34
71	62	5	51	23	14	59	19	42	4	3	66	69	17	53	36	67
73	10	27	3	53	26	56	57	68	43	5	23	58	19	45	48	60
79	50	22	42	77	7	52	65	33	15	31	71	45	60	55	24	18
83	55	46	79	59	53	51	11	37	13	34	19	66	39	70	6	22
89	68	7	55	78	19	66	41	36	75	43	15	69	47	83	8	5
97	36	63	93	10	52	87	37	55	47	67	43	64	80	75	12	26

	Numbers															
p	**66**	**67**	**68**	**69**	**70**	**71**	**72**	**73**	**74**	**75**	**76**	**77**	**78**	**79**	**80**	**81**
67	33															
71	63	47	61	41	35						Indices					
73	69	50	37	52	42	44	36									
79	73	48	29	27	41	51	14	44	23	47	40	43	39			
83	15	45	58	50	36	33	65	69	21	44	49	32	68	43	31	42
89	13	56	38	58	79	62	50	20	27	53	67	77	40	42	46	4
97	94	57	61	51	66	11	50	28	29	72	53	21	33	30	41	88

	Numbers														
p	**82**	**83**	**84**	**85**	**86**	**87**	**88**	**89**	**90**	**91**	**92**	**93**	**94**	**95**	**96**
83	41														
89	37	61	26	76	45	60	44					Indices			
97	23	17	73	90	38	83	92	54	79	56	49	20	22	82	48

Table T.4 *(Continued)*

	Indices															
p	**1**	**2**	**3**	**4**	**5**	**6**	**7**	**8**	**9**	**10**	**11**	**12**	**13**	**14**	**15**	**16**
3	2	1														
5	2	4	3	1												
7	3	2	6	4	5	1							Numbers			
11	2	4	8	5	10	9	7	3	6	1						
13	2	4	8	3	6	12	11	9	5	10	7	1				
17	3	9	10	13	5	15	11	16	14	8	7	4	12	2	6	1
19	2	4	8	16	13	7	14	9	18	17	15	11	3	6	12	5
23	5	2	10	4	20	8	17	16	11	9	22	18	21	13	19	3
29	2	4	8	16	3	6	12	24	19	9	18	7	14	28	27	25
31	3	9	27	19	26	16	17	20	29	25	13	8	24	10	30	28
37	2	4	8	16	32	27	17	34	31	25	13	26	15	30	23	9
41	6	36	11	25	27	39	29	10	19	32	28	4	24	21	3	18
43	3	9	27	38	28	41	37	25	32	10	30	4	12	36	22	23
47	5	25	31	14	23	21	11	8	40	12	13	18	43	27	41	17
53	2	4	8	16	32	11	22	44	35	17	34	15	30	7	14	28
59	2	4	8	16	32	5	10	20	40	21	42	25	50	41	23	46
61	2	4	8	16	32	3	6	12	24	48	35	9	18	36	11	22
67	2	4	8	16	32	64	61	55	43	19	38	9	18	36	5	10
71	7	49	59	58	51	2	14	27	47	45	31	4	28	54	23	19
73	5	25	52	41	59	3	15	2	10	50	31	9	45	6	30	4
79	3	9	27	2	6	18	54	4	12	36	29	8	24	72	58	16
83	2	4	8	16	32	64	45	7	14	28	56	29	58	33	66	49
89	3	9	27	81	65	17	51	64	14	42	37	22	66	20	60	2
97	5	25	28	43	21	8	40	6	30	53	71	64	29	48	46	36

	Indices																
p	**17**	**18**	**19**	**20**	**21**	**22**	**23**	**24**	**25**	**26**	**27**	**28**	**29**	**30**	**31**	**32**	**33**
19	10	1															
23	15	6	7	12	14	1							Numbers				
29	21	13	26	23	17	5	10	20	11	22	15	1					
31	22	4	12	5	15	14	11	2	6	18	23	7	21	1			
37	18	36	35	33	29	21	5	10	20	3	6	12	24	11	22	7	14
41	26	33	34	40	35	5	30	16	14	2	12	31	22	9	13	37	17
43	26	35	19	14	42	40	34	16	5	15	2	6	18	11	33	13	39
47	38	2	10	3	15	28	46	42	22	16	33	24	26	36	39	7	35
53	3	6	12	24	48	43	33	13	26	52	51	49	45	37	21	42	31
59	33	7	14	28	56	53	47	35	11	22	44	29	58	57	55	51	43
61	44	27	54	47	33	5	10	20	40	19	38	15	30	60	59	57	53
67	20	40	13	26	52	37	7	14	28	56	45	23	46	25	50	33	66
71	62	8	56	37	46	38	53	16	41	3	21	5	35	32	11	6	42
73	20	27	62	18	17	12	60	8	40	54	51	36	34	24	47	16	7
79	48	65	37	32	17	51	74	64	34	23	69	49	68	46	59	19	57
83	15	30	60	37	74	65	47	11	22	44	5	10	20	40	80	77	71
89	6	18	54	73	41	34	13	39	28	84	74	44	43	40	31	4	12
97	83	27	38	93	77	94	82	22	13	65	34	73	74	79	7	35	78

Table T.4 *(Continued)*

	Indices															
p	**34**	**35**	**36**	**37**	**38**	**39**	**40**	**41**	**42**	**43**	**44**	**45**	**46**	**47**	**48**	**49**
37	28	19	1													
41	20	38	23	15	8	7	1						Numbers			
43	31	7	21	20	17	8	24	29	1							
47	34	29	4	20	6	30	9	45	37	44	32	19	1			
53	9	18	36	19	38	23	46	39	25	50	47	41	29	5	10	20
59	27	54	49	39	19	38	17	34	9	18	36	13	26	52	45	31
61	45	29	58	55	49	37	13	26	52	43	25	50	39	17	34	7
67	65	63	59	51	35	3	6	12	24	48	29	58	49	31	62	57
71	10	70	64	22	12	13	20	69	57	44	24	26	40	67	43	17
73	35	29	72	68	48	21	32	14	70	58	71	63	23	42	64	28
79	13	39	38	35	26	78	76	70	52	77	73	61	25	75	67	43
83	59	35	70	57	31	62	41	82	81	79	75	67	51	19	38	76
89	36	19	57	82	68	26	78	56	79	59	88	86	80	62	8	24
97	2	10	50	56	86	42	16	80	12	60	9	45	31	58	96	92

	Indices															
p	**50**	**51**	**52**	**53**	**54**	**55**	**56**	**57**	**58**	**59**	**60**	**61**	**62**	**63**	**64**	**65**
53	40	27	1													
59	3	6	12	24	48	37	15	30	1					Numbers		
61	14	28	56	51	41	21	42	23	46	31	1					
67	47	27	54	41	15	30	60	53	39	11	22	44	21	42	17	34
71	48	52	9	63	15	34	25	33	18	55	30	68	50	66	36	39
73	67	43	69	53	46	11	55	56	61	13	65	33	19	22	37	39
79	50	71	55	7	21	63	31	14	42	47	62	28	5	15	45	56
83	69	55	27	54	25	50	17	34	68	53	23	46	9	18	36	72
89	72	38	25	75	47	52	67	23	69	29	87	83	71	35	16	48
97	72	69	54	76	89	57	91	67	44	26	33	68	49	51	61	14

	Indices															
p	**66**	**67**	**68**	**69**	**70**	**71**	**72**	**73**	**74**	**75**	**76**	**77**	**78**	**79**	**80**	**81**
67	1															
71	60	65	29	61	1											
73	49	26	57	66	38	44	1									
79	10	30	11	33	20	60	22	66	40	41	44	53	1			
83	61	39	78	73	63	43	3	6	12	24	48	13	26	52	21	42
89	55	76	50	61	5	15	45	46	49	58	85	77	53	70	32	7
97	70	59	4	20	3	15	75	84	32	63	24	23	18	90	67	19

	Indices															
p	**66**	**67**	**68**	**69**	**70**	**71**	**72**	**73**	**74**	**75**	**76**	**77**	**78**	**79**	**80**	**81**
83	1											Numbers				
89	21	63	11	33	10	30	1									
97	95	87	47	41	11	55	81	17	85	37	88	52	66	39	1	

Table T.4

References

1. S. B. Aiyer, "The Ganita-Sara-Sangraha of Mahaviracarya," *Mathematics Teacher* 47 (Dec. 1954), 528–533.
2. M. Andersen *et al.*, `www.library.thinkquest.org/27890/altver/biographies3/.html`.
3. T. M. Apostol, "Another GCD Problem," *Mathematics Magazine* 54 (March 1981), 86–87.
4. S. Asadulla, Problem 70.F, *Mathematical Gazette* 71 (1987), 66.
5. W. T. Bailey, "Friday-the-Thirteenth," *Mathematics Teacher* 62 (May 1969), 363–364.
6. D. W. Ballew and J. V. Bossche, "Palindromic Figurate Numbers," *J. Recreational Mathematics* 8 (1975–1976), 92–95.
7. L. Bankoff, "Curiosa on 1968," *J. Recreational Mathematics* 1 (Jan. 1968), 78.
8. P. T. Bateman *et al.*, "The New Mersenne Conjecture," *American Mathematical Monthly* 96 (Feb. 1989), 125–128.
9. J. D. Baum, "A Note on Primitive Roots," *Mathematics Magazine*, 38 (1965), 12–14.
10. J. D. Baum, "A Number-Theoretic Sum," *Mathematics Magazine* 55 (March 1982), 111–113.
11. C. Bays and R. H. Hudson, "Zeros of Dirichlet L-Functions and Irregularities in the Distribution of Primes," *Mathematics of Computation* 69 (1999), 861–866.
12. A. H. Beiler, *Recreations in the Theory of Numbers*, Dover, 2nd edition, New York, 1966.
13. J. T. Betcher and J. H. Jaroma, "An Extension of the Results of Servais and Cramer on Odd Perfect and Odd Multiply Perfect Numbers," *American Mathematical Monthly* 110 (Jan. 2003), 49–52.
14. J. L. Boal and J. H. Bevis, "Permutable Primes," *Mathematics Magazine* 55 (Jan. 1982), 38–41.
15. N. Boston, "A Taylor-Made Plug for Wiles' Proof," *College Mathematics Journal* 26 (March 1995), 100–105.
16. S. Bulman-Fleming and E. T. H. Wang, "An Application of Bertrand's Postulate," *College Mathematics Journal* 20 (May 1989), 265.

17. D. M. Burton, *Elementary Number Theory*, 5th edition, Wm. C. Brown, Dubuque, Iowa, 2002.
18. K. Chandrasekharan, *Introduction to Analytic Number Theory*, Springer-Verlag, New York, 1968.
19. M. Charosh, Problem 10, *J. Recreational Mathematics* 16 (1983–1984), 69.
20. C. Cooper and R. E. Kennedy, "Order Mod 10," *College Mathematics Journal* 27 (March 1996), 150–151.
21. A. Cuoco, "Surprising Result," *Mathematics Teacher* 87 (Nov. 1994), 640, 660.
22. R. M. Dacic, "Mersenne Numbers and Binomial Coefficients," *Mathematics Magazine* 54 (Jan. 1981), 32.
23. M. Dalezman, "From 30 to 60 Is Not Twice as Hard," *Mathematics Magazine* 73 (April 2000), 151–153.
24. R. J. M. Dawson, "Towers of Powers Modulo *m*," *College Mathematics Journal* 25 (Jan. 1994), 22–28.
25. M. J. DeLeon, "A Simple Proof of an Old Divisibility Test," *J. Recreational Mathematics* 11 (1978–1979), 186–189.
26. E. Deutsch and I. M. Gessel, Problem 1494, *Mathematics Magazine* 70 (April 1997), 143–144.
27. K. Devlin, "Move Over Fermat, Now It's Time for Beal's Problem," *Math Horizons* (Feb. 1998), 8–10.
28. L. E. Dickson, "All Integers Except 23 and 239 Are Sums of Eight Cubes," *Bulletin of the AMS* 45 (Aug. 1939), 588–591.
29. *Dictionary of Catholic Biographers*, Doubleday, New York, 1961.
30. L. E. Dixon, *History of the Theory of Numbers*, Vols. 1–3, Chelsea, New York, 1971.
31. D. Doster, "Cubic Polynomials from Curious Sums," *American Mathematical Monthly* 104 (Jan. 1997), 78.
32. H. Dubner, "Generalized Cullen Numbers," *J. Recreational Mathematics* 21 (1989), 190–194.
33. H. Dubner, "Searching for Wilson Primes," *J. Recreational Mathematics* 21 (1989), 19–20.
34. H. Dubner and H. Nelson, "Carmichael Numbers Which Are the Product of Three Carmichael Numbers," *J. Recreational Mathematics* 22 (1990), 2–6.
35. G. Duckworth, Problem 69.C, *Mathematical Gazette* 65 (1985), 302.
36. U. Dudley, *Elementary Number Theory*, W. H. Freeman and Co., New York, 1969.
37. L. R. Duffy, "The Duffinian Numbers," *J. Recreational Mathematics* 12 (1979–1980), 112–115.
38. A. F. Dunn (ed.), *Second Book of Mathematical Bafflers*, Dover, New York, 1983.
39. P. M. Dunson, "2500th Digit Out of 35,660," *Mathematics Magazine* 53 (Sept. 1980), 249.
40. C. F. Eaton, "Perfect Numbers in Terms of Triangular Numbers," *Mathematics Magazine* 69 (Oct. 1996), 308.
41. M. Eisenstein, Problem B-530, *Fibonacci Quarterly* 22 (Aug. 1984), 274.
42. M. Eisenstein, Problem B-531, *Fibonacci Quarterly* 22 (Aug. 1984), 274.

43. W. J. Ellison, "Waring's Problem," *American Mathematical Monthly* 78 (Jan. 1971), 10–36.
44. E. Emanouilidis, "Roulette and the Beastly Number," *J. Recreational Mathematics* 29 (1998), 246–247.
45. *Encyclopaedia Britannica*, Chicago, Illinois, 1973.
46. R. Euler, "A Perfect Harmonic Mean?," *J. Recreational Mathematics* 20 (1988), 153–154.
47. H. Eves, *An Introduction to the History of Mathematics*, 3rd edition, Holt, Rinehart and Winston, New York, 1969.
48. A. Filz, "Prime Circles," *J. Recreational Mathematics* 15 (1982–1983), 70–71.
49. S. Forseth and A. P. Troutman, "Using Mathematical Structures to Generate Artistic Designs," *Mathematics Teacher* 67 (May 1974), 393–398.
50. M. Gardner, *Mathematical Puzzles and Diversions*, The University of Chicago, Chicago, 1987.
51. H. Gauchman and I. Rosenholtz, Problem 1404, *Mathematics Magazine* 65 (Oct. 1992), 265.
52. G. Gilbert, Problem 1339, *Mathematics Magazine* 64 (Feb. 1991), 63.
53. C. C. Gillispie (ed.), *Dictionary of Scientific Biography*, Charles Scribner's Sons, New York, 1973.
54. J. Ginsburg, "Triplets of Equiareal Rational Triangles," *Scripta Mathematica* 20 (1954), 219.
55. J. Ginsburg, "The Wonderful Wonders of Arithmetic Progressions," *Scripta Mathematica* 9 (1943), 190.
56. S. W. Golomb, "Primitive Elements Modulo Primes and Their Squares," *American Mathematical Monthly* 104 (Jan. 1997), 71–72.
57. R. Goormaghtigh, "Identities with Digits in Reversed Order," *Scripta Mathematica* 17 (1951), 19.
58. *Grand Larousse Encylopédique*, Librairie Larousse, Paris, 1960.
59. D. Grannis, Problem 1663, *J. Recreational Mathematics* 23 (1991), 303.
60. A. Granville, "Review of BBC's Horizon Program, 'Fermat's Last Theorem,' " *Notices of the AMS* 44 (Jan. 1997), 26–28.
61. M. H. Greenblat, "Wilson's Theorem," *J. Recreational Mathematics* 4 (April 1971), 88–89.
62. R. K. Guy, *Unsolved Problems in Number Theory*, 2nd edition, Springer-Verlag, New York, 1994.
63. R. K. Guy, "What's Left?" *Math Horizons* (April 1998), 5–7.
64. G. E. Hardy and M. V. Subbarao, "A Modified Problem of Pillai and Some Related Questions," *American Mathematical Monthly* 109 (June 2002), 554–559.
65. R. J. Hendel, "Fermat's Little Theorem," *Mathematics Magazine* 62 (Oct. 1989), 281.
66. R. J. Hendel, Problem Q752, *Mathematics Magazine* 62 (Oct. 1989), 275–276.
67. H. J. Hindin, "Another Property of 1729," *J. Recreational Mathematics* 16 (1983–1984), 248–249.
68. G. L. Honaker, Problem 1817, *J. Recreational Mathematics* 23 (1991), 231.

69. R. Honsberger, *More Mathematical Morsels*, Math. Assoc. of America, Washington, D.C., 1991.
70. R. H. Hudson, "Averaging Effects on Irregularities in the Distribution of Primes in Arithmetic Progression," *Mathematics Computation* 44 (April 1985), 561–571.
71. R. H. Hudson, "A Bound for the First Occurrence of Three Consecutive Integers with Equal Quadratic Character," *Duke Mathematical Journal* 40 (March 1973), 33–39.
72. B. J. Hulbert, "Twin Primes," *Mathematical Spectrum* 33 (2000–2001), 19.
73. D. E. Iannucci, "The Neighbors of the Beast," *J. Recreational Mathematics* 31 (2002–2003), 52–55.
74. D. E. Iannucci, "The Third Largest Prime Divisor of an Odd Perfect Number Exceeds One Hundred," *Mathematics Computation* 69 (1999), 867–879.
75. R. F. Jordan, "A Triangular Generalization," *J. Recreational Mathematics* 23 (1991), 78.
76. D. R. Kaprekar, "On Kaprekar's Harshad Numbers," *J. Recreational Mathematics* 13 (1980–1981), 2–3.
77. D. R. Kaprekar and R. V. Iyer, "Identities with Digits in Reversed Order," *Scripta Mathematica* 16 (1950), 160.
78. M. Keith, "A Mental Perpetual Calendar," *J. Recreational Mathematics* 8 (1975–1976), 242–245.
79. M. Keith, "The Number 666," *J. Recreational Mathematics* 15 (1982–1983), 85–86.
80. M. Keith, "More 666 Curiosa," *J. Recreational Mathematics* 31 (2002–2003), 47–49.
81. M. Keith and T. Craver, "The Ultimate Perpetual Calendar?" *J. Recreational Mathematics* 22 (1990), 280–282.
82. S. Y. Kim, "An Elementary Proof of the Quadratic Reciprocity," *American Mathematical Monthly* 111 (Jan. 2004), 48–50.
83. C. Kimberling, "A Visual Euclidean Algorithm," *Mathematics Teacher* 76 (Feb. 1983), 108–109.
84. W. Kohnen, "A Simple Congruence Modulo *p*," *American Mathematical Monthly* 104 (May 1997), 444–445.
85. S. J. Kolpas, *The Pythagorean Theorem: Eight Classic Proofs*, Dale Seymour, Palo Alto, Calif., 1992.
86. T. Koshy, "The Ends of a Mersenne Number and an Even Perfect Number," *J. Recreational Mathematics* 29 (1998), 196–202.
87. T. Koshy, "The Digital Root of a Fermat Number," *J. Recreational Mathematics* 31 (2002–2003).
88. T. Koshy, "The Ends of a Fermat Number," *J. Recreational Mathematics* 31 (2002–2003), 183–184.
89. T. Koshy, "A Generalization of a Curious Sum," *Mathematics Gazette* 83 (March 1999), 9–13.
90. T. Koshy, "A Generalization of Euler's Theorem," *Mathematical Gazette* 82 (March 1998), 80.
91. T. Koshy, *Fibonacci and Lucas Numbers with Applications*, Wiley, New York, 2001.
92. T. Koshy, "Linear Diophantine Equations, Linear Congruences, and Matrices," *Mathematics Gazette* 82 (July 1998), 274–277.
93. T. Koshy, "Long Chains of Primes," *Pi Mu Epsilon Journal* 10 (Fall 1994), 32–33.

94. T. Koshy, "Digital Roots of Mersenne Primes and Even Perfect Numbers," *Mathematical Gazette* 89 (Nov. 2005), 464–466.
95. T. Koshy and M. Salmassi, "Parity and Primality of Catalan Numbers," *College Mathematics Journal* 37 (Jan. 2006), 52–53.
96. M. Kraitchik, "Multiplication-Table Curiosities," *Scripta Mathematica* 16 (1950), 125.
97. S. Kravitz, "Three Consecutive Abundant Numbers," *J. Recreational Mathematics* 27 (1995), 156–157.
98. L. Kuipers, Problem Q671, *Mathematics Magazine* 55 (March 1982), 116.
99. E. G. Landauer, "On Squares of Positive Integers," *Mathematics Magazine* 58 (Sept. 1985), 236.
100. L. J. Lange, "An Elegant Continued Fraction for π," *American Mathematical Monthly* 106 (May 1999), 456–458.
101. D. R. Lichtenberg, "From Geoboard to Number Theory to Complex Numbers," *Mathematics Teacher* 68 (May 1975), 370–375.
102. P. Locke, "Residue Designs," *Mathematics Teacher* 65 (March 1972), 260–263.
103. C. T. Long and J. H. Jordan, "A Limited Arithmetic on Simple Continued Fractions," *Fibonacci Quarterly* 5 (April 1967), 113–128.
104. A. R. G. MacDivitt, "The Most Recently Discovered Prime Number," *Mathematical Gazette* 63 (1979), 268–270.
105. A. A. K. Majumdar, "A Note on a Sequence Free from Powers," *Mathematical Spectrum* 29 (1996–1997), 41.
106. R. D. Mauldin, "A Generalization of Fermat's Last Theorem: The Beal Conjecture and Prize Problem," *Notices of the AMS* 44 (Dec. 1997), 1436–1437.
107. W. L. McDaniel, "Palindromic Smith Numbers," *J. Recreational Mathematics* 19 (1987), 34–37.
108. F. L. Miksa, "Table of Primitive Pythagorean Triangles Whose Areas Contain All the Digits 1, 2, 3, 4, 5, 6, 7, 8, 9 and 0," *Scripta Mathematica* 20 (1954), 231.
109. P. L. Montgomery and J. L. Selfridge, "Special Perfect Numbers," *American Mathematical Monthly* 102 (Jan. 1995), 72.
110. T. E. Moore, "A Query Regarding Euler's φ-Function," *J. Recreational Mathematics* 7 (Fall 1974), 306.
111. L. Moser, "Palindromic Primes," *Scripta Mathematica* 16 (1950), 127–128.
112. R. B. Nelsen, "Proof without Words," *Mathematics Magazine* 66 (June 1993), 180.
113. F. C. Ogg, Letter to the Editor, *The Scientific Monthly* 78 (Nov. 1951), 333.
114. R. Ondrejka, "Prime Trivia," *J. Recreational Mathematics* 14 (1918–1982), 285.
115. R. Ondrejka, "666 Again," *J. Recreational Mathematics* 16 (1983–1984), 121.
116. S. Peterburgsky, "Problem 13: Square Summands," *Math Horizons* (Feb. 1995), 33.
117. P. A. Piza, "Sums of Powers of Triangular Numbers," *Scripta Mathematica* 16 (1950), 127.
118. J. M. Pollard, "A Monte Carlo Method for Factorization," *BIT* 15 (1975), 331–334.
119. M. Polezzi, "A Geometrical Method for Finding an Explicit Formula for the Greatest Common Divisor," *American Mathematical Monthly* 104 (May 1997), 445–446.
120. C. Pomerance, "A New Primal Screen," *Focus* 22 (Nov. 2002), 4–5.

121. P. Ribenboim, *The New Book of Prime Number Records*, Springer-Verlag, New York, 1996.
122. K. H. Rosen, *Elementary Number Theory and Its Applications*, 5th edition, Addison-Wesley, Boston, Massachusetts, 2005.
123. S. M. Ruiz, "An Algebraic Identity Leading to Wilson's Theorem," *Mathematical Gazette* 80 (Nov. 1996), 579–582.
124. H. Sazegar, "A Note on Wilson's Theorem," *Mathematical Spectrum* 26 (1993–1994), 81–82.
125. N. Schaumberger, "A Diophantine Equation with Successive Powers," *College Mathematics Journal* 25 (Sept. 1994), 338.
126. J. Shallit, "Diophantine Equation, $\sigma(n) = 2^m$," *Mathematics Magazine* 63 (April 1990), 129.
127. D. Shanks, *Solved and Unsolved Problems in Number Theory*, Chelsea, New York, 1985.
128. C. L. Shedd, "Another Triplet of Equiareal Triangles," *Scripta Mathematica* 11 (1945), 273.
129. T. W. Shilgalis, "Are Most Fractions Reduced?" *Mathematics Teacher* 87 (April 1994), 236–238.
130. J. O. Silva, "Summing Powers Mod *m*," *College Mathematics Journal* 27 (Jan. 1996), 75–76.
131. D. L. Silverman, "A Prime Problem," *J. Recreational Mathematics* 1 (Oct. 1968), 236.
132. D. L. Silverman, "φ Sums," *J. Recreational Mathematics* 15 (1982–1983), 67.
133. C. Singh, "The Beast 666," *J. Recreational Mathematics* 21 (1984), 244.
134. D. Slowinski, "Searching for the 27th Mersenne Prime," *J. Recreational Mathematics* 11 (1978–1979), 258–261.
135. C. Small, "Waring's Problem," *Mathematics Magazine* 50 (Jan. 1977), 12–16.
136. H. V. Smith, "The Twenty-Fifth (Known) Perfect Number," *Mathematical Gazette* 63 (1979), 271.
137. A. P. Stevens, "Divisibility by 7," *Scripta Mathematica* 17 (1951), 146.
138. L. Talbot, "Problem 28.8," *Mathematics Spectrum* 28 (1995–1996), 45.
139. "The Art of Giant Slaying is Refined: GIMPS finds $2^{3021377} - 1$ is prime," `www.utm.edu/research/primes/notes/3021377`, 1998.
140. V. Thébault, "Number Pleasantries," *Scripta Mathematica* 12 (1946), 218.
141. "The Largest Known Primes," `www.utm.edu/research/primes/largest.html`, 2000.
142. R. D. Torre, "An Interesting Property of Prime Numbers," *Scripta Mathematica* 7 (1940), 159.
143. J. Touchard, "On Prime Numbers and Perfect Numbers," *Scripta Mathematica* 19 (1953), 35–39.
144. C. W. Trigg, "Recurrent Operations on 1968," *J. Recreational Mathematics* 1 (Oct. 1968), 243–245.
145. C. W. Trigg, "1729—More Properties," *J. Recreational Mathematics* 17 (1984–1985), 176.
146. C. W. Trigg, "That Perfectly Beastly Number," *J. Recreational Mathematics* 20 (1988), 61.

147. H. L. Umansky, "A Triangle of Pythagorean Hypotenuses," *Scripta Mathematica* 16 (1950), 128.
148. E. T. H. Wang, "Divisibility of Binomial Coefficients," *College Mathematics Journal* 26 (March 1995), 160–161.
149. S. Webb, "Mersenne Primes," *Mathematical Spectrum* 31 (Nov. 1998), 18.
150. D. Wells, *The Penguin Dictionary of Curious and Interesting Numbers*, Penguin Books, New York, 1987.
151. P. A. Weiner, "The Abundancy Ratio, a Measure of Perfection," *Mathematics Magazine* 73 (Oct. 2000), 307–310.
152. W. P. Whitlock, Jr., "An Impossible Triangle," *Scripta Mathematica* 9 (1943), 189.
153. W. P. Whitlock, Jr., "A Problem in Representations," *Scripta Mathematica* 12 (1946), 218.
154. W. P. Whitlock, Jr., "Pythagorean Triangles with Square Perimeters," *Scripta Mathematica* 14 (1948), 60.
155. W. P. Whitlock, Jr., "Rational Right Triangles with Equal Areas," *Scripta Mathematica* 9 (1943), 155–161.
156. W. P. Whitlock, Jr., "A Very Exclusive '6'," *Scripta Mathematica* 9 (1943), 189.
157. A. Wilansky, "Primitive Roots without Quadratic Reciprocity," *Mathematics Magazine* 79 (1976), 146.
158. "Wiles Receives NAS Award in Mathematics," *Notices of the AMS* 43 (July 1996), 760–763.
159. `www.groups.dcs.st-` and `.ac.uk/~history/mathematicians/Binet.html`.
160. `www.math.bmc.hu/mathhist/Mathematicians/Binet.html`.
161. `www.mathsoft.com/asolve/pwrs32/pwrs32.html`.
162. `www.mathworld.wolfram.com/Fermat-CatalanConjecture.html`
163. `www.mcs.surrey.ac.uk/personal/R.knott/Fibonacci/cfINTRO.html`
164. `www.newadvent.org/cathen/0257a.html`.
165. `www.pbs.org/wgbh/nova/proof/wiles.html`.
166. `www.prothsearch.net/fermat.html`.
167. `www.reality.sgi.com/csp/ioccc/noll/prime-press.html`.
168. `www.utm.edu/research/primes/largest.html`.
169. `www.utm.edu/research/primes/mersenne.shtml`.
170. `www.utm.edu/research/primes/notes/3021377`.
171. `www.vamri.xray.ufl.edu/proths/fermat.html`.
172. S. Yates, "Peculiar Properties of Repunits," *J. Recreational Mathematics* 2 (July 1969), 139–146.
173. S. Yates, "Smith Numbers Congruent to 4 (mod 9)," *J. Recreational Mathematics* 19 (1987), 139–141.
174. M. J. Zerger, "The 'Number of Mathematics'," *J. Recreational Mathematics* 25 (1993), 247–250.
175. M. J. Zerger, "Surprising Result II," *Mathematics Teacher* 87 (Nov. 1994), 660.

Solutions to Odd-Numbered Exercises

Chapter 1 Fundamentals

Exercises 1.1 (p. 7)

1. 1806
3. 0 or -1
5. yes, if x is an integer.
7. 370
9. Canceling $a - b = 0$ from both sides is invalid.
11. $1105 = 4^2 + 33^2 = 9^2 + 32^2 = 23^2 + 24^2 = 12^2 + 31^2$
13. Let $n^2 = 2xy89$. Then n must be a three-digit number beginning with 1 and ending in 3 or 7, so n is of the form $1t3$ or $1t7$. Since $140^2 = 19{,}600 < 2xy89 < 32{,}400 = 180^2$, $143 \le n \le 177$. Because $(10u + 3)^2 = 100u^2 + 60u + 9$, u must be 3 or 8 in order to yield 8 as the tens digit in n^2, which cannot be the case. Similarly, from $(10u + 7)^2 = 100u^2 + 140u + 49$, it follows that $u = 1$ or 6. Only $u = 6$ falls within the bounds for n, so $n = 167$. Thus, $167^2 = 27{,}889$ and $xy = 78$.
15. There are $\lfloor\sqrt{500}\rfloor = 22$ squares and $\lfloor\sqrt[3]{500}\rfloor = 7$ cubes ≤ 500. Among them, $\lfloor\sqrt[6]{500}\rfloor = 2$ are counted twice, namely 1 and 64. Thus, there are $22 + 7 - 2 = 27$ integers ≤ 500 that are not in the sequence, and 473 integers that are either nonsquares or noncubes ≤ 500. So to find the 500th term of the sequence, append 27 integers to the list $2, 3, 5, \ldots, 500$. Because we cannot use 512, it must be 528.
17. Suppose $a, b \ge 0$. Then $a + b \ge 0$, so $|a + b| = a + b = |a| + |b|$. If $a, b < 0$, then $a + b < 0$, so $|a + b| = -(a + b) = (-a) + (-b) \le a + b = |a| + |b|$. Suppose $a \ge 0$ and $b < 0$. If $|a| > |b|$, then $|a + b| = a + b \le |a| + |b|$. The other case follows similarly.
19. Let $n = 2k + 1$ for some integer k. Then $\lceil n/2 \rceil = \lceil k + 1/2 \rceil = k + 1 = (2k + 2)/2 = (n + 1)/2$.
21. Let $n = 2k + 1$ for some integer k. Then $\lceil n^2/4 \rceil = \lceil k^2 + k + 1/4 \rceil = k^2 + k + 1 = [(4k^2 + 4k + 1) + 3]/4 = (n^2 + 3)/4$.
23. Let $x = k + x'$, where $k = \lfloor x \rfloor$ and $0 < x' < 1$. Then $\lceil x \rceil = k + 1 = \lfloor x \rfloor + 1$.
25. Let $x = k + x'$, where $k = \lfloor x \rfloor$ and $0 \le x' < 1$. Then $\lceil x + n \rceil = k + 1 + n = \lceil x \rceil + n$.

27. Let $x = k + x'$, where $k = \lfloor x \rfloor$ and $0 \le x' < 1$. Then $\lfloor x/n \rfloor = \lfloor (k+x')/n \rfloor = \lfloor k/n \rfloor = \lfloor \lfloor x \rfloor / n \rfloor$.
29. $d(0, x) = |x - 0| = |x|$
31. $d(x, y) = |y - x| = |-(x - y)| = |x - y| = d(y, x)$
33. **case 1** Let $x \ge y$. Then $|x - y| = x - y$, $\max\{x, y\} = x$, and $\max\{x, y\} = y$. Thus, $\max\{x, y\} - \min\{x, y\} = x - y = |x - y|$.

    case 2 Let $x < y$. Then, $|x - y| = y - x$, $\max\{x, y\} = y$, and $\min\{x, y\} = x$. Thus, $\max\{x, y\} - \min\{x, y\} = y - x = |x - y|$.

Exercises 1.2 (p. 14)

1. 21
3. 5
5. 20
7. 21
9. 135
11. $\sum_{k=1}^{12} (2k - 1)$
13. $\sum_{k=1}^{11} k(k+1)$
15. T
17. $S = (a_{m+1} - a_m) + (a_{m+2} - a_{m+1}) + \cdots + (a_n - a_{n-1}) = a_n - a_m$
19.
$$\sum_{i=1}^{n} [(i+1)^2 - i^2] = \sum_{i=1}^{n} (i+1)^2 - \sum_{i=1}^{n} i^2$$
$$(n+1)^2 - 1 = 2\sum_{i=1}^{n} i + \sum_{i=1}^{n} 1$$
$$n^2 + n = 2\sum_{i=1}^{n} i$$
$$\sum_{i=1}^{n} i = n(n+1)/2$$
21. $[n(n+1)/2]^2$
23. 32
25. 255
27. 4420
29. 10
31. 210
33. 28
35. 6
37. 480480
39. 613
41. $a_{11} + a_{12} + a_{21} + a_{22} + a_{31} + a_{32}$
43. $2a_1 + 2a_2 + 2a_3$
45. sum $= \sum_{n=1}^{1023} [\lg(n+1) - \lg n] = \lg 1024 - \lg 1 = 10 - 0 = 10$
47. sum $= \lfloor \lg 2/1 \rfloor + \lfloor \lg 3/2 \rfloor + \lfloor \lg 4/3 \rfloor + \cdots + \lfloor \lg 1025/1024 \rfloor = 1 + 0 + \cdots + 0 = 1$
49. When $n \ge 10$, $n!$ is divisible by 100. Therefore, the tens digit in the given sum equals the tens digit in $\sum_{i=1}^{9} n!$. By direct computation, it ends in 13, so the desired tens digit is 1.
51. Given sum $= \sum_{n=0}^{\infty} \left\lfloor \frac{1000}{2^{n+1}} + \frac{1}{2} \right\rfloor$. Thus, the sum is not infinite since if $n \ge 14$, then
$$\frac{10000}{2^{n+1}} < \frac{1}{2}, \quad \text{so} \quad \left\lfloor \frac{10000}{2^{n+1}} + \frac{1}{2} \right\rfloor = 0$$
Therefore,
$$\text{given sum} = \sum_{n=0}^{13} \left\lfloor \frac{10000}{2^{n+1}} + \frac{1}{2} \right\rfloor = \sum_{n=0}^{13} \left(\left\lfloor \frac{10000}{2^n} - \frac{10000}{2^{n+1}} \right\rfloor \right),$$
because $\lfloor x + 1/2 \rfloor = \lfloor 2x \rfloor - \lfloor x \rfloor$
$$= 10{,}000 - \left\lfloor \frac{10000}{2^{14}} \right\rfloor = 10{,}000 - 0 = 10{,}000$$

Exercises 1.3 (p. 24)

1. no, because it has no least element.
3. yes
5. Assume there is an integer n between a and $a+1$: $a < n < a+1$. Then $0 < n-a < 1$. Thus exists an integer $n-a$ between 0 and 1, a contradiction.
7. **proof** (by contradiction): Assume there are no integers n such that $na \geq b$; that is, $na < b$ for every integer n. Then the set $S = \{b - na \mid n \in \mathbf{Z}^+\}$ consists of positive integers, so by the well-ordering principle, S has a least element ℓ. Then $\ell = b - ma$ for some integer m. Notice that $b - (m+1)a \in S$, but $b - (m+1)a < b - ma = \ell$, a contradiction.
9. Let $A = \{a \in \mathbf{Z} \mid a \leq n_0\}$. Then the set $B = \{n_0 - a \mid a \in A\}$ consists of nonnegative integers and hence has a least element ℓ. Thus, $n_0 - a \geq \ell$ for every element a in A; that is, $n_0 - \ell \geq a$ for every a in A. But $n_0 - \ell \leq n_0$, so $n_0 - \ell \in A$. Thus, $n_0 - \ell$ is a largest element of A.
11. 364

(*Note*: In Exercises 13 and 15, $P(n)$ denotes the given statement.)

13. **basis step:** When $n = 1$, LHS $= 1$ RHS. Therefore, $P(1)$ is true.

induction step: Assume $P(k)$ is true: $\sum_{i=1}^{k}(2i-1) = k^2$. Then

$$\sum_{i=1}^{k}(2i-1) + (2k+1) = k^2 + (2k+1) = (k+1)^2.$$

Therefore, $P(k+1)$ is true. Thus, the given result follows by PMI.

15. **basis step:** When $n = 1$, LHS $= 1 =$ RHS. Therefore, $P(1)$ is true.

induction step: Assume $P(k)$ is true: $\sum_{i=1}^{k} i^3 = \left[\frac{k(k+1)}{2}\right]^2$. Then

$$\sum_{i=1}^{k} i^3 + (k+1)^3 = \left[\frac{k(k+1)}{2}\right]^2 + (k+1)^3 = (k+1)^2\left[\frac{k^2+4k+4}{4}\right]$$

$$= \frac{(k+1)^2(k+2)^2}{4} = \left[\frac{(k+1)(k+2)}{2}\right]^2$$

Therefore, $P(k+1)$ is true. Thus, the result is true by PMI.

17. 28,335
19. $\begin{cases} n^2/4 & \text{if } n \text{ is even} \\ (n^2-1)/4 & \text{otherwise} \end{cases}$
21. $x = \sum_{i=1}^{n}(2i-1) = n^2$
23. $x = \sum_{i=1}^{n} i = \frac{n(n+1)}{2}$
25. $\frac{n(n+1)(n+2)}{6}$
27. $\frac{n(n+1)(2n+1)}{6}$
29. $(n!)^2$
31. $2^{n^2(n+1)}$
33. Recall that the kth term of the geometric sequence $a, ar, ar^2, \ldots$ is $a_k = ar^{k-1}$. So the number of grains on the last square $= n^2$th term in the sequence $1, 2, 2^2, 2^3, \ldots$. Here $a = 1$, $r = 2$, and $k = n^2$. Therefore, the number of grains on the last square is $1 \cdot (2^{n^2-1}) = 2^{n^2-1}$. (We could prove this using induction also.)

35. $P(1)$ does not imply $P(2)$.

37. $\dfrac{n(n+1)(2n+1)}{6}$

39. $\left[\dfrac{n(n+1)}{2}\right]^2$

41. 8204

43. **proof:** Suppose $q(n)$ satisfies the conditions of the second principle. Let $p(n) = q(n_0) \wedge q(n_0+1) \wedge \cdots \wedge q(n)$, where $n \geq n_0$. Because $q(n_0)$ is true, $p(n_0)$ is true. Assume $p(k)$ is true, where $k \geq n_0$. By condition (2), $q(k+1)$ is true. Therefore, $p(k+1) = q(n_0) \wedge \cdots \wedge q(k) \wedge q(k+1)$ is true. So $p(k)$ implies $p(k+1)$. So by the first principle, $p(n)$ is true for every $n \geq n_0$. Thus, $q(n)$ is true for every $n \geq n_0$.

45. $\dfrac{n(3n^4+7n^2+2)}{12}$

Exercises 1.4 (p. 31)

1. 1, 4, 7, 10

3. 1, 2, 3, 4

5. 1, 1, 2, 4

7. $a_1 = 1$
   $a_n = a_{n-1} + 3, n \geq 2$

9. $a_0 = 0$
   $a_n = 2a_{n-1} + 3, n \geq 1$

11. $a_1 = a$
    $a_n = a_{n-1} + d, n \geq 2$

13.
$$S_n = a + (a+d) + \cdots + [a + (n-2)d] + [a + (n-1)d]$$
Also, $$S_n = [a + (n-1)d] + [a + (n-2)d] + \cdots + (a+d) + a$$
Adding, $$2S_n = [2a + (n-1)d] + [2a + (n-1)d] + \cdots + [2a + (n-1)d] + [2a + (n-1)d]$$
$$= n[2a + (n-1)d]$$
Therefore, $$S_n = \frac{n}{2}[2a + (n-1)d]$$

15. $a_n = ar^{n-1}$

17. $a_1 = 1$
    $a_n = a_{n-1} + (n-1), n \geq 2$

19. By Exercise 13, sum of the numbers in row $n = \frac{n}{2}$[first term + last term]
$$= \frac{n}{2}\{[n(n-1)/2 + 1] + [n(n+1)]\}$$
$$= \frac{n(n^2+1)}{2}$$

21. $a_1 = 1$
    $a_n = a_{n-1} + n, n \geq 2$

23. $b_n = 2^n - 1$

25. 91

27. 91

29. $f(99) = f(f(110)) = f(100) = f(f(111)) = f(101) = 91$

31. Let k be the smallest integer such that $90 \leq x + 11k \leq 100$. Then $f(x) = f(f(x+11)) = f(f(f(x + 11 \cdot 2))) = \cdots = f^{k+1}(x + 11k)$. Because $90 \leq x + 11k \leq 100$ and $k \geq 1$,

$f(x+11k)=91$, by Exercise 30. Therefore, $f^{k+1}(x+11k)=f^k(91)$. Because $f(91)=91$ by Exercise 29, $f^k(91)=91$. So $f(x)=91$ for $0 \le x < 90$.

33. 3

35. 7

37. **proof** (by PMI): When $n=0$, LHS $=A(2,0)=A(1,1)=A(0,A(1,0))=1+A(1,0)-1+A(0,1)=1+2=3=$ RHS. So $P(0)$ is true.
Assume $P(k)$: $A(2,k)=3+2k$. Then

$$\begin{aligned} A(2,k+1) &= A(1,A(2,k)) = A(1,3+2k) = A(0,A(1,2+2k)) = 1+A(1,2+2k) \\ &= 1+A(0,A(1,1+2k)) = 2+A(1,1+2k) = 2+A(0,A(1,2k)) \\ &= 3+A(1,2k) = 3+(2k+2) = 3+2(k+1) \end{aligned}$$

Therefore, $P(k+1)$ is true. Thus, the result follows by induction.

39. **proof** (by PMI): When $n=0$, LHS $=A(3,0)=A(2,1)=A(1,A(2,0))=A(1,3)=2+3=2^3-3=$ RHS. So $P(0)$ is true.
Assume $P(k)$: $A(3,k)=2^{k+3}-3$. Then

$$A(3,k+1)=A(2,A(3,k))=A(2,2^{k+3}-3)=3+2(2^{k+3}-3)=2^{k+4}-3$$

Therefore, $P(k+1)$ is true. Thus, the result follows by induction.

Exercises 1.5 (p. 38)

1. The number of gifts sent on the nth day $=\sum_{i=1}^{n} i=\dfrac{n(n+1)}{2}=\dbinom{n+1}{2}$

3. 112

5. $32x^5-80x^4+80x^3-40x^2+10x-1$

7. 17,920

9. 10

11. 35

13. $\dbinom{n}{\lfloor n/2 \rfloor}$

15. 5

17. 52

19. By Exercise 18, $\dbinom{2n}{n}=2\dbinom{2n-1}{n-1}$, which is even.

21. LHS = sum of the even binomial coefficients = sum of the odd binomial coefficients = RHS

23. $(1+x)^{2n}=(1+x)^n(1+x)^n$. Because $(1+x)^{2n}=\sum_{r=0}^{2n}\dbinom{2n}{r}x^{2n-r}$, coefficient of $x^n=\dbinom{2n}{r}$. But $(1+x)^n(1+x)^n=\left[\sum_{r=0}^{n}\dbinom{n}{r}x^{n-r}\right]\left[\sum_{r=0}^{n}\dbinom{n}{n-r}x^r\right]$ so coefficient of $x^n=\sum_{r=0}^{n}\dbinom{n}{r}\dbinom{n}{n-r}$. Therefore, $\sum_{r=0}^{n}\dbinom{n}{r}\dbinom{n}{n-r}=\dbinom{2n}{n}$

25. Let $S=\sum_{r=1}^{n} r\dbinom{n}{r}=\sum_{r=1}^{n-1} r\dbinom{n}{r}+n\dbinom{n}{n}$

Reversing the sum on the RHS, $S = n\binom{n}{n} + \sum_{r=1}^{n-1}(n-r)\binom{n}{n-r}$

Adding these two equations,

$$2S = n\binom{n}{n} + \sum_{r=1}^{n-1}(n-r+r)\binom{n}{r} + n\binom{n}{n}$$

$$= n\binom{n}{0} + n\sum_{r=1}^{n-1}\binom{n}{r} + n\binom{n}{n} = n\left[\sum_{r=0}^{n}\binom{n}{r}\right] = n2^n$$

So $S = n \cdot 2^{n-1}$.

27. Suppose $C(n, r-1) < C(n, r)$. Then $\frac{n!}{(r-1)!(n-r+1)!} < \frac{n!}{r!(n-r)!}$

This yields $\frac{1}{n-r+1} < \frac{1}{r}$; that is, $n - r + 1 > r$. So $r < \frac{n+1}{2}$.

Conversely, let $r < \frac{n+1}{2}$. By retracing the steps, it can be shown that $C(n, r-1) < C(n, r)$.

29. **proof** (by PMI): Let $P(r)$ denote the given statement. When $r = 0$, LHS $= 1 =$ RHS, so $P(0)$ is true. Assume $P(k)$ is true: $\sum_{i=0}^{k}\binom{n+i}{i} = \binom{n+k+1}{k}$

Then $\sum_{i=0}^{k+1}\binom{n+i}{i} = \sum_{i=0}^{k}\binom{n+i}{i} + \binom{n+k+1}{k}$

$$= \binom{n+k+1}{k} + \binom{n+k+1}{k+1} = \binom{n+k+2}{k+1}$$

by Pascal's identity. Thus, $P(k)$ implies $P(k+1)$, so the result follows by induction.

31. **proof** (by PMI): Let $P(n)$ denote the given statement. When $n = 0$, LHS $= 1 =$ RHS. So $P(0)$ is true.

Assume $P(k)$ is true for an arbitrary $k \geq 0$: $\sum_{i=0}^{k}\binom{k}{i}^2 = \binom{2k}{k}$

Then $\sum_{i=0}^{k+1}\binom{k+1}{i}^2 = \sum_{i=0}^{k+1}\left[\binom{k}{i} + \binom{k}{i-1}\right]^2$

$$= \sum_{i=0}^{k+1}\binom{k}{i} + \sum_{i=0}^{k+1}\binom{k}{i-1} + 2\sum_{i=0}^{k+1}\binom{k}{i-1}\binom{k}{i}$$

$$= \sum_{i=0}^{k}\binom{k}{i} + \sum_{i=0}^{k}\binom{k}{i-1} + 2\sum_{i=0}^{k}\binom{k}{i-1}\binom{k}{i}$$

$$= \binom{2k}{k} + \binom{2k}{k} + 2\binom{2k}{k+1}$$

$$= 2\left[\binom{2k}{k} + \binom{2k}{k+1}\right] = 2\binom{2k+1}{k+1}$$

Therefore, $P(k)$ implies $P(k+1)$, so the result follows by induction.

33. $n(1+x)^{n-1} = \sum_{r=1}^{n} \binom{n}{r} rx^{r-1}$. Let $x = -1$. Then $0 = \sum_{r=1}^{n} r\binom{n}{r}(-1)^{r-1}$. That is,

$1\binom{n}{1} + 3\binom{n}{3} + \cdots = 2\binom{n}{2} + 4\binom{n}{4} + \cdots$ so each $= n2^{n-2}$, by Exercise 32.

35. **proof:** Let $P(n)$ denote the given statement, where $n \geq 2$. When $n = 2$, LHS $= 1 =$ RHS. So $P(2)$ is true.
Assume $P(k)$: $\sum_{i=2}^{k} \binom{i}{2} = \binom{k+1}{3}$, where $k \geq 2$
Then $\sum_{i=2}^{k+1} \binom{i}{2} = \sum_{i=2}^{k} \binom{i}{2} + \binom{k+1}{2} = \binom{k+1}{3} + \binom{k+1}{2} = \binom{k+2}{3}$
by Pascal's identity. Therefore, $P(k+1)$ is true. Thus, by induction, the result holds for every $n \geq 2$.

37. **proof:** Let $P(n)$ denote the given statement, where $n \geq 3$. When $n = 3$, LHS $= 1 =$ RHS. So $P(3)$ is true.
Assume $P(k)$: $\sum_{i=3}^{k} \binom{i}{3} = \binom{k+1}{4}$, where $k \geq 3$
Then $\sum_{i=3}^{k+1} \binom{i}{3} = \sum_{i=3}^{k} \binom{i}{3} + \binom{k+1}{3} = \binom{k+1}{4} + \binom{k+1}{3} = \binom{k+2}{4}$
by Pascal's identity. Therefore, $P(k+1)$ is true. Thus, by induction, the formula holds for every integer $n \geq 3$.

Exercises 1.6 (p. 48)

1. $t_n = t_{n-1} + n = t_{n-2} + (n-1) + n = \cdots = t_1 + 2 + 3 + \cdots + n = n(n+1)/2$

3. $s_n = s_{n-1} + (2n-1) = s_{n-2} + (2n-3) + (2n-1) = \cdots$
$= s_1 + 3 + 5 + \cdots + (2n-1) = 1 + 3 + 5 + \cdots + (2n-1) = n^2$

5. $p_1 = 1$; $p_n = p_{n-1} + 3n - 2, n \geq 2$

7. RHS $= n + 3t_{n-1} = n + 3(n-1)n/2 = (n/2)(3n-1) = p_n =$ LHS

9. RHS $= 4n(n-1)/2 + n = n(2n-1) = h_n =$ LHS

11. LHS $= \left[\frac{(n-1)n}{2}\right]^2 + \left[\frac{n(n+1)}{2}\right]^2 = \frac{n^2}{4}[(n-1)^2 + (n+1)^2] = \frac{n^2}{2}(n^2+1) = t_{n^2} =$ RHS

13. LHS $= \frac{(2n-1)(2n)}{2} = n^2 =$ RHS

15. Since $t_k = t_{k-1} + k$, the result follows by letting $k = t_n$.

17. Let $n = k(k+1)/2$ for some $k \in \mathbf{Z}^+$. Then $9n + 1 = 9k(k+1)/2 = (3k+1)(3k+2)/2$; $25n + 3 = 25k(k+1)/2 + 3 = (5k+2)(5k+3)/2$; and $49n + 6 = 49k(k+1)/2 + 6 = (7k+3)(7k+4)/2$.

19. $h_1 = 1$; $h_n = h_{n-1} + 4n - 3, n \geq 2$

21. 1, 7, 18, 34

23.
$$e_n = e_{n-1} + 5n - 4 = e_{n-2} + [5(n-1) - 4] + (5n-4)$$
$$= e_{n-2} + 5[(n-1) + n] - 2 \cdot 4$$

$$= e_{n-3} + 5[(n-2)+(n-1)+n] - 3\cdot 4$$
$$\vdots$$
$$= e_1 + 5(2+3+\cdots+n) - 4(n-1)$$
$$= 1 + 5(2+3+\cdots+n) - 4(n-1)$$
$$= 1 + 5[n(n+1)/2 - 1] - 4(n-1)$$
$$= 1 + 5n(n+1)/2 - 5 - 4n + 4 = n(5n-3)/2$$

25. $o_1 = 1$; $o_n = o_{n-1} + 6n - 5$, $n \geq 2$
27. 21, 15; 171, 105
29. 55, 66, 666
31. 41616, 1413721
33. **proof:** Notice that 1 and 36 are both triangular numbers and square numbers. Let t_n be any triangular number that is also a square. Then $4t_n$ is a square, and $8t_n + 1$ is a square by Exercise 4; so $4t_n(8t_n + 1)$ is a square. But it is also the triangular number $8t_n(8t_n + 1)/2$.
35. By Exercise 34, $\sum_{k=1}^{n} \frac{1}{t_k} = \frac{2n}{n+1}$. So as $n \to \infty$, the RHS and hence the LHS approaches 2.

Exercises 1.7 (p. 52)

1. $T_1 = 1$; $T_8 = 36$; $T_{49} = 1225$; $T_{288} = 41{,}616$
3. $S_1 = 1$; $S_n = S_{n-1} + n^2$, $n \geq 2$
5. $\sum_{k=1}^{n} p_k = \sum_{k=1}^{n} \frac{k(3k-1)}{2} = \frac{1}{2}\left[3\sum_{k=1}^{n} k^2 - \sum_{k=1}^{n} k\right]$

$$= \frac{1}{2}\left[\frac{3n(n+1)(2n+1)}{6} - \frac{n(n+1)}{2}\right] = \frac{n^2(n+1)}{2}$$

7. $P_n = P_{n-1} + \frac{n(3n-1)}{2} = P_{n-2} + \frac{(n-1)(3n-4)}{2} + \frac{n(3n-1)}{2}$

$$= \cdots$$
$$= 1 + \frac{2\cdot 5}{2} + \frac{3\cdot 8}{2} + \cdots + \frac{n(3n-1)}{2}$$
$$= \frac{1\cdot 2}{2} + \frac{2\cdot 5}{2} + \frac{3\cdot 8}{2} + \cdots + \frac{n(3n-1)}{2} = \frac{1}{2}\sum_{i=1}^{n} i(3i-1)$$
$$= \frac{1}{2}\left[\frac{3n(n+1)(2n+1)}{6} - \frac{n(n+1)}{2}\right]$$
$$= \frac{n(n+1)}{4}[(2n+1)-1] = \frac{n^2(n+1)}{2}$$

9. $H_1 = 1$; $H_n = H_{n-1} + n(2n-1)$, $n \geq 2$
11. 1, 8, 26, 60, 115

Exercises 1.8 (p. 57)

1. $C_n = \frac{(2n)!}{(n+1)!n!} = \frac{1}{n}\cdot\frac{(2n)!}{(n-1)!(n+1)!} = \frac{1}{n}\binom{2n}{n-1}$

3. $\binom{2n}{n}-\binom{2n}{n-2}=\frac{(2n)!}{n!n!}-\frac{(2n)!}{(n-2)!(n+2)!}=\frac{(2n)!}{n!(n+1)!}\left[(n+1)!-\frac{n(n-1)}{n+2}\right]$

$$=\frac{(2n)!}{n!(n+1)!}\cdot\frac{2(2n+1)}{n+2}=\frac{(2n)!(2n+1)(2n+2)}{n!(n+1)!(n+2)}$$

$$=\frac{(2n+2)!}{(n+1)!(n+2)!}=C_{n+1}$$

5. $\binom{2n-1}{n-1}-\binom{2n-1}{n-2}=\frac{(2n-1)!}{(n-1)!n!}-\frac{(2n-1)!}{(n-2)!(n+1)!}$

$$=\frac{(2n-1)!}{(n+1)!n!}[n(n+1)-n(n-1)]$$

$$=\frac{(2n-1)!}{(n+1)!n!}(2n)=\frac{(2n)!}{(n+1)!n!}=C_n$$

7. $\binom{2n+1}{n+1}-2\binom{2n}{n+1}=\frac{(2n+1)!}{(n+1)!n!}-2\frac{(2n)!}{(n+1)!(n-1)!}=(2n+1)C_n-(2n)C_n=C_n$

9. $C_6=\sum_{r=0}^{2}\binom{5}{2r}2^{5-2r}C_r=32+80+20=132$

11. By Exercise 10, $xC^2(x)-C(x)+1=0$, so $C(x)=\frac{1\pm\sqrt{1-4x}}{2}$. But every $C_n>0$; so $C(x)=\frac{1+\sqrt{1-4x}}{2}$.

Review Exercises (p. 60)

1. $\sum_{i=1}^{n}i(i+1)=\sum_{i=1}^{n}i^2+\sum_{i=1}^{n}i=\frac{n(n+1)(2n+1)}{6}+\frac{n(n+1)}{2}=\frac{n(n+1)(n+2)}{3}$

3. $\sum_{i=1}^{n}\sum_{j=1}^{n}2^i3^j=\sum_{i=1}^{n}2^i\left(\sum_{j=1}^{n}3^j\right)=e\sum_{i=1}^{n}2^i\left(\frac{3^{n+1}-1}{2}-1\right)$

$$=\frac{3^{n+1}-3}{2}\left(\sum_{i=1}^{n}2^i\right)=\frac{3^{n+1}-3}{2}\cdot\left(\frac{2^{n+1}-1}{2-1}-1\right)$$

$$=(2^n-1)(3^{n+1}-3)$$

5. $\prod_{i=1}^{n}\prod_{j=1}^{n}2^i3^j=\prod_{i=1}^{n}2^{ni}\left(\prod_{j=1}^{n}3^j\right)=\prod_{i=1}^{n}2^{ni}\cdot 3^{(\sum j)}=\prod_{i=1}^{n}2^{ni}\cdot 3^{n(n+1)/2}$

$$=3^{n^2(n+1)/2}\cdot 2^{(\sigma ni)}=3^{n^2(n+1)/2}\cdot 2^{n(\sum i)}$$

$$=3^{n^2(n+1)/2}\cdot 2^{n\cdot n(n+1)/2}=6^{n^2(n+1)/2}$$

7. $\prod_{i=1}^{n}\prod_{j=1}^{i}2^i=\prod_{i=1}^{n}2^{i^2}=2^{\sum_{i=1}^{n}i^2}=2^{n(n+1)(2n+1)/6}$

9. 2^{2^n}

11. $n\cdot n=n^2$

13. 7

15. $h(1)=0$
$h(n)=h(n-1)+4(n-1), n\geq 2$

17. **proof** (by PMI): Let $P(n)$: $h(n) = 2n(n-1)$. When $n = 1$, LHS $= 0 =$ RHS. So $P(1)$ is true. Assume $P(k)$ is true: $h(k) = 2k(k-1)$. Then $h(k+1) = h(k) + 4k = 2k(k-1) + 4k = 2k(k+1)$. Therefore, $P(k)$ implies $P(k+1)$, so the result follows by induction.
19. $a_n = 2 \cdot 3^n, n \geq 1$
21. $a_n = \sum_{i=1}^{n} (3i-1) = 3\sum_{i=1}^{n} i - n = 3 \cdot \dfrac{n(n+1)}{2} - n = \dfrac{(3n+1)}{2}, n \geq 0$
23. **proof:** Let a be any integer. Then $4a(a+1) + 1 = 4a^2 + 4a + 1 = (2a+1)^2$, a square.
25. **proof:** Clearly, $(3, 4, 5)$ is a solution: $3^2 + 4^2 = 5^2$. Let k be any integer. Then $(3k)^2 + (4k)^2 = (5k)^2$, showing that $(3k, 4k, 5k)$ is also a solution for every integer k.
27. **proof** (by PMI): When $n = 1$, LHS $= \frac{1}{3} =$ RHS, so $P(1)$ is true.

    Assume $P(k)$ is true: $\sum_{i=1}^{k} \dfrac{1}{(2i-1)(2i+1)} = \dfrac{k}{2k+1}$

$$\begin{aligned}
\text{Then } \sum_{i=1}^{k+1} \frac{1}{(2i-1)(2i+1)} &= \sum_{i=1}^{k} \frac{1}{(2i-1)(2i+1)} + \frac{1}{(2k+1)(2k+3)} \\
&= \frac{k}{2k+1} + \frac{1}{(2k+1)(2k+3)} = \frac{k(2k+3)+1}{(2k+1)(2k+3)} \\
&= \frac{(k+1)(2k+1)}{(2k+1)(2k+3)} = \frac{k+1}{2(k+1)+1}
\end{aligned}$$

    Therefore, $P(k)$ implies $P(k+1)$, so the result holds by PMI.
29. When $n = 1$, RHS $= 2 \cdot 3 = a_1 =$ LHS; so $P(1)$ is true. Assume $P(k)$ is true. Then $a_{k+1} = 3a_k = 3(2 \cdot 3^k) = 2 \cdot 3^{k+1}$. So $P(k)$ implies $P(k+1)$. Thus, the result follows by induction.
31. When $n = 0$, RHS $= 0 = a_0 =$ LHS; so $P(0)$ is true. Assume $P(k)$ is true: $a_k = \dfrac{k(3k+1)}{2}$.

    Then $a_{k+1} = a_k + (3k+2) = \dfrac{k(3k+1)}{2} + (3k+2) = \dfrac{(k+1)(3k+4)}{2}$.

    Thus, $P(k)$ implies $P(k+1)$, so the result follows by PMI.
33. Let $P(n)$ denote the given statement, where $1 \leq r \leq n$. When $n = 1$, LHS $= 1$ RHS; so $P(1)$ is true. Assume $P(k)$ is true: $\sum_{i=r}^{k} C(i, r) = C(k+1, r+1)$

$$\begin{aligned}
\text{Then } \sum_{i=r}^{k+1} C(i, r) &= \sum_{i=r}^{k} C(i, r) + C(k+1, r) = C(k+1, r+1) + C(k+1, r) \\
&= C(k+2, r+1)
\end{aligned}$$

    Thus, $P(k)$ implies $P(k+1)$, so the result follows by PMI.
35. LHS $= \left[\dfrac{n(n+1)}{2}\right]^2 - \left[\dfrac{(n-1)n}{2}\right]^2 = \dfrac{n^2}{4}[(n+1)^2 - (n-1)^2] = \dfrac{n^2}{4}(4n) = n^3$
37. 1, 3, 6, 55, 66, 171, 595, 666
39. **proof:**

$$\begin{aligned}
t_n + t_{n-1}t_{n+1} &= n(n+1)/2 + (n-1)n/2 \cdot (n+1)(n+2)/2 \\
&= [n(n+1)/2][1 + (n^2 + n - 2)/2] \\
&= n(n+1)/2 \cdot n(n+1)/2 = t_n^2
\end{aligned}$$

41. $\text{RHS} = \frac{n(n+1)}{2} + \frac{k(k+1)}{2} - (n+1)k = \frac{n^2+k^2-2nk+n-k}{2}$

$= \frac{(n-k)(n-k+1)}{2} = \text{LHS}$

43. $\text{LHS} = \frac{(n-1)n}{2} \cdot \frac{k(k+1)}{2} + \frac{n(n+1)}{2} \cdot \frac{(k-1)k)}{2} = \frac{nk}{2}(nk-1) = \text{RHS}$

45. **proof** (by PMI): When $n = 1$, the result is true. Assuming that $\frac{(nr)!}{(r!)^n}$ is an integer, $\frac{[(n+1)r]!}{(r!)^{n+1}} = \frac{(nr)!}{r!^n}(n+1)$ is also an integer. Thus, the result follows by induction.

47. $a_1 = 1 = a_2$
$a_n = a_{n-1} + a_{n-2},\ n \geq 3$

Supplementary Exercises (p. 62)

1. $(m^2-n^2)^2 + (2mn)^2 = (m^4 - 2m^2n^2 + n^4) + 4m^2n^2 = m^4 + 2m^2n^2 + n^4 = (m^2+n^2)^2$

3. $11^2 = 6+7+8+\cdots+14+15+16$; $13^2 = 7+8+9+\cdots+17+18+19$

5. $\text{RHS} = \sum_{i=n}^{3n-2} i = \sum_{i=1}^{3n-2} i - \sum_{i=1}^{n-1} i = \frac{(3n-2)(3n-1)}{2} - \frac{(n-1)n}{2}$

$= 4n^2 - 4n + 1 = (2n-1)^2 = \text{LHS}$

7. Let S denote the sum of the numbers in the first n bands of the array and S_d the sum of an arithmetic sequence in the array with common difference d, where $2 \leq d \leq n+1$. To establish the given formula, we compute the sum S in two different ways.
Notice that the common differences of the arithmetic sequences are $2, 3, 4, \ldots, (n+1), \ldots$, and their nth terms are $2n-1, 3n-2, 4n-3, \ldots, (n+1)n - n = n^2, \ldots$, respectively. The sum of the numbers B_n in the nth band is given by

$$B_n = S_{n+1} + \{(2n-1) + (3n-2) + (4n-3) + \cdots + [n \cdot n - (n-1)]\}$$

$$= \frac{n}{2}[2 \cdot 1 + (n-1)(n+1)] + \left[n\sum_{i=2}^{n} i - \sum_{i=1}^{n-1} i\right]$$

$$= \frac{(n^2+1)}{2} + n\left(\sum_{i=1}^{n} i - 1\right) - \left(\sum_{i=1}^{n} i - n\right) = \frac{n(n^2+1)}{2} + (n-1)\sum_{i=1}^{n} i$$

$$= \frac{n(n^2+1)}{2} + (n-1) \cdot \frac{n(n+1)}{2} = \frac{n(n^2+1)}{2} + \frac{n(n^2-1)}{2} = n^3$$

Thus, the sum of the numbers in each band is a cube. Adding the numbers band by band, we get $S = \sum_{i=1}^{n} B_i = \sum_{i=1}^{n} i^3$.
We can evaluate S in a different way also. Adding the sums of the arithmetic sequences row by row,

$$S = \sum_{d=2}^{n+1} S_d = \sum_{d=2}^{n+1} \left\{\frac{n}{2}[2 \cdot 1 + (n-1)d]\right\} = \frac{n}{2}\left[2n + (n-1)\sum_{d=2}^{n+1} d\right]$$

$$= \frac{n}{2}\left[2n + (n-1)\left(\sum_{d=1}^{n+1} d - 1\right)\right]$$

$$= \frac{n}{2}\left[2n + (n-1)\frac{(n+1)(n+2)}{2} - (n-1)\right]$$

$$= \frac{n}{2}\left[n + 1 + \frac{(n+2)(n^2-1)}{2}\right] = \frac{n}{4}(n^3 + 2n^2 + n) = \left[\frac{n(n+1)}{2}\right]^2$$

Thus, $\sum_{i=1}^{n} i^3 = \left[\frac{n(n+1)}{2}\right]^2$, as desired.

9. $$\begin{aligned} a_n &= a_{n-1} + (n-1) \\ &= a_{n-2} + (n-2) + (n-1) \\ &\vdots \\ &= a_1 + 1 + 2 + 3 + \cdots + (n-1) \\ &= 1 + n(n-1)/2 \end{aligned}$$

11. We shall use the following identities:

$$(1)\ n\binom{n-1}{k-1} = k\binom{n}{k} \qquad (2)\ \sum_{k=0}^{n}\binom{n}{k} = 2^n \qquad (3)\ \sum_{k=1}^{n}\binom{n}{k} = n2^{n-1}$$

$$\sum_{k=0}^{n}\binom{n}{k}k^2 = \sum_{k=0}^{n}\binom{n}{k}k \cdot k = \sum_{k=0}^{n} n\binom{n-1}{k-1}k = n\sum_{k=0}^{n}\binom{n-1}{k-1}(k-1+1)$$

$$= n\sum_{k=0}^{n}\binom{n-1}{k-1}(k-1) + n\sum_{k=0}^{n}\binom{n-1}{k-1}$$

$$= n\sum_{k=0}^{n}(n-1)\binom{n-2}{k-2} + \sum_{k=1}^{n-1}\binom{n-1}{k-1}$$

$$= n(n-1)\sum_{k=2}^{n-2}\binom{n-2}{k-2} + n\sum_{k=1}^{n-1}\binom{n-1}{k-1}$$

$$= n(n-1)\sum_{k=0}^{n-2}\binom{n-2}{k} + n\sum_{k=0}^{n-1}\binom{n-1}{k}$$

$$= n(n-1)\cdot 2^{n-2} + n\cdot 2^{n-1} = n(n+1)2^{n-2}$$

13. When $n = 3$, $\text{RHS} = (1^3 + 3^3 + 6^3) + 2(1^4 + 3^4 + 6^4) = 244 + 2756 = 3000$
$= 3(1 + 3 + 6)^3 = \text{LHS}$

When $n = 4$, $\text{RHS} = (1^3 + 3^3 + 6^3 + 10^3) + 2(1^4 + 3^4 + 6^4 + 10^4)$
$= 1244 + 22756 = 24000 = 3(1 + 3 + 6 + 10)^3 = \text{LHS}$

15. $1729 = 1^3 + 12^3 = 9^3 + 10^3$

17. $11^3 + 12^3 + 13^3 + 14^3 = 20^3$

19. Let S denote the sum of the numbers in the nth set and S_i the sum of the numbers in the first i sets. Then $S = S_n - S_{n-1}$.

Number of elements in the first n sets $= 1 + 7 + 13 + \cdots + [6(n-1)+1]$

$$= \frac{n}{2}[1 + 6(n-1) + 1] = \frac{n}{2}(6n-4) = 3n^2 - 2n$$

Then $S_n = \frac{(3n^2 - 2n)(3n^2 - 2n + 1)}{2} = \frac{9n^4 - 12n^3 + 7n^2 - 2n}{2}$

So $S_{n-1} = \frac{9(n-1)^4 - 12(n-1)^3 + 7(n-1)^2 - 2(n-1)}{2}$

$$= \frac{9n^4 - 48n^3 + 97n^2 - 88n + 30}{2}$$

Therefore,

$$S = S_n - S_{n-1} = \frac{9n^4 - 12n^3 + 7n^2 - 2n}{2} - \frac{9n^4 - 48n^3 + 97n^2 - 88n + 30}{2}$$

$$= 18n^3 - 45n^2 + 43n - 15$$

21. Let $S_n = \sum_{i=1}^{k} p_i = \sum_{i=1}^{k} \frac{i(3i-1)}{2}$, where $k = n(n+1)/2$.

Then $S_n = \frac{1}{2}\left[3\sum_{i=1}^{k} i^2 - \sum_{i=1}^{k} i\right] = \frac{k^2(k+1)}{2} = \frac{(n^2+n)^2[(n^2+n)+2]}{16}$

$$= \frac{(n^2+n)^3 + 2(n^2+n)^2}{16} = \frac{n^6 + 3n^5 + 5n^4 + 5n^3 + 2n^2}{16}$$

Changing n to $n-1$, $S_{n-1} = \frac{n^6 - 3n^5 + 5n^4 - 5n^3 + 2n^2}{16}$

Therefore, the desired sum $= S_n - S_{n-1} = \frac{6n^5 + 10n^3}{16} = \frac{n^3(3n^2+5)}{8}$

Chapter 2 Divisibility

Exercises 2.1 (p. 79)

1. 7, 1
3. −25, 0
5. 5
7. 4
9. 161
11. 2896
13. 138
15. 1695
17. T
19. F
21. F
23. T
25. F
27. $3^2 = (-3)^2$, but $3 \neq -3$.
29. $3|(4+5)$, but $3 \nmid 4$ and $3 \nmid 5$.
31. 28
33. 13/6
35. 135, 175, 315, 735
37. $f(n) = \begin{cases} 1 & \text{if } n \bmod 3 = 0 \\ 3 & \text{if } n \bmod 3 = 1 \\ 2 & \text{otherwise} \end{cases}$
39. **proof:** Because $a|b$ and $c|d$, $b = ma$ and $d = nc$ for some integers m and n. Then $bd = mn(ac)$, so $ac|bd$.
41. **proof:** Let $x = 2m+1$ and $y = 2n+1$ be any two odd integers. Then $x + y = (2m+1) + (2n+1) = 2(m+n+1)$, an even integer.

43. **proof:** Let x be an even integer and y an odd integer. Then $x = 2m$ and $y = 2n + 1$ for some integers m and n. So $x + y = (2m) + (2n + 1) = 2(m + n) + 1$, which is an odd integer.
45. **proof** (by contradiction): Let x be any integer such that x^2 is odd. Suppose x is even, so $x = 2k$ for some integer k. Then $x^2 = (2k)^2 = 2(2k^2)$, an even integer, which is a contradiction.
47. **proof:** Let $x = 4m + 1$ and $y = 4n + 1$ be any two integers of the given form. Then $x + y = (4m + 1) + (4n + 1) = 2(2m + 2n + 1)$, which is an even integer.
49. **proof:** Let $x = 3m + 1$ and $y = 3n + 1$ be any two integers of the given form. Then $x \cdot y = (3m + 1) \cdot (3n + 1) = 3(3mn + m + n) + 1$, an integer of the same form.
51. **proof:** Assume the given conclusion is false. Then both integers are odd, and hence their product is also odd. This contradicts the given hypothesis, so the result follows.
53. **proof:** Let n be an even integer, say, $n = 2m$. Then $n^2 + n = 4m^2 + 2m = 2(2m^2 + m)$, which is an even integer. On the other hand, let n be an odd integer, say, $n = 2m + 1$. Then $n^2 + n = (2m + 1)^2 + (2m + 1) = 2(2m^2 + 3m + 1)$, which is also an even integer.
55. **proof:** Let n be an even integer, say, $n = 2m$. Then $n^3 - n = 8m^3 - 2m = 2(4m^2 - m)$, which is an even integer. On the other hand, let n be an odd integer, say, $n = 2m + 1$. Then $n^3 - n = (2m + 1)^3 - (2m + 1) = 2(4m^3 + 6m^2 + 2m)$, which is also an even integer.
57. Let A, B, and C be three arbitrary finite sets. Notice that $A \cap C = C \cap A$ and $(A \cap B) \cap (A \cap C) = A \cap B \cap C$.

$$\begin{aligned}|A \cup B \cup C| &= |A \cup (B \cup C)| = |A| + |B \cup C| - |A \cap (B \cup C)| \\ &= |A| + |B \cup C| - |(A \cap B) \cup (A \cap C)| \\ &= |A| + (|B| + |C| - |B \cap C|) - [|A \cap B| + |A \cap C| - |(A \cap B) \cap (A \cap C)|] \\ &= |A| + |B| + |C| - |A \cap B| - |B \cap C| - |C \cap A| + |A \cap B \cap C|\end{aligned}$$

59. **proof:** Let n be any positive integer. Then $n(n + 1)(n + 2)(n + 3) = [n(n + 3)][(n + 1)(n + 2)] = (n^2 + 3n)(n^2 + 3n + 2) = (n^2 + 3n + 1)^2 - 1$. By Exercise 58, this cannot be a square.
61. By Exercise 60, $3|(n + 3)$, so $3|n$. Let $n = 3m$. Then $27m^3 + (3m + 1)^3 + (3m + 2)^3 = (3m + 3)^3$. Simplifying, we get $3m^3 - 2m - 1 = 0$; that is, $(m - 1)(3m^2 + 3m + 1) = 0$. Since $3m^2 + 3m + 1 = 0$ has no real solutions, $m - 1 = 0$; that is, $m = 1$, so $n = 3$.
63. **proof** (by PMI): When $n = 1$, $n^4 + 2n^3 + n^2 = 1 + 2 + 1 = 4$ is divisible by 4. So $P(1)$ is true. Assume $P(k)$ is true. Let $k^4 + 2k^3 + k^2 = 4m$ for some $m \in \mathbf{Z}^+$. Then

$$\begin{aligned}&(k + 1)^4 + 2(k + 1)^3 + (k + 1)^2 = (k^4 + 2k^3 + k^2) + 4(k^3 + 3k^2 + 2k + 1) \\ &\quad = 4m + 4(k^3 + 3k^2 + 2k + 1) = 4(m + k^3 + 3k^2 + 2k + 1)\end{aligned}$$

which is divisible by 4. Thus $P(k)$ implies $P(k + 1)$, so the result follows by PMI.

65. **proof:** $4^{2n} + 10n - 1 = (5 - 1)^{2n} + 10n - 1 = \sum_{r=0}^{2n} \binom{2n}{r} 5^{2n-r}(-1)^r + 10n - 1$

$$= \sum_{r=0}^{2n-2} \binom{2n}{r} 5^{2n-r}(-1)^r - 10n + 1 + 10n - 1$$

$$= \sum_{r=0}^{2n-2} \binom{2n}{r} 5^{2n-r}(-1)^r$$

$$= 5^{2n} - \binom{2n}{1} 5^{2n-1} + \cdots + \binom{2n}{2n-2} 5^2$$

which is clearly divisible by 25.

Exercises 2.2 (p. 87)

1. 13
3. 1022
5. 10000110100_{two}
7. 3360_{eight}
9. 15_{eight}
11. 72_{eight}
13. $1D_{\text{sixteen}}$
15. 75_{sixteen}
17. 110110_{two}
19. 1000110111_{two}
21.
$$19 = 1 + 2 + 16$$
$$19 \cdot 31 = 1 \cdot 31 + 2 \cdot 31 + 16 \cdot 31$$
$$= 31 + 62 + 496 = 589$$
23.
$$29 = 1 + 4 + 8 + 16$$
$$29 \cdot 49 = 1 \cdot 49 + 4 \cdot 49 + 8 \cdot 49 + 16 \cdot 49$$
$$= 49 + 196 + 392 + 784 = 1421$$
25.

19	9	4	2	1
31*	62*	124	248	496*

$19 \cdot 31 = 31 + 62 + 496 = 589$

27.

29	14	7	3	1
49*	98	196*	392*	784*

$29 \cdot 47 = 49 + 196 + 392 + 784 = 1421$

29.

1	2	4	8	16
19	38	76*	152*	304

$243 = 152 + 76 + \boxed{15}$
quotient $= 8 + 4 = 12$
remainder $= 15$

31.

1	2	4	8	16	32	64
35	70*	140	280	560*	1120*	2240

$1776 = 1120 + 560 + 70 + 26$
quotient $= 32 + 16 + 2 = 50$
remainder $= 26$

33. 110, 1011, 10110, 11011, 101010
35. 0; 1
37. 2
39. 5
41. If $n \geq 22$, then n^3 will contain at least 5 digits and n^4 at least 6 digits, so together they will contain at least 11 digits. If $n \leq 17$, together they will contain at most 9 digits. Therefore, $18 \leq n \leq 21$. $n = 19$, 20, or 21 does not work, but $n = 18$ works: $18^3 = 5832$ and $18^4 = 104976$.
43. 4
45. n
47. 165
49. 101
51. $2^{10} = 1024$, $2^{20} = \ldots 6$, $2^{40} = \ldots 6$. Therefore, $2^{100} = 2^{40} \cdot 2^{40} \cdot 2^{20} = \ldots 6 \times \ldots 6 \times \ldots 6 = \ldots 6$, so the ones digit is 6.
53. $10AB_{\text{twelve}}$

Exercises 2.3 (p. 97)

1.

+	0	1	2	3	4
0	0	1	2	3	4
1	1	2	3	4	10
2	2	3	4	10	11
3	3	4	10	11	12
4	4	10	11	12	13

3. 110_{two}
5. 1000_{seven}
7. $4 \le b \le 7$
9. 3174_{eight}
11. $87EC_{\text{sixteen}}$
13. 206_{seven}
15. ABB_{sixteen}
17. 10100_{two}
19. $42B5_{\text{twelve}}$

21.

×	0	1	2	3	4	5	6
0	0	0	0	0	0	0	0
1	0	1	2	3	4	5	6
2	0	2	4	6	11	13	15
3	0	3	6	12	15	21	24
4	0	4	11	15	22	26	33
5	0	5	13	21	26	34	42
6	0	6	15	24	33	42	51

23. 133102_{five}
25. $A010_{\text{twelve}}$
27. 132011_{five}
29. $B706_{\text{twelve}}$
31. 3071730_{eight}
33. $889C98_{\text{sixteen}}$

Exercises 2.4 (p. 102)

1. 21, 28
3. 51, 70
5. 56, 84
7. 13, 21
9. $16+9=25$
   $25+11=36$
11. $1+4+9+16+25=55$
    $1+4+9+16+25+36=91$
13. $1+4\cdot 6=25$
    $1+5\cdot 7=36$
15. $111111\cdot 111111=12345654321$
    $1111111\cdot 1111111=1234567654321$
17. $12345679\cdot 54=666666666$
    $12345679\cdot 63=777777777$
19. 1 1 0 0 1 1
    1 0 1 0 1 0 1
21. $1+2+3+\cdots+n=n(n+1)/2$
23. $1+2+4+\cdots+2^n=2^{n+1}-1$
25. $(n+1)^3-(n+1)=n(n+1)(n+2)$
27. $\underbrace{66\ldots 66}_{n \text{ sixes}}7\times\underbrace{66\ldots 66}_{n \text{ sixes}}7=\underbrace{44\ldots 44}_{n+1 \text{ 4s}}\underbrace{88\ldots 88}_{n \text{ 8s}}9$
29. $\underbrace{33\ldots 33}_{n \text{ threes}}4\times\underbrace{33\ldots 33}_{n \text{ threes}}4=\underbrace{11\ldots 11}_{n+1 \text{ 1s}}\underbrace{55\ldots 55}_{n \text{ 5s}}6$
31. $10^{2n}-10^n+1=\underbrace{99\ldots 99}_{n \text{ 9s}}\underbrace{00\ldots 00}_{n \text{ 0s}}1$
33. **proof** (by PMI): Let $P(n)$: $\sum_{i=1}^{n} i^2=n(n+1)(2n+1)/6$, where $n\ge 1$. Clearly, $P(1)$ is true. Assume $P(k)$ is true for any integer $k\ge 1$. Then

$$\sum_{i=1}^{k+1} i^2=\sum_{i=1}^{k} i^2+(k+1)^2=\frac{k(k+1)(2k+1)}{6}+(k+1)^2=(k+1)\frac{(2k^2+k)+6(k+1)}{6}$$
$$=\frac{(k+1)(k+2)[2(k+1)+1]}{6}$$

Because $P(k)$ implies $P(k+1)$, the result follows by PMI.

35. **proof:** $1+n(n+2)=n^2+2n+1=(n+1)^2$

37. **proof:**
$$\begin{aligned}
\text{RHS} &= 123\ldots(n-1)n(n-1)\ldots321\\
&= 1\cdot 10^{2n-2} + 2\cdot 10^{2n-3} + 3\cdot 10^{2n-4} + \cdots + 3\cdot 10^2 + 2\cdot 10 + 1\cdot 10^0\\
&= 10^{2n-2} + 10^{2n-3} + 10^{2n-4} + \cdots + 10^2 + 10 + 1\\
&\qquad + 10^{2n-3} + 10^{2n-4} + \cdots + 10^2 + 10\\
&\qquad\qquad + 10^{2n-4} + \cdots + 10^2\\
&\qquad\qquad\qquad \vdots\\
&\qquad\qquad + 10^n + 10^{n-1} + 10^{n-2}\\
&\qquad\qquad\quad + 10^{n-1}\\
&= \frac{10^{2n-1}-1}{9} + \left(\frac{10^{2n-2}-1}{9} - \frac{10-1}{9}\right) + \left(\frac{10^{2n-3}-1}{9} - \frac{10^2-1}{9}\right)\\
&\quad + \cdots + \left(\frac{10^n-1}{9} - \frac{10^{n-1}-1}{9}\right)\\
&= \frac{1}{9}\{(10^{2n-1} + 10^{2n-2} + 10^{2n-3} + \cdots + 10^n) - [(10-1) + (10^2-1)\\
&\quad + \cdots + (10^{n-1}-1)] - n\}\\
&= \frac{1}{9}\left[\left(\frac{10^{2n}-1}{9} - \frac{10^n-1}{9}\right) - (10 + 10^2 + \cdots + 10^{n-1}) + (n+1) - n\right]\\
&= \frac{1}{9}\left(\frac{10^{2n}-1}{9} - \frac{10^n-1}{9} - \frac{10^n-1}{9} + 1 + n - 1 - n\right)\\
&= \frac{1}{9}\left(\frac{10^{2n}-1}{9} - 2\cdot\frac{10^n-1}{9}\right)\\
&= \frac{10^{2n} - 2\cdot 10^n + 1}{81} = \frac{(10^n-1)(10^n+1)}{81} = \frac{10^n-1}{9}\cdot\frac{10^n+1}{9}\\
&= \underbrace{111\ldots1}_{n \text{ ones}} \times \underbrace{111\ldots1}_{n \text{ ones}} = \text{LHS}
\end{aligned}$$

39. It fails when $n = 10$.

41. 123454321, 12345654321

43. no

45. $2n - 1 = n^2 - (n-1)^2$

47.
```
     11   12   13   14   15
  16   17   18   19   20   21
```

49. $\dfrac{n(n^2+1)}{2}$

51. $n(n-1)+1, n(n+1)-1$

53.
$$\begin{aligned}
\text{RHS} &= 123\ldots(n-1)\\
&= 1\cdot 10^{n-2} + 2\cdot 10^{n-3} + 3\cdot 10^{n-4} + \cdots + (n-2)\cdot 10 + (n-1)\\
&= 10^{n-2} + 10^{n-3} + 10^{n-4} + \cdots + 10 + 1\\
&\qquad + 10^{n-3} + 10^{n-4} + \cdots + 10 + 1\\
&\qquad\qquad + 10^{n-4} + \cdots + 10^3 + 1\\
&\qquad\qquad\qquad \ddots\\
&\qquad\qquad\qquad + 10^2 + 10 + 1\\
&\qquad\qquad\qquad\qquad + 10 + 1\\
&\qquad\qquad\qquad\qquad\quad + 1
\end{aligned}$$

$$= \left(\frac{10^{n-1}-1}{9} + \frac{10^{n-2}-1}{9} + \frac{10^{n-3}-1}{9} + \cdots + \frac{10^3-1}{9} + \frac{10^2-1}{9}\right) + \frac{10-1}{9}$$

$$= \frac{1}{9}\left[(10^{n-1} + 10^{n-2} + 10^{n-3} + \cdots + 10 + 1) - (n-1) - 1\right]$$

$$= \frac{1}{9}\left(\frac{10^n-1}{9} - n\right) = \frac{10^n - 9n - 1}{81} = \text{LHS}$$

55.
$$\begin{aligned} 123\ldots n &= 1\cdot 10^{n-1} + 2\cdot 10^{n-2} + 3\cdot 10^{n-3} + \cdots + (n-1)\cdot 10 + n \\ &= 10^{n-1} + 10^{n-2} + 10^{n-3} + \cdots + 10^2 + 10 + 1 \\ &\quad + 10^{n-2} + 10^{n-3} + \cdots + 10^2 + 10 + 1 \\ &\quad\ddots \\ &\quad + 10^2 + 10 + 1 \\ &\quad + 10 + 1 \\ &\quad + 1 \\ &= \left(\frac{10^n-1}{9} + \frac{10^{n-1}-1}{9} + \cdots + \frac{10^3-1}{9} + \frac{10^2-1}{9}\right) + \frac{10-1}{9} \\ &= \frac{1}{9}\left(\frac{10^{n+1}-1}{9} - 1 - n\right) = \frac{10^{n+1} - 9n - 10}{81} \end{aligned}$$

$$\begin{aligned} \text{LHS} &= 123\ldots n \times 8 + n = \frac{8(10^{n+1} - 9n - 10)}{81} + n = \frac{80\cdot 10^n + 9n - 80}{81} \\ &= \frac{(81-1)10^n + 9n - 81 + 1}{81} \\ &= 10^n - \frac{10^n - 9n - 1}{81} - 1 = 987\ldots(10-n) \\ &= \text{RHS}, \quad \text{by Exercises 53 and 54} \end{aligned}$$

Exercises 2.5 (p. 126)

1. F
3. F
5. F
7. T
9. T
11. no, since $3|129$.
13. no, since $7|1001$.
15. 15
17. 24
19. 24, 25, 26, 27, 28
21. 362882 through 362889
23. 39916802 through 39916811
25. 5, 7
27. 7
29. 23
31. 37
33. 101
35. $n = 4$
37. $x - y = (n+1)/2 - (n-1)/2 = 1$, so $x - y$ is a trivial factor of n.
39. $1, p, q, pq, p^2, p^2q$
41. $1, p, q, pq, p^2, q^2, p^2q, pq^2, p^2q^2$
43. $2\cdot 3\cdot 7\cdot 43 + 1 = 1807 = 13\cdot 139$
45. **proof:** Let n and $n+1$ be any two consecutive integers that are primes. Because they are consecutive integers, one of them is even. Suppose it is n. Then $2|n$. But n is a prime, so $n = 2$ and hence $n + 1 = 3$. The case that $n + 1$ is even yields $n = 1$, a contradiction.

47. **proof:** Because $p^2 + 8$ is prime, p must be odd. By the division algorithm, p is of the form $3n$, $3n - 1$, or $3n + 1$. If $p = 3n \pm 1$, then $p^2 + 8 = (3n \pm 1)^2 + 8 = 9n^2 \pm 6n + 9$ is divisible by 3. So $p = 3n$, and hence $n = 1$ and $p = 3$. Then $p^3 + 4 = 31$ is a prime.
49. Since $p < \frac{p+q}{2} < q$, and p and q are successive odd primes, $r = \frac{p+q}{2}$ cannot be a prime. So r is composite.
51. Since $p^2 + 2$ is a prime, $p \geq 3$. When $p = 3$, both $p^2 + 2 = 11$ and $p^3 + 2 = 29$ are primes. Suppose $p > 3$. By the division algorithm, p is of the form $3q \pm 1$. Then $p^2 + 2 = (3q \pm 1)^2 + 2 = 9q^2 \pm 6q + 3$, so $3|(p^2 + 2)$ and hence $p^2 + 2$ is not a prime. Thus $p = 3$.
53. **proof** (of the contrapositive): If p is even, p must be 2. Then $p + 2 = 4$ is not a prime.
55. **proof:** Let p be any odd prime. By the division algorithm, p is of the form $4n$, $4n + 1$, $4n + 2$, or $4n + 3$. Since p is odd, it must be of the form $4n + 1$ or $4n + 3$.
57. **proof:** Let $n = ab$. Then $2^n - 1 = 2^{ab} - 1 = (2^a - 1)[2^{(b-1)a} + 2^{(b-2)a} + \cdots + 1]$, a composite number.
59. **proof** (by contradiction): Suppose there is a positive integer > 1 that has no prime factors. Then the set of positive integers that have no prime factors is nonempty, so by the well-ordering principle, it has a least element n. Because n has no prime factors and $n|n$, n cannot be a prime. Let $n = ab$, where $1 < a, b < n$. Since $1 < a < n$, a has a prime factor p. Since $p|a$ and $a|n$, $p|n$, which is a contradiction. Thus, every positive integer > 1 has a prime factor.
61. yes
63. yes
65. $1111 = 11 \cdot 101$ and $11111 = 41 \cdot 271$, so both R_4 and R_5 are composite.
67. Three and 5 are primes, but $R_3 = 111 = 3 \cdot 37$ and $R_5 = 11111 = 41 \cdot 271$ are not primes.
69. 3–5–7, 61–67–71
71. **proof** (by PMI): Because $E_6 = 30031 < 83521 = p_7^4$, the result is true when $n = 6$. Assume it is true for an arbitrary $n \geq 6$. Then $E_{n+1} = E_n p_{n+1} - (p_{n+1} - 1) < E_n p_{n+1} < (p_{n+1})^{n-2} p_{n+1} = (p_{n+1})^{n-1} < (p_{n+2})^{n-1}$. Thus, by induction, the result is true for all $n \geq 6$.
73. **proof:** Let $N = p_1 p_2 \cdots p_n + 1$. By Lemma 2.1, N has a prime factor p. Because $p_i \nmid N$ for $1 \leq i \leq n$, $p \geq p_{n+1}$. Since $p|N$, $p \leq N$. Therefore, $p_{n+1} \leq N$.
75. **proof:** Since $C_n = \frac{(2n)!}{(n+1)!n!}$, $(n+2)C_{n+1} = (4n+2)C_n$. Assume that C_n is a prime for some n. It follows from Segner's formula that if $n > 3$, then $\frac{n+2}{C_n} < 1$; so $C_n > n + 2$. Consequently, $C_n \mid C_{n+1}$, so $C_{n+1} = kC_n$ for some positive integer k. Then $4n + 2 = k(n + 2)$, hence $1 \leq k \leq 3$ and thus $n \leq 4$. It follows that C_2 and C_3 are the only Catalan numbers that are prime.

Exercises 2.6 (p. 137)

1.

$$\begin{aligned}
F_n &= F_{n+1} - F_{n-1} \\
F_{n-1} &= F_n - F_{n-2} \\
&\vdots \\
F_3 &= F_4 - F_2 \\
F_2 &= F_3 - F_1
\end{aligned}$$

Adding, $\sum_{i=2}^{n} F_i = (F_3 + \cdots + F_{n-1} + F_n + F_{n+1}) - (F_1 + F_2 + \cdots + F_{n-1})$

$= F_n + F_{n+1} - F_1 - F_2$

That is, $\sum_{i=1}^{n} F_i = F_n + F_{n+1} - F_2 = F_{n+2} - 1$

3. 33

5. 232

7. **proof** (by strong version of PMI): Let $P(n)$ denote the statement that $a_n = F_n - 1$, where $n \geq 1$. Because $a_0 = 0 = F_1 - 1$, $P(1)$ is true. Now assume that $P(1), P(2), \ldots,$ and $P(k)$ are true for an arbitrary integer $k \geq 1$. Since $F_{n+1} = F_n + F_{n-1}$, $a_{n+1} = a_n + a_{n-1} = (F_n - 1) + (F_{n-1} - 1) = (F_n + F_{n-1}) - 2 = F_{n+1} - 2 = a_{n+1} - 1$. Thus, $P(k+1)$ is true, so the result follows by PMI.

9. **proof:** $a_i = F_i - 1$. Therefore, $\sum_{i=1}^{n} a_i = \sum_{i=1}^{n} F_i - n = (F_{n+2} - 1) - n = F_{n+2} - 1 - n = (a_{n+2} + 1) - 1 - n = a_{n+2} - n$

11. 1; 2; 1.5; 1.66666667; 1.6; 1.625; 1.61538462; 1.61904762; 1.61764706; 1.61818182

13. $\sum_{i=1}^{n} F_{2i-1} = F_{2n}$

15. $\sum_{i=1}^{n} L_i = L_{n+2} - 3$

17. $\sum_{i=1}^{n} L_{2i} = L_{2n+1} - 1$

19. $\sum_{i=1}^{n} L_i^2 = L_n L_{n+1} - 2$

(In Exercises 20–30, $P(n)$ denotes the given predicate.)

21. **basis step** When $n = 2$, LHS $= -1 =$ RHS. So $P(2)$ is true.
induction step Assume $P(k)$ is true: $F_k^2 - F_{k-1}F_{k+1} = (-1)^{k-1}$. Then

$$\begin{aligned}F_{k+1}^2 - F_kF_{k+2} &= F_{k+1}^2 - F_k(F_k + F_{k+1}) = F_{k+1}^2 - F_k^2 - F_kF_{k+1} \\ &= F_{k+1}^2 - [F_{k-1}F_{k+1} + (-1)^{k-1}] - F_kF_{k+1} \\ &= F_{k+1}^2 - F_{k+1}(F_{k-1} + F_k) + (-1)^k \\ &= F_{k+1}^2 - F_{k+1}F_{k+1} + (-1)^k = 0 + (-1)^k = (-1)^k\end{aligned}$$

Thus, $P(k)$ implies $P(k+1)$, so the result holds for every $n \geq 1$.

23. **basis step** When $n = 1$, LHS $= 1 =$ RHS; so $P(1)$ is true.
induction step Assume $P(n)$ is true for an arbitrary integer $n \geq 1$. Then

$$\sum_{i=1}^{n+1} F_{2i-1} = \sum_{i=1}^{n} F_{2i-1} + F_{2n+1} = F_{2n} + F_{2n+1} = F_{2n+2}$$

Thus, $P(n)$ implies $P(n+1)$, so, by induction, the result is true for every $n \geq 1$.

25. **basis step** When $n = 1$, LHS $= 1 =$ RHS; so $P(1)$ is true.
induction step Assume $P(n)$ is true for an arbitrary integer $n \geq 1$. Then

$$\sum_{i=1}^{n+1} L_i = \sum_{i=1}^{n} L_i + L_{n+1} = (L_{n+2} - 3) + L_{n+1} = (L_{n+1} + L_{n+2}) - 3 = L_{n+3} - 3$$

Thus, $P(n)$ implies $P(n+1)$, so, by induction, the result is true for every $n \geq 1$.

27. **basis step** When $n = 1$, LHS $= 3 =$ RHS; so $P(1)$ is true.
induction step Assume $P(n)$ is true for an arbitrary integer $n \geq 1$. Then

$$\begin{aligned}\sum_{i=1}^{n+1} L_{2i} &= \sum_{i=1}^{n} L_{2i} + L_{2n+2} = (L_{2n+1} - 1) + L_{2n+2} \\ &= (L_{2n+1} + L_{2n+2}) - 1 = L_{2n+3} - 1\end{aligned}$$

Thus, $P(n)$ implies $P(n+1)$, so, by induction, the result is true for every $n \geq 1$.

29. **basis step** When $n = 1$, LHS $= 1 =$ RHS; so $P(1)$ is true.
induction step Assume $P(n)$ is true for an arbitrary integer $n \geq 1$. Then

$$\begin{aligned}\sum_{i=1}^{n+1} L_i^2 &= \sum_{i=1}^{n} L_i^2 + L_{n+1}^2 = (L_n L_{n+1} - 2) + L_{n+1}^2 \\ &= L_{n+1}(L_n + L_{n+1}) - 2 = L_{n+1} L_{n+2} - 2\end{aligned}$$

Thus, $P(n)$ implies $P(n+1)$, so, by induction, the result is true for every $n \geq 1$.

31. Because $|A| = -1$, $|A^n| = |A|^n = (-1)^n$. Therefore, $F_{n+1}F_{n-1} - F_n^2 = |A^n| = (-1)^n$.

33. Because α and β are solutions of the equation $x^2 = x + 1$, $\alpha + \beta = 1$. Therefore, $b_2 = \dfrac{\alpha^2 - \beta^2}{\alpha - \beta} = \alpha + \beta = 1$.

35. $u_1 = \alpha + \beta = 1$, by Exercise 33.

37. Because α and β are solutions of the equation $x^2 = x + 1$, $\alpha^2 = \alpha + 1$ and $\beta^2 = \beta + 1$. So

$$\begin{aligned}u_{n-1} + u_{n-2} &= (\alpha^{n-1} + \beta^{n-1}) + (\alpha^{n-2} + \beta^{n-2}) = \alpha^{n-2}(\alpha + 1) + \beta^{n-2}(\beta + 1) \\ &= \alpha^{n-2} \cdot \alpha^2 + \beta^{n-2} \cdot \beta^2 = \alpha^n + \beta^n = u_n\end{aligned}$$

39.
$$\begin{aligned}F_{n-1} + F_{n+1} &= \frac{\alpha^{n-1} - \beta^{n-1}}{\alpha - \beta} + \frac{\alpha^{n+1} - \beta^{n+1}}{\alpha - \beta} = \frac{\alpha^{n-1}(1 + \alpha^2) - \beta^{n-1}(1 + \beta^2)}{\alpha - \beta} \\ &= \frac{\alpha^{n-1}(\sqrt{5}\alpha) - \beta^{n-1}(-\sqrt{5}\beta)}{\alpha - \beta} = \alpha^n + \beta^n = L_n\end{aligned}$$

41.
$$\begin{aligned}L_{n-1} + L_{n+1} &= (\alpha^{n-1} + \beta^{n-1}) + (\alpha^{n+1} + \beta^{n+1}) = \alpha^{n-1}(1 + \alpha^2) + \beta^{n-1}(1 + \beta^2) \\ &= \alpha^{n-1}(\sqrt{5}\alpha) + \beta^{n-1}(-\sqrt{5}\beta) = \sqrt{5}(\alpha^n - \beta^n) = 5F_n\end{aligned}$$

43.
$$\begin{aligned}F_{n+1}^2 - F_{n-1}^2 &= \frac{(\alpha^{n+1} - \beta^{n+1})^2 - (\alpha^{n-1} - \beta^{n-1})^2}{5} \\ &= \frac{[\alpha^{2n+2} + \beta^{2n+2} - 2(\alpha\beta)^{n+1}] - [\alpha^{2n-2} + \beta^{2n-2} - 2(\alpha\beta)^{n+1}]}{5} \\ &= \frac{\alpha^{2n}(\alpha^2 - \alpha^{-2}) + \beta^{2n}(\beta^2 - \beta^{-2})}{5} \\ &= \frac{\alpha^{2n}(\alpha^2 - \beta^2) + \beta^{2n}(\beta^2 - \alpha^2)}{5} \\ &= \frac{\alpha^{2n} - \beta^{2n}}{\alpha - \beta} = F_{2n}\end{aligned}$$

45. $a_1 = 2, a_2 = 3; a_n = a_{n-1} + a_{n-2}, n \geq 3$

Exercises 2.7 (p. 142)

1. 257; 65537
3. **proof** (by PMI): Let $P(n)$ denote the statement that the ones digit in f_n is 7, where $n \geq 2$. Because $f_2 = 17$, $P(2)$ is true. Now assume $P(k)$ is true for an arbitrary integer $k \geq 2$. By Theorem 2.16, $f_{k+1} = f_k^2 - 2f_k + 2$. Because f_k ends in 7, $f_k^2 - 2f_k + 2$ ends in $9 - 4 + 2 = 7$. Thus, $P(k+1)$ is true, so, by PMI, every f_n ends in 7 for $n \geq 2$.
5. **proof** (by contradiction): Assume m is not a prime, say, $m = ab$. Then $2^{ab} - 1 = (2^b - 1)(2^{(a-1)b} + 2^{(a-2)b} + \cdots + 2^b + 1)$, so $(2^b - 1)|(2^{ab} - 1)$, a contradiction.
7. **proof:** When $k \geq 3$, $t_k = k(k+1)/2$ ends in 1, 3, 5, 6, or 8. Thus, the only Fermat number that is also triangular is 3.

9. yes; $5 \cdot 2^7 + 1 = 5 \cdot 128 + 1 = 641$.
11. no
13. yes

Review Exercises (p. 146)

1. 1666
3. 2036
5. 1024
7. 2989
9. 11110101_{two}
11. 2305_{eight}
13. 100001_{two}
15. 4233658_{sixteen}
17. 265_{eight}
19. 463_{eight}
21. $\text{B5}_{\text{sixteen}}$
23. 133_{sixteen}
25. n^2
27. $a_n = \begin{cases} n(n+2)/4 & \text{if } n \text{ is even} \\ (n+1)^2/4 & \text{otherwise} \end{cases}$
29. **proof:** Let $P(n)$ denote the given statement. Since $3|0$, $P(1)$ is true. Now assume $P(k)$ is true for an arbitrary integer $k \geq 1$: $k^3 - k = 3m$ for some integer m. Then $(k+1)^3 - (k+1) = (k^3 - k) + 3(k^2 + k) = 3m + 3(k^2 + k)$ is divisible by 3, so $P(k+1)$ is true. Thus, the result follows by PMI.
31. **proof:** When $n = 1$, LHS $= \dfrac{1}{3} =$ RHS. So $P(1)$ is true. Assume $P(k)$ is true:

$$\sum_{i=1}^{k} \frac{1}{(2i-1)(2i+1)} = \frac{k}{2k+1}. \text{ Then } \sum_{i=1}^{k+1} \frac{1}{(2i-1)(2i+1)}$$

$$= \sum_{i=1}^{k} \frac{1}{(2i-1)(2i+1)} + \frac{1}{(2k+1)(2k+3)} = \frac{k}{2k+1} + \frac{1}{(2k+1)(2k+3)}$$

$$= \frac{k(2k+3)+1}{(2k+1)(2k+3)} = \frac{(k+1)(2k+1)}{(2k+1)(2k+3)} = \frac{k+1}{2(k+1)+1}.$$

Thus, $P(k)$ implies $P(k+1)$, so the result follows by PMI.

33. **proof:** Let $P(n)$: We must select at least $2n+1$ socks to ensure n matching pairs. By the PHP, $P(1)$ is true. Assume $P(k)$ is true, that is, we must select at least $2k+1$ socks to ensure k matching pairs. Add one matching pair. This ensures $k+1$ matching pairs by selecting $(2k+1)+2 = 2(k+1)+1$ socks. Therefore, $P(k)$ implies $P(k+1)$. Thus, $P(n)$ is true for every integer $n \geq 1$.

35. $777777 \cdot 999999 = 777776222223$
$7777777 \cdot 9999999 = 77777762222223$
37. composite, since $3|327$.
39. prime
41. 32
43. 35
45. 5042 through 5047
47. 87, 178, 291, 202 through 87, 178, 291, 214

$$p_1 = 2 = 2 \cdot 1 \qquad p_2 = 3 = 1 + 2 \qquad p_3 = 5 = 1 - 2 + 2 \cdot 3$$

$$p_4 = 7 = 1 - 2 + 3 + 5 \qquad p_5 = 11 = 1 - 2 + 3 - 5 + 2 \cdot 7$$

$$p_6 = 13 = 1 + 2 - 3 - 5 + 7 + 11 \qquad p_7 = 17 = -1 - 2 + 3 - 5 + 7 - 11 + 2 \cdot 13$$

$$p_8 = 19 = 1 - 2 - 3 - 5 + 9 - 11 + 13 + 17$$

51. Let $p = 29$. Then $2^{29} + 1 = 536{,}870{,}913 = 3 \cdot 59 \cdot 3{,}033{,}169$. So $q = 59 \cdot 3{,}033{,}169$ is not a prime.
53. **proof:** $n^4 - n^2 = n^2(n^2 - 1) = n(n-1)n(n+1)$ is divisible by 3.
55. **proof:** By the division algorithm, every odd integer n is of the form $4k + 1$ or $4k + 3$. If $n = 4k + 1$, then $n^2 = (4k + 1)^2 = 16k^2 + 8k + 1 = 8(2k^2 + k) + 1$. If $n = 4k + 3$, then $n^2 = (4k + 3)^2 = 16k^2 + 24k + 1 = 8(2k^2 + 3k) + 1$. Thus, in both cases, n^2 is of the form $8m + 1$.
57. Since $b^{3n} \pm 1 = (b^n)^3 \pm 1^3 = (b^n \pm 1)(b^{2n} \mp 1)$, $b^{3n} \pm 1$ is composite for $n \geq 1$.
59. **proof** (by R. L. Patton): Let N be the given number. Then $N = (3655^n - 3482^n) - (2012^n - 1848^n)$. Since $3655 - 3482 = 173 = 2021 - 1848$ and $a - b|a^n - b^n$, $173|N$. N is the difference of two odd numbers, $2|N$. Therefore, $2 \cdot 173 = 346|N$.
61. $F_2 = 1$, not a prime; $F_{19} = 4181 = 37 \cdot 113$, not a prime.
63. yes
65. yes
67. **proof:** Let $P(n)$ denote the given statement. When $n = 1$, LHS $= 1 =$ RHS, so $P(1)$ is true. Assume $P(k)$ is true: $F_k = (\alpha^k - \beta^k)/(\alpha - \beta)$. Because $x^n = xF_n + F_{n-1}$, $x^{n+1} = xF_{n+1} + F_n$. Therefore, $\alpha^{n+1} = \alpha F_{n+1} + F_n$ and $\beta^{n+1} = \beta F_{n+1} + F_n$. Then $\alpha^{n+1} - \beta^{n+1} = (\alpha - \beta)F_{k+1}$, so $P(k + 1)$ is true. Thus, the result follows by induction.
69. **proof** (by PMI): Let $P(n)$ be the statement that F_{3n-2} and F_{3n-1} are odd, and F_{3n} is even. Because F_1 and F_2 are odd, and F_3 is even, $P(1)$ is true. Now assume $P(k)$ is true: F_{3k-2} and F_{3k-1} are odd, and F_{3k} is even. Because $F_{3(k+1)} = F_{3k+3} = F_{3k+2} + F_{3k+1} = (F_{3k+1} + F_{3k}) + (F_{3k} + F_{3k-1}) = 2F_{3k} + (F_{3k+1} + F_{3k-1})$ is even by the inductive hypothesis. Thus, the result follows by PMI.

Supplementary Exercises (p. 148)

1. $x = 40$
3. 6174
5. $K^1(1968) = 8172$; $K^2(1968) = K(8172) = 7443$; $K^3(1968) = K(7443) = 3996$; $K^4(1968) = K(3996) = 6264$; $K^5(1968) = K(6264) = 4176$; $K^6(1968) = K(4176) = 6174$.
7. 113, 131, 311, 199, 919, 991, 337, 373, 733
9. 3779, 9377, 7937,7793

11. If a cyclic prime N contains 0, 2, 4, 6, or 8, then some permutation of N will end in 0, 2, 4, 6, or 8; the resulting number would not be a prime. Similarly, it cannot contain 5.
13. 2, 3, 5, 7, 11, 13, 17, 31, 37, 71, 73, 79, 97
15. 337 is a reversible prime, but not palindromic.
17. Because the ith step leads to a new lucky number for every positive integer i, there are infinitely many lucky numbers.
19. even; even
21. $S_1 = 1$; $S_n = [(b^n - b) + (b-1)]/(b-1) = bS_{n-1} + 1, n \geq 2$
23. **proof:** When $b = a^2$, $S_n = (a^{2n} - 1)/(a^2 - 1) = (a^n - 1)(a^n + 1)/(a^2 - 1)$. When n is even and > 2, $a^2 - 1 | a^n - 1$. When n is odd and > 2, $a - 1 | a^n - 1$ and $a + 1 | a^n + 1$. In both cases, S_n is composite.
25. $$\begin{aligned}2n(3n^4 + 13n^2 + 8) &= 6n^5 + 26n^3 + 16n = (4n^3 + 12n)n^2 + (2n^4 + 14n^2 + 16)n \\ &= n^2(n^4 + 2n^3 + 7n^2 + 6n + 8) + n(n^4 + 2n^3 + 7n^2 + 6n + 8) \\ &\quad - n^2(n^4 - 2n^3 + 7n^2 - 6n + 8) + n(n^4 - 2n^3 + 7n^2 - 6n + 8) \\ &= (n^2 + n)(n^4 + 2n^3 + 7n^2 + 6n + 8) - (n^2 - n)(n^4 - 2n^3 + 7n^2 - 6n + 8) \\ &= (n^2 + n)(n^2 + n + 2)(n^2 + n + 4) - (n^2 - n)(n^2 - n + 2)(n^2 - n + 4)\end{aligned}$$

    Let $n^2 + n = n(n+1) = 2m$. Then $K = (n^2 + n)(n^2 + n + 2)(n^2 + n + 4) = 2m(2m + 2)(2m + 4) = 8m(m+1)(m+2)$. Because $m(m+1)(m+2)$ is the product of three consecutive integers, it is divisible by 6. Therefore, $48|K$. Similarly, $48|L$, where $L = (n^2 - n)(n^2 - n + 2)(n^2 - n + 4)$. So $48|(K - L)$. Thus $24|[n(3n^4 + 13n^2 + 8)]$.
27. $A = 1, B = 27721$; $A = 1, B = 11! + 1$. (If $A = B$, then $A + n | B + n$.)
29. $g_1 = 2, g_2 = 3$

    $g_n = g_{n-1} + g_{n-2}, n \geq 3$
31. $f(n, 1) = 1$

    $f(n, n) = \begin{cases} 1 & \text{if } n = 0 \text{ or } 1 \\ 0 & \text{otherwise} \end{cases}$

    $f(n, k) = f(n-2, k-1) + f(n-1, k)$
33. $a_n = \begin{cases} 1 & \text{if } n = 0 \\ (1 + a_{n-1})/k & \text{if } n \geq 1 \end{cases}$
35. **proof** (by PMI): Let $P(n)$ be the statement that $a_n = (k^n + k - 2)/[k^n(k-1)]$. Clearly, $P(0)$ is true. Assume $P(n)$ is true for an arbitrary integer $n \geq 0$. Then

    $$\begin{aligned}a_{n+1} &= (1 + a_n)/k = \{1 + (k^n + k - 2)/[k^n(k-1)]\}/k \\ &= [k^n(k-1) + k^n + k - 2]/k^{n+1}(k-1) = (k^{n+1} + k - 2)/k^{n+1}(k-1)\end{aligned}$$

    Thus $P(k+1)$ is true. Hence, the result follows by PMI.
37. Let $P(k)$ denote the sum of the first k Fibonacci numbers. Then $P(k) = \sum_{i=1}^{k} F_i = F_{k+2} - 1$.

    So $S_n = P[n(n+1)/2] - P[n(n-1)/2] = \sum_{i=1}^{n(n+1)/2} F_i - \sum_{i=1}^{n(n-1)/2} F_i$

    $= F_{n(n+1)/2+2} - F_{n(n-1)/2+2}$

Chapter 3 Greatest Common Divisors

Exercises 3.1 (p. 165)

1. T
3. F
5. F
7. T
9. F
11. F
13. $4 = (-1) \cdot 24 + 1 \cdot 28$
15. $1 = 5 \cdot 21 + (-4) \cdot 26$
17. 12
19. 8
21. n
23. 1
25. 1
27. a
29. a
31. $a + b$
33. $a^2 - b^2$
35. $3 = (-1) \cdot 15 + 1 \cdot 18 + 0 \cdot 24$
37. $1 = (-1) \cdot 15 + 2 \cdot 18 + (-1) \cdot 20 + 0 \cdot 28$
39. $(15, 24, 28, 45) = ((15, 24, 28), 45) = (((15, 24), 28), 45) = ((3, 28), 45) = (1, 45) = 1$
41. $(18, 24, 36, 63) = ((18, 24, 36), 63) = (((18, 24), 36), 63) = ((6, 36), 63) = (6, 63) = 3$
43. $(a^2b^2, ab^3, a^2b^3, a^3b^4, a^4b^4) = ((a^2b^2, ab^3, a^2b^3, a^3b^4), a^4b^4) = (((a^2b^2, ab^3, a^2b^3), a^3b^4), a^4b^4) = ((((a^2b^2, ab^3), a^2b^3), a^3b^4), a^4b^4) = (((ab^2, a^2b^3), a^3b^4), a^4b^4) = ((ab^2, a^3b^4), a^4b^4) = (ab^2, a^4b^4) = ab^2$
45. Let $a = 4$, $b = 6$, and $c = 8$
47. **proof:** Let $d = (a, b)$ and $d' = (a, -b)$. Because $d|b$, $d| - b$. Thus $d|d'$. Similarly, $d'|d$. Thus $d = d'$.
49. **proof:** By Exercises 47 and 48, $(-a, -b) = (a, -b) = (a, b)$.
51. **proof:** Let $d = (a, b)$, so $d = \alpha a + \beta b$ for some integers α and β. Then $dc = \alpha(ac) + \beta(bc)$. Thus $dc = (ac, bc)$; that is, $c(a, b) = (ac, bc)$.
53. **proof:** Let $d = (p, a)$ and $d > 1$. Then $d|p$ and $d|a$. Because $d|p$, $d = p$, so $p|a$, a contradiction. Thus $(p, a) = 1$.
55. **proof** (by PMI): Let $P(n)$ denote the given statement. Clearly, $P(2)$ is true. Assume $P(k)$ is true for an arbitrary integer $k \geq 2$. Because $d = (a_1, a_2, \ldots, a_{k+1}) = ((a_1, a_2, \ldots, a_k), a_{k+1})$, $d|(a_1, a_2, \ldots, a_k)$ and $d|a_{k+1}$. By the inductive hypothesis, $d|a_i$ for every i, where $1 \leq i \leq k$. So $d|a_i$ for every i, where $1 \leq i \leq k + 1$. Thus, the result follows by PMI.
57. **proof:** By Theorem 3.4, $(a, a - b) = (a, b)$. Thus, $(a, a - b) = 1$ if and only if $(a, b) = 1$.
59. **proof:** Since $(a, b) = 1 = (a, c)$, $ra + sb = 1 = xa + yc$ for some integers r, s, x, and y. Then $(ra + sb)(xa + yc) = 1$; that is, $\alpha a + \beta bc = 1$, where $\alpha = rxa + yc + sxb$ and $\beta = sy$.
61. **proof:** Let $(a^2 + b^2, a^2 + 2ab) = 1$ or 5. Let $d = (a, b)$. Then $a = \alpha d$ and $b = \beta d$, where $(\alpha, \beta) = 1$. Then $a^2 + b^2 = (\alpha^2 + \beta^2)d^2$ and $a^2 + 2ab = (\alpha^2 + 2\alpha\beta)d^2$; so $d^2|(a^2 + b^2, a^2 + 2ab)$. Thus, $d^2|1$ and $d^2|5$. In either case, $d = 1$. Thus $(a, b) = 1$.
63. **proof:** Let $d = (a^n - 1, a^m + 1)$. Then $a^n = rd + 1$ and $a^m = sd - 1$ for some integers r and s. Then, by the binomial theorem

$$a^{mn} = (a^n)^m = (rd + 1)^m = \sum_{i=0}^{m} \binom{m}{i} (rd)^i = 1 + Ad$$

and

$$a^{mn} = (a^m)^n = (sd - 1)^n = (-1)^n \sum_{j=0}^{n} \binom{n}{j} (-sd)^j = -\sum_{j=0}^{n} \binom{n}{j} (-sd)^j = Bd - 1$$

for some integers A and B. Thus, $1 + Ad = Bd - 1$, where $A \neq B$. Then $(B - A)d = 2$, so $d|2$. Thus, $d = 1$ or 2.

65. **proof:** $\dfrac{a(a+b-1)!}{a!b!} = \dbinom{a+b-1}{a-1}$ and $\dfrac{b(a+b-1)!}{a!b!} = \dbinom{a+b-1}{b-1}$. Then

$$(a,b)\frac{(a+b-1)!}{a!b!} = \left(\frac{a(a+b-1)!}{a!b!}, \frac{b(a+b-1)!}{a!b!}\right) = \left(\binom{a+b-1}{a-1}, \binom{a+b-1}{b-1}\right)$$

is an integer. Thus, $a!b!|(a,b)(a+b-1)!$.

67. **proof:** We have $(F_m, F_n) = F_{(m,n)}$. Suppose $F_m|F_n$. Then $(F_m, F_n) = F_m = F_{(m,n)}$, so $m = (m, n)$. This implies, $m|n$.
Conversely, let $m|n$. Then $(F_m, F_n) = F_{(m,n)} = F_m$, so $F_m|F_n$.

69. **proof:** $\dfrac{f_n - 2}{f_m} = \dfrac{a^k - 1}{a+1}$, where $a = 2^{2^m}$ and $k = 2^{n-m} = a^{k-1} - a^{k-2} + \cdots - 1$, so $f_m|(f_n - 2)$.

71. **proof:** Let $P(n)$: Every f_n contributes a new prime factor. Clearly, $P(0)$ is true. Assume $P(n)$ is true for $n \le k-1$. f_k has a prime factor p. Suppose $p|f_i$ for some i, where $0 \le i \le k-1$. Then $p|f_i$ and $p|f_k$, which is a contradiction by Theorem 3.2. So p is different from the prime factors of $f_0, f_1, \ldots,$ and f_{k-1}. Thus, $P(k)$ is true, so the result follows by PMI.

73. **proof:** Let $d = (m, n)$. Then, by Theorem 3.5, there are integers A and B such that $d = Am + Bn$. Multiplying both sides by $\dbinom{m}{n}$, this yields $d\dbinom{m}{n} = Am\dbinom{m}{n} + Bn\dbinom{m}{n} = m\left[A\dbinom{m}{n} + B\dbinom{m-1}{n-1}\right] = mC$ where C is an integer. So $\dfrac{m}{d}\Big|\dbinom{m}{n}$. That is, $\dfrac{m}{(m,n)}\Big|\dbinom{m}{n}$.

Exercises 3.2 (p. 173)

1. 8
3. 4
5. 8
7. 4
9. $8 = 42 \cdot 1024 - 43 \cdot 1000$
11. $4 = (-85) \cdot 2076 + 164 \cdot 1076$
13. $(-71) \cdot 1976 + 79 \cdot 1776$
15. $(-97) \cdot 3076 + 151 \cdot 1976$
17. **proof:** By the division algorithm, $a = bq + r$ for some integer q. Since $d'|b$ and $d'|r$, $d'|a$. Thus, $d'|a$ and $d'|b$. Thus, $d'|(a,b)$; that is, $d'|d$.
19. **proof** (by PMI): Let $P(n)$ denote the given statement.
basis step When $n = 2$, $F_2 = 1 < (1+\sqrt{5})/2 = \alpha$. So $P(2)$ is true.
induction step Assume $P(2), P(3), \ldots,$ and $P(k)$ are true, that is, $F_i < \alpha^{i-1}, 2 \le i \le k$. Then $\alpha^k = \alpha^{k-1} + \alpha^{k-2} > F_k + F_{k-1} = F_{k+1}$.
Thus, $P(k)$ implies $P(k+1)$, so the result holds for every $n \ge 2$.

Exercises 3.3 (p. 182)

1. $3 \cdot 11 \cdot 59$
3. $3^4 \cdot 23$
5. $2^{27} + 1 = (2^9)^3 = 1^3 = (2^9+1)(2^{18} - 2^9 + 1) = (2^3+1)(2^6 - 2^3 + 1)(2^{18} - 2^9 + 1)$
$= 3^2 \cdot 57 \cdot 261633 = 3^4 \cdot 19 \cdot 87211$
7. Let $x = 100$. Then $10,510,100,501 = 1 \cdot x^5 + 5x^4 + 10x^3 + 10x^2 + 5x + 1 = (x+1)^5 = 101^5$.
9. $1, p$
11. $1, p, q, pq$
13. 1, 2, 3, 4, 6, 8, 12, 16, 24, 48

15. 1, 2, 3, 5, 6, 7, 10, 14, 15, 21, 30, 35, 42, 70, 105, 210
17. 24
19. 149
21. 240; 241; 242; 243; 244
23. 30
25. pq^2
27. 6
29. 1
31. 2
33. 4
35. largest integer b such that $2^b|b$
37. $333 + 111 + 37 + 12 + 4 + 1 = 498$
39. $143 + 20 + 2 = 165$
41. $b = n - e = 1001 - 994 = 7$
43. $b = n - e = 3076 - 3073 = 3$
45. **proof:** Because $p|a \cdot a$, $p|a$ by Lemma 3.3.
47. **proof** (by PMI): Clearly, the statement is true when $n = 1$, so assume it is true for an arbitrary integer $n \geq 1$. Then $\prod_{i=1}^{n+1} (4k_i + 1) = \left[\prod_{i=1}^{n} (4k_i + 1)\right](4k_{n+1} + 1) = (4m + 1)(4k_{n+1} + 1)$, which is of the form $4k + 1$ by Lemma 3.5.
49. **proof** (by contradiction): Assume $(a, b) > 1$. Let p be a prime factor of (a, b). Then $p|a$ and $p|b$. Then $p|a^n$ and $p|b^n$, so $p|(a^n, b^n)$; that is, $p|1$, a contradiction.
51. **proof:** Using Dirichlet's theorem with $a = 5$ and $b = 8$, there are infinitely primes of the form $8n + 5$.
53. **proof** (by strong version of PMI): The statement is true when $n = 1$. So assume it is true for every positive integer 1 through an arbitrary positive integer k. Consider the integer $k + 1$. If $k + 1$ is even, $k + 1 = 2m$ for some integer m. Then m is of the form 2^a5^bc, by the inductive hypothesis, so $k + 1 = 2(2^a5^bc) = 2^{a+1}5^bc$, where c is not divisible by 2 or 5. Suppose $k + 1$ is odd. If $k + 1$ is divisible by 5, then $k + 1 = 5x$ for some integer x. Then x is of the form 2^a5^bc by the inductive hypothesis, so $k + 1 = 5x = 5(2^a5^bc) = 2^a5^{b+1}c$, where c is not divisible by 2 or 5. Finally, assume $k + 1$ is not divisible by 5. Then $k + 1 = 2^05^0c$, where $c = k + 1$ is not divisible by 2 or 5.
55. 4
57. 9
59. $\dfrac{p^{i+1} - 1}{p - 1}$
61. $\dfrac{(p^{i+1} - 1)(q^{j+1} - 1)}{(p - 1)(q - 1)}$
63. **proof:** Assume n is square-free and its prime-factorization contains a factor p^k, where $k \geq 2$. Then $p^2|n$, so n is not square-free, a contradiction.
    Conversely, let $n = p_1p_2 \cdots p_k$, where the primes p_i are distinct. Assume n is not square-free. Then $p^2|n$ for some prime p, so $p^2|p_1p_2 \cdots p_k$, which is a contradiction.
65. 4, 8, 9
67. **proof:** Let $p^a \parallel m$ and $p^b \parallel n$, so $m = p^aA$ and $n = p^bB$, where $p \nmid A$ and $p \nmid B$. Then $mn = (p^aA)(p^bB) = p^{a+b}(AB)$, where $p \nmid AB$. So $p^{a+b} \parallel mn$.
69. **proof:** Let $p^a \parallel m$ and $p^b \parallel n$, so $m = p^aA$ and $n = p^bB$, where $p \nmid A$ and $p \nmid B$. Suppose $a \leq b$. Then $m + n = p^aA + p^bB = p^a(A + p^{b-a}B)$. Because $p \nmid A$, it follows that $p^a \parallel (m + n)$; that is, $p^{\min(a,b)} \parallel (m + n)$. The case that $a > b$ is quite similar.

Exercises 3.4 (p. 188)

1. F
3. T
5. T
7. F
9. F
11. 2310
13. b
15. b
17. ab
19. 231
21. 11

23. 20, 840; 40, 420; 60, 280; 120, 140

25. $a = b$

27. $[15, 18, 24, 30] = [[15, 18, 24], 30] = [[15, 18], 24, 30] = [[90, 24], 30] = [360, 30] = 360$

29. $[12, 15, 18, 25, 30] = [[12, 15, 18, 25], 30] = [[[12, 15, 18], 25], 30] = [[[12, 15], 18], 25, 30] = [[[60, 18], 25], 30] = [[180, 25], 30] = [900, 30] = 900$

31. Let $n = \prod_i p_i^{e_i}$. Because n is a square, $2|e_i$; similarly, $3|e_i$ and $5|e_i$. Therefore, $[2, 3, 5]|e_i$; that is, $30|e_i$ for every i. Thus, the least possible exponent is 30 and the least possible base is 2, so $n = 2^{30} = 1, 073, 741, 824$.

33. In the proof of Theorem 3.17 we found that $(a, b) \cdot [a, b] = ab$, so $(a, b)|[a, b]$.

35. **proof:** Let $\ell = [a, b]$. Then $m = q\ell + r$, where $0 \le r \le \ell$. Then $r = m - q\ell$. Because both m and 1 are common multiples of a and b, r is also a common multiple of a and b. But this is impossible since $r < \ell$. Therefore, $r = 0$. Thus $\ell|m$.

37. **proof** (by PMI): Let $P(n)$ denote the given statement. By Exercise 34, $P(2)$ is true. Now assume $P(t)$ is true for $t \ge 2$: $[ka_1, ka_2, \dots, ka_t] = k[a_1, a_2, \dots, a_t]$. Then $[ka_1, ka_2, \dots, ka_{t+1}] = [[ka_1, ka_2, \dots, ka_t], ka_{t+1}] = [k[a_1, a_2, \dots, a_t], ka_{t+1}] = k[[a_1, a_2, \dots, a_t], a_{t+1}] = k[a_1, a_2, \dots, a_t, a_{t+1}]$, by the inductive hypothesis, and Exercises 34 and 36.

39. **proof:** Let $x \le y \le z$. Then LHS $= z - x + x + y + x = x + y + z =$ RHS. The other five cases can be handled similarly.

41. Let $a = \prod_i p_i^{a_i}$, $b = \prod_i p_i^{b_i}$, and $c = \prod_i p_i^{c_i}$. Suppose $a_i \le b_i \le c_i$. Then $[b, c] = \prod_i p_i^{c_i}$, $(a, [b, c]) = \prod_i p_i^{a_i}$, $(a, b) = \prod_i p_i^{a_i}$, and $(a, c) = \prod_i p_i^{a_i}$. So $[(a, b), (a, c)] = \prod_i p_i^{a_i} = (a, [b, c])$. Similarly, it is true in the other five cases.

Exercises 3.5 (p. 203)

1. 84
3. no
5. yes
7. no
9. $x = 3 + 4t, y = -1 - 3t$
11. $x = -3 + 7t, y = 4 - 5t$
13. $x = 41001 - 247t, y = -36849 + 222t$
15. $x = 2 - 3t, y = 2t$
17. $x = -1 + 13t, y = 2 - 12t$
19. $x = 1 - 13t, y = 1 + 4t$
21. 18 mangoes
23. The clerk has made a computational error.
25. 3 apples
27. By Cassini's formula, $F_{13}F_{11} - F^2 12 = 1$; that is, $233 \cdot 89 - 144^2 = 1$. So $233 \cdot (19 \cdot 89) - 144(19 \cdot 144) = 19$. Thus, the general solution is $x = 19 \cdot 89 - 144t = 1681 - 144t$, $y = 19 \cdot -233t = 2736 - 233t$.
29. yes
31. yes
33. $x = 3t, y = 0, z = 2 - t$
35. $x = -22 - 5t + 2t', y = -t', z = -11 - 2t$
37. The general solution is: $n = 8 + 12t$, $d = 6 - t$, $q = 12 - 2t$. The unique solution is 8 nickels, 6 dimes, and 12 quarters, corresponding to $t = 0$.
39. The general solution is $h = 1535 - 4t$, $c = 1600 - 4t$, $z = -1600 + 5t$, where $356 \le t \le 383$. The possible combinations are: 111, 176, 180; 107, 172, 185; …; 7, 72, 310; and 3, 68, 315.

41. **proof** (by PMI): The statement is true when $n = 2$, by Theorem 3.19, so assume it is true for an arbitrary integer $k \geq 2$. Assume $\sum_{i=1}^{k+1} a_i x_i = \sum_{i=1}^{k} a_i x_i + a_{k+1}x_{k+1} = c$ is solvable. By the inductive hypothesis, $\sum_{i=1}^{k} a_i x_i = y$ is solvable, so $(a_1, \ldots, a_k)|y$, since $y + a_{k+1}x_{k+1} = c$, $(y, a_{k+1})|c$. Since $(a_1, \ldots, a_k, a_{k+1}) = ((a_1, \ldots, a_k), a_{k+1})$, it follows that $(a_1, \ldots, a_k, a_{k+1})|c$.
Conversely, suppose $(a_1, \ldots, a_n)|c$. Then $c = d \cdot (a_1, \ldots, a_n)$. But $(a_1, \ldots, a_n) = a_1 y_1 + \cdots + a_n y_n$ for some integers y_i. Then $c = a_1(dy_1) + \cdots + a_n(dy_n)$, so the LDE $a_1x_1 + \cdots + a_nx_n$ is solvable.

Review Exercises (p. 207)

1. 90
3. 16
5. 4
7. 3
9. $(16, 20, 36, 48) = ((16, 20, 36), 48) = (((16, 20), 36), 48) = ((4, 36), 48) = (4, 48) = 4$
11. $(28, 48, 68, 78) = ((28, 48, 68), 78) = (((28, 48), 68), 78) = ((4, 68), 78) = (4, 78) = 2$
13. $2 = 4 \cdot 14 + (-3) \cdot 18$
15. $2 = 0 \cdot 12 + (-1) \cdot 18 + 1 \cdot 20$
17. 1, 2, 7, 14, 49, 98
19. $1, p, \ldots, p^i, q, pq, \ldots, p^i q$
21. $2^4 \cdot 5^3$
23. $7 \cdot 11 \cdot 23$
25. $52 + 10 + 2 = 64$
27. $280 + 56 + 11 + 2 = 349$
29. 0
31. 0
33. 2
35. 0
37. 630
39. 528
41. 28, 29
43. 192
45. 630
47. $[16, 20, 36, 48] = [[16, 20, 36], 48] = [[[16, 20], 36], 48] = [[80, 36], 48] = [720, 48] = 720$
49. $[28, 48, 68, 78] = [[28, 48, 68], 78] = [[[28, 48], 68], 78] = [[336, 68], 78] = [5712, 78] = 74,256$
51. **proof** (by PMI): Let $P(n)$ denote the given statement. Clearly, $P(1)$ is true. Assume $P(k)$ is true: If $p|a^k$, then $p^k|a^k$. Because $a^{k+1} = a^k \cdot a$, $p^k \cdot p|a^{k+1}$, by the inductive hypothesis. That is, $p^{k+1}|a^{k+1}$. Thus, by PMI, the result is true for every $n \geq 1$.
53. **proof:** The product of any three consecutive integers is divisible by 2 and that of any three consecutive integers is divisible by 3. Therefore, by Review Exercise 52, it is divisible by $[2, 3] = 6$.
55. **proof:** Let $(p, p + 2) = d$ and $d > 1$. Then $d|p$ and $d|(p + 2)$, so $d|2$ and hence $d = 2$. Then $p + 2 = 4$ is not a prime. Therefore, $d = 1 = (p, p + 2)$.
57. **proof:** Let $(a, c) = t$. Then $a = mt$ and $c = nt$, where $(m, n) = 1$. Because $a|b$, $b = \alpha t = \alpha mt$ for some integer α. Likewise, $d = \beta c = \beta nt$ for some integer β. Therefore, $(b, d) = (\beta nt, \alpha mt) = (\beta n, \alpha m)t$, so $t|(b, d)$; that is, $(a, c)|(b, d)$.
59. **proof:** Let $x = 4i + 3$ and $y = 4j + 3$ be any two integers of the given form. Then $xy = (4i + 3)(4j + 3) = 4(4ij + i + j + 2) + 1 = 4n + 1$, where $n = 4ij + i + j$. Thus, xy is of the desired form.
61. **proof:** Let $n = \prod_i p_i^{e_i}$. By the division algorithm, $e_i = 3q_i + r_i$, where $0 \leq r_i < 3$. Then

$$n = \prod_i p_i^{e_i} = \prod_i p_i^{3q_i+r_i} = \left(\prod_i p_i^{3q_i}\right)\left(\prod_i p_i^{r_i}\right)$$

Thus, n is a cube if and only if $r_i = 0$ for every i, that is, if and only if every exponent in the canonical decomposition of n is divisible by 3.

63. Using Dirichlet's theorem with $a = 4$ and $b = 7$, there is an infinite number of primes of the form $7n + 4$.

65. **proof:** Suppose the power of p in the canonical decomposition of n is e. By Theorem 2.5, there are exactly $\lfloor n/p^i \rfloor$ positive integers that are less than or equal to n and divisible by p^i, where $i \geq 1$. Thus, the number of positive integers $\leq n$ and divisible by p is $\sum_{i \geq 1} \lfloor n/p^i \rfloor$; that is, $e = \sum_{i \geq 1} \lfloor n/p^i \rfloor$.

67. $a = \prod_i p_i^{a_i}$, $b = \prod_i p_i^{b_i}$, and $c = \prod_i p_i^{c_i}$. Suppose $a_i \leq b_i \leq c_i$. Then $[a, b] = \prod_i p_i^{b_i}$, $[b, c] = \prod_i p_i^{c_i}$, $[c, a] = \prod_i p_i^{c_i}$, $(a, b) = \prod_i p_i^{a_i}$, $(b, c) = \prod_i p_i^{b_i}$, and $(c, a) = \prod_i p_i^{a_i}$. So

$$([a, b], [b, c], [c, a]) = \prod_i p_i^{b_i} = [(a, b), (b, c), (c, a)]$$

The other five cases follow similarly.

69. **proof** (by PMI): The result is true when $n = 0$. Assume it is true for an arbitrary integer $n \geq 0$. Since $6(n+1) | (3n+1)(3n+2)(3n+3)$, it follows that $6(n+1)6^n n! | (3n)!(3n+1)(3n+2)(3n+3)$; that is, $6^{n+1}(n+1)! | [3(n+1)!]$. Thus, the result follows by PMI.

71. $15! = 2^{11} \cdot 3^6 \cdot 5^3 \cdot 7^2 \cdot 11 \cdot 13$

73. $23! = 2^{19} \cdot 3^9 \cdot 5^4 \cdot 7^3 \cdot 11^2 \cdot 13 \cdot 17 \cdot 19 \cdot 23$

75. Since $(76, 176) = 4$ and $4|276$, the LDE is solvable.

77. By trial and error, $x_0 = -1$, $y_0 = 2$ is a particular solution. The general solution is $x = -1 + (176/4)t = -1 + 44t$, $y = 2 - (76/4)t = 2 - 19t$.

79. $76x + 176y = 276$, so $x = \dfrac{276 - 176y}{76} = 3 - 2y + \dfrac{48 - 24y}{76} = 3 - 2y + u$, where $u = \dfrac{48 - 24y}{76}$. Then $24y + 76u = 48$, so $y = 2 - \dfrac{19u}{6}$. Since y is an integer, $u = 6t$ for some integer t. Then $y = 2 - 19t$ and $x = 3 - 2(2 - 19t) + 6t = -1 + 44t$. Thus, the general solution is $x = -1 + 4t$, $y = 2 - 19t$.

81. By Exercise 38 in Section 3.5, the LDE is $1024n - 15625z = 8404$. Then

$$n = \frac{15625z + 8404}{1024} = 15z + 8 + \frac{265z + 212}{1024}$$

Let $a = \dfrac{265z + 212}{1024}$, so $265z = 1024a - 212$; that is, $z = 3a + \dfrac{229a - 212}{265}$. Let $b = \dfrac{229a - 212}{265}$, so $229a = 265b + 212$, so $a = b + \dfrac{36b + 212}{229}$. Let $c = \dfrac{36b + 212}{229}$, so $b = 6c - 5 + \dfrac{13c - 32}{36}$. Let $d = \dfrac{13c - 32}{36}$, so $c = 2d + 2 + \dfrac{10d + 6}{13}$. Let $e = \dfrac{10d + 6}{13}$. Then $d = e + \dfrac{3e - 6}{10}$. Now we let $f = \dfrac{3e - 6}{10}$, so $e = 3f + 2 + \dfrac{f}{3}$. Since e is an integer, $f = 3t$ for some integer t. Then

$$e = 3f + 2 + t = 10t + 2, \quad d = e + f = (10t + 2) + 3t = 13t + 2$$

$$c = 2d + 2 + e = 2(13t + 2) + 2 + 10t + 2 = 36t + 8$$

$$b = 6c - 5 + d = 6(36t + 8) - 5 + (13t + 2) = 229t + 45$$

and $a = b + c = (229t + 45) + (36t + 8) = 265t + 5$

$$z = 3a + b = 3(265t + 53) + (229t + 45) = 1024t + 204$$

So $n = 15z + 8 + a = 15(1024t + 204) + 8 + (265t + 53) = 15625t + 3121$

83. Let a denote the number of apples, e the number of oranges, and p the number of pears. Then $a + e + p = 12$ (1) and $75a + 30e + 60p = 630$ (2). Substituting for p in (2), we get $15a - 30e = -90$; that is, $a - 2e = -6$. So $a_0 = 0$, $e_0 = 3$ is a particular solution of this LDE. The general solution is $a = -2t$, $e = 3 - t$, $p = 9 + 3t$. Since $a > 0$, $t < 0$; since $p > 0$, $-3 < t$. Thus, $-3 < t < 0$. $t = -1$ implies $a = 2$, $e = 4$, and $p = 6$; and $t = -2$ yields $a = 4$, $e = 5$, and $p = 3$.

Supplementary Exercises (p. 209)

1. $[1, 2, 3, 4, 5] = 60$; $[1, 2, 3, 4, 5, 6] = 60$; $[1, 2, 3, 4, 5, 6, 7] = 420$; $[1, 2, 3, 4, 5, 6, 7, 8] = 840$
3. $9^\star = 3(8^\star) = 3 \cdot 840 = 2520$
5. **proof:** Let $2 \le i \le n + 1$. Then $i|(n + 1)^\star$. Therefore, $i|[(n + 1)^\star + i]$ for $2 \le i \le n + 1$. Thus, $(n + 1)^\star + 2, (n + 1)^\star + 3, \dots, (n + 1)^\star + (n + 1)$ are n consecutive composite numbers.
7. $2, 3, 7, 13, 61, 61$, and 421; they are all primes. Therefore, $n^\star + 1$ is a prime for every $n \ge 1$.
9. $5, 11, 59, 419, 419$, and 839; they are primes. Therefore, $n^\star - 1$ is a prime for every $n \ge 3$.
11. **proof** (by J. S. Shipman): Suppose $(a, b) = 1$. Then $M = ab$ and $(c, m) = (c, ab)$. Let $(c, a) = d$, where $d > 1$. Then $c = c'd$ and $a = a'd$ for some integers c' and a'. So $b = c - a = d(c' - a')$, and hence $d|b$. Thus, $d|a$ and $d|b$, which contradicts the hypothesis that $(a, b) = 1$; so $(c, a) = 1$. Similarly, $(c, b) = 1$. Therefore, $(c, ab) = 1$. Thus, $(c, m) = (c, ab) = 1 = (a, b)$.
    Suppose $(a, b) = d > 1$. Let $a = a'd$ and $b = b'd$ for some integers a' and b', where $(a', b') = 1$. Then $c = a + b = (a' + b')d = c'd$, where $c' = a' + b'$. Since
    $$m = \frac{ab}{(a, b)} = \frac{a'b'd^2}{d} = a'b'd$$
    it follows that $(c, m) = (c'd, a'b'd) = d(c', a'b') = d = (a, b)$
13. **proof:**
    $$a^n - b^n = (a - b)\left(\sum_{k=0}^{n-1} a^k b^{n-1-k}\right)$$
    $$\begin{aligned}\frac{a^n - b^n}{a - b} &= \sum_{k=0}^{n-1} a^k b^{n-1-k} = \sum_{k=1}^{n-1} [(a^k - b^k)b^{n-1-k} + b^{n-1}] + b^{n-1} \\ &= \sum_{k=1}^{n-1} [(a^k - b^k)b^{n-1-k}] + (n - 1)b^{n-1} + b^{n-1} \\ &= (a - b)Q(a, b) + nb^{n-1} \qquad (1)\end{aligned}$$

where $Q(a,b)$ is a polynomial in a and b with integral coefficients. Let $d = ((a^n - b^n)/(a-b), a-b)$ and $e = (n(a,b)^{n-1}, a-b)$. From equation (1), $d|nb^{n-1}$. Similarly, $d|na^{n-1}$, so $d|(na^{n-1}, nb^{n-1})$. But $(na^{n-1}, nb^{n-1}) = n(a^{n-1}, b^{n-1}) = n(a,b)^{n-1}$, hence $d|e$.

Because $e|n(a,b)^{n-1}$ and $n(a,b)^{n-1} = (na^{n-1}, nb^{n-1})$, $e|nb^{n-1}$. Also $e|a-b$, so by equation (1), $e|((a^n - b^n)/(a-b))$. Thus, $e|d$. Thus, $d|e$ and $e|d$, so $e = d$.

15. **proof:** Let $d = (m,n)$, $e = a^d - b^d$, and $f = (a^m - b^m, a^n - b^n)$. Because $d|m$, $a^d - b^d|a^m - b^m$; that is, $e|a^m - b^m$. Similarly, $e|a^n - b^n$, so $e|f$.

    To show that $f|e$, because $d = (m,n)$, $d = mx - ny$ for some integers x and y. Then $a^{mx} = a^{ny+d} = a^{ny}(b^d + e)$. Therefore, $a^{mx} - b^{mx} = a^{ny}(b^d + e) - b^{mx} = b^d(a^{ny} - b^{ny}) + ea^{ny}$ (1). Because $f|(a^m - b^m)$ and $(a^m - b^m)|(a^{mx} - b^{mx})$, $f|(a^{mx} - b^{mx})$. Likewise, $f|(a^{ny} - b^{ny})$. So, by equation (1), $f|ea^{ny}$.

    To show that $(f,a) = 1$, let $(f,a) = d'$. Then $d'|a$, so $d'|a^m$ and $d'|(a^m - b^m)$; so $d'|[a^m - (a^m - b^m)]$; that is, $d'|b^m$. Thus, $d'|a^m$ and $d'|b^m$, but $(a^m, b^m) = 1$ because $(a,b) = 1$. So $d' = 1$. Thus, $(f,a) = 1$. Since $f|ea^{ny}$, this implies $f|e$. Thus, $e|f$ and $f|e$, so $f = e$.

Chapter 4 Congruences

Exercises 4.1 (p. 228)

1. T 3. T 5. T 7. T 9. T 11. T
13. T 15. F 17. F 19. T 21. F 23. T
25. $n \equiv 1 \pmod 2$ 27. $n \equiv 0 \pmod 5$ 29. Friday 31. 1 P.M.
33. $3^2 \equiv 2^2 \pmod 5$, but $3 \not\equiv 2 \pmod 5$.
35. Because $k! \equiv 0 \pmod{10}$ if $k \geq 5$, $1! + 2! + \cdots + 1000! \equiv 1! + 2! + 3! + 4! \equiv 3 \pmod{10}$, so the remainder is 3.
37. Because $k! \equiv 0 \pmod{12}$ if $k \geq 4$, $1! + 2! + \cdots + 1000! \equiv 1! + 2! + 3! \equiv 9 \pmod{12}$, so the remainder is 9.
39. $2^3 \equiv 1 \pmod 7$, so $2^{35} = (2^3)^{11} \cdot 4 \equiv 1 \cdot 4 \equiv 4 \pmod 7$. Thus, the remainder is 4.
41. $23^{1001} \equiv 6^{1001} = (6^{16})^{62} \cdot 6^8 \cdot 6 \equiv 1^{62} \cdot (-1) \cdot 6 \equiv -6 \equiv 11 \pmod{17}$, so the remainder is 11.
43. $2^{97} \equiv 2^{64} \cdot 2^{32} \cdot 2 \equiv 3 \cdot 9 \cdot 2 \equiv 2 \pmod{13}$. Thus, the desired remainder is 2.
45. $13^{218} \equiv 13^{128} \cdot 13^{64} \cdot 13^{16} \cdot 13^8 \cdot 13^2 \equiv 1 \cdot 1 \cdot 1 \cdot 1 \cdot (-1) \equiv 16 \pmod{17}$. So the remainder is 16.
47. $1776 \equiv 6 \pmod{10}$ and $6^n \equiv 6 \pmod{10}$ for every $n \geq 1$. Thus, $1776^n \equiv 6^n \equiv 6 \pmod{10}$ for $n \geq 1$, so the desired units digit is 6.
49. $1077 \equiv 7 \pmod{10}$. By Example 4.12, $7^a \equiv 1 \pmod{10}$ if $a \equiv 0 \pmod 4$; $7 \pmod{10}$ if $a \equiv 1 \pmod 4$; $9 \pmod{10}$ if $a \equiv 2 \pmod 4$; and $3 \pmod{10}$ if $a \equiv 3 \pmod 4$. Because $1177 \equiv 1 \pmod 4$, $1177^n \equiv 1 \pmod 4$ for every $n \geq 1$. Because $a \equiv 1 \pmod 4$ corresponds to $7 \pmod{10}$, the ones digit is 7.
51. $1776 \equiv 76 \pmod{100}$. Since $76^n \equiv 76 \pmod{100}$ for every $n \geq 1$, the desired number is 76.

53. r
55. 1, 5
57. 1, 3, 5, 7
59. at least 2 for $m \geq 3$
61. a
63. 2
65. 6
67. $-1 \pmod p$
69. **proof:** Because $a \equiv b \pmod m$ and $c \equiv c \pmod m$, the result follows by Theorem 4.4.
71. **proof:** Because $a \equiv b \pmod m$ and $c \equiv c \pmod m$, the result follows by Theorem 4.4.
73. **proof:** Because $ac \equiv bc \pmod p$, $p|(ac - bc)$; that is, $p|c(a-b)$. But $p \nmid c$, so $p|(a-b)$; that is, $a \equiv b \pmod p$.
75. **proof:** Let $f(x) = \sum_{i=0}^{n} c_i x^i$. Because $a \equiv b \pmod m$, $ca^i \equiv cb^i \pmod m$, so $\sum_{i=0}^{n} c_i a^i \equiv \sum_{i=0}^{n} c_i b^i$. That is, $f(a) \equiv f(b) \pmod m$.
77. **proof:** By the division algorithm, every odd integer is of the form $4k+1$ or $4k+3$; so it is congruent to 1 or 3 modulo 4.
79. **proof:** Let $n = 2k+1$ be any odd integer. Then $n^2 = 4(k^2+k)+1 \equiv 1 \pmod 4$.
81. **proof:** Because $2a \equiv 0 \pmod p$, $p|2a$. But $p \nmid 2$, so $p|a$; that is, $a \equiv 0 \pmod p$.
83. **proof** (by PMI): The given statement is true when $n = 0$. So assume $P(k)$ is true for an arbitrary integer $k \geq 0$. Then $(k+1)^4 + 2(k+1)^3 + (k+1)^2 = (k^4 + 2k^3 + k^2) + 4(3k^2 + 3k + 1) \equiv 0 \pmod 4$, by the inductive hypothesis, so the statement is true when $n = k+1$. Thus, the result is true for every $n \geq 0$.
85. **proof:** Let m be a positive integer. Because $m(m+1) \equiv 0, 2, 6 \pmod{10}$, $m(m+1)/2 \equiv 0, 1, 3 \pmod 5$. Therefore, every triangular number t_m ends in 0, 1, 3, 5, 6, or 8.
    When $n \geq 2$, the Fermat number $f_n \equiv 7 \pmod{10}$; that is, when $n \geq 2$, f_n ends in 7. Therefore, when $n \geq 2$, no f_n can be a triangular number. The Fermat numbers $f_1 = 5$ and $f_2 = 17$ are not triangular, but $f_0 = 3$ is.
87. **proof** (by PMI): The result is true when $n = 3$. Assume it is true for all integers k, where $3 \leq k < n$: $5^{2^k} \equiv 5^{2^{k-1}} \pmod{10^{k+1}}$; so $5^{2^k} = 5^{2^{k-1}} + t \cdot 10^{k+1}$ for some integer t. Then $5^{2^{k+1}} = 5^{2^k} + t^2 \cdot 10^{2k+2} + 2t \cdot 5^{2^{k-1}} \cdot 10^{k+1} = 5^{2^k} + 10^{k+2}(t^2 \cdot 10^k + 5^{2^{k-1}-1}) \equiv 5^{2^k} \pmod{10^{k+2}}$. Thus, the result follows by the strong version of PMI.
89. $\binom{2p}{r} = \dfrac{(2p)(2p-1)\cdots[2p-(p-1)]}{r!} \equiv 0 \pmod p$, since $0 < r < p$.
91. **proof:** Clearly, p must be odd. By the division algorithm, $p \equiv 0$, 1, or $2 \pmod 3$. Since $p \nmid d$, $d \equiv \pm 1 \pmod 3$.
    case 1 Suppose $p \equiv 0 \pmod 3$. Since p is a prime, $p = 3$.
    case 2 Suppose $p \equiv 1 \pmod 3$. If $d \equiv 1 \pmod 3$, then $p + 2d \equiv 1 + 2 \equiv 0 \pmod 3$; so $p + 2d$ is composite. Hence $d \not\equiv 1 \pmod 3$. If $d \equiv -1 \pmod 3$, then likewise $p + 4d$ is composite. Thus, $p \not\equiv 1 \pmod 3$.
    case 3 Suppose $p \equiv -1 \pmod 3$. If $d \equiv 1 \pmod 3$, then $p + 4d$ is composite. If $d \equiv -1 \pmod 3$, then likewise $p + 2d$ is composite. Thus, $d \not\equiv \pm 1 \pmod 3$; so $p \not\equiv -1 \pmod 3$. Thus $p = 3$.
93. Since $10^{10^{100}} \equiv 3^{10^{100}} \pmod 7$ and $10^{100} \equiv 4^{100} \equiv 4 \pmod 6$, the desired remainder is 4.

Exercises 4.2 (p. 237)

1. yes
3. yes
5. no
7. 3
9. 1
11. 12

13. $ax_0 \equiv b \pmod{m}$. Then $a[x_0 + (m/d)t] \equiv ax_0 + (am/d)t \equiv b + m(a/d)t \equiv b + 0 \equiv b \pmod{m}$.
15. The incongruent solutions are given by $1 + (91/7)t = 1 + 13t$, where $0 \le t \le 6$. They are 0, 14, 27, 40, 53, 66, and 79.
17. The incongruent solutions are given by $1 + (28/7)t = 1 + 4t$, where $0 \le t \le 6$. They are 1, 5, 9, 13, 17, 21, and 25.
19. The incongruent solutions are given by $2 + (1076/4)t = 2 + 269t$, where $0 \le t \le 3$. They are 2, 271, 540, and 809.
21. $3x \equiv 5 \pmod{4}$, so $x \equiv 3 \pmod{4}$; that is, $x = 3 + 4t$. Then $3(3 + 4t) + 4y = 5$, so $y = -1 - 3t$. Thus, the general solution of the LDE is $x = 3 + 4t$, $y = -1 - 3t$.
23. Because $15x \equiv 39 \pmod{21}$, $x \equiv 4 \pmod{7}$. Thus, $x = 4 + 7t$. Then $15(4 + 7t) + 21y = 39$, so $y = -1 - 5t$. Thus, the general solution is $x = 4 + 7t$, $y = -1 - 5t$.
25. $48x \equiv 144 \pmod{84}$, so $4x \equiv 12 \pmod{7}$; that is, $x \equiv 3 \pmod{7}$ and $x = 3 + 7t$. Then $4(3 + 7t) = 12$, so $y = -4t$. Thus, the general solution of the LDE is $x = 3 + 7t$, $y = -4t$.
27. 1, 2, 3, 4
29. 1, 6
31. Because $5x \equiv 3 \pmod{6}$ and $5^{-1} \equiv 5 \pmod{6}$, $5 \cdot 5x \equiv 5 \cdot 3 \pmod{6}$; that is, $x \equiv 3 \pmod{6}$.
33. $19x \equiv 29 \pmod{16}$, so $3x \equiv 13 \pmod{16}$. Because $3^{-1} \equiv 11 \pmod{16}$, $11 \cdot 3x \equiv 11 \cdot 13 \pmod{16}$; that is, $x \equiv 15 \pmod{16}$.
35. Because $ab = ba \equiv 1 \pmod{m}$, a is an inverse of b modulo m.
37. **proof:** Suppose a is invertible modulo m. Then $a^2 \equiv 1 \pmod{m}$; that is, $(a - 1)(a + 1) = a^2 - 1 \equiv 0 \pmod{p}$. Thus, $a \equiv \pm 1 \pmod{p}$. Conversely, let $a \equiv \pm 1 \pmod{p}$. Then $a^2 - 1 = (a - 1)(a + 1) \equiv 0 \pmod{p}$; that is, $a^2 \equiv 1 \pmod{p}$. Thus, a is self-invertible modulo p.
39. Because $19^{10} \equiv 1 \pmod{100}$, $19^{1991} = (19^{10})^{199} \cdot 19 \equiv 1 \cdot 19 \pmod{100}$, the last two-digit number is 19.
41.
$$\begin{aligned}17^{1776} &= (17^5)^{355} \cdot 17 \equiv (-143)^{355} \cdot 17 \equiv -(143^3)^{118} \cdot 143 \cdot 17 \equiv -207^{118} \cdot 431\\ &\equiv -(207^2)^{59} \cdot 431 \equiv -(-151)^{59} \cdot 431 \equiv 151^{59} \cdot 431 \equiv (151^3)^{19} \cdot 151^2 \cdot 431\\ &\equiv (-49)^{19} \cdot 231 \equiv -(49^4)^4 \cdot 49^3 \cdot 231 \equiv -(-199)^4 \cdot 919 \equiv -199^3 \cdot 199 \cdot 919\\ &\equiv -599 \cdot (-119) \equiv 281 \pmod{1000}\end{aligned}$$

43. By Exercise 18 in Section 3.3, 376! has 93 trailing zeros, so the last nonzero digit d in 376! is the ones digit in $376!/10^{93}$. Because d is even, $d = 2, 4, 6$, or 8.

$$\frac{375!}{5^{93}} = \frac{375!}{5^{75} \cdot 75!} \cdot \frac{75!}{5^{15} \cdot 15!} \cdot \frac{15!}{5^3} \equiv (-1)^{75} \cdot (-1)^{15} \cdot (-1)^3 \equiv -1 \equiv 4 \pmod{5}$$

so $\dfrac{376!}{5^{93}} \equiv 4 \pmod{5}$. Therefore,

$$\frac{376!}{10^{93}} \equiv 4 \cdot 2^{-93} \equiv 4 \cdot (2^4)^{-23} \cdot 2^{-1} \equiv 4 \cdot 1 \cdot 3 \equiv 2 \pmod{5}$$

Thus, $d = 2$.

45. By Exercise 20 in Section 3.3, 1010! has 251 trailing zeros, so the last nonzero digit d in 1010! is the ones digit in $1010!/10^{251}$. Because d is even, $d = 2, 4, 6$, or 8.

$$\frac{1010!}{5^{251}} = \frac{1010!}{5^{202} \cdot 202!} \cdot \frac{202!}{5^{40} \cdot 40!} \cdot \frac{40!}{5^{8} \cdot 8!} \cdot \frac{8!}{5}$$
$$\equiv (-1)^{202} \cdot [(-1)^{40} \cdot 1 \cdot 2] \cdot (-1)^{8} \cdot [(-1) \cdot 1 \cdot 2 \cdot 3] \equiv 3 \pmod 5$$

Because $2^{-4} \equiv 1 \pmod 5$ and $2^{-3} \equiv 2 \pmod 5$,

$$\frac{1010!}{10^{251}} = \frac{1010! \cdot 2^{-251}}{5^{251}} \equiv 3 \cdot 2^{-251} \equiv 3 \cdot 2^{-248} \cdot 2^{-3} \equiv 3 \cdot 1 \cdot 2 \equiv 1 \pmod 5$$

so $d = 6$.

47. $x = -3337 + 148t$

Exercises 4.3 (p. 240)

1. We have the sequence 2, 5, 26, 677, ... Since $(x_2 - x_1, 3199) = 7$, $7|3199$. Hence, $3199 = 7 \cdot 457$.
3. We have the sequence 2, 5, 26, 677, 8772, 5156, 1777, 1051, 253, 5372, 8489, 6793, 6517, 7605, 9185, 3690, 2312, ... Since $(x_{16} - x_8, 9773) = 29$, $19|9773$; so $9773 = 29 \cdot 337$.
5. We have 2, 5, 26, 677, 18363, 25060, 10704, 24209, 39638, ... Since $(x_6 - x_3, 39997) = 37$, $37|39997$.
7. We have 3, 8, 3968, 26202, 36295, 25829, 27277, 10534, 13477, 3151, 9544, 14766, 11108, ... Then $(x_{12} - x_6, 39997) = 851$, so $851|39997$.

Review Exercises (p. 241)

1. Thursday
3. 5:30 A.M.
5. $(6, 9) = (3, 9) = 3$, but $6 \not\equiv 3 \pmod 9$.
7. $p = 5$
9. no
11. 5
13. 8
15. The incongruent solutions are $x = 7 + 13t$, where $0 \le t \le 11$. They are 7, 20, 33, 46, 59, 72, 85, 98, 111, 124, 137, and 150.
17. Because $36x \equiv 96 \pmod{156}$, by Exercise 15, $x = 7 + 13t$. Then $y = -1 - 3t$.
19. 1, 17
21. $k! \equiv 0 \pmod{13}$ when $k \ge 13$. So $1! + 2! + \cdots + 300! \equiv 1! + 2! + \cdots + 12! \equiv 9 \pmod{13}$
23. 0 or 1
25. $5^{103} = 5^{64} \cdot 5^{32} \cdot 5^{4} \cdot 5^{2} \cdot 5 \equiv 1 \cdot 1 \cdot 1 \cdot (-1) \cdot 5 \equiv -5 \equiv 8 \pmod{13}$
27. $3^{100} = (3^8)^{12} \cdot 3^4 \equiv 9^{12} \cdot 9^2 = (9^4)^3 \cdot 9^2 \equiv 9^5 \equiv 81 \pmod{91}$
29. $2^{100} = (2^5)^{20} \equiv 1^{20} \equiv 1 \pmod{11}$ and $3^{123} = (3^5)^{24} \cdot 3^3 \equiv 1^{24} \cdot 5 \equiv 5 \pmod{11}$. Therefore, $2^{10} + 3^{123} \equiv 1 + 5 \equiv 6 \pmod{11}$.
31. Since $13 \equiv -4 \pmod{17}$, $(-4)^2 \equiv -1 \pmod{17}$, $(-1)^3 \equiv -1 \pmod{17}$, and $(-1)^4 \equiv 1 \pmod{17}$, the desired remainder is 1.
33. $3^{3434} = (3^{10})^{343} \cdot 3^4 \equiv 49^{343} \cdot 81 \equiv (49^2)^{171} \cdot 49 \cdot 81 \equiv 1^{171} \cdot 69 \equiv 69 \pmod{100}$
35. $3^{3003} = (3^{10})^{30} \cdot 3^3 \equiv 49^{300} \cdot 27 \equiv (49^{10})^{30} \cdot 27 \equiv 1^{30} \cdot 27 \equiv 027 \pmod{1000}$

37. $1024 \equiv 4 \pmod{10}$. Let $n \geq 1$. If n is odd, then $4^n \equiv 4 \pmod{10}$; otherwise, $4^n \equiv 6 \pmod{10}$. Since $k = 1025^n$ is always odd, $4^k \equiv 4 \pmod{10}$, so the desired ones digit is 4.
39. When $k \geq 4$, $k! \equiv 0 \pmod{12}$. Therefore, $1! + 2! + 3! + \cdots + 100! \equiv 1! + 2! + 3! \equiv 9 \pmod{12}$. Thus, the desired remainder is 9.
41. When $k \geq 4$, $k! \equiv 0 \pmod{24}$. Using a 24-hour clock, $15 + \sum_{k=1}^{100} k! \equiv 15 + 1! + 2! + 3! \equiv 0 \pmod{24}$. So the time will be midnight.
43. Let $q = p + 2$. When $p = 3$, $q = 5$ and $pq - 2 = 13$; they are all primes. Suppose $p > 3$. Since p is prime, $p \equiv \pm 1 \pmod 3$. If $p \equiv 1 \pmod 3$, then $q \equiv 0 \pmod 3$, which is impossible. If $p \equiv -1 \pmod 3$, then $q \equiv 1 \pmod 3$; then $pq - 2 \equiv 0 \pmod 3$, which is also impossible. Thus, $p = 3$ is the only such value.
45. Let $n = 7429$. With $x_0 = 2$ and $f(x) = x_2 + 1$, we have the sequence 5, 26, 677, 5161, 2957, 7346, 6890, 791, 1646, ... Then $(x_2 - x_1, n) = (x_4 - x_2, n) = 1$, but $(x_6 - x_3, n) = 19$. So $19|7429$. Thus, $7429 = 19 \cdot 391 = 19 \cdot 17 \cdot 23 = 17 \cdot 19 \cdot 23$.
47. **proof:** Because $a^2 \equiv b^2 \pmod p$, $p|(a-b)(a+b)$. So either $p|a-b$ or $p|a+b$; that is, either $a \equiv b \pmod p$ or $a \equiv -b \pmod p$; that is, $a \equiv \pm b \pmod p$.
49. **proof:** Because $n^3 - n = (n-1)n(n+1)$ is the product of three consecutive integers, it is divisible by 3. So $n^3 - n \equiv 0 \pmod 3$; that is, $n^3 \equiv n \pmod 3$.
51. **proof** (by PMI): Because $1 + 0 \equiv 1 \pmod{25}$, the statement is true when $n = 0$. Assume it is true for an arbitrary integer $k \geq 0$: $4^{2k} + 10k \equiv 1 \pmod{25}$. Then,

$$\begin{aligned} 4^{2(k+1)} + 10(k+1) &= 16 \cdot 4^{2k} + 10k + 10 \equiv 16(1 - 10k) + 10k + 10 \\ &= 26 - 150k \equiv 1 \pmod{25} \end{aligned}$$

Thus, the result follows by PMI.
53. **proof:** Because $a \equiv b \pmod p$, $p|a-b$. Similarly, $q|a-b$. So $[p,q]|a-b$; that is, $pq|a-b$. In other words, $a \equiv b \pmod{pq}$.
55. **proof** (by U. Dudley): Let $N = 12 \cdot 900^n + 1$, where $n \geq 1$. By the division algorithm, let $n = 3k + r$, where $0 \leq r < 3$. When $r = 0$, $N \equiv 0 \pmod{13}$; when $r = 1$, $N \equiv 0 \pmod 7$; and when $r = 2$, $N \equiv 0 \pmod{19}$. Thus, N is composite when $n \geq 1$. So $n = 0$ and hence $N = 13$, a twin prime.
57. **proof** (by R. S. Stacy): Let $N = 2^p + 3^p$. Since p is odd, $N = 5S$, where

$$S = \sum_{k=0}^{p-1} 2^{p-1-k} 3^k \equiv \sum_{k=0}^{p-1} 2^{p-1-k}(-2)^k \equiv \sum_{k=0}^{p-1} 2^{p-1} \equiv p2^{p-1} \pmod 5$$

Suppose $p \neq 5$. Then $p \equiv \pm 1, \pm 2 \pmod 5$, so $S \equiv k \not\equiv 0 \pmod 5$. Thus $5 \nmid S$, so $5^2 \nmid N$. Consequently, N is not a perfect power. When $p = 5$, $N = 275$, again not a perfect power.

Supplementary Exercises (p. 243)

1. Because $1287xy6 \equiv 0 \pmod{72}$,

$$1287xy6 \equiv 0 \pmod 8 \quad \text{and} \tag{1}$$

$$1287xy6 \equiv 0 \pmod 9 \tag{2}$$

Congruence (1) implies $y+1 \equiv 0 \pmod 2$, so y is odd. Thus, $y = 1, 3, 5, 7$, or 9. The two congruences yield

$$2x + y + 3 \equiv 0 \pmod 4 \quad \text{and} \tag{3}$$

$$x + y + 6 \equiv 0 \pmod 9 \tag{4}$$

If $y = 1$, then $2x+4 \equiv 0 \pmod 4$; that is, $x = 0, 2, 4, 6$, or 8, of which only 2 satisfies (4), so $(2, 1)$ is a solution. If $y = 3$, then $2x + 6 \equiv 0 \pmod 4$ by (3), so $x + 1 \equiv 0 \pmod 2$. Thus, $x = 1, 3, 5, 7$, or 9, of which only $x = 9$ satisfies (4). The corresponding solution is $(9, 3)$. Similarly, we get one more solution, $(5, 7)$. Thus, there are three solutions: $(2, 1)$, $(5, 7)$, and $(9, 3)$.

3. Let k be the largest factor of A_n. Since $A_1 = -1995 = -3 \cdot 5 \cdot 7 \cdot 19$, $k|1995$. Because $A_n \equiv 2^n - 2^n - 1^n + 1^n \equiv 0 \pmod 3$, $3|A_n$. Likewise, $5|A_n$, $7|A_n$, and $19|A_n$. Therefore, $[3, 5, 7, 19] = 1995|A_n$. Thus, $k = 1995$.
5. Because

$$\sum_{i=1001}^{2000} i^{1999} = \sum_{j=1}^{1000} (j+1000)^{1999} \equiv \sum_{j=1}^{1000} (j-1001)^{1999} \equiv \sum_{i=1}^{1000} (-i)^{1999} \pmod{2001}$$

it follows that

$$\sum_{i=1}^{2000} i^{1999} = \sum_{i=1}^{1000} i^{1999} + \sum_{i=1001}^{2000} i^{1999} \equiv \sum_{i=1}^{1000} i^{1999} + \sum_{i=1}^{1000} (-i)^{1999} \equiv 0 \pmod{2001}$$

7. **proof:** Let d be the ones digit in a. Then $a \equiv d \pmod{10}$ and $a^2 \equiv d^2 \pmod{10}$. Because a is automorphic, $d^2 \equiv d \pmod{10}$. The only solutions of this congruence are 0, 1, 5 and 6.
9. (by B. D. Beasly) Notice that x, y, and z are relatively prime. Thus, $2 \le x < y < z$. The given congruences can be combined to get $xy + yz + zx - 1 \equiv 0 \pmod{x, y, \text{ and } z}$. Since x, y, and z are relatively prime, this implies $xy + yz + zx \equiv 1 \pmod{xyz}$. Therefore, $xy + yz + zx = 1 + k(xyz)$ for some positive integer k. This yields $1/x + 1/y + 1/z = k + 1/xyz > 1$. Because $2 \le x < y < z$, this implies $(x, y, z) = (2, 3, 4)$ or $(2, 3, 5)$. But $(2, 4) \neq 1$, so the unique solution of the given congruences is $(2, 3, 5)$.
11. (by S. Fernandez) Let (x, y, z) be a solution. Then, by symmetry, a permutation of (x, y, z) is also a solution. The entries are distinct and are ≥ 3. Since $xy \bmod z = 2$, $xy - 2 = kz$ for some positive integer k. Since $y < z$, $ky < xy - 2$, so $k < x$.
   Since $zx \bmod y = 2$, $-2(x+k) \equiv (xy-2)x - 2k \equiv k(zx-2) \equiv 0 \pmod y$. So $y|2(x+k)$; that is, $2(x+k) = \ell y$ for some positive integer ℓ. Then $\ell y = 2(x+k) < 4x < 4y$; so $1 \le \ell \le 3$.
   Since $yz \bmod x = 2$, $-2(2+\ell) \equiv 2(xy-2) - 2\ell \equiv 2kz - 2\ell \equiv 2(x+k)z - 2\ell \equiv \ell(yz-2) \equiv 0 \pmod x$. Therefore, $x|2(2+\ell)$.
   Thus, $y = \dfrac{2(x+k)}{\ell}$ and $z = \dfrac{xy-2}{k}$, where $1 \le \ell \le 3 \le x$, $1 \le k < x$, $x|2(2+\ell)$, and $\ell|2(x+k)$. Of the 22 triplets (ℓ, x, k) that satisfy these conditions, only six yield valid solutions: $(1, 3, 1)$, $(1, 3, 2)$, $(2, 4, 1)$, $(2, 4, 2)$, $(1, 6, 1)$, and $(1, 6, 5)$. The corresponding solutions (x, y, z) are $(3, 8, 22)$, $(3, 10, 14)$, $(4, 5, 18)$, $(4, 6, 11)$, $(6, 14, 82)$, and $(6, 22, 26)$.

Chapter 5 Congruence Applications

Exercises 5.1 (p. 252)

1. F
3. T
5. T
7. By 2 and 4
9. By 2, 4, and 8.
11. Not by 3 or 9.
13. Not by 3 or 9.
15. yes
17. yes
19. no
21. yes
23. definitely wrong
25. definitely wrong
27. definitely wrong
29. 8
31. 1
33. $4(10^3)+a(10^2)+b(10)+8\equiv 0 \pmod 8$ yields $2a+b\equiv 0 \pmod 4$. Likewise, $a+b\equiv 6 \pmod 9$. These two congruences yield $a=2$ and $b=4$. Thus, the desired number is 4248.
35. Let n be the number with the given properties. By the division algorithm, $n=(i+1)q_i+i$ for some q_i. Then $n+1=(i+1)q_i+(i+1)$, so $(i+1)|(n+1)$; that is, 2, 3, ..., and 9 are factors of $n+1$. Thus, $n+1=[2,3,\dots,9,10]=2520$, so $n=2519$.
37. Let $n=n_k.\ 10^k+\cdots+n_1\cdot 10+n_0$. Since $10^3\equiv 1 \pmod{37}$, $n\equiv n_2\cdot 10^2+n_1\cdot 10+n_0\equiv n_2n_1n_0 \pmod{37}$. Thus, n is divisible by 37 if and only if the three-digit number $n_2n_1n_0$ is divisible by 37.
39. Definitely wrong.
41. $16429058\equiv 8 \pmod 9$, so the digital root is 8.
43. $1776^{1776}\equiv 3^{1776}\equiv (3^2)^{888}\equiv 0^{888}\equiv 0 \pmod 9$, so the digital root is 9.
45. **proof:** Let d be the digital root of n, so $n\equiv d\equiv 0 \pmod 9$. Then $mn\equiv 0 \pmod 9$, so the digital root of mn is also 9.
47. Let n be any integer and d its digital root. Then $n^2\equiv d^2 \pmod 9$. Since $0^2\equiv 0 \pmod 9$, $(\pm 1)^2\equiv 1 \pmod 9$, $(\pm 2)^2\equiv 4 \pmod 9$, $(\pm 3)^2\equiv 0 \pmod 9$, and $(\pm 4)^2\equiv 7 \pmod 9$, it follows that the digital root of a square must be 1, 4, 7, or 9.

    Since the digital root of 61,194,858,376 is 6, it follows that it cannot be a square.
49. **proof:** Let N be a triangular number. Then $N=n(n+1)/2$ for some positive integer n.

    case 1 Let n be even, say, $n=2k$. Then $N=k(2k+1)$. Let d denote the units digit in k. Then $N\equiv d(2d+1) \pmod{10}$. Because $0\le d\le 9$, the residues of $d(2d+1)$ modulo 10 are 0, 3, 0, 1, 6, 5, 8, 5, 6, and 1, so the possible units digits are 0, 1, 3, 5, 6, and 8.

    case 2 Let n be odd, say, $n=2k+1$. Then $N=(k+1)(2k+1)$. Let d denote the units digit in k. Then $N\equiv (d+1)(2d+1) \pmod{10}$. Since $0\le d\le 9$, the residues of $(d+1)(2d+1)$ modulo 10 are 3, 0, 1, 6, 5, 8, 5, 6, 1, and 0, so the possible units digits are 0, 1, 3, 5, 6, and 8.
51. **proof:** Because $R_n=(10^n-1)/9$, $d|R_n$ implies $10^n\equiv 1 \pmod d$. Because $d|a_{n-1}\dots a_1a_0$, $a_{n-1}10^{n-1}+a_{n-2}10^{n-2}+\cdots+a_1 10+a_0\equiv 0 \pmod d$. Therefore, $a_{n-1}10^n+a_{n-2}10^{n-1}+\cdots+a_1 10^2+a_0 10\equiv 0 \pmod d$; that is, $a_{n-2}10^{n-1}+a_{n-3}10^{n-2}+\cdots+a_0 10+a_{n-1}\equiv 0 \pmod d$, so $d|a_{n-2}\dots a_0a_{n-1}$. Similarly, d divides the other cyclic permutations also.
53. **proof:** Let $N=2^{2000}+2^{2001}+2^{2003}+\cdots+2^{2007}$. Since $2^3\equiv -1 \pmod 9$, $N=(2^3)^{666}\cdot 2^2+(2^3)^{667}+(2^3)^{667}+(2^3)^{667}\cdot 2^2+(2^3)^{668}+(2^3)^{668}\cdot 2+(2^3)^{668}\cdot 2^2+(2^3)^{669}\equiv(-1)4+(-1)+(-1)2+(-1)4+1+2+4-1\equiv 3 \pmod 9$. So the digital root of N is 3. But the digital root of a square is 1, 4, 7, or 9. Consequently, N cannot be a square.

Exercises 5.2 (p. 257)

1. 

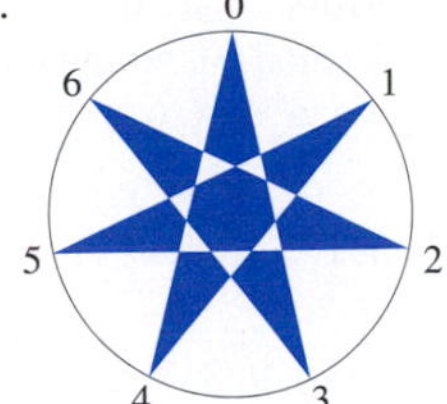

3.

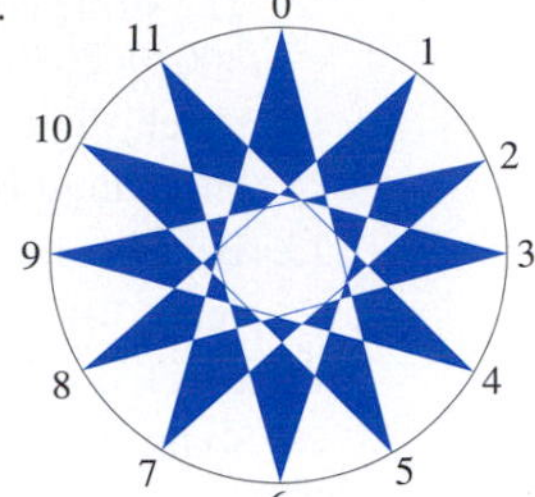

5.

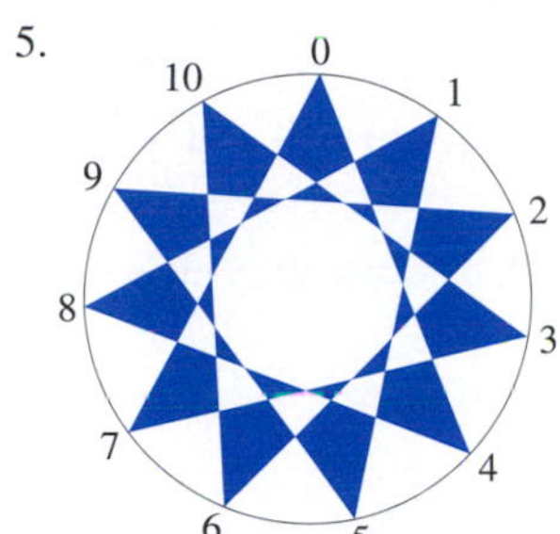

7.

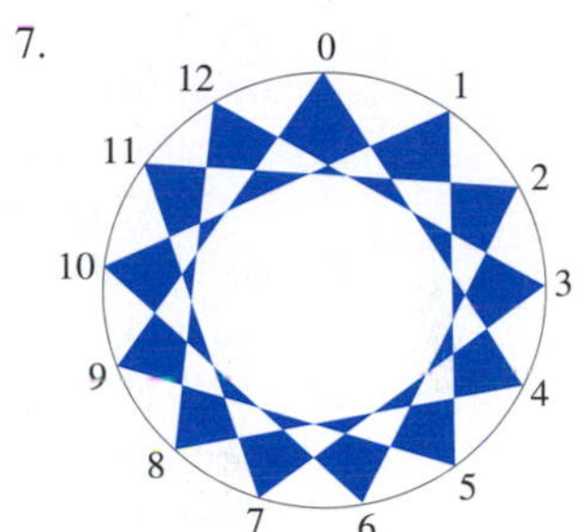

9.

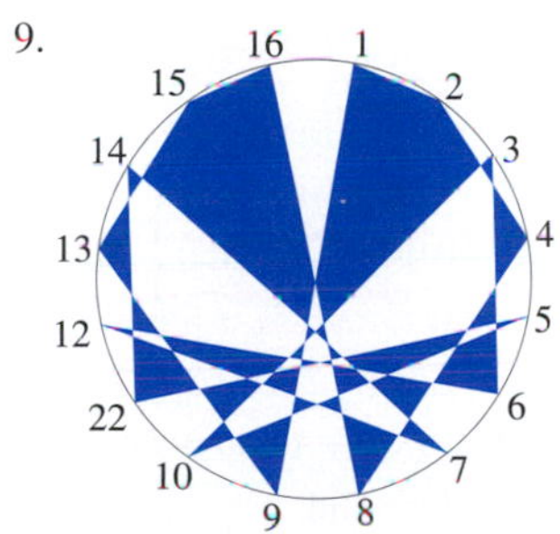

11.

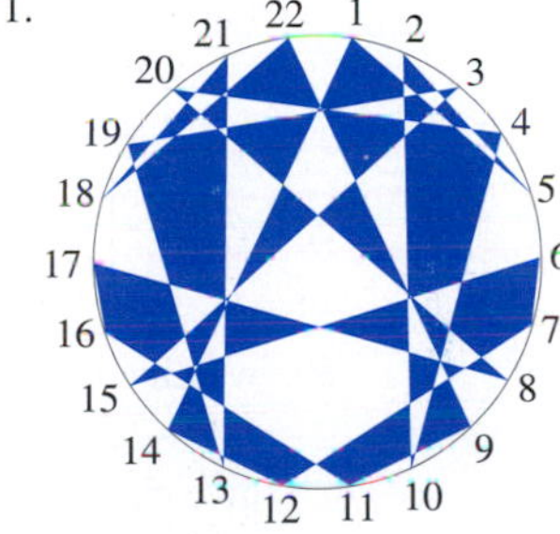

13.

15.

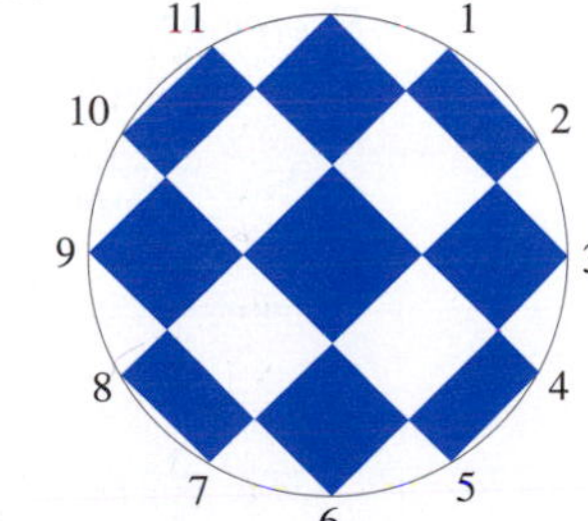

Exercises 5.3 (p. 272)

1. 0
3. 0
5. The string contains an even number of errors.
7. 0
9. $300706202013 = 6 \pmod 9$ and $-6 \equiv 3 \pmod 9$, so the check digit is 3.
11. check digit $\equiv (7,9,0,0,2,9,6,6) \cdot (7,3,9,7,3,9,7,3) \equiv 223 \equiv 3 \pmod{10}$, so the check digit is 3.
13. Because the check digit is 3 and $(3,3,1,3,d,4,4,7) \cdot (7,3,9,7,3,9,7,3) \equiv 3d+5 \pmod{10}$, $3d+5 \equiv 2 \pmod{10}$, so $d = 6$.
15. check digit $\equiv -(0,8,7,6,2,0,3,2,1) \cdot (10,9,8,7,6,5,4,3,2) \equiv -202 \equiv 7 \pmod{11}$, so the check digit is 7.
17. yes, because check digit $= 1 \equiv -(0,2,0,1,5,7,6,0,3) \cdot (10,9,8,7,6,5,4,3,2) = -120 \equiv 1 \pmod{11}$.
19. check digit $\equiv -(0,1,6,0,0,0,6,6,6,1,0) \cdot (3,1,3,1,3,1,3,1,3,1,3) \equiv -62 \equiv 8 \pmod{10}$, so it is 8.
21. no, since check digit $= 8 \not\equiv -(0,7,0,7,3,4,0,6,3,1,0) \cdot (3,1,3,1,3,1,3,1,3,1,3) = -43 \equiv 7 \pmod{10}$.
23. no, because $d_{10} \equiv -(0,6,5,4,6,3,3,0,7) \cdot (3,7,6,1,8,9,4,5,2) = -177 \equiv 10 \pmod{11}$, but d_{10} is given to be 0.
25. check digit $\equiv -(1,2,3,4,5,6,7) \cdot (8,7,6,5,4,3,2) \equiv -112 \equiv 9 \pmod{11}$, so the check digit is 9.
27. check digit $\equiv (3,1,5,7,4,0,6) \cdot (7,3,1,7,3,1,7) \equiv 132 \equiv 2 \pmod{10}$. Therefore, check digit $= 2$.
29. check digit $\equiv (9,8,7,6,5,4,3,2) \cdot (1,4,9,2,1,9,9,4) = 192 \equiv 2 \pmod{10}$, so it is 2.
31. Let $x \equiv -(0,2,4,3,5,7,9) \cdot (2,7,6,5,4,3,2) = -112 \equiv 9 \pmod{11}$. So the check digit is $x = 9$.
33. Using Exercise 31, the check digit is still 9.
35. $d_7 \equiv -\left\{\left[\sum_{i=1}^{3}(2d_{2i} \text{ div } 10) + \lfloor 2d_i/10 \rfloor\right] + \sum_{i=1}^{3} d_{2i-1}\right\} \pmod{10}$
37. 6
39. weighted sum $= 8\cdot 2 + 7\cdot 8 + 6\cdot 7 + 5\cdot 5 + 4\cdot 2 + 3\cdot 1 + 2\cdot 6 + 10\cdot 5 + 9\cdot 1 + 8\cdot 8 + 7\cdot 5 + 6\cdot 4 + 5\cdot 1 + 4\cdot 8 + 3\cdot 7 + 2\cdot 3 \equiv 1 \pmod{11}$. So check digit $= 1$, same as the give check digit. Therefore, the VIN is valid.
41. yes

Exercises 5.5 (p. 282)

1. $g_n = 0 + [1 + 2 + \cdots + (n-1)] = n(n-1)/2$
3.

Round \ Team	1	2	3	4	5	6
1	5	4	6	2	1	3
2	6	5	4	3	2	1
3	2	1	5	6	3	4
4	3	6	1	5	4	2
5	4	3	2	1	6	5

5.

Round \ Team	1	2	3	4	5	6	7	8	9
1	9	8	7	6	bye	4	3	2	1
2	bye	9	8	7	6	5	4	3	2
3	2	1	9	8	7	bye	5	4	3
4	3	bye	1	9	8	7	6	5	4
5	4	3	2	1	9	8	bye	6	5
6	5	4	bye	2	1	9	8	7	6
7	6	5	4	3	2	1	9	bye	7
8	7	6	5	bye	3	2	1	9	8
9	8	7	6	5	4	3	2	1	bye

7. No, because the pairing is cyclic. For example, in round 1, team 1 plays team 2, but team 2 plays team 3.

Exercises 5.6 (p. 287)

1. Thursday
3. Monday
5. T
7. T
9. Because there are 20 years in the range 2000–2019, of which five are leap years, January 1, 2020, falls on day $20 + 5 \equiv 4 \pmod 7$, that is, Wednesday.
11. Because $d \equiv 17 + \lfloor 2.6 \times 11 - 0.2 \rfloor - 34 + 5 + \lfloor 17/4 \rfloor + \lfloor 5/4 \rfloor \equiv 0 \pmod 7$, the desired day is Sunday.
13. $d \equiv 4 + \lfloor 2.6 \times 5 - 0.2 \rfloor - 34 + 76 + \lfloor 17/4 \rfloor + \lfloor 76/4 \rfloor \equiv 4 \pmod 7$, so the desired day is Thursday.
15. $d \equiv 7 + \lfloor 2.6 \times 1 - 0.2 \rfloor - 36 + 76 + \lfloor 18/4 \rfloor + \lfloor 76/4 \rfloor \equiv 2 \pmod 7$. So the day is Tuesday.
17. $d \equiv 16 + \lfloor 2.6 \times 5 - 0.2 \rfloor - 38 + 45 + \lfloor 19/4 \rfloor + \lfloor 45/4 \rfloor \equiv 1 \pmod 7$. Thus the day is Monday.
19. $d \equiv 12 + \lfloor 2.6 \times 2 - 0.2 \rfloor - 38 + 61 + \lfloor 19/4 \rfloor + \lfloor 61/4 \rfloor \equiv 3 \pmod 7$, so the day is Wednesday.
21. $x \equiv 2000 + \lfloor 1999/4 \rfloor - \lfloor 1999/100 \rfloor + \lfloor 1999/400 \rfloor \equiv 6 \pmod 7$, so it falls on a Saturday.
23. $x \equiv 2076 + \lfloor 2075/4 \rfloor - \lfloor 2075/100 \rfloor + \lfloor 2075/400 \rfloor \equiv 3 \pmod 7$. So Jan. 1, 2076 is a Wednesday.
25. one; October 13
27. two; March 13, November 13
29. April 7
31. April 26
33. $d \equiv \lfloor 23/9 \rfloor + 17 + 4 + 1706 + 426 - 17 + 4 - 0 \equiv 0 \pmod 7$, so the day is Sunday.
35. $d \equiv \lfloor 23 \times 7/9 \rfloor + 4 + 4 + 1776 + 444 - 17 + 4 - 2 \equiv 4 \pmod 7$, thus the day is Thursday.
37. $d \equiv \lfloor 23 \times 3/9 \rfloor + 7 + 4 + 1876 + 469 - 18 + 4 - 2 \equiv 2 \pmod 7$. Thus the day is Tuesday.
39. $d \equiv \lfloor 23 \times 7/9 \rfloor + 16 + 4 + 1945 + 486 - 19 + 4 - 2 \equiv 1 \pmod 7$. So it is Monday.
41. $d \equiv \lfloor 23 \times 4/9 \rfloor + 12 + 4 + 1961 + 490 - 19 + 4 - 2 \equiv 3 \pmod 7$. Thus it is Wednesday.
43. **proof:** Because $D < 100$, $D/400 < 0.25$. Let $C = 4q + r$, where $0 \le r \le 3$, so $C/4 = q + r/4 < q + 0.75$. Then $C/4 + D/400 < q + 1$, so $\lfloor C/4 + D/400 \rfloor = q = \lfloor C/4 \rfloor$.

Review Exercises (p. 289)

1. 3
3. yes
5. no
7. 1, 3, 5, 7, or 9
9. 9
11. definitely wrong
13. definitely wrong
15. 0
17. 1
19. 9
21. $5555^{1122} \equiv 2^{1122} \equiv (2^3)^{374} \equiv (-1)^{374} \equiv 1 \pmod 9$, so the digital root is 1.
23. The digital root of the given number is 3. But the digital root of a square is 1, 4, 7, or 9, so it cannot be a square.
25. The digital root of the given number is 8. Because the digital root of a cube is 1, 8, or 9, it can be a cube. (In fact, it is not a cube.)
27. $d \equiv -(0, 1, 3, 1, 3, 0, 0, 3, 6, 2, 2) \cdot (3, 1, 3, 1, 3, 1, 3, 1, 3, 1, 3) \equiv -9 \equiv 1 \pmod{10}$, so the check digit is 1. But the given check digit is 4, so it is an invalid UPC number.
29. Since $d \equiv -(7, 5, 1, 7, 4, 6, 0, 3, 3, 6, 1) \cdot (3, 1, 3, 1, 3, 1, 3, 1, 3, 1, 3) \equiv -5 \equiv 5 \pmod{10}$, the UPC number is 7-51746-03361-5.
31. check digit $\equiv -(3, 3, 0, 1, 4, 0, 0, 0, 9, 9, 0, 7, 3) \cdot (2, 1, 2, 1, 2, 1, 2, 1, 2, 1, 2, 1, 2) - 1 \equiv -(6+3+1+8+18+9+0+7+6) - 1 \equiv -9 \equiv 1 \pmod{10}$, so the check digit is 1.
33. check digit $\equiv -[\rho(10) + \rho(4) + \rho(0) + \rho(2) + \rho(8) + \rho(10) + \rho(14) + \rho(12) + (3 + 9+4+9+2+3+3)] \equiv -[(1+4+0+2+8+1+5+3)+33] \equiv -57 \equiv 3 \pmod{10}$, so it is 3.
35. $d \equiv 1 + \lfloor 2.6 \times 12 - 0.2 \rfloor - 34 + 90 + \lfloor 17/4 \rfloor + \lfloor 90/4 \rfloor \equiv 2 \pmod 7$. So the desired day is a Tuesday.
37. 7
39. $f(1) \star f^2(2) \star f^3(6) \star f^4(2) \star f^5(4) \star f^6(4) \star f^7(1) \star f^8(2) \star f^9(9) \star f^{10}(8) \star s_{11} = 0$; this yields $s_{11} = 1$.

Supplementary Exercises (p. 291)

1. **proof:** Because $0 < p(n) \le 9$, $\rho(\rho(n)) = \rho(n)$.
3. **proof:** Let $m = \sum_i m_i 10^i$ and $n = \sum_i n_i 10^i$. Then $mn = \left(\sum_i m_i 10^i\right)\left(\sum_i n_i 10^i\right)$, so $mn \equiv \left(\sum_i m_i\right)\left(\sum_i n_i\right) \pmod 9$. That is, $\rho(mn) = \rho(m)\rho(n)$.
5. **proof:** Because $d|b-1$, $b \equiv 1 \pmod d$. Then $n = \sum_i n_i b^i \equiv \sum_i n_i \pmod d$. Thus, n is divisible by d if and only if $\sum_i n_i$ is divisible by d.
7. Using $a = 1$, $3|n$ if and only if $3|q+r$.
9. With $a = 1$, $9|n$ if and only if $9|q+r$.
11. With $a = 4$, $13|n$ if and only if $3|q+4r$.
13. Using $a = 2$, $3|n$ if and only if $3|q+2r$.
15. By a repeated application of the test in Exercise 10, 11|28303.
17. By a repeated application of the test in Exercise 13, 19|61731.
19. **proof:** Let $N = 100a + 10b + c \equiv 2a + 3b + c \equiv 0 \pmod 7$. Then $N' = (100c + 10b + a) - (c - a) \equiv 2a + 3b + c \equiv 0 \pmod 7$.
21. **proof:** If $p = 2$, then $n = 6$, so its digital root is 6. Assume $p > 2$. Then p is odd and $2^{p-1} \equiv (-1)^{p-1} \equiv 1 \pmod 3$, so $2^{p-1} = 3k + 1$ for some integer k. Then $n = (3k + 1)(6k + 1) = 18k^2 + 9k + 1 \equiv 1 \pmod 9$, so the digital root of n is 1.

Chapter 6 Systems of Linear Congruences

Exercises 6.1 (p. 302)

1. yes
3. no
5. $x = 2 + 5t_1$. Then $2 + 5t_1 \equiv 3 \pmod 7$; that is, $t_1 \equiv 3 \pmod 7$. So $t_1 = 3 + 7t$. Thus, $x = 17 + 35t$.
7. $x \equiv 1 \pmod 2$ yields $x = 1 + 2t_1$. Then $1 + 2t_1 \equiv 2 \pmod 3$, so $t_1 \equiv 2 \pmod 3$; that is, $t_1 = 2 + 3t_2$. Therefore, $x = 5 + 6t_2$. Then $5 + 6t_2 \equiv 3 \pmod 5$; that is, $t_2 \equiv 3 \pmod 5$. So $t_2 = 3 + 5t$. Thus, $x = 23 + 30t$.
9. $x \equiv 1 \pmod 3$ yields $x = 1 + 3t_1$. Then $1 + 3t_1 \equiv 3 \pmod 4$; that is, $t_1 = 2 + 4t_2$, so $x = 7 + 12t_2$. Then $7 + 12t_2 \equiv 4 \pmod 7$, so $t_2 = 5 + 7t_3$. Therefore, $x = 67 + 84t_3$. Then $67 + 84t_3 \equiv 7 \pmod{11}$; that is, $t_3 = 4 + 11t$. Thus, $x = 403 + 924t$.
11. $M_1 = 7$ and $M_2 = 5$. $M_1y_1 \equiv 1 \pmod{m_1}$ yields $7y_1 \equiv 1 \pmod 5$; that is, $y_1 \equiv 3 \pmod 5$. Similarly, $y_2 \equiv 3 \pmod 7$. Thus, $x \equiv \sum_i a_iM_iy_i \equiv 2 \cdot 7 \cdot 3 + 3 \cdot 5 \cdot 3 \equiv 17 \pmod{35}$. Thus, $x = 17 + 35t$.
13. $M_1 = 15$, $M_2 = 10$, $M_3 = 6$, $y_1 \equiv 1 \pmod 2$, $y_2 \equiv 1 \pmod 3$, and $y_3 \equiv 1 \pmod 5$. Then $x \equiv 1 \cdot 15 \cdot 1 + 2 \cdot 10 \cdot 1 + 3 \cdot 6 \cdot 1 \equiv 23 \pmod{30}$, so $x = 23 + 30t$.
15. $M_1 = 308$, $M_2 = 231$, $M_3 = 132$, $M_4 = 84$, $y_1 \equiv 2 \pmod 3$, $y_2 \equiv 3 \pmod 4$, $y_3 \equiv 6 \pmod 7$, and $y_4 \equiv 8 \pmod{11}$. Then $x \equiv 1 \cdot 308 \cdot 2 + 3 \cdot 231 \cdot 3 + 4 \cdot 132 \cdot 6 + 7 \cdot 84 \cdot 8 \equiv 403 \pmod{924}$, so $x = 403 + 924t$.
17. Because $x = 2 + 4t$, $2 + 4t \equiv 3 \pmod 6$; that is, $4t \equiv 1 \pmod 6$ which is not solvable because $(4, 6) \neq 1$.
19. Let x be the desired integer. Then $x \equiv 3 \pmod 7$, $x \equiv 4 \pmod 9$, and $x \equiv 8 \pmod{11}$. Using the CRT with $M_1 = 99$, $M_2 = 77$, $M_3 = 63$, $y_1 \equiv 1 \pmod 7$, $y_2 \equiv 2 \pmod 9$, and $y_3 \equiv 7 \pmod{11}$, $x \equiv 3 \cdot 99 \cdot 1 + 4 \cdot 77 \cdot 2 + 8 \cdot 63 \cdot 7 \equiv 283 \pmod{693}$, so $x = 283$.
21. Let x be the desired integer. Then $x \equiv 0 \pmod 7$, $x \equiv 2 \pmod 5$, $x \equiv 3 \pmod 6$, and $x \equiv 5 \pmod{11}$. Using the CRT with $M_2 = 462$, $M_3 = 385$, $M_4 = 210$, $y_2 \equiv 3 \pmod 7$, $y_3 \equiv 1 \pmod 6$, and $y_4 \equiv 1 \pmod{11}$, $x \equiv 0 \cdot M_1 \cdot y_1 + 2 \cdot 462 \cdot 3 + 3 \cdot 385 \cdot 1 + 5 \cdot 210 \cdot 1 \equiv 357 \pmod{2310}$, so $x = 357$.
23. We have $n \equiv 0 \pmod 2$, $n \equiv 2 \pmod 3$, $n \equiv 3 \pmod 5$, $n \equiv 4 \pmod 7$, and $n \equiv 7 \pmod{11}$. Using the CRT with $M_2 = 770$, $M_3 = 462$, $M_4 = 330$, $M_5 = 210$, $y_2 \equiv 2 \pmod 3$, $y_3 \equiv 3 \pmod 5$, $y_4 \equiv 1 \pmod 7$, and $y_5 \equiv 1 \pmod{11}$, $n \equiv 0 \cdot M_1 \cdot y_1 + 2 \cdot 770 \cdot 2 + 3 \cdot 462 \cdot 3 + 4 \cdot 330 \cdot 1 + 7 \cdot 210 \cdot 1 \equiv 788 \pmod{2310}$, so $n = 788$.
25. $n \equiv 0 \pmod 9$, $n \equiv -1 \pmod{16}$, and $n \equiv -2 \pmod{25}$. Using the CRT with $M_2 = 225$, $M_3 = 144$, $y_2 \equiv 1 \pmod{16}$, and $y_3 \equiv 4 \pmod{25}$, $n \equiv 0 \cdot M_1 \cdot y_1 + (-1) \cdot 225 \cdot 1 + (-2) \cdot 144 \cdot 4 \equiv 2223 \pmod{3600}$, so $n = 2223$.
27. Let n be an integer that leaves the given remainders. Then $n \equiv 0 \pmod 3$, $n \equiv 2 \pmod 5$, $n \equiv 3 \pmod 7$, and $n \equiv 5 \pmod{13}$. Using the CRT with $M_2 = 273$, $M_3 = 195$, $M_4 = 105$, $y_2 \equiv 2 \pmod 5$, $y_3 \equiv -1 \pmod 7$, and $y_4 \equiv 1 \pmod{13}$, $n \equiv 0 \cdot M_1 \cdot y_1 + 2 \cdot 273 \cdot 2 + 3 \cdot 195 \cdot (-1) + 5 \cdot 105 \cdot 1 \equiv 1032 \pmod{1365}$, so $n = 1032 + 1365t$. So the largest such integer < 6000 is $n = 5127$.
29. Because $4^3 | n + 2$, $2^3 | n + 2$. Thus $2^3 | n$ and $2^3 | n + 2$; so $2^3 | 2$, which is a contradiction. Thus, no such integer exists.

Exercises 6.2 (p. 306)

1. yes, because $(10, 5) = 5|(2 - 5)$
3. no, because $(9, 18) = 9 \nmid (4 + 2)$
5. yes
7. yes
9. $66 + [8, 9, 10]t = 66 + 360t$
11. Because $(12, 15) = 3$ and $3|(10 - 4)$, the system is solvable. $x \equiv 10 \pmod{12}$ yields $x = 10 + 12t_1 \equiv 4 \pmod 5$; that is, $t_1 = 2 + 5t$. Thus, $x = 10 + 12(2 + 5t) = 34 + 60t$.
13. By Theorem 6.3, the linear system is solvable. Since $x = 1 + 10t_1 \equiv 5 \pmod{12}$, $t_1 = 4 + 6t$, so $x = 1 + 10(4 + 6t) = 41 + 60t$. But $41 + 60t \equiv -4 \pmod{15}$ is true for all values of t, so $x = 41 + 60t$.
15. Because $x = 2 + 6t_1 \equiv 5 \pmod 9$, $t_1 = 2 + 3t_2$. Then $x = 2 + 6(2 + 3t_2) = 14 + 18t_2 \equiv 8 \pmod{11}$, so $t_2 = 7 + 11t_3$. So $x = 14 + 18(7 + 11t_3) = 140 + 198t_3 \equiv 11 \pmod{15}$, so $t_3 = 2 + 5t$. Thus, $x = 140 + 198(2 + 5t) = 536 + 990t$.
17. By Theorem 4.8, $x \equiv a \pmod{[m_1, m_2, \dots, m_k]}$, so $x = a + [m_1, m_2, \dots, m_k]t$.
19. Let n denote the number of fruits. Then $n \equiv 5 \pmod{12}$, $n \equiv 9 \pmod{16}$, and $n \equiv 11 \pmod{18}$. Thus, $n = 5 + 12t_1 \equiv 9 \pmod{16}$, so $t_1 = 3 + 4t_2$. Therefore, $n = 5 + 12(3 + 4t_2) = 41 = 48t_2 \equiv 11 \pmod{18}$, so $t_2 = 2 + 3t$. Thus, $n = 41 + 48(2 + 3t) = 137 + 144t$. Because $n > 500$, $n = 137 + 144 \cdot 3 = 569$.
21. We have $n \equiv 0 \pmod 3$, $n \equiv -1 \pmod 4$, $n \equiv -2 \pmod 5$, $n \equiv -3 \pmod 6$, and $n \equiv -4 \pmod 7$. Then $n = 3t_1 \equiv -1 \pmod 4$, so $t_1 = 1 + 4t_2$. Therefore, $n = 3(1 + 4t_2) = 3 + 12t_2 \equiv -2 \pmod 5$, so $t_2 = 5t_3$. Then $n = 3 + 60t_3 \equiv -3 \pmod 6$, which is true for all t_3, so $n = 3 + 60t_3 \equiv -4 \pmod 7$. This yields $t_3 = 7t$. Therefore, $n = 3 + 420t$. Because $n \geq 4$, $n = 423$.
23. Let n denote the number of eggs in the basket. Then $n \equiv 1 \pmod 2$, $n \equiv 2 \pmod 3$, $n \equiv 3 \pmod 4$, $n \equiv 5 \pmod 6$, and $n \equiv 0 \pmod 7$. Because $n \equiv 1 \pmod 2$, $n = 1 + 2t_1 \equiv 2 \pmod 3$, so $t_1 = 2 + 3t_2$. Therefore, $n = 1 + 2(2 + 3t_2) = 5 + 6t_2 \equiv 3 \pmod 4$, so $t_2 = 1 + 2t_3$. Therefore, $n = 5 + 6(1 + 2t_3) = 11 + 12t_3 \equiv 4 \pmod 5$, so $t_3 = 4 + 5t_4$. Then $n = 11 + 12(4 + 5t_4) = 59 + 60t_4 \equiv 5 \pmod 6$, which is true for all t_4. Therefore, $n = 59 + 60t_4 \equiv 0 \pmod 7$, so $t_4 = 1 + t$. Thus, $n = 59 + 60(1 + 7t) = 119 + 420t$, so there must have been at least 119 eggs in the basket.
25. We have $n \equiv 0 \pmod 9$, $n \equiv -1 \pmod{16}$, $n \equiv -2 \pmod{25}$, and $n \equiv -3 \pmod{49}$. Then $n = 9t_1 \equiv -1 \pmod{16}$, so $t_1 = 7 + 16t_2$. Therefore, $n = 63 + 144t_2 \equiv -2 \pmod{25}$, so $t_2 = 15 + 25t_3$. Then $n = 63 + 144(15 + 25t_3) = 2223 + 3600t_3 \equiv -3 \pmod{49}$, so $t_3 = 14 + 49t$. Therefore, $n = 2223 + 3600(14 + 49t) = 52623 + 176400t$, so the desired smallest positive integer is 52,623.

Exercises 6.3 (p. 312)

1. $\Delta \equiv 6 \pmod 7$ and $(6, 7) = 1$, so the system is solvable.
3. $\Delta \equiv 7 \pmod{13}$ and $(7, 13) = 1$, so the system is solvable.
5. $\Delta \equiv 3 \pmod{11}$ and $(3, 11) = 1$, so the system is solvable.
7. $x \equiv 6 \pmod 7$, $y \equiv 2 \pmod 7$.
9. $x \equiv 5 \pmod{13}$, $y \equiv 4 \pmod{13}$.
11. $x \equiv 4 \pmod{11}$, $y \equiv 7 \pmod{11}$.
13. $\Delta \equiv 6 \pmod 7$ and $\Delta^{-1} \equiv 6 \pmod 7$. Therefore, $x \equiv 6(25 - 24) \equiv 6 \pmod 7$ and $y \equiv 6(18 - 20) \equiv 2 \pmod 7$.

15. $\Delta \equiv 7 \pmod{13}$ and $\Delta^{-1} \equiv 2 \pmod{13}$. Therefore, $x \equiv 2(-70-12) \equiv 5 \pmod{13}$ and $y \equiv 2(10-60) \equiv 4 \pmod 7$.
17. $\Delta \equiv 8 \pmod{11}$ and $\Delta^{-1} \equiv 7 \pmod{11}$. $x \equiv 7(3-15) \equiv 4 \pmod{11}$ and $y \equiv 7(5-15) \equiv 7 \pmod{11}$.
19. $\Delta \equiv 8 \pmod{13}$ and $\Delta^{-1} \equiv 5 \pmod{13}$. Therefore, $x \equiv 5(40-99) \equiv 4 \pmod{13}$ and $y \equiv 5(45-88) \equiv 6 \pmod{13}$.
21. $\Delta \equiv 1 \pmod{18}$ and $\Delta^{-1} \equiv 1 \pmod{18}$. Therefore, $x \equiv (-12)\cdot 12 - (-11)\cdot 2 \equiv 4 \pmod{18}$ and $y \equiv 7\cdot 2 - 11\cdot 12 \equiv 8 \pmod{18}$.
23. $x \equiv 1 \pmod{11}$, $y \equiv 8 \pmod{11}$, $z \equiv 1 \pmod{11}$.
25. Because $\Delta \equiv 3 \pmod{11}$ and $(3, 11) = 1$, the system has a unique solution modulo 11.
27. Because $\Delta \equiv 14 \pmod{13}$ and $(14, 13) = 1$, the system has a unique solution modulo 13.
29. $\Delta \equiv 3 \pmod{11}$ and $\Delta^{-1} \equiv 4 \pmod{11}$. Then $x \equiv 4\cdot 3 \equiv 1 \pmod{11}$, $y \equiv 4\cdot 2 \equiv 8 \pmod{11}$, $z \equiv 4\cdot 3 \equiv 1 \pmod{11}$.
31. $\Delta = 14 \equiv 1 \pmod{13}$ and $\Delta^{-1} \equiv 1 \pmod{13}$. Then $x \equiv 1\cdot 68 \equiv 3 \pmod{13}$, $y \equiv 1\cdot(-48) \equiv 4 \pmod{13}$, $z \equiv 1\cdot(-112) \equiv 5 \pmod{13}$.

Review Exercises (p. 314)

1. Because $x = 3 + 7t_1 \equiv 5 \pmod{10}$, $t_1 \equiv 6 + 10t$, so $x = 3 + 7(6+10t) = 45 + 70t$.
3. Because $x = 2 + 5t_1 \equiv 3 \pmod 7$, $t_1 = 3 + 7t_2$. Then $x = 2 + 5(3 + 7t_2) = 17 + 35t_2 \equiv 5 \pmod 8$, so $t_2 = 4 + 8t$. Thus, $x = 17 + 35(4+8t) = 157 + 280t$.
5. Using the CRT with $M_1 = 10$, $M_2 = 7$, $y_1 \equiv 5 \pmod 7$, and $y_2 \equiv 3 \pmod{10}$, $x \equiv 3\cdot 10\cdot 5 + 5\cdot 7\cdot 3 = 255 \equiv 45 \pmod{70}$, so $x = 45 + 70t$.
7. Using the CRT with $M_1 = 56$, $M_2 = 40$, $M_3 = 35$, $y_1 \equiv 1 \pmod 5$, $y_2 \equiv 3 \pmod 7$, and $y_3 \equiv 3 \pmod 8$, $x \equiv 2\cdot 56\cdot 1 + 3\cdot 40\cdot 3 + 5\cdot 35\cdot 3 = 997 \equiv 157 \pmod{280}$, so $x = 157 + 280t$.
9. Let n be such an integer. Then $n \equiv 1 \pmod 7$, $n \equiv 7 \pmod{11}$, and $n \equiv 11 \pmod{15}$. Because $n = 1 + 7t_1 \equiv 7 \pmod{11}$, $t_1 = 4 + 11t_2$. Therefore, $n = 1 + 7(4 + 11t_2) = 29 + 77t_2 \equiv 11 \pmod{15}$, so $t_2 = 6 + 15t$. Thus, $n = 29 + 77(6 + 15t) = 491 + 1155t$, so the least such positive integer is 491.
11. Let n be such an integer. Then $n \equiv 2 \pmod 5$, $n \equiv 5 \pmod{11}$, $n \equiv 0 \pmod{12}$, and $n \equiv 1 \pmod{13}$. Because $n \equiv 0 \pmod{12}$, $n = 12t_1 \equiv 2 \pmod 5$, $t_1 = 1 + 5t_2$. Therefore, $n = 12(1 + 5t_2) = 12 + 60t_2 \equiv 5 \pmod{11}$, so $t_2 = 3 + 11t_3$. Then $n = 12 + 60(3 + 11t_3) = 192 + 660t_3 \equiv 1 \pmod{13}$, so $t_3 = 3 + 13t$. Thus, $n = 192 + 660(3 + 13t) = 2172 + 8580t$. So the largest integer < 15, with the desired remainders is $2172 + 8580 = 10{,}752$.
13. We have $n \equiv 0 \pmod 3$, $n \equiv -1 \pmod 4$, $n \equiv -2 \pmod 5$, $n \equiv -3 \pmod 7$, and $n \equiv -4 \pmod{11}$. Using the CRT with $M_2 = 1155$, $M_3 = 924$, $M_4 = 660$, $M_5 = 420$, $y_2 \equiv 3 \pmod 4$, $y_3 \equiv 4 \pmod 5$, $y_4 \equiv 4 \pmod 7$, and $y_5 \equiv 6 \pmod{11}$, $x \equiv 0\cdot M_1\cdot y_1 + (-1)\cdot 1155\cdot 3 + (-2)\cdot 924\cdot 4 + (-3)\cdot 660\cdot 4 + (-4)\cdot 420\cdot 6 = -28,857 \equiv 3483 \pmod{4620}$, so the smallest desired integer is 3483.
15. Let n denote the number of pennies in the jar. Then $n \equiv 3 \pmod 5$, $n \equiv 5 \pmod 6$, $n \equiv 4 \pmod 7$, and $n \equiv 8 \pmod{11}$. Using the CRT with $M_1 = 462$, $M_2 = 385$, $M_3 = 330$, $M_4 = 210$, $y_1 \equiv 3 \pmod 5$, $y_2 \equiv 1 \pmod 6$, $y_3 \equiv 1 \pmod 7$, and $y_4 \equiv 1 \pmod{11}$, $x \equiv 3\cdot 462\cdot 3 + 5\cdot 385\cdot 1 + 4\cdot 330\cdot 1 + 8\cdot 210\cdot 1 = 9083 \equiv 2153 \pmod{2310}$. Thus, the least number of coins in the jar is 2153.

17. Because $\Delta \equiv 5 \cdot 5 - 6 \cdot 7 \equiv 1 \pmod 9$ and $(\Delta, 9) = 1$, the linear system is solvable.
19. Because $\Delta \equiv 8 \cdot 9 - 7 \cdot 11 \equiv 8 \pmod{13}$ and $(\Delta, 13) = 1$, the linear system is solvable.
21. $5x + 7y \equiv 3 \pmod 9$ (1) and $6x + 5y \equiv 4 \pmod 9$ (2). Multiply (1) by 6 and (2) by 5; subtract one from the other. Then $17y \equiv -2 \pmod 9$; that is, $y \equiv 2 \pmod 9$, so $x \equiv 5 \pmod 9$. Thus, $x \equiv 5 \pmod 9$ and $y \equiv 2 \pmod 9$.
23. $8x + 5y \equiv 4 \pmod{15}$ (1) and $3x + 11y \equiv 7 \pmod{15}$ (2). Multiply (1) by 3 and (2) by 8. Subtracting one from the other, $73y \equiv 44 \pmod{15}$; that is, $y \equiv 8 \pmod{15}$, so $x \equiv 3 \pmod{15}$. Thus, $x \equiv 3 \pmod{15}$ and $y \equiv 8 \pmod{15}$.
25. $x - y - z \equiv 5 \pmod{17}$ (1), $x + 2y + z \equiv 2 \pmod{17}$ (2), and $2x - 3y - z \equiv 0 \pmod{17}$ (3). Adding (1) and (2) yields $2x + y \equiv 7 \pmod{17}$ (4). Adding (2) and (3) yields $3x - y \equiv 2 \pmod{17}$ (5). Now add (4) and (5): $5x \equiv 9 \pmod{17}$; that is, $x \equiv 12 \pmod{17}$. Therefore, $y \equiv 7 - 24 \equiv 0 \pmod{17}$. Then from (2), $z \equiv 2 - 12 - 0 \equiv 7 \pmod{17}$. Thus, $x \equiv 12 \pmod{17}$, $y \equiv 0 \pmod{17}$, and $z \equiv 7 \pmod{17}$.
27. $\Delta = 25 - 42 \equiv 1 \pmod 9$ and $\Delta^{-1} \equiv 1 \pmod 9$. So $x \equiv 5 \cdot 3 - 7 \cdot 4 \equiv 5 \pmod 9$ and $y \equiv 5 \cdot 4 - 6 \cdot 3 \equiv 2 \pmod 9$.
29. $\Delta = 88 - 15 \equiv 13 \pmod{15}$ and $\Delta^{-1} \equiv 7 \pmod{15}$. Therefore, $x \equiv 7(11 \cdot 4 - 5 \cdot 7) \equiv 3 \pmod{15}$ and $y \equiv 7(8 \cdot 7 - 3 \cdot 4) \equiv 8 \pmod{15}$.
31. $\Delta = \begin{vmatrix} 1 & -1 & -1 \\ 1 & 2 & 1 \\ 2 & -3 & -1 \end{vmatrix} \equiv 5 \pmod{17}$, $\Delta^{-1} \equiv 7 \pmod{17}$, $D_x \equiv 9 \pmod{17}$, $D_y \equiv 0 \pmod{17}$, $D_z \equiv 1 \pmod{17}$. Then $x \equiv \Delta^{-1} \cdot D_x \equiv 7 \cdot 9 \equiv 12 \pmod{17}$. Similarly, $y \equiv 7 \cdot 0 \equiv 0 \pmod{17}$ and $z \equiv 7 \cdot 1 \equiv 7 \pmod{17}$.
33. Let x denote the number of one-bedroom apartments and y that of two-bedroom apartments. Then $69x + 66y \equiv 54 \pmod{101}$ and $70x + 67y \equiv 53 \pmod{101}$. Because $\Delta \equiv 3 \pmod{101}$ and $\Delta^{-1} \equiv 34 \pmod{101}$, $x \equiv 34(67 \cdot 54 - 66 \cdot 53) \equiv 40 \pmod{101}$ and $y \equiv 34(69 \cdot 53 - 70 \cdot 54) \equiv 60 \pmod{101}$. Thus, there are 40 one-bedroom and 60 two-bedroom apartments.

Supplementary Exercises (p. 316)

1. $$\begin{bmatrix} 3 & 5 \\ 4 & 7 \end{bmatrix}\begin{bmatrix} 7 & 6 \\ 7 & 3 \end{bmatrix} = \begin{bmatrix} 56 & 33 \\ 77 & 45 \end{bmatrix} \equiv \begin{bmatrix} 1 & 0 \\ 0 & 1 \end{bmatrix} \pmod{11} \text{ and}$$
$$\begin{bmatrix} 7 & 6 \\ 7 & 3 \end{bmatrix}\begin{bmatrix} 3 & 5 \\ 4 & 7 \end{bmatrix} = \begin{bmatrix} 45 & 77 \\ 33 & 56 \end{bmatrix} \equiv \begin{bmatrix} 1 & 0 \\ 0 & 1 \end{bmatrix} \pmod{11}$$
Therefore, they are inverses of each other modulo 11.

3. $$\begin{bmatrix} 3 & 5 \\ 4 & 7 \end{bmatrix}\begin{bmatrix} x \\ y \end{bmatrix} \equiv \begin{bmatrix} 7 \\ 9 \end{bmatrix} \pmod{11}$$

5. $$\begin{bmatrix} 3 & 5 \\ 4 & 7 \end{bmatrix}\begin{bmatrix} x \\ y \end{bmatrix} \equiv \begin{bmatrix} 7 \\ 9 \end{bmatrix} \pmod{11}. \text{ Therefore, by Exercise 2,}$$
$$\begin{bmatrix} x \\ y \end{bmatrix} \equiv \begin{bmatrix} 7 & 6 \\ 7 & 3 \end{bmatrix}\begin{bmatrix} 7 \\ 9 \end{bmatrix} = \begin{bmatrix} 103 \\ 76 \end{bmatrix} \equiv \begin{bmatrix} 4 \\ 10 \end{bmatrix} \pmod{13}$$

Thus, $x \equiv 4 \pmod{13}$ and $y \equiv 10 \pmod{13}$.

7. **proof:** $\begin{bmatrix} a & b \\ c & d \end{bmatrix} \cdot \Delta^{-1} \begin{bmatrix} d & -b \\ -c & a \end{bmatrix} = \Delta^{-1} \begin{bmatrix} \Delta & 0 \\ 0 & \Delta \end{bmatrix} = I_2$. Also,

$$\Delta^{-1} \begin{bmatrix} d & -b \\ -c & a \end{bmatrix} \begin{bmatrix} a & b \\ c & d \end{bmatrix} = \Delta^{-1} \begin{bmatrix} \Delta & 0 \\ 0 & \Delta \end{bmatrix} = I_2.$$

Thus, one is the inverse of the other.

9. $\Delta \equiv 5 \cdot 7 - 3 \cdot 8 \equiv 11 \pmod{13}$ and $\Delta^{-1} \equiv 6 \pmod{13}$.

$$A^{-1} \equiv 6 \begin{bmatrix} 7 & -8 \\ -3 & 5 \end{bmatrix} = \begin{bmatrix} 42 & -48 \\ -18 & 30 \end{bmatrix} \equiv \begin{bmatrix} 3 & 4 \\ 8 & 4 \end{bmatrix} \pmod{13}$$

11. $\begin{bmatrix} 3 & 4 \\ 5 & 6 \end{bmatrix} \begin{bmatrix} x \\ y \end{bmatrix} \equiv \begin{bmatrix} 2 \\ 3 \end{bmatrix} \pmod{7}$. Therefore,

$$\begin{bmatrix} x \\ y \end{bmatrix} \equiv \begin{bmatrix} 4 & 2 \\ 6 & 2 \end{bmatrix} \begin{bmatrix} 2 \\ 3 \end{bmatrix} = \begin{bmatrix} 14 \\ 18 \end{bmatrix} \equiv \begin{bmatrix} 0 \\ 4 \end{bmatrix} \pmod{7}, \text{ by Exercise 8.}$$

13. $\begin{bmatrix} 8 & 13 \\ 10 & 11 \end{bmatrix} \begin{bmatrix} x \\ y \end{bmatrix} \equiv \begin{bmatrix} 9 \\ 8 \end{bmatrix} \pmod{17}$. Therefore,

$$\begin{bmatrix} x \\ y \end{bmatrix} \equiv \begin{bmatrix} 5 & 8 \\ 14 & 16 \end{bmatrix} \begin{bmatrix} 9 \\ 8 \end{bmatrix} = \begin{bmatrix} 109 \\ 254 \end{bmatrix} \equiv \begin{bmatrix} 7 \\ 16 \end{bmatrix} \pmod{17}$$

by Exercise 10.

Chapter 7 Three Classical Milestones

Exercises 7.1 (p. 326)

1. 1, 6   3. 1, 18   5. 1, 5   7. 1, 5, 7, 11

9. The congruence $x^2 \equiv 1 \pmod{6}$ has exactly two solutions, but $m = 6$ is not a prime.

11. Because $a^2 \equiv 1 \pmod{m}$, $(m-a)^2 = m^2 - 2am + a^2 \equiv a^2 \equiv 1 \pmod{m}$, so $m - a$ is also a solution.

13. With $p = 5$, $(p-1)! = 4! = 24 \equiv -1 \pmod{5}$.

15. With $p = 13$, $(p-1)! = 12! = (1 \cdot 12)(2 \cdot 6)(3 \cdot 4)(5 \cdot 8)(7 \cdot 11)(9 \cdot 10) \equiv (-1)^5 \equiv -1 \pmod{13}$.

17. **proof:** $(p-1)(p-2)\cdots(p-k) \equiv (-1)(-2)\cdots(-k) \equiv (-1)^k k! \pmod{p}$.

19. **proof:** By Wilson's theorem, $(p-1)! \equiv -1 \pmod{p}$, so $[(p-1)!]^2 \equiv 1 \pmod{p}$; that is, $[1^2 \cdot 3^2 \cdots (p-2)^2][2^2 \cdot 4^2 \cdots (p-1)^2] \equiv 1 \pmod{p}$. Then, by Exercise 18, $(-1)^{(p+1)/2}[2^2 \cdot 4^2 \cdots (p-1)^2] \equiv 1 \pmod{p}$. Therefore, $2^2 \cdot 4^2 \cdots (p-1)^2 \equiv (-1)^{(p+1)/2} \pmod{p}$.

21. **proof:** By Wilson's theorem, $-1 \equiv (p-1)! = (p-r-1)![(p-r)\cdots(p-1)] \equiv (p-r-1)!(-1)^r r! \equiv (p-r-1)!(-1)^r(-1)^r \equiv (p-r-1)! \pmod{p}$. Thus, $(p-r-1)! \equiv -1 \pmod{p}$.

23. **proof** (by PMI): By Wilson's theorem, the result is true when $r = 0$. Now assume it is true for an arbitrary integer r, where $0 \le r \le p-1$. Then $0 \equiv r![p-(r+1)! + (-1)^r \equiv r![p-(r+2)]![p-(r+1)] + (-1)^r \equiv r!(p-r-2)!p - (r+1)!(p-r-2)! + (-1)^r \equiv 0 - (r+1)!(p-r-2)! + (-1)^r \equiv (r+1)!(p-r-2)! + (-1)^r \pmod{p}$. Thus, the result follows by PMI.
25. **proof:** Since $\binom{np}{p} = n\binom{np-1}{p-1}, n\left[\binom{np-1}{p-1} - 1\right] = \binom{np}{p} - n$. So, by Exercise 24, $n\left[\binom{np-1}{p-1} - 1\right] \equiv \pmod{p}$. Since $(n,p) = 1$, this yields the desired result.
27. **proof** (by The CPPPS group): $[m! + (m+1)!][n! + (n+1)!] = m![1+(m+1)]n![1+(n+1)] = m!n!(m+2)(n+2) = m!n!(mn+m+n+p+1) \equiv m!n!(mn+m+n+1) \equiv m!n!(m+1)(n+1) \equiv (m+1)!(n+1)! \pmod{p}$ (1). Since $m+1 = p-(n+2)$, $(m+1)! = [p-(n+2)]! = [p-(n+2)][p-(n+3)]\cdots[p-(p-1)] \equiv (-1)^{m+1}(n+2)(n+3)\cdots(p-1) \pmod{p}$. Substituting for $(m+1)!$ in (1), $(m+1)!(n+1)! \equiv (n+1)!(-1)^{m+1}(n+2)(n+3)\cdots(p-1) \equiv (-1)^{m+1}(p-1)! \pmod{p}$. By Wilson's theorem, this yields the desired result.
29. With $p = 5$, $[(p-1)/2]! = 2$, so $[((p-1)/2)!]^2 = 4 \equiv -1 \pmod{5}$.
31. With $p = 17$, $[(p-1)/2]! = 40320 \equiv 13 \pmod{17}$, so $[(p-1)/2)!]^2 \equiv 13^2 \equiv -1 \pmod{17}$.
33. If p is a prime $\equiv 1 \pmod{4}$, then $[(p-1)/2)!]^2 \equiv -1 \pmod{p}$.
35. **proof:** Because $p - i \equiv -i \pmod{p}$, $(p-1)! = [1\cdot 2\cdots(p-1)/2][(p+1)/2)\cdots(p-1)] = [(p-1)/2]!\{[-(p-1)/2]\cdots(-2)(-1)\} = [(p-1)/2]![(p-1)/2)]!(-1)^{(p-1)/2} = (-1)^{(p-1)/2}\{[(p-1)/2]!\}^2$. Therefore, by Wilson's theorem, $(-1)^{(p-1)/2}\{[(p-1)/2]!\}^2 \equiv -1 \pmod{p}$; that is, $\{[(p-1)/2]!\}^2 \equiv (-1)^{(p+1)/2} \pmod{p}$. Since $p \equiv 1 \pmod{4}$, $(p+1)/2$ is odd, so $\{[(p-1)/2]!\}^2 \equiv -1 \pmod{p}$.

Exercises 7.2 (p. 336)

1. $7^{1001} = (7^{16})^{62} \cdot 7^8 \cdot 7 \equiv 1^{62} \cdot (-1) \cdot 7 \equiv -7 \equiv 10 \pmod{17}$.
3. $15^{1976} \equiv (15^{22})^{89} \cdot 15^{18} \equiv 1^{89} \cdot (15^4)^4 \cdot 15^2 \equiv 2^4 \cdot (-5) \equiv 12 \pmod{23}$.
5. $5^{101} = (5^6)^{16} \cdot 5^5 \equiv 1^{16} \cdot 3 \equiv 3 \pmod{7}$.
7. Because $29 \equiv 1 \pmod{7}$, $29^{2076} \equiv 1 \pmod{7}$.
9. Because $8^{-1} \equiv 7 \pmod{11}$, $7 \cdot 8x \equiv 7 \cdot 3 \pmod{11}$; that is, $x \equiv 10 \pmod{11}$.
11. Because $26x \equiv 12 \pmod{17}$, $9x \equiv 12 \pmod{17}$; that is, $3x \equiv 4 \pmod{17}$. Then $6(3x) \equiv 6 \cdot 4 \pmod{17}$; that is, $x \equiv 7 \pmod{17}$.
13. Because $2^{10} \equiv 1 \pmod{11}$ and $2^{30} \equiv 1 \pmod{31}$, $2^{340} = (2^{10})^{34} \equiv 1 \pmod{11}$ and $2^{340} = (2^{30})^{11} \cdot 2^{10} \equiv 1 \cdot 1 \equiv 1 \pmod{31}$. Therefore, by Theorem 4.8, $2^{340} \equiv 1 \pmod{[11, 31]}$; that is, $2^{340} \equiv 1 \pmod{341}$.
15. Notice that $247 \equiv 13 \cdot 19$. By Fermat's Little Theorem $13^{18} \equiv 1 \pmod{19}$ and $19^{12} \equiv 1 \pmod{13}$. Therefore, $13^{18} + 19^{12} \equiv 1 + 0 \equiv 1 \pmod{19}$ and $13^{18} + 19^{12} \equiv 0 + 1 \equiv 1 \pmod{13}$. Then $13^{18} + 19^{12} \equiv 1 \pmod{[13, 19]} \equiv 1 \pmod{247}$.
17. By Theorem 7.7, $(16+21)^{23} = 37^{23} \equiv 37 = 16 + 21 \equiv 16^{23} + 21^{23} \pmod{23}$.
19. **proof:** $a^p \equiv a \pmod{p}$, so $a^{pq} \equiv (a^p)^q \equiv a^q \equiv a \pmod{p}$. Similarly, $a^{pq} \equiv a \pmod{q}$. Then, by Theorem 4.8, $a^{pq} \equiv a \pmod{[p, q]}$; that is, $a^{pq} \equiv a \pmod{pq}$.

21. **proof:** By Theorem 7.7, $(a-b)^p \equiv a-b \pmod{p}$. But, by Fermat's little theorem, $a-b \equiv a^p - b^p \equiv 0 \pmod{p}$. Therefore, $(a-b)^p \equiv 0 \pmod{p}$; that is, $a \equiv b \pmod{p}$.
23. **proof:** By Fermat's little theorem, $p^{q-1} + q^{p-1} \equiv 1 + 0 \equiv 1 \pmod{q}$ and $p^{q-1} + q^{p-1} \equiv 0 + 1 \equiv 1 \pmod{p}$. Therefore, by Theorem 4.8, $p^{q-1} + q^{p-1} \equiv 1 \pmod{[p,q]} \equiv 1 \pmod{pq}$.
25. Since $3!|(n-1)n(n+1)$, $n^5 \equiv n \pmod{6}$. By Theorem 7.7, $n^5 - n \equiv 0 \pmod{5}$. Thus, $n^5 - n \equiv \pmod{[6,5]}$; that is, $n^5 \equiv n \pmod{30}$.
27. **proof:** Because $r^{p-1} \equiv 1 \pmod{p}$, where $0 < r < p$, $1^{p-1} + 2^{p-1} + \cdots + (p-1)^{p-1} \equiv p - 1 \equiv -1 \pmod{p}$.
29. **proof:** Let $0 < r < p$. Then, by Theorem 7.7, $(a+r)^p \equiv a + r \pmod{p}$. Because p is odd,

$$\sum_{r=1}^{p-1} (a+r)^p \equiv \sum_{r=1}^{p-1} (a+r) = (p-1)a + \frac{(p-1)p}{2} 2 \equiv -a \pmod{p}$$

31. **proof:** By Theorem 7.7, $a^p \equiv a \pmod{p}$. Since $p-1$ is even, $a^{p-1} = 1 = (a^{(p-1)/2} - 1)[a^{(p-1)/2} + 1] = (a-1)[a^{(p-3)/2} + \cdots + 1](a+1)[a^{(p-3)/2} - \cdots - 1]$. So $a|a^p - a$, $a - 1|a^p - a$, and $a + 1|a^p - a$, so $a^p \equiv a \pmod{6}$. Thus, by Theorem 4.8, $a^p \equiv a \pmod{6p}$.
33. By the binomial theorem,

$$(a+b)^p = \sum_{k=0}^{p} \binom{p}{k} a^{p-k} b^k = a^p + \sum_{k=1}^{p-1} \binom{p}{k} a^{p-k} b^k + b^p \equiv a^p + b^p \pmod{p}$$

by Exercise 30.

35. **proof** (by PMI): The statement is clearly true when $x = 0$, so assume it is true for an arbitrary integer $r \geq 0$: $(r+1)^p \equiv r^p + 1 \pmod{p}$. Then, by Exercise 30,

$$\begin{aligned}(r+2)^p = [(r+1)+1]^p = \sum_{k=0}^{p} \binom{p}{k} (r+1)^k &= 1 + \sum_{k=1}^{p-1} \binom{p}{k} (r+1)^k + (r+1)^p \\ &\equiv 1 + (r+1)^p \equiv 1 + (r^p + 1) = r^p + 2 \equiv (r+1)^p + 1\end{aligned}$$

by the IH. Therefore, by PMI, the statement is true for every integer $x \geq 0$.

37. **proof:** By the binomial theorem and Exercise 35,

$$\sum_{i=0}^{pa} \binom{pa}{i} x^i = (x+1)^{pa} = [(x+1)^p]^a \equiv (x^p + 1)^a \equiv \sum_{j=0}^{a} \binom{a}{j} x^{jp} \pmod{p}$$

The result now follows by equating the coefficients of x^{bp} from both sides.

39. Since $2^{5!} \equiv 196^{6!} \equiv 1861 \pmod{7967}$ and $(1860, 7967) = 31$, $31|7967$.

Exercises 7.3 (p. 340)

1. Because $10 = 2 \cdot 5$, $2^2 - 1 = 3$ and $2^5 - 1 = 31$ are factors of $2^{10} - 1$. Therefore, $2^{10} - 1 = 3 \cdot 341 = 3 \cdot 31 \cdot 11 = 3 \cdot 11 \cdot 31$.
3. We have $15 = 3 \cdot 5$, so both $2^3 - 1 = 7$ and $2^5 - 1 = 31$ are factors of $2^{15} - 1$. Therefore, $2^{15} - 1 = 7 \cdot 4681 = 7 \cdot 31 \cdot 151$.

5. Notice that $561 = 3 \cdot 11 \cdot 17$. Because $2^{561} \equiv (-1)^{561} \equiv -1 \equiv 2 \pmod 3$, $2^{561} = (2^5)^{112} \cdot 2 \equiv (-1)^{112} \cdot 2 \equiv 1 \cdot 2 \equiv 2 \pmod{11}$, and $2^{561} = (2^4)^{140} \cdot 2 \equiv (-1)^{140} \cdot 2 \equiv 1 \cdot 2 \equiv 2 \pmod{17}$, by Theorem 4.8, $2^{561} \equiv 2 \pmod{[3, 11, 17]} \equiv 2 \pmod{561}$.
7. Notice that $1105 = 5 \cdot 13 \cdot 17$. Since $2^{1105} = (2^4)^{276} \cdot 2 \equiv 1^{276} \cdot 2 \equiv 2 \pmod 5$, $2^{1105} = (2^{12})^{92} \cdot 2 \equiv 1^{92} \cdot 2 \equiv 2 \pmod{13}$, and $2^{1105} = (2^{16})^{69} \cdot 2 \equiv 1^{69} \cdot 2 \equiv 2 \pmod{17}$, by Theorem 4.8, $2^{1105} \equiv 2 \pmod{[5, 13, 17]} \equiv 2 \pmod{1105}$.
9. $2^{10} \equiv 4 \pmod{340}$, $2^{30} \equiv 2^6 \pmod{340}$, $2^{40} \equiv 2^8 \pmod{340}$, and $2^{300} \equiv 2^4 \pmod{340}$. Therefore, $2^{340} = 2^{300} \cdot 2^{40} \equiv 2^4 \cdot 2^8 \equiv 2^{10} \cdot 4 \equiv 4 \cdot 4 \equiv 16 \pmod{340}$. Thus $2^{340} \not\equiv 2 \pmod{340}$.
11. $4^{14} \equiv (4^2)^7 \equiv 1^7 \equiv 1 \pmod{15}$.
13. First, notice that $35 = 5 \cdot 7$. Because $6^{34} \equiv 1^{34} \equiv 1 \pmod 5$ and $6^{34} \equiv (-1)^{34} \equiv 1 \pmod 7$, by Theorem 4.8, $6^{34} \equiv 1 \pmod{35}$.
15. Because $91 = 7 \cdot 13$, we compute 2^{90} modulo 7 and then modulo 13: $2^{90} = (2^6)^{15} \equiv 1^{15} \equiv 1 \pmod 7$, but $2^{90} = (2^{12})^7 \cdot 2^6 \equiv 1^7 \cdot (-1) \not\equiv 1 \pmod{13}$. Therefore, by Theorem 4.8, $2^{90} \not\equiv 1 \pmod{91}$.
17. Because $1105 = 5 \cdot 13 \cdot 17$, it suffices to show that $a^{1104} \equiv 1 \pmod{1105}$ for all positive integers a, where $(a, 1105) = 1$. By Fermat's Little Theorem, $a^4 \equiv 1 \pmod 5$, $a^{12} \equiv 1 \pmod{13}$, and $a^{16} \equiv 1 \pmod{17}$. Therefore, $a^{1104} = (a^4)^{276} \equiv 1^{276} \equiv 1 \pmod 5$, $a^{1104} = (a^{12})^{92} \equiv 1^{92} \equiv 1 \pmod{13}$, and $a^{1104} = (a^{16})^{69} \equiv 1^{69} \equiv 1 \pmod{17}$. Therefore, by Theorem 4.8, $a^{1104} \equiv 1 \pmod{1105}$.
19. We have $2465 = 5 \cdot 17 \cdot 29$. Let a be any positive integer with $(a, 2465) = 1$. Then, by Fermat's Little Theorem, $a^4 \equiv 1 \pmod 5$, $a^{16} \equiv 1 \pmod{17}$, and $a^{28} \equiv 1 \pmod{29}$. Therefore, $a^{2464} = (a^4)^{616} \equiv 1^{616} \equiv 1 \pmod 5$, $a^{2464} = (a^{16})^{154} \equiv 1^{154} \equiv 1 \pmod{17}$, and $a^{2464} = (a^{28})^{88} \equiv 1^{88} \equiv 1 \pmod{29}$. Therefore, by Theorem 4.8, $a^{2464} \equiv 1 \pmod{2465}$.
21. $7^4 = 49^2 \equiv (-1)^2 \equiv 1 \pmod{5^2}$.
23. $19^6 = (19^3)^2 \equiv (-1)^3 \equiv 1 \pmod{7^3}$.
25. $239^{12} = (239^4)^3 \equiv 1^3 \equiv 1 \pmod{13^4}$.

Exercises 7.4 (p. 347)

1. $\varphi(8) = \varphi(2^3) = 2^3 - 2^2 = 8 - 4 = 4$.
3. $\varphi(21) = \varphi(3 \cdot 7) = \varphi(3)\varphi(7) = 2 \cdot 6 = 12$.
5. 1, 2, 4, 7, 8, 11, 13, and 14.
7. 1, 3, 5, 9, 11, 13, 15, 17, 19, 23, 25, and 27.
9. $1^6 \equiv 1 \pmod{18}$; $5^6 \equiv (5^3)^2 \equiv (-1)^2 \equiv 1 \pmod{18}$; $7^6 = (7^3)^2 \equiv 1^2 \equiv 1 \pmod{18}$; $11^6 = (11^3)^2 \equiv (-1)^2 \equiv 1 \pmod{18}$; $13^6 = (13^3)^2 \equiv 1^2 \equiv 1 \pmod{18}$; and $17^6 \equiv (-1)^6 \equiv 1 \pmod{18}$.
11. Because $\varphi(6) = 2$ and $(a, 6) = 1$, $a = 1$ or 5. Then $1^2 \equiv 1 \equiv (-1)^2 \equiv 5^2 \pmod 6$.
13. We have $\varphi(15) = 8$ and $a = 1, 2, 4, 7, 8, 11, 13,$ and 14. Then, $1^8 \equiv 1 \equiv (-1)^8 \equiv 14^8 \pmod{15}$; $2^8 = (2^4)^2 \equiv 1^2 \equiv 1 \equiv (-2)^8 \equiv 13^8 \pmod{15}$; $4^8 = (4^2)^4 \equiv 1^4 \equiv 1 \equiv (-4)^8 \equiv 11^8 \pmod{15}$; and $7^8 = (7^4)^2 \equiv 1^2 \equiv 1 \equiv (-7)^8 \equiv 8^8 \pmod{15}$.
15. Because $\varphi(15) = 8$, by Euler's theorem, $7^8 \equiv 1 \pmod{15}$. Therefore, $7^{1020} = (7^8)^{127} \cdot 7^4 \equiv 1^{127} \cdot 1 \equiv 1 \pmod{15}$.
17. Notice that $\varphi(24) = 8$. Therefore, $79^{1776} \equiv 7^{1776} \equiv (7^8)^{222} \equiv 1^{222} \equiv 1 \pmod{24}$.
19. We have $\varphi(10) = 4$ and $17 \equiv 7 \pmod{10}$. So, by Euler's theorem, $7^4 \equiv 1 \pmod{10}$. Therefore, $17^{6666} \equiv (7^4)^{1666} \cdot 7^2 \equiv 1^{1666} \cdot 9 \equiv 9 \pmod{10}$, so the ones digit is 9.

21. Because $\varphi(16) = 8$, by Euler's theorem $7^8 \equiv 1 \pmod{16}$. Therefore, $7^{1030} \equiv (7^8)^{128} \cdot (7^2)^3 \equiv 1^{128} \cdot 1^3 \equiv 1 \pmod{16}$, so the desired ones digit is 1.
23. Because $\varphi(10) = 4$, multiply both sides by $7^3 \equiv 3 \pmod{10}$: $3(7x) \equiv 3 \cdot 8 \pmod{10}$; that is, $x \equiv 4 \pmod{10}$.
25. We have $7x \equiv 13 \pmod{18}$. Because $\varphi(18) = 6$, multiply both sides by $7^5 \equiv 13 \pmod{18}$: $13(7x) \equiv 13 \cdot 13 \pmod{18}$; that is, $x \equiv 7 \pmod{18}$.
27. We have $3x \equiv 7 \pmod{20}$, $\varphi(20) = 8$, and $3^7 \equiv 7 \pmod{20}$. Therefore, $7(3x) \equiv 7 \cdot 7 \pmod{20}$; that is, $x \equiv 9 \pmod{20}$.
29. $\varphi(15) = \varphi(3 \cdot 5) = \varphi(3)\varphi(5) = 2 \cdot 4 = 8$.
31. $\varphi(105) = \varphi(3 \cdot 5 \cdot 7) = \varphi(3)\varphi(5)\varphi(7) = 2 \cdot 4 \cdot 6 = 48$.
33. $\sum_{d|7} \varphi(d) = \varphi(1) + \varphi(7) = 1 + 6 = 7$.
35. $\sum_{d|12} \varphi(d) = \varphi(1) + \varphi(2) + \varphi(3) + \varphi(4) + \varphi(6) + \varphi(12) = 1 + 1 + 2 + 2 + 2 + 4 = 12$
37. $\sum_{d|n} \varphi(d) = n$.
39. $$\sum_{d|10} (-1)^{10/d}\varphi(d) = (-1)^{10/1}\varphi(1) + (-1)^{10/2}\varphi(2) + (-1)^{10/5}\varphi(5) + (-1)^{10/10}\varphi(10) = 1 - 1 + 4 - 4 = 0$$
41. $\sum_{d|17} (-1)^{17/d}\varphi(d) = (-1)^{17/1}\varphi(1) + (-1)^{17/17}\varphi(17) = -1 - 16 = -17$.
43. Let $S = 1 + 9 + \cdots + 9^{23} = \dfrac{9^{24} - 1}{9 - 1}$. Therefore, $8S = 9^{24} - 1 \equiv 0 \pmod{35}$ by Euler's theorem. But $(8, 35) = 1$, so $S \equiv 0 \pmod{35}$.
45. Let $a = 4$ and $b = 7$. Then $\varphi(4, 7) = \varphi(1) = 1 \neq 2 = (2, 6) = (\varphi(4), \varphi(7))$.
47. **proof:** By Euler's theorem, $a^{\varphi(m)} \equiv 1 \pmod{m}$; that is, $a \cdot a^{\varphi(m)-1} \equiv 1 \pmod{m}$. Therefore, $a^{\varphi(m)-1}$ is an inverse of a modulo m.
49. **proof:** By Euler's theorem $a^{\varphi(b)} \equiv 1 \pmod{b}$ and $b^{\varphi(a)} \equiv 1 \pmod{a}$. Therefore, $a^{\varphi(b)} + b^{\varphi(a)} \equiv 1 + 0 \equiv 1 \pmod{b}$ and $a^{\varphi(b)} + b^{\varphi(a)} \equiv 0 + 1 \equiv 1 \pmod{a}$. Thus, by Theorem 4.8, $a^{\varphi(b)} + b^{\varphi(a)} \equiv 1 \pmod{[a, b]} \equiv 1 \pmod{ab}$.
51. **proof:** By Euler's theorem, $(a - 1)(a^{\varphi(m)-1} + \cdots + a + 1) = a^{\varphi(m)} \equiv 0 \pmod{m}$. But $(a - 1, m) = 1$, so $a^{\varphi(m)-1} + \cdots + a + 1 \equiv 0 \pmod{m}$.
53. There are four positive integers $\leq 2^3$ and relatively prime to it, namely, 1, 3, 5, and 7. Therefore, $\varphi(2^3) = 4 = 2^3 - 2^2$.
55. There are 18 positive integers $\leq 3^4$ and relatively prime to it, namely, 1, 2, 4, 5, 7, 8, 10, 11, 13, 14, 16, 17, 19, 20, 22, 23, 25, and 26. Therefore, $\varphi(3^4) = 18 = 3^3 - 3^2$.
57. $\varphi(p^n) = p^n - p^{n-1}$.
59. **proof:** Let $M_k = [\varphi(m_1), \varphi(m_2), \ldots, \varphi(m_k)]$. Because $\varphi(m_i)|M_k$, $M_k/\varphi(m_i)$ is a positive integer, where $1 \leq i \leq k$. Then $a^{M_k} = [a^{\varphi(m_i)}]^{M_k/\varphi(m_i)} \equiv 1^{M_k/\varphi(m_i)} \equiv 1 \pmod{m_i}$. Therefore, by Theorem 4.8, we have the desired conclusion, $a^{M_k} \equiv 1 \pmod{[m_1, m_2, \ldots, m_k]}$.
61. Because the moduli m_i are pairwise relatively prime, $(M_i, m_i) = 1$ for all i, so by Euler's theorem $M_i^{\varphi(m_i)} \equiv 1 \pmod{m_i}$. When $j \neq i$, $M_j \equiv 0 \pmod{m_i}$. Because $x \equiv \sum_{j \neq i} a_j M_j^{\varphi(m_j)} + a_i M_i^{\varphi(m_i)} \equiv 0 + a_i \equiv \pmod{m_i}$ for every i, it follows that x is a solution of the linear system. By the CRT, the solution is unique.

63. $M_1 = 9$, $M_2 = 4$, $\varphi(m_1) = \varphi(4) = 2$, and $\varphi(m_2) = \varphi(9) = 6$. Therefore, $x \equiv 3 \cdot 9^2 + 5 \cdot 4^6 \equiv 23 \pmod{36}$.
65. $M_1 = 35$, $M_2 = 21$, $M_3 = 15$, $\varphi(3) = 2$, $\varphi(5) = 4$, and $\varphi(7) = 6$. Therefore, $x \equiv 2 \cdot 35^2 + 4 \cdot 21^4 + 5 \cdot 15^6 \equiv 89 \pmod{105}$.
67. $M_1 = 585$, $M_2 = 468$, $M_3 = 260$, $M_4 = 180$, $\varphi(4) = 2$, $\varphi(5) = 4$, $\varphi(9) = 6$, and $\varphi(13) = 12$. Therefore, $x \equiv 2 \cdot 585^2 + 3 \cdot 468^4 + 4 \cdot 260^6 + 5 \cdot 180^{12} \equiv 1318 \pmod{2340}$.

Review Exercises (p. 350)

1. $18! = (1 \cdot 18)(2 \cdot 10)(3 \cdot 13)(4 \cdot 5)(6 \cdot 16)(7 \cdot 11)(8 \cdot 12)(9 \cdot 17)(14 \cdot 15)$
$\equiv (-1) \cdot 1 \cdot 1 \cdot 1 \cdot 1 \cdot 1 \cdot 1 \cdot 1 \cdot 1 \equiv -1 \pmod{19}$.
3. $11! \equiv 0 \pmod 3$ and $11! \equiv 0 \pmod 4$, so $11! \equiv 0 \pmod{12}$. Thus $11! \not\equiv -1 \pmod{12}$.
5. $19^{22} \equiv (-4)^{22} \equiv 4^{22} \equiv (4^6)^3 \cdot 4^4 \equiv 2^3 \cdot 3 \equiv 1 \pmod{23}$.
7. $18^{4567} \equiv 5^{4567} \equiv (5^2)^{2283} \cdot 5 \equiv (-1)^{2283} \cdot 5 \equiv -5 \equiv 8 \pmod{13}$.
9. $55^{1876} \equiv 1^{1876} \equiv 1 \pmod 3$; $55^{1876} \equiv (-1)^{1876} \equiv 1^{1876} \equiv 1 \pmod 4$. Therefore, by Theorem 4.8, $55^{1876} \equiv 1 \pmod{[3,4]} \equiv 1 \pmod{12}$.
11. By Exercise 50 in Section 7.4, $13^{16} + 17^{12} \equiv 1 \pmod{17 \cdot 13} \equiv 1 \pmod{221}$.
13. By Exercise 26 in Section 7.2, $11^{19} + 19^{11} \equiv 11 + 19 \equiv 30 \pmod{11 \cdot 19} \equiv 30 \pmod{209}$.
15. $18^{20} = (18^5)^4 \equiv 18^4 \equiv 126 \pmod{450}$ and $25^6 \equiv 325 \pmod{450}$. Therefore, $18^{20} + 25^6 \equiv 126 + 325 \equiv 1 \pmod{450}$.
17. $15^{1942} \equiv 4^{1942} \equiv (4^5)^{388} \cdot 4^2 \equiv 1^{388} \cdot 4^2 \equiv 1^{388} \cdot 5 \equiv 1 \cdot 5 \equiv 5 \pmod{11}$, so the ones digit is 5.
19. Since $2^{4!} \equiv 460 \pmod{5899}$ and $(459, 5899) = 17$, $17 | 5899$.
21. $\varphi(16) = \varphi(2^4) = 2^4 - 2^3 = 8$.
23. $\varphi(200) = \varphi(2^3 \cdot 5^2) = \varphi(2^3)\varphi(5^2) = (2^3 - 2^2)(5^2 - 5) = 4 \cdot 20 = 80$.
25. $\varphi(17) = 16$
27. $\varphi(25) = 20$
29. There are $\varphi(16) = 8$ positive integers ≤ 16 and relatively prime to it. They are 1, 3, 5, 7, 9, 11, 13, and 15. Then, $1^8 \equiv 1 \equiv (-1)^8 \equiv 15^8 \pmod{16}$; $3^8 = (3^4)^2 \equiv 1^2 \equiv 1 \equiv (-1)^8 \equiv 13^8 \pmod{16}$; $5^8 = (5^4)^2 \equiv 1^2 \equiv 1 \equiv (-1)^8 \equiv 11^8 \pmod{16}$; $7^8 = (7^2)^4 \equiv 1^4 \equiv 1 \equiv (-1)^8 \equiv 9^8 \pmod{16}$.
31. $273^{1961} \equiv 73^{1961} = (73^5)^{392} \cdot 73 \equiv (-7)^{392} \cdot 73 \equiv 7^{392} \cdot 73 \equiv (7^4)^{98} \cdot 73 \equiv 1^{98} \cdot 73 \equiv 73 \pmod{100}$, so the last two-digit number is 73.
33. The given congruence yields $7x \equiv 10 \pmod{13}$. Because $7^{11} \equiv 2 \pmod{13}$, multiply both sides by 2: $2(7x) \equiv 20 \pmod{13}$; that is, $x \equiv 7 \pmod{13}$.
35. Because $65x \equiv 27 \pmod{18}$, we have $11x \equiv 9 \pmod{18}$. Multiply both sides by $11^{\varphi(18)-1} = 11^5 \equiv 5 \pmod{18}$: $5(11x) \equiv 5 \cdot 9 \pmod{18}$; that is, $x \equiv 9 \pmod{18}$.
37. $2^{2047} = (2^{11})^{186} \cdot 2 \equiv 1^{186} \cdot 2 \equiv 1 \cdot 2 \equiv 2 \pmod{2047}$, so 2047 is a pseudoprime.
39. Let $S = 1 + 7 + \cdots + 7^{17} = \dfrac{7^{18} - 1}{18}$. Therefore, $18S = 7^{18} - 1 \equiv 0 \pmod{19}$ by Fermat's Little Theorem. But $(18, 19) = 1$, so $S \equiv 0 \pmod{19}$.
41. $12^{65} \equiv 2^{65} = (2^4)^{16} \cdot 2 \equiv 1^{16} \cdot 2 \equiv 1 \cdot 2 \equiv 2 \equiv 12 \pmod 5$ and $12^{65} \equiv (-1)^{65} \equiv -1 \equiv 12 \pmod{13}$. Therefore, by Theorem 4.8, $12^{65} \equiv 12 \pmod{[5, 13]} \equiv 12 \pmod{65}$.
43. $28^{87} \equiv 1^{87} \equiv 1 \equiv 28 \pmod 3$ and $28^{87} \equiv (-1)^{87} \equiv -1 \equiv 28 \pmod{29}$. Therefore, by Theorem 4.8, $28^{87} \equiv 28 \pmod{[3, 29]} \equiv \pmod{87}$.

45. $38^{16} = (38^2)^8 \equiv (-1)^8 \equiv 1 \pmod{17^2}$.
47. $\sum_{d|8} \varphi(d) = \varphi(1) + \varphi(2) + \varphi(4) + \varphi(8) = 1 + 1 + 3 + 3 = 8$.
49. $\sum_{d|18} \varphi(d) = \varphi(1) + \varphi(2) + \varphi(3) + \varphi(6) + \varphi(9) + \varphi(18) = 1 + 1 + 2 + 2 + 6 + 6 = 18$.
51. $$\begin{aligned}\sum_{d|8} (-1)^{n/d}\varphi(d) &= (-1)^{8/1}\varphi(1) + (-1)^{8/2}\varphi(2) + (-1)^{8/4}\varphi(4) + (-1)^{8/8}\varphi(8)\\ &= 1 \cdot 1 + 1 \cdot 1 + 1 \cdot 2 + (-1) \cdot 4 = 0.\end{aligned}$$
53. $$\begin{aligned}\sum_{d|18} (-1)^{18/d}\varphi(d) &= (-1)^{18/1}\varphi(1) + (-1)^{18/2}\varphi(2) + (-1)^{18/3}\varphi(3) + (-1)^{18/6}\varphi(6)\\ &\quad + (-1)^{18/9}\varphi(9) + (-1)^{18/18}\varphi(18) = 1 - 1 + 2 - 2 + 6 - 6 = 0.\end{aligned}$$
55. Let a be any positive integer with $(a, 8911) = 1$. By Fermat's Little Theorem, $a^6 \equiv 1 \pmod 7$, $a^{18} \equiv 1 \pmod{19}$, and $a^{66} \equiv 1 \pmod{67}$. Therefore, $a^{8910} = (a^6)^{1485} \equiv 1^{1485} \equiv 1 \pmod 7$, $a^{8910} = (a^{18})^{495} \equiv 1^{495} \equiv 1 \pmod{19}$, and $a^{8910} = (a^{66})^{135} \equiv 1^{135} \equiv 1 \pmod{67}$. Therefore, by Theorem 4.8, $a^{8910} \equiv 1 \pmod{[7, 19, 67]} \equiv 1 \pmod{8911}$. Thus, 8911 is a Carmichael number.
57. $\sum_{i=1}^{p-1} ai = a\left(\sum_{i=1}^{p-1} i\right) = a\dfrac{(p-1)p}{2} \equiv 0 \pmod p$, since p is odd.
59. **proof:** Because $p - i \equiv -i \pmod p$,

$$\begin{aligned}(p-1)! &= [1 \cdot 2 \cdots (p-1)/2][(p+1)/2) \cdots (p-1)]\\ &= [(p-1)/2]!\{[-(p-1)/2] \cdots (-2)(-1)\}\\ &= [(p-1)/2]![(p-1)/2)]!(-1)^{(p-1)/2} = (-1)^{(p-1)/2}\{[(p-1)/2]!\}^2\end{aligned}$$

Therefore, by Wilson's theorem, $(-1)^{(p-1)/2}\{[(p-1)/2]!\}^2 \equiv -1 \pmod p$; that is, $\{[(p-1)/2]!\}^2 \equiv (-1)^{(p+1)/2} \pmod p$. Since $p \equiv 3 \pmod 4$, $(p+1)/2$ is even, so $\{[(p-1)/2]!\}^2 \equiv -1 \pmod p$. Thus, $[(p-1)/2]! \equiv \pm 1 \pmod p$.

Supplementary Exercises (p. 351)

1. $W(5) = \dfrac{4! + 1}{5} = \dfrac{25}{5} = 5 \equiv 0 \pmod 5$, so $W(5)$ is a Wilson prime. Likewise, $W(13) = \dfrac{12! + 1}{13} = \dfrac{479001601}{13} = 36846277 \equiv 0 \pmod{13}$, so 13 is also a Wilson prime.
3. (a) $202 = 2 \cdot 101$. Sum of the digits in $202 = 2 + 0 + 2 = 4 = 2 + 1 + 0 + 1 =$ sum of the digits in the prime factors of 202. So, 202 is a Smith number.
   (b) $265 = 5 \cdot 53$. Sum of the digits in $265 = 2 + 6 + 5 = 13 = 5 + 5 + 3 =$ sum of the digits in the prime factors of 265. So, 265 is a Smith number.
   (c) $666 = 2 \cdot 3 \cdot 3 \cdot 37$. Sum of the digits in $666 = 6 + 6 + 6 = 18 = 2 + 3 + 3 + 3 + 7 =$ sum of the digits in the prime factors of 666. So, 666 is a Smith number.
   (d) $1111 = 11 \cdot 101$. Sum of the digits in $1111 = 1 + 1 + 1 + 1 = 4 = 2 + 2 =$ sum of the digits in the prime factors of 1111. So, 1111 is a Smith number.
5. $$\phi(1) = 1 = \phi(2); \qquad \phi(3) = 2 = \phi(4)$$
$$\phi(15) = \phi(3)\phi(5) = 2 \cdot 4 = 8 = 2^4 - 2^3 = \phi(16)$$

$$\phi(104)=\phi(8)\phi(13)=4\cdot 12=48=2\cdot 4\cdot 6=\phi(3)\phi(5)\phi(7)=\phi(105)$$

$$\phi(164)=\phi(4)\phi(41)=2\cdot 40=2\cdot 4\cdot 10=\phi(3)\phi(5)\phi(11)=\phi(165)$$

$$\phi(194)=\phi(2)\phi(97)=1\cdot 96=96=2\cdot 4\cdot 12=\phi(3)\phi(5)\phi(13)=\phi(195)$$

$$\phi(255)=\phi(3)\phi(5)\phi(17)=2\cdot 4\cdot 16=128=2^8-2^7=\phi(2^8)=\phi(256)$$

$$\phi(495)=\phi(3^2)\phi(5)\phi(11)=6\cdot 4\cdot 10=240=8\cdot 30=\phi(16)\phi(31)=\phi(496) \quad \text{and}$$

$$\phi(65535)=\phi(3)\phi(5)\phi(17)\phi(257)=2\cdot 4\cdot 16\cdot 256=32768=2^{16}-2^{15}$$
$$=\phi(2^{16})=\phi(65536)$$

Thus 1, 3, 15, 104, 164, 194, 255, 495, and 65535 are solutions of the equation.

7. **proof:** Since $\phi(2^k)=2^k-2^{k-1}=2^{k-1}$, $f(2^k)=2^{k-1}+2^{k-2}+\cdots+1+\phi(1)=(2^{k-1}+2^{k-2}+\cdots+1)+1=1+(2^k-1)/(2-1)=2^k$

9. **proof** (by PMI): The statement is true when $n=1$, by virtue of Wilson's theorem. Assume it is true for an arbitrary positive integer k: $\dfrac{(kp-1)!}{(k-1)!p^{k-1}}\equiv(-1)^k \pmod p$. We have $(kp+1)\cdots[(k+1)p-1]\equiv(p-1)!\equiv -1 \pmod p$, by Wilson's theorem. Then

$$\frac{[(k+1)p-1]!}{k!p^k}=\frac{(kp-1)!}{(k-1)!p^{k-1}}\cdot\frac{(kp)(kp+1)\cdots[(k+1)p-1]}{kp}$$
$$=\frac{(kp-1)!}{(k-1)!p^{k-1}}\cdot(kp)(kp+1)\cdots[(k+1)p-1]\equiv(-1)^k(p-1)!\equiv(-1)^{k+1}$$

by the inductive hypothesis and Wilson's theorem. Thus, the statement is true when $n=k+1$, so the result follows by PMI.

Chapter 8 Multiplicative Functions

Exercises 8.1 (p. 364)

1. Because $f(mn)=0=0\cdot 0=f(m)f(n)$, f is multiplicative.
3. $\varphi(341)=\varphi(11\cdot 31)=\varphi(11)\varphi(31)=10\cdot 30=300$.
5. $\varphi(1105)=\varphi(5\cdot 13\cdot 17)=\varphi(5)\varphi(13)\varphi(17)=4\cdot 12\cdot 16=768$.
7. $\varphi(6860)=\varphi(2^2\cdot 5\cdot 7^3)=\varphi(2^2)\varphi(5)\varphi(7^3)=2\cdot 4\cdot 294=2352$.
9. $\varphi(183920)=\varphi(2^4\cdot 5\cdot 11^2\cdot 19)=\varphi(2^4)\varphi(5)\varphi(11^2)\varphi(19)=8\cdot 4\cdot 110\cdot 18=63{,}360$.
11. $\varphi(3!)=\varphi(6)=2$.
13. $\varphi(7!)=\varphi(2^4\cdot 3^2\cdot 5\cdot 7)=8\cdot 6\cdot 4\cdot 6=1152$.
15. 1
17. 7, 9, 14, 18
19. $\varphi(pq)=\varphi(p)\varphi(q)=(p-1)(q-1)=(p-1)(p+1)=p^2-1$.
21. Let $p<q$. By Exercise 19, $\varphi(pq)=p^2-1=288$. Then $p^2=289$, so $p=17$. Hence, $q=19$.
23. No, since $\varphi(n)\le n$ for every positive integer n.
25. $\varphi(n)=\varphi(2^k)=2^k-2^{k-1}=2^{k-1}=2^k/2=n/2$.
27. Since n is odd, $(4,n)=1$. Therefore, $\varphi(4n)=\varphi(4)\varphi(n)=2\varphi(n)$.
29. $\varphi(n)=\varphi(2^j)=2^{j-1}=n/2$.

31. $\varphi(n) = \varphi(2^j 3^k) = \varphi(2^j)\varphi(3^k) = 2^{j-1}3^{k-1} \cdot 2 = 2^j 3^{k-1} = n/3$, so $3\varphi(n) = n$.
33. **proof:** $\varphi(2^{2k+1}) = 2^{2k} = (2^k)^2$ is a square.
35. **proof:** Because $p \nmid n$, $(p, n) = 1$. Therefore, $\varphi(pn) = \varphi(p)\varphi(n) = (p-1)\varphi(n)$.
37. $\varphi(p^e) = p^{e-1}(p-1)$, so $\varphi(\varphi(p^e)) = \varphi(p^e(p-1)) = \varphi(p^{e-1})\varphi(p-1) = p^{e-2}(p-1)$ $\varphi(p-1) = p^{e-2}\varphi(p)\varphi(p-1) = p^{e-2}\varphi(p(p-1))$
39. **proof:** Let $m = p^i a$ and $n = pb$, where $(a, p) = (b, p) = 1$, and $i \geq 1$. Then $\varphi(mn) = \varphi(p^{i+1}ab) = \varphi(p^{i+1})\varphi(ab) = p^i(p-1)\varphi(ab)$. So

$$\varphi(m)\varphi(n) = \varphi(p^i a)\varphi(pb) = p^{i-1}(p-1)^2\varphi(a)\varphi(b) = (p-1)\varphi(mn)/p$$

41. **proof:** Let $m = da$ and $n = db$, where $(a, b) = 1$. Let $d = \prod_i p_i^{e_i}$. Then $mn = \prod_i p_i^{2e_i} \cdot ab$, so $\varphi(mn) = \prod_i \varphi(p_i^{2e_i})\varphi(ab) = \prod_i (p_i^{2e_i-1})(p_i - 1)\varphi(ab)$. $\varphi(m)\varphi(n) = \prod_i [p_i^{2e_i-1}(p_i - 1)$ $\varphi(a)\varphi(b)(p_i - 1)/p_i] = \varphi(mn)\prod_i (1 - 1/p_i) = \varphi(mn)\varphi(d)/d$
43. Let $m = n$. Then $d = n$. Then, by Exercise 41, $\varphi(n^2) = n/\varphi(n) \cdot \varphi(n) \cdot \varphi(n) = n\varphi(n)$.
45. **proof** (by PMI): The statement is clearly true when $e = 1$, so assume it is true for an arbitrary positive integer $e = k$. Then, using Exercise 41 with $m = n^k$ and $d = n$, $\varphi(n^{k+1}) = n/\varphi(n) \cdot \varphi(n^k)\varphi(n) = n \cdot n^{k-1}\varphi(n) = n^k\varphi(n)$, by the IH. Therefore, by PMI, the result is true for every exponent $e \geq 1$.
47. $\varphi(48) = \varphi(6 \cdot 8) = 2/\varphi(2) \cdot \varphi(6)\varphi(8) = 2 \cdot 2 \cdot 4 = 16$.
49. $\varphi(375) = \varphi(15 \cdot 25) = 5/\varphi(5) \cdot \varphi(15)\varphi(25) = 5/4 \cdot 8 \cdot 20 = 200$.
51. $\varphi(16) = \varphi(2^4) = 2^3\varphi(2) = 2^3 \cdot 1 = 8$.
53. $\varphi(2401) = \varphi(7^4) = 7^3\varphi(7) = 7^3 \cdot 6 = 2058$.
55. 6
57. 27
59. $n\varphi(n)/2$
61. **proof** (by D. A. Breault): Clearly, $n > 4$; so $2|n$ and $3|n$, and hence $6|n$. Thus $n = 6m$ for some positive integer m. Then $\varphi(n) = \varphi(6)\varphi(m) = 2\varphi(m)$; so $3\varphi(n) = 6\varphi(m) = 6m\prod_{p|m}(1 - 1/p) = n\prod_{p|m}(1 - 1/p) \leq n$, as desired.
63. **proof:** We have $\varphi(n)/n = \prod_i (1 - 1/p)$. Let $n = 2^a 3^b$, where $a, b > 0$. Then $\varphi(n)/n = 1/3$. Since a and b are arbitrary, this implies that there are infinitely many positive integers n such that $\varphi(n)/n = 1/3$. On the other hand, let $\varphi(n)/n = 1/4$. Then $\prod_{p|m}(1 - 1/p) = 1/4$; that is, $4\prod_{p|m}(p-1) = \prod_{p|m} p$; so 4|RHS, which is impossible.

Exercises 8.2 (p. 372)

1. Because 43 is a prime, $\tau(43) = 2$.
3. $2187 = 3^7$, so $\tau(2187) = 7 + 1 = 8$.
5. Because the only positive factors of 43 are 1 and 43, $\sigma(43) = 1 + 43 = 44$.
7. $2187 = 3^7$, so $\sigma(2187) = \sigma(3^7) = (3^8 - 1)/2 = 3280$.
9. $1, p, q$, and pq.
11. $1, p, q, pq, p^2$, and p^2q.
13. $1 + p + q + pq$
15. $(1 + q)(1 + p + p^2)$
17. $1, p, \ldots, p^i, q, pq, \ldots, p^i q, \ldots, q^j, \ldots, p^i q^j$
19. They are primes.
21. $\tau(n) = 2k$ and $\sigma(n) = (p_1 + 1)(p_2 + 1) \cdots (p_k + 1)$.

23. Product $= 1 \cdot 2 \cdot 2^2 \cdot 2^3 \cdots 2^{2^e} = 2^{1+2+3+\cdots+2^e} = 2^{2^e(2^e+1)/2} = 2^{2^{e-1}(2^e+1)}$.

25. Product $= \prod_{\substack{0\le i\le a\\ 0\le j\le b}} p^i q^j = \prod_j p^{a(a+1)/2} q^{(a+1)j} = p^{(b+1)a(a+1)/2} q^{(a+1)b(b+1)/2}$

$= (p^a q^b)^{(a+1)(b+1)/2} = n^{\tau(n)/2}$

27. $\tau(n) = 2p$

29. $\sigma(6) = 12$

31. $\sigma(496) = \sigma(2^4 \cdot 31) = 31 \cdot 32 = 992$.

33. Let $n = 2^{p-1}(2^p - 1)$, where p and $2^p - 1$ are primes. Then $\sigma(n) = 2n$.

35. $\sigma(p+2) = 1 + (p+2) = (1+p) + 2 = \sigma(p) + 2$.

37. For $\sigma(p) = 1 + p$ to be odd, p must be even. So $p = 2$.

39. RHS $= 2 \cdot \sigma(332) = 2 \cdot 588 = 1176 =$ LHS.

41. LHS $= 432 = 2 \cdot 216 = 2\varphi(666) =$ RHS.

43. $\varphi(p) + \sigma(p) = (p-1) + (p+1) = 2p$.

45. **proof:** Let $n = \prod_i p_i^{e_i}$. Then $\tau(n) = \prod_i (e_i + 1)$. If $\tau(n)$ is odd, e_i must be even, say, $e_i = 2k_i$. Then $n = \prod_i p_i^{2k_i} = (\prod_i p_i^{k_i})^2$ is a square.

47. **proof:** Let $n = \prod_i p_i^{e_i}$. Then $\tau(n) = \prod_i (e_i + 1)$. If $\tau(n)$ is a prime, then $i = 1$ and either $e_1 = 1$ or e_1 must be an even integer $2e$. If $e_1 = 1$, then n is a prime. If $e_1 = 2e$, then $n = p^{2e}$.

49. **proof:** $1 + p + \cdots + p^k = 1 + k$ odd numbers. This sum is odd if and only if k is even.

51. **proof:** Let $n = \prod_i p_i^{e_i}$. Then $\sigma(n) = \prod_i (p_i^{e_i} - 1)/(p_i - 1)$. Let $\sigma(n)$ be odd. If every p_i is odd, then $(p_i^{e_i} - 1)/(p_i - 1)$ is odd for every i. So e_i is even for every i. Therefore, n is a square. Suppose $p_1 = 2$. Then $\sigma(n) = (2^{e_{i+1}} - 1) \prod_{i>1} (p_i^{e_i+1} - 1)/(p_i - 1)$ is odd, so both factors are odd. Again, all e_i are even as before. If $e_i = 0$, then n is of the form m^2, a square. If $e_i > 0$, then 2^{e_i} is even, so n is of the form $2^e m^2$. Thus, in both cases, n is of the form $2^e m^2$, where $e \ge 0$.

53. **proof:** Since every divisor of m is also a divisor of n, $\sum_{d|m} \frac{1}{d} \le \sum_{d|n} \frac{1}{d}$. The desired result follows by Exercise 42.

55. **proof:** Let $n = \prod_i p_i^{e_i}$ be the canonical decomposition of n. Then $\sigma(n) = \prod_i \sigma(p_i^{e_i})$. So

$$\frac{\sigma(n)}{n} = \prod_i \frac{\sigma(p_i^{e_i})}{p_i^{e_i}} < \prod_i \frac{p_i}{p_i - 1} = \prod_{p|n} \frac{p}{p-1}$$

57. $\sigma_2(18) = \sum_{d|18} d^2 = 1^2 + 2^2 + 3^2 + 6^2 + 9^2 + 18^2 = 455$.

59. $\sigma_3(28) = \sum_{d|28} d^3 = 1^3 + 2^3 + 4^3 + 7^3 + 14^3 + 28^2 = 25{,}112$.

61. $\sigma_k(p^e) = \sum_{d|p^e} d^k = \sum_{i=0}^{e} (p^i)^k = \sum_{i=0}^{e} p^{ki} = \dfrac{p^{(e+1)k} - 1}{p^k - 1}$

63. **proof:** By Theorem 8.6, $F(n) = \sum_{d|n} f(d)$ is multiplicative. Let $f(n) = n^k$. Then $\sigma_k = F(n) = \sum_{d|n} d^k$ is multiplicative, since f is multiplicative.

65. $\sigma_k(n) = \prod_i \sigma_k(p_i^{e_i}) = \prod_i \left[(p_i^{(e_i+1)k} - 1)/(p_i - 1)\right]$

67. $\sigma_3(18) = \sigma_3(2 \cdot 3^2) = (2^6 - 1)/1 \cdot (3^9 - 1)/2 = 619{,}983$
69. $\sigma_4(84) = \sigma_4(2^2 \cdot 3 \cdot 7) = (2^{12} - 1)/1 \cdot (3^8 - 1)/2 \cdot (7^8 - 1)/6 = 12{,}905{,}081{,}280{,}000$

Exercises 8.3 (p. 379)

1. Because $2p$ is a perfect number, $\sigma(2p) = 4p$; that is, $1 + 2 + p + 2p = 4p$, so $p = 3$.
3. $\varphi(n) = \varphi(2^{p-1})\varphi(2^p - 1) = 2^{p-2}(2^p - 2) = 2^{2p-2} - 2^{p-1} = 2^{2p-2} + n - 2^{2p-1} = n - 2^{2p-2}$.
5. Let $m = 2^p - 1$. Then $t_m = m(m+1)/2 = 2^{p-1}(2^p - 1)$ is a triangular number.
7. Because p and q are distinct odd primes, $\sigma(pq) = 1 + p + q + pq$. Also $(p - q)(q - 1) > 2$; that is, $1 - p - q + pq > 2$, so $1 + p + q < pq$. Thus, $\sigma(pq) < 2pq$, so pq is not a perfect number.
9. Notice that 1 and 2 are not solutions, but 3 is a solution. We will now show that the equation is not solvable if $n > 3$. Let $n = 3 + k$, where $k \ge 1$. Then RHS $= 2(3+k) + 3 = 2k + 9$ and LHS $= (3+k)^{k+2} = (3+k)^k(3+k)^2 = (3+k)^k(9 + 6k + k^2) > 2k + 9 =$ RHS. Thus, if $n > 3$, then the equation is not solvable.
11. Because the sum of the cubes of the first n odd integers $= n^2(2n^2 - 1)$, the sum of the cubes of the first $2^{(p-1)/2}$ odd numbers equals $(2^{(p-1)/2})^2[2 \cdot (2^{(p-1)/2})^2 - 1] = 2^{p-1}(2^p - 1)$, a perfect number.
13. **proof:** Let $m = 2^{p-1}(2^p - 1)$ and $n = 2^{q-1}(2^q - 1)$ be even perfect numbers. Then $mn = 2^{p+q-1}(2^p - 1)(2^q - 1)$ is not a perfect number since $(2^p - 1)(2^q - 1)$ is not a prime.
15. **proof:** Since $q = 2^p - 1$ is a prime, every factor d of n is of the form $2^i q^j$, where $0 \le i \le p - 1$ and $0 \le j \le 1$. So

$$\prod_{d|n} d = \prod_{i,j} 2^i q^j = \left(\prod_{i=0}^{p-1} 2^i\right)\left(\prod_{i=0}^{p-1} 2^i q\right) = [2^{p(p-1)/2}][2^{p(p-1)/2} q^p] = 2^{p(p-1)} q^p = n^p.$$

17. **proof:** Since p and e are odd, let $p = 2r + 1$ and $e = 2s + 1$. Then $n = (2r+1)^{2s+1} m^2$. Since n is odd perfect, $2 | \sigma(n)$, but $4 \nmid \sigma(n)$. But $\sigma(n) = \sigma(2r+1)^{2s+1} \cdot \sigma(m^2)$. So, $4 \nmid \sigma(2r+1)^{2s+1}$.

    Now,

$$(2r+1)^i = \sum_{j=0}^{i} \binom{i}{j} (2r)^j \equiv 1 + 2ir \pmod 4.$$

    Therefore,

$$\begin{aligned} \sigma(2r+1)^{2s+1} &\equiv \sum_{i=0}^{2s+1} (1 + 2ir) \equiv (2s+2) + 2r \cdot \frac{(2s+1)(2s+2)}{2} \\ &\equiv (2s+2)(1 + r + 2rs) \pmod 4 \equiv (r+1)(s+1) \pmod 2 \qquad (1) \end{aligned}$$

    Since $4 \nmid \sigma(2r+1)^{2s+1}$, it follows from (1) that $(r+1)(s+1) \not\equiv 0 \pmod 2$; so $(r+1)(s+1) \equiv 1 \pmod 2$. Consequently, $r \equiv s \equiv 0 \pmod 2$; that is, both r and s are even, say, $r = 2a$ and $s = 2b$. Then $p = 4a + 1$ and $e = 4b + 1$, so $p \equiv 1 \equiv a \pmod 4$.
19. **proof:** We have $n = p^e m^2$, where $p \equiv 1 \pmod 4$ and m is odd. Since m is odd, $m \equiv \pm 1 \pmod 4$, so $m^2 \equiv 1 \pmod 4$. Thus, $n \equiv 1^e \cdot 1 \equiv 1 \pmod 4$.
21. $\sigma(88) = 180 > 2 \cdot 88$, so 88 is abundant.
23. Since $\sigma(315) = 624 < 2 \cdot 315$, 315 is deficient.

25. Let $n = 2^{22}M_{23}$. Since $M_{23} = 2^{23} - 1 = 47.\ 178{,}481$ is composite, $\sigma(M_{23}) > 1 + M_{23} = 2^{23}$. Therefore, $\sigma(n) = \sigma(2^{22})\sigma(M_{23}) = (2^{23} - 1)\sigma(M_{23}) > M_{23} \cdot 2^{23} = 2n$. Thus, n is abundant.
27. Let $n = 2^{k-1}(2^k - 1)$. Since $2^k - 1$ is composite, $\sigma(2^k - 1) > (2^k - 1) + 1 = 2^k$. Therefore, $\sigma(n) = (2^k - 1)\sigma(2^k - 1) > 2^k(2^k - 1) = 2n$, so n is abundant.
29. Because $\sigma(30240) = \sigma(2^5 \cdot 3^3 \cdot 5 \cdot 7) = 63 \cdot 40 \cdot 6 \cdot 8 = 120{,}960$, 30,240 is 4-perfect.
31. $2620 = 2^2 \cdot 5 \cdot 131$ and $2924 = 2^2 \cdot 17 \cdot 43$. Because $\sigma(2620) = 7 \cdot 6 \cdot 132 = 5544 = 7 \cdot 18 \cdot 44 = \sigma(2924) = 2924 + 2620$, 2620 and 2924 are amicable numbers.
33. When $n = 2$, $2^n ab = 4(3 \cdot 4 - 1)(3 \cdot 2 - 1) = 4 \cdot 11 \cdot 5 = 220$ and $2^n c = 4(9 \cdot 8 - 1) = 284$ are amicable numbers; likewise, $n = 4$ yields the amicable numbers 17,296 and 18,416.
35. Sum of the first 17 primes $= 2 + 3 + 5 + 7 + 11 + 13 + 17 + 19 + 23 + 29 + 31 + 37 + 41 + 43 + 47 + 53 + 59 = 440 = 2 \cdot 220$; sum of their squares $= 16756 = 59 \cdot 284$.
37. $\nu(20) = 1 \cdot 2 \cdot 4 \cdot 5 \cdot 10 = 20^2$
39. $\nu(24) = 1 \cdot 2 \cdot 3 \cdot 4 \cdot 6 \cdot 8 \cdot 12 = 24^3$
41. $\nu(pq) = 1 \cdot p \cdot q = pq$
43. $\sigma(p) = 1 + p < 2p$, so p is deficient
45. $\sigma(p^e) = (p^{e+1} - 1)/(p - 1) < 2p^e$, so p^e is deficient.
47. **proof:** Because p is prime and $2^e p$ is abundant, $\sigma(2^e p) = (2^{e+1} - 1)(p + 1) > 2 \cdot 2^e p$. That is, $p2^{e+1} + 2^{e+1} - p - 1 > p2^{e+1}$. So $2^{e+1} > p + 1$; that is, $e + 1 \geq \lg(p + 1)$. Thus, $e \geq \lceil \lg 2(p + 1) \rceil - 1$.
49. **proof:** $\sigma(\sigma(2^k)) = \sigma(2^{k+1} - 1) = 2^{k+1} = 2 \cdot 2^k$. Therefore, 2^k is superperfect.
51. **proof:** Because m and n are amicable, $\sigma(m) = m + n = \sigma(n)$. $\sigma(m) = \sum_{d|m} \frac{m}{d} = m \sum_{d|m} \frac{1}{d}$.

    Therefore, $\left(\sum_{d|m} \frac{1}{d}\right)^{-1} = \frac{m}{\sigma(m)}$. Similarly, $\left(\sum_{d|n} \frac{1}{d}\right)^{-1} = \frac{n}{\sigma(n)}$. Therefore,

$$\left(\sum_{d|m} \frac{1}{d}\right)^{-1} + \left(\sum_{d|n} \frac{1}{d}\right)^{-1} = \frac{m}{\sigma(m)} + \frac{n}{\sigma(n)} = \frac{m+n}{\sigma(m)} = 1$$

53. **proof:** $\nu(n) = \prod_{i=0}^{e-1} p^i = p^{e(e-1)/2} = n^{(e-1)/2}$. Thus, $\nu(n)$ is a power of n if and only if $(e - 1)/2$ is an integer; that is, if and only if e is odd and $e > 1$.
55. **proof:** Let $\nu(n) = n^k$. Then $n^{\tau(n)/2} = n^k$, so $\tau(n) = 2(k + 1)$. Thus, $\tau(n)$ is even and is ≥ 4. Conversely, let $\tau(n)$ be even and ≥ 4, say, $2k$, where $k \geq 2$. Then $\nu(n) = n^{k-1}$, which is a power of n.
57. $$\frac{1}{h(n)} = \frac{1}{\tau(n)}\left(\sum_{d|n} \frac{1}{d}\right) = \frac{1}{\tau(n)}\left(\sum_{d|n} \frac{1}{n/d}\right) = \frac{1}{\tau(n)}\left(\sum_{d|n} \frac{d}{n}\right) = \frac{1}{n\tau(n)} \sum_{d|n} d = \frac{\sigma(n)}{n\tau(n)}$$

    Therefore, $h(n) = \frac{n\tau(n)}{\sigma(n)}$.

Exercises 8.4 (p. 397)

1. 1, 11, 111, 1111, 11111.
3. Number of digits in $M_{2281} = \lceil 2281 \log 2 \rceil = 687$.
5. Number of digits in $M_{110503} = \lceil 110503 \log 2 \rceil = 33{,}265$.

7. Because $4^k \equiv 4 \pmod{10}$ if k is odd, $2^{127} = (2^2)^{63} \cdot 2 = 4^{63} \cdot 2 = 4 \cdot 2 \equiv 8 \pmod{10}$, $M_{127} \equiv 7 \pmod{10}$, so M_{127} ends in 7.
9. When k is even, $4^k \equiv 6 \pmod{10}$. So $2^{11213} = (2^2)^{5606} \cdot 2 = 4^{5606} \cdot 2 \equiv 6 \cdot 2 \equiv 2 \pmod{10}$. Then $M_{11213} \equiv 1 \pmod{10}$, so M_{11213} ends in 1.
11. Because $24^k \equiv 76 \pmod{100}$ when k is even, $2^{127} = (2^{10})^{12} \cdot 2^7 = 24^{12} \cdot 28 \equiv 76 \cdot 28 \equiv 28 \pmod{100}$. Thus, $M_{127} \equiv 27 \pmod{100}$, so M_{127} ends in 27.
13. Because $24^k \equiv 76 \pmod{100}$ when k is even, $2^{9941} = (2^{10})^{994} \cdot 2 = 24^{994} \cdot 2 \equiv 76 \cdot 2 \equiv 52 \pmod{100}$. Thus, $M_{9941} \equiv 51 \pmod{100}$, so M_{9941} ends in 51.
15. Because $2^{103} \equiv 8 \pmod{1000}$, $2^{1279} = (2^{103})^{12} \cdot 2^{25} \cdot 2^{18} \equiv 8^{12} \cdot 432 \cdot 144 \equiv 088 \pmod{1000}$, $M_{1279} \equiv 087 \pmod{1000}$. Thus, M_{1279} ends in 087.
17. Because $2^{103} \equiv 8 \pmod{1000}$, $2^{110503} = (2^{103})^{1072} \cdot 2^{87} \equiv 8^{1072} \cdot 2^{87} = 2^{3303} = (2^{103})^{32} \cdot 2^7 \equiv 8^{32} \cdot 2^7 = 2^{103} \equiv 008 \pmod{1000}$. Thus, $M_{110503} \equiv 007 \pmod{1000}$, so M_{110503} ends in 007.
19. **proof:** When k is even, $4^k \equiv 6 \pmod{10}$. Because $n \geq 2$, $2^{2^n} = 2^{2 \cdot 2^{n-1}} = 4^{2^{n-1}} \equiv 6 \pmod{10}$, $f_n \equiv 7 \pmod{10}$. Thus, if $n \geq 2, f_n$ ends in 7.
21. By Exercise 20, the number of digits in f_{13} is given by $\lceil 2^{13} \log 2 \rceil = 2467$.
23. By Exercise 20, f_{23} contains $\lceil 2^{23} \log 2 \rceil = 2{,}525{,}223$ digits.

25. 1 followed by $2^n - 1$ zeros and then a 1. 27. p ones followed by $p - 1$ zeros.

29. There are $\varphi(10) = 4$ positive integers ≤ 10 and relatively prime to 10, namely, 1, 3, 7, and 9. Because $k|4$, $k = 1, 2$, or 4. Clearly, $k \neq 1$. Since $3^2 \not\equiv 1 \pmod{10}$, $k \neq 2$. So $k = 4$.
31. There are $\varphi(18) = 6$ positive integers ≤ 18 and relatively prime to 18, namely, 1, 5, 7, 11, 13, and 17. Because $k|6$, $k = 1, 2, 3$, or 6. Clearly, $k \neq 1$. Since $5^2 \not\equiv 1 \pmod{18}$, $k \neq 2$. Besides, $5^3 \not\equiv 1 \pmod{18}$, so $k \neq 3$. Thus, $k = 6$.
33. $M_7 = 2^7 - 1 = 127$. By Theorem 8.12, every factor of M_7 is of the form $14k + 1$. Because there are no primes of the form $14k + 1$ that are $\leq \lfloor\sqrt{127}\rfloor = 11$, M_7 is a prime.
35. If $M_{17} = 2^{17} - 1 = 131071$ is composite, it must have a prime factor $\leq \lfloor\sqrt{131071}\rfloor = 362$, which is of the form $34k + 1$. Such primes are 103, 137, 239, and 307. But none of them is a factor of M_{17}, so it is a prime.
37. Every prime factor of M_{29} is of the form $58k + 1$. When $k = 4$, $58k + 1 = 233$ and $233|M_{29}$, so M_{29} is composite.
39. The prime factors of M_{43} are of the form $86k + 1$. When $k = 5$, $86k + 1 = 431$ and $431|M_{43}$, so M_{43} is composite.
41. **proof:** Let $m = 2^{n/2}$.

$$\text{Sum of the cubes of the first } m \text{ odd positive integers} = \sum_{i=1}^{m} (2i-1)^3$$

$$= 8\sum_{i=1}^{m} i^3 - 12\sum_{i=1}^{m} i^2 + 6\sum_{i=1}^{m} i - \sum_{i=1}^{m} 1$$

$$= 8[m(m+1)/2]^2 - 12[m(m+1)(2m+1)/6] + 6[m(m+1)/2] - m$$

$$= 2m^2(m+1)^2 - 2m(m+1)(2m+1) + 3m(m+1) - m$$

$$= m(m+1)[(2m^2 + 2m) - (4m+2)] - m$$

$$= m(m+1)(2m^2 - 2m + 1) - m = m[(m+1)(2m^2 - 2m + 1) - 1]$$
$$= m(2m^3 - m) = m^2(2m^2 - 1) = 2^n(2^{n+1} - 1)$$

43. **proof:** Let $n = 2k$. Then, by Exercise 41,

$$S = 2^{2k}(2^{2k+1} - 1) = 4^k(2 \cdot 4^k - 1)$$
$$\equiv \begin{cases} 4(2 \cdot 4 - 1) \equiv 8 \pmod{10} & \text{if } k = n/2 \text{ is odd} \\ 6(2 \cdot 6 - 1) \equiv 6 \pmod{10} & \text{otherwise} \end{cases}$$

Now consider the case $n/2 = 2j + 1$ is odd. Then $S = 2^{4j+2}(2^{4j+3} - 1) = 4 \cdot 16^j(8 \cdot 16^j - 1)$. But $16^i \equiv 76 \pmod{100}$ if $i \equiv 0 \pmod 5$; $16^i \equiv 16 \pmod{100}$ if $i \equiv 1 \pmod 5$; $16^i \equiv 56 \pmod{100}$ if $i \equiv 2 \pmod 5$; $16^i \equiv 96 \pmod{100}$ if $i \equiv 3 \pmod 5$; and $16^i \equiv 36 \pmod{100}$ if $i \equiv 4 \pmod 5$. In each case, it can be shown that $S \equiv 28 \pmod{100}$. Thus, S ends in 6 or 28.

45. Clearly, the digital root of M_2 is 3 and that of M_3 is 7. Suppose $p > 3$. Then p is of the form $6k + 1$ or $6k + 5$. When $p = 6k + 1$, $M_p \equiv 1 \pmod 9$; when $p = 6k + 5$, $M_p \equiv 4 \pmod 9$. Thus, the digital root of M_p is 1 if $p \equiv 1 \pmod 6$ and 4 if $p \equiv 5 \pmod 6$.

47. It follows from Segner's formula that $C_n = 2(C_0C_{n-1} + C_1C_{n-2} + \cdots + C_{\frac{n-3}{2}}C_{\frac{n+1}{2}}) + C^2_{\frac{n-1}{2}}$. Consequently, for $n > 0$, C_n is odd if and only if both n and $C_{\frac{n-1}{2}}$ are odd. The same argument implies that C_n is odd if and only if $\frac{n-1}{2}$ and $C_{\frac{n-3}{4}}$ are both odd or $\frac{n-1}{2} = 0$. Continuing this finite descent, it follows that C_n is odd if and only if $C_{\frac{n-(2^m-1)}{2^m}}$ is odd, where $m \geq 1$. But the least value of k for which C_k is odd is $k = 0$. Thus, the sequence of these *if and only if* statements terminates when $\frac{n - (2^m - 1)}{2^m} = 0$; that is, when $n = 2^m - 1$, a Mersenne number.

Exercises 8.5 (p. 405)

1. 101 is prime, so $\mu(101) = -1$.
3. Because $2047 = 23 \cdot 89$, $\mu(2047) = (-1)^2 = 1$.
5. $\mu(p) = -1$.
7. Because $2^2 | 2^{p-1}(2^p - 1)$, $\mu(2^{p-1}(2^p - 1)) = 0$.
9. $\sum_{d|5} \mu(d)\tau(5/d) = \mu(1)\tau(5) + \mu(5)\tau(1) = 1 \cdot 2 + (-1) \cdot 1 = 1$.
11. $$\sum_{d|10} \mu(d)\tau(10/d) = \mu(1)\tau(10) + \mu(2)\tau(5) + \mu(5)\tau(2) + \mu(10)\tau(1)$$
$$= 1 \cdot 4 + (-1) \cdot 2 + (-1) \cdot 2 + (-1)^2 \cdot 1 = 4 - 2 - 2 + 1 = 1$$
13. $\sum_{d|5} \mu(d)\sigma(5/d) = \mu(1)\sigma(5) + \mu(5)\sigma(1) = 1 \cdot 6 + (-1) \cdot 1 = 6 - 1 = 5$
15. $$\sum_{d|10} \mu(d)\sigma(10/d) = \mu(1)\sigma(10) + \mu(2)\sigma(5) + \mu(5)\sigma(2) + \mu(10)\sigma(1)$$
$$= 1 \cdot 18 + (-1) \cdot 6 + (-1) \cdot 3 + (-1)^2 \cdot 1 = 18 - 6 - 3 + 1 = 10$$
17. $\varphi(23) = 23 \sum_{d|23} \frac{\mu(d)}{d} = 23\left[\frac{\mu(d)}{1} + \frac{\mu(d)}{23}\right] = 23(1 - 1/23) = 22.$

19. $$\begin{aligned}\varphi(36) &= 36\sum_{d|36}\frac{\mu(d)}{d}\\ &= 36\left[\frac{\mu(1)}{1}+\frac{\mu(2)}{2}+\frac{\mu(3)}{3}+\frac{\mu(4)}{4}+\frac{\mu(6)}{6}+\frac{\mu(9)}{9}+\frac{\mu(18)}{18}+\frac{\mu(36)}{36}\right]\\ &= 36(1-1/2-1/3+0/4+1/6+0/9+0/18+0/36)=12\end{aligned}$$

21. $\mu(pq)=(-1)^2=(-1)(-1)=\mu(p)\mu(q)$

23. $\mu(p^2qr)=0=0\cdot(-1)\cdot(-1)=\mu(p^2)\mu(q)\mu(r)$

25. $$\begin{aligned}\sum_{d|pq}\mu(d)\tau(d) &= \mu(1)\tau(1)+\mu(p)\tau(p)+\mu(q)\tau(q)+\mu(pq)\tau(pq)\\ &= 1\cdot 1+(-1)\cdot 2+(-1)\cdot 2+(-1)^2\cdot 4=1-2-2+4=1\end{aligned}$$

27. $$\begin{aligned}\sum_{d|p^2qr}\mu(d)\tau(d) &= \mu(1)\tau(1)+\mu(p)\tau(p)+\mu(q)\tau(q)+\mu(r)\tau(r)+\mu(p^2)\tau(p^2)\\ &\quad+\mu(pq)\tau(pq)+\mu(pr)\tau(pr)+\mu(qr)\tau(qr)+\mu(p^2q)\tau(p^2q)\\ &\quad+\mu(p^2r)\tau(p^2r)+\mu(pqr)\tau(pqr)+\mu(p^2qr)\tau(p^2qr)\\ &= 1-2-2-2+0+4+4+4+0+0-8+0=-1\end{aligned}$$

29. $\sum_{d|pq}\mu(d)\tau(d)=\mu(1)\sigma(1)+\mu(p)\sigma(p)=1\cdot 1+(-1)\cdot(p+1)=1-p-1=-p$

31. $$\begin{aligned}\sum_{d|p^2q}\mu(d)\sigma(d) &= \mu(1)\sigma(1)+\mu(p)\sigma(p)+\mu(q)\tau(q)+\mu(p^2)\sigma(p^2)+\mu(pq)\tau(pq)\\ &\quad+\mu(p^2q)\tau(p^2q)\\ &= 1-p-1-q-1+1+p+q+pq=pq\end{aligned}$$

33. $\sum_{d|p}\mu(d)\tau(d)=-p=(-1)^1p$; $\sum_{d|pq}\mu(d)\tau(d)=pq=(-1)^2pq$; $\sum_{d|pqr}\mu(d)\tau(d)=-pqr=(-1)^3pqr$. Thus, we conjecture that $\sum_{d|n}\mu(d)\tau(d)=(-1)^kn$, where n is a product of k distinct primes.

35. Because $104=2^3\cdot 3$, $\lambda(104)=(-1)^{3+1}=1$.

37. $3024=2^4\cdot 3^3\cdot 7$, so $\lambda(3024)=(-1)^{4+3+1}=1$

39. $\sum_{d|12}\lambda(d)=\lambda(1)+\lambda(2)+\lambda(3)+\lambda(4)+\lambda(6)+\lambda(12)=1-1-1+1+1-1=0$

41. $\sum_{d|28}\lambda(d)=\lambda(1)+\lambda(2)+\lambda(4)+\lambda(7)+\lambda(14)+\lambda(28)=1-1+1-1+1-1=0$

43. **proof:** By Theorem 8.5, $n=\sum_{d|n}\varphi(d)$. Therefore, by Möbius inversion formula, $\varphi(n)=\sum_{d|n}\mu(d)(n/d)$.

45. **proof:** $\sum_{d|n}\frac{\mu(d)}{d}=\sum_{i,j}\frac{\mu(d)}{p^iq^j}=\left(\sum_i\frac{\mu(d)}{p^i}\right)\left(\sum_j\frac{\mu(d)}{q^j}\right)=(1-1/p)(1-1/q)$

47. **proof:** By Theorem 8.18,

$$\varphi(p^e)=\sum_{d|p^e}\mu(d)(p^e/d)=\sum_{i=0}^{e}\mu(d)(p^e/p^i)=\mu(1)p^e+\mu(p)(p^e/p)=p^e-p^{e-1}$$

49. **proof:** Let $m=\prod_i p_i^{e_i}$ and $n=\prod_j q_j^{f_j}$, where $(p_i,q_j)=1$. Then $mn=\prod_{i,j}p_i^{e_i}q_j^{f_j}$, so

$$\lambda(mn)=(-1)^{\sum e_i+\sum f_j}=(-1)^{\sum e_i}(-1)^{\sum f_j}=\lambda(m)\lambda(n)$$

Thus, λ is multiplicative.

51. **proof:** Let $f(n) = \sum_{d|n} \lambda(d)$. Then $f(p^e) = \sum_i \lambda(p^i) = \lambda(1) + \lambda(p) + \cdots + \lambda(p^e) = 1$ if and only if e is even. Let $n = \prod_{i=1}^{k} p_i^{e_i}$. Since λ is multiplicative, so is f. Therefore, $f(n) = \prod_i f(p_i^{e_i}) = 1^k$ if and only if each e_i is even; that is, if and only if n is a square.

53. **proof:** Let $(m, n) = 1$. Every divisor d of mn can be written as $d = xy$, where $x|m$, $y|n$, and $(x, y) = 1$. So, by the inversion formula,

$$f(mn) = \sum_{d|mn} \mu(d)F(mn/d) = \sum_{\substack{x|m\\ y|n}} \mu(xy)F(mn/xy) = \sum_{\substack{x|m\\ y|n}} \mu(x)\mu(y)F(m/x)F(n/y)$$

$$= \sum_{x|m} \mu(x)F(m/x) \sum_{y|n} \mu(y)F(m/y) = f(m)f(n) \quad \text{so } f \text{ is multiplicative.}$$

Review Exercises (p. 408)

1. sum $= \varphi(2020) = 800$.

3. $\sigma(945) = \sigma(3^3)\sigma(5)\sigma(7) = 40 \cdot 6 \cdot 8 = 1920 > 1890 = 2 \cdot 945$, so 945 is abundant. $\sigma(45{,}045) = \sigma(3^2)\sigma(5)\sigma(7)\sigma(11)\sigma(13) = 13 \cdot 6 \cdot 8 \cdot 12 \cdot 14 = 104{,}832 > 90{,}090 = 2 \cdot 45{,}045$, so 45,045 is also abundant.

5. $\sigma(12{,}285) = \sigma(3^3)\sigma(5)\sigma(7)\sigma(13) = 40 \cdot 6 \cdot 8 \cdot 14 = 26{,}880$; $\sigma(14{,}595) = \sigma(3)\sigma(5)\sigma(7)$ $\sigma(139) = 4 \cdot 6 \cdot 8 \cdot 140 = 26{,}880 = 12{,}285 + 14{,}595$. Therefore, they are amicable.

7. Every factor of $M_{53} = 2^{53} - 1$ is of the form $106k + 1$. When $k = 60$, $106k + 1 = 6361$ and $6361|M_{53}$.

9. $\sum_{i=0}^{n} \sigma(2^i) = \sum_{i=0}^{n} (2^{i+1} - 1) = \sum_{i=0}^{n} 2^{i+1} - (n+1) = 2^{n+2} - n - 2$.

11. σ and τ are multiplicative. But $(\sigma + \tau)(4 \cdot 7) = \sigma(28) + \tau(28) = 56 + 6 = 62 \neq 100 = 10 \cdot 10 = (\sigma + \tau)(4) \cdot (\sigma + \tau)(7)$, so $\sigma + \tau$ is not multiplicative.

13. **proof:** Let $(m, n) = 1$. Then $(fg)(mn) = f(mn) \cdot g(mn) = f(m)f(n) \cdot g(m)g(n) = (fg)(m) \cdot (fg)(n)$. Thus, fg is multiplicative.

15. **proof:** Every hexagonal number is of the form $h_n = n(2n - 1)$. Let $n = 2^{p-1}$. Then $h_n = 2^{p-1}(2^p - 1)$, an even perfect number.

17. **proof:** Let $n = \prod_i p_i$, where $p_i = 2^{q_i} - 1$ is a Mersenne prime. Then $\sigma(n) = \prod_i 2^{q_i} = 2^e$, where $e = \sum_i q_i$, so $\sigma(n)$ is a power of 2.

19. **proof:** $\sigma(f_n) - \varphi(f_n) = (1 + f_n) - (f_n - 1) = 2$.

21. **proof** (by induction on e): When $e = 1$, $a^{\varphi(n)} = a^{p-1} \equiv 1 \pmod{p}$, which is true by Fermat's Little Theorem. So assume it is true for an arbitrary positive exponent e: $a^{\varphi(p^e)} \equiv 1 \pmod{p^e}$; that is, $a^{\varphi(p^e)} = 1 + kp^e$ for some positive integer k. Then

$$a^{\varphi(p^{e+1})} = a^{p\varphi(p^e)} = [a^{\varphi(p^e)}]^p = (1 + kp^e)^p = \sum_{r=0}^{p} \binom{p}{r} (kp^e)^r$$

$$= 1 + \sum_{r=0}^{p} \binom{p}{r} (k^r p^{er}) \equiv 1 \pmod{p^{e+1}}$$

since $p\left|\binom{p}{r}\right.$ when $0 < r < p$. Thus, by PMI, the result is true for every positive exponent e.

23. **proof:** Let $N = 2^{p-1}(2^p - 1)$, where both p and $2^{p-1} - 1$ are primes. When $p = 2$, $N = 6$ and when $p = 5$, $N = 496$, and both end in 6. So assume $p \geq 7$. Then $p = r \pmod{10}$, where $r = 1, 3, 7$, or 9. Let $p = 10k + r$. Then $N = 2^{10k+r-1}(2^{10k+r} - 1)$. Since $2^{10} \equiv 4 \pmod{10}$, $2^{10} \equiv 4^k \pmod{10}$. Therefore, $N = 4^k 2^{r-1}(4^k 2^r - 1)$. But $4^k \equiv 4$ or $6 \pmod{10}$.

   case 1 Let $4^k \equiv 4 \pmod{10}$. Then $N \equiv 4 \cdot 2^{r-1}(4 \cdot 2^r - 1) \pmod{10}$. If $r = 1$, then $N \equiv 4(8-1) \equiv 8 \pmod{10}$; if $r = 3$, $N \equiv 4 \cdot 4(4 \cdot 8 - 1) \equiv 6 \cdot 1 \equiv 6 \pmod{10}$; if $r = 7$, $N \equiv 4 \cdot 2^6(4 \cdot 2^7 - 1) \equiv 6(4 \cdot 8 - 1) \equiv 6 \pmod{10}$; and if $r = 9$, $N \equiv 4 \cdot 2^8(4 \cdot 2^9 - 1) \equiv 4(4 \cdot 2 - 1) \equiv 8 \pmod{10}$.

   case 2 Let $4^k \equiv 6 \pmod{10}$. Then $N \equiv 6 \cdot 2^{r-1}(6 \cdot 2^r - 1) \pmod{10}$. This case is quite similar. Thus, in both cases, N ends in 6 or 8.

25. **proof:** Let $n = 2^k \cdot 3 \cdot p$. Since $\sigma(n) = 3n$, $(2^{k+1} - 1) \cdot 4 \cdot (p+1) = 3 \cdot 2^k \cdot 3p$. Because $p+1$ is even, let $p + 1 = 2^\ell q$, where q is odd, $\ell > 1$, and $(p, q) = 1$. Then $(2^{k+1} - 1)2^{\ell+2} \cdot q = 3^2 \cdot 2^k \cdot p$. Clearly, $\ell + 2 = k$, so $k \geq 3$. Therefore,

$$(2^{\ell+3} - 1)q = 9p \tag{1}$$

   Because $(p, q) = 1$, $q|9$. Therefore, $q = 1, 3$, or 9.

   case 1 Let $q = 1$. Then $p + 1 = 2^\ell$; that is, $p = 2^\ell - 1$. Therefore, by equation (1), $2^{\ell+3} - 1 = 9(2^\ell - 1)$; that is, $8 \cdot 2^\ell - 1 = 9 \cdot 2^\ell - 9$. Therefore, $2^\ell = 8$, so $\ell = 3$, $k = 5$, and $p = 2^3 - 1 = 7$. Thus $n = 2^5 \cdot 3 \cdot 7 = 672$.

   case 2 Let $q = 3$. Then $p + 1 = 3 \cdot 2^\ell$; that is, $p = 3 \cdot 2^\ell - 1$. Therefore, by equation (1), $(8 \cdot 2^{\ell+3} - 1) \cdot 3 = 9(3 \cdot 2^\ell - 1)$; that is, $8 \cdot 2^\ell - 1 = 9 \cdot 2^\ell - 3$. Therefore, $2^\ell = 2$, so $\ell = 1 \cdot k = 3$, and $p = 3 \cdot 2^1 - 1 = 5$. Thus, $n = 2^3 \cdot 3 \cdot 5 = 120$.

   case 3 Let $q = 9$. Then $p + 1 = 9 \cdot 2^\ell$; that is, $p = 9 \cdot 2^\ell - 1$. Therefore, by equation (1), $9(8 \cdot 2^{\ell+3} - 1) \cdot 3 = 9(9 \cdot 2^\ell - 1)$; that is, $8 \cdot 2^\ell - 1 = 9 \cdot 2^\ell$, which is impossible.

   Thus, 120 and 672 are the only 3-perfect numbers of the form $2^k \cdot 3 \cdot p$.

Supplementary Exercises (p. 409)

1. 5775, 5776.
3. **proof:** $\sigma(M) = \sigma(d)(a+1)(b+1)$ and $\sigma(N) = \sigma(d)(c+1)$. Because $\sigma(M) = \sigma(N)$, $\sigma(d)(a+1)(b+1) = \sigma(d)(c+1)$; that is, $(a+1)(b+1) = (c+1)$. Because $\sigma(M) = \sigma(d)(a+1)(b+1)$ and $\sigma(M) = M + N = d(ab+c)$, $\sigma(d)(a+1)(b+1) = d(ab+c)$.
5. **proof:** Because $\sigma(2^n) = 2^{n+1} - 1$, by equation (3), $(2^{n+1} - 1)(ab + a + b + 1) = 2^{n+1}ab + 2^n a + 2^n b$; that is, $ab - 2^n a - 2^n b + a + b - 2^{n+1} + 1 = 0$. Adding 2^{2n} to both sides, this yields, $a[b - (2^n - 1)] - b(2^n - 1) + (2^n - 1)^2 = 2^{2n}$; that is, $[a - (2^n - 1)][b - (2^n - 1)] = 2^{2n}$.
7. Use $m = 1$, $a = 3 \cdot 2^n - 1$, $b = 3 \cdot 2^{n-1} - 1$ and $c = 9 \cdot 2^{2n-1} - 1$.

9. **proof:** $\sigma(d)/d = \sigma(3^2 \cdot 7 \cdot 13)/9 \cdot 7 \cdot 13 = 1456/9 \cdot 7 \cdot 13 = 16/9$. Therefore, By Exercise 8, $2 - (a+b+2)/(a+1)(b+1) = 16/9$; that is, $2(ab+a+b+1) = 9(a+b+2)$. Then $4ab - 14a - 14b + 49$; that is, $(2a-7)(2b-7) = 81$.
11. When $a = 5$, $b = 17$, and $c = 107$, $M = 3^2 \cdot 7 \cdot 13 \cdot 5 \cdot 17 = 69{,}615$ and $N = 3^2 \cdot 7 \cdot 13 \cdot 107 = 87{,}633$.
13. **proof** (by W. D. Blair): Because $(m_i, m_j) = 1$ for $i \neq j$, by Euler's theorem, $m_i^{\varphi(m_i)} \equiv 1 \pmod{m_j}$. Because φ is multiplicative, $\varphi(m) = \varphi(m_1)\varphi(m_2)\cdots\varphi(m_n)$. Hence, $m_i^{\varphi(m)/\varphi(m_i)} \equiv 1 \pmod{m_j}$ for all $i \neq j$. Also $m_j^{\varphi(m)/\varphi(m_j)} \equiv 0 \pmod{m_j}$. Thus for every j, $1 \le j \le n$, $\sum_{i=1}^{n} m_i^{\varphi(m)/\varphi(m_j)} \equiv n - 1 \pmod{m_j}$. Because the m_j's are relatively prime, the result now follows.
15. $\sigma(2) = 3$, $\sigma(3) = 4$, $\sigma(4) = 7$, and $\sigma(5) = 6$.
17. **proof:** Let $f(n) = \sum_{d|n} \mu(d)\tau(d)$. Since μ and τ are multiplicative, so is f. We have $f(p^e) = \sum_{d|n} \mu(d)\tau(d) = \mu(1)\tau(1) + \mu(p)\tau(p) + 0 = 1 - 2 = -1$. So
$$F(n) = \prod_{i=1}^{k} f(p_i^{e_i}) = \prod_{i=1}^{k} (-1) = (-1)^k$$
19. **proof:** Let $f(n) = \sum_{d|n} \mu(d)\varphi(d)$. Since μ and φ are multiplicative, so is f. We have $f(p^e) = \sum_{d|n} \mu(d)\varphi(d) = \mu(1)\varphi(1) + \mu(p)\tau(p) + 0 = 1 - (p-1) = 2 - p$. So
$$F(n) = \prod_{i=1}^{k} f(p_i^{e_i}) = \prod_{i=1}^{k} (2 - p_i)$$
21. Using $f = \tau$, $\sum_{d|n} \mu(d)\tau(d) = \prod_{i=1}^{k} [1 - \tau(p_i)] = \prod_{i=1}^{k} (1-2) = (-1)^k$
23. Using $f = \varphi$, $\sum_{d|n} \mu(d)\varphi(d) = \prod_{i=1}^{k} [1 - \varphi(p_i)] = \prod_{i=1}^{k} [1 - (p_i - 1)] = \prod_{i=1}^{k} (2 - p_i)$
25. Using $f = 1/d$, $\sum_{d|n} \frac{\mu(d)}{d} = \prod_{i=1}^{k} (1 - 1/p_i)$
27. $\sigma(16) = \sigma(2^4) = 2^5 - 1 = 31 = 2 \cdot 16 - 1$, so 16 is near-perfect.

Chapter 9 Cryptology

Exercises 9.1 (p. 424)

1. DOOLV ZHOOW KDWHQ GVZHO O
3. NECESSITY IS THE MOTHER OF INVENTION
5. YZWPR LNJTD DZRCP LELDS ZYPDE J
7. TIME IS THE BEST MEDICINE
9. HMCFU ZXWKT HPMBF JHIXB WXGTS TG
11. KNHOX AETRM KUNWO GKVEW TEZMB MZ
13. THE BUCK STOPS HERE

15. none
17. D, Q
19. A FOOL SEES NOT THE SAME TREE THAT A WISE MAN SEES
21. $\varphi(26) \cdot 26 = 12 \cdot 26 = 312$
23. IMPDG YZMGY PMK
25. WE ARE ALWAYS GETTING READY TO LIVE BUT NEVER LIVING
27. OIIOID CBXJWZ UBWLHF QZPUHK JMZLCK QBWLWT KMCJIJ
29. LIFE WONT WAIT

Exercises 9.2 (p. 430)

1. JV BF NA OI HO AQ
3. SEND MORE MONEY
5. ZRO RWV CAJ DJW ZWL ABN NDE QTY
7. A HUNGRY MAN IS NOT A FREE MAN
9. Solving the linear system $\begin{bmatrix} 5 & 13 \\ 3 & 18 \end{bmatrix} \begin{bmatrix} x \\ y \end{bmatrix} \equiv \begin{bmatrix} x \\ y \end{bmatrix}$ (mod 26) yields $x = 0 = y$. Thus, the block left fixed is AA.
11. The enciphering key for the product cipher is $\begin{bmatrix} 7 & 15 \\ 3 & 4 \end{bmatrix} \begin{bmatrix} 2 & 11 \\ 5 & 13 \end{bmatrix} \equiv \begin{bmatrix} 11 & 12 \\ 0 & 7 \end{bmatrix}$ (mod 26), so the cipher text is XA FE DM DA GP.
13. By Exercise 11, the enciphering matrix for the product cipher is $\begin{bmatrix} 11 & 12 \\ 0 & 7 \end{bmatrix}$ (mod 26). Let $\begin{bmatrix} x \\ y \end{bmatrix}$ be a fixed block. Then the matrix congruence $\begin{bmatrix} 11 & 12 \\ 0 & 7 \end{bmatrix} \begin{bmatrix} x \\ y \end{bmatrix} \equiv \begin{bmatrix} x \\ y \end{bmatrix}$ (mod 26) yields the linear system

$$10x + 12y \equiv 0 \pmod{26}$$
$$6y \equiv 0 \pmod{26}$$

Solving this system, we get $x = y = 0$; $x = 0, y = 13$; $x = 13, y = 0$; and $x = 13 = y$. Thus, there are four fixed blocks: AA, AN, NA, and NN.
15. A LABOR OF LOVE INDEED

Exercises 9.3 (p. 433)

1. 6
3. Because $AB = 0001 = P$, $C = P^e$ (mod p), and $1^e \equiv 1$ (mod p), it follows that $C = 0001 = P$.
5. 1443 0748 1917 1557 2620 0351
7. 0388 1048 2080 1093 1224 1404 0335 14xx
9. PANCAKE
11. FIREBRAND
13. $e = 2^{7 \cdot 17} \equiv 30 \pmod{131}$
15. Because $83 \cdot 30 \equiv 1 \pmod{131}$, the common deciphering key is 83.

Exercises 9.4 (p. 442)

1. 1466 0248 1575 0315
3. 1836 1683 1719 2466
5. With $d = 251$, the plaintext is MORNING.
7. With $d = 2273$, the plaintext is MORNING.
9. Because $\varphi(n) = \varphi(pq) = \varphi(p)\varphi(q) = (p-1)(q-1) = pq - p - q + 1 = n - p - q + 1$, $p + q = n - \varphi(n) + 1$.
11. Using Exercises 9 and 10,

$$p = [(p+q)+(p-q)]/2 = \left\{n - \varphi(n) + 1 + \sqrt{[n-\varphi(n)+1]^2 - 4n}\right\}/2$$

and $$q = [(p+q)-(p-q)]/2 = \left\{n - \varphi(n) + 1 - \sqrt{[n-\varphi(n)+1]^2 - 4n}\right\}/2$$

13. By Exercises 9 and 10, $p+q = 3953 - 3828 + 1 = 126$ and $p - q = \sqrt{126^2 - 4 \cdot 3953} = 8$. Therefore, $p = 67$ and $q = 59$.
15. We have $e_1 = 13$, $d_1 = 2437$, $e_2 = 17$, and $d_2 = 1553$. Using the coding schemes $A = D_2(P) \equiv P^{d_2} \pmod{n}$ and $S = E_1(A) = A^{e_1} \pmod{n}$, the signed message sent is 0464 1719 1322.
17. We have $e_1 = 13$, $d_1 = 2437$, $e_2 = 17$, and $d_2 = 1553$. Let $A = D_2(P) \equiv P^{d_2} \pmod{2747}$ and $S = E_1(A) \equiv A^{e_1} \pmod{2747}$, Betsey's signed message. Then $D_1(S) = A^{e_1 d_1} \equiv A \pmod{2747}$ and $P \equiv A^{e_2 d_2} \pmod{2747}$. Using these schemes, Betsey's plaintext message is MONEY.

Exercises 9.5 (p. 448)

1. No
3. (1, 1, 1, 0, 1)
5. (1, 0, 1, 0, 1)
7. 16, 32, 11, 22
9. 38 22 22 43 49 00 11 70 65
11. 60 49 54 54 00 54 54
13. SEND MORE
15. OPEN BOOK

Review Exercises (p. 450)

1. YDSDX FRZCU DHCLC UDSYW
3. TOP SECRET
5. ETCETRA
7. XSTJEE WBZOBQ
9. CUA HNO MQD
11. HONEY MOON
13. Solving the congruence $\begin{bmatrix} 2 & 11 & 5 \\ 7 & 0 & 4 \\ 9 & 3 & 8 \end{bmatrix}\begin{bmatrix} x \\ y \\ z \end{bmatrix} \equiv \begin{bmatrix} x \\ y \\ z \end{bmatrix} \pmod{26}$, we get $x = y = z = 0$, or $x = 13 = y$ and $z = 0$. Thus, the blocks left fixed are AAA and NNA.
15. 2524 2599 0546 2338
17. 2470 2652 0996
19. 2737 0696 2524 3514 3372
21. GOOD JOB
23. 1532 1426
25. FAT
27. No
29. (1, 1, 0, 1, 0)
31. 37 80 71 46 37 00 37 85 43
33. MARKET OPEN

Supplementary Exercises (p. 451)

1. $C \equiv c(aP + b) + d \equiv acP + bc + d \pmod{26}$

3. The matrix congruence $\begin{bmatrix} a & b \\ c & d \end{bmatrix}\begin{bmatrix} x \\ y \end{bmatrix} \equiv \begin{bmatrix} x \\ y \end{bmatrix}$ (mod 26) yields the linear system

$$(a-1)x + by \equiv 0 \pmod{26}$$

$$cx + (d-1)y \equiv 0 \pmod{26}$$

Eliminating y, $[(a-1)(d-1) - bc]x \equiv 0 \pmod{26}$; that is, $(ad - bc - a - d + 1)x \equiv 0 \pmod{26}$. This congruence has $k = (ad - bc - a - d + 1, 26)$ solutions. Thus, k blocks are left fixed by the enciphering matrix.

5. No, because matrix multiplication is not commutative.
7. By Exercise 6, the probability equals $1/p + 1/q - 1/pq < 10^{-99} + 10^{-99} - 10^{-9801} < 2 \cdot 10^{-99}$.
9. **proof:** Let $a_{i+1} > 2a_i$. Then $\sum_{i=1}^{j} a_{i+1} > 2\sum_{i=1}^{j} a_i$; that is, $\sum_{i=2}^{j} a_i + a_{j+1} > 2\sum_{i=2}^{j} a_i + 2a_1$. Then $a_{i+1} > \sum_{i=2}^{j} a_i + 2a_1 = \sum_{i=1}^{j} a_i + a_1$. Thus, $a_{i+1} > \sum_{i=1}^{j} a_i$, so the sequence is superincreasing.

Chapter 10 Primitive Roots and Indices

Exercises 10.1 (p. 463)

1. 6
3. 5
5. Because $\text{ord}_{23} 9 = 11$, $9^{11} \equiv 1 \pmod{23}$; that is, $(-14)^{11} \equiv 1 \pmod{23}$. Then $14^{11} \equiv -1 \pmod{23}$, so $14^{22} \equiv 1 \pmod{23}$. Since $14^2 \not\equiv 1 \pmod{23}$, this implies $\text{ord}_{23} 14 = 22$.
7. Because $5^8 \equiv 1 \pmod{13}$, $\text{ord}_{13} 5 = 1, 2, 4$, or 8. But $5^1 \not\equiv 1 \pmod{13}$ and $5^2 \not\equiv 1 \pmod{13}$. Since $5^4 \equiv 1 \pmod{13}$, $\text{ord}_{13} 5 = 4$.
9. Because $\text{ord}_{13} 4 = 6$, $\text{ord}_{13} 4^5 = 6$ by Corollary 10.3.
11. By Theorem 10.2, $\text{ord}_{11} 7^4 = 10/(10, 4) = 10/2 = 5$.
13. $\text{ord}_{13} 4 = 6$, whereas $\text{ord}_{13}(-4) = \text{ord}_{13} 9 = 1$.
15. $3^4 + 3^3 + 3^2 + 3 + 1 = (3^5 - 1)/(3 - 1) = (3^5 - 1)/2 \equiv 6(3^5 - 1) \equiv 6 \cdot 0 \equiv 0 \pmod{11}$.
17. $4^5 + 4^4 + 4^3 + 4^2 + 4 + 1 = (4^6 - 1)/(4 - 1) = (4^6 - 1)/3 \equiv 9(4^6 - 1) \equiv 9 \cdot 0 \equiv 0 \pmod{13}$.
19. Because $\text{ord}_m 5 \le \varphi(m) \le m - 1$, $22 \le m - 1$; so $m \ge 23$. But $5^{22} \equiv 1 \pmod{23}$, so $m = 23$.
21. Because $\text{ord}_m 7 \le \varphi(m) \le m - 1$, $10 \le m - 1$; so $m \ge 11$. But $7^{10} \equiv 1 \pmod{11}$, so $m = 11$.
23. Let $p = 7$ and $a = 2$. Then $\text{ord}_7 2 = 3 \ne 6 = \text{ord}_7 5 = \text{ord}_7(7 - 2)$.
25. $p = 3$ has exactly one primitive root, namely, 2.
27. Because $5^4 \equiv 1 \pmod{13}$, $5^{1001} = (5^4)^{250} \cdot 5 \equiv 1 \cdot 5 \equiv 5 \pmod{13}$, so the desired remainder is 5.
29. We have $7^3 \equiv 1 \pmod 9$ and $8^3 \equiv 1 \pmod 9$. Therefore, $(7 \cdot 8)^6 = (7^3)^2 \cdot (8^2)^3 \equiv 1 \pmod 9$. Thus, $\text{ord}_9 56 = 1, 2, 3$, or 6. Because $56^1 \equiv 2 \not\equiv 1 \pmod 9$, $56^2 \equiv 4 \not\equiv 1 \pmod 9$, and $56^3 \equiv 8 \not\equiv 1 \pmod 9$, it follows that $\text{ord}_9 56 = 6$.

31. $\varphi(10) = 4$

33. $\varphi(28) = \varphi(4 \cdot 7) = \varphi(4)\varphi(7) = 12$

35. 3, 5

37. 2, 6, 7, 11

39. The product of the incongruent primitive roots is congruent to 1 modulo p.

41. There are four positive integers ≤ 12 and relatively prime to 12: 1, 5, 7, and 11. Because $1^2 \equiv 5^2 \equiv 7^2 \equiv 11^2 \equiv 1 \pmod{12}$, $\text{ord}_{12} 1 = \text{ord}_{12} 5 = \text{ord}_{12} 7 = \text{ord}_{12} 11 = 2 \neq 4 = \varphi(12)$. Therefore, 12 has no primitive roots.

43. The prime factors of R_5 are of the form $q = 2kp + 1$, where $p \geq 5$ and $k \geq 1$. When $p = 5$, $q = 10k + 1$. The numbers 11 and 31 of the form $10k + 1$ are not factors of R_5; but $41|R_5$. Similarly, $239|R_7$, $53|R_{13}$, and $83|R_{41}$.

45. **proof:** Let $\text{ord}_m a = h$ and $\text{ord}_m(a^{-1}) = k$. Because $a^h \equiv 1 \pmod{m}$, $(a^{-1})^h = a^{-h} \equiv 1 \pmod{m}$. So $k|h$. Similarly, $h|k$. Thus, $h = k$.

47. **proof:** Assume m is not a prime. Then $\varphi(m) < m - 1$. Because $\text{ord}_m a|\varphi(m)$, $\text{ord}_m a \leq \varphi(m) < m - 1$. Because this contradicts the hypothesis, m is a prime.

49. **proof:** Let $p - 1 = 4k$, where $k \geq 1$. Let a be a primitive root modulo p, so $\text{ord}_p a = \varphi(p) = p - 1$. Then $(p - a)^{\varphi(p)} \equiv (-a)^{4k} \equiv a^{4k} \equiv a^{p-1} \equiv 1 \pmod{p}$, so $\text{ord}_p(p - a)|\varphi(p)$. Let $\text{ord}_p(p - a) = h < p - 1$. Then $(p - a)^h \equiv 1 \pmod{p}$; that is, $(-a)^h \equiv 1 \pmod{p}$, so $a^{2h} \equiv 1 \pmod{p}$. Then $\text{ord}_p a|2h$; that is, $p - 1|h$. Since $(p - 1)/2|h$ and $h < p - 1$, $(p - 1)/2 = h = 2k$. Thus, h is even. This implies $a^h \equiv (-a)^h \equiv 1 \pmod{p}$, where $h < p - 1$, which is a contradiction.
Conversely, let $p - a$ be a primitive root modulo p. Then $a^{p-1} \equiv (-a)^{p-1} \equiv 1 \pmod{p}$, so $\text{ord}_p a|p - 1$. Suppose $\text{ord}_p a = r$, where $r|p - 1$, but $r < p - 1$. Since $(p - a)^{2r} \equiv (-a)^{2r} \equiv (a^r)^2 \equiv 1 \pmod{p}$, $p - 1|2r$. So $(p - 1)/2|r$. Because $(p - 1)/2$ is even, this implies r is even. Therefore, $(p - a)^r \equiv (-a)^r \equiv a^r \equiv 1 \pmod{p}$, where $r < p - 1$. Because this is a contradiction, $\text{ord}_p a = p - 1$.

51. **proof:** Because $\text{ord}_m a = hk$, by Theorem 10.2, $\text{ord}_m(a^h) = hk/(hk, h) = hk/h = k$.

53. **proof:** Because $q|a^p - 1$, $a^p \equiv 1 \pmod{q}$; so $\text{ord}_m a = 1$ or p. If $\text{ord}_m a = 1$, then $q|a - 1$. If $\text{ord}_m a = p$, then $p|\varphi(q)$; that is, $p|q - 1$, so $q = 2kp + 1$, since $q - 1$ is even.

55. **proof:** $R_p = (10^p - 1)/9$, so $10^p - 1 = 9 \cdot R_p$. Since $q|R_p$, this implies $q|10^p - 1$. Therefore, $10^p \equiv 1 \pmod{q}$. Thus, $p|q - 1$, so $q = 2kp + 1$, since $q - 1$ is even.

57. **proof:** Let p be an odd prime factor of $n^4 + 1$. Then $n^4 \equiv -1 \pmod{p}$, so $n^8 \equiv 1 \pmod{p}$. Thus, $8|\varphi(p) = p - 1$; so p is of the form $8k + 1$.

59. **proof:** Assume that there are only finitely many primes, p_1 through p_r, of the form $8k + 1$. Let $N = (2p_1 \cdots p_r)^4 + 1$; it is of the form $n^4 + 1$. By Exercise 57, it has an odd prime factor p of the form $8k + 1$. Then $p \neq p_i$ for every i, which contradicts the hypothesis.

Exercises 10.2 (p. 466)

1. $2^{10} = (2^{20})^{10} \equiv 14^{10} \equiv 1 \pmod{101}$. The prime factors of $100 = 2^2 \cdot 5^2$ are 2 and 5. When $q = 2, 2^{100/2} = 2^{50} \equiv (2^{10})^5 \equiv 14^5 \equiv -1 \pmod{101}$; when $q = 5, 2^{100/5} = 2^{20} \equiv (2^{10})^2 \equiv 14^2 \equiv 95 \pmod{101}$. Thus, $2^{100/q} \not\equiv 1 \pmod{101}$ for all prime factors q of 100. Therefore, by Lucas' theorem, 101 is a prime.

3. $3^{772} = (3^{100})^7 \cdot 3^{60} \cdot 3^{12} \equiv 38^7 \cdot 160^3 \cdot 390 \equiv 733 \cdot 646 \cdot 390 \equiv 1 \pmod{773}$. The prime factors of $772 = 2^2 \cdot 193$ are 2 and 193. When $q = 2, 3^{772/2} = 3^{386} \equiv (3^{100})^3 \cdot 3^{80} \cdot 3^6 \equiv 38^5 \cdot 160^4 \cdot 3^6 \equiv 762 \cdot 551 \cdot 729 \equiv -1 \pmod{773}$; when $q = 193, 3^{772/193} = 3^4 \equiv 81$

(mod 773). Thus, $3^{772/q} \not\equiv 1$ (mod 773) for all prime factors q of 773. Therefore, by Lucas' theorem, 773 is a prime.

5. First, $3^{(127-1)/2} = 3^{63} = (3^{20})^3 \cdot 3^3 \equiv 36^3 \cdot 27 \equiv -1 \pmod{127}$. Because $126 = 2 \cdot 3^2 \cdot 7$, the odd prime factors q of 126 are 3 and 7. When $q = 3$, $3^{42} = (3^{10})^4 \cdot 3^2 \equiv 6^4 \cdot 9 \equiv 107 \pmod{127}$; when $q = 7$, $3^{18} = 3^{10} \cdot 3^8 \equiv (-6)(84) \equiv 4 \pmod{127}$. Thus, in both cases, $3^{(127-1)/q} \not\equiv 1 \pmod{127}$, so 127 is a prime.
7. First, notice that $5^{576/2} = 5^{288} = (5^{100})^2 \cdot (5^{40})^2 \cdot 5^8 \equiv 162^2 \cdot 130^2 \cdot 573 \equiv -1 \pmod{577}$. Because $576 = 2^6 \cdot 9$, its only odd prime factor q is 9. When $q = 9$, $3^{(577-1)/9} = 5^{64} \equiv 335 \not\equiv 1 \pmod{577}$; so 577 is prime.
9. Because there is a positive integer x such that $x^{f_n - 1} \equiv 1 \pmod{f_n}$ and $x^{(f_n-1)/2} \not\equiv 1 \pmod{f_n}$, it follows that f_n is a prime.

Exercises 10.3 (p. 473)

1. 2, 5
3. 2, 4
5. none
7. 5, 10
9. $x - 1$, 1; $x^2 - 1$, 1, 6; $x^3 - 1$, 1, 2, 4; $x^6 - 1$, 1 through 6.
11. $f(x) = (x^4 - 10x^3 + 35x^2 - 50x + 24) - x^4 + 1 = -10x^3 + 35x^2 - 50x + 25$. Clearly, 5 is a factor of every coefficient.
13. $\varphi(6) = 2$
15. $\varphi(48) = 16$
17. no
19. $3^5 + 3^4 + \cdots + 1 = (3^6 - 1)/(3 - 1) = (3^6 - 1)/2 \equiv 4(3^6 - 1) \equiv 4 \cdot 0 \equiv 0 \pmod 7$.
21. 2 and 6 are primitive roots modulo 11. But $2 \cdot 6 \equiv 1 \pmod{11}$ is not, since $\text{ord}_{11} 1 = 1 \neq 10$.
23. 3 and 5 are primitive roots modulo 17, and so is $3 + 5 \equiv 8 \pmod{17}$.
25. $\text{ord}_{11}(-2) = \text{ord}_{11} 9 = 5$; $\text{ord}_{11}(-6) = \text{ord}_{11} 5 = 5$; $\text{ord}_{11}(-7) = \text{ord}_{11} 4 = 5$; and $\text{ord}_{11}(-8) = \text{ord}_{11} 3 = 5$.
27. yes, because $\text{ord}_5(-2) = \text{ord}_5 3 = 4 = \varphi(5)$.
29. yes, because $\text{ord}_{17}(-5) = \text{ord}_{17} 12 = 16 = \varphi(17)$.
31. $p \equiv 1 \pmod 4$
33. The product of incongruent primitive roots modulo an odd prime p is congruent to 1 modulo p.
35. $\sum_{d|12} \psi(d) = \psi(1) + \psi(2) + \psi(3) + \psi(6) + \psi(12) = 1 + 1 + 2 + 2 + 2 + 4 = 12 = 13 - 1$
37. $d|16$ implies $d = 1, 2, 4, 8$, or 16. $\psi(1) = 1 = \psi(2)$, $\psi(4) = 2$, $\psi(8) = 4$, and $\psi(16) = 8$.
39. $\psi(1) = 1 = \varphi(1)$, $\psi(2) = 1 = \varphi(2)$, $\psi(4) = 2 = \varphi(4)$, $\psi(8) = 4 = \varphi(8)$, and $\psi(16) = 8 = \varphi(16)$.
41. $d = 1, 2, 11$, or 22. The only least residue of order 1 modulo 23 is 1; the residue of order 2 modulo 23 is 22; there are 10 residues of order 11: 2, 3, 4, 6, 8, 9, 12, 13, 16, and 18; and there are 10 least residues of order 22: 5, 7, 10, 11, 14, 15, 17, 19, 20, and 21.
43. The $\varphi(16) = 8$ primitive roots modulo 17 are given by 3^k modulo 17, where $(k, 16) = 1$. They are $3^1, 3^3, 3^5, 3^7, 3^9, 3^{11}, 3^{13}$, and 3^{15} modulo 17, that is, 3, 5, 6, 7, 10, 11, 12, and 14.
45. The $\varphi(30) = 8$ primitive roots modulo 31 are given by 3^k modulo 31, where $(k, 30) = 1$. They are $3^1, 3^7, 3^{11}, 3^{13}, 3^{17}, 3^{19}, 3^{23}$, and 3^{29} modulo 31, that is, 3, 11, 12, 13, 17, 21, 22, and 24.

47. **proof:** $(\alpha-1)(\alpha^{p-2}+\alpha^{p-3}+\cdots+\alpha+1)=\alpha^{p-1}-1\equiv 0 \pmod{p}$. But $\alpha \not\equiv 1 \pmod{p}$, so $\alpha^{p-2}+\alpha^{p-3}+\cdots+\alpha+1\equiv 0 \pmod{p}$.
49. Because α is a primitive root modulo p, $\alpha^{p-1}\equiv 1 \pmod{p}$, so $\alpha^{(p-1)/2}\equiv -1 \pmod{p}$. Then $[\alpha^{(p-1)/4}]^2+1=\alpha^{(p-1)/2}+1\equiv -1+1\equiv 0 \pmod{p}$. Thus, $\alpha^{(p-1)/4}$ satisfies the congruence $x^2+1\equiv 0 \pmod{p}$.
51. **proof:** Let $p=4k+3$, where $k\geq 1$. Because α is a primitive root modulo p, $\alpha^{2k+1}=\alpha^{(p-1)/2}\equiv -1 \pmod{p}$. Then $(-\alpha)^{(p-1)/2}=-\alpha^{2k+1}\equiv 1 \pmod{p}$. Therefore, $\text{ord}_p(-\alpha)|2k+1$. Let $\text{ord}_p(-\alpha)=t$, where $t<2k+1$. Then $(-\alpha)^t\equiv 1 \pmod{p}$; that is, $\alpha^t\equiv(-1)^t \pmod{p}$, so $\alpha^{2t}\equiv 1 \pmod{p}$. Then $p-1|2t$, so $(p-1)/2|t$, where $t<2k+1=(p-1)/2$. Because this is impossible, $t=2k+1=(p-1)/2$; that is, $\text{ord}_p(-\alpha)=(p-1)/2$.
53. **proof:** Because α is a primitive root modulo p^j, $\alpha^{\varphi(p^j)}\equiv 1 \pmod{p^j}$ and $\alpha^t\not\equiv 1 \pmod{p^j}$ for $t<\varphi(p^j)$. Since $\varphi(2p^j)=\varphi(p^j)$, it follows that $\alpha^{\varphi(2p^j)}\equiv 1 \pmod{p^j}$. Because α is odd, $\alpha^{\varphi(2p^j)}\equiv 1 \pmod{2}$. Therefore, $\alpha^{\varphi(2p^j)}\equiv 1 \pmod{2p^j}$.
    Suppose $\alpha^e\equiv 1 \pmod{2p^j}$, where $e<\varphi(2p^j)$. This implies $\alpha^e\equiv 1 \pmod{p^j}$, where $e<\varphi(p^j)$. This contradicts the assumption that α is a primitive root modulo p^j. Thus, α is also a primitive root modulo $2p^j$.

Exercises 10.4 (p. 482)

1. First, $3^{\varphi(25)}=3^{20}\equiv 1 \pmod{25}$. Because $3^e\not\equiv 1 \pmod{25}$, when $e=1, 2, 4, 5$, or 10, 3 is a primitive root modulo 5^2.
3. We have $3^{\varphi(625)}=3^{500}\equiv 1 \pmod{625}$. Because $3^e\not\equiv 1 \pmod{625}$ for $e=1, 2, 4, 5, 10, 20$, or 100, it follows that 3 is a primitive root modulo 5^4.
5. We have $3^{\varphi(49)}=3^{42}\equiv 1 \pmod{49}$. Because $3^e\not\equiv 1 \pmod{49}$ for $e=1, 2, 3, 6, 7, 14$, or 21, it follows that 3 is a primitive root modulo 7^2.
7. 2 is a primitive root modulo both 11 and 11^2.
9. 3 is a primitive root modulo both 17 and 17^2.
11. 2 is a primitive root modulo both 3 and 3^4.
13. 3 is a primitive root modulo both 7 and 7^3.
15. yes
17. yes
19. Because $n=2p$ and $\alpha=3$ is odd, by Theorem 10.12, 3 is also a primitive root modulo 10.
21. Because $n=486=2\cdot 3^5$ and $\alpha=2$ is even, by Theorem 10.12, $\alpha+p^j=2+3^5=245$ is a primitive root modulo 486.
23. yes
25. no
27. yes
29. no
31. 7, 13, 17, 19
33. 2, 5, 11, 14, 20, 23
35. **proof:** number of primitive roots modulo $2p^k=\varphi(\varphi(2p^k))=\varphi(\varphi(2)\varphi(p^k))=\varphi(\varphi(p^k))=$ number of primitive roots modulo p^k.

Exercises 10.5 (p. 487)

1. $\text{ind}_\alpha\,\alpha=1$
3. $\text{ind}_2 8\equiv\text{ind}_2 5+6\equiv 9+6\equiv 3 \pmod{12}$, so $\text{ind}_2 8=3$.
5. $\text{ind}_3 6\equiv\text{ind}_3 11+8\equiv 7+8\equiv 15 \pmod{17}$, so $\text{ind}_3 6=15$.

7.

a	1	2	3	4	5	6
$\text{ind}_3\, a$	6	2	1	4	5	3

9. $\text{ind}_2\, 8 = 12 - \text{ind}_2\, 5 = 12 - 9 = 3$

11. $\text{ind}_2\, 12 = 18 - \text{ind}_2\, 8 = 18 - 3 = 15$

13. $\text{ind}_2\, 11 = 7 = 12 - 5 = \varphi(13) - \text{ind}_2\, 6$, so $11 \equiv 6^{-1} \pmod{13}$.

15. $\text{ind}_2\, 13 = 5 \neq 18 - 6 = \varphi(19) - \text{ind}_2\, 5$, so $13 \not\equiv 5^{-1} \pmod{19}$.

17.
$$\begin{aligned}\text{ind}_6(8x^5) &\equiv \text{ind}_6\, 3 \pmod{12}\\ \text{ind}_6\, 8 + 5\,\text{ind}_6\, x &\equiv \text{ind}_6\, 3 \pmod{12}\\ 3 + 5\,\text{ind}_6\, x &\equiv 1 \pmod{12}\\ x &\equiv 6 \pmod{13}\end{aligned}$$

19.
$$\begin{aligned}\text{ind}_{11}(8x^5) &\equiv \text{ind}_{11}\, 3 \pmod{12}\\ \text{ind}_{11}\, 8 + 5\,\text{ind}_{11}\, x &\equiv \text{ind}_{11}\, 3 \pmod{12}\\ 9 + 5\,\text{ind}_{11}\, x &\equiv 4 \pmod{12}\\ \text{ind}_{11}\, x &\equiv 12 \pmod{12}\\ x &\equiv 6 \pmod{13}\end{aligned}$$

21.
$$\begin{aligned}\text{ind}_5(7x) &\equiv \text{ind}_5\, 13 \pmod{6}\\ \text{ind}_6\, 7 + \text{ind}_5\, x &\equiv \text{ind}_5\, 13 \pmod{6}\\ 2 + \text{ind}_5\, x &\equiv 4 \pmod{6}\\ \text{ind}_6\, x &\equiv 2 \pmod{6}\\ x &\equiv 7 \pmod{18}\end{aligned}$$

23.
$$\begin{aligned}\text{ind}_2(2x^4) &\equiv \text{ind}_2\, 5 \pmod{12}\\ \text{ind}_2\, 2 + 4\,\text{ind}_2\, x &\equiv \text{ind}_2\, 5 \pmod{12}\\ 1 + 4\,\text{ind}_2\, x &\equiv 9 \pmod{12}\\ 4\,\text{ind}_2\, x &\equiv 8 \pmod{12}\\ \text{ind}_2\, x &\equiv 2 \pmod{3}\\ \text{ind}_2\, x &\equiv 2, 5, 8, 11 \pmod{12}\\ x &\equiv 4, 6, 7, 9 \pmod{13}\end{aligned}$$

25.
$$\begin{aligned}\text{ind}_3(4x^3) &\equiv \text{ind}_3\, 5 \pmod{16}\\ \text{ind}_3\, 4 + 3\,\text{ind}_3\, x &\equiv \text{ind}_3\, 5 \pmod{16}\\ 12 + 3\,\text{ind}_3\, x &\equiv 5 \pmod{16}\\ 3\,\text{ind}_3\, x &\equiv 9 \pmod{16}\\ \text{ind}_3\, x &\equiv 3 \pmod{16}\\ x &\equiv 10 \pmod{17}\end{aligned}$$

27.
$$\begin{aligned}\text{ind}_2(7^{5x-1}) &\equiv \text{ind}_2\, 5 \pmod{12}\\ (5x-1)\,\text{ind}_2\, 7 &\equiv \text{ind}_2\, 5 \pmod{12}\\ 11(5x-1) &\equiv 9 \pmod{12}\\ 5x &\equiv 4 \pmod{12}\\ x &\equiv 8 \pmod{12}\\ x &\equiv 8 \pmod{13}\end{aligned}$$

29. Let $x = 23^{1001} \equiv 10^{1001} \pmod{13}$. Then $\text{ind}_2 x \equiv 1001 \cdot \text{ind}_2\, 10 \equiv 1001 \cdot 10 \equiv 2 \pmod{12}$, so $x \equiv 4 \pmod{13}$.

31. Let $x = 5^{17} \cdot 7^{19} \pmod{13}$. Then $\text{ind}_2 x \equiv 17 \cdot \text{ind}_2\, 5 + 19 \cdot \text{ind}_2\, 7 \equiv 17 \cdot 9 + 19 \cdot 11 \equiv 2 \pmod{12}$, so $x \equiv 4 \pmod{13}$.

33. **proof:** Because $\alpha^{\varphi(m)} \equiv 1 \pmod{m}$, $\alpha^{\varphi(m)/2} \equiv -1 \pmod{m}$. Therefore, $\text{ind}_\alpha(m-1) = \varphi(m)/2$.

35. **proof:** Because α is a primitive root modulo m, $\alpha^{\varphi(m)} \equiv 1 \pmod{m}$. Then $\alpha^{\varphi(m)/2} \equiv -1 \pmod{m}$ and $m - a \equiv a \cdot \alpha^{\varphi(m)/2}$. Therefore, $\text{ind}_\alpha(m-a) \equiv \text{ind}_\alpha\, a + \text{ind}_\alpha\, \alpha^{\varphi(m)/2} \pmod{\varphi(m)}$. That is, $\text{ind}_\alpha(m-a) \equiv \text{ind}_\alpha\, a + \varphi(m)/2 \pmod{\varphi(m)}$. Thus, $\text{ind}_\alpha(m-a) \equiv \text{ind}_u\, a + \varphi(m)/2$.

37. Follows by Exercise 36 with $m = p$.

39. **proof** (by contradiction): Because $p_i \equiv 1 \pmod{4}$, $q = (p_1 \cdots p_n)^2 + 1 \equiv 1 + 1 \equiv 2 \pmod{4}$. Therefore, $q = 4x + 2 = 2(2x+1)$ for some integer x. Since q has an odd factor, it has an odd prime factor p. So, $q \equiv 0 \pmod{p}$, and hence, $(p_1 \cdots p_n)^2 \equiv -1 \pmod{p}$. Therefore, by Exercise 38, p is of the form $4k+1$. If $p = p_i$ for some i, then $p|1$, which is a contradiction. So, $p \neq p_i$ for every i. Thus, q has a prime factor of the form $4k+1$ different from every p_i; this is again a contradiction.

41. **proof** (by contradiction): Because $p_i \equiv 1 \pmod{8}$, $p_1 \cdots p_n \equiv 1 \pmod{8}$. Therefore, $q = (p_1 \cdots p_n)^4 + 1 \equiv 1 + 1 \equiv 2 \pmod{8}$. Then $q = 8x + 2 = 2(4x+1)$ has an odd prime factor and hence $(p_1 \cdots p_n)^4 \equiv -1 \pmod{p}$. Thus, by Exercise 40, p is of the form $8k+1$. If

$p = p_i$ for some i, then $p|1$, which is a contradiction. So $p \neq p_i$ for every i. Thus, q has a prime factor of the form $8k+1$, different from every p_i; this is again a contradiction.

43. Follows from Exercise 42 with $m = p$.

Review Exercises (p. 491)

1. 3
3. 5
5. $\text{ord}_{13}(5^7) = 4/(4,7) = 4/1 = 4$
7. $3 \cdot 4 = 12$
9. 0
11. 0
13. yes, because 1723 is a prime.
15. yes, because $48778 = 2 \cdot 29^3$.
17. α^i is a primitive root modulo 50 if and only if $(i, \varphi(50)) = (i, 20) = 1$. So, the remaining primitive roots modulo 50 are given by $3^3, 3^7, 3^9, 3^{11}, 3^{13}, 3^{17}$, and 3^{19} modulo 50, that is, 13, 17, 23, 27, 33, 37, and 47.
19. $\beta = \alpha$ if α is odd; otherwise, $\beta = \alpha + p^j$.
21. 5
23. $2 + 5^3 = 127$
25. 3, 13, 17, 23, 27, 33, 37, and 47.
27. 2, 5, 8, 11, 14, 17, 20, 23, 26, 32, 35, 41, 44, 47, 50, 53, 56, 59, 62, 68, 71, and 77.
29. 2, 6, 7, 8, 17, 18, 19, 24, 28, 29, 30, 35, 39, 41, 46, 50, 51, 52, 57, 62, 63, 68, 72, 73, 74, 76, 79, 83, 84, 85, 87, 90, 94, 95, 96, 101, 105, 107, 116, and 117.
31. 5, 11, 23, 29, 41,47, 59, 65, 77, 83, 95, 101, 113, 119, 131, 137, 149, and 155.
33. $6^{108} \equiv 1 \pmod{109}$, so

$$424^{2076} \equiv (6^4)^{2076} \equiv 6^{8304} \equiv (6^{108})^{76} \cdot 6^{96} \equiv 1 \cdot 6^{96} = (6^6)^{16} \equiv 4^{16} \equiv 75 \pmod{109}$$

35. 4
37. 3
39. $\text{ind}_3(13 \cdot 47) \equiv \text{ind}_3 13 + \text{ind}_3 47 \equiv 17 + 11 \equiv 28 \pmod{50}$.
41. $\text{ind}_3 9 \equiv \text{ind}_3 5 + \varphi(14)/2 \pmod{\varphi(14)} \equiv 5 + 3 \equiv 2 \pmod 6$.
43.

$$\begin{aligned} \text{ind}_2(3x^7) &\equiv \text{ind}_2 4 \pmod{10} \\ \text{ind}_2 3 + 7\,\text{ind}_2 x &\equiv 2\,\text{ind}_2 2 \pmod{10} \\ 8 + 7\,\text{ind}_2 x &\equiv 2 \pmod{10} \\ 7\,\text{ind}_2 x &\equiv 4 \pmod{10} \\ \text{ind}_3 x &\equiv 2 \pmod{10} \\ x &\equiv 4 \pmod{11} \end{aligned}$$

45.

$$\begin{aligned} \text{ind}_3(3^{4x-1}) &\equiv \text{ind}_3 11 \pmod{16} \\ (4x-1)\,\text{ind}_3 5 &\equiv \text{ind}_3 11 \pmod{16} \\ 5(4x-1) &\equiv 7 \pmod{16} \\ 5x &\equiv 3 \pmod 4 \\ x &\equiv 3 \pmod 4 \\ x &\equiv 3, 7, 10, 14 \pmod{17} \end{aligned}$$

47. Let $x = 50^{1976} \equiv 11^{1976} \pmod{13}$. Then $\text{ind}_2 x \equiv 1976 \cdot \text{ind}_2 11 \equiv 1976 \cdot 7 \equiv 8 \pmod{12}$, so $x \equiv 2^8 \equiv 9 \pmod{13}$.
49. $3^{136} \equiv (3^{20})^6 \cdot 3^{16} \equiv 4^6 \cdot 88 \equiv 123 \cdot 88 \equiv 1 \pmod{137}$. The prime factors q of $136 = 2^3 \cdot 17$ are 2 and 17. When $q = 2, 3^{68} = (3^{20})^3 \cdot 3^8 \equiv 4^3 \cdot 122 \equiv -1 \not\equiv 1 \pmod{137}$; when $q = 17, 3^8 = 122 \not\equiv 1 \pmod{137}$. Therefore, by Lucas' theorem, 137 is a prime.
51. As in Exercise 49, $3^{136} \equiv 1 \pmod{137}$. The only odd prime factor q of 136 is 17. When $q = 17, 3^8 = 122 \not\equiv 1 \pmod{137}$. Therefore, by Corollary 10.8, 137 is a prime.

Supplementary Exercises (p. 492)

1. **proof:** $f_n = 2^{2^n} + 1 \equiv 0 \pmod{f_n}$, so $2^{2^n} \equiv -1 \pmod{f_n}$. Then $2^{2^{n+1}} \equiv 1 \pmod{f_n}$, so $\text{ord}_{f_n} 2 | 2^{n+1}$.

3. **proof:** Because p is a prime factor of $f_n, f_n = 2^{2^n} + 1 \equiv 0 \pmod{p}$. Then $2^{2^n} \equiv -1 \pmod{p}$, so $2^{2^{n+1}} \equiv 1 \pmod{p}$. Therefore, $\text{ord}_p 2 | 2^{n+1}$. Let $\text{ord}_p 2 = 2^k$, where $k \le n$. Then $2^{2^k} \equiv 1 \pmod{p}$, so $2^{2^k \cdot 2^{n-k}} \equiv 2^{2^n} \equiv 1 \pmod{p}$. Because this is impossible, $k = n+1$. Thus, $\text{ord}_p 2 = 2^{n+1}$.
5. Because $\text{ord}_{19} 7 = 3$, $7^3 \equiv 1 \pmod{19}$. Then $8^3 \equiv (7+1)^3 = 7^3 + 3 \cdot 7^2 + 3 \cdot 7 + 1 \equiv 1 + 14 + 2 + 1 \equiv -1 \pmod{19}$, so $8^6 \equiv 1 \pmod{19}$. Thus, $\text{ord}_{19} 8 = 6$.
7. **proof:** Because p is odd and $\text{ord}_p a = 3$, $p > 3$. Because $a^3 \equiv 1 \pmod{p}$ and $a \not\equiv 1 \pmod{p}$, $a^2 + a + 1 \equiv 0 \pmod{p}$. Then $(a+1)^6 = a^6 + 6a^5 + 15a^4 + 20a^3 + 15a^2 + 6a + 1 \equiv 1 + 6a^5 + 15a^4 + 20a^3 + 15a^2 + 6a + 1 \equiv 1 + 21(a^2 + a + 1) \equiv 1 \pmod{p}$. Therefore, $\text{ord}_p(a+1) = 1, 2, 3,$ or 6.

   case 1 If $\text{ord}_p(a+1) = 1$, then $a \equiv 0 \pmod{p}$. This is impossible.

   case 2 If $\text{ord}_p(a+1) = 2$, then $(a+1)^2 \equiv 1 \pmod{p}$. This implies $a^2 + 2a \equiv 0 \pmod{p}$, so $p|a$ or $p|a+2$. But $p \nmid a$. So suppose $p|a+2$. Because $a^2 + 2a \equiv 0 \pmod{p}$, $a^2 + 2 \equiv 2(a^2 + a + 1) \equiv 0 \pmod{p}$. Therefore, $(a^2 + a + 1) + 3 \equiv (a^2 + 2) + (a + 2) \pmod{p}$; that is, $0 + 3 \equiv 0 + 0 \pmod{p}$. Thus, $3 \equiv 0 \pmod{p}$. This is also impossible.

   case 3 Let $\text{ord}_p(a+1) = 3$. Then $(a+1)^2 \equiv 1 \pmod{p}$, so $a^3 + 3a^2 + 3a \equiv 0 \pmod{p}$; that is, $3a(a^2 + a + 1) \equiv 2a^3 \pmod{p}$. This implies $2a^3 \equiv 0 \pmod{p}$. Therefore, $p|2$ or $p|a$; both are impossible. Thus, $\text{ord}_p(a+1) = 6$.
9. **proof** (by C. Vitale): Let $(m,n) = (m,10) = (n,10) = 1$. Let $d = [\text{ord}_m 10, \text{ord}_n 10]$, so $d = x \cdot \text{ord}_p 10 = y \cdot \text{ord}_p 10$ for some positive integers x and y. Then $10^{x \cdot \text{ord}_m 10} \equiv 10^d \equiv 1 \pmod{m}$ and $10^{y \cdot \text{ord}_n 10} \equiv 10^d \equiv 1 \pmod{n}$. This implies $10^d \equiv 1 \pmod{mn}$, because $(m,n) = 1$. Therefore, $\text{ord}_{mn} 10 | d$. But, by definition, $10^{\text{ord}_{mn} 10} \equiv 1 \pmod{mn}$, so $10^{\text{ord}_{mn} 10} \equiv 1 \pmod{m}$ and $10^{\text{ord}_{mn} 10} \equiv 1 \pmod{n}$. This implies $\text{ord}_m 10 | \text{ord}_{mn} 10$ and $\text{ord}_n 10 | \text{ord}_{mn} 10$. Therefore, $d | \text{ord}_{mn} 10$. Thus $d = [\text{ord}_m 10, \text{ord}_n 10] = \text{ord}_{mn} 10$.
11. $p = 29$, $g = 14$.

Chapter 11 Quadratic Congruences

Exercises 11.1 (p. 500)

1. 1, 5
3. 1, 5, 7, 11
5. none
7. 2, 5
9. none
11. $(4x+3)^2 \equiv 1 \pmod{7}$
    $4x + 3 \equiv 1, 6 \pmod{7}$
    $x \equiv 3, 6 \pmod{7}$
13. $(2x-5)^2 \equiv 4 \pmod{13}$
    $2x - 5 \equiv \pm 2 \pmod{13}$
    $2x \equiv 3, 7 \pmod{13}$
    $x \equiv 8, 10 \pmod{13}$
15. 1
17. 50
19. 1, 4, 7, 9, 10, 13, 16
21. 1, 2, 3, 4, 6, 8, 9, 12, 13, 16, 18
23. Because $6^2 \equiv 7 \pmod{29}$, 7 is a quadratic residue of 29.
25. Because $12^2 \equiv 3 \pmod{47}$, 3 is a quadratic residue of 47.
27. $x^2 \equiv 1, 3, 4, 5,$ or $9 \pmod{11}$ for every nonzero residue modulo 11. Therefore, $x^2 \equiv 7 \pmod{11}$ and $x^2 \equiv 10 \pmod{11}$ are not solvable. But $x^2 \equiv 7 \cdot 10 \equiv 4 \pmod{11}$ has two solutions, 2 and 9.

29. **proof:** Let α be a primitive root modulo p. Then $\alpha^{p-1} \equiv 1 \pmod{p}$, so $[\alpha^{(p-1)/2} - 1][\alpha^{(p-1)/2} + 1] \equiv 0 \pmod{p}$. But $\alpha^{(p-1)/2} \not\equiv 1 \pmod{p}$, so α is a quadratic nonresidue of p.
31. **proof:** Because α is a quadratic residue of $p, \alpha^{(p-1)/2} \equiv 1 \pmod{p}$. Then $(p - a)^{(p-1)/2} \equiv (-a)^{(p-1)/2} = a^{(p-1)/2}(-1)^{(p-1)/2} \equiv (-1)^{(p-1)/2} \pmod{p}$. This is congruent to 1 (mod p) if and only if $p \equiv 1 \pmod{4}$.
33. **proof:** Let a and b be quadratic residues of p. Then $x^2 \equiv a \pmod{p}$ and $y^2 \equiv a \pmod{p}$ are solvable. So $(xy)^2 = x^2y^2 \equiv ab \pmod{p}$. Thus, ab is also a quadratic residue of p.
35. **proof:** Let a be a quadratic nonresidue of p. Then $a^{(p-1)/2} \equiv -1 \pmod{p}$, so $(a^2)^{(p-1)/2} = [a^{(p-1)/2}]^2 \equiv (-1)^2 \equiv 1 \pmod{p}$. Thus, a^2 is a quadratic residue of p. (This result follows trivially from Exercise 34.)
37. **proof** (by contradiction): Let a be a quadratic residue of p and $b \equiv a^{-1} \pmod{p}$. Suppose b is a quadratic nonresidue. Then, by Exercise 36, $1 \equiv ab \pmod{p}$ is a quadratic nonresidue, which is a contradiction.
39. **proof:** Suppose a is a quadratic residue of M_p. Then $a^{(M_p-1)/2} \equiv 1 \pmod{p}$, so $(-a)^{(M_p-1)/2} = -a^{(M_p-1)/2} \equiv -1 \pmod{M_p}$. Thus, $p - a \equiv -a$ is also a quadratic nonresidue of M_p. The converse follows by reversing the steps.
41. **proof:** Suppose not all three congruences are solvable.

    case 1 Suppose none are solvable. Then both a and b are quadratic nonresidues, so, by Exercise 34, ab is a quadratic residue. Hence $x^2 \equiv ab$ is solvable, which is a contradiction.

    case 2 If exactly one congruence is solvable, then we are done.

    case 3 Suppose exactly two congruences, say, $x^2 \equiv a \pmod{p}$ and $x^2 \equiv b \pmod{p}$, are solvable.

    Then a and b, and hence ab are quadratic residues by Exercises 33. In other words, $x^2 \equiv ab \pmod{p}$ is solvable, again a contradiction.

Exercises 11.2 (p. 513)

1. -1
3. -1
5. $(16/17) = (-1/7) = 1$
7. $(-1/29) = 1$
9. $(128/23) = (2^7/23) = (2/23)^7 = 1^7 = 1$
11. $(600/23) = (2/23)^3(3/23)(5/23)^2 = 1^3 \cdot 1 \cdot (-1)^2 = 1$
13. $(27/19) = (3/19)^3 = (-1)^3 = -1$
15. $(147/19) = (3/19)(7/19)^2 = (-1)(-1)^2 = -1$
17. There are two residues > 6.5, namely, 9 and 12; so $v = 2$.
19. There are four residues > 8.5, namely, 11, 12, 15, and 16; so $v = 4$.
21. $(5/13) = (-1)^3 = -1$
23. $(7/19) = (-1)^4 = 1$
25. $(2/19) = -1$
27. $(2/41) = 1$
29. $(13/31) = -1$
31. $(41/43) = (-2/43) = (-1/43)(2/43) = (-1)(-1) = 1$
33. 13, 29, 53, 149, 173
35. $(12/19) = (4/12)(3/12) = 1 \cdot (-1) = -1$
37. $(35/47) = (-12/47) = (-1/47)(3/47)(4/47) = (-1) \cdot 1 \cdot 1 = -1$
39. $2^2, 2^4, 2^6, 2^8, 2^{10}, 2^{12}$ modulo 13, that is, 1, 3, 4, 9, 10, 12.

41. Because $1913 \equiv 1 \pmod 8$, $(2/1913) = 1$. Therefore, $2^{956} \equiv 1 \pmod{1913}$; that is, $1913 | 2^{956} - 1$.
43. **proof:** Because $a \equiv b \pmod p$, $(a/p) = (b/p)$. Therefore, either both a and b are quadratic residues or both are quadratic nonresidues.
45. **proof:** Because a is a quadratic residue of p, $(a/p) = 1 \equiv a^{(p-1)/2} \pmod p$. Then $(p - a)^{(p-1)/2} \equiv (-a)^{(p-1)/2} \equiv a^{(p-1)/2} \equiv 1 \pmod p$, because $p \equiv 1 \pmod 8$. So $p - a$ is also a quadratic residue of p.
47. **proof:** Because $(2a/p) = (2/p)(a/p) = 1 \cdot 1 = 1$, $2a$ is a quadratic residue of p.
49. **proof:** Given sum $= [(p-1)/2] \cdot 1 + [(p-1)/2] \cdot (-1) = 0$.
51. **proof:** Because $q | M_p$, $2^p \equiv 1 \pmod q$. Let $\text{ord}_q 2 = e$. Then $e | p$, so $e = 1$ or $e = p$. But $e \neq 1$, so $e = p$. By Euler's criterion, $(2/q) \equiv 2^{(q-1)/2} \pmod q$. So $(2/q) = 1$ if and only if $2^{(q-1)/2} \equiv 1 \pmod q$; that is, if and only if $p | (q-1)/2$. Thus, 2 is a quadratic residue of q if and only if q is of the form $2kp + 1$.
53. **proof:** Because $(a/p)(b/p) = (ab/p) = (1/p) = 1$, $(a/p) = (b/p)$.
55. **proof:** The quadratic residues of p are congruent to $1^2, 2^2, \ldots, [(p-1)/2]^2$ modulo p. Let S be their sum. Then

$$6S = 6 \sum_{i=1}^{(p-1)/2} i^2 \equiv [(p-1)/2] \cdot [(p+1)/2] \equiv 0 \pmod p$$

But $6 \nmid p$. Therefore, $S \equiv 0 \pmod p$.
57. **proof:** Because $\alpha^{p-1} \equiv 1 \pmod p$, $(a/p) = (\alpha^k/p) = (\alpha/p)^k \equiv [\alpha^{(p-1)/2}]^k \equiv (-1)^k \pmod p$. Thus, $(a/p) = 1$ if and only if k is even.
59. **proof** (by PMI): When $n = 1$, $a = p_1^{e_1}$. Because $(a/p) = (p_1^{e_1}/p) = (p_1/p)^{e_1}$, the result is true when $n = 1$. Now assume it is true for an arbitrary positive integer k. Let $a = \prod_{i=1}^{k+1} p_i^{e_i}$.

$$\text{Then } (a/p) = \left(\prod_{i=1}^{k+1} p_i^{e_i} \cdot p_{k+1}^{e_{k+1}}/p\right) = \left(\prod_{i=1}^{k+1} p_i^{e_i}/p\right)(p_{k+1}^{e_{k+1}}/p)$$
$$= \prod_{i=1}^{k+1} (p_i^{e_i}/p) \cdot (p_{k+1}^{e_{k+1}}/p) = \prod_{i=1}^{k+1} (p_k/p)^{e_i}$$

Thus, by PMI, the result is true for every $n \geq 1$.
61. **proof:** Let $ab \equiv 1 \pmod p$. As a runs over the integers 1 through $p - 2$, $b + 1$ runs over the residues 2 through $p - 1$. Therefore, by Exercise 54,

$$\sum_{a=1}^{p-2} (a(a+1)/p) = \sum_{b=2}^{p-1} ((b+1)/p) = \sum_{i=2}^{p-1} (i/p) = \sum_{i=1}^{p-1} (i/p) - (1/p)$$
$$= 0 - 1 = -1$$

63. **proof:** Let α be a quadratic nonresidue of f_n. Then $\alpha^{2^{2^n-1}} \equiv -1 \pmod{f_n}$, so $\alpha^{2^{2^n}} \equiv 1 \pmod{f_n}$. Thus, $\text{ord}_{f_n} \alpha | 2^{2^n}$.

Let $\text{ord}_{f_n} \alpha = e$. Then $e = 2^k$, where $k \leq 2^n$. Suppose $k < 2^n$. Then $\alpha^e = 2^{2^k} \equiv 1 \pmod{f_n}$, so $2^{2^n-1} = (\alpha^{2^k})^{2^{2^n-k-1}} \equiv 1 \pmod{f_n}$, which is a contradiction. Thus, $\text{ord}_{f_n} \alpha = 2^{2^n}$ and α is a primitive root modulo f_n.

Exercises 11.3 (p. 526)

1. $$\sum_{k=1}^{2} \lfloor 11k/5 \rfloor = \lfloor 11/5 \rfloor + \lfloor 22/5 \rfloor = 2+4=6 \quad \text{and}$$
$$\sum_{k=1}^{5} \lfloor 5k/11 \rfloor = \lfloor 5/11 \rfloor + \lfloor 10/11 \rfloor + \lfloor 15/11 \rfloor + \lfloor 20/11 \rfloor + \lfloor 25/11 \rfloor$$
$$= 0+0+1+1+2=4$$

   Therefore, LHS $= 6+4 = 10 = 2 \cdot 5 =$ RHS.

3. $(261/47) = (26/47) = (2/47)(13/47) = 1 \cdot (47/13) = (8/13) = (2/13)^3 = (-1)^3 = -1$
5. $(176/241) = (2^4/241)(11/241) = (11/241) = (-1/11) = -1$
7. $(-1776/1013) = (250/1013) = (2/1013)(5^3/1013) = -(5/1013)^3 = -(1013/5)^3$
   $= -(3/5)^3 = -1^3 = -1$
9. $(1428/2411) = (2^2/2411)(3/2411)(7/2411)(17/2411)$
   $= 1 \cdot -(2411/3) \cdot -(2411/7) \cdot (2411/17)$
   $= (2/3)(3/7)(-3/17) = (-1) \cdot -(7/3)(-1/17)(3/17)$
   $= (1/3)(-1/17)(17/3) = 1 \cdot 1 \cdot (2/3) = -1$
11. $17 = 4 \cdot 3 + 5$, so $(3/17) = (3/5) = (-2/5) = (-1/5)(2/5) = 1 \cdot (-1) = -1$
13. $191 = 4 \cdot 43 + 19$, so $(43/191) = (43/19) = (5/19) = (19/5) = (-1/5) = 1$
15. $x \equiv \pm 13^{(23+1)/4} \equiv \pm 13^6 \equiv \pm 6 \pmod{23}$, so $x \equiv 6, 17 \pmod{23}$.
17. Pepin's test cannot be applied, since $n = 0$; nonetheless, f_0 is a prime.
19. Because $3^{(f_2-1)/2} = 3^{128} \equiv -1 \pmod{17}$, 17 is a prime.
21. By the law of quadratic reciprocity,

$$(5/p) = (p/5) \begin{cases} (1/5) = 1 & \text{if } p \equiv 1 \pmod 5 \\ (2/5) = -1 & \text{if } p \equiv 2 \pmod 5 \\ (3/5) = -1 & \text{if } p \equiv 3 \pmod 5 \\ (4/5) = 1 & \text{if } p \equiv 4 \pmod 5 \end{cases} = \begin{cases} 1 & \text{if } p \equiv \pm 1 \pmod 5 \\ -1 & \text{if } p \equiv \pm 2 \pmod 5 \end{cases}$$

23. Because $79 \equiv -1 \pmod 5$, $F_{79} \equiv 1 \pmod{79}$.
25. Because $97 \equiv 2 \pmod 5$, $F_{97} \equiv -1 \pmod{97}$.
27. $817 = 19 \cdot 43$. The solutions of $x^2 \equiv 17 \pmod{19}$ are $x = 6, 13$; and those of $x^2 \equiv 17 \pmod{43}$ are $x = 19, 24$. Using the CRT, the solution of the system $x \equiv 6 \pmod{19}$ and $x \equiv 19 \pmod{43}$ is $x \equiv 6 \cdot 43 \cdot 4 + 19 \cdot 19 \cdot 34 \equiv 234 \pmod{817}$; similarly we get three more solutions: $x \equiv 196, 621, 583$ modulo 817.
29. **proof:** Let $m = 5s+1$ and $n = 5t+1$. Then $mn = (5s+1)(5t+1) = 5(5st+s+t)+1$.
31. **proof:** Notice that $p \equiv 3 \equiv q \pmod 4$. Because $p = q + 4a$, $(4a/p) = ((p-q)/p) = ((-q)/p) = ((-1/p)(q/p)) = (-1)(-1)(p/q) = (p/q)$; that is, $(a/p) = (p/q)$. Besides, $(4a/q) = ((p-q)/q) = (p/q)$; that is, $(a/q) = (p/q)$. Thus $(a/p) = (a/q)$.
33. **proof:** Using Exercise 32,

$$(-3/p) = (-1/p)(3/p) = \begin{cases} (-1/p) & \text{if } p \equiv \pm 1 \pmod{12} \\ -(-1/p) & \text{if } p \equiv \pm 5 \pmod{12} \end{cases}$$

   When $p \equiv 1 \pmod{12}$, $p \equiv 1 \pmod 6$; when $p \equiv -1 \pmod{12}$, $p \equiv 5 \pmod 6$; and similarly the other two cases. Combining the four cases, we get the desired result.

35. **proof** (by PMI): Because $f_1 = 5 \equiv 1 \pmod 4$, the result is true when $n = 1$. Assume it is true for an arbitrary positive integer n. Then $f_{n+1} = 2^{2^{n+1}} + 1 = 2^{2 \cdot 2^n} + 1 = 4^{2^n} + 1 \equiv 1 \pmod 4$. Thus, the result follows by induction.
37. **proof:** When $n > 0$, $f_n \equiv 1$, (mod 4) by Exercise 35, and when $p > 2$, $M_p \equiv 3 \pmod 4$. Thus, by the law of quadratic reciprocity $(f_n/M_p) = (M_p/f_n)$.
39. **proof:** If $M_t = 2^t - 1 > 3$, $t > 2$. If $t \equiv 1 \pmod 4$, then $M_t = 2^{4t+t} - 1 = 2(2^4)^t - 1 \equiv 2 \cdot 0 - 1 \equiv 3 \pmod 4$. If $t \equiv 3 \pmod 4$, then $M_t = 2^{4t+3} - 1 = 8(2^4)^t - 1 \equiv 0 - 1 \equiv 3 \pmod 4$. Thus, if t is an odd prime, $M_t \equiv 3 \pmod 4$. Consequently, because $M_p \equiv 3 \equiv M_q \pmod 4$, $(M_p/M_q) = -(M_q/M_p)$, by Corollary 11.7.
41. **proof** (by contradiction): Assume that there are only finitely many such primes $p_1, \ldots, p_k$. Let $N = (2p_1 \cdots p_k)^2 + 3$. It must have an odd prime factor p, so $(2p_1 \cdots p_k)^2 \equiv -3 \pmod p$ and $(-3/p) = 1$. By Exercise 33, this implies $p \equiv 1 \pmod 6$. Suppose $p = p_i$ for some i. Then $p|p_i$ and $p|N$, so $p = 3$, a contradiction. Thus $p \neq p_i$ for every i, which is also a contradiction.
43. By the law of quadratic reciprocity, $(7/p) = \begin{cases} (p/7) & \text{if } p \equiv 1 \pmod 4 \\ -(p/7) & \text{if } p \equiv 3 \pmod 4 \end{cases}$

    If $p \equiv 1 \pmod 4$, then, by the CRT, $(7/p) = \begin{cases} 1 & \text{if } p \equiv 1, 9, 25 \pmod{28} \\ -1 & \text{if } p \equiv 5, 13, 17 \pmod{28} \end{cases}$

    If $p \equiv 3 \pmod 4$, then, by a similar argument, $(7/p) = \begin{cases} 1 & \text{if } p \equiv 11, 15, 23 \pmod{28} \\ -1 & \text{if } p \equiv 3, 9, 27 \pmod{28} \end{cases}$

    Thus, $(7/p) = \begin{cases} 1 & \text{if } p \equiv 1, 3, 9, 19, 25, 27 \pmod{28} \\ -1 & \text{if } p \equiv 5, 11, 13, 15, 17, 23 \pmod{28} \end{cases}$

Exercises 11.4 (p. 535)

1. 1, 3, 4, 9, 12, 15, 16, 22, 25, 27, 31
3. $(3/35) = (3/5)(3/7) = (-1) \cdot -(7/3) = (1/3) = 1$
5. $(23/65) = (23/5)(23/13) = (3/5)(-3/13) = (-1) \cdot (-1/3) \cdot (13/3) = -(1/3) = -1$
7. $(442/385) = (442/5)(442/7)(442/11) = (2/5)(1/7)(2/11) = (-1) \cdot 1 \cdot (-1) = 1$
9. $(-198/2873) = (-198/17)(-198/13^2) = (6/7)(-198/13)^2 = (2/7)(3/17) \cdot 1$
   $= 1 \cdot (-1) \cdot 1 = -1$
11. $(17/33) = (17/3)(17/11) = (2/3)(6/11) = (-1) \cdot (2/11)(3/11) = -(-1) \cdot 1 = 1$
13. By Exercise 1, 17 is not a quadratic residue of 33. So the congruence is not solvable.
15. $(3/5^3 \cdot 7^5) = (3/5)^3(3/7)^5 = (-1)^3(-1)^5 = 1$
17. $(3/5 \cdot 7^3 \cdot 13^6) = (3/5)(3/7)^3(3/13)^6 = (-1)(-1)^3 \cdot 1 = 1$
19. $(3/m) = (3/p)^a(3/q)^b(3/r)^c(3/s)^d = (-1)^a(-1)^b(-1)^c(1)^d = (-1)^{a+b+c}$, so $(3/m) = 1$ if and only if $a + b + c$ is even.
21. **proof:** Let $m = \prod_i p_i^{e_i}$, where each p_i is odd. If $m \equiv 1 \pmod 4$, then every factor p_i is congruent to 1 (mod 4), or m contains an even number e_j of prime factors $p_j \equiv 3 \pmod 4$, where $j \neq i$. Then, by Corollary 11.2,

$$(-1/m) = \prod_i (-1/p_i)^{e_i} \cdot \prod_{j \neq i} (-1/p_j)^{e_j} = \prod_i 1^{e_i} \cdot \prod_{j \neq i} (-1)^{e_j} = 1 \cdot 1 = 1$$

If $m \equiv -1 \pmod 4$, then there must be an odd number of prime factors of m that are congruent to $-1 \pmod 4$; since $(-1/p) = -1$, by Corollary 11.2 for each such factor, it follows that $(-1/m) = -1$. Thus, $(-1/m) = 1$ if $m \equiv 1 \pmod 4$ and -1 otherwise.

23. **proof:** By the generalized law of quadratic reciprocity, $(m/n)(n/m) = (-1)^{(m-1)/2 \cdot (n-1)/2}$. If $m \equiv 1 \pmod 4$, then $(m-1)/2$ is even; if $n \equiv 1 \pmod 4$, then $(n-1)/2$ is even. In both cases, $(m-1)/2 \cdot (n-1)/2$ is even, so $(m/n)(n/m) = 1$. Therefore, $(n/m) = (m/n)$.
If $m \equiv 3 \equiv n \pmod 4$, then both $(m-1)/2$ and $(n-1)/2$ are odd, so $(m/n)(n/m) = -1$. Therefore, $(n/m) = -(m/n)$.

Exercises 11.5 (p. 542)

1. If $x^2 \equiv 4 \pmod 5$, then $x \equiv 2, 3 \pmod 5$. If $x^2 \equiv 4 \pmod 7$, then $x \equiv 2, 5 \pmod 7$. Then, by the CRT, $x \equiv 2, 12, 23, 33 \pmod{35}$.
3. If $x^2 \equiv 43 \equiv 4 \pmod{13}$, then $x \equiv 2, 11 \pmod{13}$. If $x^2 \equiv 43 \equiv 9 \pmod{17}$, then $x \equiv 3, 14 \pmod{17}$. Then, by the CRT, $x \equiv 37, 54, 167, 184 \pmod{221}$.
5. Because 108 is a solution of $x^2 \equiv 3 \pmod{13^2}$, $108^2 = 3 + 13^2 k$, where $k = 69$. Now look for a solution of the form $108 + 13^2\ell$. Then $(108 + 13^2\ell)^2 \equiv 108^2 + 216 \cdot 13^2\ell \equiv (3 + 69 \cdot 13^2) + 216 \cdot 13^2\ell \equiv 3 + 13^2(69 + 216\ell) \pmod{13^3}$. Set $69 + 216\ell \equiv 0 \pmod{13}$; this yields $\ell = 6$. Then $108 + 13^2\ell = 1122$ is a solution of $x^2 \equiv 3 \pmod{13^3}$. The remaining solution is $-1122 \equiv 1075 \pmod{13^3}$.
7. By Example 11.30, 87 is a solution of $x^2 \equiv 23 \pmod{7^3}$. Then $87^2 = 23 + 7^3 k$, where $k = 22$. Now look for a solution of the form $87 + 7^3\ell$: $(87 + 7^3\ell)^2 \equiv 87^2 + 174 \cdot 7^3\ell \equiv (23 + 22 \cdot 7^3) + 174 \cdot 7^3\ell \equiv 23 + 7^3(22 + 174\ell) \pmod{7^4}$. Solving $22 + 174\ell \equiv 0 \pmod 7$ yields $\ell = 1$. Therefore, $87 + 7^3 \cdot 1 = 430$ is a solution of $x^2 \equiv 23 \pmod{7^4}$. The remaining solution is $-430 \equiv 1971 \pmod{7^4}$.
9. Because 6 is a solution of $x^2 \equiv 10 \pmod{13}$, $6^2 = 10 + 13k$, where $k = 2$. Now look for a solution of the form $6 + 13\ell$: $(6 + 13\ell)^2 \equiv 6^2 + 12 \cdot 13\ell \equiv (10 + 2 \cdot 13) + 12. 13\ell \equiv 10 + 13(2 + 12\ell) \pmod{13^2}$. Solving $2 + 12\ell \equiv 0 \pmod{13}$ yields $\ell = 2$. Then $6 + 13\ell = 32$ is a solution of $x^2 \equiv 10 \pmod{13^2}$; so the other solution is $-32 \equiv 137 \pmod{13^2}$.
11. Because 4 is a solution of $x^2 \equiv 5 \pmod{11}$, $4^2 = 5 + 11k$, where $k = 1$. Now look for a solution of the form $4 + 11\ell$: $(4 + 11\ell)^2 \equiv 4^2 + 8 \cdot 11\ell \equiv (5 + 1 \cdot 11) + 8 \cdot 11\ell \equiv 5 + 11(1 + 8\ell) \pmod{11^2}$. Solving $1 + 8\ell \equiv 0 \pmod{11}$ yields $\ell = 4$. Therefore, $4 + 11\ell = 48$ is a solution of $x^2 \equiv 5 \pmod{11^2}$; the remaining solution is $-48 \equiv 73 \pmod{11^2}$.
13. Because 8 is a solution of $x^2 \equiv 13 \pmod{17}$, $8^2 = 13 + 17k$, where $3 = 1$. Now look for a solution of the form $8 + 17\ell$: $(8 + 17\ell)^2 \equiv 8^2 + 16 \cdot 17\ell \equiv (13 + 3 \cdot 17) + 16 \cdot 17\ell \equiv 13 + 17(3 + 16\ell) \pmod{17^2}$. Solving $3 + 16\ell \equiv 0 \pmod{17}$ yields $\ell = 3$. Therefore, $8 + 17\ell = 59$ is a solution of $x^2 \equiv 13 \pmod{17^2}$; the remaining solution is $-59 \equiv 230 \pmod{17^2}$.
15. Using step 4 in Example 11.32, $\alpha = 9$ is a solution of $x^2 \equiv 17 \pmod{32}$. Then $9^2 = 17 + 32i$, where $i = 2$. Now choose j such that $i + \alpha j = 2 + 9j \equiv 0 \pmod 2$: $j = 0$. Then $\alpha + j2^{k-1} = 9 + 0 \cdot 2^4 = 9$ is a solution of $x^2 \equiv 17 \pmod{64}$. The remaining solutions are $-9 \equiv 55$ and $2^5 \pm 9$ modulo 64, that is, 23, 41, and 55 modulo 64.

17. $\alpha = 1$ is a solution of $x^2 \equiv 25 \equiv 1 \pmod 8$, so $1^2 = 25 + 8i$, where $i = -3$. Now choose j such that $i + \alpha j = -3 + 1j \equiv 0 \pmod 2$: $j = 1$. Then $\alpha + j2^{k-1} = 1 + 0 \cdot 2^2 = 5$ is a solution of $x^2 \equiv 25 \pmod{16}$, so $5^2 = 25 + 16i$, where $i = 0$. Now choose j such that $i + \alpha j = 0 + 5j \equiv 0 \pmod 2$: $j = 0$. Then $\alpha + j2^{k-1} = 5 + 0 \cdot 2^3 = 5$ is a solution of $x^2 \equiv 25 \pmod{32}$. The remaining solutions are -5 and $2^4 \pm 5$ modulo 32, that is, 11, 21, and 27 modulo 32.
19. Because $\alpha = 1$ is a solution of $x^2 \equiv 33 \equiv 1 \pmod 8$, $1^2 = 33 + 8i$, where $i = -4$. Now choose j such that $i + \alpha j = -4 + 1j \equiv 0 \pmod 2$: $j = 0$. Then $\alpha + j2^{k-1} = 1 + 0 \cdot 2^2 = 1$ is a solution of $x^2 \equiv 33 \pmod{16}$, so $1^2 = 33 + 16i$, where $i = -2$. This implies $\alpha + j2^{k-1} = 1 + 0 \cdot 2^3 = 1$ is a solution of $x^2 \equiv 33 \pmod{32}$, so $1^2 = 33 + 32i$, where $i = -1$. This yields $\alpha + j2^{k-1} = 1 + 1 \cdot 2^4 = 17$ as a solution of $x^2 \equiv 33 \pmod{64}$. The remaining solutions are -17 and $2^5 \pm 17$ modulo 64, that is, 15, 47, and 49 modulo 64.
21. $\alpha = 1$ is a solution of $x^2 \equiv 41 \equiv 1 \pmod 8$. Then $\alpha + j2^{k-1} = 1 + 1 \cdot 2^2 = 5$ is a solution of $x^2 \equiv 41 \pmod{16}$. This yields $\alpha + j2^{k-1} = 5 + 1 \cdot 2^3 = 13$ is a solution of $x^2 \equiv 41 \pmod{32}$. The remaining solutions are $-13 \equiv 19$, $2^4 - 13 \equiv 3$, and $2^4 + 13 \equiv 29$ modulo 32.
23. By Exercise 22, $\alpha = 77$ is a solution of $x^2 \equiv 41 \equiv 1 \pmod{256}$. So $\alpha + j2^{k-1} = 77 + (-1) \cdot 2^7 = -51$ is a solution of $x^2 \equiv 41 \pmod{512}$. This yields $\alpha + j2^{k-1} = -51 + 1 \cdot 2^8 = 205$ is a solution of $x^2 \equiv 41 \pmod{1024}$. The remaining solutions are $-205 \equiv 819$, $2^9 - 205 \equiv 307$, and $2^9 + 205 \equiv 717$ modulo 1024.
25. The given congruence can be written as $y^2 \equiv 4 \pmod{5^2}$, where $y = 2x + 1$. Clearly, $y = \pm 2$ are solutions of this congruence. When $y = 2$, $2x + 1 \equiv 2 \pmod{25}$, so $x \equiv 13 \pmod{25}$. The other solution is $x \equiv 11 \pmod{25}$, corresponding to $y \equiv -2 \pmod{25}$.
27. $9(2x^2 + 1) \equiv 0 \pmod{17}$; that is, $x^2 \equiv 8 \pmod{17}$. Therefore, $x \equiv 5, 12 \pmod{17}$.
29. $13(3x^2 + 1) \equiv 0 \pmod{19}$; that is, $x^2 \equiv 6 \pmod{19}$. Therefore, $x \equiv 5, 14 \pmod{19}$.
31. Because $23 \not\equiv 1 \pmod 4$, $x^2 \equiv 23 \pmod 4$ is not solvable by Theorem 11.15. Therefore, the given congruence is not solvable.
33. We have $x^2 \equiv 13 \pmod 4$ and $x^2 \equiv 13 \pmod{17^2}$. The solutions of $x^2 \equiv 13 \pmod 4$ are $1, 3 \pmod 4$ and those of the other are $59, 230 \pmod{17^2}$. When $x \equiv 1 \pmod 4$ and $x \equiv 59 \pmod{17^2}$, by the CRT, $x \equiv 1 \cdot 17^2 \cdot 1 + 59 \cdot 4 \cdot 217 \equiv 637 \pmod{1156}$. Similarly, we get three more solutions: 59, 519, and 1097 modulo 1156.
35. We have $x^2 \equiv 17 \pmod 8$ and $x^2 \equiv 17 \pmod{19^2}$. The solutions of $x^2 \equiv 17 \equiv 1 \pmod 8$ are 1, 3, 5, and 7 (mod 8); $x = 6$ is a solution of $x^2 \equiv 17 \pmod{19}$, so the two solutions of $x^2 \equiv 17 \pmod{19^2}$ are $6 + 19 \cdot 11 \equiv 215 \pmod{19^2}$ and $-215 \equiv 146 \pmod{19^2}$. When $x \equiv 1 \pmod 8$ and $x \equiv 215 \pmod{19^2}$, by the CRT, $x \equiv 1 \cdot 19^2 \cdot 1 + 215 \cdot 8 \cdot 316 \equiv 215 \pmod{2888}$; when $x \equiv 1 \pmod 8$ and $x \equiv 146 \pmod{19^2}$, $x \equiv 1 \cdot 19^2 \cdot 1 + 146 \cdot 8 \cdot 316 \equiv 2673 \pmod{2888}$; Similarly, we get six more solutions: 507, 937, 1229, 1659, 1951, and 2381 modulo 2888.

37. 2 39. 8 41. 32 43. 2^k 45. 2^{k+2}

47. **proof:** Because $ab \equiv 1 \pmod p$, $(ab/p) = (1/p) = 1$. That is, $(a/p)(b/p) = 1$, so $(a/p) = (b/p)$.
49. **proof:** $3x^2 + 1 \equiv 0 \pmod p$ is solvable if and only if $x^2 \equiv -3^{-1} \pmod p$ is solvable; that is, if and only if $(-3^{-1}/p) = 1$. But $(-3^{-1}/p) = 1$ if and only if $(-3/p) = 1$; that is, if and only if $p \equiv 1 \pmod 6$, by Exercise 33 in Section 11.3.

51. **proof:** Let α be a solution. Then, by Exercise 50, $p^n - \alpha$ is also a solution. Suppose $\alpha \equiv p^n - \alpha \pmod{p^n}$. Then $2\alpha \equiv 0 \pmod{p^n}$, so $\alpha \equiv 0 \pmod{p^n}$, which is impossible. Therefore, α and $p^n - \alpha$ are two incongruent solutions.
    Let β be any solution of the congruence. Then $\alpha^2 \equiv a \equiv \beta^2 \pmod{p^n}$, so $(\alpha + \beta)(\alpha - \beta) = \alpha^2 - \beta^2 \equiv 0 \pmod{p^n}$. Because $n \geq 2$, this implies either $p^n | \alpha + \beta$, $p^n | \alpha - \beta$, or $p | \alpha + \beta$ and $p | \alpha - \beta$. If $p^n | \alpha + \beta$, then $\beta = p^n - \alpha$. If $p^n | \alpha - \beta$, then $\alpha = \beta$. In the last case, $p | 2\alpha$, so $p | \alpha$ since p is odd. Let $\alpha = mp$. Then $m^2p^2 \equiv a \pmod{p^n}$, which implies $p | a$, which is a contradiction.
    Thus, the given congruence has exactly two incongruent solutions, α and $p^n - \alpha$.
53. **proof:** Because $(2^n - \alpha)^2 \equiv (-\alpha)^2 \equiv a \pmod{p^n}$ and $(2^{n-1} \pm \alpha)^2 = 2^{2n-2} \pm 2^n\alpha + \alpha^2 \equiv 0 \pm 0 \cdot \alpha \equiv a \pmod{p^n}$, the result follows.

Review Exercises (p. 546)

1. 8, 9
3. $(116/73) = (43/73) = (73/43) = (40/43) = (10/43) = (2/43)(3/5) = -(-1) = 1$
5. $(3749/1739) = (271/1739) = (271/37)(271/47) = (12/37)(36/47)$
    $= (3/37) = (37/3) = (1/3) = 1$
7. $(1/M_p) = 1$
9. Because $M_p \equiv 3 \pmod 4$, $(-1/M_p) = -1$.
11. Because $M_p = 2^p - 1 \equiv (-1)^p - 1 \equiv 1 \pmod 3$, $(3/M_p) = -(M_p/3) = -(1/3) = -1$.
13. Because $f_n \equiv 1 \pmod 4$ and $f_n \equiv 2 \pmod 3$, $(3/f_n) = (f_n/3) = (2/3) = -1$.
15. Because $n \geq 5$, $p = n! + 1 \equiv 0 + 1 \equiv 1 \pmod 5$. Therefore, $(5/p) = (p/5) = (1/5) = 1$.
17. Because n is even, $p = 4^k + 1 \equiv 4 + 1 \equiv 5 \pmod{12}$; so $(3/p) = -1$.
19. Because $p = 16^n + 1 \equiv 1 + 1 \equiv 2 \pmod 5$, $(5/p) = -1$.
21. Because the product of two quadratic nonresidues is a quadratic residue, $13 \cdot 29 \equiv 12 \pmod{73}$ is a quadratic residue.
23. $3989 \equiv 5 \pmod 8$, so $(2/3989) = -1$. Then $2^{1994} \equiv -1 \pmod{3989}$.
25. Because $4793 \equiv 5 \pmod{12}$, $(3/4793) = -1$; that is, $3^{2396} \equiv -1 \pmod{4793}$.
27. Because 14 is a solution of $x^2 \equiv 9 \pmod{17}$, it is also a solution of $(x/2)^2 \equiv 9 \pmod{17}$. Therefore, $x/2 \equiv 14 \pmod{17}$. Thus, $x \equiv 28 \equiv 11 \pmod{17}$.
29. $167 | 2^{83} - 1$
31. $2027 | 2^{1013} - 1$
33. Let $p | n^2 + 1$. Then $n^2 \equiv -1 \pmod p$, so $(-1/p) = (n^2/p) = 1$. But $(-1/p) = 1$ if and only if $p \equiv 1 \pmod 4$. Thus, $p | n^2 + 1$ if and only if $p \equiv 1 \pmod 4$.
35. The given congruence can be written as $y^2 \equiv 114 \pmod{11^2}$, where $y \equiv 4x + 1 \pmod{11^2}$. Because 2 is a solution of $y^2 \equiv 114 \equiv 4 \pmod{11}$, $2^2 \equiv 114 + 11i$, where $i = -10$. Then $2 + 11j = 2 + 11 \cdot 8 = 90$ is a solution of $y^2 \equiv 114 \pmod{11^2}$, and so is $-90 \equiv 31 \pmod{11^2}$. When $y = 90$, $4x + 1 \equiv 90 \pmod{11^2}$, so $x \equiv 112 \pmod{11^2}$. When $y = 31$, $4x + 1 \equiv 31 \pmod{11^2}$, so $x \equiv 68 \pmod{11^2}$.
37. Because $2431 = 11 \cdot 13 \cdot 17$, $x^2 \equiv 53 \pmod{2431}$ implies $x^2 \equiv 9 \pmod{11}$, $x^2 \equiv 1 \pmod{13}$, and $x^2 \equiv 2 \pmod{17}$. Solving them, we get $x \equiv 3, 8 \pmod{11}$, $x \equiv 1, 12 \pmod{13}$, and $x \equiv 6, 11 \pmod{17}$, respectively. Then, by the CRT, $x \equiv 261, 300, 844, 1026, 1405, 1587, 2131, 2170 \pmod{2431}$.
39. Because $9724 = 4 \cdot 11 \cdot 13 \cdot 17$, the given congruence implies $x^2 \equiv 169 \pmod 4$, $x^2 \equiv 169 \pmod{11}$, $x^2 \equiv 169 \pmod{13}$, and $x^2 \equiv 169 \pmod{17}$. The only solution of $x^2 \equiv 169$

(mod 13) is 0; that is, $x \equiv 13 \pmod{13}$. The various solutions of the given congruence are 13, 871, 3991, 4849, 4875, 5733, 8853, and 9711 modulo 9724.

41. 6 is a solution of $x^2 \equiv 226 \equiv 17 \pmod{19}$. Then $6 + 19j = 6 + 19 \cdot 4 = 82$ is a solution of $x^2 \equiv 226 \pmod{19^2}$. Consequently, $82 + 19^2 j = 82 + 19^2 \cdot 8 = 2970$ is a solution of $x^2 \equiv 226 \pmod{19^3}$. The other solution is $-2970 \equiv 3889 \pmod{19^3}$.

43. $p = 1{,}000{,}151 \equiv 3 \pmod 4$ and $q = 2p + 1 = 2{,}000{,}303$ are primes. Therefore, $2000303 | M_{1000151}$.

45. **proof:** $(4a/p) = (4/p)(a/p) = (a/p)$. Thus, $(4a/p) = 1$ if and only if $(a/p) = 1$. In other words, the congruence $x^2 \equiv 4a \pmod p$ is solvable if and only if $x^2 \equiv a \pmod p$ is solvable.

47. **proof:** Because $q \equiv -1 \pmod p$, $(q/p) = (-1/p) = 1$. Then $(p/q) = (q/p) = 1$.

49. **proof:** Because $p \equiv \pm 1 \pmod 8$, $(2/p) = 1$; that is, $2^{(p-1)/2} \equiv 1 \pmod p$, so $p | 2^{(p-1)/2} - 1$.

51. **proof:** Let α be a quadratic nonresidue of f_n. Then $\alpha^{(f_n - 1)/2} \equiv -1 \pmod{f_n}$. Therefore, $\alpha^{f_n - 1} \equiv 1 \pmod{f_n}$. Thus α is a primitive root modulo f_n, because $\text{ord}_{f_n} \alpha \nless f_n$.

53. **proof:** Because $p \equiv 1 \pmod 4$, $q \equiv 3 \pmod 8$; so $(2/q) = -1$. Then, by Euler's criterion, $2^{(q-1)/2} \equiv -1 \pmod q$; that is, $2^{2p} \equiv 1 \pmod q$. Consequently, $\text{ord}_q 2 = 1, 2, p$, or $2p$. Clearly, $\text{ord}_q 2 \neq 1$. If $\text{ord}_q 2 = 2$, then $2^2 \equiv 1 \pmod q$; that is, $q|3$, which is impossible. Because $2^p = 2^{(q-1)/2} \not\equiv 1 \pmod q$, $\text{ord}_q 2 \neq p$. Thus $\text{ord}_q 2 = 2p = \varphi(q)$, and hence 2 is a primitive root modulo q.
On the other hand, let $p \equiv 3 \pmod 4$. Then $q \equiv -1 \pmod 8$, so $(-1/q) = -1$ and $(2/q) = 1$. Thus $(-2/q) = -1 \equiv (-2)^p \pmod q$; that is, $(-2)^{2p} \equiv 1 \pmod q$. As earlier, it can be shown that -2 is a primitive root modulo q.

55. **proof** (by contradiction): Assume that there are only a finite number of such primes, $p_1, \ldots, p_k$. Let $N = (p_1 \cdots p_k)^2 + 2$. Notice that $N \equiv 3 \pmod 8$. Let p be an odd prime factor of N. Because $p|N$, $(p_1 \cdots p_k)^2 \equiv -2 \pmod p$, so $(-2/p) = 1$. Therefore, by Exercise 50 in Section 11.2, $p \equiv 1$ or $3 \pmod 8$. If every prime factor of N is $\equiv 1 \pmod 8$, then N also would be $\equiv 1 \pmod 8$. Because $N \equiv 3 \pmod 8$, it must have a prime factor $q \equiv 3 \pmod 8$.
If $q = p_i$ for some i, then $q|2$, which is impossible. Therefore, $q \neq p_i$ for every i, which is a contradiction.

57. **proof:** Because $p \equiv (-1)^n + 1 \pmod 3$, $p \equiv 2 \pmod 3$ if n is even and $0 \pmod 3$ otherwise. Therefore, n must be even for p to be a prime; so $(p/3) = (2/3) = -1$; that is, $3^{(p-1)/2} \equiv 3^{2^{n-1}} \equiv -1 \pmod p$. Then $3^{2^n} \equiv 1 \pmod p$.
Let $\text{ord}_p 3 = e$. Then $e | 2^n$, so $e = 2^k$, where $k \le n$. If $k < n$, then $1 = (3^{2^k})^{2^{n-k-1}} = 3^{2^{n-1}} \equiv -1 \pmod p$, which is a contradiction. Thus $\text{ord}_p 3 = 2^n$, and 3 is a primitive root modulo p.

59. **proof:** Let $m \equiv 1 \pmod 4$. Then $(m-1)/2$ is even, so $m^* = m$. Therefore, $(m^*/n) = (m/n) = (n/m)$, by Theorem 11.13.
On the other hand, let $m \equiv 3 \pmod 4$. Then $(m-1)/2$ is odd. Therefore, $(m^*/n) = (-m/n) = (-1/n)(m/n)$. If $n \equiv 1 \pmod 4$, then $(-1/n) = 1$ and $(m/n) = (n/m)$; thus, $(m^*/n) = 1$. $(n/m) = (n/m)$. If $n \equiv 3 \pmod 4$, then $(-1/n) \equiv -1$ and $(m/n) = -(n/m)$; thus $(m^*/n) = (-1)(-(n/m)) = (n/m)$.

Supplementary Exercises (p. 548)

1. **proof:** $f(ab) = (ab/p) = (a/p)(b/p) = f(a) \cdot f(b)$.
3. $K =$ set of quadratic residues of p
5. 1, 2, 3, 4; 1, 6
7. **proof:** Let a be a biquadratic residue of p, so $x^4 \equiv a \pmod p$ is solvable. Then $(x^2)^2 \equiv a \pmod p$, so $y^2 \equiv a \pmod p$ is solvable. Thus, a is also a quadratic residue of p.
9. Because $8^4 \equiv -1 \pmod 7$, -1 is a biquadratic residue of 17. Since the biquadratic residues of 13 are 1, 3, and 9, -1 is a biquadratic nonresidue of 13.
11. $(85/2) = -1$, since $85 \equiv 5 \pmod 8$.
13. $$\begin{aligned}(85/3861) &= (5\cdot 17/3)^3(5\cdot 17/11)(5\cdot 17/13) = (5\cdot 17/3)(5\cdot 17/11)(5\cdot 17/13)\\ &= (4/3)(8/11)(7/13) = (1/3)(2/11)\cdot[-(13/7)]\\ &= 1\cdot(-1)\cdot[-(-1/7)] = (-1/7) = -1\end{aligned}$$
15. Let $m = \prod_i p_i^{e_i}$ and $n = \prod_i q_i^{f_i}$ be the prime factorizations of m and n. Then $$(a/mn) = \prod_i (a/p_i)^{e_i} \cdot \prod_i (a/q_i)^{f_i} = (a/m)(a/n)$$

Chapter 12 Continued Fractions

Exercises 12.1 (p. 564)

1. [2; 2, 11]
3. [−3; 2, 8]
5. $\dfrac{113}{77}$
7. $\dfrac{-441}{157}$
9. The 21×13 rectangle contains one 13×13 square, one 8×8 square, one 5×5 square, one 3×3 square, one 2×2 square, and two 1×1 square. So the desired continued fraction is [1; 1, 1, 1, 1, 2].
11. It follows from the table

n	1	2	3	4	5	6
a_n	1	1	1	1	1	1
p_n	2	3	5	8	13	21
q_n	1	2	3	5	8	13

that $[1; 1, 1, 1, 1, 1, 1] = \dfrac{p_6}{q_6} = \dfrac{21}{13}$.

13. $[3; 1, 4, 2, 7] = \dfrac{p_4}{q_4} = \dfrac{313}{82}$
15. $c_4 = \dfrac{p_4}{q_4} = \dfrac{a_4p_3 + p_2}{a_4q_3 + q_2} = \dfrac{5\cdot 43 + 10}{5\cdot 30 + 7} = \dfrac{225}{157}$
17. $c_6 = \dfrac{p_6}{q_6} = \dfrac{a_6p_5 + p_4}{a_6q_5 + q_4} = \dfrac{1\cdot 13 + 8}{1\cdot 8 + 5} = \dfrac{21}{13}$
19. **proof** (by PMI): The result is clearly true when $n = 1$ and $n = 2$. Now assume it is true for all positive integers $\le k$. Then, by Theorem 12.3, $$c_{k+1} = \frac{p_{k+1}}{q_{k+1}} = \frac{a_{k+1}p_k + p_{k-1}}{a_{k+1}q_k + q_{k-1}} = \frac{1\cdot F_{k+2} + F_{k+1}}{1\cdot F_{k+1} + F_k} = \frac{F_{k+3}}{F_{k+2}}$$ Thus, the result follows by strong induction.
21. [5; 4, 3, 2, 1]
23. [5; 4, 3, 2]
25. $\dfrac{p_k}{p_{k-1}} = [a_k; a_{k-1}, \ldots, a_1, a_0]$
27. [0; 2, 2, 11]
29. $\dfrac{1}{r} = [0; a_0, a_1, \ldots, a_n]$
31. Since $(12, 13) = 1$, the LDE is solvable. Since $\dfrac{12}{13} = [0; 1, 12]$ and $p_2q_1 - q_2p_1 = -1$, $12\cdot 14 + 13(-14) = 14$; so the general solution is $x = 14 + 13t$, $y = -14 - 12t$.

33. Since $(1776, 1976) = 8$ and $8|4152$, the LDE is solvable. The LDE can be rewritten as $222x + 247y = 519$. Using the continued fraction $\frac{222}{247} = [0; 1, 8, 1, 7, 3]$, $p_5q_4 - q_5p_4 = 1$. This yields $x = 41001 + 247t$, $y = -36849 - 222t$.

Exercises 12.2 (p. 575)

1. $[1; \overline{2}]$
3. $[2; \overline{4}]$
5. $[3; 7, 15, 1, 292, 1, 1, 2, 1, 3, 1, \dots]$
7. It follows from the table

n	1	2	3	4	5
a_n	2	2	2	2	2
p_n	3	7	17	41	99
q_n	2	5	12	29	70

that $c_1 = 3/2$, $c_2 = 7/5$, $c_3 = 17/12$, $c_4 = 41/29$, and $c_5 = 99/70$.

9. We have $\pi = [3; 7, 15, 1, 292, 1, 1, 2, 1, 3, 1, \dots]$. It follows from the table

n	1	2	3	4	5
a_n	7	15	1	292	1
p_n	22	333	355	103993	104348
q_n	7	106	113	33102	33215

that $c_1 = 22/7$, $c_2 = 333/106$, $c_3 = 355/113$, $c_4 = 103993/33102$, and $c_5 = 104348/33215$.

11. By Exercise 9, $c_4 = \frac{103993}{33102} \approx 3.1415926530$, and $c_5 = \frac{104348}{33215} \approx 3.1415926539$; so $\pi \approx 3.14159265$.
13. Follows by Exercise 19 in Section 12.1.
15. Let $x = [F_n; F_n, F_n, \dots]$. Then $x = F_n + \frac{1}{x}$. Solving this, $x = \frac{F_n \pm \sqrt{F_n^2 + 4}}{2}$. But $x > 0$, so $x = \frac{F_n + \sqrt{F_n^2 + 4}}{2}$.

Review Exercises (p. 576)

1. $[2; 2, 9]$
3. $\frac{225}{43}$
5. $c_5 = \frac{p_5}{q_5} = \frac{a_5p_4 + p_3}{a_5q_4 + q_3} = \frac{9 \cdot 115 + 16}{9 \cdot 151 + 21} = \frac{1051}{1380}$
7. $[0; 1, 2, 1, 1, 3]$
9. Since $(43, 23) = 1$, the LDE is solvable. Using the continued fraction $\frac{43}{23} = [1; 1, 6, 1, 2]$, the general solution is $x = -264 + 23t$, $y = 495 - 43t$.
11. Since $(76, 176) = 4$ and $4|276$, the LDE is solvable. Dividing the LDE by 4, it becomes $19x + 44y = 69$. Using the continued fraction $\frac{19}{44} = [0; 2, 3, 6]$, the general solution is $x = 483 + 44t$, $y = -207 - 19t$.
13. Using the convergents $c_9 = \frac{p_9}{q_9} = \frac{F_{11}}{F_{10}} = \frac{89}{55}$ and $c_{10} = \frac{p_{10}}{q_{10}} = \frac{F_{12}}{F_{11}} = \frac{144}{89}$, the general solution is $x = -1265 + 89t$, $y = 2047 - 144t$.
15. $[2; 1, 1, 1, 3, 0, 1, 1, 1, \dots]$
17. $[1; 1, 1, 1, 5, 1, 1, 9, 1, 1, 13, 1, 1, \dots]$

19. $c_5 = \dfrac{15541}{33630} = 0.462117157\ldots$ and $c_6 = \dfrac{342762}{741721} = 0.462117157\ldots$. So $\dfrac{e-1}{e+1} \approx 0.46211715$.

21. **proof:** Since $r < 1, r = [0; a_1, a_2, \ldots, a_n] = \dfrac{1}{[a_1; a_2, \ldots, a_n]}$. So $\dfrac{1}{r} = [a_1; a_2, \ldots, a_n]$.

Supplementary Exercises (p. 578)

1. Since $p_k = a_k p_{k-1} + p_{k-2}$, $\dfrac{p_k}{p_{k-1}} = a_k + \dfrac{p_{k-2}}{p_{k-1}} = a_k + \frac{1}{\frac{p_{k-1}}{p_{k-2}}}$. Continuing like this, we get:

$$\frac{p_k}{p_{k-1}} = a_k + \cfrac{1}{a_{k-1} + \cfrac{1}{p_{k-2}/p_{k-3}}} = a_k + \cfrac{1}{a_{k-1} + \cfrac{1}{a_{k-2} + \cfrac{1}{\ddots + \cfrac{1}{a_1}}}}$$

$$= [a_k; a_{k-1}, \ldots, a_1, a_0]$$

3. **proof** (by PMI): Since $q_2 = a_2 a_1 + 1 \geq 2$ and $q_3 = a_3 q_2 + q_1 \geq q_2 + q_1 > 2q_1 \geq 2$, the result is true when $k = 2$ and $k = 3$. Now assume that $q_i \geq 2^{i/2}$ for all integers i, where $2 \leq i \leq k-1$. Then $q_k = a_k q_{k-1} + q_{k-2} \geq 2q_{k-2} \geq 2 \cdot 2^{(k-2)/2}$; that is, $q_k \geq 2^{k/2}$. Thus, the result follows by strong induction, where $k \geq 2$.

Chapter 13 Miscellaneous Nonlinear Diophantine Equations

Exercises 13.1 (p. 589)

1. $m \not\equiv n \pmod 2$

3. Let x–$(x+1)$–$(x+2)$ be a Pythagorean triple consisting of consecutive positive integers. Then $x^2 + (x+1)^2 = (x+2)^2$. This yields $(x-3)(x+1) = 0$. Because $x > 0$, $x = 3$. Thus, 3–4–5 is the only Pythagorean triple consecutive positive integers; clearly, it is primitive.

5. 2000001 2000002000000 2000002000001
   20000001 200000020000000 200000020000001

7. 44–117–125, 88–105–137, 132–85–157, 176–57–185, 220–21–221

9. $13^2 + 84^2 = 85^2$; $15^2 + 112^2 = 113^2$

11. The conjecture in Exercise 10 is $(2n+1)^2 + [2n(n+1)]^2 = [2n(n+1)+1]^2$, where $n \geq 1$.
    proof: RHS $= [2n(n+1)+1]^2 = [2n(n+1)]^2 + 2[2n(n+1)] + 1 = [2n(n+1)]^2 + (4n^2 + 4n + 1) = 2n(n+1)^2 + (2n+1)^2 =$ LHS.

13. $(4n)^2 + (4n^2 - 1)^2 = (4n^2 + 1)^2, n \geq 2$.

15. **proof:** If $3|m$ or $3|n$, then $3|x$ or $3|y$. Suppose $3 \nmid m$ and $3 \nmid n$. Then, by Fermat's Little Theorem, $m^2 \equiv 1 \equiv n^2 \pmod 3$. Therefore, $y = m^2 - n^2 \equiv 0 \pmod 3$, so $3|y$.

17. **proof:** If $2|m$, then $4|x$ and we are done. If $2 \nmid m$, then m is odd. Therefore, n has to be even and hence $2|n$. Then also $4|x$. Thus, at least one of the numbers x, y, and z is divisible by 4.

19. **proof:** By Exercise 18, $12|xyz$. By Exercise 16, 5 divides x, y, or z; so $5|xyz$. Thus, $60|xyz$.

21. **proof:** Because $2 = z - y = (m^2 + n^2) - (m^2 - n^2)$, $2n^2 = 2$, so $n = 1$. Then $x = 2m$, $y = m^2 - 1$, and $z = m^2 + 1$.
23. Area = (base)(height)/2 = $(2mn)(m^2 - n^2)/2 = mn(m^2 - n^2)$.
25. No, because if it does, then $m^2 - n^2 + 1 = m^2 + n^2$. This yields $2n^2 = 1$; so n is an irrational number, which is a contradiction.
27. Area = (base)(height)/2 = $(k2mn) \cdot k(m^2 - n^2)/2 = k^2mn(m^2 - n^2)$, where k is an arbitrary positive integer.
29. Area = $xy/2 = (1294)(693)/2 = 666{,}666$.
31. **proof:** First, notice that $z_n^2 = x_n^2 + (x_n + 1)^2 = 2x_n^2 + 2x_n + 1$. Also, $y_{n+1} + 1 = x_{n+1} + 1$. Next we show that $x_{n+1}^2 + y_{n+1}^2 = z_{n+1}^2$:

$$\begin{aligned}
x_{n+1}^2 + y_{n+1}^2 &= (3x_n + 2z_n + 1)^2 + (3x_n + 2z_n + 2)^2 \\
&= 18x_n^2 + 8z_n^2 + 24z_nx_n + 34x_n + 12z + 5 \\
&= 18x_n^2 + 8(2x_n^2 + 2x_n + 1) + 24z_nx_n + 18x_n + 12z_n + 5 \\
&= 34x_n^2 + 24z_nx_n + 34x_n + 12z_n + 13 \\
z_{n+1}^2 &= (4x_n + 3z_n + 2)^2 = 16x_n^2 + 9z_n^2 + 24z_nx_n + 16x_n + 12z_n + 4 \\
&= 16x_n^2 + 9(2x_n^2 - 2x_n + 1) + 24z_nx_n + 16x_n + 12z_n + 4 \\
&= 34x_n^2 + 24z_nx_n + 34x_n + 12z_n + 13 = x_{n+1}^2 + y_{n+1}^2
\end{aligned}$$

Finally, let p be a prime factor of $(x_{n+1}, y_{n+1}, z_{n+1})$. Then $p|x_{n+1}$ and $p|y_{n+1}$, so $p|(y_{n+1} - x_{n+1})$. That is, $p|1$, which is a contradiction. Therefore, $(x_{n+1}, y_{n+1}, z_{n+1}) = 1$. Thus x_{n+1}–y_{n+1}–z_{n+1} is also a primitive Pythagorean triple.
33. $x = 15$, $y = 20$, $z = 12$.

Exercises 13.2 (p. 601)

1. **proof:** Suppose $x^2 + y^2 = r^2$ and $x^2 - y^2 = s^2$ for some integers r and s. Then $x^4 - y^4 = (x^2 + y^2)(x^2 - y^2) = (rs)^2$. Because this contradicts Theorem 13.3, the result follows.
3. **proof:** Area of a Pythagorean triangle $= xy/2$. Suppose the area equals $2u^2$ for some integer u. Then $xy = 4u^2$. This implies $(x + y)^2 = z^2 + 8u^2$ and $(x - y)^2 = z^2 - 8u^2$, so $(x + y)^2(x - y)^2 = (z^2 + 8u^2)(z^2 - 8u^2)$. That is, $(x^2 - y^2)^2 = z^4 - (4u)^4$. Because this negates Theorem 13.3, the result follows.
5. **proof:** If p divides x, y, or z, then $p|xyz$. Otherwise, by Fermat's little theorem, $x^{p-1} \equiv y^{p-1} \equiv z^{p-1} \equiv 1 \pmod{p}$. Since $x^{p-1} + y^{p-1} = z^{p-1}$, this implies $1 + 1 \equiv 1 \pmod{p}$, which is a contradiction.
7. 1–1–1
9. We have $z^2 = x^2 + y^2$. Suppose $x = r^2$ and $y = s^2$ for some integers r and s. Then $r^4 + s^4 = z^2$, which is impossible by Theorem 13.2. On the other hand, let x and z be squares. Let $x = r^2$ and $z = t^2$ for some integers r and t. Then $t^4 - r^4 = y^2$, which is impossible by Theorem 13.3. By symmetry, it is impossible for y and z also to be squares.
11. 2–2

13. Because 3–4–5–6 is a solution by Exercise 12, $3n$–$4n$–$5n$–$6n$ is also a solution for every positive integer. Thus, the equation has infinitely many solutions.
15. When $a = 2$ and $b = 1$, the formula yields $9^3 + 12^3 + 15^3 = 18^3$, so 9–12–15–18 is a solution. Similarly, when $a = 3$ and $b = 1$, the formula yields the solution 28–53–75–84.
17. **proof** (by contradiction): Assume that $x^4 - y^4 = z^2$ has integral solutions and x–y–z is a solution with the least positive value of x. Then $z^2 + y^4 = x^4$; that is, $z^2 + (y^2)^2 = (x^2)^2$. Then z–y^2–x^2 is a primitive Pythagorean triple.

    case 1 Let y^2 be odd. Then there exist positive integers m and n such that $z = 2mn$, $y^2 = m^2 - n^2$, and $x^2 = m^2 + n^2$, where $m > n$ and $(m, n) = 1$. This implies $m^4 - n^4 = (m^2 + n^2)(m^2 - n^2) = x^2y^2 = (xy)^2$, so m–n–xy is also a solution of the given equation. Because $0 < m^2 < m^2 + n^2 = x^2$, it follows that m–n–xy is smaller than the least solution, which is a contradiction.

    case 2 Let y^2 be even. Then there exist positive integers m and n such that $y^2 = 2mn$, $z^2 = m^2 - n^2$, and $x^2 = m^2 + n^2$, where $m > n$, $(m, n) = 1$, m is even, and n is odd. Because $(2m, n) = 1$ and $y^2 = 2mn$, $2m = r^2$ and $n = s^2$ for some integers r and s. Because r is even, this implies $m = 2t^2$ for some integer t. Therefore, $x^2 = 4t^4 + s^4$, so $2t^2$–s^2–x is a primitive Pythagorean triple. Consequently, there are positive integers u and v such that $2t^2 = 2uv$, $s^2 = u^2 - v^2$, and $x = u^2 + v^2$, where $u > v$ and $(u, v) = 1$. Since $t^2 = uv$, $u = a^2$ and $v = b^2$ for some integers a and b. Then $s^2 = a^4 - b^4$, so a–b–s is also a solution of the given equation $x^4 - y^4 = z^2$. Because $0 < a = \sqrt{u} < u^2 + v^2 = x$, this yields a contradiction. Thus, the given equation has no positive integral solutions.

Exercises 13.3 (p. 612)

1. $13 = 2^2 + 3^2$
3. $41 = 4^2 + 5^2$
5. $k^2n = k^2(x^2 + y^2) = (kx)^2 + (ky)^2$
7. $13 = 2^2 + 3^2$ and $17 = 1^2 + 4^2$. Therefore, $13 \cdot 17 = (2 \cdot 1 + 3 \cdot 4)^2 + (2 \cdot 4 - 3 \cdot 1)^2 = 14^2 + 5^2$.
9. yes, because 101 is a prime and $101 \equiv 1 \pmod 4$.
11. yes, since 233 is a prime and $233 \equiv 1 \pmod 4$.
13. $149 = 10^2 + 7^2$
15. $200 = 14^2 + 2^2$
17. Although $r \equiv 3 \pmod 4$, since the exponent of r is odd, the given number is not expressible as the sum of two squares.
19. Because $2n(n + 1) + 1 = 2n^2 + 2n + 1 = n^2 + (n^2 + 2n + 1) = n^2 + (n + 1)^2$, the result follows.
21. Let $x^2 + y^2 + z^2 \equiv 0 \pmod 4$. If all the integers are odd, then $x^2 + y^2 + z^2 \equiv 1 + 1 + 1 \equiv 3 \pmod 4$, which is a contradiction. Suppose two of them, say, x and y, are odd. Then $x^2 + y^2 + z^2 \equiv 1 + 1 + 0 \equiv 2 \pmod 4$, again a contradiction. Similarly, even if one of them is odd, we will again get a contradiction. Thus, $x \equiv y \equiv z \equiv 0 \pmod 2$.
23. Follows by letting $c = d = g = h = 0$ in Lemma 13.11.
25. $$\begin{aligned} 217 = 7 \cdot 31 &= (2^2 + 1^2 + 1^2 + 1^2)(5^2 + 2^2 + 1^2 + 1^2) \\ &= (2 \cdot 5 + 1 \cdot 2 + 1 \cdot 1 + 1 \cdot 1)^2 + (2 \cdot 2 - 1 \cdot 5 + 1 \cdot 1 - 1 \cdot 1)^2 \\ &\quad + (2 \cdot 1 - 1 \cdot 1 - 1 \cdot 5 + 1 \cdot 2)^2 + (2 \cdot 1 + 1 \cdot 1 - 1 \cdot 2 - 1 \cdot 5)^2 \\ &= 14^2 + 1^2 + 2^2 + 4^2 \end{aligned}$$

27. $29{,}791 = 31 \cdot 31^2 = (5^2 + 2^2 + 1^2 + 1^2) \cdot 31^2 = 155^2 + 62^2 + 31^2 + 31^2$
29. $89 = 8^2 + 5^2 + 0^2 + 0^2$
31. $349 = 18^2 + 5^2 + 0^2 + 0^2$
33. $840 = 2^3 \cdot 7 \cdot 15 = 2^2(2 \cdot 105) = 2^2(1^2 + 1^2 + 0^2 + 0^2)(10^2 + 2^2 + 1^2 + 0^2)$
$= 2^2(14^2 + 3^2 + 2^2 + 1^2) = 28^2 + 6^2 + 4^2 + 2^2$
35. Because $64{,}125 = 3^3 \cdot 5^3 \cdot 19$, $3 \cdot 5 = 3^2 + 2^2 + 1^2 + 1^2$, and $19 = 4^2 + 1^2 + 1^2 + 1^2$,

$$64{,}125 = 3^2 \cdot 5^2(3 \cdot 5 \cdot 19) = 15^2(3^2 + 2^2 + 1^2 + 1^2)(4^2 + 1^2 + 1^2 + 1^2)$$
$$= 15^2(14^2 + 9^2 + 2^2 + 2^2) = 210^2 + 135^2 + 30^2 + 30^2$$

37. $23 = 2^3 + 2^3 + 1^3 + 1^3 + 1^3 + 1^3 + 1^3 + 1^3 + 1^3$

Exercises 13.4 (p. 620)

1. 9, 4; 161, 72 3. 19, 6; 721, 228 5. 4, 1 7. 15, 4
9. Let $y = 2v$. Then $u^2 - 2y^2 = 1$; 17, 12 is a solution of this equation. Therefore, $2v = 12$ and $v = 6$. Thus, $u = 17$, $v = 6$ is a solution.
11. **proof:** LHS $= a^2c^2 + N^2b^2d^2 - N(a^2d^2 + b^2c^2)$
$= (ac + Nbd)^2 - N(2abcd + a^2d^2 + b^2c^2)$
$= (ac + Nbd)^2 - N(ad + bc)^2 =$ RHS
13. When $n = 2$, Theorem 13.7 yields $x = [(5 + 2\sqrt{6})^2 + (5 - 2\sqrt{6})^2]/2 = 49$ and $y = [(5 + 2\sqrt{6})^2 - (5 - 2\sqrt{6})^2]/2\sqrt{6} = 20$; when $n = 3$, Theorem 13.7 yields $x = [(5 + 2\sqrt{6})^3 + (5 - 2\sqrt{6})^3]/2 = 485$ and $y = [(5 + 2\sqrt{6})^3 - (5 - 2\sqrt{6})^3]/2\sqrt{6} = 198$.
15. **proof** (by PMI): The statement is clearly true when $n = 1$. Assume it is true for an arbitrary positive integer $n = k$. Then $r^{k+1} = r^k \cdot r$ also yields a solution by Lemma 12.14. Thus, by induction, the statement is true for every $n \geq 1$.
17.
$$\begin{aligned} x_n^2 - Ny_n^2 &= (\alpha x_{n-1} + \beta N y_{n-1})^2 - N(\beta x_{n-1} + \alpha y_{n-1})^2 \\ &= (\alpha^2 x_{n-1}^2 + \beta^2 N^2 y_{n-1}^2 + 2\alpha\beta N x_{n-1} y_{n-1}) \\ &\quad - N(\beta^2 x_{n-1}^2 + \alpha^2 y_{n-1}^2 + 2\alpha\beta x_{n-1} y_{n-1}) \\ &= \alpha^2(x_{n-1}^2 - Ny_{n-1}^2) - N\beta^2(x_{n-1}^2 - Ny_{n-1}^2) \\ &= \alpha^2 \cdot 1 - N\beta^2 \cdot 1 = \alpha^2 - N\beta^2 = 1. \end{aligned}$$

Thus, x_n, y_n is a solution.
19. $\alpha = 3$, $\beta = 2 = N$.

(a) $x_1 = \alpha x_0 + \beta N y_0 = \alpha^2 + 2\beta^2 = 9 + 2 \cdot 4 = 17$, $y_1 = \beta x_0 + \alpha y_0 = \beta\alpha + \alpha\beta = 2\alpha\beta = 2 \cdot 3 \cdot 2 = 12$.

(b) $x_2 = \alpha x_1 + \beta N y_1 = 3 \cdot 17 + 2 \cdot 2 \cdot 12 = 99$, $y_2 = \beta x_1 + \alpha y_1 = 2 \cdot 17 + 3 \cdot 12 = 70$.

21. **proof** (by contradiction): Suppose there is a solution x, y. Then $x^2 + 1 = 3y^2 \equiv 0 \pmod 3$. But $x^2 + 1 \equiv 1$ or $2 \pmod 3$ for all integers x. Because this is a contradiction, no solutions exist.
23. **proof:** $x^2 - Ny^2 = -1 = \alpha^2 - N\beta^2 = (\alpha^2 - N\beta^2)^{2n-1}$; that is, $(x + y\sqrt{N})(x - y\sqrt{N}) = (\alpha + \beta\sqrt{N})^{2n-1}(\alpha - \beta\sqrt{N})^{2n-1}$. Now equate factors with the same sign: $x + y\sqrt{N} =$

$(\alpha + \beta\sqrt{N})^{2n-1}$ and $x - y\sqrt{N} = (\alpha - \beta\sqrt{N})^{2n-1}$. Solving, $x = [(\alpha + \beta\sqrt{N})^{2n-1} + (\alpha - \beta\sqrt{N})^{2n-1}]/2$ and $y = [(\alpha + \beta\sqrt{N})^{2n-1} - (\alpha - \beta\sqrt{N})^{2n-1}]/2\sqrt{N}$.

25. $x_n = 3x_{n-1} + 4y_{n-1}$, $y_n = 2x_{n-1} + 3y_{n-1}$, where $x_1 = 1 = y_1$ and $n \geq 2$.
27. The solutions of the equation $x^2 - 2y^2 = 1$ are given by $x = [(3+2\sqrt{2})^n + (3-2\sqrt{2})^n]/2$ and $y = [(3+2\sqrt{2})^n - (3-2\sqrt{2})^n]/2\sqrt{2}$. Since $3+2\sqrt{2} = (1+\sqrt{2})^2$, these equations can be rewritten as

$$x = \left[(1+\sqrt{2})^{2n} + (1-\sqrt{2})^{2n}\right]/2, \quad y = \left[(1+\sqrt{2})^{2n} - (1-\sqrt{2})^{2n}\right]/2\sqrt{2}$$

The solutions of the equation $x^2 - 2y^2 = -1$ are given by

$$x = \left[(1+\sqrt{2})^{2n-1} + (1-\sqrt{2})^{2n-1}\right]/2, \quad y = \left[(1+\sqrt{2})^{2n-1} - (1-\sqrt{2})^{2n-1}\right]/2\sqrt{2}$$

Combining these two cases, we get the solution:

$$x = \left[(1+\sqrt{2})^k + (1-\sqrt{2})^k\right]/2, \quad y = \left[(1+\sqrt{2})^k - (1-\sqrt{2})^k\right]/2\sqrt{2}$$

where $k \geq 1$.

Review Exercises (p. 623)

1. $600009^2 + 20000600000^2 = 20000600009^2$
   $6000009^2 + 2000006000000^2 = 2000006000009^2$
3. LHS $= [3(2n+3)]^2 + [2n(n+3)]^2 = (36n^2 + 108n + 81) + (4n^4 + 24n^3 + 36n^2)$.
   $= 4n^4 + 24n^3 + 72n^2 + 108n + 81 = (2n^2 + 6n + 9)^2$
   $= [2n(n+3) + 9]^2 =$ RHS where $n = 10^m$
5. If the generators of a Pythagorean triple are consecutive triangular numbers, then the length of a leg of the triangle is a cube.
7. Because x, y, and z are in arithmetic progression, we can assume that $x = a - n$, $y = a$, and $z = a + n$ for some positive integers a and n. Because they form a Pythagorean triangle, $(a-n)^2 + a^2 = (a+n)^2$. Simplifying, we get $a = 4n$. Thus, x–y–z = $3n$–$4n$–$5n$.
9. The corresponding Pythagorean triples are 23408–9945–25433, 10640–21879–24329, 40698–5720–41098, and 17290–13464–21914.
11. **proof:** $x = 2ab = 2(m^2 + mn)(3n^2 - mn) = 2mn(m+n)(m-n)$ and $y = a^2 - b^2 = (m^2 + mn)^2 - (3n^2 - mn)^2 = (m^2 + 3n^2)(m + 3n)(m - n)$. Therefore, area $= xy/2 = mn(m+n)(3n-m)(m^2+3n^2)(m+3n)(m-n) = mn(m^2-n^2)(m^2-9n^2)(m^2+3n^2)$. In the second case, $x = 2cd = 2(m^2 - mn)(3n^2 + mn) = 2mn(m-n)(m+3n)$ and $y = c^2 - d^2 = (m^2 - mn)^2 - (3n^2 + mn)^2 = (m^2 + 3n^2)(m - 3n)(m + n)$. Therefore, area $= xy/2 = mn(m-n)(m+3n)(m^2+3n^2)(m-3n)(m+n) = mn(m^2-n^2)(m^2-9n^2)(m^2+3n^2)$. Thus, the areas are equal.
13. 50 53 58 65 74 85
    65 68 73 80 89 100 113
15. We have $x = 2mn$, $y = m^2 - n^2$, and $z = m^2 + n^2$, where $m > n$, $(m, n) = 1$, and $m \not\equiv n \pmod 2$. Then $P = 2m^2 + 2mn = 2m(m+n)$.
17. By Exercise 15, perimeter $P = 2m(m+n)$. Because $(m, n) = 1$, $(m, m+n) = 1$. Because $m+n$ is odd, it follows that $(2m, m+n) = 1$. Because P is a square, $2m = u^2$ and $m+n = v^2$ for some integers u and v. Therefore, m is even and hence n is odd.

19. no

21. no

23. $289 = 17^2 + 0^2$

25. $449 = 20^2 + 7^2$

27. $37 \cdot 41 = (6^2 + 1^2)(5^2 + 4^2) = (6 \cdot 5 + 1 \cdot 4)^2 + (6 \cdot 4 - 1 \cdot 5)^2 = 34^2 + 19^2$

29. $179 = 13^2 + 3^2 + 1^2 + 0^2$

31. $337 = 18^2 + 3^2 + 2^2 + 0^2$

33. $13 \cdot 17 = (3^2 + 2^2 + 0^2 + 0^2)(4^2 + 1^2 + 0^2 + 0^2) = 14^2 + 5^2 + 0^2 + 0^2$

35. $$\begin{aligned} 1360 &= 2^4 \cdot 5 \cdot 17 = 2^4 \cdot (5 \cdot 17) = 2^4(2^2 + 1^2 + 0^2 + 0^2)(4^2 + 1^2 + 0^2 + 0^2) \\ &= 4^2(8^2 + 4^2 + 2^2 + 1^2) = 32^2 + 16^2 + 8^2 + 4^2 \end{aligned}$$

37. Because $9^3 + 10^3 = 12^3 + 1^3$, 9–10–12–1 is a solution; so is $9n$–$10n$–$12n$–n for every positive integer n. Thus, the equation has infinitely many integral solutions.

39. Because 59–158–133–134 is a solution, $5n$–$158n$–$133n$–$134n$ is also a solution for every integer n; so the given equation has infinitely many integral solutions.

41. 9, 2

43. Because $r = 23 + 4\sqrt{33}$ is a solution, $r^2 = 1057 + 184\sqrt{33}$ also yields a solution. That is, 1057, 184 is also a solution.

45. $\alpha = x_1 = 48$, $\beta = y_1 = 7$, and $N = 47$. Then $x_2 = \alpha x_1 + N\beta y_1 = 48^2 + 47 \cdot 7 = 4607$ and $y_2 = \beta x_1 + \alpha y_1 = 7 \cdot 48 + 48 \cdot 7 = 672$. Therefore, 4607, 672 is also a solution.

47. $\alpha = 2$, $\beta = 1$, and $N = 5$. When $n = 2$, Theorem 13.7 yields $x = [(2 + \sqrt{5})^3 + (2 - \sqrt{5})^3]/2 = 38$ and $y = [(2 + \sqrt{5})^3 - (2 - \sqrt{5})^3]/2\sqrt{5} = 17$. When $n = 3$, the theorem yields $x = [(2 + \sqrt{5})^5 + (2 - \sqrt{5})^5]/2 = 682$ and $y = [(2 + \sqrt{5})^5 - (2 - \sqrt{5})^5]/2\sqrt{5} = 305$.

Thus, 38, 17; and 682, 305 are two new solutions.

Supplementary Exercises (p. 626)

1. The legs of the corresponding primitive Pythagorean triangle are 31360, 99999; 71492, 70965; 53452, 96795; and 106684, 56763. Their areas are 1567984320, 2536714890, 2586943170, and 3027851946, respectively.

3. $336700 = 33^3 + 67^3 + 00^3$
   $336701 = 33^3 + 67^3 + 01^3$

5. $$\begin{array}{rclrcl} 370 &=& 3^3 + 7^3 + 0^3 & 371 &=& 3^3 + 7^3 + 1^3 \\ 336700 &=& 33^3 + 67^3 + 00^3 & 336701 &=& 33^3 + 67^3 + 01^3 \\ 333667000 &=& 333^3 + 667^3 + 000^3 & 333667001 &=& 333^3 + 667^3 + 001^3 \\ 333366670000 &=& 3333^3 + 6667^3 + 0000^3 & 333366670001 &=& 3333^3 + 6667^3 + 0001^3 \end{array}$$

7. $25^2 + 26^2 + 27^2$

9. $55^2 + 56^2 + 57^2 + 58^2 + 59^2 + 60^2 = 61^2 + 62^2 + 63^2 + 64^2 + 65^2$

11. 77^2

13. The positive factors of 30 are 1, 2, 3, 5, 6, 10, 15, and 30; they have 1, 2, 2, 2, 4, 4, 4, and 8 positive factors, respectively. Then $(1 + 2 + 2 + 2 + 4 + 4 + 4 + 8)^2 = 729 = 1^3 + 2^3 + 2^3 + 2^3 + 4^3 + 4^3 + 4^3 + 8^3$

15. Let t_k denote the kth triangular number. Then $3\left(\sum_{k=1}^{n} t_k\right)^3 = \sum_{k=1}^{n} t_k^3 + 2\sum_{k=1}^{n} t_k^4$.

17. Let $x - 1$, x, $x + 1$, and $x + 2$ be the four consecutive integers. Then $(x - 1)^3 + x^3 + (x + 1)^3 = (x + 2)^3$. Simplifying, we get $(x - 4)(x^2 + x + 1) = 0$. Because $x^2 + x + 1 = 0$ has no integral solutions, $x = 4$. Therefore, $w = x - 1 = 3$.

19. $14^2 + 36^2 + 58^2 + 90^2 = 09^2 + 85^2 + 63^2 + 41^2 = 12{,}956$
$16^2 + 38^2 + 50^2 + 94^2 = 49^2 + 05^2 + 83^2 + 61^2 = 13{,}036$

21. $26^2 + 35^2 + 84^2$

23. $12^2 + 43^2 + 65^2 + 78^2$

25. $11^3 + 12^3 + 13^3 + 14^3 = 20^3$

27. $666 = 2^2 + 3^2 + 5^2 + 7^2 + 11^2 + 13^2 + 17^2$

Appendix A.1 (p. 637)

1. yes

3. yes

5. Let x and y be any two even integers. Then $x = 2m$ and $y = 2n$ for some integers m and n. Then $x + y = 2m + 2n = 2(m + n)$, which is also an even integer.

7. Let $x = 2m$ be any even integer. Then $x^2 = (2m)^2 = 2(2m^2)$ is also an even integer.

9. Let x be any odd integer. Then $x = 2m + 1$ for some integer m. Then $x^2 = (2m + 1)^2 = 2(2m^2 + 2m) + 1$, which is an odd integer.

11. Let x be any even integer and y any odd integer. Then $x = 2m$ and $y = 2n + 1$ for some integers m and n. Then $xy = (2m)(2n + 1) = 2(2mn + m)$, an even integer.

13. Let $x = 4k + 1$. Then $x^2 = (4k + 1)^2 = 16k^2 + 8k + 1 = 4(4k^2 + 2k) + 1 = 4m + 1$, where $m = 4k^2 + 2k$ is an even integer. Therefore, x^2 is also an integer of the same form.

15. Let x be any integer. Assume it is not even, so it is odd and is of the form $2m + 1$. Then $x^2 = (2m + 1)^2 = 4m^2 + 4m + 1 = 2(2m^2 + 2m) + 1$, which is an odd integer. So, the given hypothesis is false and the result follows.

17. Let x and y be any two integers. Assume that the given conclusion is false; that is, assume that both x and y are odd integers. Then $x = 2m + 1$ and $y = 2n + 1$ for some integers m and n, so $xy = (2m + 1)(2n + 1) = 2(2mn + m + n) + 1$, an odd integer. This negates the given hypothesis and so the result follows.

19. Suppose $\sqrt{2}$ is not an irrational number; that is, $\sqrt{2}$ is a rational number. Let $\sqrt{2} = a/b$, where a and b have no positive common factors except 1. Then $(a/b)^2 = 2$ or $a^2 = 2b^2$. Therefore, 2 is a factor of a^2 and hence of a. Then $a = 2m$ for some integer m. So $(2m)^2 = 2b^2$ or $b^2 = 2m^2$. Then 2 is a factor of b^2 and hence of b. Consequently, 2 is a common factor of a and b, which is a contradiction.

21. Assume $\sqrt{p}$ is a rational number a/b, where a and b have no positive common factors except 1. Then $\sqrt{p} = a/b$, $(a/b)^2 = p$, or $a^2 = pb^2$. Consequently, p is a factor of a^2 and hence of a. So let $a = mp$. Then $(mp)^2 = pb^2$ or $b^2 = pm^2$. As before, this shows p is a factor of b. Thus, p is a common factor of a and b, a contradiction.

23. **proof: case 1** Let n be an even integer, say, $2m$. Then $n^2 + n = (2m)^2 + 2m = 2(2m^2 + m)$, an even integer.

    case 2 Let n be an odd integer, say, $2m + 1$. Then $n^2 + n = (2m + 1)^2 + (2m + 1) = (4m^2 + 4m + 1) + (2m + 1) = 2(2m^2 + 3m + 1)$, again, an even integer.

25. **proof:** Every integer n is of the form $3k$, $3k + 1$, or $3k + 2$.

    case 1 Let $n = 3k$. Then $n^3 - n = (3k)^3 - (3k) = 27k^3 - 3k = 3(9k^3 - k)$.

    case 2 Let $n = 3k + 1$. Then $n^3 - n = (3k + 1)^3 - (3k + 1) = (27k^3 + 27k^2 + 9k + 1) - (3k + 1) = 27k^3 + 27k^2 + 6k = 3(9k^3 + 9k^2 + 2k)$.

    case 3 Let $n = 3k + 2$. Then $n^3 - n = (3k + 2)^3 - (3k + 2) = (27k^3 + 54k^2 + 36k + 8) - (3k + 2) = 27k^3 + 54k^2 + 33k + 6 = 3(9k^3 + 18k^2 + 11k + 2)$

    Thus, in every case, $n^3 - n$ is divisible by 3.

27. $x = 3$
29. **proof:** Because 3–4–5 is a solution, $3n$–$4n$–$5n$ is also a solution for every integer n.
31. 0, because 0^2 is not positive.
33. January has 31 days.
35. **proof:** Because $a < b$, there is a positive real number x such that $a + x = b$. Therefore, $(a + x) + c = b + c$; that is, $(a + c) + x = b + c$. Therefore, $a + c < b + c$.
37. **proof:** Let $a \cdot b = 0$ and $a \neq 0$. Because $a \cdot 0 = 0$, $a \cdot b = a \cdot 0$. Canceling a from both sides, we get $b = 0$.
39. 11, because $2047 = 23 \cdot 89$.

Credits

TEXT AND ART

p. 42 Figure 1.18. M. J. Zerger, "Proof without Words: Sums of Triangular Numbers" from *Mathematics Magazine* 63 (December 1990), 314. Reprinted with the permission of M. J. Zerger, Adams State College.

p. 46 Figure 1.25. R. B. Nelsen, "Proof without Words: The Sum of the Squares of Consecutive Triangular Numbers Is Triangular" from *Mathematics Magazine* 70 (1997), 130. Reprinted with the permission of Roger B. Nelson, Lewis & Clark College, Portland.

p. 47 Figure 1.26. E. G. Landauer, "Proof without words: On Squares of Positive Integers" from E. G. Landauer, "Proof without Words," *Mathematics Magazine* 58 (September 1985), 203. Reprinted with the permission of Edwin G. Landauer, Clackamas Community College, Oregon City, Oregon.

p. 108 Figure 2.26. Clive Thompson, "Outsider Math" from *The New York Times Magazine* (December 15, 2002), 107. Reprinted with permission.

pp. 119–120 Figure 2.27. Barry Cipra, "How Number Theory Got the Best of the Pentium Chip" from *Science* 267 (January 13, 1995). Copyright © 1995 American Association for the Advancement of Science. Reprinted with permission.

p. 123 Figure 2.28. "Publishers Offer Prize for Proof of Goldbach's Conjecture" from *MAA Focus* (May/June 2000), 11. Copyright © 2000 by The Mathematical Association of America. Reprinted with permission.

p. 130 Figure 2.29b. Text from H. E. Huntley, *The Divine Proportion* (Mineola, New York: Dover Publications, 1970), p. 165. Reprinted with the permission of the publishers.

p. 255 Figures 5.2a, b. "The (19, 9) residue design," and "The (21, 10) design" from P. Locke, "Residue Designs," *Mathematics Teacher* 65 (March 1972), 262. Copyright © 1972 by the National Council of Teachers of Mathematics. All rights reserved.

pp. 257–258 Figures 5.7, 5.8, 5.9, 5.10, and 5.12. Designs from S. Forseth and A. P. Troutman, "Using Mathematical Structures to Generate Artistic Designs," *Mathematics Teacher* 67 (May 1974). Copyright © 1974 by the National Council of Teachers of Mathematics. All rights reserved.

p. 374 "The Perfect Eye Chart" from "Reader Reflections," *Mathematics Teacher* 89 (November 1996), 624. Reprinted with the permission of Ron Ruemmler and Bob Urbanski, Middlesex County College, Edison, NJ.

p. 383 Figure 8.3. Postage meter stamp. Reprinted with the permission of the University of Illinois.

p. 383 Figure 8.4. T. J. W. Research Center envelope marking the 1971 discovery of a prime number by B. Tuckerman (IBM). Reprinted with permission from IBM's T. J. Watson Research Laboratory.

p. 384 Figure 8.5. Excerpt from Michael Stroh, "Worldwide Number Search Inspires Prime Competition" from *The Baltimore Sun*. Reprinted with permission.

p. 386 Figure 8.6. Reuters, "Prime Number Is Largest Ever." Copyright © 2001 by Reuters. Reprinted with permission.

p. 395 Figure 8.8. Pascal's Binary Triangle. K. M. Shannon, Salisbury University, Salisbury, MD.

p. 415 Figure 9.1. Sabra Chartrand, excerpt from "A New Encryption System Would Protect a Coveted Digital Data Stream Music on the Web" from *The New York Times* (July 3, 2000). Copyright © 2000 by The New York Times Company. Reprinted with permission.

pp. 435–436 Figure 9.5. Simson Garfinkel, "A Prime Argument in Patent Debate" from *The Boston Globe* (April 6, 1995). Copyright © 1996 by Simson L. Garfinkel. Reprinted with permission.

p. 437 Figure 9.6. Laurence Zuckerman, excerpt from "I.B.M. Researchers Develop a New Encryption Formula" from *The New York Times* (May 7, 1997). Copyright © 1997 by The New York Times Company. Reprinted with permission.

p. 440 Figure 9.7. Barnaby J. Feder, excerpt from "E-Signing Law Seen as a Boon to E-Business" from *The New York Times* (July 4, 2000). Copyright © 2000 by The New York Times Company. Reprinted with permission.

p. 441 Figure 9.8. Ivars Peterson, "Cracking Huge Numbers" from *Science News* (May 7, 1994). Reprinted with permission from *Science News*, the weekly newsmagazine of science. Copyright © 1994 by Science Service Inc.

pp. 595–596 Figure 13.3. Gina Kolata, excerpt from "At Last, Shout of 'Eureka!' An Age-Old Math Mystery" from *The New York Times* (June 24, 1993). Copyright © 1993 by The New York Times Company. Reprinted with permission.

p. 596 Figure 13.4. Gina Kolata, excerpt from "How a Gap in the Fermat Proof Was Bridged" from *The New York Times* (January 31, 1995). Copyright © 1995 by The New York Times Company. Reprinted with permission.

p. 598 Figure 13.5. "Fermat's Last Theorem: The Musical" from *MAA Focus* (November 2000). Copyright © 2000 by The Mathematical Association of America. Reprinted with permission.

p. 599 Figure 13.6. Greg Nemec, illustration, "Beal's Problem" from *Math Horizons* (February 1998), 9. Copyright © 1998. Reprinted with permission.

PHOTOS

p. 4 Leopold Kronecker. Courtesy David Eugene Smith Collection, Rare Book and Manuscript Library, Columbia University.

p. 9 Joseph Louis Lagrange. Courtesy Academy of Sciences, France.

p. 32 John McCarthy. Courtesy Roger Ressmeyer/Corbis.

p. 39 Blaise Pascal. Courtesy Academy of Sciences, France.

p. 53 Eugene Charles Catalan. Courtesy David Eugene Smith Collection, Rare Book and Manuscript Library, Columbia University.

p. 73 Gustav Peter Lejeune Dirichlet. Courtesy Academy of Sciences, France.

p. 112 Jacques Hadamard. Courtesy Academy of Sciences, France.

p. 113 Charles-Jean-Gustave-Nicholas de la Valleé-Poussin. Courtesy of the Archives, California Institute of Technology.

p. 124 Joseph Louis François Bertrand. Courtesy Academy of Sciences, France.

p. 129 Leonardo Fibonacci. Courtesy David Eugene Smith Collection, Rare Book and Manuscript Library, Columbia University.

p. 130 Figure 2.29a. Sunflower head. Courtesy Runk/Schoenberger/Grant Heilman Photography, Inc.

p. 134 Giovanni Domenico Cassini. Courtesy Bettmann/Corbis.

p. 136 Jacques Philippe Marie Binet. Courtesy Academy of Sciences, France.

p. 137 Gabriel Lamé. Courtesy Academy of Sciences, France.

p. 139 Pierre de Fermat. Courtesy David Eugene Smith Collection, Rare Book and Manuscript Library, Columbia University.

p. 322 Edward Waring. Courtesy David Eugene Smith Collection, Rare Book and Manuscript Library, Columbia University.

p. 339 Robert Daniel Carmichael. Courtesy The Mathematical Association of America.

p. 342 Leonhard Euler. Courtesy David Eugene Smith Collection, Rare Book and Manuscript Library, Columbia University.

p. 382 Marin Mersenne. Courtesy Academy of Sciences, France.

p. 392 Derrick Henry Lehmer. Courtesy Math Department, University of California at Berkeley.

p. 398 August Ferdinand Möbius. Courtesy David Eugene Smith Collection, Rare Book and Manuscript Library, Columbia University.

p. 414 Godfrey Harold Hardy. Courtesy Academy of Sciences, France.

p. 502 Adrien-Marie Legendre. Courtesy David Eugene Smith Collection, Rare Book and Manuscript Library, Columbia University.

p. 517 Ferdinand Gotthold Eisenstein. Courtesy David Eugene Smith Collection, Rare Book and Manuscript Library, Columbia University.

p. 528 Karl Gustav Jacob Jacobi. Courtesy David Eugene Smith Collection, Rare Book and Manuscript Library, Columbia University.

p. 567 Srinivasa Aiyangar Ramanujan.

p. 580 Figure 13.1. Babylonian Tablet. Plimpton 322, Rare Book and Manuscript Library, Columbia University.

p. 588 Charles Luttwidge Dodgson (Lewis Carroll). Courtesy David Eugene Smith Collection, Rare Book and Manuscript Library, Columbia University.

p. 594 Ernst Eduard Kummer. Courtesy David Eugene Smith Collection, Rare Book and Manuscript Library, Columbia University.

p. 595 Andrew J. Wiles. Courtesy Andrew Wiles/Princeton University.

p. 600 Andrew Beal. Courtesy D. Andrew Beal.

Index

H

I

M

N

U

V

W

Y

Z